INTEGRATED ENVIRONMENTAL MODELING

Pollutant Transport, Fate, and Risk in the Environment

Anu Ramaswami

Jana B. Milford

Mitchell J. Small

WILEY

JOHN WILEY & SONS, INC.

This book is printed on acid-free paper. ♾

Published by John Wiley & Sons, Inc., Hoboken, New Jersey
Published simultaneously in Canada

Library of Congress Cataloging-in-Publication Data:

Ramaswami, Anu.
Integrated environmental modeling : pollutant transport, fate, and risk in the environment / Anu Ramaswami, Jana B. Milford, Mitchell J. Small.
p. cm.
Includes index.
1. Pollution—Mathematical models. 2. Pollution—Environmental aspects. 3. Pollutants—Environmental aspects. 4. Cross-media pollution—Mathematical models. 5. Environmental risk assessment. I. Milford, Jana B. II. Small, Mitchell J. III. Title.
TD174 .R36 2005
628.5'2'015118—dc22 2004011078

Printed in the United States of America

10 9 8 7 6 5 4 3 2

CONTENTS

This book is dedicated to:
My parents, Padma and Ramaswami; and my son, Varun
My husband, Rick; and parents, Ken and Betty Milford
My wife, Mindy; parents, Ida and Martin Small;
and my children, Ammi, Yehudit, Dovid,
Devorah, Levi, and Yossi

ACKNOWLEDGMENTS

We thank our teachers who introduced the world of environmental modeling to us: Francis McMichael, Richard Shane, William Darby, Edward Rubin, Richard Luthy, Greg McRae, Cliff Davidson, Armistead (Ted) Russell, Granger Morgan and Indira Nair at Carnegie Mellon University; Dominic Di Toro, Robert Thomann, John Mueller, and the late Donald O'Connor at Manhattan College, Hydroscience and Hydroqual, Inc.; and Perry Samson, Walter Weber, Jonathan Bulkley and Raymond Canale at the University of Michigan.

We acknowledge our many colleagues whose seminal pedagogical works, research insights and field case studies have been important in our synthesis of environmental modeling methods in this book: Julian Andelman, Mary Anderson, Robert Borden, Philip Bedient, Deborah Bennett, Gordon Bennett, Steve Chapra, Yoram Cohen, Bernard (Jack) Cosby, Francis DiGiano, David Dzombak, Paul Fischbeck, George Hornberger, William Jury, Don Mackay, Tom McKone, Shelley Miller, James Morgan, Wayne Ott, Spyros Pandis, Marina Pantazidou, Prabhakar Clement, Kenneth Reckhow, Peter Reichert, Paul Roberts, Robert Runkel, Mark Schervish, Jerry Schnoor, Rene Schwarzenbach, John Seinfeld, Laura Sigg, Werner Stumm, Tim Sullivan, Louis Thibodeaux, and Chunmiao Zheng.

We thank Tom McKone, Edward Butler, Nicole Rowan, and Mike Sohn for sections contributed to Chapters 13, 10, and 9, respectively, as well as Glen McConville for the original cartoons that appear in this book. Mark Milke, Robert Harley, and Mike Hannigan reviewed various chapters of this book and provided useful insights and feedback.

We thank several of our current and former students who contributed figures and example problems, and participated in many rounds of review and revisions over the years that this book was in development. In particular, we are grateful to Cindy Beeler, Dominic Boccelli, Kevin Brand, Grant Bromhal, Joseph DeCarolis, Geoff Germano, Nicholas Giardino, Amit Goyal, Lori Gross, Patrick Gurian, Mehmet Isleyen, Jae Gun Jung, Daniel Kovacs, Sylvia Myhre, Smita Narayan, Anand Patwardhan, Mark Pitterle, Donna Riley, Mary Schoen, Tanarit Sakulyanontvittaya, Martin Schultz, Rajarishi Sinha, Neil Stiber, Satoshi Takahama, Win Trivitayanurak, Jason West, Charles Wilkes, and Sonia Yeh.

Finally, each of us would like to thank our families for their patience and support, and each other for the collaborative spirit without which this book would not have come to be.

PREFACE

Environmental engineers and scientists are increasingly required to be familiar with contaminant behavior in all three environmental media—air, water, and soil/subsurface systems. There is also increased awareness of the need to understand pollution impacts on the biosphere (humans, flora, and fauna) by assessing pollution of air, water, and soil systems in an integrated manner, rather than within media-specific boundaries. The primary reasons for focusing on multimedia contaminant behavior are:

- Contaminant transfer often occurs between multiple media—for example, washout of air pollutants by precipitation (wet deposition) creates a source of water pollution, as in the acid rain problem. Likewise, gasoline spills can result in contamination of air as well as water and soil. An understanding of the multimedia transport and fate of pollutants is required to assess the overall phase distribution and space–time location of the pollutant.
- Health risks to humans arise from exposure to contaminants through multiple pathways that encompass the different environmental media—air, water, and soil. Inhalation exposure to pollutants occurs through the medium of air, while ingestion of and dermal contact with pollutants can occur through contaminated soils, water and food. Estimates of contaminant concentrations in air, water, soil and flora are required for human health risk assessment.
- Laws for hazardous waste management and new chemical regulation are founded on an integrated assessment of human health risks posed by contaminants in multiple media. In addition, screening and implementing remediation technologies at hazardous waste sites requires consideration of the multimedia transport and fate of pollutants.

Although the need exists to learn the principles underlying contaminant transport in all three environmental systems in a unified manner, most textbooks and environmental engineering curricula tend to focus on individual media. Furthermore, few books address the use of environmental models in determining pollution impacts on the living biosphere, encompassing humans, flora, and fauna. Therefore, a student typically must take three or four courses to learn about pollutant transport in all three environmental systems and to evaluate the impact of such contaminant transport on the biosphere. This textbook offers a unique, comprehensive, one-semester treatment of contaminant behavior in all three environmental systems, and resulting impacts on the biosphere, by focusing on the fundamental processes underlying environmental modeling.

Fundamental processes underlying the transport and transformation of environmental pollutants are essentially the same in the different environmental media. Furthermore, mathematical models for transport and transformation processes, and techniques for linking these process models together, are also similar for the different environmental systems. Building contaminant transport models may therefore be viewed as putting together different process models and testing the overall output against real-world data. The process of building, implementing, and testing environmental models is thus a generalized process, not media-specific in nature. However, because of the media-specific approach of traditional environmental science and engineering curricula, students tend to focus on media-specific issues, remaining unaware of common threads and principles that underlie contaminant behavior in the total environment. The primary goal of this book is to teach model development, model implementation, and model testing skills in a unified manner, cross-cutting the three environmental systems by focusing on parallels and similarities between them.

The objective of this book is to provide broad-based training in the development of pollutant transport and fate models in air, water, and soil, with focus on five essential components:

1. Understanding of the process principles that govern contaminant transport and transformations in multimedia environments, with emphasis on the parallels and linkages between different media
2. Developing more complex and realistic models by coupling various process modules appropriate for the system being considered
3. Knowledge of data sources (including compiled data and site-specific lab or field tests) and parameter estimation techniques from which input parameters for fate and transport models may be estimated
4. Practice with computer-aided implementation and evaluation of fate and transport models, using self-guided example problems and case studies
5. Application of fate and transport models to evaluate pollutant interactions with the biosphere, particularly in the context of human exposure modeling and health risk assessment

The book is unique in that it provides in-depth coverage of all five of the above components, making it an instructive textbook, as well as a comprehensive reference for professionals in environmental science, engineering, and management. The textbook is designed for use by graduate students interested in developing a broad knowledge base of contaminant transport and transformation processes in the natural environment. Balanced treatment of issues pertaining to atmospheric, surface water, and groundwater systems is achieved by the coordinated efforts of three co-authors with primary specialization in the different systems. Pollutant interactions with the biosphere are discussed through physically based uptake and partition models, as well as physiologically based pharmacokinetic models used for human health risk assessment. The text is intended for use in a one-semester course, with lectures covering the process fundamentals and basic concepts underlying environmental contaminant transport and fate models, and an accompanying computer laboratory to provide students the opportunity to apply the models using real-world case studies. Guidance is provided for different sequences of chapters that can be utilized for courses emphasizing either environmental engineering and science or risk assessment and environmental management.

Chapter 1 introduces the reader to the realm of environmental pollution modeling, summarizing the basic principles and the applications of multimedia contaminant fate and transport models. A framework for formulating models is presented that leads the reader through the sequence of steps required to describe contaminant movement from a source location to a receptor location. The nature of environmental pollutants is discussed in Chapter 2. The reader is encouraged to classify pollutants using a variety of attributes including physical state, chemical structure, toxicity, health effects, and legislation. Parameters pertaining to the physical state of the contaminant, for example, melting and boiling point, solubility, and vapor pressure, are discussed. Chemical parameters related to reaction rate constants and equilibria are presented. The reader is introduced to toxicity indices and to the broad range of pollutants regulated under various environmental laws. Thus, Chapter 2 explores the physical, chemical, toxicological, and regulatory aspects of pollutant characterization.

Equilibrium processes that govern interphase transfer of contaminants between the different media are discussed in Chapter 3. Equilibrium partitioning constants between different pairs of media are presented along with parameter estimation techniques. Physically based partition coefficients that describe the equilibrium interaction between contaminants and biota are also presented. The kinetics of interphase mass transfer is discussed in Chapter 4. Interphase mass-transfer rate coefficients are introduced, and the reader is provided with a comprehensive resource list for estimating these coefficients for different pairs of environmental media, including air–water, soil–water, and nonaqueous liquid–water transfer. The impact of intermedia exchanges on contaminant transport is illustrated by means of examples drawn from all three environmental systems. The reader is introduced to simultaneous multimedia compartment models that incorporate interphase partitioning including their application to evaluate chemical persistence and long-range transport potential in the global environment.

The fundamental transport processes: advection, dispersion, and diffusion, which act within individual environmental media, are discussed in Chapter 5. Basic concepts underlying advection, dispersion, and diffusion are presented along with a generalized modeling framework for representing each of the processes in environmental systems. Advection, dispersion, and diffusion parameters are defined, and quantitative estimates of these parameters are provided for all three environmental media, enabling comparison of the three systems. A control volume mass balance is shown to result in the generalized advective–dispersive transport equation, which is common to all three environmental systems. The transition from the limiting case of the completely mixed reactor to the plug flow reactor is examined as a simple means of characterizing the advective–dispersive transport model.

Numerical methods for computer-aided implementation of environmental models are presented in Chapters 6 and 7, motivated by a discussion of the need for numerical models to address spatial and temporal heterogeneities in environmental systems. Chapter 6 presents methods for discretization of spatial and temporal modeling domains, along with associated solution techniques. Chapter 7 presents an overview of probability concepts and methods used in stochastic modeling and Monte Carlo simulations.

In Chapters 8 to 10, the generalized advective–dispersive equation is adapted to address contaminant transport in the atmosphere, groundwater, and surface water systems, respectively. Analytical models are presented that describe contaminant transport in the individual systems primarily for two idealized situations: an instantaneous release of contaminant and a continuous steady release. Techniques for estimating the advection, dispersion, and diffusion model parameters, which are highly media specific in nature, are presented at the end of each of the chapters. Case studies are provided to introduce numerical models such as the Industrial Source Complex (ISC) model for air quality modeling, a three-dimensional subsurface contaminant transport model (MT3D), and the One-dimensional Transport with Inflow and Storage (OTIS) model for water quality in streams. A review of currently available fate and transport modeling software is presented in which the advantages, disadvantages, and unique features of the various packages are discussed.

Chemical transformation processes that alter the structure of contaminants in atmospheric and aqueous systems are discussed in Chapters 11 and 12, respectively. Chapter 11 focuses on gas-phase photochemical reactions, phase transformations, and aerosol growth, which are especially important in atmospheric systems. Processes that are important in aqueous systems, including acid–base equilibria, mineral weathering kinetics, and biodegradation kinetics, are presented in Chapter 12.

Multimedia contaminant exposure models and risk calculations are presented in Chapter 13, along with a case study using the CalTOX model. Techniques for evaluating model performance are presented in Chapter 14. Statistical methods are presented for comparing model results with data, incorporating both model uncertainty and parameter variability. Case studies provide opportunities to gain experience with sensitivity and uncertainty analysis tools.

1 Introduction to Modeling the Transport and Transformation of Contaminants in the Environment

1.1 DEFINITION AND ROLE OF CONTAMINANT TRANSPORT AND FATE MODELS

The air, water, soil, and biota that comprise Earth's environment form a dynamic and evolving tapestry. The biosphere is sustained by energy, material, and information flows remarkable in their complexity and coherence. The human role in the environment is likewise complex, though not always as coherent. We are sustained by the energy, material, and information flows of the biosphere but also play a major role in affecting them. These impacts are not all positive for human health or for the long-term sustainability of the environment.

Figure 1.1 presents some of the key components and processes of environmental fate, transport, exposure, and impact. Pollutants are transported across the air, water, soil, and biological compartments and undergo physical, chemical, and biological transformations within these compartments. Humans and other living receptors are exposed to pollutants through a variety of pathways, with impacts that depend on the magnitude and temporal and spatial distribution of the exposure.

To better understand the workings of the environment and the impacts of human activity on it, scientists and engineers from a variety of disciplines have developed conceptual frameworks and corresponding mathematical models to describe environmental processes such as those depicted in Figure 1.1. Mathematical models provide a framework for understanding the physical, chemical, and biological processes that determine the cycling of elements and compounds through the environment. They provide a basis for relating human activities and environmental impacts and thus for predicting the changes that might occur in response to alterations in the activities. In this mode, they can assist in making decisions about environmental strategies and programs.

Environmental models are developed both for scientific purposes and as applied tools for policy development, implementation, and management. In policy or management applications, models are used to address questions such as:

- By how much must nutrient loads to a lake or reservoir be reduced to reverse chronic eutrophication? What is the relative contribution of industrial versus

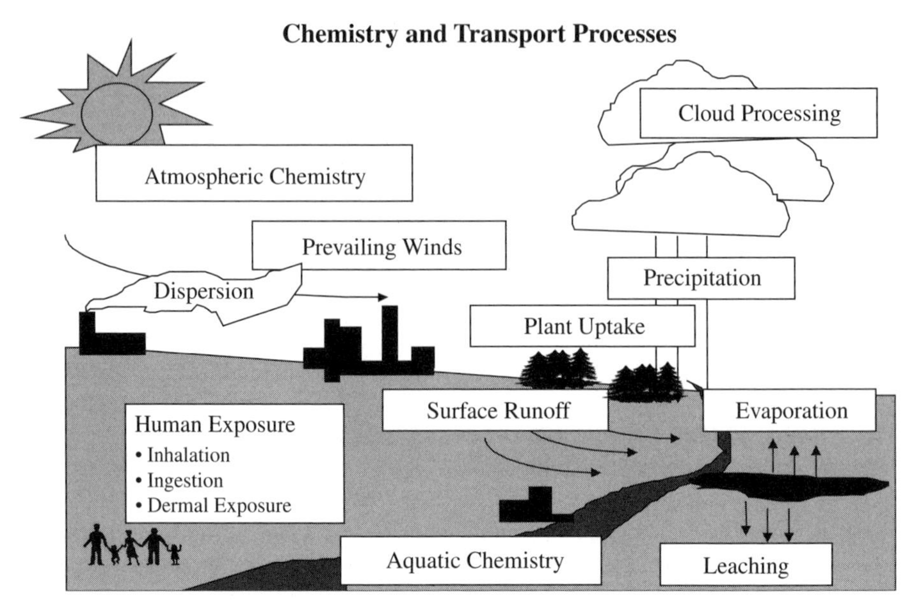

Figure 1.1 Transport and transformation processes for contaminants in the environment. Understanding these pathways and processes is critical for predicting human exposure, health effects, and ecosystem impacts.

agricultural sources to the eutrophication problem, and how should requirements for reducing nutrient loads be allocated?

- What is the likely effect of reductions in sulfur emissions to the atmosphere on ambient sulfate, fine particulate matter, and associated human exposure and health impacts? What effects will these reductions have on the acidification of lakes and streams? What effects will they have on visibility?
- How long will a groundwater aquifer remain contaminated by pesticides? What are the implications for human health risks? What remediation strategies would lead to the greatest risk reductions?
- Which of a number of alternative products or product designs results in the lowest environmental impact over its life cycle?

These are all questions asked by industry, citizens, and regulators seeking to determine whether and how to implement environmental programs and controls. Environmental models can play a central role in answering these questions. Along with laboratory and field studies, models can provide improved understanding leading to better management of environmental quality and sustainability.

In addition to providing a basis for environmental management, mathematical models serve a fundamental scientific purpose, providing a framework for organizing and integrating scientific understanding of environmental processes and interactions. In this context, a model can be viewed as a testable hypothesis, refutable with laboratory and field observations. Ideally, models and observations are used to advance understanding in an iterative manner, with models used to direct the collection of environmental data that in turn are used to challenge and refine the models. As with other scientific theories, models once viewed as "state-of-the-art" may later be discarded as new evidence and insights are collected and considered.

1.2 ENVIRONMENTAL POLLUTANTS

Environmental pollutants are identified as such because they produce or pose risks of undesirable effects on the health of humans, animals, plants, or ecosystems, because they degrade aesthetic qualities of the environment such as odor or visual range, or because they pose economic costs by damaging materials or property. Lists of environmental pollutants are not fixed but rather evolve as new concerns are recognized. Though pollutants are associated with human activity by definition, many chemicals characterized as pollutants have natural sources as well.

Table 1.1 lists a variety of pollutants that are currently of concern to environmental scientists, engineers, and policymakers. For each pollutant listed, the table shows the primary physical state of the contaminant in the environment (at standard temperature and pressure), together with some of the effects of concern. As indicated next to their sources, the pollutants listed in Table 1.1 are categorized as either primary, meaning that they are directly emitted into the environment, or secondary, meaning that they are formed in the environment from the reactions of other pollutants.

Although the predominant associations of the contaminant phases in Table 1.1 are gas phase with air, liquid phase with water and solid phase with soil, all three phases are present in all media, including the biota. Very fine particles of solid or liquid contaminants can remain suspended in the atmosphere for long periods of time. Cloud and rain droplets and sleet and snow are additional examples of condensed matter in the air medium. Similarly, surface water contains suspended solid particles as well as gas bubbles. The soil vadose zone from the surface to the groundwater table includes both gases and liquids in soil pores. In contrast, soil pores in the underlying saturated zone are entirely filled with liquid.

While the natural physical state of a chemical in the environment is governed by its melting and boiling points relative to the ambient temperature, most chemicals are present in the environment in more than one physical state or phase and consequently in more than one medium. For example, volatile liquids having a significant vapor pressure at ambient conditions will tend to maintain equilibrium between gas and liquid phases. Airborne gases that are soluble in water will also be present in both media. Substances that naturally occur in a liquid or solid state may dissolve in water and also exist as a separate phase when saturation of the aqueous phase occurs. Thus, interphase mass-transfer processes determine the distribution of a chemical in the different environmental media.

In some cases, chemical reactions affect the distribution of pollutants between different media. For example, oxidized forms of mercury (Hg^{2+}), including methyl mercury, are much more soluble in water than the elemental form and thus more easily removed from the atmosphere by precipitation scavenging. In other cases, intermedia transfers can affect the reactivity of pollutants. For example, the fuel oxygenate

TABLE 1.1 Examples of Environmental Contaminants, Their Sources, and Effects[a]

Pollutant	Phase	Sources	Effects
Methane	Gas	Swamps, landfills, livestock (P)	Explosive, greenhouse gas
Carbon dioxide	Gas	Deforestation, fossil fuel combustion (P)	Greenhouse gas
Carbon monoxide	Gas	Incomplete combustion (P)	Acute health effects: binds with hemoglobin
Sulfur dioxide	Gas	Fossil fuel combustion (P)	Respiratory irritant
Fine particulate matter	Solid or liquid (aerosol)	Incomplete combustion, windblown soil (P), oxidation of ambient SO_2 and NO_2 (S)	Acute and chronic respiratory and pulmonary health effects, visibility impairment
Sulfate ion	Liquid	Sea salt aerosols, mineral dissolution in groundwater (P), oxidation of ambient SO_2 (S)	Acidification of surface waters, materials damage
Nitrate ion	Liquid	Oxidation of NO_2 and NH_3 fertilizer (S)	Acidification and eutrophication of surface waters, reduced to nitrite in body
Nitrite ion	Liquid	Meat preservative (P), oxidation of NH_3 fertilizer (S)	Methemoglobinemia, aquatic toxicity, possible carcinogen
Trihalomethanes	Liquid	Disinfection by-products (S)	Carcinogen
Microbial pathogens	Solid (suspended)	Human and animal wastes, indoor ventilation systems (P)	Infectious diseases
Ozone	Gas	Photochemical reactions (S)	Respiratory irritant
Plutonium	Solid	Nuclear fuel and weapons processing (P)	Carcinogen

(*continued*)

TABLE 1.1 ***(Continued)***

Pollutant	Phase	Sources	Effects
Radon	Gas	Natural uranium deposits, weapons production (S)	Carcinogen
Arsenic	Solid/liquid ionic and organic compounds	Natural soil deposits, inorganic pesticides, wood treatment (P)	Arsenic poisoning/ gangrene/carcinogen
Lead	Solid	Gasoline additive, smelting (P)	Central nervous system damage
Perchlorate	Liquid (dissolved ion)	Dissolution of salts used as solid propellant for weapons, fireworks, air bags (S)	Thyroid inhibition, fetal impairment, possible thyroid carcinoma
Elemental mercury	Liquid (volatile)	Oceans (P)	Oxidized to methyl mercury
Methyl mercury	Liquid (volatile)	Coal and waste combustion (P), oxidation of Hg^0 (S)	Central nervous system damage
Methylene chloride	Liquid (volatile)	Solvents, furniture stripper, hairspray (P)	Acute toxin, probable carcinogen
Benzene	Liquid (volatile)	Gasoline (P)	Carcinogen
Trichloroethylene	Liquid (volatile)	Solvent (P)	Probable carcinogen
Benzo(*a*)pyrene	Liquid (semivolatile)	Combustion (P)	Probable carcinogen
Polychlorinated biphenyls	Liquid (semivolatile)	Electric power transformers, environmental reservoirs (P)	Dermal chloracne, bioaccumulation, possible reproductive effects, probable carcinogen

[a]P refers to a primary pollutant directly emitted into the environment, while S refers to a secondary pollutant formed by transformations of primary pollutants. Some environmental contaminants are both primary and secondary pollutants, depending upon their sources.

methyl-tert-butyl ether (MTBE) undergoes relatively rapid photodegradation in air but is nonreactive in aqueous environments over time scales of several months. Ultimately, for pollutants that reside in more than one medium, the possibility of multiple human exposure pathways must be considered. For example, humans are exposed to elemental lead through inhalation of airborne particles and through ingestion of lead in food, water, and soil. Modeling the movement and transformation of chemicals in multiple media provides a holistic representation of human exposure to pollutants and associated health risks.

1.3 MASS BALANCES

The foundation of environmental fate and transport models is a mass balance, which accounts for the production, loss, and accumulation of the contaminant within a specified control volume. Transport phenomena and physical, chemical, and biological transformations are represented within the framework of this fundamental concept.

Mass balances underlie back of the envelope calculations as well as complex global circulation models. The control volume around which a mass balance is calculated can be a water droplet, your bloodstream, a lake, the global atmosphere, or a mathematical construct representing an infinitesimally small parcel of air, soil, or water. The mass-balance principle holds that within a well-defined control volume, as illustrated in Figure 1.2, the rate of accumulation of mass must balance the rates of production within the control volume, input from outside, loss across the boundaries, and loss by reaction. The principle is expressed mathematically as a time-dependent differential equation:

$$\frac{dM}{dt} = \frac{dC\,V}{dt} = \dot{M}_{\text{in}}(C,t) - \dot{M}_{\text{out}}(C,t) + S(C,t) \pm \text{Rxn}(C,t) \tag{1.1}$$

in which M is the mass of the contaminant within the control volume, V is its volume, $C(= M/V)$ is the concentration, t is time, $\dot{M}_{\text{in}}$ and $\dot{M}_{\text{out}}$ are the transport rates across the boundaries of the control volume from and to the surrounding environment, respectively, S is the source emission rate, and Rxn is the rate of internal reactions that may either produce or consume the contaminant. Concentration units are preferred to mass units in environmental modeling because the former is an intensive property and thus is easier to measure and more directly related to environmental impacts.

The transport and reaction rates in Eq. (1.1) generally depend on the concentration within (and for transport, around) the control volume at a particular time. Source mechanisms may include direct emissions into the medium of interest or interphase mass transfer from another medium. Direct source terms (emissions rates) are usually independent of C, while interphase mass-transport processes can be concentration dependent. For many problems, the reaction processes are assumed to be first-order losses, for example, chemical decay with rates directly proportional to the concentration of the pollutant. However, more complex chemical transformation processes can involve higher-order reactions among multiple pollutants. Interphase mass transfer

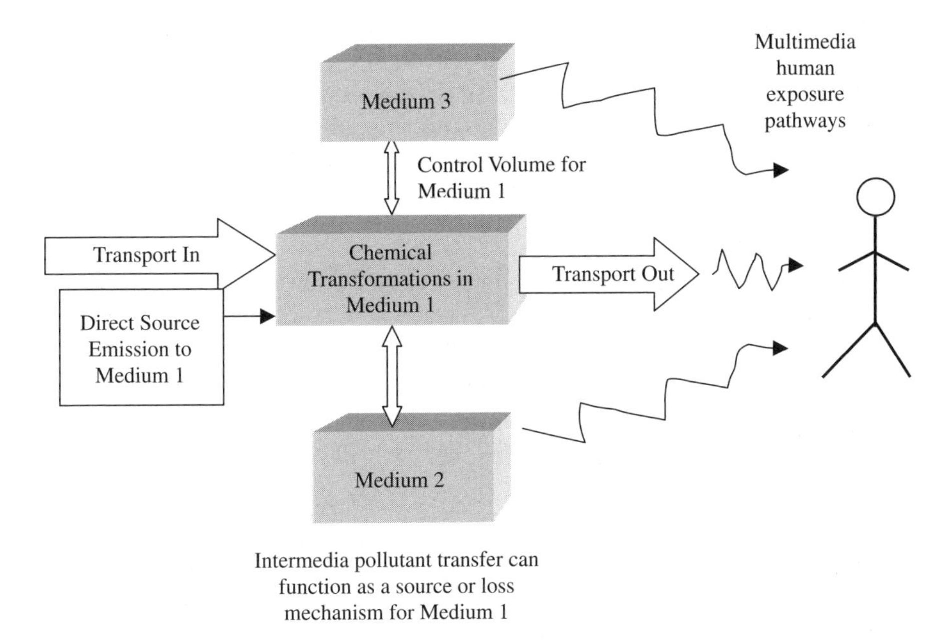

Figure 1.2 Emission, transport, intermedia mass transfer, and chemical transformations in an environmental medium, leading to multimedia routes of human exposure to a chemical pollutant.

out from the media of interest can also result in a loss of the pollutant from the control volume, and this is also often modeled as a first-order process. The various source, transport, mass-transfer and reaction mechanisms are schematically represented in Figure 1.2. Pollutant transport, accumulations, and transformations in different media eventually result in different pathways of human exposure, including inhalation, ingestion, and dermal contact.

1.4 COMPONENTS OF AN ENVIRONMENTAL MODEL

Comprehensive contaminant transport and fate models couple source mechanisms with transport, transformation, and removal processes in order to predict the spatial and temporal distribution of pollutants. This section introduces each of these components of an environmental model.

1.4.1 Source Mechanisms

The source of contaminant discharge to the environment is typically characterized by a loading or emission rate, given as the mass of the pollutant discharged to the environment per unit time. For many problems, such as those involving industrial discharges to the air or surface waters, direct emission rates are determined from monitoring data and given as an exogenous input to the environmental model. With more complicated source terms, the source characterization itself can involve a detailed and sophisticated modeling effort. Engineering process models may be utilized to estimate the emissions to various environmental media, considering industrial inputs and operations and incorporating the same mass-balance concepts that are used to predict contaminant fate and transport in the natural environment. Examples include:

- Models for electric power plant operations and pollution generation that evaluate fuel, process, and emissions control options, and their impact on plant efficiency, costs, and environmental discharges (e.g., Berkenpas et al., 1996)
- Models for wastewater treatment plant operations that determine the effectiveness of different unit operations and the distribution of influent contaminants to the effluent water, solid residuals, and fugitive air emissions (e.g., Namkung and Rittmann, 1987)

Other models consider the overall emissions from multiple sources. For example, mobile source models predict air pollutant emissions from the vehicular fleet of a metropolitan area, considering vehicle technology and use, fuel composition, and driving conditions (Sawyer et al., 2000).

A second type of source characterization involves chemical releases that do not occur directly into the medium of interest but instead involve complex interaction with the surrounding environment, including intermedia mass transfers. Such situations include:

- Non-point-source runoff into waterways from forest, agricultural, and urban areas, requiring tracking of contaminant accumulation and mobilization from the land surface and transport through natural and engineered conveyance (e.g., Nikolaidis et al., 1993; Cassell and Clausen, 1993).
- Dynamic source zone concentrations and emission rates in the medium of interest, derived from intermedia mass-transfer processes. For example, dissolution of chemicals to groundwater from multicomponent non-aqueous-phase liquids (NAPLs) in the subsurface (Peters et al., 1999).

Source characterization includes selection of the appropriate temporal and spatial resolution for the source term. Sources may be instantaneous (e.g., from a spill), intermittent (e.g., from wet deposition, rainfall–runoff events, or intermittent activities in the home or workplace), or continuous, exhibiting constant behavior, periodic variation, or a long-term trend. The degree of temporal resolution required for the source term will depend on the time scale of the problem and the dynamic response characteristics of the contaminant in the environment.

The spatial resolution of the source term also depends on the properties of the contaminant and the receiving environment. Emissions may be characterized as occurring from a point, line, area, or volumetric source (Fig. 1.3). Industrial stacks, discharge pipes, and underground storage tanks are often characterized as point sources.

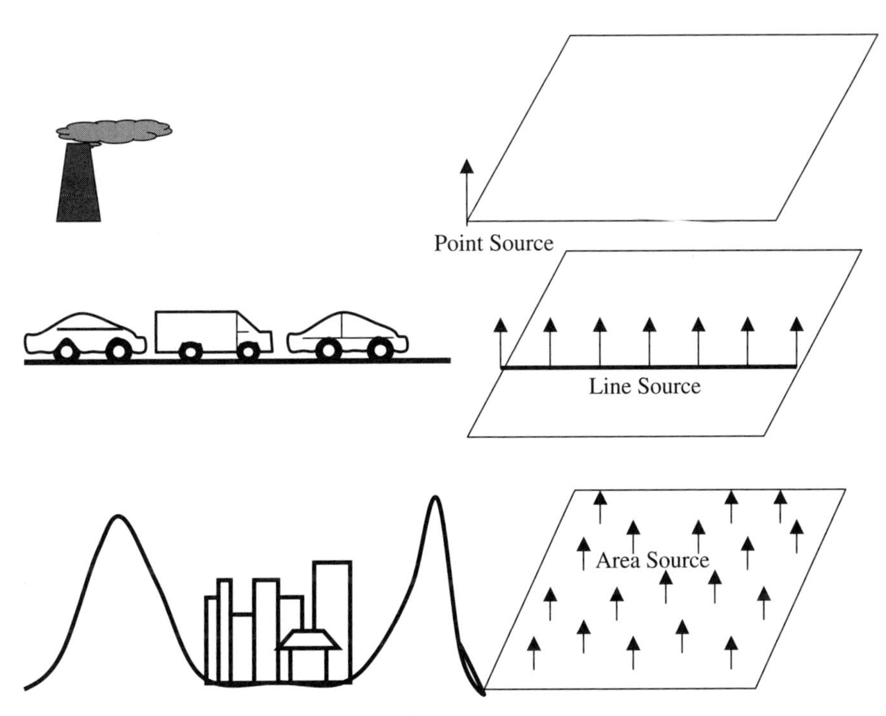

Figure 1.3 Source geometry is often idealized to simplify the spatial modeling of pollutant transport from source to receptor. The arrows represent idealized pollutant emissions. A single smoke stack might be represented as a point source with emissions in units of mass per time being of interest. A highway with cars emitting pollutants might be viewed as a line source with an emission rate per unit length of highway. In a regional-scale model, the consolidated emissions from an entire urban area might be viewed as an area source with an emissions rate per unit area.

Motor vehicle emissions from a highway, runoff to surface water, and dissolution from a broadly distributed spill are examples of distributed sources. If the area of environmental impact is large relative to the dimensions of the source, and the point of concern is not too close to the source, distributed or multiple point sources may be represented as occurring at a single effective location. The appropriate level of spatial aggregation thus depends on the spatial extent of impact and the need to differentiate concentrations and exposures at different locations.

1.4.2 Transport Mechanisms

Once a pollutant is released into an environmental medium, pollutant transport occurs through three mechanisms: advection, diffusion, and dispersion, which are common to surface water, groundwater, and air. Advection is the transport of contaminants along with the mean or bulk flow of the air or water. Diffusion is the mixing of contaminants that is driven by gradients in contaminant concentration and is usually viewed as occurring in two ways. First, molecular diffusion occurs due to the random motion of molecules within the fluid. Second, turbulent diffusion, which is mixing due to turbulent motions in the fluid, occurs as a scale-dependent phenomenon. If fluid motions are considered at a sufficiently fine scale, this process is recognized as advection. Shear dispersion is defined as mixing due to velocity gradients in the fluid. Both turbulent diffusion and shear dispersion are often represented in environmental models as analogous to molecular diffusion, that is, having rates that are proportional to the gradient of the contaminant concentration. As discussed in Chapter 5, the overall effects of turbulent diffusion and shear dispersion are generally combined into a net dispersion or diffusion coefficient.[1] This relationship is expressed by Fick's law (shown for a one-dimensional case):

$$J_x = -D_x \frac{\partial C}{\partial x} \tag{1.2}$$

where J_x is the mass flux of the contaminant per unit area in the x direction ($M\ L^{-2}\ T^{-1}$) and D_x is an overall dispersion coefficient ($L^2\ T^{-1}$) representing the strength of molecular, turbulent, and/or shear forces for dispersive transport.

The dimensionality of pollutant transport in environmental systems depends on whether complete mixing can be assumed across various dimensions of the system, as shown in Figure 1.4. Transport through natural systems is often represented using idealized reactor types from chemical engineering. The simplest of these reactors is the completely stirred tank reactor (CSTR), which is assumed to be well-mixed across its volume, that is, across x, y, and z dimensions. As a result, the CSTR represents a zero-dimensional transport system, with advection in and out but no concentration gradients within the tank. It thus represents a case of "infinite" dispersion and is often used as a model for well-mixed ponds, lakes, or rooms in a building. A well-mixed CSTR box model is also employed in multimedia models that focus on the overall distribution of pollutants in the environment, ignoring the spatial variations that occur within each medium.

The simplest model representing spatial variation in pollutant concentrations is the plug flow reactor (PFR), which represents the case of purely advective transport in one dimension, with no dispersion occurring in that dimension. A PFR is often used for simple, steady-state water quality models for streams and rivers and is occasionally used as an initial model for contaminant plumes in the atmosphere or groundwater. One-dimensional transport systems that include both advection and dispersion (because neither can be ignored) are often represented by a plug flow with dispersion reactor (PFDR) model. In both the PFR and PFDR models, the cross-sectional area perpendicular to flow is assumed to be completely mixed, so that concentration gradients in these ancillary dimensions are ignored. The primary direction of flow (advection) is, by convention, modeled to occur along the positive

[1] As is common in groundwater and surface water modeling, we use the term *dispersion* to represent the net, overall mixing resulting from these processes. Atmospheric modelers prefer the term *diffusion* to represent the overall mixing, reserving dispersion for the shear component only.

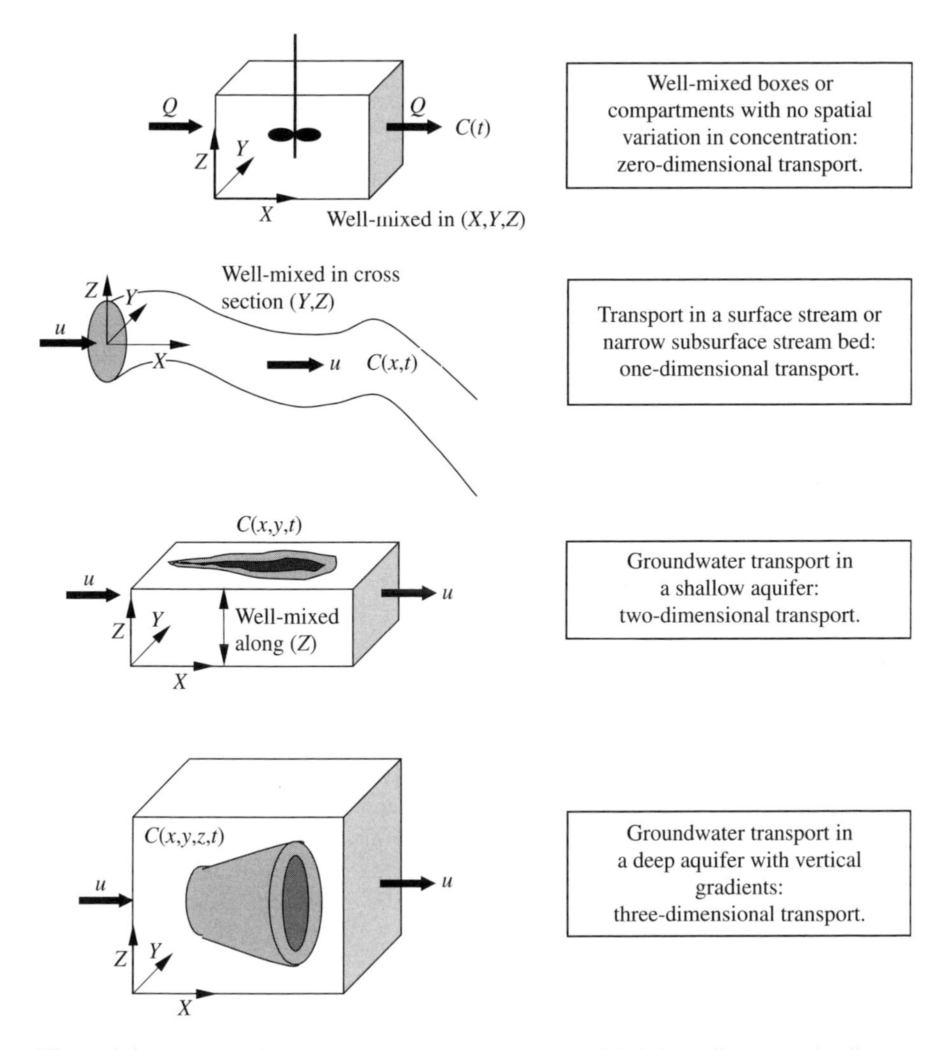

Figure 1.4 Transport from source to receptor can be simplified depending upon the dimensions over which uniformly mixed conditions can be assumed.

x axis. When incomplete mixing and concentration gradients occur in the ancillary dimensions, dispersion in multiple dimensions is incorporated into more realistic modeling of environmental systems, for example, the two-dimensional (2D) and three-dimensional (3D) aquifers shown in Figure 1.4.

The CSTR, PFR, and PFDR models, as well as idealized, homogeneous 2D or 3D systems, are frequently used because they give rise to mass-balance equations with analytical solutions. However, heterogeneous systems with multidimensional variations in loadings, transport, and reaction generally require more sophisticated numerical models, though these are often constructed by piecewise aggregation of multiple idealized reactor units for local areas or zones. More on the use of these reactor types and their aggregation is presented in Chapters 5 and 6 and in the individual chapters on air, groundwater, and surface water modeling. Examples of simple CSTR and PFR models are presented later in this chapter (Section 1.6).

In general, the mathematical representation of the effect of transport on a contaminant mass balance is given by the three-dimensional advection–dispersion equation:

$$\frac{\partial C(\mathbf{x}, t)}{\partial t} + \mathbf{u} \cdot \nabla C = \nabla \cdot D \nabla C \pm \mathrm{Rxn}(C) + S(\mathbf{x}, t) \tag{1.3}$$

In Eq. (1.3), $\mathbf{x}$ is a vector giving the location in Cartesian coordinates; $\mathbf{u}$ is the fluid velocity vector; ∇ is the gradient operator; D is the dispersivity tensor; and Rxn and S are reaction and source terms, respectively. The first term in the equation represents the rate of accumulation of the contaminant. The second term represents advective transport. The first term on the right side of the equation represents dispersive transport. Production or loss through transfers to or from other media are generally treated as boundary conditions for Eq. (1.3). They can also be treated as source terms, as in the case of wet deposition loadings of a chemical to the surface of a lake, or reaction processes, as in the case of oxygen transfer to a stream to reduce a dissolved oxygen deficit (see the example in Section 1.6.2). The advection–dispersion equation is equally applicable as a description of contaminant transport in air, water, or groundwater.

Solving the advection–dispersion equation is often difficult because the fluid flows affecting environmental transport are often unsteady and extremely complex. Examples include the case of turbulent atmospheric motions near Earth's surface, fresh and saline water circulation in tidal rivers, estuaries, and bays, and groundwater flow in complex heterogeneous subsurface systems. Various simplifying assumptions concerning spatial and temporal variations and mixing are used to make these cases tractable. Transport processes and the models used to describe them are discussed in more detail in Chapters 5, 8, 9, and 10.

1.4.3 Interphase Transfers

The transport processes mentioned above occur within a given medium. However, transport or mass transfer also occurs across media and pollutant phases (Fig. 1.2). Examples include dissolution of sulfur dioxide from the air into a rain drop, adsorption of a pesticide onto a soil particle, or volatilization of xylene into the air from a liquid gasoline spill. In environmental models that focus on a single medium, interphase transfers play the role of a source or sink term for pollutant accumulation within that medium. In contrast, multimedia models treat these processes as exchanges between two or more active compartments.

At equilibrium, the balance in concentrations between two media, or compartments i and j, is described by a partition coefficient:

$$K_{ij} = \frac{C_i}{C_j} \tag{1.4}$$

Partition coefficients that are commonly used in environmental modeling are those for air–water, soil–water, and octanol–water exchange. The octanol–water partition coefficient is significant because it is correlated with the tendency of compounds to adsorb on soil, as well as the tendency to partition to fatty tissues in plants or animals. The rate of interphase transfer of pollutants is typically represented as a first-order process driven by the concentration gradient across the compartmental boundaries of the system. Partition coefficients and interphase transfers are discussed in detail in Chapters 3 and 4.

1.4.4 Chemical Transformations

In addition to phase changes and partitioning, many contaminants participate in chemical or biological reactions in the environment. In some cases these reactions are viewed as benign or beneficial, removing contaminants from the environment by degrading them. In other cases, the products of the reactions may be more harmful than the original reactants. As noted in Table 1.1, pollutants formed as products of reactions that occur in the environment are referred to as secondary pollutants.

Both kinetic and equilibrium analyses of pollutant transformations are widely used in environmental modeling. The two are related by the definition of chemical equilibrium. For an elementary reversible reaction involving compounds, A, B, C, and D:

$$p\mathrm{A} + q\mathrm{B} \underset{k_r}{\overset{k_f}{\rightleftarrows}} r\mathrm{C} + s\mathrm{D}$$

the mass law rate equation is

$$\frac{d[\mathrm{A}]}{p\,dt} = \frac{d[\mathrm{B}]}{q\,dt} = -\frac{d[\mathrm{C}]}{r\,dt} = -\frac{d[\mathrm{D}]}{s\,dt} = -k_f[\mathrm{A}]^p[\mathrm{B}]^q + k_r[\mathrm{C}]^r[\mathrm{D}]^s \qquad (1.5)$$

where the brackets denote compound concentrations; the lowercase letters represent stoichiometric coefficients in the reaction; k_f is the rate constant for the forward reaction and k_r the rate constant for the reverse reaction. At equilibrium, the concentrations are constant, so:

$$K_{\mathrm{eq}} = \frac{k_f}{k_r} = \frac{[\mathrm{C}]^r[\mathrm{D}]^s}{[\mathrm{A}]^p[\mathrm{B}]^q} \qquad (1.6)$$

where K_{eq} is the equilibrium constant for the reaction. Equilibrium analyses are appropriate for reversible reactions when the time required for equilibration is short compared to the time scales for mass transfer, transport, and other reaction processes. For example, acid–base reactions in aquatic systems occur very quickly compared to mass-transfer processes in water droplets, so equilibrium treatment of chemistry in cloud droplets is usually appropriate. In contrast, explicit treatment of kinetics is required for slow or irreversible reactions. Irreversible reaction kinetics are modeled according to the law of mass action, which states that the rate of a reaction is proportional to the molar concentrations of the reactants. When only the forward reaction rate in Eq. (1.5) is considered, we obtain

$$\frac{d[\mathrm{A}]}{p\,dt} = \frac{d[\mathrm{B}]}{q\,dt} = -\frac{d[\mathrm{C}]}{r\,dt} = -\frac{d[\mathrm{D}]}{s\,dt} = -k_f[\mathrm{A}]^p[\mathrm{B}]^q \tag{1.7}$$

1.4.5 Significance of Transport and Transformation Processes

The distance over which a pollutant can migrate from its source and produce significant effects determines its spatial scale (Scheringer, 1997; Van Pul et al., 1998; Beyer, et al., 2000; Hertwich and McKone, 2001). Large-scale circulation in the atmosphere or oceans disperses some contaminants on a global scale ($\sim$10,000 km). Regional-scale transport ($\sim$100 to 1000 km) occurs in coastal zones of the oceans, large lakes, and river systems as well as the atmosphere. Transport on the local scale ($<$100 km) and site or microscale ($<$100 m) occurs in all media. Temporal scales of impact depend on the residence time of the pollutant in the system, as well as the rate of pollutant transport. In turn, the residence time depends on removal processes such as chemical or biological decay, dry or wet deposition from the atmosphere, or volatilization from surface water. For example, fine particle sulfate remains in the lower troposphere for up to 1 to 2 weeks before removal by dry or wet deposition and can thus be transported from source regions in the midwestern United States to the maritime provinces in Canada (and similarly from western Europe to Scandinavia, and from China to Japan). In contrast, certain nonreactive polycyclic aromatic hydrocarbons (PAH) and polychlorinated biphenyls (PCBs) can persist in groundwater for several decades, yet migrate only a few hundred meters or less due to slow groundwater flow rates and adsorption onto the aquifer soil matrix.

The relative rates at which various transport, interphase transfer, and chemical or biological transformation processes occur dictates whether and how these processes are represented in fate and transport models. For example, chemical transformations that occur at time scales that are much longer than the time required to transport a pollutant from its source to a critical receptor can reasonably be neglected. Thus, the oxidation of CO to CO_2 in the atmosphere is appropriately left out of air quality models used to predict CO concentrations in urban areas that are due to local sources but must be considered in long-term global CO_2 inventory and climate models. In contrast, as noted above, very fast reactions can be treated using an equilibrium model, rather than a kinetic representation. Since acid–base protonation/deprotonation reactions in surface waters occur much more quickly than the other processes that affect surface water pH (such as the dissolution/precipitation of carbonate minerals and the exchange of CO_2 with the atmosphere), acid–base speciation is usually treated as an instantaneous, equilibrium-controlled process in models for stream or lake acidity. Chemical transformations are introduced in Chapter 2 and subsequently incorporated with transport processes in Chapters 11 and 12.

1.5 MODELING APPROACHES

This book focuses on process models of environmental contaminant transport and transformation that are based on mass-balance equations incorporating mathematical descriptions of the relevant physical, chemical, or biological processes. However, there are two other categories of environmental models that should be recognized. First, significant advances in understanding environmental systems have been made using physical-scale models to try to reproduce environmental processes in the laboratory. Examples of physical modeling tools with environmental applications include wind tunnels that have been used to investigate pollutant dispersion and deposition, scale-model waterways for flow and circulation studies, and bench-scale soil columns that have been used to study the transport of pollutants in soils and groundwater. The second category is empirical models that estimate statistical relationships between observed pollutant concentrations and a variety of potential explanatory factors. Statistical analysis techniques that have been fruitfully applied to pollution problems include kriging, for interpolation of spatially varying data (Venkatram, 1988; Conradsen, et al., 1992; Casada, et al., 1994), time-series analysis for real-time pollution forecasts (Robeson and Steyn, 1990; Box et al., 1994), and factor analysis for identifying sources that contribute to existing pollution levels (Anderson et al., 2001; Hopke, 1985; i Salau et al., 1997; Polissar et al., 2001). While extremely valuable for identifying existing or historical relationships, the predictive capability of empirical models is limited to the conditions under which they were developed. Without a theoretical model to extrapolate beyond those conditions, there is no general basis for expecting the observed relationships to hold when conditions change. Still, these methods provide a useful basis for characterizing data, and thus support the development and validation of models. Further discussion of these and other statistical methods is provided in Chapter 7.

1.5.1 Characteristics of Fate and Transport Models

Contaminant transport and transformation models can be classified according to a number of contrasting characteristics, including:

- *Single-Media versus Multimedia Models* Single-media models for air, surface water, groundwater, and soil pollution have been traditionally utilized by different disciplines to address media-specific problems. These models generally provide a more detailed description of the space–time distribution of a pollutant within a single medium and are presented in Chapters 5 and 8 to 12. While these models often incorporate mass transfer from (or to) other media as sources (or sinks) or boundary conditions, they lack the ability to characterize the total environmental impact of a contaminant release. Multimedia fate, transport, and exposure models have been developed in recent years. These either link a number of single-media models for multiple exposure pathways within a single computing environment, or model the simultaneous partitioning of chemicals among multiple compartments representing each medium in a simplified "unit world." This latter approach is introduced in Chapters 3 and 4, while comprehensive multimedia environmental fate, transport, exposure, and risk models are reviewed in Chapter 13.

- *Dynamic or Time Variable versus Steady State* Dynamic models predict changes in ambient concentrations over time and allow consideration of temporal variations in source release and other model inputs. Steady-state models apply to conditions of assumed constant input and where the reaction, transport, and mass-transfer terms have achieved a balance. In this case constant (time-invariant) concentrations are determined.
- *Distributed versus Lumped Models* Distributed models represent the spatial variation of environmental processes in an explicit manner, with spatial variations in transport and reaction terms represented, to the extent possible, as they actually occur and are measured across the environmental landscape. Lumped models employ a greater degree of spatial aggregation, using average or effective parameters that may not be directly observable in the environment but are meaningful in the context of the model predictions.
- *Lagrangian versus Eulerian Frameworks* For distributed models, there are two distinct reference frames from which to view pollutant transport. The first reference frame is Eulerian, in which the coordinate system is fixed with respect to Earth's surface. In that case, a succession of fluid parcels is viewed as moving past a stationary observer. The second reference frame is Lagrangian, moving with the fluid as it flows and carrying the observer along with it. Lagrangian models have been developed for long-range air pollutant transport and for models of stream and groundwater quality.
- *Analytical versus Numerical Models* Another useful demarcation of mathematical models is whether they have analytical (closed-form) solutions or require approximate, computer-based solutions employing numerical discretization. Models that have analytical solutions generally require significant simplifying assumptions (e.g., steady-state conditions and spatial homogeneity or complete mixing). The assumptions may be well justified from direct observations or simply adequate for the intended purpose of the model. Numerical solution techniques are generally required for models that account for significant spatial heterogeneity, temporal variability, multicomponent chemistry, or nonlinear transformation processes. These techniques solve differential equations describing the pollutant mass balance over discrete cells in space and/or discrete steps in time. Numerical techniques for solving environmental models are discussed in Chapter 6.
- *Deterministic versus Stochastic Models* Deterministic models calculate a single concentration for each location and time in the model. Stochastic models produce a range or distribution of values for each prediction. This distribution may represent variability—reflecting the temporal or spatial variation that occurs in the environment—or uncertainty—reflecting the imperfect knowledge

and prediction achievable on the part of the model and modeler. Methods for implementing stochastic models are introduced in Chapter 7.

1.5.2 Model Evaluation

As discussed by Oreskes et al. (1994), environmental models can never be "verified," meaning established as true, because they are necessarily imperfect representations of the real world and, thus, strictly speaking, untrue. Nevertheless, a critical element of model development is evaluation, through benchmarking, comparison with observations, and sensitivity and uncertainty analysis. Benchmarking entails establishing that the solution for a model matches the known solutions for idealized cases in which model inputs and parameter values are given. Performance evaluation involves careful comparison of model results with field observations for historical cases in which source terms and environmental conditions can be specified with an accuracy that is commensurate with that expected from the model. For many models, performance evaluation is preceded by a calibration step. In this stage, the model is calibrated against observations to estimate parameters that cannot be measured directly. This may be done in an informal, trial-and-error manner or more formally using parameter estimation methods that optimize the match between the model predictions and observed data. Using the estimated parameter values, the model is subsequently evaluated with an independent data set. Finally, because model inputs and parameters can never be known exactly, analysis of the effect of uncertainty in these variables is essential if model predictions are to be used for decision making. Requirements and techniques for model evaluation are presented in Chapter 14.

1.6 EXAMPLES OF SIMPLE ENVIRONMENTAL MODELS

To illustrate the formulation and application of environmental process models, two simple examples are presented that describe pollutant transport and transformations in individual media: air and water. The first uses a CSTR representation for indoor air quality in a ventilated room and applies the model to estimate carbon monoxide concentrations resulting from cigarette smoking. The second uses a PFR representation for a stream and computes the classic biochemical oxygen demand (BOD)—dissolved oxygen (DO) concentration profiles resulting from an organic wastewater discharge.

1.6.1 Completely Mixed Systems

Figure 1.5 shows a CSTR control volume used to compute the mass balance of carbon monoxide concentrations in a room with several burning cigarettes. The control volume is defined as the room, which has a specified volume V. The carbon monoxide concentration outside the room, C_{amb}, is assumed to be constant. Inside the room, the cigarettes produce CO at a fixed production rate, S. Carbon monoxide is assumed to

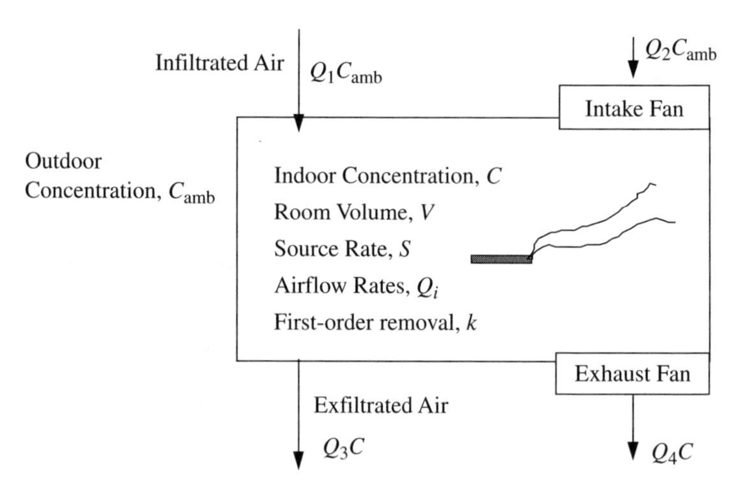

Figure 1.5 Control volume for a completely mixed model of indoor air.

undergo removal due to deposition or chemical decay at a rate that is proportional to its concentration, with rate constant k. Air enters the room through infiltration at volumetric flow rate Q_1 and through an intake fan at flow rate Q_2. Air exits the room through exfiltration at volumetric flow rate Q_3 and through an exhaust fan at flow rate Q_4. Assuming that CO is uniformly distributed throughout the room, the time dependence of the indoor concentration, $C(t)$, can be determined from a mass-balance equation that accounts for the production of CO from the cigarette, the rate of input with the incoming air, and the rate of loss due to deposition/reaction and outflow with the exfiltration and exhaust fan air.

For this particular case, with $Q_{\text{in}} = Q_1 + Q_2$ and $Q_{\text{out}} = Q_3 + Q_4$, the mass-balance equation is

$$V\frac{dC}{dt} = (Q_{\text{in}})\,C_{\text{amb}} - (Q_{\text{out}})\,C + S - kVC \tag{1.8}$$

The steady-state solution for this system, when $dC/dt = 0$, is

$$C_{\text{ss}} = \frac{(Q_{\text{in}})C_{\text{amb}} + S}{(Q_{\text{out}}) + kV} \tag{1.9}$$

For example, consider a poorly ventilated room with $V = 80$ m^3, and $Q_{\text{in}} = Q_{\text{out}} = 40$ m^3/h, in which four card players have been smoking for several hours, each at a rate of four cigarettes per hour. Each cigarette produces 125 mg of CO. The CO decays with a rate constant $k = 0.1$ h^{-1}. The ambient concentration of CO is negligible. In this case, the steady-state concentration of CO in the room is

$$C_{\text{ss}} = \frac{(4 \times 4 \times 125)\text{mg/h}}{40\ \text{m}^3/\text{h} + (0.1/\text{h} \times 80\ \text{m}^3)} = 41.7\ \text{mg/m}^3$$

Now consider how quickly CO concentrations buildup. If the card party and smoking start at 10 p.m., what is the CO concentration at 10:30 p.m.? The time-dependent solution to Eq. (1.8), with a step function loading S beginning at time $t = 0$ and initial concentration $C(t = 0) = 0$, is

$$C(t) = C_{ss}\left\{1 - \exp\left[-\left(\frac{Q_{out}}{V} + k\right)t\right]\right\} \tag{1.10}$$

The solution for this case, with $t = 0$ corresponding to 10 p.m., is plotted in Figure 1.6. The solution shows that the concentration at 10:30 p.m. is 10.8 mg/m^3.

Note that the overall removal of CO from the room is controlled by the parameter $\alpha = k + (Q_{out}/V)$ (T^{-1}). In indoor air pollution, the ratio Q/V is referred to as the air exchange rate and is usually expressed in terms of air changes per hour (ACH). In other applications, attention is focused on the inverse ratio, V/Q, which is referred to as the retention time for the contaminant-carrying fluid moving through the control volume. Correspondingly, $1/k$ is known as the lifetime with respect to decay within the control volume. For a case with a nonzero initial concentration, $C(t = 0) = C_0$, and no source term, the solution to Eq. (1.8) is the familiar exponential decay equation:

$$C(t) = C_0 \exp(-\alpha t) \tag{1.11}$$

The residence time of the contaminant is the time τ required for the concentration to fall to C_0/e. By inspection of Eq. (1.11), $\tau = 1/\alpha$.

As discussed above, Eq. (1.8) is described as a zero-dimensional or CSTR model because no spatial variations are considered. Use of the complete mixing assumption leads to little error in calculating contaminant concentrations if the characteristic time for mixing in the compartment being modeled is short compared to the time scales for decay and for transport out of the system.

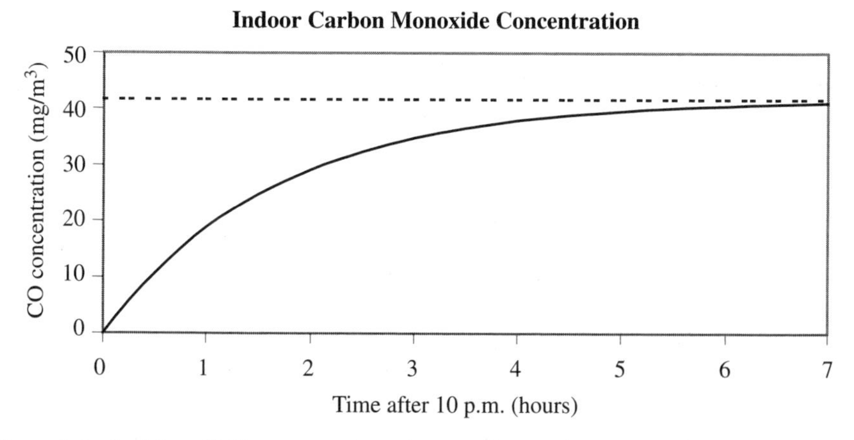

Figure 1.6 Buildup of CO concentration after smoking commences at 10 p.m. The steady-state concentration is shown as the dotted line.

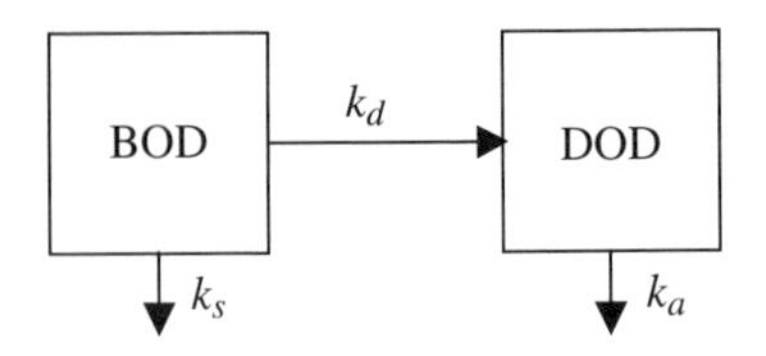

Figure 1.7 Consecutive reaction sequence for removal of waste biochemical oxygen demand (BOD) by deoxygenation with rate constant k_d to form dissolved oxygen deficit (DOD), and by settling with rate constant k_s. The DOD is removed by reaeration, with rate constant k_a.

1.6.2 Plug Flow Systems

One of the earliest models of surface water quality was published by Streeter and Phelps (1925) and describes the effect of biodegradable organic wastes such as sewage or food processing wastes on dissolved oxygen (DO) levels in a stream or river. The oxygen required to decompose the wastes under aerobic conditions is called biochemical oxygen demand (BOD). The Streeter–Phelps model treats the competing effects of oxygen consumption by the decomposing wastes and reaeration by transfer of oxygen from the air to the water. It is formulated as two mass balances. The first is for the decomposing waste, which can be removed from the stream by settling as well as decomposition. The second mass balance is for the dissolved oxygen deficit (DOD), which is the difference between the saturation concentration of oxygen in water at equilibrium with the air, and the actual oxygen concentration at a given point in the stream. The reaction sequence is shown in Figure 1.7.

Consider an industry that continuously discharges a biodegradable waste with discharge rate $S = 0.2$ kg BOD/s into a stream with water flow rate $Q = 20$ m^3/s, such that the in-stream concentration at the discharge point is $C_0 = S/Q = 10$ mg/L. The stream is assumed to be well mixed in its cross section but not along the flow direction (x). In this case, instead of considering the entire length of the stream to be completely mixed, it is modeled as a plug flow reactor (PFR). In a PFR, parcels of water flow sequentially downstream with no mixing between them. Figure 1.8 illustrates the control volume for a mass-balance model written for the amount of waste, $C(x)$, remaining at a distance x below the discharge point. The control volume for the mass balance in this case has a differential length, Δx, and the same cross section as the channel.

If u represents the stream velocity ($= Q/A$), the flux of waste [mass-transport rate per unit area ($M\ T^{-1}\ L^{-2}$)] carried into the element from upstream is $uC(x)$,

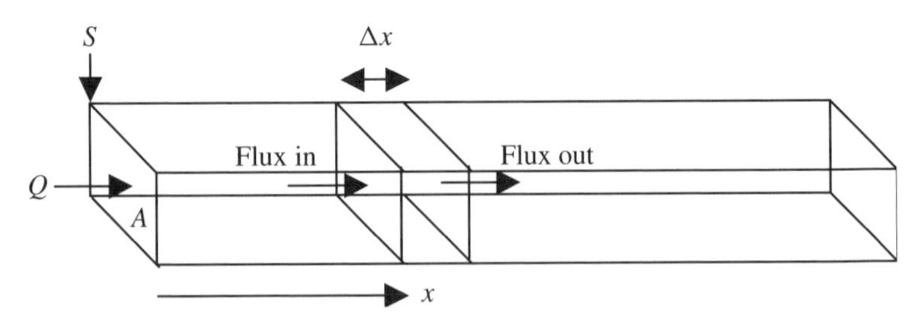

Figure 1.8 Control volume for a plug flow reactor. A mass balance is performed for the cell with infinitesimal length, Δx and channel cross section, A.

and the flux carried out of the element at the downstream boundary is $u\{C(x) + [dC(x)/dx]\Delta x\}$. Assuming that the waste undergoes first-order decay, with a total rate constant k, the mass balance for the waste in the control volume is

$$\frac{\partial C}{\partial t} = -u\frac{\partial C}{\partial x} - kC \tag{1.12}$$

where k is the sum of the rate constants for settling, k_s, and biochemical decomposition, k_d, shown in Figure 1.7. At steady state, Eq. (1.12) becomes

$$\frac{dC}{dx} = -\frac{k}{u}C \tag{1.13}$$

which can be solved to give

$$C(x) = C_0 \exp\left(\frac{-kx}{u}\right) \tag{1.14}$$

The ratio x/u is the *travel time* required for the wastes to be carried from the discharge point to position x. The concentration of waste in the stream thus decays exponentially with distance downstream of the discharge point, or equivalently, with travel time. Figure 1.9 shows the result for the example, with $u = 0.1$ m/s, $k_d = 0.6$ day^{-1} and $k_s = 0.2$ day^{-1} so that $k = 0.8$ day^{-1}.

As waste is decomposed at a rate $k_d C(x)$, the oxygen dissolved in the stream is consumed in the reaction to produce a DOD that is defined as the difference between the actual dissolved oxygen concentration (DO) and the saturation concentration (DO_{sat}):

$$\mathrm{DOD} = \mathrm{DO_{sat}} - \mathrm{DO} \tag{1.15}$$

However, the effect of deoxygenation from waste decomposition is offset by reaeration, the transfer of oxygen from the atmosphere. The rate at which this process

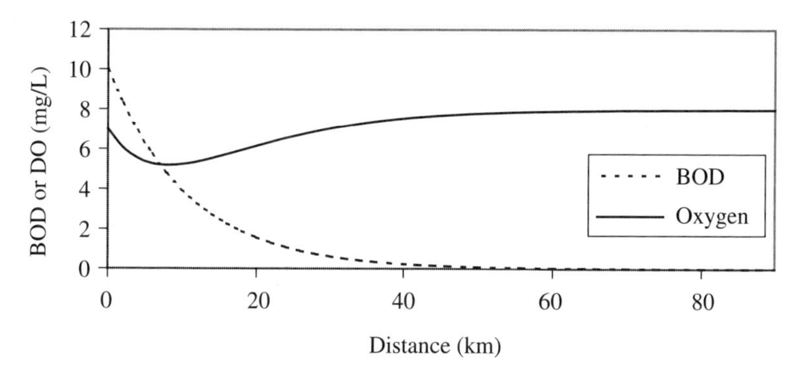

Figure 1.9 BOD and DO concentrations downstream from an industrial discharge point.

occurs is proportional to the oxygen deficit, DOD. A mass balance for the deficit of oxygen in the stream control volume can thus be written as:

$$\frac{\partial \text{DOD}}{\partial t} = -u\frac{\partial \text{DOD}}{\partial x} + k_d C(x,t) - k_a \text{DOD} \tag{1.16}$$

in which k_a is the reaeration rate constant. For steady-state conditions, with the result from Eq. (1.14) substituted for $C(x,t)$, Eq. (1.16) becomes

$$\frac{d\text{DOD}}{dx} = \frac{k_d}{u}C_0 \exp\left(\frac{-kx}{u}\right) - \frac{k_a}{u}\text{DOD} \tag{1.17}$$

With an initial deficit, DOD($x = 0$) $=$ DOD_0, the solution to this equation is the Streeter–Phelps oxygen sag curve:

$$\text{DOD}(x) = \frac{k_d C_0}{k_a - k}\left[\exp\left(\frac{-kx}{u}\right) - \exp\left(\frac{-k_a x}{u}\right)\right] + \text{DOD}_0 \exp\left(\frac{-k_a x}{u}\right) \tag{1.18}$$

Figure 1.9 shows the dissolved oxygen concentration (DO_{sat} − DOD) as a function of distance in a stream with the waste discharge and stream parameters given above and with $\text{DO}_{\text{sat}} = 8$ mg/L, $\text{DOD}_0 = 1$ mg/L, and $k_a = 1.0$ day^{-1}.

The Streeter–Phelps sag curve is formulated as a plug flow model, with no mixing in the longitudinal (x) direction and with instantaneous mixing in the vertical and lateral directions. The model is one dimensional because spatial variations are captured only in the downstream direction.

1.6.3 Challenges in Modeling Multimedia Transport and Fate of Contaminants

The indoor air pollution and biochemical oxygen demand examples presented above provide relatively simple illustrations of the mass-balance principle. However, most environmental contamination problems are not so straightforward, due to one or more of the following characteristics:

- Spatial or temporal variability in environmental characteristics or processes
- Unknown or poorly characterized properties or behavior of the medium containing the contaminant
- Unknown, poorly characterized or variable contaminant sources
- Physical or chemical interactions between contaminants or between contaminants and natural components of the environment
- Multimedia exchanges between air, water, soil, and groundwater
- Feedback effects through which the presence of the contaminant changes the behavior of the environment

Most of this book is devoted to dealing with these characteristics, as is most of the current effort that goes into research, development, and application of environmental models.

1.7 ROLE OF CONTAMINANT TRANSPORT MODELS IN ENVIRONMENTAL MANAGEMENT

As discussed in the introduction, mathematical models are used for scientific research and environmental management. In management applications, the objective of fate and transport modeling is usually to assess the impact of proposed source modifications on contaminant concentrations in the environment. Fate and transport model results are then linked to information about the potential impacts of the predicted contaminant levels. This occurs explicitly in the case of an integrated environmental health risk assessment (EHRA) or an ecological risk assessment, in which exposures, dose, risk, and health endpoints are predicted. This link is implicit in the case of modeling performed to test compliance with regulatory standards or target levels for ambient concentrations or environmental fluxes chosen based on health or ecosystem risk criteria.[2]

1.7.1 Design of Control, Treatment, and Remediation Systems

An important engineering application of environmental models is the design or analysis of engineered systems for pollutant control, treatment, or remediation. The optimal design problem is commonly expressed either as:

1. Minimize the cost of controls or remediation required to meet an environmental standard or
2. Minimize pollutant concentrations in the environment subject to a budget constraint on total pollution control or remediation expenditures

Either way, environmental models are needed to relate the engineering activities of control, treatment, or remediation to environmental concentrations. With a more complete model of environmental risk, the constraints or design objectives could also be expressed in terms of predicted exposure, risk, or population impacts.

Because there are no practical means for cleaning up air pollutants once they have been released into the atmosphere, design activities for air pollution focus on source control. Identification of cost-effective control strategies is an especially intriguing design problem for secondary pollutants that are formed in the atmosphere through a complex set of reactants and reaction pathways (Trijonis, 1974; Wang and Milford, 2001). For surface water and especially groundwater, both source control and in situ treatment may be options for reducing environmental contamination. In groundwater applications, mathematical optimization with contaminant transport models has been used to determine the optimal placement and operating rates for pumping and

[2] Ambient water quality standards are often based on drinking water standards known as maximum contaminant levels (MCLs), or ecological criteria. Health-based ambient air quality standards are established for criteria air pollutants as part of the National Ambient Air Quality Standards (NAAQS). Further discussion of these standards is presented later in this chapter and in Chapter 2. For an example of standards based on environmental fluxes, see discussion of "critical loads" developed in Europe to protect against ecological damage from acid deposition (Skeffington, 1999).

injection wells and related site remediation activities (e.g., Gorelick et al., 1984; Culver and Shoemaker, 1992; McKinney and Lin, 1995).

1.7.2 Regulatory Frameworks

Environmental models play a significant role in the regulatory frameworks set up by several U.S. laws and international agreements designed to prevent, control, or clean up pollution. For example, modeling is often an element of the environmental impact statements required for federal projects by the U.S. National Environmental Policy Act of 1970. Models are used for total maximum daily load (TMDL) calculation and waste load allocations under the Clean Water Act and amendments of 1972 and 1987, for determining the status of hazardous wastes listed under the Resource Conservation and Recovery Act (RCRA) and subsequent amendments, for risk assessment purposes under the Comprehensive Environmental Response, Compensation and Liability Act (CERCLA), for target load determination for atmospheric deposition in Europe by the European Union (EU) and the United Nations Economic Commission for Europe (UNECE), and to identify persistent organic pollutants (POPs) for international treaties aimed to control the emissions and global transport of these compounds (e.g., Rodan et al., 1999).

1.7.3 Risk Assessment

Our ultimate concern about pollutants derives from their impact upon human health and the environment. For such impacts to occur, human or ecological receptors must be exposed to pollutants at doses sufficient to result in a health or environmental effect. The integrated study of contaminant release, transport, fate, exposure, dose, and response is the basis for the environmental health risk assessment paradigm, illustrated in Figure 1.10. The relationship between pollutant emissions and ambient concentrations, which is the focus of most of this book, is only the first step in a health risk assessment. As discussed in Chapter 13, ambient concentrations are related to exposure through the use of an exposure model or an exposure assessment. In the case of human health, principal routes of exposure include inhalation of ambient air (both outdoor and indoor), ingestion (e.g., of drinking water, contaminated food, or contaminated soil), and dermal exposure to the air, soil, and domestic or recreational

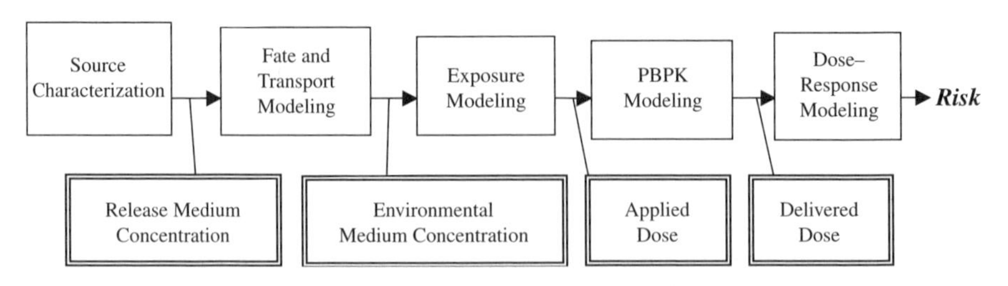

Figure 1.10 Integrated environmental health risk assessment model.

waters. These are typically computed and summed along multiple exposure pathways for a pollutant of concern.

The exposure for each pathway is determined by the product of the pollutant concentration in the contact medium and an exposure factor, representing, for example, the breathing rate, water ingestion rate, soil ingestion rate, or dermal contact rate. Exposure concentrations and exposure factors for an individual are dependent upon their time–activity patterns, that is, where the individual spends his or her time and in what type of activity. Pollutant exposures may be computed over short periods, such as a day for acute pollutant impacts, or over a lifetime for chronic impacts such as cancer. The exposure received by an individual may be thought of as a potential dose; the actual effective dose received by an individual depends upon how much of the pollutant is retained in the body and how much of it reaches target organs or cells. This calculation, which involves the transformation from exposure (or potential dose) to effective dose is sometimes made invoking simple statistical or conservative assumptions but can also be made using sophisticated physiologically based pharmacokinetic (PBPK) models. PBPK models predict the movement and accumulation of contaminants in humans or animals, through their respiratory, digestive, and circulatory systems to internal compartments, target organs, and cells, in much the same way that a multimedia environmental model predicts pollutant transport through the ambient environment. An example of a PBPK model is presented in Chapter 13.

The final stage of the health risk assessment is to relate dose to health impact through the dose–response function. Functional relationships can range from a simple linear, no-threshold relationship defined solely by a "slope factor," to more sophisticated relationships that account for multiple stresses over time to target cells in individuals with varying susceptibility to the health endpoint. These relationships can be estimated from past human exposures to spills or high occupational doses, animal experiments, or epidemiological studies. However they are estimated, they typically involve a high degree of uncertainty. The characterization of variability (e.g., from individual to individual within a target population) and uncertainty (in individual or population risks) are thus key steps in the environmental health risk assessment.

1.8 DEVELOPING CONTAMINANT TRANSPORT AND TRANSFORMATION MODELS

The chapters that follow present an integrated discussion of mathematical models of the fate and transport of chemical pollutants in air, water, groundwater, soil, and biota. The text emphasizes common modeling techniques and parallels among the processes occurring in the three fluid compartments, as well as transfers that occur between the media. This approach is motivated by the fact that health and ecosystem risks arise from exposures through multiple pathways encompassing the different environmental media.

To preview the understanding you should develop from the remainder of this text, key issues that guide the formulation of contaminant transport and transformation models are identified in Figure 1.11.

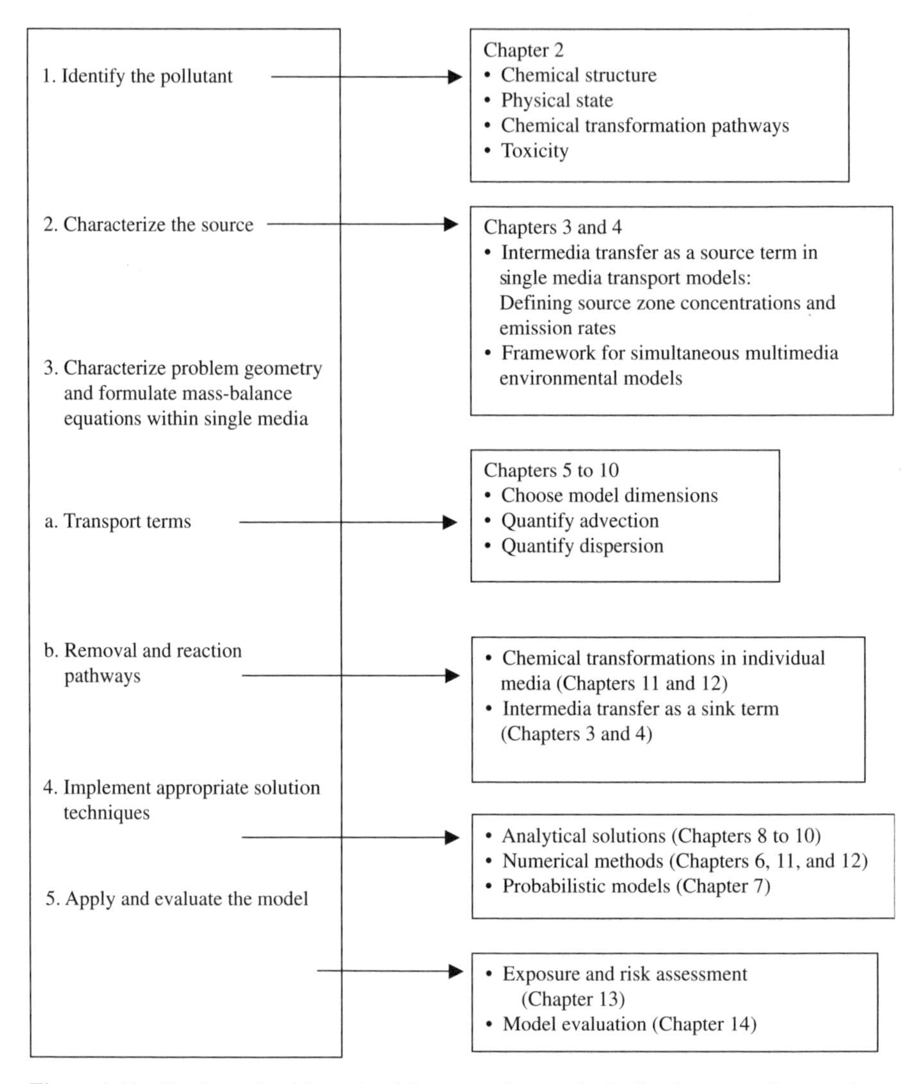

Figure 1.11 Roadmap for this textbook in terms of stages in the development of contaminant transport, transformation and risk assessment models.

APPENDIX 1A CONCENTRATION UNITS

Environmental models usually describe pollutant quantities in terms of concentrations in air, water, or soil. Conventional concentration units differ across environmental media and pollutant phase. The mass of contaminant per volume is used for pollutants in both air and water (e.g., mg/L, $\mu g/m^3$). The mass of contaminant per mass is used for water and soil [e.g., mg/kg, parts per million (ppm)]. In general, the mass in the denominator represents the total mass of the carrier medium and the contaminant (e.g., water plus contaminant). However, contaminant concentrations in environmental media are typically small, so the contaminant mass can often be neglected in the denominator term. In air, the concentration of a gaseous pollutant is often expressed as a mixing ratio, denoting the number of moles per mole of air or equivalently the volume of contaminant per volume of air [e.g., ppm(v) = volume of contaminant per 10^6 volumes of air]. The ideal gas law gives the temperature and pressure dependence of the conversion from mixing ratio units to mass per volume units:

$$P_{\text{atm}} V = nRT_{\text{env}} \tag{A.1}$$

in which P_{atm} represents the absolute ambient pressure, V the volume, n the number of moles of the gas, R the ideal gas constant, and T_{env} the absolute environmental temperature. Care should be taken to use the value of R that corresponds to the selected units of P, V, and T. For ideal gases in air, the mole and volume fractions and partial pressure ratios for contaminant A are equivalent:

$$Y^{\text{A}} = \frac{n^{\text{A}}}{n} = \frac{V^{\text{A}}}{V} = \frac{P^{\text{A}}}{P_{\text{atm}}} \tag{A.2}$$

The mass concentration for contaminant A is thus related to its mole fraction, Y^{A} by:

$$\frac{M^{\text{A}}}{V} = \frac{P_{\text{atm}} Y^{\text{A}} \text{MW}^{\text{A}}}{RT_{\text{env}}} \tag{A.3}$$

in which MW^{A} is the molecular weight of the contaminant. For atmospheric pressure and $T = 298$ K, Eq. (A.3) gives

$$\frac{M^{\text{A}}}{V}(\mu\text{g/m}^{-3}) = 40.9C^{\text{A}}(\text{ppm})\text{MW}^{\text{A}} \tag{A.4}$$

Due to the difference in common usage of mass fraction units in soil and water but volume or mole fraction units in air, particular attention to units is warranted in applications involving intermedia transfers.

Homework problems for this chapter are provided at www.wiley.com/college/ramaswami.

2 Nature of Environmental Pollutants

Modern industrial societies pursue a philosophy embodied by the advertising slogan introduced by DuPont in the 1930s: "Better Living Through Chemistry"![1] More than 1000 new chemicals are synthesized each year, and about 60,000 chemicals are in daily use, in products ranging from drugs, food preservatives, and additives to pesticides, household cleaners, soaps, detergents, and cosmetics (Heinsohn and Kabel, 1999). Complete toxicity and hazard information is available for only a small percentage of the synthetic chemicals produced by industry and consumed by society. The full impact of chemical releases on human health and the environment remains unknown since assessment of chemical toxicity and reactivity in the environment requires observational data collected over long periods of time. Thus, it was that the use of DDT (dichloro-diphenyl-trichloroethane) and other pesticides was widespread for several years before harmful effects on the environment were discovered (Carson, 1962). Many naturally occurring substances can also pose a health hazard to humans when ingested in large doses. For example, the leaching of naturally occurring arsenic into groundwater has resulted in arsenic poisoning of tens of millions of people in parts of India and Bangladesh, with similar problems now found in Vietnam and China (Mandal et al., 1996; Christen, 2001).

Chemical pollutants vary widely in their physical state, chemical structure, and their effects on human health and ecosystems. Before trying to solve an environmental problem resulting from a chemical release, it is useful to first fully understand the many different facets and properties of the chemical itself. The molecular structure of a chemical defines its identity. The size and structure of a chemical determine fundamental properties such as its natural physical state, aqueous solubility, vapor pressure, and toxicity/reactivity. The toxicity and reactivity of a chemical determine its potential risk to human health and safety. This chapter begins with a discussion of the chemical structure of various environmental pollutants. Subsequent sections focus on linking chemical structure with important properties shown in Figure 2.1. The final sections of this chapter briefly cover laws and regulations that have been developed to address the harmful effects of environmental pollutants.

[1] See, for example, http://heritage.dupont.com/touchpoints/tp_1939/overview.shtml; William L. Bird's "Better Living:" Advertising, Media, and the New Vocabulary of Business Leadership, 1935–1955 (Evanston, 1999); http://pubs.acs.org/cen/125th/pdf/7913baker.txt.pdf; http://www.nrdc.org/air/transportation/hleadgas.asp; and http://www.monitor.net/rachel/r152.html.

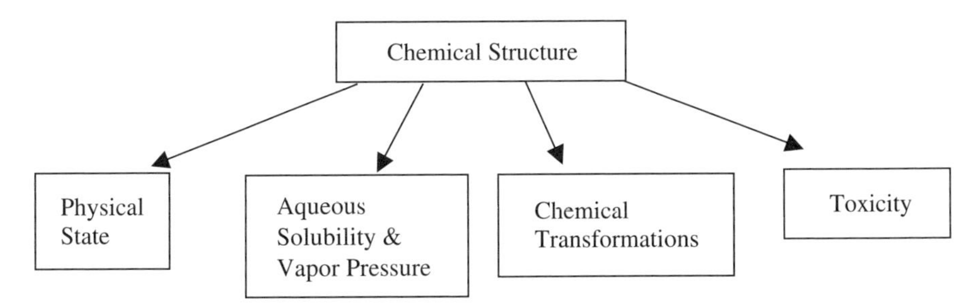

Figure 2.1 Impact of chemical structure on pollutant properties and behavior.

2.1 CHEMICAL STRUCTURE

Chemical pollutants may be broadly divided into two categories: organic and inorganic and then further classified as shown in Figure 2.2. Organic chemicals contain tetravalent carbon bonded primarily with hydrogen atoms (such compounds are called hydrocarbons) and to a lesser extent with oxygen, nitrogen, sulfur, and the halogenic elements, such as chlorine and bromine. The tetravalent carbon in organic compounds forms covalent bonds with other elements by means of a shared electron pair. As a result, organic compounds are typically nonpolar in nature, that is, equal sharing of the electron pair results in an even charge distribution in the molecule. Inorganic compounds do not contain the tetravalent carbon atom. They are formed by various combinations of the elements seen in the periodic table (Table 2.1), including carbon without tetravalent bonds.

2.1.1 Inorganic Pollutants

Inorganic elements may broadly be classified as metals, nonmetals, and transition group elements. Much can be learned about the nature of inorganic atoms by studying the periodic table of the elements (Table 2.1). Representative elements are those found in groups IA through VIIA of the periodic table with the group number representing the number of electrons in the outermost or valence shell of these elements. Representative elements form molecules by lending, borrowing, or sharing electrons in order to achieve a stable electronic configuration of two or eight electrons in their outermost shell.[2] Among the representative elements, the group IA through IIIA elements are the strongly electropositive metals, that is, these elements readily donate electrons to form positively charged species (cations) within molecules. The magnitude of the charge is determined by the number of electrons donated from their outermost shell, easily identified by the group numbers on the periodic table. Thus, the IA metals such as sodium form cations with a +1 charge, the IIA elements such

[2] The group VIII (aka group 0) elements are inert gases and do not participate in reactions since they already have a stable electronic configuration.

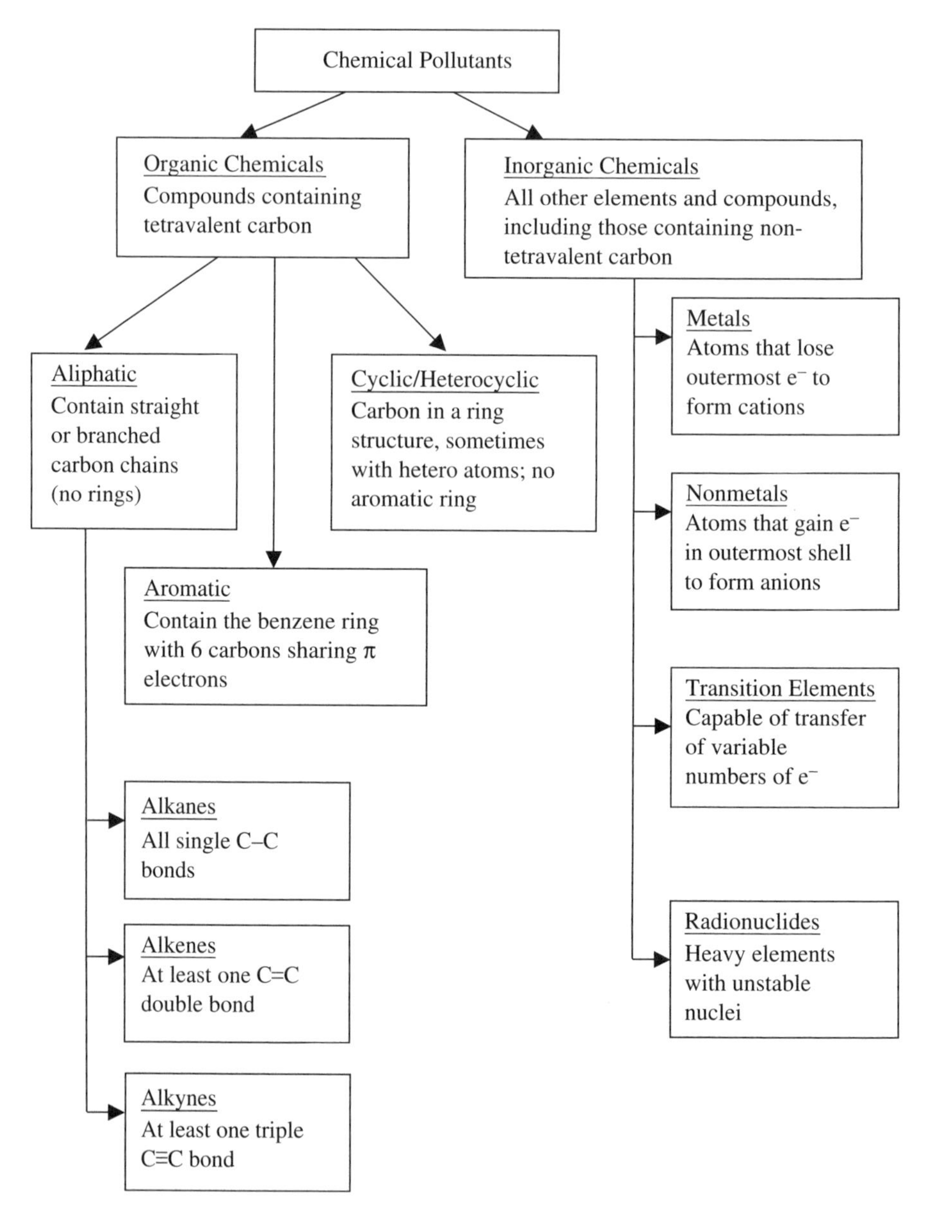

Figure 2.2 Classification of pollutants by chemical structure. (e^- represents electrons).

as magnesium form cations with a +2 ionic charge, and so forth. Representative nonmetals are grouped toward the right side of the periodic table, and, during molecule formation, they "borrow" electrons from other atoms to form anions with a stable number of electrons (two or eight) in their outermost shells. Thus, the group VII elements (halogens) form anions with a -1 charge since they require one additional

TABLE 2.1 Periodic Table of the Elements (From Snoeyink and Jenkins, 1980)

PERIODIC TABLE OF THE ELEMENTS

Key: Atomic No. / Symbol / Atomic Wt. (oxidation numbers shown before the atomic number)

Representative Elements *s* block (IA, IIA); Transition Elements *d* block (IIIB–IIB); Representative Elements *p* block (IIIA–VIIA); Noble gases (O)

Periods	IA	IIA	IIIB	IVB	VB	VIB	VIIB	VIII	VIII	VIII	IB	IIB	IIIA	IVA	VA	VIA	VIIA	O
1	+1 1 H 1.0079																	2 He 4.003
2	+1 3 Li 6.941	+2 4 Be 9.012											+3 5 B 10.81	+4 6 C 12.011	+5 +3 7 N −3 14.007	8 O −2 15.999	9 F −1 18.998	10 Ne 20.18
3	+1 11 Na 22.99	+2 12 Mg 24.30											+3 13 Al 26.98	+4 14 Si 28.08	+5 +3 15 P −3 30.97	+6 +4 16 S −2 32.06	+7 +5 +3 +1 17 Cl −1 35.45	18 Ar 39.95
4	+1 19 K 39.10	+2 20 Ca 40.08	21 Sc 44.96	22 Ti 47.90	23 V 50.94	+6 +3 24 Cr 52.00	+7 +4 +2 25 Mn 54.94	+3 +2 26 Fe 55.85	+3 +2 27 Co 58.93	+3 +2 28 Ni 58.71	+2 +1 29 Cu 63.55	+2 30 Zn 65.38	+3 31 Ga 69.72	+4 +2 32 Ge 72.59	+5 +3 33 As 74.92	+6 +4 34 Se −2 78.96	+7 +5 +3 +1 35 Br −1 79.90	36 Kr 83.80
5	+1 37 Rb 85.47	+2 38 Sr 87.62	39 Y 88.91	40 Zr 91.22	41 Nb 92.91	42 Mo 95.94	43 Tc 98.91	44 Ru 101.07	45 Rh 102.91	46 Pd 106.4	+1 47 Ag 107.87	+2 48 Cd 112.40	+3 49 In 114.82	+4 +2 50 Sn 118.69	+5 +3 51 Sb 121.75	+6 +4 52 Te −2 127.60	+7 +5 +3 +1 53 I −1 126.90	54 Xe 131.30
6	+1 55 Cs 132.91	+2 56 Ba 137.34	57 La 138.91	72 Hf 178.49	73 Ta 180.95	74 W 183.85	75 Re 186.2	76 Os 190.2	77 Ir 192.22	78 Pt 195.09	79 Au 196.97	+2 +1 80 Hg 200.6	+3 +1 81 Tl 204.4	+4 +2 82 Pb 207.2	+5 +3 83 Bi 209.0	+6 +4 84 Po (210)	85 At (210)	86 Rn (222)
7	+1 87 Fr (223)	+2 88 Ra 226.0	89 Ac (227)	104 Ku* 	105 Ha* 													

Inner Transition Elements *f* block

Series														
Lanthanum Series	58 Ce 140.12	59 Pr 140.9	60 Nd 144.24	61 Pm (147)	62 Sm 150.4	63 Eu 151.96	64 Gd 157.2	65 Tb 158.93	66 Dy 162.50	67 Ho 164.93	68 Er 167.26	69 Tm 168.93	70 Yb 173.04	71 Lu 174.97
Actinium Series	90 Th 232.0	91 Pa 231.0	92 U 238.0	93 Np 237.0	94 Pu (242)	95 Am (243)	96 Cm (247)	97 Bk (247)	98 Cf (247)	99 Es (254)	100 Fm (253)	101 Md (256)	102 No (254)	103 Lr (257)

Mass numbers of the most stable or most abundant isotopes are shown in parentheses.
The elements to the right of the bold lines are called the nonmetals and the elements to the left of the bold line are called the metals.
Common oxidation numbers are given for the representative elements and some transition elements.
*Kurchatovium and Hahnium are tentative names for these elements.

electron to maintain stable valence configuration. The group VI elements (oxygen, sulfur, etc.) typically form anions with a −2 charge since two electrons must be borrowed to attain stable configurations. The strongly electropositive metals and electronegative nonmetals typically combine together to form white or colorless ionic salts (e.g., NaCl) or else they combine with OH^- and H^+ ions, respectively, to form strong bases and acids (for example, hydrochloric acid, HCl, and sodium hydroxide, NaOH). Elemental forms of these atoms, for example, pure sodium metal or pure chlorine, are extremely reactive and must be stored and handled carefully.

Among the group A representative elements described above, the outermost electronic shell is being filled sequentially as one moves across a period (row). As a result, group number, which represents the number of electrons in this outermost *valence* shell, fixes the valency of the group A elements. In contrast, the group B transition elements have a partially filled outermost electronic shell, but their penultimate shell is also incomplete and is being filled sequentially with electrons as one moves from left to right across a period. Because both the outermost and the penultimate electronic shells are incomplete, the transition group metals can form many different types of bonds with other elements resulting in colored salts and complexes. They can engage in ionic bonds by lending of an electron, pure covalent bonds by equal sharing of an electron pair, polar covalent bonds by unequal sharing of an electron pair, or coordination bonds by attracting an entire electron pair. As a result, the transition elements grouped in the middle of the periodic table (IB to VIIIB) can exhibit multiple valencies, or more appropriately, multiple stable oxidation states (denoted above each element in the periodic table). Oxidation state, usually denoted by Roman numerals, represents the number of electrons lost, gained, or shared by an atom during the formation of ionic, covalent, polar covalent, or coordination bonds. In some cases, numerous stable oxidation states are possible enabling the atom to appear in cationic as well as anionic form by forming coordination bonds with various entities. For example, cationic forms of chromium in aqueous solution include Cr^{2+} and Cr^{3+}. But, chromium can also form coordination bonds with oxygen and appear in aqueous solution in the form of a variety of anionic species such as chromite [$(CrO_2)^-$], chromate [$(CrO_4)^{2-}$], and dichromate [$(Cr_2O_7)^{2-}$]. Speciation is thus very important when dealing with inorganic compounds, particularly since the toxicity and the mobility of different chemical species can vary widely. For example, the chromate ion with chromium in the +6 oxidation state is more toxic and also much more mobile in groundwater than the chromite ion, which has chromium in the +3 oxidation state [Cr(III)].

Many of the metals that pose serious environmental hazards are high-molecular-weight metals and radionuclides that are located in the lower portion of the periodic table. Radionuclides are heavy elements with unstable nuclear structure; their nuclei undergo radioactive decay to attain a more stable structure, resulting in the formation of a different *daughter* or *progeny* atom.

The process of radioactive decay releases ionizing radiation in the form of α or β particles and γ rays. Of these, the α particles (He^{2+}) are the heaviest and possess the greatest ionizing strength but do not penetrate very deeply into solid surfaces or across human skin. However, α particles can still be harmful when inhaled or ingested. Beta particles are lighter than α particles and carry a lower charge. Consequently, they have less ionization power but can penetrate further into solid surfaces. Gamma rays, being massless, have the greatest penetration power and can also produce ionization in the materials that they impact.

The effect of ionizing radiation on living cells is to cause cell death at high doses and also to cause cell mutations that become apparent at low doses. Severe health effects have been observed when humans have been exposed to radiation as a result of nuclear weapons explosions. Such effects have also occurred during accidental exposures at nuclear power plants, while taking or administering nuclear medicine, and in mining operations. Radionuclides such as uranium and plutonium, which are used in the manufacture of weapons and in power generation, are high-molecular-weight heavy metals grouped in the actinide series at the bottom of the periodic table. Most elements in the lanthanide series are produced from nuclear reactions. Radon, which is a daughter product of uranium decay, is itself a radioactive gas and an important indoor air pollutant. Radium and strontium are heavy alkali earth metals (group II) used widely in nuclear medicine.

Heavy metals that are nonradioactive, such as mercury, zinc, cadmium, and lead, can also pose serious health hazards. These metals are sparingly soluble in water and are also referred to as trace metals. However, even at low levels, they have been associated with serious health effects. For example, Minamata disease, which affects the central nervous system, is associated with mercury poisoning (www.nimd.go.jp), and Itai-Itai disease, which affects kidney function, is caused by cadmium (Purves, 1985). Heavy metals may also bioaccumulate and become biomagnified through the food chain, leading to fish advisories and causing potential damage to sensitive higher-order predators (Roelke et al., 1991; Lange et al., 1993; Spalding et al., 1994; Facemire, 1995; Wiener, 1995).

In summary, inorganic pollutants can occur in purely elemental form, such as elemental lead or zinc released to air from metal smelting operations. More frequently, however, metals and nonmetals form cations and anions and a variety of other complex species depending upon their electronic configuration and oxidation state. The various species of an inorganic pollutant often carry charge and hence are strongly impacted by pH and redox conditions in the environment. Chemical toxicity, in turn, is strongly affected by such speciation. Because of these complex phenomena, the chemistry of inorganic pollutants is briefly discussed in Section 2.4 but is largely reserved for more detailed treatment in Chapter 12. Chapters 3 to 5 and 8 to 10 focus primarily on the intermedia transfer and transport of nonpolar organic chemicals, many of which are frequently encountered at hazardous waste sites. The structure and classification of organic chemicals is discussed next.

2.1.2 Organic Pollutants

Organic chemicals are named so because many of them were initially discovered or derived from naturally occurring organic matter. All organic compounds contain the tetravalent carbon atom that forms four covalent bonds with other carbon atoms or with hydrogen atoms. Industrial chemical technologies developed in the past century have produced a wide variety of synthetic organic chemicals including pharmaceuticals, petrochemicals, pesticides, herbicides, and plastics.

Organic chemicals are broadly subdivided into three categories based on chemical structure—aliphatic compounds, aromatic ring compounds, and cyclic or heterocyclic ring compounds. Aliphatic compounds are hydrocarbons in which carbon atoms are attached together to form an open chain, that is, the chain does not loop around to form a closed ring. The chain may be straight or branched as shown in Figure 2.3. The bonds between the carbon atoms may be exclusively single bonds leading to the formation of alkanes. Alkenes contain at least one double bond, while the formation of alkynes includes at least one triple bond between adjacent carbon atoms. Alkanes are named with an *ane* suffix, alkenes with an *ene* suffix, and alkynes with a *yne* suffix. Alkanes, alkenes, and alkynes form a homologous series with nomenclature as shown in Figure 2.3. Hydrogen atoms in the parent aliphatic compound may be replaced by various functional groups, for example, the –OH group yielding alcohols, the –Cl group yielding chloro compounds, or the $–NH_2$ group yielding amines. Some chlorinated aliphatic compounds of environmental significance are TCE (trichloroethylene), chloroform ($CHCl_3$), and DCE (dichloroethylene). In some cases, a hydrogen atom in the parent aliphatic compound may be substituted by another "alkane"—the incoming alkane is termed an "alkyl" group and named with an *yl* suffix: methyl, ethyl, propyl, butyl, and so on, with the general formula C_nH_{2n+1}. In the case of substituted compounds, the location of substitution is denoted by the numbers assigned to the carbon chain, as illustrated in Figure 2.3.

Aromatic ring compounds contain the benzene ring structure shown in Figure 2.4. Benzene contains six carbon atoms attached in a ring, each carbon also being attached to a single hydrogen atom. Every alternate carbon atom is double bonded with its neighbor to satisfy the tetravalent nature of carbon atoms. In reality, all six carbon atoms share an electron cloud represented by the ring at the center of the molecule. The hydrogen atoms in the benzene ring can be replaced by different atoms, forming numerous compounds, for example, phenol (–OH replaces H), toluene ($–CH_3$ replaces H), ethyl-benzene (C_2H_5 replaces –H), and xylene (two $–CH_3$ groups replace two –H atoms). Gasoline spills often result in the release of benzene, toluene, ethylbenzene, and xylene, which are collectively termed the BTEX compounds. Polyaromatic hydrocarbons, or PAH, compounds are composed of multiple aromatic rings. For example, naphthalene is composed of two rings and phenanthrene has three rings (see Fig. 2.4). Again, replacement of the hydrogen atom can result in the formation of substituted PAH compounds.

When a benzene ring is joined to another group, it is called the *phenyl-* functional group. When two phenyl groups are attached to each other, a biphenyl molecule is formed (Fig. 2.5). If biphenyl is chlorinated, it yields a mixture of molecules containing 1 to 10 chlorine atoms. These compounds are called polychlorinated biphenyls, or PCBs. There are 10 different homologs of PCBs, corresponding to the number of chlorine atoms in the molecule (molecules in a given homolog have the same chemical formula), but 209 different congeners, that is, molecules with different degrees of chlorination *or* structure. Different congeners within the same homolog group have the same chemical formula, but the chlorine atoms are bonded at different

# of C Atoms	Alkane	Alkene	Alkyne	Some Substituted Alkanes	
(n)	(C_nH_{2n+2})	(C_nH_{2n})	(C_nH_{2n-2})	Chlorinated Compounds	Alcohols
1	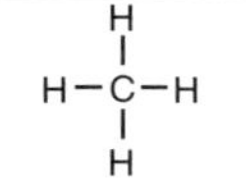 Methane (CH_4)	—	—	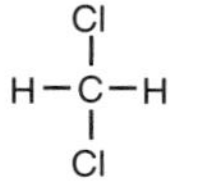 Dichloromethane (CH_2Cl_2)	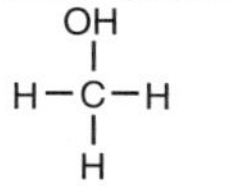 Methanol (CH_3OH)
2	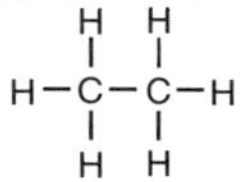 Ethane (C_2H_6)	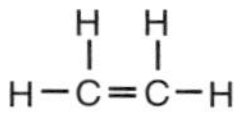 Ethene (C_2H_4)	 Ethyne (C_2H_2)	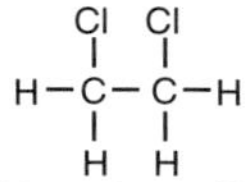 1,2-Dichloroethane ($C_2H_4Cl_2$) 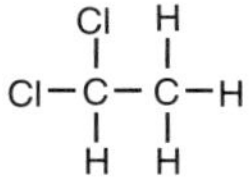 1,1-Dichloroethane ($C_2H_4Cl_2$)	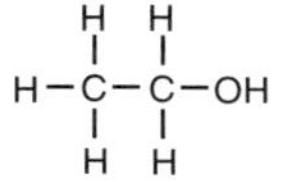 Ethanol (C_2H_5OH)
3	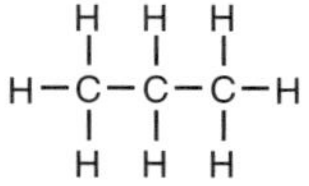 Propane (C_3H_8)	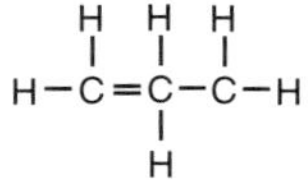 Propene (C_3H_6)	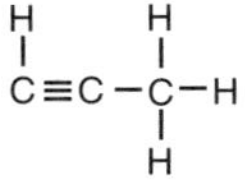 Propyne (C_3H_4)	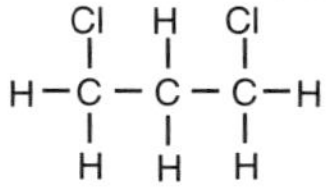 1,3-Dichloropropane (C_3H6Cl_2)	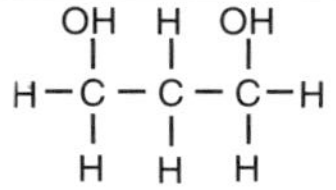 Propane-di-ol ($C_3H_6(OH)_2$)
4	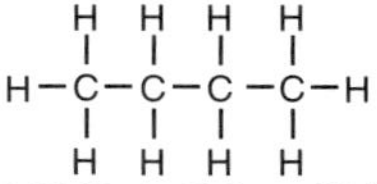 Straight Chain Butane (C_4H_{10}) 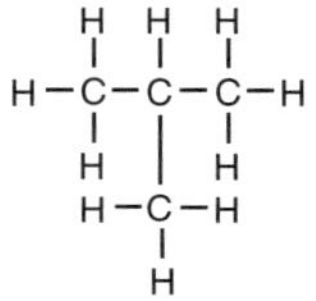 Branched-chain butane (C_4H_{10})	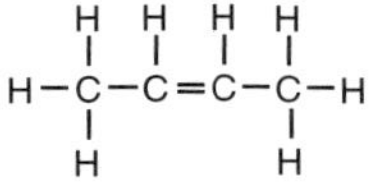 But-2-ene (C_4H_8)	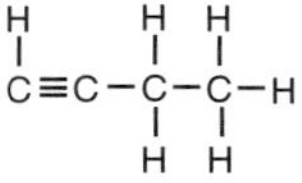 But-1-yne (C_4H_6)		
5	Pentane (C_5H_{12})	Pentene (C_5H_{10})	Pentyne (C_5H_8)		
6	Hexane (C_6H_{14})	Hexene (C_6H_{12})	Hexyne (C_6H_{10})		

Figure 2.3 Aliphatic compounds are open chain (straight or branched) compounds containing primarily tetravalent carbon and hydrogen. Alkanes contain carbon-to-carbon single bonds (C–C), alkenes contain at least one double bond (C=C), and alkynes contain at least one triple bond (C≡C).

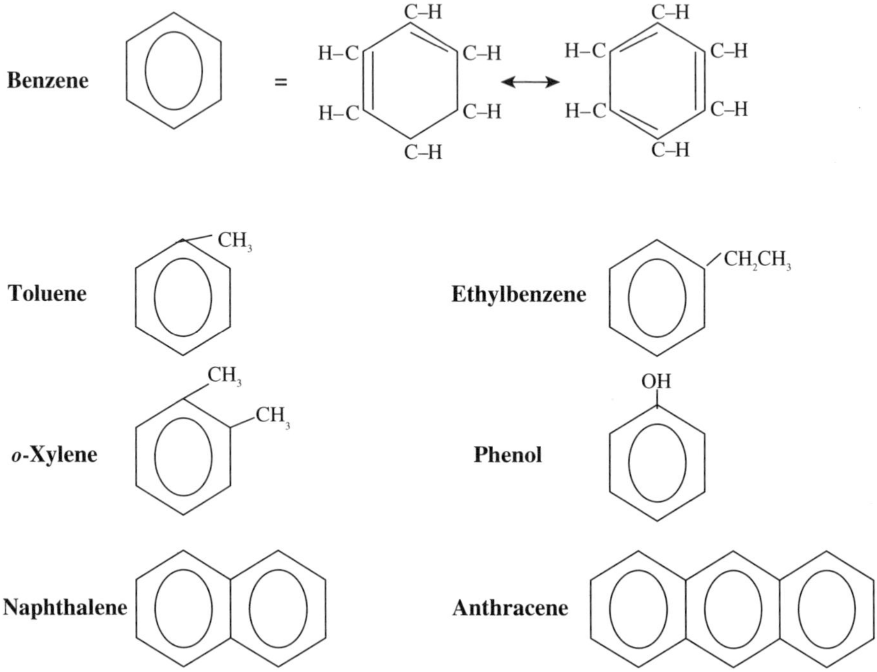

Figure 2.4 Structure of some aromatic and polycyclic aromatic hydrocarbons.

locations on the biphenyl rings, which can affect the behavior of the compound. PCBs are quite resistant to chemical, thermal, or biological degradation and tend to persist in the environment (especially those with a higher degree of chlorination), with adverse effects on the endocrine and reproductive systems of biota. Pesticides like DDT (Fig. 2.5), 2,4-D, and 2,4,5-T are examples of complex organic pollutants containing phenyl and chlorinated phenyl groups.

Cyclic compounds are those in which carbon atoms form a ring structure without the floating shared electron cloud denoted in the benzene ring, for example, cyclopentane, cyclohexane, and the like. In heterocyclic compounds, one of the carbon atoms is replaced by a different (*hetero*) atom, such as oxygen, nitrogen, or sulfur. Some chlorinated cyclic compounds such as benzene hexachloride are of environmental significance due to their toxic effects. The reader is referred to a classic organic chemistry textbook for further details of the nomenclature and structure of the organic compounds.

2.1.3 Individual Chemicals versus Aggregate Measures of Pollution

In this chapter and in the rest of the book, individual chemical pollutants are generically represented by the superscript A. Thus, A may represent any individual chemical compound such as benzene, toluene, or trichloroethylene (TCE). This book is primarily about the transport, fate, and risk posed by the individual chemical species.

Biphenyl

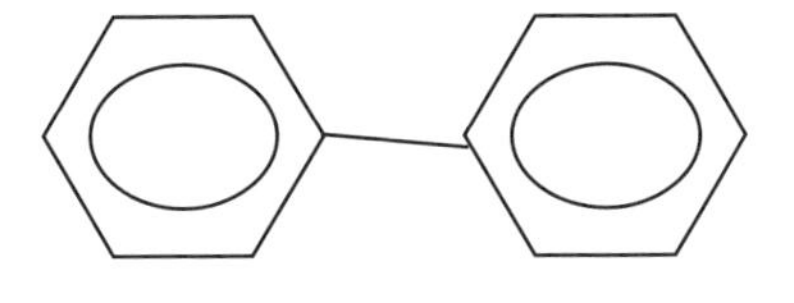

PCB: Polychlorinated biphenyl

(Up to 10 chlorine atoms can bond at the numbered locations)

DDT: Dichloro-diphenyl-trichloroethane

Figure 2.5 Some phenyl compounds.

In many cases, however, it is convenient to report the quality of a sample of air, water, or soil by reporting the aggregate concentration of various groups of chemicals, the categorization or grouping being based upon similarity in chemical structure and properties. Methods for chemical analysis of pollutants in air, water, and soil are specified by the EPA for these specific groups of chemicals based on their similar physical and chemical properties. For example, a test for total petroleum hydrocarbons (TPH) includes all the organic pollutants that are part of petroleum. Volatile organic compounds (VOCs) are a group of hydrocarbons, with generally fewer than about 12 carbon atoms, that readily volatilize to the air. Semivolatiles (SVOCs) refer to less volatile compounds, including pesticides such as DDT, dieldrin, and endrin. Lists of chemicals classified as VOCs and SVOCs may be found at http://www.frtr.gov/matrix2.

As noted in Chapter 1, the release of organic wastes to water, through municipal sewage, industrial waste, or natural runoff, results in the consumption of dissolved oxygen in the water. This impact is reported by measuring the aggregate oxygen consumption by all the organic pollution present in a water sample—reported either as biochemical oxygen demand (BOD), chemical oxygen demand (COD), or theoretical oxygen demand (ThOD). Similar aggregation occurs when reporting inorganic pollution. For example, hardness refers to the sum of precipitable cations in a water sample, typically the total of the calcium, magnesium, and iron concentrations dissolved in water. Total dissolved solids (TDS) is a measure of all of the inorganic salts dissolved in a water sample. Particulate matter (PM) suspended in the atmosphere is measured and reported on a mass concentration basis, often with a size cut-off. For example, PM_{10} refers to particulate matter less than 10 μm in size but composed of many different liquid or solid-phase chemical constituents, including elemental and organic carbon, nitrates, sulfates, and other compounds.

Chemical pollutants may also be aggregated by the nature of their source. In many cases, the individual chemical may be released from a complex matrix composed of many different chemicals. A primary example is a non-aqueous- (organic) phase liquid (NAPL) released to the environment. NAPLs are typically complex multicomponent mixtures of several different organic compounds. Gasoline, diesel fuel, and coal tar are examples of multicomponent NAPLs. Gasoline and diesel fuel are lighter than water and are therefore referred to as LNAPLs, while coal tar and several chlorinated solvent mixtures are typically denser than water and are called DNAPLs.

Table 2.2 shows the chemical composition of a coal tar sample obtained from a field site. Contaminants of concern in NAPL coal tar are typically the BTEX compounds: benzene, toluene, ethylbenzene, and xylene, as well as polyaromatic compounds such as naphthalene and benzo(*a*)pyrene. Although the aggregate NAPL is essentially immiscible in water and exists as a separate phase because of the low aqueous solubility of its organic constituents, individual constituents may dissolve in water or volatilize to air at concentrations high enough to cause environmental

TABLE 2.2 Major constituents of sample of coal tar obtained from gas plant site

Compound	Molecular Weight (g/mol)	Weight %
	Volatile Organic Compounds	
Benzene	78	0.100
Ethylbenzene	92	0.340
Styrene	106	0.044
Total xylenes	106	0.250
	Polycyclic Aromatic Hydrocarbons	
Naphthalene	128	10.000
2-Methylnaphthalene	142	5.300
Acenaphthylene	152	0.370
Acenaphthene	154	1.300
Fluorene	166	0.180
Phenanthrene	178	0.160
Anthracene	178	2.000
Fluoranthene	202	0.550
Pyrene	202	0.320
Benz(*a*)anthracene	228	1.000
Chrysene	228	0.360
Benzo(*b*)fluoranthene	252	0.400
Benzo(*k*)fluoranthene	252	0.160
Benzo(*a*)pyrene	252	0.360
Dibenz(*a*,*h*)anthracene	252	0.040

Source: Ghoshal et al. (1996).

concern. Due to the slow dissolution of NAPL components in groundwater, trapped NAPL lenses can function as long-term sources of soil and groundwater contamination (Hunt et al., 1988). Surface spills of NAPLs give rise to air and surface water pollution, such as in the Exxon *Valdez* oil spill in Alaska (Wells et al., 1995). NAPLs thus represent the liquid waste matrix from which many individual chemicals are released into ambient air, water, and soil.

2.2 PROPERTIES OF CHEMICALS

2.2.1 Physical State

The natural physical state of a pollutant determines which medium is immediately and more readily polluted by the chemical. For example, release of a gaseous pollutant will cause an immediate air pollution problem, although some proportion of the gaseous chemical may later transfer to water or soil. The natural physical state of a pollutant depends upon its melting point temperature, T_m, its boiling point temperature, T_b, and the temperature of the environment at large, represented as T_{env}. The melting and boiling points of selected organic compounds are shown in Table 2.3. Unless otherwise stated, T_{env} in this book is taken as 25°C or 298 K.

- If $T_{\text{env}} < T_m$, the pollutant's natural physical state at T_{env} is solid.
- If $T_m < T_{\text{env}} < T_b$, the pollutant's natural physical state at T_{env} is liquid.
- If $T_{\text{env}} > T_b$, the pollutant's natural physical state at T_{env} is gaseous.

The melting point, boiling point, and hence the natural physical state of a pollutant are determined by the size and molecular structure of the chemical. Smaller, nonpolar organic compounds are typically gases at ambient temperatures, while larger or more polar compounds tend to be liquids or solids. The smaller compounds are lighter and readily transfer to the gas phase; the nonpolar nature indicates that there are no polar attractions or bonds to bind the molecules together in liquid form. In the series of alkanes, for example, methane, ethane, propane, and butane are gases at ambient temperatures, pentane through dodecane are liquids, and larger straight-chain alkanes are solids at ambient temperatures. Polarity introduced into the lighter alkanes, for example, by substituting an –OH group as in the alcohols, changes the natural physical state of these now polar chemicals. Thus, methanol and ethanol are liquids due to the polarity introduced by the –OH group.

EXAMPLE 2.1

What are the natural physical states of the following compounds at T_{env} of 25°C: benzene, methane, trichloroethene (TCE), and naphthalene? Refer to Table 2.3.

Solution 2.1

The natural state of benzene is liquid since $T_m (= 5.5°C) < T_{env} < T_b (= 80°C)$.
The natural state of methane is gaseous since $T_{env} > T_b (= -164°C)$.
The natural state of TCE is liquid since $T_m (= -73°C) < T_{env} < T_b (= 87°C)$.
The natural state of naphthalene is solid since $T_{env} < T_m (= 81°C)$.

The physical state of compounds when present in complex mixtures such as NAPLs may be affected by chemical interactions. Consider the case of coal tar, which is a NAPL composed of several hundred organic compounds, including naphthalene. At 25°C, pure naphthalene is a solid; yet when dissolved in coal tar, naphthalene exists as a liquid at 25°C. In this case, naphthalene in coal tar is referred to as a subcooled liquid, that is, the naphthalene exists as a liquid at ambient environmental temperatures due to chemical interaction effects within the NAPL, even though it has been cooled below its melting point. If the liquid naphthalene present in coal tar could be isolated in pure liquid form and maintained at T_{env} = 25°C, this would represent pure hypothetical subcooled liquid naphthalene. Pure subcooled liquid naphthalene is a hypothetical entity because it is the solvation energy of the coal tar mixture that provides the energy for liquefying the naphthalene; yet, this provides an important theoretical construct used later in describing contaminant equilibria in the presence of NAPLs.

2.2.2 Fundamental Properties of Pure Chemicals Present in Abundance

Like its physical state, the tendency of a chemical pollutant to enter air or water is also governed by its structure. These tendencies are characterized by two fundamental properties of the pure chemical: aqueous solubility and vapor pressure. Note that both fundamental properties refer to the ability of a chemical to saturate water and air, when present in *great abundance in its pure and natural state*. A great abundance indicates that upon fully saturating all the environmental phases present, some amount of the chemical will still remain in its pure and natural state. "Pure" implies that the chemical is not present in a mixture with any other compound. Natural state represents the physical state of the pollutant, as determined by its boiling point (T_b), melting point (T_m), and the environmental temperature (T_{env}).

2.2.3 Aqueous Solubility

When a chemical, present in great abundance in its pure and natural state, is contacted with water for an extended period of time, the chemical will enter the aqueous phase until the water is fully saturated with that chemical. The maximum saturation concentration of the chemical A in water is called its aqueous solubility, $C_{w,\text{sat}}^{\text{A}}$, where the superscript represents the identity of the chemical:

$$C_{w,\text{sat}}^{\text{A}} = \text{Maximum saturation concentration of A in water (mol/L or g/L)} \qquad (2.1)$$

TABLE 2.3 Some Physical and Chemical Properties of Various Chemicals.[a] Log units represent base 10.

Chemical	Molecular Weight (g/mol)	T_m (°C)	T_b (°C)	Density (g/cm^3)	Aqueous Solubility $-\log C_{w,sat}^{A}$ (mol/L)	Vapor Pressure $-\log p_v^{A}$ (atm)	Subcooled/Superheated Aqueous Solubility $-\log C_{w,sat}^{A(L)}$ (mol/L)	Subcooled/Superheated Vapor Pressure $-\log p_v^{A(L)}$ (atm)	Henry's Law Constant $\log K_H^A$ (L atm/mol)	Octanol Water Coef. $\log K_{ow}^A$ (L_{wat}/L_{oct})
Aliphatic Compounds and Halogenated Aliphatic Compounds										
n-Butane	58.1	−138.4	−0.5		2.98		2.59	−0.39	2.98	2.89
Carbon tetrachloride	153.8	−22.9	77.0	1.59	2.20	0.82			1.38	2.73
Chloroform (trichloromethane)	119.4	−63.5	61.7	1.48	1.19	0.59			0.60	1.93
1,1-Dichloroethane	99.0	−97.0	57.5	1.18	1.30	0.52			0.78	1.79
1,2-Dichloroethane	99.0	−35.4	83.5	1.24	1.07	1.04			0.03	1.47
Ethane	30.1	−183.3	−88.6		2.69		1.09	−1.60	2.69	1.81
Ethanol	46.0	−114.0	78.0	0.79	∞	.078			−0.31	
Hexane	86.2	−95.0	69.0	0.66	3.83	0.69			3.14	4.11
Methane	16.0	−182.5	−164.0		2.82		0.38	−2.44	2.82	1.09
Methylene chloride	84.9	−95.1	39.7	1.33	0.64	0.23			0.41	1.15
n-Octane	114.2	−56.8	125.7	0.70	5.20	1.73			3.47	5.18
n-Pentane	72.2	−129.7	36.1	0.63	3.25	0.16			3.09	3.62
Propane	44.1	−189.7	−42.1		2.85		1.88	−0.97	2.85	2.36
1,1,1-Trichloroethane (TCA)	133.4	−30.4	74.1	1.34	2.07	0.78			1.29	2.48
Trichloroethene (TCE)	131.4	−73.0	87.0	1.46	2.04	1.01			1.03	2.42
Vinyl chloride (chloroethene)	62.5	−153.8	−13.4		1.35		0.76	−0.59	1.35	0.60
Aromatic Compounds										
Aniline	93.1	−6.3	184.0	1.08	0.41	2.89			−2.48	0.90
Benzene	78.1	5.5	80.1	0.88	1.64	0.90			0.74	2.13
Benz(*a*)anthracene	228.3	159.8	435.0		7.31	9.55	5.96	8.20	−2.24	5.91
Benzo(*a*)pyrene	252.3	176.5			8.22	11.14	6.71	9.63	−2.92	6.50
n-Butylbenzene	134.2	−88.0	183.0	0.86	3.97	2.86			1.13	4.28

(continued)

TABLE 2.3 (*Continued*)

Chemical	Molecular Weight (g/mol)	T_m (°C)	T_b (°C)	Density (g/cm^3)	Aqueous Solubility $-\log C_{w,sat}^{A}$ (mol/L)	Vapor Pressure $-\log p_v^{A}$ (atm)	Subcooled/Superheated Aqueous Solubility $-\log C_{w,sat}^{A(L)}$ (mol/L)	Subcooled/Superheated Vapor Pressure $-\log p_v^{A(L)}$ (atm)	Henry's Law Constant $\log K_H^{A}$ (L atm/mol)	Octanol Water Coef. $\log K_{ow}^{A}$ (L_{wat}/L_{oct})
Aromatic Compounds (continued)										
Chlorobenzene	112.6	−45.6	132.0	1.11	2.35	1.80			0.55	2.92
Ethylbenzene	106.2	−95.0	136.2	0.87	2.80	1.90			0.90	3.15
Naphthalene	128.2	80.6	217.9	1.03	3.61	3.98	3.06	3.43	−0.37	3.36
n-Pentylbenzene	148.3	−75.0	205.4		4.59	3.36			1.23	4.90
Phenanthrene	178.2	99.5	340.2	0.98	5.20	6.79	4.46	6.05	−1.59	4.57
Phenol	94.1	43.0	181.7	1.07	0.20	3.59	0.02	3.41	−3.39	1.45
n-Propylbenzene	120.2	−101.6	159.2	0.86	3.34	2.35			0.99	3.63
Toluene (methyl-benzene)	92.1	−95.0	110.6	0.87	2.25	1.42			0.83	2.69
Cyclic and Heterocyclic Compounds										
Cyclohexane	84.16	6.6	80.7	0.78	3.15	0.9			2.25	3.44
Lindane	290.90	112.9			4.59	7.08	3.71	6.20	−2.49	3.78
Inorganic Gases										
Nitrogen	28.01	−187.9	−195.8						[b]	
Oxygen	32.00	−218.8	−182.9						[b]	

[a]Densities are reported for all liquids and selected solid chemicals. Densities were measured between 15.5 and 22°C, except for phenanthrene at 4°C. Aqueous solubility ($C_{w,sat}^{A}$), vapor pressure (p_v^{A}), Henry's law constant (K_H^{A}) and octanol–water partition coefficient (K_{ow}^{A}) are for 25°C at an ambient pressure of 1 atm, unless otherwise noted. Properties of subcooled and superheated chemicals are represented as: aqueous solubility ($C_{w,sat}^{A(L)}$), vapor pressure ($p_v^{A(L)}$).

[b]The Henry's constant, K_H, for inorganic gases is defined as the inverse of that for organic chemicals. At 25°C, $\log K_H$ for oxygen and nitrogen are −2.9 and −3.19, respectively, in units of mol/L atm.

Sources: Adapted from CRC (2001) and Schwarzenbach et al. (1993).

A common-place illustration of aqueous solubility involves dissolving sugar in water. Upon sequentially adding spoonfuls of sugar to a cup of water, saturation conditions will be attained when any excess sugar added remains undissolved. Further experimentation with sugar reveals (what all cooks and tea drinkers know) that the aqueous solubility of sugar increases as the temperature of water is increased. Indeed, an increase in $C^{\mathrm{A}}_{w,\mathrm{sat}}$ with T_{env} occurs for all solid chemicals since solids require heat energy to melt into the liquid state prior to dissolving in water. Conversely, an increase in T_{env} provides greater energy for gaseous chemicals to escape from the liquid-phase water to the air (gas) phase. Hence, the aqueous solubility of gaseous chemicals decreases as T_{env} increases. In the case of liquid chemicals, the impact of temperature on aqueous solubility is not as significant as for solid and gaseous chemicals.

The aqueous solubility of an organic chemical is strongly dependent on its chemical structure. Larger nonpolar organic contaminants tend to be more hydrophobic (water avoiding), exhibiting lower aqueous solubility when compared to their smaller homologs within a certain compound class. Thus, in the benzene family of compounds, increased alkylation (addition of larger alkyl groups to the benzene ring) results in the larger compounds exhibiting a lower aqueous solubility, with the order of decreasing aqueous solubility being: benzene ($10^{-1.64}$ mol/L), toluene, that is, methylbenzene ($10^{-2.25}$ mol/L), ethylbenzene ($10^{-2.80}$ mol/L), propylbenzene ($10^{-3.34}$ mol/L), butylbenzene ($10^{-3.97}$ mol/L), and pentylbenzene ($10^{-4.59}$ mol/L). On the other hand, inclusion of polar groups on the benzene ring, such as the hydroxyl group (–OH) in phenol, and the amine group ($-NH_2$) in aniline, results in enhanced solubility of the polar compounds in water. Following the "like-dissolves-like" principle, the more polar compounds are attracted to the polar nature of water and dissolve more readily in it. Thus, the polar compounds, phenol and aniline, exhibit aqueous solubilities of $10^{-0.20}$ and $10^{-0.41}$ mol/L, respectively, compared to the solubility of $10^{-1.64}$ mol/L for nonpolar benzene.

Qualitative trends in aqueous solubility, such as those discussed above, enable the environmental modeler to develop an intuitive "feel" for the behavior of organic chemicals in water. Quantitative data on aqueous solubility and other compound properties are presented for selected chemicals in Table 2.3.

2.2.4 Vapor Pressure

Vapor pressure is the analog of aqueous solubility for the air system. Consider a great abundance of a pure liquid chemical, A, exposed to air at ambient pressure and temperature for an extended period of time. Some molecules of chemical A, with sufficient kinetic energy to escape the liquid phase, will enter the air phase (see Fig. 2.6). The kinetic energy of the molecules depends upon the ambient temperature, T_{env}, as well as the size (molecular weight) of chemical A. When equilibrium has been attained between air and the abundant pure liquid chemical, the air is saturated with the maximum number of vapor-phase molecules of A energetically capable of escaping the liquid phase at the given temperature. The pressure exerted by the vapors of a pure liquid chemical at saturation conditions at a certain environmental temperature, is termed its vapor pressure:

$$p_v^{\mathrm{A}} = \text{Saturation vapor pressure of A in air (atm)} \tag{2.2}$$

From the universal gas law, since the pressure exerted by a gas is proportional to the number of moles of that gas, the mole fraction of A in air, $Y_{\mathrm{air}}^{\mathrm{A}}$, is given by:

$$Y_{\mathrm{air}}^{\mathrm{A}} = \frac{\text{Moles of A}}{\text{Moles of air}} = \frac{n^{\mathrm{A}}}{n^{\mathrm{air}}} = \frac{p^{\mathrm{A}}}{P_{\mathrm{atm}}} \tag{2.3a}$$

where p^{A} is the partial pressure of A in air, that is, the pressure exerted by the molecules of A in air at any given condition (i.e., not necessarily saturation), and P_{atm} is the ambient atmospheric pressure (usually taken to be 1 atm). For the specific condition when vapors of A have saturated air, we get

$$Y_{\mathrm{air,sat}}^{\mathrm{A}} = \text{Saturation mole fraction of A in air} = \frac{p_v^{\mathrm{A}}}{P_{\mathrm{atm}}} \tag{2.3b}$$

Note, for those chemicals that are solids or liquids at T_{env}, the vapor pressure is always less than 1 atm. As the ambient temperature ($T_{\mathrm{env.}}$) is increased, p_v^{A} and $Y_{\mathrm{air,sat}}^{\mathrm{A}}$ increase. When the ambient temperature reaches the boiling point, T_b, a pure gas is released above the boiling liquid such that $p_v^{\mathrm{A}} = P_{\mathrm{atm}}$; $Y_{\mathrm{air,sat}}^{\mathrm{A}} = 1$. When T_{env} exceeds T_b, the natural physical state of chemical A changes to the gaseous phase, with vapor pressure now in excess of the ambient pressure. Because the vapor pressure of a gas exceeds ambient pressure, gases must be stored in pressurized containers to prevent their release and dissipation in the atmosphere.

The effect of temperature on vapor pressure can be estimated using the Antoine equation:

$$\log p_v^{\mathrm{A}} = \frac{-b}{2.303T_{\mathrm{env}}} + a \tag{2.4}$$

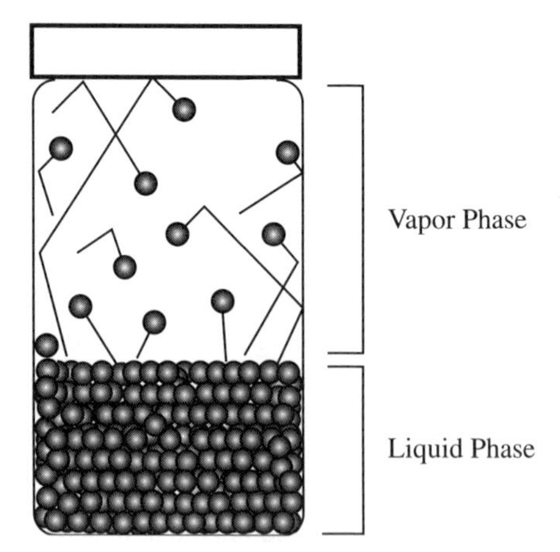

Figure 2.6 Visualizing vapor pressure: The vapor phase consists of molecules with more energy than those in the liquid phase. Vapor pressure is caused by the more energetic molecules "bouncing" against the walls and lid of the container.

where p_v^A represents the vapor pressure of chemical A at the ambient temperature, T_{env}, expressed in degrees Kelvin, and a and b are constants. The values of these constants have been tabulated for individual compounds (CRC, 2001–02). Vapor pressure can change by more than 2 orders of magnitude for a 40°C change in ambient temperature. As a result, temperature corrections of vapor pressures are required at locations with significant variation in ambient temperatures.

In addition to temperature, chemical structure also affects the vapor pressure of organic compounds. Smaller nonpolar compounds are lighter and more readily escape to the vapor phase, compared to large, nonpolar organic compounds in the same chemical series. For example, the lightest alkane, methane, is a gas at T_{env} = 25°C and exhibits a vapor pressure (as a superheated liquid) of $10^{2.44}$ atm. Higher alkanes in the series, ethane, propane, butane, and pentane, are also gases at T_{env} of 25°C but with vapor pressures decreasing from $10^{1.60}$ atm for ethane to $10^{0.39}$ atm for pentane. Alkanes larger than pentane, for example, hexane ($p_v = 10^{-0.16}$ atm), are all liquids at 25°C, with vapor pressures less than one and decreasing as compound size increases. The inclusion of polar groups, such as the alcohol group (–OH), increases intramolecular attractive forces (such as hydrogen bonds and dipole–dipole interactions) between the molecules of the polar chemical, such that the escaping tendency to the vapor phase is diminished, resulting in reduced vapor pressure. The impact of polarity on vapor pressure is illustrated by comparing the alcohol of ethane, namely, ethanol, with the parent compound. Ethanol is a liquid at $T_{env} = 25$°C with vapor pressure of $10^{-0.078}$ atm, much less than the vapor pressure of ethane, which is a gas at $T_{env} = 25$°C.

Table 2.3 provides a compilation of vapor pressure data at T_{env} = 25°C, for selected chemicals. The data may be utilized to assess the saturation vapor pressure [Eq. (2.2)] as well as the saturation mole fraction [Eq. (2.3)] of pure chemical A in air, provided a great excess of the compound A is available to saturate the air. In some situations, such as in modeling inhalation exposure to a pollutant, knowledge of pollutant concentration in air may be desired in units of mass of the chemical per unit volume of air inhaled. In such situations, the universal gas law is employed to determine saturation pollutant concentrations in air on a mass per unit volume basis:

$$C_{\text{air,sat}}^{\text{A}} = \text{Saturation conc. of A in air (g/L)} = \frac{Y_{\text{air,sat}}^{\text{A}} P_{\text{atm}} \text{MW}^{\text{A}}}{RT_{\text{env}}} = \frac{p_v^{\text{A}} \text{MW}^{\text{A}}}{RT_{\text{env}}} \tag{2.5}$$

where RT_{env}/P_{atm} is the volume occupied by one mole of air at ambient temperature and pressure, and MW^A is the molecular weight of the chemical. The reader is cautioned to use appropriate units for the universal gas constant, R, as discussed in Appendix 1A. Note, because saturation conditions represent the availability of a great abundance of pure chemical, saturation vapor pressure [Eq. (2.2)], saturation mole fraction [Eq. (2.3b)], and saturation chemical concentration [Eq. (2.5)] represent the upper limit value of these parameters, which cannot be exceeded at a given T_{env}.

EXAMPLE 2.2

A10-L waste container is half-filled with air and half with a benzene-based solvent, which is conservatively assumed to be pure liquid benzene. What is the maximum possible concentration of benzene in air in the container? Express concentration both as a mole fraction and in units of mass per volume. What is the maximum possible mass of benzene present in the air of the container? Assume standard temperature and pressure.

Air(V_{air} = 5 L)
Pure liquid benzene (5 L)

Solution 2.2 We can assume a great excess of benzene in the container (this is verified below). Maximum possible pollutant concentrations correspond to equilibrium conditions with saturation of the air phase at the vapor pressure of benzene since a great excess of the pure compound is available.

Equilibrium partial pressure of benzene in air = Saturation vapor pressure

$$= p_{air,eq}^{benzene} = p_v^{benzene} = 10^{-0.9} \text{ atm} = 0.125 \text{ atm}$$

$$\text{Mole fraction of benzene in air} = Y_{air,eq}^{benzene} = \frac{P_v^{benzene}}{P_{atm}} = \frac{0.125 \text{ atm}}{1 \text{ atm}}$$

$$= 0.125 \frac{\text{mol benzene}}{\text{mol air}}$$

$$\text{Benzene concentration in air} = C_{air,eq}^{benzene} = \frac{P_{air,eq}^{benzene} \text{MW}^{benzene}}{RT_{env}}$$

$$= \frac{0.125 \text{ atm} \times 78 \text{ g/mol}}{0.0821 \text{ L atm/mol K} \times 298 \text{ K}} = 0.4 \frac{\text{g benzene}}{\text{L air}}$$

$$\text{Mass of benzene in air} = M_{air}^{benzene} = C_{air,eq}^{benzene} \times V_{air} = 0.4 \frac{\text{g benzene}}{\text{L air}} \times 5 \text{ L}_{air}$$

$$= 2 \text{ g benzene}$$

The 2g of benzene in air is negligible compared to the 4400 g of benzene that was introduced into the container as 5L of pure liquid benzene (use the density of benzene from Table 2.3). The initial 5 L of benzene is thus more than sufficient to ensure headspace saturation in the container.

2.2.5 Aqueous Solubility and Vapor Pressure of Subcooled Liquids

It is implicitly understood that aqueous solubility and vapor pressure apply to the compound in its pure and natural state. The concept of a subcooled liquid was introduced in Section 2.2.1 in the context of multicomponent NAPLs. A subcooled liquid compound is one that would normally be a solid in its pure and natural state (at T_{env}) but exists as a liquid when dissolved within a multicomponent NAPL mixture. Although a pure subcooled liquid is a hypothetical entity, its properties can be determined from thermodynamic considerations (Schwarzenbach et al., 1993) and employed to predict the behavior of the subcooled liquid chemical by accounting for mixture effects in the NAPL (see Chapter 3).

The aqueous solubility and vapor pressure of selected pure subcooled liquid compounds are compiled in Table 2.3, with the notation A(L) representing the properties of solid compounds when present in pure liquid form. It is important to note that the properties of the hypothetical pure subcooled liquid are quite different from those of the pure solid compound. This is due to the fact that the subcooled liquid already exists in liquid form and hence more readily dissolves in water or volatilizes to air at a given temperature. In the case of the solid contaminant, "energy costs" to liquify the contaminant must first be paid before the compound can dissolve in water or volatilize to air. Example 2.3 illustrates these phenomena.

EXAMPLE 2.3

First consider moth balls (pure solid naphthalene) at 25°C, and then also consider a situation in which the moth balls have somehow been melted to a hypothetical subcooled liquid state at 25°C. Compare the aqueous solubility and vapor pressure of solid naphthalene with that of the hypothetical subcooled liquid. In which case is the naphthalene concentration in air and water greater? Why?

Solution 2.3 Properties of pure solid naphthalene are

$$C_{w,\text{sat}}^{\text{naphthalene}} = 10^{-3.61} \text{ mol/L} \times 128 \text{ g/mol} = 0.03 \text{ g naphthalene/L water}$$

$$p_v^{\text{naphthalene}} = 10^{-3.98} \text{ atm}$$

Properties of hypothetical pure subcooled liquid naphthalene are

$$C_{w,\text{sat}}^{\text{naphthalene (L)}} = 10^{-3.06} \text{mol/L} \times 128 \text{ g/mol} = 0.11 \text{ g naphthalene/L water}$$

$$p_v^{\text{naphthalene (L)}} = 10^{-3.43} \text{ atm}$$

Aqueous solubility and vapor pressure are greater in the case of the hypothetical subcooled liquid naphthalene since energy previously required to melt the pure solid naphthalene can now be employed to enhance the aqueous solubility and vapor pressure of the liquid naphthalene.

2.3 CHEMICAL TRANSFORMATIONS

Once a chemical enters air, water, or soil, it can undergo chemical transformations. Chemical transformations are important either for eliminating a pollutant by conversion to benign products or for producing new, secondary pollutants. Many chemical reactions occur due to the presence of light—these are termed photochemical reactions. When chemical transformations require enzymes produced by living organisms (plants and microbes), such transformations are called biologically mediated transformations or biochemical transformations. In contrast, abiotic transformations proceed in the absence of living organisms or their enzymes.

Chemical transformations are broadly classified as irreversible and reversible reactions (Table 2.4). Irreversible reactions go from reactants to products in the forward direction only. For example, the degradation of organic pollutants to carbon dioxide and water often occurs through irreversible biochemical reaction pathways, a few of which are illustrated in Table 2.4. These irreversible reaction pathways can ultimately result in the destruction of organic pollutants to form carbon dioxide, a process called mineralization.[3]

The chemistry of most inorganic chemicals, for example, metals, is quite different —no final natural "sink" exists for metals. They are transformed into different chemical species that are transported or sequestered variously in different environmental compartments but are not "degraded" in the sense that organic pollutants can be degraded to carbon dioxide. Typically, the solution to pollution problems involving metals and other inorganics is to transform them into immobile and/or inert species. Such speciation of inorganic pollutants is often governed by reversible reactions that can go either in the forward or backward direction, depending upon environmental conditions. Pairs of reversible reactions that are important in determining the speciation of inorganic chemicals in the environment are shown in Table 2.4. Mathematical models that describe the progress of irreversible and reversible reactions are described briefly next.

2.3.1 Irreversible Chemical Transformations

An irreversible chemical transformation involving pollutant A, may be written as:

$$pA + qB \rightarrow rC + sD \tag{2.6}$$

where A and B are the reactants, and C and D are the products. In a batch reactor, the reactants are consumed to exhaustion to yield the products. According to the law of mass action, the rate of the reaction is proportional to the concentration of the reactants, expressed in molar units. Thus, the rate of consumption of A by the forward reaction may be written as:

[3] Mineralization is ultimately reversed when CO_2 in the form of inorganic carbon in the carbon cycle is fixed by plants via photosynthesis to yield natural organic matter. However, photosynthesis is separate and independent (spatially, temporally, and mechanistically) from the biodegradation/mineralization of organic chemicals.

TABLE 2.4 Overview of Chemical Transformations Occurring in Environment

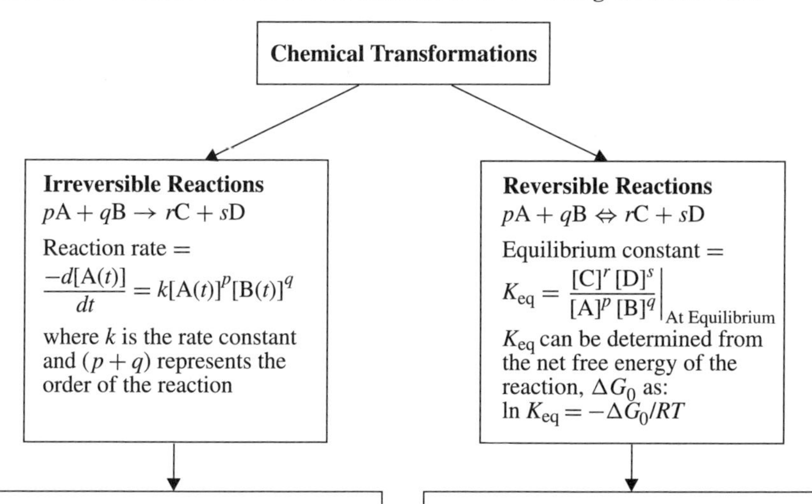

Some Irreversible Reactions of Organic Pollutants

Nucleophilic reactions (water with OH^- *group is a ubiquitous nucleophile):*

1. Hydrolysis (nucleophillic substitution):

   $R\text{-}X + OH^- \rightarrow R\text{-}OH + X^-$

2. Hydrogen–halogen elimination

$X\text{-}R\text{-}R'\text{-}H + OH^- \rightarrow R{=}R' + X^- + H_2O$

Electrophilic reactions with electron-deficient oxygen

1. Biochemical oxidation of benzene to catechol precedes ring cleavage:

Benzene $\xrightarrow[+O_2]{\text{enzymes}}$ Catechol (OH, OH) $\rightarrow$ Ring Cleavage

2. Biochemical oxidation of alkanes to alcohols, aldehydes, and acids

$R\text{-}CH_3 \rightarrow RCH_2OH \rightarrow RCH_2O$
$\rightarrow RCOOH \rightarrow CO_2$
Alkane → Alcohol → Aldehyde → Acid
→ CO_2

Photochemically driven atmospheric oxidation of alkanes by free radicals:

$CH_4 + HO\cdot \rightarrow CH_3\cdot + H_2O$
$CH_3\cdot + O_2 + M \rightarrow CH_3O_2\cdot + M$

Some Reversible Reactions of Inorganic Chemicals

Dissociation of weak acids and bases:

$HA \Leftrightarrow H^+ + A^- \quad K_a = \frac{[H^+][A^-]}{[HA]}$

Both species, HA and A^-, coexist in water. Acid dissociation is favored and A^- is dominant when pH $\gg$ pK_a. HA is dominant when pH $\ll$ pK_a

Mineral dissolution and precipitation:

$MA_{2(solid)} \Leftrightarrow M^{2+} + 2A^-$

$$K_{sp} = \frac{[M^{2+}][A^-]^2}{[\text{Solid activity} = 1]}$$

Dissolution is favored when $[M^{2+}][A^-]^2 < K_{sp}$, while precipitation occurs when $[M^{2+}][A^-]^2 > K_{sp}$

Reduction and oxidation of metals:

$M^{2+} + e^- \Leftrightarrow M^+ \quad K = \frac{[M^+]}{[M^{2+}][e^-]}$ or

$$\log[K] + \log[e^-] = \log\frac{[M^+]}{[M^{2+}]}$$

When pe_0 is defined as log[K], and $-\log[e^-]$ is defined as pe, the above equation becomes

$$pe = pe_0 - \log\frac{[M^+]}{[M^{2+}]}$$

Reduction is favored when pe $\ll$ pe_0

$$\frac{-d\,[\mathrm{A}(t)]}{dt} = k_f\,[\mathrm{A}(t)]^p\,[\mathrm{B}(t)]^q \tag{2.7}$$

where k_f is the rate constant for the forward reaction,[4] $[\mathrm{A}(t)]$ and $[\mathrm{B}(t)]$ represent the molar concentrations of the reactants at time t, and the sum of the exponents $[p+q]$ is the order of the reaction. A first-order reaction represents a unimolecular transformation (A $\rightarrow$ products), the rate of which is proportional to the concentration of the reactant, A, raised to the power $p = 1$ ($q = 0$). Radioactive decay of a chemical is a good example of a first-order reaction. In many environmental situations, we also consider pseudo-first-order reactions. In these instances, although A is involved in bimolecular or multimolecular reactions (A + B + C $\rightarrow$ products), the concentrations of the other reactants, B and C, are sufficiently large that they are assumed to be unvarying and can be lumped together into the rate constant. Certain hydrolysis reactions can effectively be represented as first-order processes because water functions as the second reactant, B, and has a large unvarying concentration in aqueous environments.

The mass-law expression for the first-order or pseudo-first-order removal of chemical A is written as:

$$\frac{-d\,[\mathrm{A}(t)]}{dt} = k\,[\mathrm{A}(t)] \tag{2.8}$$

and solved to yield the exponential decay equation:

$$[\mathrm{A}(t)] = [\mathrm{A}(0)]\,e^{-kt} \tag{2.9a}$$

where k is the reaction rate constant and $[\mathrm{A}(0)]$ is the molar concentration of the pollutant A at time $t = 0$. Note the rate constant k with units of [1/time] is related to the chemical half-life of the pollutant $t_{1/2}$ as:

$$t_{1/2} = \frac{-\ln\left(\frac{1}{2}\right)}{k} = \frac{0.693}{k} \tag{2.9b}$$

where the half-life denotes the time it takes for the chemical concentration to be reduced to one-half of its original value. The reaction rate equation and solution in Eqs. (2.8) and (2.9) are traditionally represented with concentration in [mass/volume] units, that is, $C^{\mathrm{A}}(t)$. We maintain molar concentration units in this chapter to be consistent with the notation used to describe reversible chemical reactions.

Chemical transformation half-lives for many organic chemicals are compiled in the *Handbook of Environmental Degradation Rates* (Howard et al., 1991). Where a full matrix of information is available, hydrolysis and photolysis half-lives are reported in air and in water; degradation half-lives (loosely referring to microbial degradation) are noted for soil and aqueous systems. Since microbial reaction rates

[4] The forward rate constant, k_f, written here includes the stoichiometric factor, p, shown on the lhs of Eqn. (1.5).

are sensitive to environmental conditions and the suite of microbes present at a site, actual half-lives can in fact be quite variable from one system to another.

As expected, the structure of an organic pollutant governs the ease and speed with which it is transformed. Most abiotic transformations in the aqueous phase occur by means of nucleophilic attacks on the organic molecule. A nucleophile is an electron-rich species (e.g., OH^-) that is attracted to and bonds with positively charged entities within the organic molecule, typically the electron-deficient carbon participating in polar covalent bonds within the parent molecule. Since water molecules with the hydroxide ions (OH^-) are the most ubiquitous nucleophiles in the aquatic environment, hydrolysis is a common abiotic transformation pathway for polar organic molecules. When there are no steric limitations, the incoming OH^- ion displaces the electronegative element, typically a halogen, X, that was participating in the polar bond within the parent molecule. When steric limitations prevent the OH^- ion from fully displacing the halogen, elimination of both a hydrogen and a halogen ion occurs, as shown in Table 2.4.

Abiotic transformations of gaseous pollutants in air are frequently initiated by photochemically driven reactions involving free radical species such as the hydroxyl radical (HO·). Free radicals are first produced in the atmosphere when precursor molecules are broken down by sunlight. Free radicals do not carry an ionic charge but contain unpaired electrons that make them very reactive. (The unpaired electron is represented by a dot adjacent to the molecular formula of the free radical.) As shown in Table 2.4, the hydroxyl radical (HO·) attacks alkanes and other molecules with C–H bonds by abstracting a hydrogen atom to yield water vapor along with a hydrocarbon radical. With alkenes and other molecules containing C=C bonds, the hydroxyl radical can also initiate decomposition by adding to the double bond, again creating a new radical. In either case, the free radical products can continue in a chain of reactions that produce secondary pollutants, including ozone and nitric acid (see Chapter 11). A third body molecule, for example, nitrogen, represented as *M* in Table 2.4, often absorbs some of the excess energy of the free-radical reactions and stabilizes the products.

The other major category of reactions that result in the breakdown of organic chemicals in the environment is biochemical transformation. Biochemical transformation pathways often involve reactions of organic molecules with electrophilic species, wherein the organic chemical loses electrons and hence is oxidized.[5] Under abiotic conditions, most organic compounds are oxidized very slowly, if at all, by diatomic oxygen (O_2) in air at room temperature. However, they are quite susceptible to attack by electrophilic oxygen, which is an electron-deficient form of oxygen that "likes" and seeks out electrons. Enzymes derived from microbes, such as the dioxygenase and monooxygenase enzymes, carry the electron-deficient form of oxygen and greatly enhance its reactive power. In biodegradation reactions, attack with the

[5] In some cases, reduction of organic chemicals is pertinent. For example, reductive dehalogenation of chlorinated compounds occurs in the presence of electron-rich species, for example, iron, both in biotic and abiotic systems, for example, the reduction of chloroform to methyl chloride:

$$CHCl_3 + Fe^0 + 2H^+ \rightarrow CH_3Cl + Fe^{2+} + 2Cl^-$$

electrophilic oxygen occurs at electron-rich sites within an organic molecule, such as double- or triple-bond sites or the π-electron cloud of a benzene ring. Functional groups that contribute to electron-rich centers, such as the –OH group in phenol, promote attack by electrophilic oxygen and consequently biodegrade rapidly. In contrast, chemicals that offer steric limitations or contain functional groups that withdraw electrons from the parent molecule suppress oxidation and biodegrade very slowly. Typical biodegradation pathways for aromatic compounds and alkanes are shown in Table 2.4. As shown in the table, cleavage of the benzene ring typically begins with the formation of catechol, which is then readily metabolized by microbes. Alkanes are likewise transformed to alcohols, aldehydes, acids, and ultimately mineralized to CO_2 in the presence of oxygen. More information on biodegradation pathways of organic molecules can be found in Schwarzenbach et al. (1993).

EXAMPLE 2.4

Explain which of the following chemicals will more readily undergo biochemical oxidation: ethylbenzene, benzene, phenol, pentachlorophenol, and dinitrophenol. Compare with literature estimates of biodegradation half-lives for these chemicals.

Solution 2.4 We expect functional groups that tend to contribute a negative charge to the benzene ring, such as the ethyl group in ethyl benzene and the hydroxyl group in phenol, to promote aerobic biodegradation rates. In contrast, the chloro groups have a lesser tendency to donate electrons to the benzene ring, while the nitro groups withdraw electrons, making pentachlorophenol and dinitrophenol quite recalcitrant to aerobic oxidation processes. Consistent with these expectations, the aerobic half-lives (low range values) reported for these chemicals by Howard et al. (1991) are 6 h, 3 days, 5 days, 23 days, and 2.25 months for phenol, ethylbenzene, benzene, pentachlorophenol, and dinitrophenol, respectively.

The rate of biodegradation of organic chemicals is represented by two different kinetic models:

- The Michaelis–Menten model assumes no significant microbial growth occurs from utilization of the substrate and that the rate of substrate utilization is controlled by the availability of enzymes capable of cleaving the molecule.
- The Monod model is based on the assumption that the utilization of the organic substrate causes significant growth of the microbial population such that the rate of disappearance of the substrate is closely related to the growth of the microbial population.

Although the underlying processes are different, both models have the same mathematical structure and, under certain conditions, can be simplified to yield a pseudo-first-order rate expression for biodegradation (see Homework Problem 2.4). Biodegradation modeling is discussed in more detail in Chapter 12. Note that electrophilic oxygen functions as a strong oxidizing agent in the environment and leads to aerobic biodegradation of chemicals. In the absence of oxygen, other electron acceptors such as NO_3^-, SO_4^{2-}, and even CO_2 can function as oxidizing agents in anaerobic biodegradation reactions wherein the organic chemical is oxidized, that is, loses electrons, while the electron acceptors are themselves reduced. Rates of anaerobic oxidation of organic chemicals are typically slower than aerobic degradation rates but still proceed irreversibly in the forward direction. In contrast, oxidation–reduction reactions involving inorganic species often occur reversibly in the environment, as discussed next.

2.3.2 Reversible Reactions

Consider a reversible reaction involving an inorganic chemical A:

$$pA + qB \Leftrightarrow rC + sD \tag{2.10}$$

Chemical A is consumed by the forward reaction but also produced when the reverse reaction occurs. Both reactions proceed very rapidly, for example, half-lives for many forward and reverse dissociation reactions occurring in water are of the order of fractions of seconds (Stumm and Morgan, 1981). However, since the forward and reverse reactions occur in opposition, a balance between these rates determines the overall progress of the reaction. Hence equilibrium considerations can be applied.

According to the law of mass action, the rate of consumption of A by the forward reaction is given by Eq. (2.7), while the rate of formation of A by the reverse reaction is given by:

$$\frac{d\,[A(t)]}{dt} = k_r\,[C(t)]^r\,[D(t)]^s \tag{2.11}$$

Equilibrium is attained when the rate of the forward reaction is equal to the rate of the reverse reaction such that the concentration of A, and all the other reactants and products, remains constant. At equilibrium, an equilibrium constant, K_{eq}, is defined as the ratio of the forward and backward *rate constants*, k_f and k_r, which is also mathematically equal to the ratio of the equilibrium product and reactant molar concentrations, raised to the power of their respective stoichiometric coefficients, as shown in Eq. (2.12). Note that the time dependence of molar reactant and product concentrations is omitted in Eq. (2.12) to represent invariant equilibrium conditions:

$$K_{eq} = \frac{k_f}{k_r} = \frac{[C]^r\,[D]^s}{[A]^p\,[B]^q} \tag{2.12}$$

When the numerical value of K_{eq} is much greater than 1, the reaction proceeds spontaneously in the forward direction. The equilibrium constant K_{eq} is thermodynamically related to the free energy change of the reaction, ΔG_0, as:

$$\ln K_{eq} = \frac{-\Delta G_0}{RT} \tag{2.13}$$

Equation (2.13) indicates that large negative values of ΔG_0 will promote the feasibility and spontaneity of the reaction in the forward mode. The free energy change, ΔG_0, may be computed from the deficit between the free energy needed to create the products and that created by breakdown of the reactants and reflects the impact of chemical structure and bonding on reversible reactions. For a particular reversible reaction, departures from equilibrium can occur due to changes in environmental conditions. In such cases the reaction will proceed, in forward or in reverse mode, as needed to regain equilibrium. The direction in which the reaction proceeds can be determined by computing a reaction quotient, RQ, which is then compared with the equilibrium constant K_{eq}.

$$\mathrm{RQ} = \frac{[\mathrm{C}(t)]^r\,[\mathrm{D}(t)]^s}{[\mathrm{A}(t)]^p\,[\mathrm{B}(t)]^q} \tag{2.14}$$

When RQ $> K_{eq}$, the reaction proceeds to the left (reverse) to regain equilibrium.

When RQ $< K_{eq}$, the reaction proceeds to the right (forward) to regain equilibrium.

Reversible reactions such as acid–base dissociation, mineral dissolution–precipitation, metal complex formation and destabilization, and oxidation and reduction together govern the species of inorganic chemicals that are dominant in the environment.

Dissociation of Acids and Bases An acid (more specifically, a Lewis acid) is defined as a chemical that dissociates in water to release protons, that is, hydrogen ions (H^+ ions). Acids may be monoprotic, such as hydrochloric acid, which can only release one proton, or diprotic (e.g., carbonic acid, H_2CO_3), or triprotic (phosphoric acid, H_3PO_4), releasing two or three protons, respectively. In contrast, a base is a chemical that releases the hydroxyl ion (OH^-). Water itself is composed of hydrogen and hydroxyl ions. The dissociation of a water molecule may be written as:

$$\mathrm{H_2O} \Leftrightarrow \mathrm{H^+} + \mathrm{OH^-} \qquad K = \frac{[\mathrm{H^+}][\mathrm{OH^-}]}{\text{Activity of } \mathrm{H_2O} = 1} \tag{2.15}$$

Since the activity of water is unity by convention, the equilibrium constant for the dissociation of water is called the ionic product of water, K_w, defined as:

$$K_w = [\mathrm{H^+}][\mathrm{OH^-}] \quad \text{or} \quad -\log[K_w] = \mathrm{pH} + \mathrm{pOH} \tag{2.16}$$

where log[ ] denotes the base 10 logarithm, and K_w is approximately 10^{-14} at a temperature of 25°C, and is slightly sensitive to environmental temperatures. Equation (2.16) indicates that the sum of the pH of water, defined as the negative log (base 10) of the hydrogen ion concentration (pH = $-\log[H^+]$), and the pOH of water (pOH = $-\log[OH^-]$) is 14. Thus, the pH of water can, at most, vary from 0 to 14. Most natural water bodies have pH values ranging from 5 to 9, below and beyond which aquatic life can be threatened. The pH of deionized water is called the neutral pH, where pH = pOH $\sim$ 7, with no other ions being present in solution.

A weak acid is one that can exist in undissociated form within the pH range of waters. The dissociation of an acid is written as:

$$\mathrm{HA} \Leftrightarrow \mathrm{H}^+ + \mathrm{A}^- \qquad K_a = \frac{[\mathrm{H}^+][\mathrm{A}^-]}{[\mathrm{HA}]} \tag{2.17}$$

where K_a is the equilibrium constant, called the acid dissociation constant. Strong acids are characterized by large values of $K_a(\gg 1)$, that promote the forward reaction, that is, addition to water of strong acids such as hydrochloric acid and sulfuric acid results in the irreversible and complete dissociation of these acids to release equivalent amounts of H^+ and A^- ions. In contrast, weak acids, with K_a in the range of 1 to 10^{-14}, exhibit reversible dissociation of acid molecules within the pH range of water. Monoprotic weak acids such as acetic acid can exist in water in both dissociated form (deprotonated to A^-, after release of H^+), as well as the undissociated form (HA), the dissociation being readily reversible and strongly dependent upon the pH of water. The total molar concentration of species A in water, $C_{T,\mathrm{A}}$, is the sum of the molar concentrations of the protonated form [HA] and the deprotonated form $[A^-]$, yielding a simple mass balance:

$$C_{T,\mathrm{A}} = [\mathrm{HA}] + [\mathrm{A}^-] \tag{2.18}$$

The fraction of $C_{T,\mathrm{A}}$ that exists in the form of HA and A^- can be determined by combining Eq. (2.17) with (2.18) to yield:

$$\begin{aligned} \text{Undissociated acid fraction} = \alpha_0 &= \frac{[\mathrm{HA}]}{C_{T,\mathrm{A}}} = \left[1 + \frac{K_a}{[\mathrm{H}^+]}\right]^{-1} \\ \text{First dissociated acid fraction} = \alpha_1 &= \frac{[\mathrm{A}^-]}{C_{T,\mathrm{A}}} = \left[\frac{[\mathrm{H}^+]}{K_a} + 1\right]^{-1} \end{aligned} \tag{2.19}$$

Equation (2.19) indicates that when $H^+ \gg K_a$, that is, pH $\ll$ pK_a, the acid is largely undissociated and $C_{T,\mathrm{A}} \sim$ [HA]. In contrast, when the pH is large (pH $\gg$ pK_a), the H^+ concentration is very small and the dissociation reaction [Eq. (2.17)] is promoted in the forward direction, resulting in dominance of the A^- species, such that $C_{T,\mathrm{A}}$ $\sim [A^-]$.

The relationships in Eq. (2.19) can be plotted on a graph of log[concentration] versus pH, as shown in Figure 2.7. A charge balance, or electroneutrality condition,

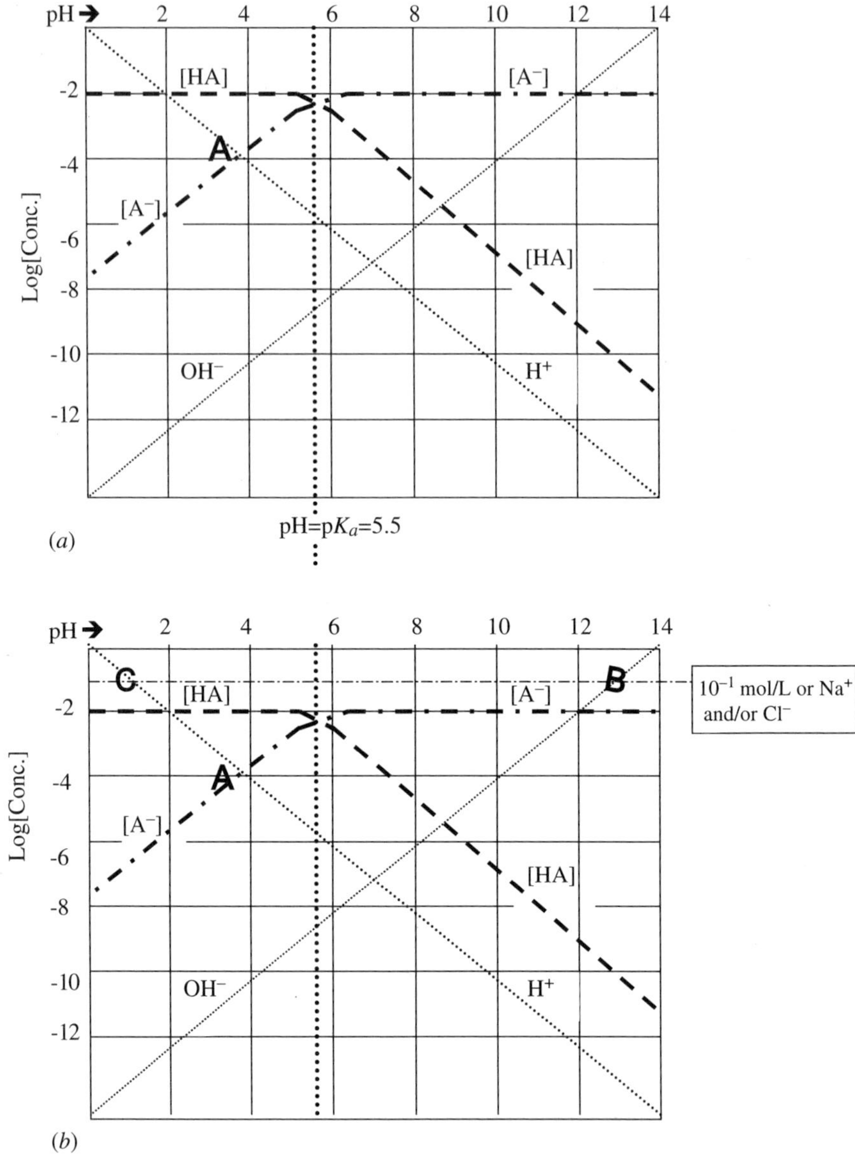

Figure 2.7 (*a*) Log[concentration] versus pH graph for the equilibrium dissociation of a monoprotic acid, HA, with $pK_a = 5.5$. The slopes of the log[HA] and log[A^-] curves can be determined by inspection of Eq. (2.19) after applying $\log_{10}$ transformations to both sides of the equation. (*b*) Graphical solution technique to determine equilibrium species concentrations with various additions to the HA system.

- Mass Balance
- Charge Balance
- Mass Action

is then imposed along with the coupled mass balance and dissociation equations [Eq. (2.19)] to yield a system of three nonlinear equations. Assuming that HA is added to deionized water (water containing no other ions), the charge balance yields:

$$[H^+] = [OH^-] + [A^-] \tag{2.20}$$

The dissociation equilibrium equation [Eq. (2.17)], mass-balance equation [Eq. (2.18)], and charge-balance equation [Eq. (2.20)] are solved simultaneously to determine the three unknown species concentrations in water, namely, [HA], $[A^-]$, and $[H^+]$. Numerical methods described in Chapter 6 can be used to determine the species concentrations, or, as is illustrated in Example 2.5, an elegant graphical solution technique may be employed. Note that OH^- concentrations can readily be determined from Equation (2.16) once the equilibrium pH is determined. Changes in equilibrium pH due to additions of strong acids and bases to the system can also be determined by the same techniques. Example 2.5 illustrates how the dominant weak acid–base species change in aqueous systems as the pH varies.

EXAMPLE 2.5

Use Figure 2.7*a* to determine the equilibrium condition (pH, HA, and A^- concentrations) when 10^{-2} mol of HA are added to 1 L of water. The pK_a for HA is 5.5. How does the equilibrium pH shift when, in addition to the HA, the following are added? (a) 10^{-6} mol of NaOH (a strong base); (b) 10^{-1} mol of NaOH; (c) 10^{-1} mol of HCl (a strong acid); (d) 10^{-1} mol each of NaOH and HCl. What is the impact of changing pH on species dominance in the system?

Solution 2.5 In the graphical method, we plot the coupled mass-balance and dissociation equilibria relationships shown in Eq. (2.19) on a graph of log[concentration] versus pH. Next, we apply appropriate charge-balance conditions, which yield solutions as shown in Figure 2.7*b*.

Base Case When 10^{-2} mol/L HA is added to water, the electroneutrality condition becomes $[A^-] + [OH^-] = [H^+]$, which is satisfied at point A in Figure 2.7*a*, at a pH of approximately 3.8. At this pH, the HA and A^- concentrations can be read off the graph and are found to be $\sim 10^{-2}$ and $10^{-3.8}$ mol/L, respectively. At this pH, HA is the dominant species. The charge balance is satisfied since $[A^-] = [H^+] = 10^{-3.8}$ mol/L, with $[OH^-]$ being much smaller at $10^{-10.2}$ mol/L and hence contributing negligibly to the charge balance.

Addition of other chemicals yields the following equilibrium conditions:

- Addition of 10^{-6} mol of NaOH (a strong base): The Na^+ concentration in water is 10^{-6} mol/L since NaOH is a strong base and dissociates completely. This yields the following charge balance: $[Na^+] + [H^+] = [A^-] + [OH^-]$

(electroneutrality). The $[Na^+]$ concentration is so much smaller than the $[H^+]$ concentration at point A, ($10^{-6} \ll 10^{-3.8}$) that it does not significantly change the equilibrium condition achieved at point A. Thus, the addition of strong base was at insufficient quantities to change the pH in a measurable manner.

- Addition of 10^{-1} mol of NaOH: Now a significantly larger amount of Na^+ enters the charge balance, which effectively becomes $[Na^+] \sim [OH^-]$ since $[A^-]$ is an order of magnitude smaller than $[Na^+]$ and can never satisfy the constraint. The equilibrium condition is at point B, at a pH of approximately 13. (Very basic!) At this condition, pH $\sim$ 13, pOH $\sim$ 1, $[Na^+] = 10^{-1}$, and $[A^-] \sim C_{T,A} = 10^{-2}$ mol/L. A^- is thus the dominant species.
- When 10^{-1} mol of HCl is added, it dissociates irreversibly yielding 10^{-1} mol/L of Cl^- ions in aqueous solution. The electroneutrality condition becomes $[H^+] = [A^-] + [Cl^-] + [OH^-]$, which can only be solved when $[H^+] \sim [Cl^-]$ at point C in Figure 2.7*b*, at a pH of 1. At this pH both $[A^-]$ and $[OH^-]$ are at a much lower concentration than 10^{-1} mol/L, and hence can be neglected from the electroneutrality condition. An equilibrium pH of 1 indicates very acidic conditions due to the high concentration of strong acid in water. At this pH the reverse acid dissociation reaction is promoted to compensate for the excess H^+ ions in solution and HA becomes the dominant species.
- 10^{-1} mol each of NaOH and HCl will dissociate completely such that the 10^{-1} mol/L of Na^+ and Cl^- in solution neutralize each other. The electroneutrality condition—$[Na^+] + [H^+] = [A^-] + [Cl^-] + [OH^-]$—reverts back to the base case with equilibrium at point A.

Naturally occurring weak acids such as acetic acid and carbonic acid provide a great deal of flexibility to aqueous chemical systems. When pH is lowered by the addition of foreign agents such as a strong acid, the weak acid–base system is able to compensate by reversing the acid dissociation reactions [i.e., consuming the excess H^+ ions by reversing Eq. (2.17)]. On the other hand, when the pH is increased, resulting in a drop in H^+ ion concentrations, the acid dissociation proceeds in the forward direction producing more H^+ ions. The reversible nature of weak acid–base reactions thus provides water with the buffering capacity to mitigate large pH changes, providing a stable environment for the support of aquatic life.

Dissolution–Precipitation of Minerals In addition to naturally occurring weak acids and bases, water quality is also affected by the nature and quantities of mineral salts that are dissolved in it. Consider the dissolution–precipitation equilibrium for a generic mineral MA_2, where each mineral molecule dissolves to release one cation (M^{2+}) and two anions (A^-) in solution.

$$MA_{2(\text{solid})} \Leftrightarrow M^{2+} + 2A^- \qquad K_{sp} = \frac{[M^+][A^-]^2}{[\text{Unit solid activity} = 1]} \tag{2.21}$$

Because the activity of the solid is set to unity by convention, the equilibrium constant for the dissolution–precipitation of mineral MA_2 becomes the product of the equilibrium molar aqueous concentrations of the constituent cations and anions in solution raised to the power of their stoichiometric coefficients. The equilibrium constant, K_{sp}, is hence called the solubility product, and K_{sp} sets the upper limit for the concentrations of dissolved mineral species in water. The solubility S of the mineral MA_2 may thus be obtained from K_{sp} data, as illustrated in Example 2.6. Once the solubility is computed, the aqueous concentrations of the dissolved cations and anions can be obtained from stoichiometry, assuming these ions are released into water only by the dissolution of a single mineral. When multiple sources release the same cation or anion to water, a "common ion" effect is observed. In such cases the total ion concentration in water (derived from all sources) represents the ionic activity in water; and the ion activity impacts Eq. (2.21) by lowering mineral solubility to compensate for the common ion occurring from another source. This effect is illustrated in Example 2.6.

EXAMPLE 2.6

The K_{sp} for $PbCl_2$ is $10^{-4.8}$. What is the solubility S of this mineral in deionized water? How would the solubility change if the mineral came into contact with water that already contained 0.6 mol/L chloride in solution from some other source?

Solution 2.6 The dissolution of $PbCl_2$ follows the same stoichiometry represented in Equation 2.21. The dissolution of S moles of the mineral will result in S moles of cations and $2S$ moles of anions. Hence, from Equation (2.21), solubility S can be computed from $[S]\,[2S]^2 = K_{sp}$; $[S]\,[2S]^2 = 10^{-4.8}$, which yields $S =$ 0.016 mol/L, $[Pb^{2+}] = 0.016$ mol/L, and $[Cl^-] = 0.032$ mol/L.

When 0.6 mol/L of chloride is already present in the water, the new mineral solubility becomes $[S]\,[0.6 + 2S]^2$, which yields $S = [Pb^{2+}] \sim 4.4 \times 10^{-5}$ mol/L. Thus, independent increases in chloride (anion) concentrations result in a decrease in the aqueous solubility of lead chloride and the corresponding aqueous lead ion concentration.

Mineral dissolution often sets the upper limit for aqueous-phase metal concentrations in the source zone, that is, at mineral-enriched matrices in contact with surface water and groundwater. Note that aqueous solutions are not always instantaneously at equilibrium with associated solid phases. In some cases, when the product of the aqueous cation and anion concentrations, that is, the RQ, is less than the K_{sp}, waters are undersaturated with the mineral species and dissolution is favored. When the product exceeds the solubility product, a supersaturated solution is formed, eventually leading to mineral precipitation to maintain equilibrium conditions. In some cases, minerals can also be transformed from one solid form to another, in a phenomenon called incongruent dissolution. However, in general, minerals largely undergo congruent dissolution, that is, the solid phase dissolves to release mineral cations and anions to water.

Formation and Destabilization of Metal Complexes The chemistry of complex formation can significantly affect the behavior of dissolved metals in water. Typically, metal ions function as the electron-accepting groups in solution, being attracted to negatively charged ions in water to form a "complex" via a coordination bond. Both electrons in the coordination bond are donated by the negatively charged species, which is called a ligand; the electron-seeking (electrophilic) metal ion functions as the central atom in the complex. The central atom is characterized by a coordination number, denoting the number of electrons needed to achieve a stable configuration. Likewise, ligands are further classified as monodentate, bidentate, tridentate, and so forth, based on the number of sites at which coordination bonds can be formed. Complex formation of a central metal atom with a multidentate ligand is also called chelation, in which case the mulidentate ligand is also referred to as a chelating agent and the complex formed is called a chelate. Chelates have complex structures based on the bonding sites in the ligand and central atom, and chelation can often be accompanied by a color change in aqueous solution due to a shift in the electrons participating in the coordination bond. Some chelating agents are produced naturally by plants and microbes; others are synthetic, such as EDTA (ethylenediaminetetraacetic acid), which is used in metal assays to extract metals from solution. Complexation by naturally occurring chelating agents plays a very important role in determining metal transport and toxicity in the environment. Complexation increases the apparent solubility of the metal in water. Furthermore, metals in complexed form are, in general, less toxic to aquatic organisms when compared with their corresponding free ionic species.

A metal complex can be destabilized to yield its component metals and ligands in solution. Thus, complex formation can be viewed as an equilibrium process, in which the equilibrium constant, referred to as the stability constant, K_s, provides a measure of the stability of the complexed metal species in water:

$$\mathrm{M^{2+} + Li^{2-}\ (ligand) \Leftrightarrow MLi\ (complex)}$$
$$K_s = \frac{[\mathrm{MLi}]}{[\mathrm{M^{2+}}][\mathrm{Li^{2-}}]} \tag{2.22}$$

Complex formation is favored when the reaction quotient, RQ, for Equation (2.22) is less than the stability constant.

Oxidation and Reduction In addition to complex formation, the mobility and toxicity of metals in the environment are strongly affected by oxidation and reduction reactions. Oxidation is defined as a reaction involving loss of electrons, while reduction involves a gain of electrons. An oxidation half reaction is normalized to the loss of one electron and is coupled with a reduction half reaction that is also normalized stoichiometrically for a gain of one electron, to yield a redox reaction pair.

TABLE 2.5 Equilibrium Constants for Various Reducing Half Reactions at 25°C[a]

Reaction	$pe_0 = \log K_{eq}$	ΔG_0, calories ($= -RT \ln K_{eq}$)
$\frac{1}{4}O_2(g) + H^+ + e = \frac{1}{2}H_2O$	+20.75	28191
$\frac{1}{2}NO_3^- + H^+ + e = \frac{1}{2}NO_2^- + \frac{1}{2}H_2O$	+14.15	19224
$\frac{1}{8}NO_3^- + 5/4H^+ + e = \frac{1}{8}NH_4^+ + \frac{3}{8}H_2O$	+14.90	20243
$\frac{1}{6}SO_4^{2-} + 4/3H^+ + e = \frac{1}{6}S(s) + \frac{1}{4}H_2O$	+6.03	8192
$\frac{1}{2}S(s) + H^+ + e = \frac{1}{2}H_2S(g)$	+2.89	3926
$\frac{1}{8}CO_2(g) + H^+ + e = \frac{1}{8}CH_4(g) + \frac{1}{4}H_2O$	+2.87	3899
$\frac{1}{4}CO_2(g) + H^+ + e = \frac{1}{4}CH_2O + \frac{1}{4}H_2O$	−1.20	−1630

[a]Oxidizing agents are being reduced in the first six tabulated half reactions; the preference for oxidizing agents is determined by the magnitude of log K_{eq} ($O_2 > NO_3^- > SO_4^{2-} > S > CO_2$).
Source: Compiled from Stumm and Morgan (1983).

Conventionally, all half reactions are written in the form of reducing half reactions, that is, as a gain of electrons. Equilibrium constants for some important half reactions are tabulated in Table 2.5. Various combinations of reducing reactions such as those summarized in Table 2.5 can be coupled with oxidation reactions (reduction reactions written in reverse form) to form a redox reaction pair. A redox reaction will occur spontaneously if the combination of the reducing and oxidizing half reactions yields a net negative value of ΔG_0.

Several oxidizing agents are available in the environment. These chemicals themselves get reduced while they oxidize the chemical of interest. The strength of these oxidizing agents is determined by the reduction reaction that yields the greatest free energy change (most negative value of ΔG_0, or equivalently most positive value of $\log K_{eq}$). From Table 2.5, we see that the oxidizing agents can be ranked in order of decreasing strength as: $O_2 > NO_3^- > SO_4^{2-} > S > CO_2$. The last reducing half reaction shown in Table 2.5 is infeasible as written since its $K_{eq} < 1$. This reaction in fact occurs as an oxidizing reaction in reverse and describes the oxidation of CH_2O. A redox pair is a feasible combination of reducing and oxidizing half reactions yielding a net negative ΔG_0 (or large net values of $K_{eq} > 1$).

EXAMPLE 2.7

Write a redox reaction pair representing the oxidation of CH_2O by O_2. What are K_{eq} and ΔG_0 for the net reaction? Is CH_2O more spontaneously oxidized by sulfur (S) than by O_2?

Solution 2.7 The redox pair representing oxidation of CH_2O by O_2 is:

Reduction of $O_2(g)$: $\frac{1}{4}O_2(g) + H^+ + e = \frac{1}{2}H_2O$ $\quad K_{eq} = 10^{20.75}$

Oxidation of CH_2O: $\frac{1}{4}CH_2O + \frac{1}{4}H_2O = \frac{1}{4}CO_2(g) + H^+ + e$ $\quad K_{eq} = 10^{1.20}$

Net redox reaction: $\frac{1}{4}CH_2O + \frac{1}{4}O_2 = \frac{1}{4}CO_2 + \frac{1}{4}H_2O$ $\quad K_{eq} = 10^{21.95}$

The redox pair representing oxidation of CH_2O by S is:

Reduction of S: $\frac{1}{2}S(s) + H^+ + e = \frac{1}{2}H_2S(g)$ $\quad K_{eq} = 10^{2.89}$

Oxidation of CH_2O: $\frac{1}{4}CH_2O + \frac{1}{4}H_2O = \frac{1}{4}CO_2(g) + H^+ + e$ $\quad K_{eq} = 10^{1.20}$

Net redox reaction: $\frac{1}{4}CH_2O + \frac{1}{2}S + \frac{1}{4}H_2O = \frac{1}{4}CO_2 + \frac{1}{2}H_2S$ $\quad K_{eq} = 10^{4.09}$

The K_{eq} for the net reaction is obtained as a product of the equilibrium constants of the two constituent half reactions. From the magnitudes of the net K_{eq}, we see that oxidation by O_2 is favored, that is, O_2 is a stronger oxidizing agent.

ΔG_0 for oxidation of CH_2O by O_2 is computed from Equation (2.13) as:

$$\Delta G_0 = -RT \ln K_{eq} = 1.98 \text{ cal/equiv. K} \times 298 \text{ K} \times \ln[10^{21.95}] = 29{,}820 \text{ cal}$$

Equilibrium relationships for the reducing half reactions are constructed in the usual way, that is, the ratio of product to reactant concentrations raised to their respective stoichiometric coefficients. The activity of free electrons in solution is denoted by $[e^-]$, with $pe = -\log_{10}[e^-]$. Thus, the equilibrium relationship for reduction of a metal ion may be written as:

$$M^{2+} + e^- \Leftrightarrow M^+ \qquad K = \frac{[M^+]}{[M^{2+}][e^-]} \qquad \text{or}$$
$$\log[K] + \log[e^-] = \log\frac{[M^+]}{[M^{2+}]} \tag{2.23}$$

When pe^0 is defined as log[K], and pe as –log[e^-], the above equation becomes

$$pe = pe^0 - \log\frac{[M^+]}{[M^{2+}]} \tag{2.24}$$

Values for pe^0 for various half reactions are shown in Table 2.5. One can view pe as analogous to pH in an acid–base equilibrium system, with pe^0 corresponding to pK_a. When $pe \gg pe^0$, the number of free electrons in solution is very small and the oxidation reaction [reverse of Eq. (2.23)] is promoted resulting in predominance of M^{2+} ions in solution with a corresponding release of free electrons. Conversely, when $pe \ll pe^0$, there is a large supply of free electrons and the reduction reaction is favored. Thus, on a plot of log[conc.] versus pe, the trend in species dominance is from reduced to oxidized species as pe increases beyond pe^0. Thus, low pe conditions represent reducing conditions, while high pe conditions promote oxidation. Environmental conditions, characterized by pe, control the predominance of reduced and oxidized pollutant species in the environment.

2.4 TOXICITY

Thus far, we have focused on the spontaneity and rate with which a chemical, A, is transformed in the environment to yield various products. Similar transformations occurring within the human body can be expected to control the ultimate toxicity of the chemical to humans. However, because of the interplay between complex enzymatic systems active in living beings, it is quite difficult to make a priori predictions of chemical toxicity based on reaction pathways. While a few generalizations have been made on the transformations that pollutants can undergo in mammalian systems (e.g., Reeves, 1981; Vogel et al., 1987; Lu, 1991), toxicity information is largely obtained from observational data.

The toxic effects of a chemical can manifest themselves in the human body in many different ways. A chemical is termed acutely toxic if short-term exposure to the chemical results in illness. Carbon monoxide poisoning is an example of this type of an effect. Exposure to high concentrations of carbon monoxide in air can rapidly produce symptoms beginning with headache and dizziness and eventually lead to asphyxiation and death. Chronic toxicity, on the other hand, refers to manifestation of ill effects after long periods of exposure to the chemical.

A chemical is known as a mutagen if it can alter the cellular genetic material [deoxyribonucleic acid (DNA)] in ways that can be transmitted during cell division. Cancer-causing chemicals and agents are termed carcinogens. Carcinogens include genotoxic agents, that is, agents that initiate the cell DNA transformation process, as well as epigenetic agents, such as promoters and co-carcinogens that enable and promote the formation of tumors. Not all cell mutations result in cancer—thus, a mutagen is not necessarily a carcinogen. A teratogen is a chemical that can adversely affect a developing fetus while not necessarily harming the mother.

The harmful human health effects of chemicals are quantified via dose–response curves. Such curves, illustrated in Figure 2.8, plot the level of adverse health effects on the ordinate versus increasing doses of the chemical on the abscissa. Two types of idealized dose–response curves are shown in Figure 2.8: a threshold limit value curve and a no-threshold curve. The dotted line represents a threshold limit value chemical; the threshold represents a "safe" dose below which no adverse effects are observed. Carcinogens are often assumed to exhibit no-threshold dose, as suggested by the solid line in Figure 2.8. There is no safe dose for such a chemical. In both scenarios, the slope of the dose–response curve (above the threshold) provides a measure of the expected incremental increase in adverse effects caused by an incremental increase in dose.

In practice, the toxicity of a chemical is first evaluated from epidemiological studies, animal studies, or cell assays. Epidemiological studies examine health effects data in human populations, comparing populations that are exposed versus those not

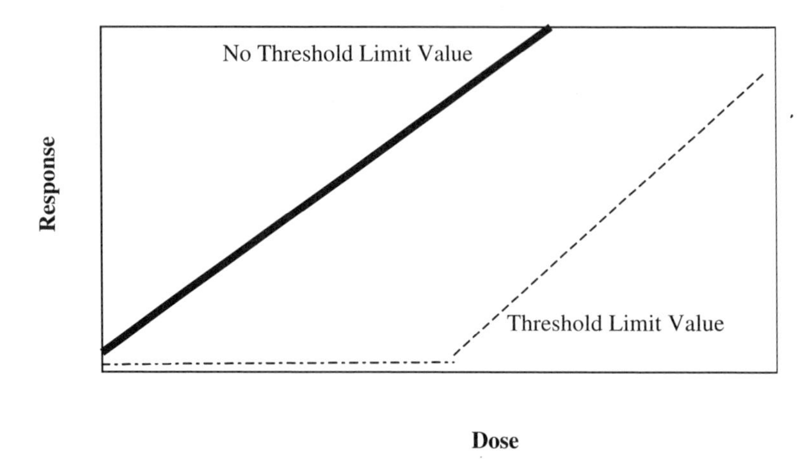

Figure 2.8 Dose–response curves for carcinogens (solid line) are assumed to have no threshold: that is, any exposure produces some chance of causing cancer. Many noncarcinogens are modeled with a threshold limit value (dotted line).

exposed to a chemical in cohort studies, or comparing exposures in people with and without a disease in case control studies. The studies may look at past data (retrospective) or at the behavior of controlled groups in food and drug test trials that go forward in time (prospective). Epidemiological studies are the best indicator of a chemical's toxic effects on humans but are limited by the lack of reliable retrospective data, lack of adequate control groups, small population sizes, and confounding parameters, for example, variation in diet and personal habits such as cigarette smoking.

Most chemicals are tested for toxicity in relatively quicker and better controlled animal studies. Animal tests can be short-term, acute tests where a single species is subjected to very high doses of a chemical. If the incidence of adverse effects is greater in the exposed population than that in the unexposed population, longer-term chronic tests at lower doses are conducted with more animals. While the animal tests can be conducted with larger population sizes and better controls, results from these tests are not directly applicable to humans. This is because of cross-species differences in metabolism and susceptibility, and also the need to administer high doses to the animals in order to observe effects in the small numbers tested.

Several indices of toxicity are obtained from dose–response curves developed from animal studies. The acute toxicity tests that are administered at high doses often result in death in the exposed animal populations. These tests can yield information on the lethal dose (LD) of a chemical. Typically, the LD_{50} is reported, that is, the dose at which half the exposed population perishes, usually in a 14-day test period. For air pollutants a lethal concentration (LC_{50}) is reported on the same basis. Figure 2.9*a* shows the LD_{50} for three chemicals. Note that chemicals with higher LD_{50} values are less toxic since a higher dose is required to elicit the same 50% death response. The low-dose chronic exposure tests are conducted with animals administered doses well below the LD_{50} to elicit information on the threshold limit value, or "safe" dose for

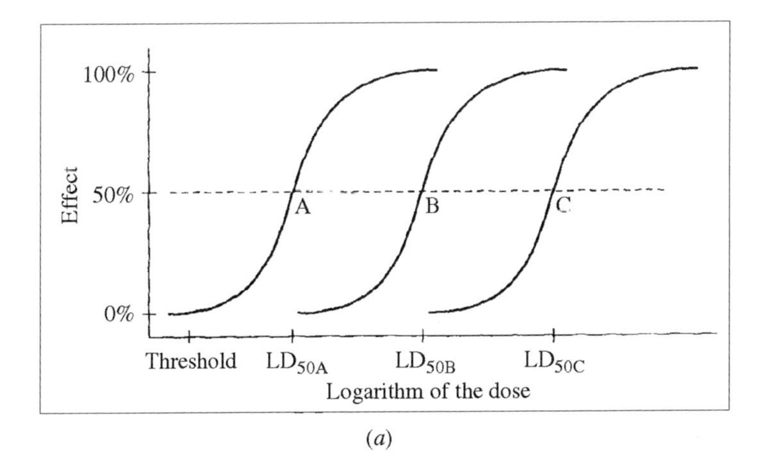

(*a*)

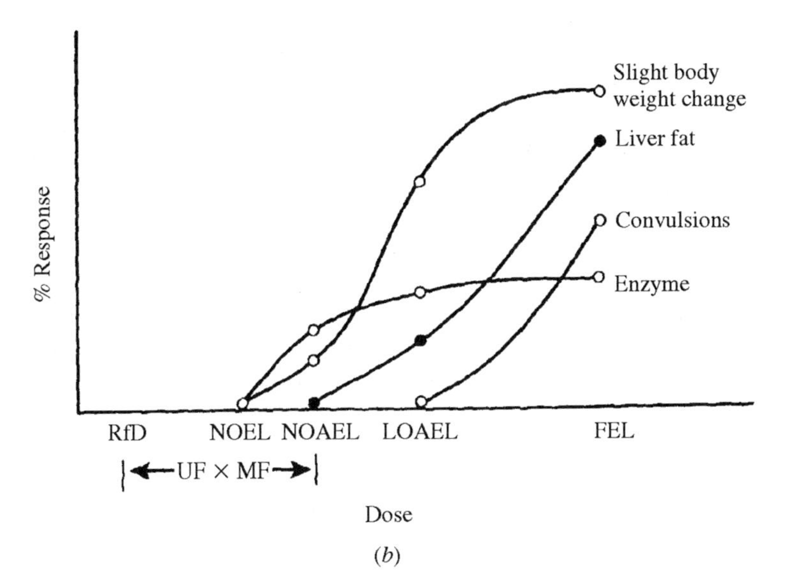

(*b*)

Figure 2.9 (*a*) By plotting cumulative dose–response curves (log dose), one can identify those doses of a toxicant or toxicants that affect a given percentage of the exposed population. A comparison of the values of LD_{50A}, LD_{50B}, and LD_{50C} ranks the toxicants according to relative potency for the response monitored (from Ostler et al., 1996). (*b*) Use of uncertainty factors (UF) and modifying factors (MF) to calculate RfD from an experimental NOAEL. At the lower dose, the NOEL (no-observed-effect level), no changes are observed. The next higher dose (NOAEL, no-observed-adverse-effect level) elicits response(s) such as slight body weight change, but none considered to be adverse. Changes at the LOAEL (lowest-observed-adverse-effect level) are deleterious. The FEL (frank-effect level) is a dose that induces very serious or life-threatening effects. (© 1993 From Hazard Assessment of Chemicals, 8th ed., by C. O. Abernathy, C. Cantilli, J. T. Du, and O. A. Levander. Reproduced by permission of Routledge Inc., part of The Taylor & Francis Group.)

noncarcinogens (see Fig. 2.8). Beginning at low doses, a no observable effect level (NOEL), and a no observable adverse effect level (NOAEL) are noted (see Fig. 2.9*b*). Beyond the lowest observed adverse effect level (LOAEL) one observes deleterious effects on animal health, while doses much above this level elicit serious or life-threatening effects and are termed the frank effect level (FEL). The safe reference dose (RfD) for humans is computed from the NOAEL obtained from animal studies by applying uncertainty factors and cross-species extrapolation factors. Chapter 13 discusses these calculations in more detail, as well as further scientific and ethical issues involved in toxicity assessment.

The assessment of carcinogenicity through animal studies typically requires both acute and chronic exposure tests with multiple species. The longer-term animal studies may be very expensive, requiring more than a year to obtain results. A combination of human epidemiological evidence and animal tests results is used to positively identify a chemical as a human carcinogen. Due to uncertainty in our knowledge of the carcinogenic nature of many chemicals, the U.S. EPA adopted a weight-of-evidence framework in 1986 that classified chemicals into categories A, B1, B2, C, D, and E, ranging from known (group A), probable (group B), and possible (group C) human carcinogens to unclassified chemicals (group D) and those with evidence of non-carcinogenicity (group E). Based on updated information, EPA more recently proposed replacing the original 6 categories with three broad categories of carcinogens—known/likely, cannot be determined, and not likely—with a total of 13 qualitative subdescriptors across the three categories (U.S. EPA, 1997c). As of the date of publication of this book, these new categories have not yet been widely adopted.

In addition to the epidemiological and animal studies described above, rapid-screening in vitro tissue culture tests can be used to determine whether a chemical can induce mutagenesis. Cell mutagenesis tests, such as the Ames test, provide quick initial inferences about the mutagenicity of a chemical. For example, the Ames test employs a strain of the bacteria *Salmonella typhimurium* that has been altered such that it is highly sensitive to mutagens and unable to repair the damage done to its DNA. However, since mutagenesis does not always result in carcinogenesis, such tests provide no conclusive evidence of the carcinogenicity of a chemical. Furthermore, they use bacterial cells that do not respond to metals and organ-specific chemicals the way mammalian cells do. To overcome these difficulties, other cell assays have been developed that employ mammalian cells. Although these tests offer a quick screening assessment, they are all based on the assumption that all mutations lead to cancer and hence do not provide conclusive evidence of carcinogenicity.

The reactivity and toxicity of chemical pollutants are both closely related to chemical structure, however, the structure–activity relationships for toxicity effects are not fully and clearly understood. No universal rule has been found to determine what it is that makes a chemical toxic to humans. This is probably due to the complexity of the human body's metabolic and immune systems, variations across gender and age, the vast array of natural and synthetic chemicals ingested, and the often confounding synergistic or antagonistic effects of chemicals on the human body.

In recent years, quantitative structure–activity relationships (QSARs), have been developed to relate chemical structure to the toxicity and reactivity of a chemical.

Physicochemical parameters such as molecular weight, density, diffusion coefficient, hydrophobicity, vapor pressure, and heat of fusion are correlated with indices that describe the toxicity of the pollutant (Cronin and Dearden, 1995; Karcher and Devillers, 1990). Various correlations are used to represent different interactions of the chemical with biota, including human organ systems. For example, baseline toxicity (also called nonpolar narcosis) of simple nonreactive hydrocarbons is caused by their physical ability to be present within cell material, which is correlated with their hydrophobicity. More complex models consider the physiology of the organism and its mode of intake, digestion, metabolism, and excretion of the chemical. These models are called physiologically based pharmacokinetic (PBPK) models and are discussed in detail in Chapter 13.

2.5 REGULATION OF ENVIRONMENTAL POLLUTANTS

The U.S. EPA currently administers 10 comprehensive environmental protection laws, as illustrated in Figure 2.10. An overview of the main environmental laws applicable to air, water, and land systems is provided below.

Air The Clean Air Act sets standards and goals for air quality in the United States. The 1990 Clean Air Act amendments also address transboundary air pollution problems of acid rain and stratospheric ozone depletion. Under the Clean Air Act, the EPA has established National Ambient Air Quality Standards (NAAQS) for six criteria pollutants. These six pollutants are carbon monoxide (CO), lead (Pb), nitrogen dioxide (NO_2), ozone (O_3), sulfur dioxide (SO_2), and particulate matter (PM). Standards for particulate matter have now been expanded to include particles less than 2.5 μm in

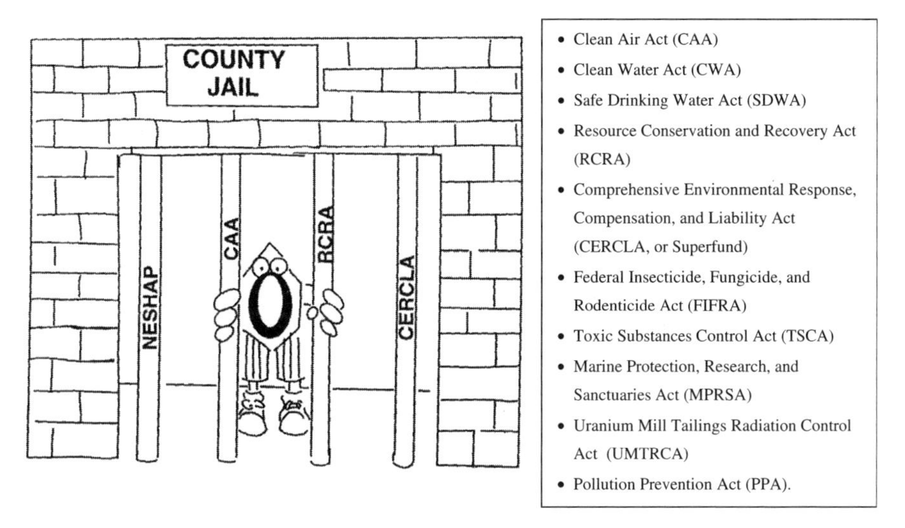

Figure 2.10 Major environmental regulations.

diameter (PM-2.5). Two types of standards are in place: primary standards designed to protect human health, and secondary standards intended to protect public welfare (vegetation, materials, and visibility). The current primary and secondary standards for the six criteria pollutants can be found at http://www.epa.gov/ttnamti1/files/ambient/criteria/reldocs/naaqs699.pdf.

Individual states, through their state implementation plans (SIPs), monitor and enforce the control of air pollutant emissions within their regions to ensure that the NAAQS are met. Air quality modeling is used in the development of SIPs to determine areawide emission limits needed to meet the NAAQS. Modeling is also used to set limits for individual pollution sources that emit the six criteria pollutants. The 1990 Clean Air Act Amendments added provisions for limiting human health risk posed by 189 toxic or hazardous air pollutants (HAPs). Hazardous air pollutants are those pollutants that are known or suspected to cause cancer or other serious health effects, including damage to the respiratory or nervous systems, birth defects, and reproductive effects. The HAPs are listed at http://www.epa.gov/ttn/atw/orig189.html. Technology-based emission reductions, to be followed up with residual risk determinations, are specified for 175 source categories that emit one or more of the HAPs.

Water The two major federal statutes governing water quality are the Safe Drinking Water Act (SDWA) and the Clean Water Act (CWA). The SDWA is concerned with public health associated with safe drinking water while the CWA has a broader goal of clean, fishable, and swimmable surface waters. The 1972 Clean Water Act established a national priority of ending the discharge of pollutants into waterways. Discharges into waterways started being regulated by the National Pollution Discharge Elimination System (NPDES) using permits for limiting chemical discharges from point sources such as industrial facilities and municipal wastewater treatment plants (publicly owned treatment works, or POTWs). However, point source effluent control alone was not successful in improving water quality in many locations in the United States to a "fishable and swimmable" level due to continued non-point-source loadings and the persistance of contaminants in sediments. Given this, focus has shifted to ambient-based water quality standards and a wasteload evaluation and allocation process known as the total maximum daily load (TMDL) method.

The TMDLs define the total maximum daily load (point + nonpoint source) that can occur while ensuring compliance with a water quality standard that supports a designated use for the waterbody. Water quality models play a critical role in this process, both in identifying the TMDL for a waterbody and in identifying the allocation of discharges that can achieve it (e.g., U.S. EPA, 1991, 1997a,b; Hession et al., 1996; Chen et al., 1999). The National Research Council report on *Assessing the TMDL Approach to Water Quality Management* (NRC, 2001) notes the particular need to better integrate water quality models for TMDLs with monitoring data and statistical approaches for uncertainty analysis. Methods in support of such an approach are presented in Chapters 10 and 14.

Under the authority of the Safe Drinking Water Act (SDWA), EPA sets standards for approximately 90 contaminants in drinking water. For each of these contaminants, EPA sets a legally enforceable standard limit, called a maximum contaminant level (MCL), established through the National Primary Drinking Water Regulations

(NPDWRs). Water that meets these standards is safe to drink, although people with severely compromised immune systems and children may have special needs. The MCLs are periodically reviewed based upon new information on pollutant health effects and risk and are listed at http://www.epa.gov/safewater/sdwa/sdwa.html.

National Secondary Drinking Water Regulations (NSDWRs) are nonenforceable guidelines regulating contaminants that may cause cosmetic effects (such as skin or tooth discoloration) or aesthetic effects (such as taste, odor, or color) in drinking water. EPA recommends secondary standards (SMCLs), but does not require water systems to comply. However, states may choose to adopt them as enforceable standards.

Hazardous Waste and Contaminated Land The Resource Conservation and Recovery Act (RCRA) and the Comprehensive Environmental Response, Compensation and Liability Act (CERCLA) work together to manage hazardous wastes. A waste is designated as hazardous if it exhibits one or more of the following characteristics: ignitability, corrosivity, reactivity, and toxicity. See http://www.epa.gov/epaoswer/hazwaste/id/index.htm.

Unless specifically exempt, wastes exhibiting hazardous characteristics are subject to EPA's Subtitle C hazardous waste regulations. Subtitle C contains RCRA regulations governing hazardous materials that are in current industrial use. RCRA requires a cradle-to-grave manifest that tracks hazardous chemicals from the generator to the transporter to the final treatment, storage, and disposal facility (TSDF), thereby ensuring that hazardous chemicals are not improperly dumped or disposed of during their current-use cycle. In contrast to RCRA, CERCLA looks backwards in time and ensures cleanup of sites contaminated by past incidents of hazardous waste releases. CERCLA assigns liability and collects monetary compensation for removal and remedial actions required at the site. This law, commonly known as Superfund, was passed by Congress in 1980 and was amended in 1986 by the Superfund Amendments and Reauthorization Act (SARA). The Superfund refers to the trust fund that finances cleanup actions that is replenished on a "polluter pays" basis.

In the CERCLA process, a site found to be contaminated is investigated and evaluated for risk based upon a Hazard Ranking System (HRS). The HRS generates a score that assesses the potential for chemicals to enter air, water, and soil and create human exposure pathways via inhalation, ingestion, and dermal contact. Scores above 28.50 on a scale of 0 to 100 result in listing of the contaminated site on the National Priorities List, that is, as a Superfund site. Since 1991, the EPA has, on average, added 30 sites per year to the NPL (http://www.epa.gov/oerrpage/superfund/resources/hrstrain/htmain/index.htm). Cleanup at a Superfund site is specified by regulations to occur in distinct phases, including site assessment, remedial investigation/feasibility study (RI/FS), followed by record of decision (ROD) and then remedial action (RA). For all contaminated sites, whether Superfund or otherwise, knowledge of the multimedia transport and fate of pollutants enables quantitative assessment of current contamination levels, planning of remedial actions based upon remediation goals, and estimation of human exposure and risk before and after site cleanup. Most often remediation goals are based upon "safe" contaminant concentrations in air and water, as specified by standards set in the Clean Air Act and the Safe Drinking Water Act. Soil cleanup goals are more difficult to set since no universal "clean soil" standards have been formulated owing to the natural heterogeneity in pollutant concentrations and bioavailability in soils. In such cases, an approach termed risk-based cleanup action (RBCA) is employed, where the cleanup goal is set based upon minimizing risk to acceptable levels rather than achieving the infeasible goal of "zero" pollution levels. Chemical fate and transport modeling plays an important role in risk assessment and management at hazardous waste sites.

International Environmental Initiatives When the spatial scale of environmental impacts is so large that it crosses national boundaries, international accords on controlling pollution may be required. Three such issues have been at the forefront of recent international negotiations: global warming, stratospheric ozone depletion, and control of persistent organic pollutants (POPs).

Global warming is largely associated with elevated levels of carbon dioxide in Earth's atmosphere. Greenhouse gases,[6] such as CO_2, absorb terrestrial radiation and radiate it back to Earth, contributing to the so-called greenhouse effect that warms Earth's surface. The sharply increased rate of anthropogenic carbon dioxide production due to fossil fuel combustion is associated with the very rapid increase in CO_2 concentration that has been observed in Earth's atmosphere since 1900. Mathematical models have been used to understand the various equilibrium and dynamic processes that together contribute to climate change. Predictive models must address uncertainty in projections of global fossil fuel consumption, CO_2 removal mechanisms in the atmosphere, sinks such as forests and plankton, feedback mechanisms such as increased cloud cover that can cause an opposing "cooling" effect, as well as the effect of thermal mixing in shallow and deep ocean waters. Impact models must then predict the effects of global warming on future weather patterns, agricultural production, forest cover, and sea-level rise. Despite uncertainty in model predictions of climate change[7] and its impact (IPCC, 1996), there is growing international consensus that preventive measures must be taken to curb global warming, or at the least, to ensure against the risks imposed by such warming. To this end, the Kyoto Accord was established in 1997 with about 180 signatory nations endorsing a goal to reduce CO_2 emissions, with developed nations having a goal of about a 30% reduction relative to projected uncontrolled emissions by the year 2010, that is, about 10% below emissions in the year 2000. The objective is to slow the continued increase in atmospheric CO_2 concentrations so that they stabilize at or near double preindustrial levels ($\sim$550 ppmv).[8]

The Kyoto Accord was set for concerted implementation after ratification by at least a subset of industrialized nations that together contribute 55% of anthropogenic CO_2 emissions in the developed world. However, the United States (which is estimated to contribute 30% of the global man-made CO_2 emissions) withdrew from the Kyoto agreement in 2001, citing disagreements on implementation strategies and source reduction allocations for developing nations (C&EN, 2000). Despite withdrawal by the United States, 15 European Union nations and Japan proceeded to ratify

[6] Greenhouse gases include methane, N_2O, chlorofluorocarbons, SF_6, as well as CO_2, which is dominant.
[7] Most models predict global warming ranging from 1 to 3.5°C over the next century.
[8] Preindustrial values of ambient CO_2 are about 280 ppmv, with current (2000) values of $\sim$380 ppmv.

the treaty in early 2002. The Kyoto Accord is expected to enter the implementation phase by the end of 2004 when Russia ratifies the treaty, thereby achieving threshold participation for implementation. Notwithstanding these developments, the Kyoto Accord will not be fully effective in the long term unless the United States, along with highly populated developing countries such as China and India, join the concerted global effort. The Kyoto protocol and follow-on negotiations highlight the challenges facing international and intergenerational policymaking against a backdrop of scientific and modeling uncertainties (Grubler, 2000).

The Montreal Protocol on Substances that Deplete the Ozone Layer is an example of successful international action to address a global-scale pollution problem. Ozone in the stratosphere shields Earth by absorbing harmful cancer-causing ultraviolet radiation. Based on theoretical considerations, Molina and Rowland (1974) predicted that stratospheric ozone could be irreversibly consumed by chloroflorocarbons (CFCs), a group of stable organic chemicals widely used as aerosol propellants, solvents and refrigerants, which remain unreacted in the atmosphere for several decades after release. The scientific theory was confirmed in 1985 by the detection of an ozone hole over Antarctica, a dramatic drop-off in stratospheric ozone concentrations at high latitudes (Farman et al., 1985). Confronted with the reality of ozone depletion, the Montreal Protocol was adopted by 29 governments in 1987 (www.unep.org/ozone). The protocol aims to reduce and eventually eliminate the emissions of man-made ozone depleting substances. Its control provisions have been strengthened through four adjustments adopted in London (1990), Copenhagen (1992), Vienna (1995), Montreal (1997), and Beijing (1999). Much success has already been observed with rapid phase out of CFCs accomplished in most developed nations by 1996.

Building upon the success of the Montreal Protocol, the Stockholm Accord was adopted by 122 nations in Sweden in 2001 and aims to minimize and ultimately eliminate the use of POPs around the world. Control measures will immediately be applied to a list of 12 chemicals, most of which will be banned from production, with an exception for DDT in developing countries (ES&T, 1999). Multimedia environmental models have been used to prioritize POPs based on their potential for long-range transport (e.g., Rodan et al., 1999; Bennet et al., 1999; Beyer et al., 2000). Thus, environmental modeling emerges as an important component of all three major international environmental initiatives.

Subsequent chapters in this book describe physical, chemical, and biological processes that control chemical movement, transformations, and risk in the environment. The groundwork for such a treatment has been provided in this chapter with an overview of the physical properties, chemical transformations, and health effects of pollutants, presented in the overall context of the structure of organic and inorganic pollutants. Some data resources that provide physical properties, chemical transformation rates, and toxicity indices of various pollutants are provided in the supplemental information for chapter 2 at www.wiley.com/college/ramaswami.

Homework problems for this chapter are provided at www.wiley.com/college/ramaswami.

3 Intermedia Contaminant Transfer: Equilibrium Analysis

Pollutant transfer between different environmental media is discussed in this chapter and the next, employing a "closed-system" framework shown in Figure 3.1. Figure 3.1 assumes no bulk movement of the four principal environmental media (also referred to as environmental compartments)—air, water, soil, and non-aqueous-phase liquid (NAPL). Furthermore, chemical or biochemical transformations of pollutants are also ignored at this stage. The primary focus of this chapter is on quantifying the equilibrium distribution of pollutants in the four environmental compartments after long contact time periods have been established between the pollutant and the media. A special topics section (Section 3.4) describes the interaction of pollutants with living organisms (plants and fish) employing physical equilibrium process models, that is, living tissues are modeled as additional passive environmental compartments. The kinetics of intermedia pollutant transfer is discussed in Chapter 4. Pollutant transport within individual media due to fluid flow and mixing is incorporated into the modeling framework in Chapters 5, 8, 9, and 10.

Consider a nonreactive pollutant, A, that is introduced into the closed system represented in Figure 3.1. After a long period of time has elapsed, an equilibrium is achieved when all possible exchanges of A between the different compartments have stabilized. Equilibrium conditions are often *not* attained in the environment due to slow rates of mass transfer between media and the effects of advection, dilution, and chemical transformations within individual media. However, equilibrium calculations are relatively straightforward to make and provide a good first estimate of the overall distribution of a chemical in the different media, indicating where in the environment the chemical is most likely to be. This chapter primarily describes the equilibrium behavior of a generic *nonpolar* organic contaminant, referred to as pollutant A. Section 3.3 discusses modifications to the analysis that arise when charged pollutant species (e.g., metals) or polar organic compounds are considered.

3.1 PROPERTIES OF PURE CHEMICALS PRESENT IN ABUNDANCE

Before studying the equilibrium distribution of chemicals in different environmental compartments, it is useful to recall from Chapter 2 the saturation conditions that

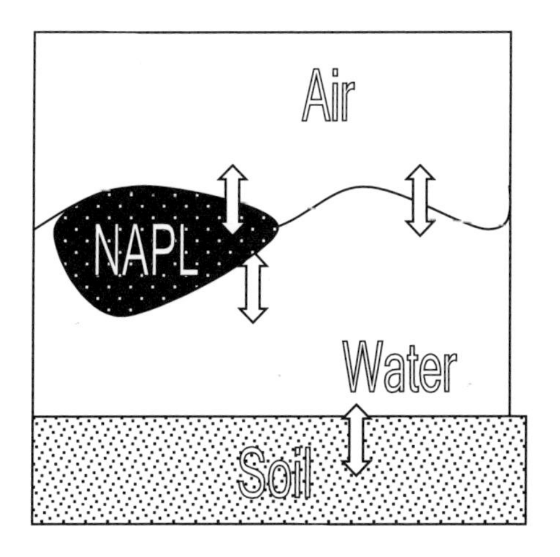

Figure 3.1 Conceptual closed-system framework for treatment of intermedia contaminant transfer. No bulk flow of individual media into or out of the system is considered. No chemical reactions and transformations are considered. Intermedia exchanges are indicated by arrows, representing air–water, NAPL–water, NAPL–air, and soil–water transfer of a pollutant.

develop in water and in air when contacted with a great abundance of a chemical in its pure and natural state:

$$\text{Aqueous solubility} = C^{\text{A}}_{w,\text{sat}} = \text{saturation conc. of A in water (mol/L or g/L)} \quad (3.1)$$

$$\text{Vapor pressure} = p^{\text{A}}_{v} = \text{pressure exerted by vapors of A when saturating air} \quad (3.2)$$

$$\text{Saturation mole fraction of A in air} = Y^{\text{A}}_{\text{air,sat}} = \frac{p^{\text{A}}_{v}}{P_{\text{atm}}} \quad (3.3)$$

where P_{atm} is the ambient atmospheric pressure, usually taken to be 1 atm. Vapor pressure and aqueous solubility data (Table 2.3) can be utilized to compute saturation chemical concentrations in air and in water, representing the upper limit of these concentrations at equilibrium conditions that cannot be exceeded at a given T_{env} (unless supersaturation occurs due to kinetic limitations on intermedia pollutant transfer). The use of aqueous solubility and vapor pressure data is demonstrated in Example 3.1.

EXAMPLE 3.1

A closed waste container is found in a laboratory. The container is filled with 5 L water, 5 L air and 5 L of a benzene-based solvent. Assume that the solvent is composed of pure benzene. Determine the equilibrium concentration and mass of benzene in the water and in the air in the container.

Solution 3.1 Since 5 L of benzene (approximately 4.5 kg) represents a great abundance of benzene in the container, saturation concentrations of benzene in the air and the water are expected to be achieved at equilibrium. Therefore,

$$C_{w,\mathrm{eq}}^{\mathrm{benzene}} = C_{w,\mathrm{sat}}^{\mathrm{benzene}} = 10^{-1.64}\frac{\text{mol benzene}}{\text{L water}} \times 78\frac{\text{g benzene}}{\text{mole}} = 1.8\frac{\text{g benzene}}{\text{L water}}$$

Air ($V_{\mathrm{air}} = 5$ L)
Pure Liquid Benzene (5 L)
Water ($V_w = 5$ L)

The mass of benzene in water is:

$$M_{\mathrm{water}}^{\mathrm{benzene}} = C_{w,\mathrm{eq}}^{\mathrm{benzene}} \times V_{\mathrm{water}} = 1.8\frac{\text{g benzene}}{\text{L water}} \times 5\ \mathrm{L_{water}} = 9 \text{ g benzene}$$

At saturation, the partial pressure of benzene in air will equal the vapor pressure. Therefore, applying Eqn. (3.2) and Eqn. (3.3):

$$p_{\mathrm{air,eq}}^{\mathrm{benzene}} = p_v^{\mathrm{benzene}} = 10^{-0.9}\text{atm} = 0.126 \text{ atm}$$

$$\text{Mole fraction of benzene in air} = Y_{\mathrm{air,eq}}^{\mathrm{benzene}} = \frac{P_v^{\mathrm{benzene}}}{P_{\mathrm{total}}} = \frac{0.126 \text{ atm}}{1 \text{ atm}} = 0.126$$

The benzene concentration in mass per volume units in computed from Eqn. (2.5).

$$\text{Benzene concentration in air} = C_{\mathrm{air,eq}}^{\mathrm{benzene}} = \frac{p_{\mathrm{air,eq}}^{\mathrm{benzene}}\, MW^{\mathrm{benzene}}}{RT_{\mathrm{env}}}$$

$$= \frac{0.126 \text{ atm} \times 78 \text{ g/mol}}{0.0821 \text{ L atm/mol K} \times 298 \text{ K}} = 0.4\frac{\text{g benzene}}{\text{L air}}$$

$$\text{Mass of benzene in air} = M_{\mathrm{air}}^{\mathrm{benzene}} = C_{\mathrm{air,eq}}^{\mathrm{benzene}} \times V_{\mathrm{air}} = 0.4\frac{\text{g benzene}}{\text{L air}} \times 5\ \mathrm{L_{air}}$$

$$= 2 \text{ g benzene}$$

The 9 g of benzene present in water and the 2 g of benzene present in air, together, represent a very small portion of the 4.5 kg of benzene present in the container. Thus, the assumption that a great excess of benzene is present is valid, enabling the use of the saturation aqueous solubility concentration and the vapor pressure for equilibrium conditions.

3.2 ENVIRONMENTAL PARTITIONING

Aqueous solubility and vapor pressure are attained when a great abundance of the pollutant in its pure and natural state is made available to the environment. However, large quantities of pure chemicals are rarely discharged into the environment. (These would be valuable raw materials for the chemical industry!) Instead, it is more common that mixtures of chemicals, or small quantities of a single chemical, are released to the environment and distributed among the different environmental media. In these situations, the environmental media are not saturated with the contaminant. Instead, the contaminant distributes itself between the different media in proportion to its relative affinity for the various compartments.

The concept of environmental partitioning was developed to describe the distribution of a contaminant between pairs of media. A simple example of partitioning is presented in Figure 3.2. Consider a saturated solution of sugar in water to which a globule of glycerin is added (glycerin may be considered to behave as a NAPL, being largely immiscible in water). No excess amount of sugar granules is available in the system. Upon addition of glycerin to water, some of the sugar originally dissolved in water will enter the glycerin phase. As a result, the sugar concentration in the water will decline, while the sugar concentration in the glycerin will increase. After a long

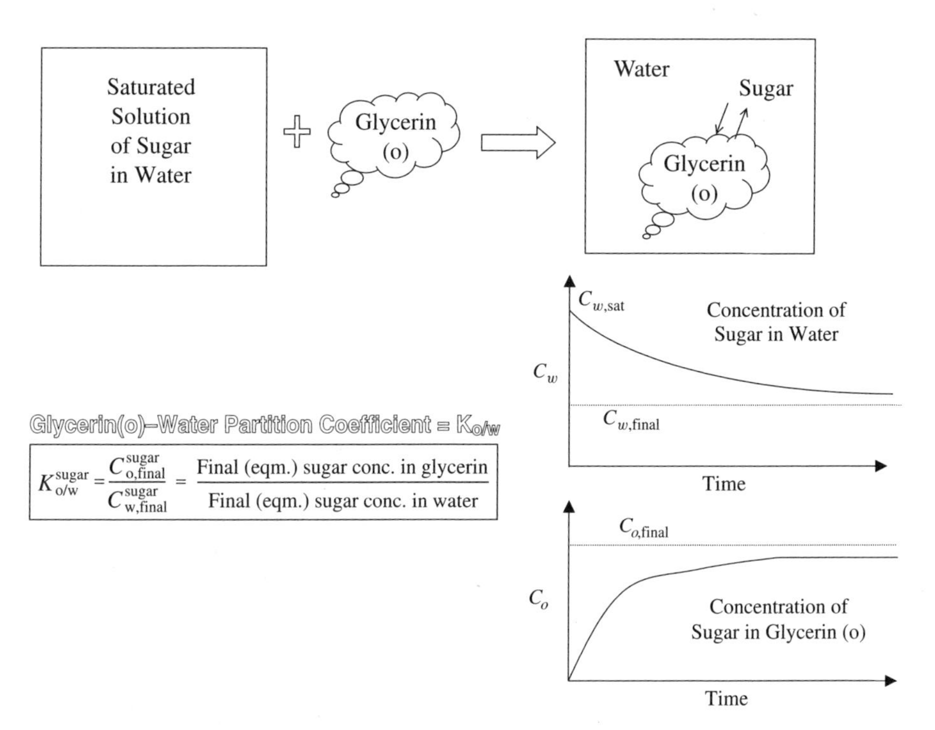

Figure 3.2 Schematic illustrating partitioning of sugar between glycerin and water.

period of time, the sugar concentrations in glycerin (NAPL) and in water will stabilize at their respective equilibrium concentrations. The ratio of the equilibrium sugar concentration in the organic phase glycerin to that in the aqueous phase is defined as the partition coefficient for sugar between these two phases.

An equilibrium partition coefficient thus represents the ratio of the equilibrium concentration of a chemical between pairs of environmental compartments. Three equilibrium partition coefficients that are commonly reported and frequently used are the air–water partition coefficient (K_{aw}), the octanol–water partition coefficient (K_{ow}), and the soil–water partition coefficient (K_d). The three coefficients are defined in Equations (3.4) to (3.6):

$$K_{\mathrm{aw}}^{\mathrm{A}} = \frac{C_{\mathrm{air,eq}}^{\mathrm{A}}}{C_{w,\mathrm{eq}}^{\mathrm{A}}} = \frac{\text{Equilibrium conc. of A in air (mass of A/L air)}}{\text{Equilibrium conc. of A in water (mass of A/L water)}} \tag{3.4}$$

$$K_{\mathrm{ow}}^{\mathrm{A}} = \frac{C_{o,\mathrm{eq}}^{\mathrm{A}}}{C_{w,\mathrm{eq}}^{\mathrm{A}}} = \frac{\text{Equilibrium conc. of A in octanol (mass of A/L octanol)}}{\text{Equilibrium conc. of A in water (mass of A/L water)}} \tag{3.5}$$

$$K_{d}^{\mathrm{A}} = \frac{C_{\mathrm{soil,eq}}^{\mathrm{A}}}{C_{w,\mathrm{eq}}^{\mathrm{A}}} = \frac{\text{Equilibrium conc. of A in soil (mass of A/kg soil)}}{\text{Equilibrium conc. of A in water (mass of A/L water)}} \tag{3.6}$$

Units and dimensions in Equations (3.4) to (3.6) are important. The air–water and octanol–water coefficients employ pollutant concentrations in identical units in the two pairs of media; hence, the air–water coefficient (K_{aw}) and the octanol–water coefficient (K_{ow}) are dimensionless. On the other hand, contaminant concentrations in soil are commonly expressed on a mass/mass basis, while contaminant concentrations in water are typically expressed on a mass/volume basis. As a result, the soil–water partition coefficient, K_d, has units of L water/kg soil. Each of these partition coefficients, and those derived from these primary coefficients, are described in more detail in the following sections.

3.2.1 Air–Water Partitioning

The dimensionless air–water partition coefficient K_{aw} represents the ratio of equilibrium pollutant concentrations in air and water, when expressed in identical units. Often, contaminant concentrations in air are represented in terms of partial pressure (atmospheres), while aqueous concentrations remain in mass/volume (mol/L or M L^{-3}) units. In this case, the ratio of equilibrium contaminant concentrations in air and water is the constant traditionally reported as the Henry's law constant, K_H, which has units of L atm/mol. The universal gas law is used to convert the Henry's law constant to the dimensionless K_{aw}:

$$K_{\mathrm{aw}}^{\mathrm{A}} = \frac{K_H^{\mathrm{A}}}{RT_{\mathrm{env}}} \quad \text{where} \quad K_H^{\mathrm{A}} = \frac{p_{\mathrm{air,eq}}^{\mathrm{A}}}{C_{w,\mathrm{eq}}^{\mathrm{A}}} = \frac{\text{equilibrium partial pressure of A in air (atm)}}{\text{equilibrium conc. of A in water (mol/L)}} \tag{3.7}$$

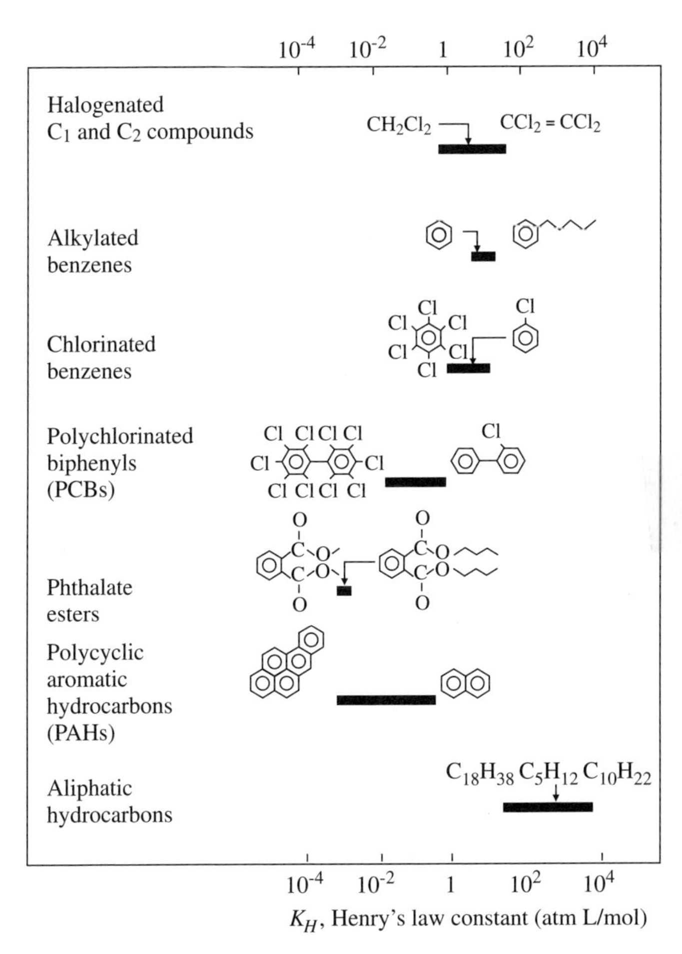

Figure 3.3 Ranges in Henry's law constants (K_H) for some important organic compounds. (Reprinted by permission from Schwarzenbach et al., 1993.)

In either form, the magnitude of the air–water partition coefficient depends upon the relative affinity of the contaminant for air versus water. In general, lighter (smaller) and more nonpolar organic molecules will prefer air to water. Figure 3.3 depicts the decreasing trend in K_H as the size of the molecule increases within each compound class. Measured K_H parameter values for selected organic and inorganic chemicals[1] are tabulated in Table 2.3. The use of the air–water partition coefficient is demonstrated in Example 3.2.

[1] K_H for inorganic gases such as O_2 and CO_2 is defined by convention as the inverse of K_H for organic chemicals, that is, the ratio of the molar equilibrium concentration of the dissolved gas in water to its partial pressure in air.

EXAMPLE 3.2

a. First consider a closed container at 25°C with 10 L of water (no air) to which 0.1 g benzene is added. What is the equilibrium concentration of benzene in water? Is the water saturated with respect to dissolved benzene?

b. Next, consider that the waste container in (a) has 10 L of air in addition to the 10 L of water, with the same 0.1 g of benzene added. What is the equilibrium concentration of benzene in water and in air? Compare with part (a).

Solution 3.2a

Closed tank with
10 L water (no air)
+
0.1 g benzene

$$C_{w,\text{eq}}^{\text{benzene}} = \frac{0.1 \text{ g benzene}}{10 \text{ L water}} = \frac{0.010 \text{ g benzene}}{\text{L water}}$$

Since $C_{w,\text{eq}}^{\text{benzene}}$ is less than $C_{w,\text{sat}}^{\text{benzene}}$ (1.8 g/L computed in Solution 3.1), water is undersaturated with respect to benzene and no pure phase benzene remains.

Solution 3.2b

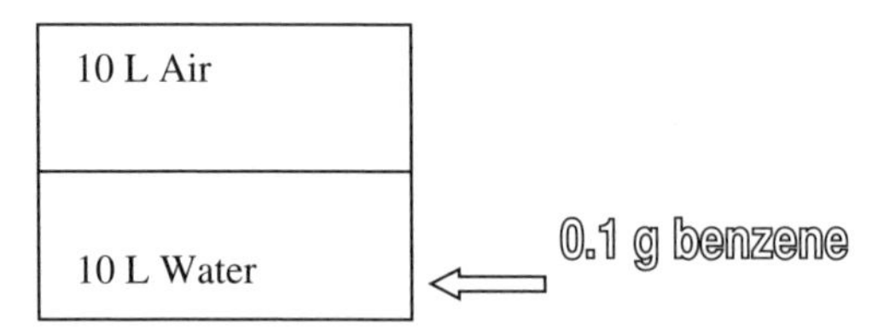

This example is similar to case 3.2(a), except that now 10 L of air will compete with the water for benzene. The use of partition coefficients is indicated since a limited quantity of benzene is available and hence both air- and aqueous-phase benzene concentrations are unknown, the system being unsaturated with benzene. A system with two unknowns requires two equations in order to obtain a solution. The first equation is a benzene mass balance in the system: The total mass of benzene in the container (0.1 g) equals the sum of benzene mass in air and the benzene mass in water.

$$M_{\text{total}}^{\text{benzene}} = M_{\text{air}}^{\text{benzene}} + M_{\text{water}}^{\text{benzene}}$$

$$0.1 \text{ g} = C_{\text{air,eq}}^{\text{benzene}} \times V_{\text{air}} + C_{w,\text{eq}}^{\text{benzene}} \times V_w \tag{1}$$

The second equation relates the unknown concentrations, $C_{\text{air,eq}}^{\text{benzene}}$ and $C_{w,\text{eq}}^{\text{benzene}}$, via the dimensionless air–water partition coefficient:

$$K_{\text{aw}}^{\text{benzene}} \text{ (dimensionless)} = \frac{C_{\text{air,eq}}^{\text{benzene}}}{C_{w,\text{eq}}^{\text{benzene}}} \tag{2}$$

Rearranging Eq. (2) for the equilibrium concentration of benzene in air, and substituting into Eq. (1):

$$0.1 \text{ g benzene} = M_{\text{total}}^{\text{benzene}} = K_{\text{aw}}^{\text{benzene}} \times C_{w,\text{eq}}^{\text{benzene}} \times V_{\text{air}} + C_{w,\text{eq}}^{\text{benzene}} \times V_{\text{water}}$$

so that,

$$C_{w,\text{eq}}^{\text{benzene}} = \frac{M_{\text{total}}^{\text{benzene}}}{\left(V_{\text{air}} \times K_{\text{aw}}^{\text{benzene}}\right) + V_{\text{water}}}$$

From Table 2.3,

$$K_H^{\text{benzene}} = 10^{0.74} = 5.495 \frac{\text{atm} \cdot \text{L}}{\text{mol}}$$

$$\Rightarrow K_{\text{aw}}^{\text{benzene}} = \frac{K_H^{\text{benzene}}}{RT} = \frac{5.495 \text{ atm} \cdot \text{L/mol}}{(0.0821 \text{ L} \cdot \text{atm/mol} \cdot \text{K})(273 + 25) \text{ K}} = 0.225$$

Therefore,

$$C_{w,\text{eq}}^{\text{benzene}} = \frac{M_{\text{total}}^{\text{benzene}}}{\left(V_{\text{air}} \times K_{\text{aw}}^{\text{benzene}}\right) + V_{\text{water}}} = \frac{0.1 \text{ g benzene}}{(10 \text{ L} \times 0.225) + 10 \text{ L}}$$

$$= 0.008 \text{ g benzene/L water}$$

Compared with case 3.2(a), the benzene concentration in water has decreased since some benzene has gone to the air. Equation (2) may next be employed to determine the equilibrium benzene concentration in air, equal to 0.002 g/L. The mass-balance equation is verified to ensure that the mass of benzene in air (0.02 g) and the mass of benzene in water (0.08 g) add up to the total of 0.1 g benzene that is placed in the container.

3.2.2 Octanol–Water Partitioning

The octanol–water partition coefficient defined in Eq. (3.5) represents the degree to which a contaminant prefers organic material to water. Large organic compounds, as well as highly nonpolar compounds, will follow the principle of "like attracts like" and exhibit a high degree of hydrophobicity. Such hydrophobic compounds prefer to partition to octanol relative to water and hence have a correspondingly large value of K_{ow}. Conversely, smaller or more polar molecules will be less hydrophobic and exhibit smaller values of K_{ow}. These trends are demonstrated in Figure 3.4 in which the K_{ow} spans more than 5 orders of magnitude for a variety of organic chemicals.

It should be noted that octanol serves as a generalized surrogate for organic media (e.g., NAPLs, animal lipid tissue, etc.). The reason for using octanol as a representative organic phase is historical: During the early years of pharmaceutical research, researchers found that octanol served as an inexpensive surrogate for human tissue (Overton, 1899; Myers, 1899). As a result, pharmaceutical studies often involved partitioning tests using octanol as an index for drug uptake by organisms, and K_{ow}

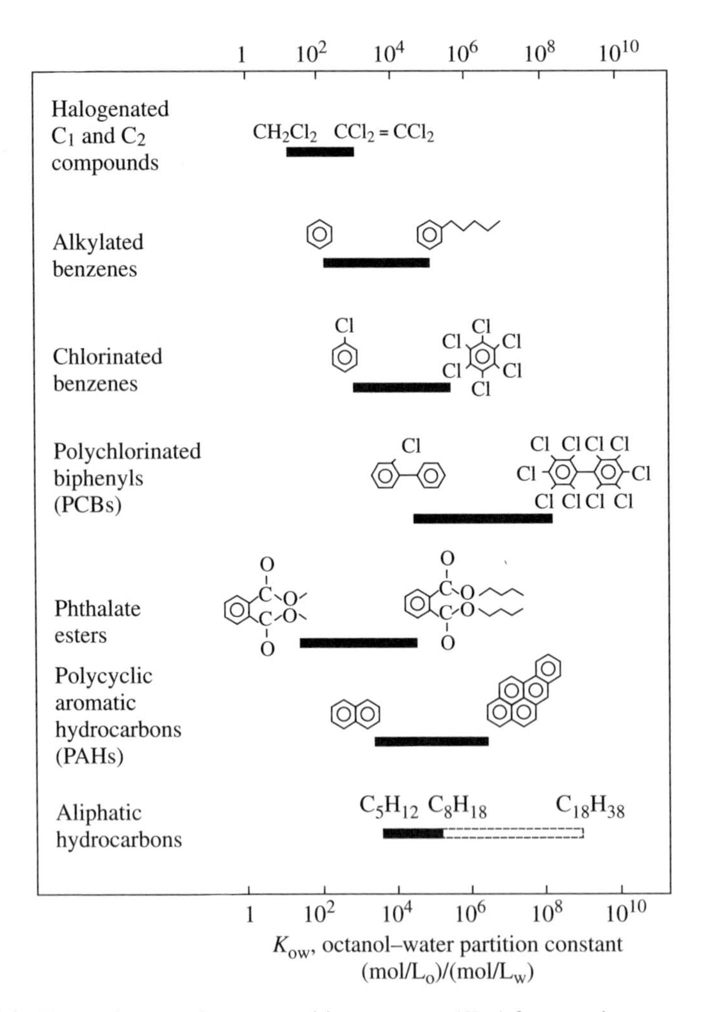

Figure 3.4 Ranges in octanol–water partition constants (K_{ow}) for some important classes of organic compounds. (Reprinted by permission from Schwarzenbach et al., 1993.)

is now commonly reported for most synthetic organic chemicals. Although octanol is considered a good representative for organic matter, the behavior of octanol does not exactly mimic the behavior of the wide variety of NAPLs and organic tissues present in the environment. As a result, statistical correlations are used to relate the behavior of the organic phase of interest with that of octanol. One particular organic phase of environmental interest is the organic matter present in soil. Partitioning of organic chemicals to soil organic matter, and the impact of K_{ow} on such partitioning, is discussed in the following section. Partitioning of pollutants to biota is also related to the K_{ow} of the pollutant and is discussed as an advanced topic in Section 3.4.

3.2.3 Partitioning Between Water and Soil Organic Matter

Soil may be described as composed of a mineral phase (typically inorganic, i.e., oxides of silica, potassium, aluminum, etc.) coated with naturally occurring organic material derived from plant and animal detritus. The fraction of natural organic matter present in soil is often used as a criterion in classifying various types of soils. For example, clean sand is basically a pure mineral phase (silica) with little or no organic matter. On the other hand, clayey soils are composed of minerals such as kaolinite and bentonite and also typically contain a larger amount of organic matter.

The mass fraction of natural organic material in soil is represented using two different notations: f_{OM} represents the mass fraction of organic matter (OM) present in soil, while f_{OC} represents the mass fraction of organic carbon (OC) in soil. Mass fraction f_{OM} considers the mass of the entire organic molecule (e.g., cell material $C_5H_7O_2N$ has a molecular weight of 113 g/mol), while f_{OC} considers only the mass of carbon present in the organic matter (e.g., only the five carbon atoms in the cell material, with total organic carbon mass of 60 g/mol). Using the rule of thumb that the weight of organic matter is roughly double the weight of organic carbon, a useful approximation is $f_{OM} \sim 2f_{OC}$.

Following the qualitative rule of organic chemistry, like attracts like, it is primarily the organic portion of soil that attracts organic contaminants. Sorption to the mineral phase in soil appears to provide a significant relative portion of sorption only when soil f_{OC} is much less than 0.1% (Schwarzenbach and Westall, 1981; Banerjee et al., 1985). Here we limit discussion to soils with f_{OC} larger than 0.1%, for which sorption to soil organic material is the dominant mechanism. The degree to which a certain contaminant is attracted to soil OM may be quantified using a partition coefficient between water and soil organic matter, represented by K_{OM}:

$$K_{OM}^{A} = \frac{C_{OM,eq}^{A}}{C_{w,eq}^{A}} = \frac{\text{Equilibrium conc. of A in soil OM (mass of A/kg OM)}}{\text{Equilibrium conc. of A in water (mass of A/L water)}} \quad (3.8)$$

A partition coefficient, K_{OC}, between water and soil organic carbon (OC) is also defined in an analogous manner as:

$$K_{OC}^{A} = \frac{C_{OC,eq}^{A}}{C_{w,eq}^{A}} = \frac{\text{Equilibrium conc. of A in soil OC (mass of A/kg OC)}}{\text{Equilibrium conc. of A in water (mass of A/L water)}} \quad (3.9)$$

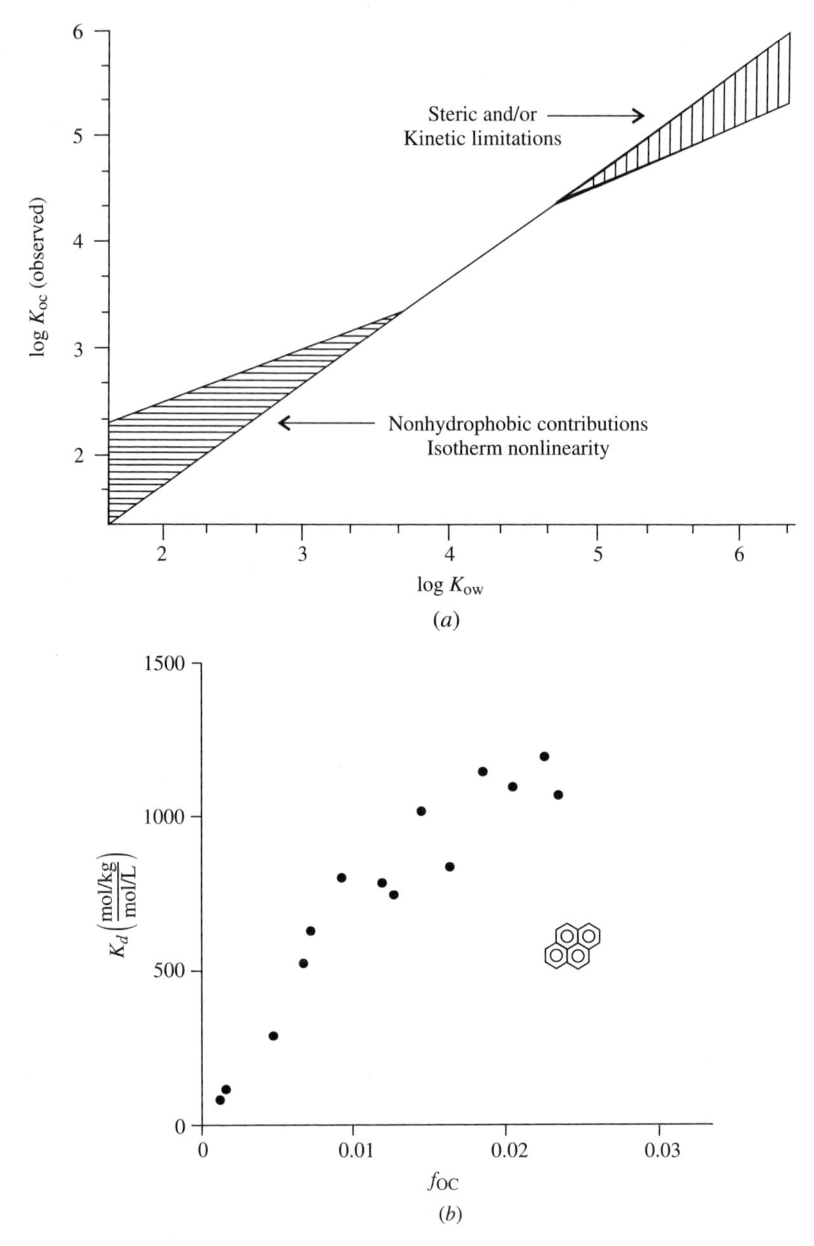

Figure 3.5 Soil–water partitioning: (*a*) Observed relationship between compound K_{oc} (organic carbon–water coefficient) and compound K_{ow} (octanol–water coefficient). A linear relationship is employed for nonionic, hydrophobic organic compounds. Nonlinearities are caused by steric factors and nonhydrophobic interactions (*Source:* Karickhoff, 1983). (*b*) Observed increase in solid–water partition coefficient, K_d, for hydrophobic compound, pyrene, as a function of the organic matter content of the solid (measured as organic carbon $f_{OC} = \frac{1}{2} f_{OM}$) in a variety of soils and sediments. (Reprinted by permission from Schwarzenbach et al., 1993.)

TABLE 3.1 Empirical Regression Models for Linear Free Energy Relationships (LFERs) between (a) $\log_{10} K_{OM}$ (L_{water}/kg_{OM}) and $\log_{10} K_{OW}$ ($L_{water}/L_{octanol}$) and (b) $\log_{10} K_{OC}$ (L_{water}/kg_{OC}) and $\log_{10} K_{OW}$

(a)	$\log_{10} K_{OM} = a \cdot \log_{10} K_{OW} + b$		
		r^2	
$a = 1.01$	$b = -0.72$	(0.99)	Aromatic compounds
0.88	−0.27	(0.97)	Chlorinated compounds
0.37	+1.15	(0.93)	Chloro-*S*-triazines
1.12	+0.15	(0.93)	Phenyl ureas
(b)	$\log_{10} K_{OC} = c \cdot \log_{10} K_{OW} + d$		
		r^2	
$c = 0.937$	$d = -0.006$	(0.95)	Aromatic compounds ($n = 19$)
1.00	−0.21	(1.00)	Mostly aromatic, two chlorinated ($n = 10$)
0.94	+0.02	NA	Chloro-*S*-triazines ($n = 9$)
0.52	+0.85	(0.84)	Phenylureas ($n = 30$)

NA = Not available.
n = number of chemicals used in regression analysis.
Sources: (a) Reprinted by permission from Schwarzenbach et al. (1993) and (b) from Lyman et al. (1982).

The greater the hydrophobicity of the contaminant, the more it is expected to "dislike" water and prefer the organic material in soil. Therefore, compounds with larger K_{OW}(more hydrophobic), exhibit larger values of K_{OM} and K_{OC}. Studies with a variety of organic contaminants have generally shown a linear log–log correlation between K_{OW} and K_{OM} or K_{OC}, with nonlinearities being caused by sorption to the mineral portion of soils and steric limitations (Fig. 3.5*a*). The pioneering work of Karickhoff (1984) yielded a general rule of thumb, $K_{OC} = 0.63 \times K_{OW}$, based on observations of a large variety of organic chemicals. Table 3.1 shows more refined correlations that have since been developed for different classes of organic compounds, for example, halogenated compounds, aromatic compounds, and so forth. The correlations in Table 3.1 enable estimation of K_{OM} and K_{OC} from K_{OW} values that are readily available in the literature (see Table 2.3). Example 3.3 illustrates the use of these correlations in estimating equilibrium pollutant concentrations in soil organic matter.

EXAMPLE 3.3

If 100 g organic matter (OM) is included in the waste container of Example 3.1 as shown below, what is the maximum concentration of benzene in the OM? What is the maximum mass of benzene present in the OM?

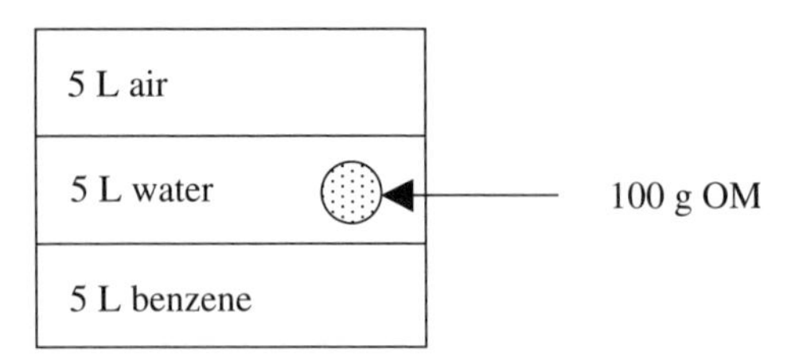

Solution 3.3 As in Example 3.1, a great abundance of benzene (~4.5 kg) is available so that the equilibrium benzene concentration in water is found to be the same as the aqueous solubility: $C_{w,\text{eq}}^{\text{benzene}} = C_{w,\text{sat}}^{\text{benzene}} = 1.8$ g/L. The benzene concentration in air is similarly obtained from its saturation vapor pressure (as in Solution 3.1) but is not used in further calculations since partitioning of benzene between OM and water is assumed for this problem.

The benzene in the OM will equilibrate with that in the water such that [from Eq. (3.8)] is valid. Note, unlike Example 3.2, a great excess of benzene is available, and the OM does not compete with the water for benzene. Instead, all three compartments, that is, air, water, and OM, are saturated with their respective maximum quantities of benzene, determined by the vapor pressure, aqueous solubility, and K_{OM} values for benzene. To determine K_{OM}, we use available correlations with K_{OW} (See Table 3.1). For aromatic compounds such as benzene,

$$\log_{10} K_{\text{OM}} = c \times \log_{10} K_{\text{OW}} + d \quad \text{where } c = 1.01 \text{ and } d = -0.72$$

$$\log_{10} K_{\text{OM}} = 1.01 \times (2.13) - 0.72 = 1.43$$

$$K_{\text{OM}}^{\text{benzene}} = 10^{1.43} = 27 \frac{\text{g benzene/kg OM}}{\text{g benzene/L water}}$$

Therefore, applying Eq. (3.8),

$$C_{\text{OM,eq}}^{\text{benzene}} = K_{\text{OM}}^{\text{benzene}} \times C_{w,\text{eq}}^{\text{benzene}} = 27 \frac{\text{L water}}{\text{kg OM}} \times \frac{1.8 \text{ g benzene}}{\text{L water}} = 48.6 \frac{\text{g benzene}}{\text{kg OM}}$$

and the benzene mass in the OM is

$$M_{\text{OM}}^{\text{benzene}} = M_{\text{OM}} \times C_{\text{OM}}^{\text{benzene}} = 0.1 \text{ kg OM} \times 48.6 \frac{\text{g benzene}}{\text{kg OM}} = 5 \text{ g benzene}$$

The additional 5 g of benzene in OM brings the total benzene mass in the three compartments, air, water, and OM, to $2 + 9 + 5 = 16$ g, which is still much less than the 4.5 kg benzene available in the pure state, enabling the use of saturation concentrations.

3.2.4 Soil–Water Partitioning

The soil organic matter partition coefficient, K_{OM}, enables equilibrium assessment of contaminant concentrations in the organic matter fraction of soil. It is more practical, however, to assess equilibrium contaminant concentrations in the aggregate soil mass, composed of both organic and mineral components. The soil–water partition coefficient, K_d, defined in Eq. (3.6), represents the ratio of the equilibrium contaminant concentration in soil to that in water. The coefficient, K_d, may be estimated by incorporating the contaminant's intrinsic affinity for soil organic matter, represented by K_{OM}, with the fraction of organic matter present in soil, represented by the soil f_{OM}. Both factors are combined in a simple multiplicative relationship, yielding:

$$K_d^{A}\left[\frac{\text{mass of A per kg soil}}{\text{mass of A per L water}}\right] = K_{OM}^{A}\left[\frac{\text{mass of A per kg OM}}{\text{mass of A per L water}}\right] \times f_{OM}\left[\frac{\text{mass of OM}}{\text{kg soil}}\right] \tag{3.10}$$

If soil organic carbon is being measured in lieu of soil organic matter, the soil–water partition coefficient is estimated as:

$$K_d^{A}\left[\frac{\text{mass of A per kg soil}}{\text{mass of A per L water}}\right] = K_{OC}^{A}\left[\frac{\text{mass of A per kg OC}}{\text{mass of A per L water}}\right] \times f_{OC}\left[\frac{\text{mass of OC}}{\text{kg soil}}\right] \tag{3.11}$$

The soil–water partition coefficient thus depends on the properties of the (soil) media represented by f_{OM} (or f_{OC}), as well as the properties of the contaminant, represented by K_{OM} (or K_{OC}). Equations (3.10) and (3.11) indicate that more hydrophobic pollutants have a greater affinity for soil. Furthermore, an organic pollutant's affinity for soil increases as the organic content of soil increases, as shown in Figure 3.5*b* for pyrene. Example 3.4 illustrates the use of the soil–water coefficient.

EXAMPLE 3.4

Add 10 kg soil with 1% OM to the container in Example 3.2(b) in which 10 L air, 10 L water, and 0.1 g benzene are present. What is the equilibrium concentration of benzene in air, water, and soil? Compare with case 3.2(b).

Solution 3.4 This example is similar to case 3.2(b), except that now 10 kg of soil as well as 10 L of air will compete for benzene with the 10 L of water. The use of partition coefficients is indicated since a limited quantity of benzene is available.

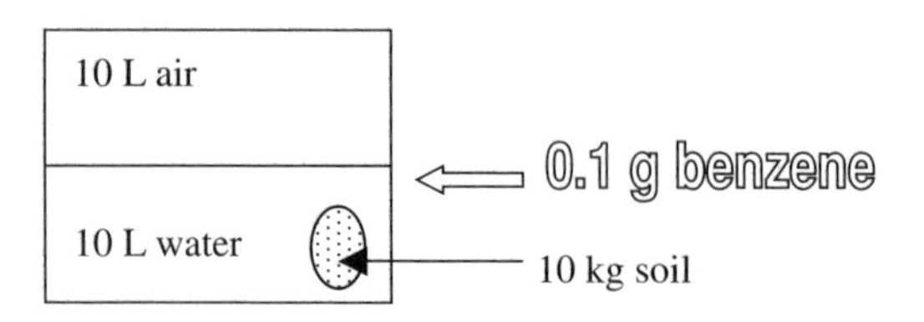

The total mass of benzene in the container (0.1 g) equals the sum of the benzene mass in air, the benzene mass in water, and the benzene mass in soil:

$$\begin{aligned} M_{\text{total}}^{\text{benzene}} &= M_{\text{air}}^{\text{benzene}} + M_{\text{water}}^{\text{benzene}} + M_{\text{soil}}^{\text{benzene}} \\ 0.1 \text{ g} &= C_{\text{air,eq}}^{\text{benzene}} \times V_{\text{air}} + C_{w,\text{eq}}^{\text{benzene}} \times V_w + C_{\text{soil,eq}}^{\text{benzene}} \times M_{\text{soil}} \end{aligned} \tag{1}$$

where benzene concentrations in air and water are expressed on a mass/volume basis, while benzene concentrations in soil are expressed on a mass/mass basis. The volumes of air (V_{air}) and water (V_w), as well as the mass of soil (M_{soil}) are known. But equilibrium benzene concentrations in air, water, and soil are not known and need to be determined. The equilibrium concentrations in the three media are interrelated by the partition coefficients:

$$K_{\text{aw}}^{\text{benzene}} \text{ (dimensionless)} = \frac{C_{\text{air,eq}}^{\text{benzene}}}{C_{w,\text{eq}}^{\text{benzene}}} \tag{2}$$

$$K_d^{\text{benzene}} \left[\frac{\text{L water}}{\text{kg soil}} \right] = \frac{C_{\text{soil,eq}}^{\text{benzene}}}{C_{w,\text{eq}}^{\text{benzene}}} \tag{3}$$

where $K_{\text{aw}} (= 0.225)$ has been estimated for benzene in a previous example. Using Eq. (3.10), the K_d for benzene may be estimated from the f_{OM} of soil and the K_{OM} for benzene computed in Example 3.3 ($K_{\text{OM}} = 27$ L water/kg OM] as:

$$K_d^{\text{benzene}} = f_{\text{OM}} \times K_{\text{OM}}^{\text{benzene}} = \frac{0.01 \text{ kg OM}}{1 \text{ kg Soil}} \times \frac{27 \text{ L water}}{\text{kg OM}} = \frac{0.27 \text{ L water}}{\text{kg Soil}}$$

Solving for benzene concentrations in air and in soil, from Eqs. (2) and (3), respectively, and substituting into Eq. (1), yields a benzene mass balance equation:

$$0.1 \text{ g} = \left(K_{\text{aw}}^{\text{benzene}} C_{w,\text{eq}}^{\text{benzene}} V_{\text{air}}\right) + \left(C_{w,\text{eq}}^{\text{benzene}} V_w\right) + \left(K_d^{\text{benzene}} C_{w,\text{eq}}^{\text{benzene}} M_{\text{soil}}\right)$$

$$C_{w,\text{eq}}^{\text{benzene}} = \frac{M_{\text{total}}^{\text{benzene}}}{\left(V_{\text{air}} K_{\text{aw}}^{\text{benzene}}\right) + V_w + \left(K_d^{\text{benzene}} M_{\text{soil}}\right)} = 0.0067 \text{ g benzene/L water}$$

Equilibrium benzene concentrations in air and in soil are then computed from (2) and (3) to be 0.0016 g/L and 0.0018 g/kg, respectively. Compared with Example

3.2(b), benzene concentrations in water and in air have decreased since some of the benzene has gone to the soil. The mass-balance equation should be verified by the reader to ensure that the mass of benzene in air, water, and soil add up to the 0.1 g total benzene present in the container.

3.2.5 NAPL–Water Partitioning

As discussed in Chapter 2, NAPLs can serve as long-term sources of contamination to the surrounding environment. Hence, equilibrium transfers of contaminants from NAPL to water and from NAPL to air are of interest. Since most NAPLs are multicomponent mixtures, it is useful to describe the partitioning behavior of a single constituent, A, when present within a multicomponent NAPL mixture, and to compare this partitioning behavior with that of a pure NAPL composed exclusively of chemical A. Consider these two situations illustrated in Figure 3.6.

In the case of a single-component NAPL composed of chemical A (in pure liquid form), the NAPL–water interface is composed exclusively of chemical A. The equilibrium concentration of A in the water adjacent to the NAPL is given by the aqueous solubility of the pure liquid A, such that: $C_{w,\mathrm{eq}}^{\mathrm{A}} = C_{w,\mathrm{sat}}^{\mathrm{A(L)}}$. However, when the NAPL

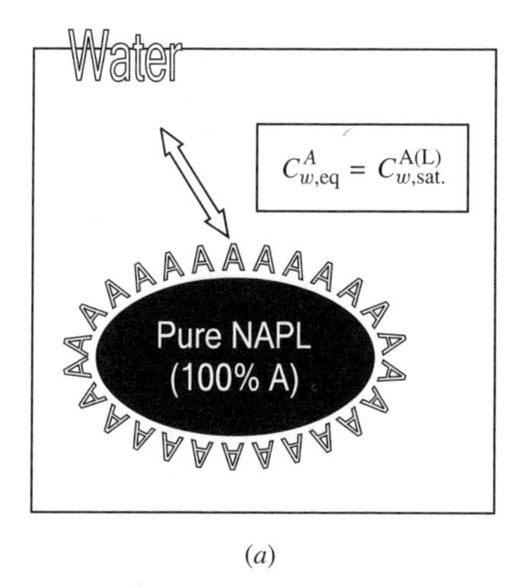

(*a*)

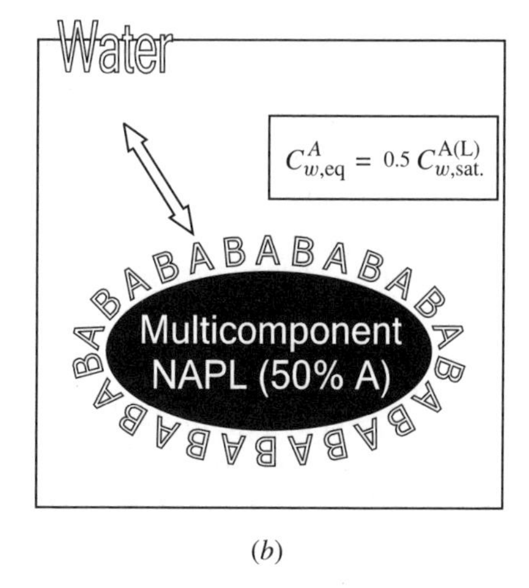

(*b*)

Figure 3.6 Impact of NAPL composition and surface molecular coverage on NAPL–water equilibria. (*a*) Pure NAPL composed exclusively of chemical A. The NAPL–water interface is covered fully with molecules of A, and the equilibrium aqueous-phase concentration of chemical A is the same as the saturation solubility of compound A in water. (*b*) A multicomponent ideal NAPL composed of 50% mole fraction of two chemicals, A and B. The NAPL–water interface is now half covered with molecules of chemical A and half with chemical B. The equilibrium aqueous-phase concentration of chemical A is now 50% that of its saturation solubility, because of surface competition by the other chemical(s) present in the ideal NAPL mixture.

is a multicomponent mixture consisting of, for example, 50% A mixed with other constituents, then the NAPL–water interface looks much different. In order to subsequently dissolve in water, all the other NAPL constituents will now compete to occupy a position at the NAPL–water interface. The degree of dissolution to water is thermodynamically related to the degree of NAPL surface coverage offered by molecules of A. In an ideal NAPL mixture, the fraction of surface coverage provided by molecules of A is the same as the mole fraction of A in the mixture. Hence, the equilibrium concentration in water in contact with a multicomponent NAPL, $C_{w,\text{eq}}^{\text{A}}$, is given by the aqueous solubility of pure liquid A in water, $C_{w,\text{sat}}^{\text{A}}$, scaled down by the mole fraction of compound A in the NAPL, $X_{\text{NAPL}}^{\text{A}}$:

$$C_{w,\text{eq}}^{\text{A}} = X_{\text{NAPL}}^{\text{A}} \, C_{w,\text{sat}}^{\text{A}} \tag{3.12}$$

For compounds that are solids in pure form, but exist as "dissolved" liquids within the NAPL mixture, Eq. (3.12) must be modified such that the saturation solubility of the hypothetical pure subcooled liquid,[2] $C_{w,\text{sat}}^{\text{A(L)}}$, is used in place of the solubility of the pure compound, $C_{w,\text{sat}}^{\text{A}}$. Equation (3.12) is valid only for ideal NAPL mixtures and is known as Raoult's law. For the case of nonideal mixtures, an additional term is required to account for the nonideality of the mixture:

$$C_{w,\text{eq}}^{\text{A}} = \gamma_{\text{NAPL}}^{\text{A}} \, X_{\text{NAPL}}^{\text{A}} \, C_{w,\text{sat}}^{\text{A}} \tag{3.13}$$

where $\gamma_{\text{NAPL}}^{\text{A}}$ is the activity coefficient of A in the NAPL, numerically set to unity for the case of an ideal NAPL mixture. Mixtures are termed as ideal if the fraction of total volume and total surface area occupied by any one constituent is given by the mole fraction of that constituent in the NAPL. Typically, NAPLs composed of similar constituents behave ideally or near-ideally; NAPLs composed of disparate compounds are expected to exhibit nonideal behavior. For several NAPLs of environmental concern, such as gasoline, diesel fuel, and coal tar, experiments have shown that the assumption of an ideal NAPL results in errors of the magnitude of 10% in estimating equilibrium aqueous solubility (Mackay et al., 1991; Lee et al., 1992; Ramaswami and Luthy, 1997). Errors of this magnitude are often acceptable in comparison to measurement errors or variability in transport parameters that can result in uncertainties of one or more orders of magnitude. Thus, the assumption of an ideal NAPL yields a reasonable first estimate of the equilibrium aqueous concentration of a contaminant released from a multicomponent NAPL. If nonidealities must be considered, the activity coefficient of the contaminant in the organic phase may be estimated and employed in Eq. (3.13) (Peters et al., 1998).

Raoult's law with the ideal NAPL assumption can also be used to describe dynamic changes in NAPL–water equilibria associated with water flushing. As a NAPL blob or pool is flushed with water, the more soluble constituents are depleted from the NAPL, thereby altering the chemical mole fraction in the NAPL, $X_{\text{NAPL}}^{\text{A}}$. By tracking dynamic changes in the mole fraction, $X_{\text{NAPL}}^{\text{A}}(t)$, the corresponding changes in NAPL–water

[2] See Chapter 2, Section 2.2.1.

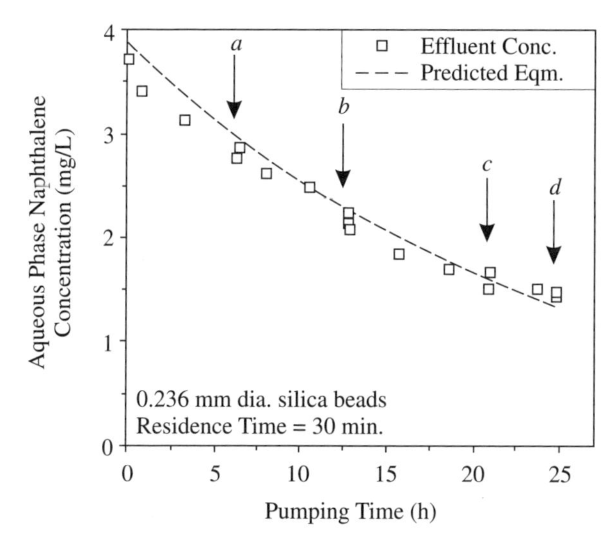

Figure 3.7 Dynamic changes in equilibrium aqueous naphthalene concentrations in response to water pumping in a coal tar (NAPL)–water system (Ramaswami et al., 1997). Points *a*–*d* indicate times when pumping was discontinued to confirm that the reactor contents were at equilibrium. Changes in equilibrium aqueous-phase naphthalene concentrations were consistent with dynamic equilibrium predictions based on naphthalene depletion from coal tar.

equilibria may also be predicted using Eq. (3.12). Experimental evaluations of naphthalene depletion from NAPL coal tar (Ramaswami and Luthy, 1997) have shown the utility of Raoult's law in predicting dynamic NAPL–water equilibria by tracking the mass of naphthalene flushed from the NAPL, and thereby updating the naphthalene mole fraction remaining within the NAPL (see Fig. 3.7).

3.2.6 NAPL–Air Partitioning

Equilibrium partitioning of a contaminant from a multicomponent NAPL mixture to air follows the same rules as those outlined above for NAPL–water partitioning. Thus, the equilibrium partial pressure of contaminant A in air in contact with a multicomponent NAPL, $p^{\mathrm{A}}_{\mathrm{air,eq}}$, is given by the vapor pressure of pure compound A, p^{A}_{v}, scaled down by the mole fraction of A in the NAPL, $X^{\mathrm{A}}_{\mathrm{NAPL}}$, as well as any corrections required to address the nonideality of the NAPL:

$$p^{\mathrm{A}}_{\mathrm{air,eq}} = \gamma^{\mathrm{A}}_{\mathrm{NAPL}}\, X^{\mathrm{A}}_{\mathrm{NAPL}}\, p^{\mathrm{A}}_{v} \tag{3.14}$$

For compounds that are solids in pure form, but exist as "dissolved" liquids within the NAPL mixture, Eq. (3.14) must be modified by using the vapor pressure of the hypothetical pure subcooled liquid, $p^{\mathrm{A(L)}}_{v}$ in place of the vapor pressure of the pure chemical in its natural state, p^{A}_{v}.

EXAMPLE 3.5

Consider the waste solvent in the container of Example 3.1 to be a multicomponent NAPL containing 10% naphthalene on a mole fraction basis. What is the maximum possible concentration of naphthalene in air and in water in contact with the NAPL?

Solution 3.5 This example emphasizes the need to use subcooled liquid properties (See Table 2.3) when describing the NAPL–water and NAPL–air partitioning of contaminants present as solids in their pure, natural state. Note that the use of pure naphthalene properties would result in an error of over a factor of 4. Assume an ideal NAPL with $\gamma_{\text{NAPL}}^{\text{naphthalene}} = 1.0$. Applying Raoult's law, using the properties of subcooled liquid naphthalene:

Air and Water
NAPL with 10% naphthalene

$$p_{\text{air}}^{\text{naphthalene}} = X_{\text{NAPL}}^{\text{naphthalene}} \times p_v^{\text{naphthalene(L)}} = 0.1 \times 10^{-3.43} \text{ atm}$$
$$= 3.7 \times 10^{-5} \text{ atm} \tag{3.15a}$$

$$C_{w,\text{eq}}^{\text{naphthalene}} = X_{\text{NAPL}}^{\text{naphthalene}} \times C_{w,\text{sat}}^{\text{naphthalene(L)}} = 0.1 \times 10^{-3.06} \text{ mol/L}$$
$$\times\ 128 \text{ g naphthalene/mol} = 0.011 \text{ g naphthalene/L water} \tag{3.15b}$$

3.3 PARAMETERS REQUIRED FOR INTERMEDIA EQUILIBRIUM COMPUTATIONS

Examples 3.1 to 3.5 illustrate the different classes of environmental problems that may be solved using the equilibrium concepts presented in Sections 3.1 and 3.2. Figure 3.8 identifies the three different categories of environmental partitioning problems and the parameters required to address these situations. The first category of problems addresses situations in which a great abundance of a chemical is present in its pure and natural state, requiring the use of aqueous solubility and vapor pressure parameters. In this case, both water and air are saturated with the pollutant; the pollutant concentration in soil is determined by the soil–water partition coefficient. In the second category of problems, a large abundance of the pollutant chemical is present, but within a NAPL mixture. This category of problems requires the use of Raoult's law, with pure compound solubility and vapor pressure parameters scaled down by the mole fraction (and activity) of the chemical in the multicomponent NAPL, to yield equilibrium aqueous-phase and air-phase pollutant concentrations, respectively. The

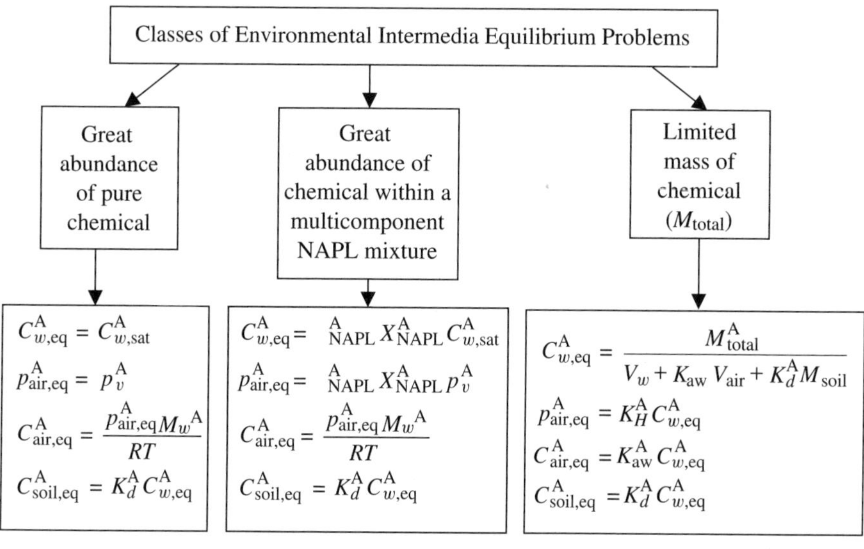

Parameters Required

Pure compound properties: $C^{A}_{w,\text{sat}}$ = Aqueous solubility
p_v = Vapor pressure
$M_w{}^{A}$ = Molecular Mass

NAPL properties: X^{A}_{NAPL} = Mole fraction of A in NAPL
${}^{A}_{\text{NAPL}}$ = Activity of A in NAPL

Partition coefficients: K_H = Henry's law constant
K_{aw} = Dimensionless air–water coefficient
K_d = Soil–water partition coefficient

Media compartment sizes: V_{air} = Volume of air
V = Volume of water
${}_{\text{soil}}$ = Mass of soil

Figure 3.8 Schematic showing different categories of equilibrium computations and the parameters required for intermedia equilibrium computations.

pollutant concentration in soil is again determined by the soil–water partition coefficient. In the third category of problems, sufficient pollutant is not available to saturate the media. Instead, there is a limited mass of the pollutant that distributes itself between air, water, and soil, in a manner proportional to the air–water and soil–water partition coefficients. The volumes of the air, water, and soil compartments, as well as the magnitudes of the air–water and soil–water partition coefficients, are important parameters that govern the distribution of the pollutant between the different media in this case.

3.3.1 Parameter Estimation Methods

The pure compound properties and the equilibrium partitioning parameters listed in Figure 3.8 may be estimated in three ways:

1. *From values reported in the literature with suitable correction factors* Data sources for aqueous solubility, vapor pressure, octanol–water partition coefficient, and the Henry's law constant for a variety of organic compounds are summarized in the supplementary information for Chapter 3 at www.wiley.com/college/ramaswami.

   When using reported values for the parameters from another study, care should be taken to correct for any changes in temperature and/or other environmental factors (e.g., pH and ionic strength) that may affect the parameters of interest. Factors that affect aqueous solubility, vapor pressure, and partition coefficients are summarized in Table 3.2. The qualitative reasoning as well as the quantitative correction factors used to account for these impacts are described in Table 3.2. Of particular importance is the effect of speciation on aqueous solubility and other water-based partition coefficients. Several polar molecules, for example, organic and inorganic

TABLE 3.2 Environmental Factors That Affect Aqueous Solubility, Vapor Pressure, and Environmental Partitioning of Organic Pollutants

Environmental Factor	Impact on Transfer to Water: $C_{w,\text{sat}}$	Impact on Transfer to Air: p_v and K_H
Increase in ambient temperature, T_{env}	Thermal energy promotes melting of solids to liquid state, enhancing $C_{w,\text{sat}}$ of solids. Thermal energy promotes escape of gas molecules from water, decreasing $C_{w,\text{sat}}$ of gases.	Thermal energy increases kinetic energy of pollutant molecules, resulting in greater tendency to escape to the vapor phase and hence higher vapor pressures. See Antoine equation [Eq. (2.4)]
Increase in aqueous salt conc., e.g., in seawater	Salt ions are attracted to polar water molecules and "sequester" them such that less water is available to dissolve nonpolar organic pollutants. Aqueous solubility in salt water decreases: $\left[\dfrac{C_{w,\text{sat,salt}}}{C_{w,\text{sat}}}\right] = 10^{-K_s \cdot M_{\text{salt}}}$ M_{salt} is the total molar salt conc. in water; K_s is the Setschenow or salting constant.	Transfer of organic pollutants from seawater to air appears to be enhanced because of the decreased affinity of organic pollutants to saline water. K_H appears to increase correspondingly.

(*continued*)

TABLE 3.2 (*Continued*)

Environmental Factor	Impact on Transfer to Water: $C_{w,\text{sat}}$	Impact on Transfer to Air: p_v and K_H
Presence of co-dissolved organic chemicals in water	Organic chemicals are attracted to other organic chemicals dissolved in water: Aqueous solubility appears to be enhanced by this co-solvent effect. Quantitative estimation of the co-solvent effect is described in Yalkowsky et al., 1976; Morris et al., 1988; Baum, 1998.	Organic co-solvents in water have variable effects on transfer of the pollutant to air, depending upon the relative volatility of the co-solvents.
pH, charge, and speciation [α_0 is the fraction of total acid species ($HA + A^-$) in water that remains undissociated as HA]. See Eq. (2.19)	Charged species, such as dissociated organic acid A^-, have a strong affinity for water (which is polar) hence dissolve readily in water. The undissociated species HA is limited by its (nonpolar) aqueous solubility, $C_{w,\text{sat}}^{\text{HA}}$. Aqueous solubility of combined HA and A^- species (total HA) is enhanced at pH > pK_a when dissociation to A^- is favored. $C_{w,\text{sat}}^{\text{Tot.HA}} = C_{w,\text{sat}}^{\text{HA}}/\alpha_0$	The charged species, A^-, has a great affinity for water and hence does not readily partition to air. When pH > pK_a, dissociation to A^- is favored, the total acid species (Tot. HA = $HA + A^-$) manifests a lowered affinity for air driven by the behavior of A^-. $K_H^{\text{Tot.HA}} = \alpha_0 \cdot K_H^{\text{HA}}$

Soil–Water Partitioning: The impact of environmental parameters on the soil–water partition coefficient, K_d, is the inverse of that on aqueous solubility, i.e., factors that result in an increase in $C_{w,\text{sat}}$ will cause a corresponding decrease in K_d.

acids and bases, can exist in water in undissociated form, as well as in dissociated forms that typically carry an electronic charge. The charged species have a strong affinity for polar water molecules compared to the undissociated species, causing dramatic changes in pollutant partitioning behavior with pH changes that control dissociation (see Fig. 3.9). Partitioning behavior of metals, particularly sorption to soil, is affected both by chemical speciation in water and chemical reactions occurring at the soil–water interface. Chemistry-based, mechanistic models that describe the sorption of metals and other inorganic species to soil are presented in Chapter 12. However, the bulk of this book continues to focus on the transport and fate of a nonpolar organic chemical, employing primarily physical process principles.

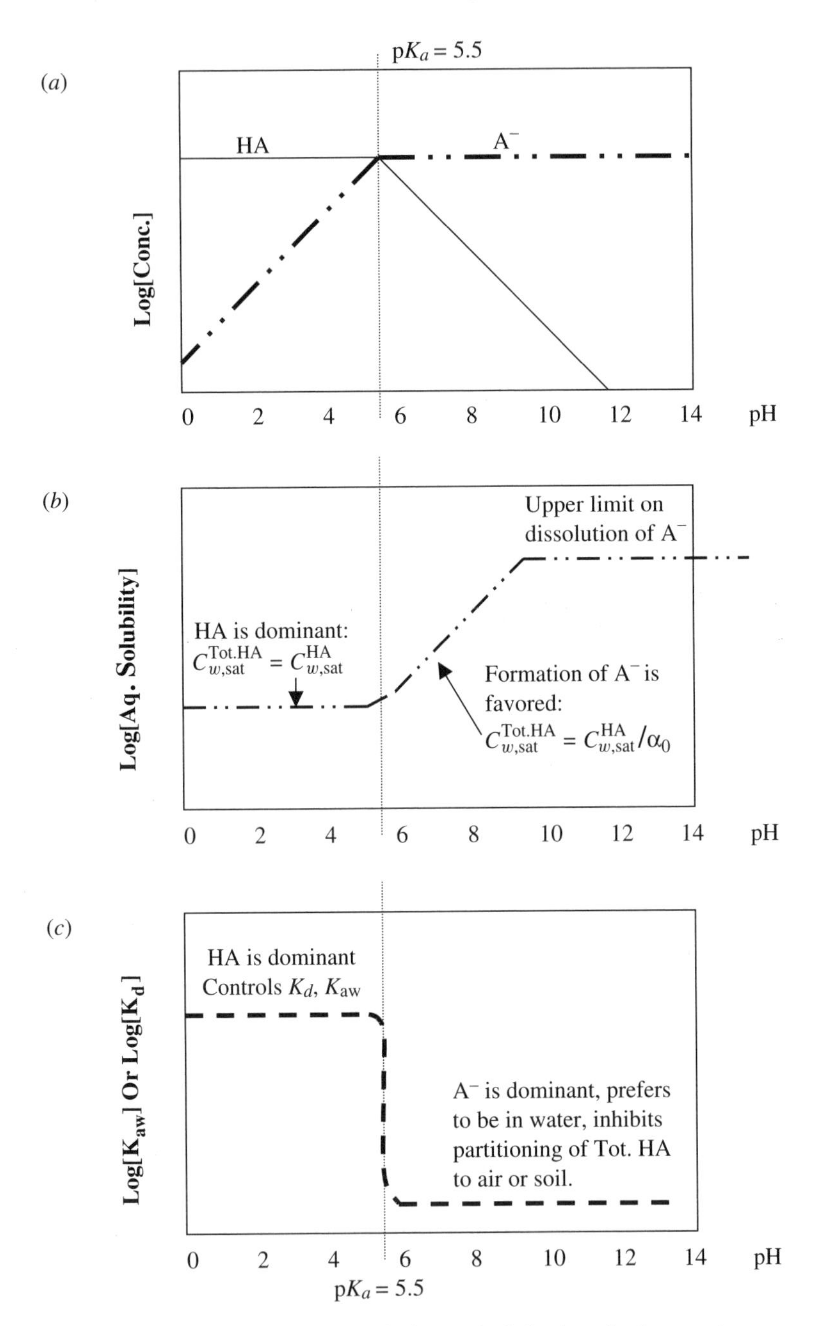

Figure 3.9 Impact of solution pH and speciation on the behavior of polar organic compounds in water. (*a*) Dissociation of HA to A^- when a fixed amount of Tot. HA = [HA]+[A^-] is present in water. (*b*) Dissolution of HA into water when an infinite supply of pure HA is available. Aqueous solubility of Tot. HA, $C_{w,sat}^{Tot.HA}$, increases dramatically when pH > pK_a. (*c*) Partition coefficients, K_{aw} or K_d, relative to water. Partition coefficients decrease dramatically at pH > pK_a when formation of A^- is favored. Note the scales on the *Y* axes are all different.

2. *Estimation of physical partitioning parameters from compound structure* The molecular size and structure (polarity) of chemicals can significantly affect properties such as aqueous solubility, vapor pressure, and air–water and octanol–water partition coefficients. Solubility and vapor pressure are often estimated through correlation with other compound properties such as melting and boiling point temperatures, which are in turn dependent upon molecular structure (Lyman et al., 1982; Baum, 1998). In the case of the octanol–water partition coefficient and the air–water partition coefficient, the contribution of each chemical structural unit and chemical bond in an organic compound can be quantified in the form of structural unit contributions. Contributions to K_{aw} and K_{ow} by each of the constituent chemical bonds and chemical units in the molecule are then tallied, to yield estimates of K_{aw} and K_{ow} for the aggregate chemical (Hine and Mookerjee, 1975; Hansch and Leo, 1979). The reader is referred to Lyman et al. (1992) and Baum (1998) for a comprehensive discussion of different techniques for estimating pure compound properties and partition coefficients from chemical structure information. An online resource, CHEMEST (Lyman and Potts, 1987) is also available for chemical property estimation. UNIFAC predictions of aqueous and nonaqueous solubilities of some chemicals of environmental interest are reported by Kan and Tomson (1996).

3. *Direct experimental measurement* Experimental measurements are typically much more expensive but also more accurate than parameter estimation using the two methods described above. Although experimental measurements of aqueous solubility and vapor pressure and the partition coefficients, K_H and K_{ow}, are conceptually simple, the measurement of extremely low pollutant concentrations in air and water can pose unique challenges. In general, the chemical concentration in one medium is known and measured accurately. Known volumes of the competing medium are then equilibrated with the primary medium in a closed container. Chemical concentrations in the competing medium, if too low for detection, are often determined from mass-balance considerations based on measured changes in chemical concentration occurring in the primary medium. Equilibrium chemical concentrations in one medium are plotted against the other. The slope of such a graph yields the equilibrium concentration ratio for the two media, that is, the partition coefficient of interest.

Soil–water partitioning is measured using similar principles. A fixed mass of soil is contacted with a fixed volume of water in several closed vials containing varying amounts of the contaminant. Equilibration is allowed to take place and the final pollutant concentrations in water and in soil are determined in all of the vials. Vials with a larger contaminant mass input will have larger water– and soil–contaminant concentrations at equilibrium. A graph of $C_{\text{soil,eq}}$ versus $C_{w,\text{eq}}$ is constructed; this function is referred to as the sorption isotherm since it is developed for a given temperature. If the function is linear, the slope of the graph yields the soil–water partition coefficient, K_d. Linear isotherms occur when there are insignificant intermolecular interactions on the soil surface, so that the fraction of the sorption sites used by the solute is well below capacity. Nonlinear isotherms, such as the

Langmuir and Freundlich isotherms,[3] are needed when these conditions do not apply, especially at higher solute concentrations. For most organic pollutants in the environment, aqueous-phase concentrations are low enough that a linear sorption isotherm with a slope of K_d provides an acceptable representation.

3.4 POLLUTANT INTERACTIONS WITH BIOTA

3.4.1 Aquatic Organisms

Pollutant uptake and absorption by aquatic organisms depends upon physiological processes such as ingestion and excretion of the pollutant, as well as the passive physical exchange of the pollutant between the lipid content of the organism and the surrounding waters (see Fig. 3.10). A bioconcentration factor (BCF), developed primarily for fish, focuses on pollutant uptake into fish tissue by nondietary routes, that is, only by the process of physical exchange between the fish tissue and the surrounding waters. A related parameter, termed the bioaccumulation factor (BAF), applicable for a variety of organisms including fish, considers uptake of a chemical into the organism by all exposure routes, that is, ingestion, respiration, and direct physical exchange. Both parameters represent the ratio of the pollutant concentration in fish to that in the surrounding medium (water), with the implicit assumption that equilibrium or steady-state conditions have been attained. The primary difference between the two parameters is in the mechanisms of pollutant intake considered.

Models that consider the dynamics of physiological processes such as ingestion, digestion, respiration, and excretion are called physiologically based pharmacokinetic (PBPK) models and are described in detail in Chapter 13. PBPK models incorporate specific physiological features of the organism that influence the rate of pollutant absorption, such as residence time of the pollutant in the digestive tract, pH of gastric juices (stomach fluids), and so forth. While PBPK models provide a more accurate representation of the movement of the pollutant through the organism, these models require knowledge of several physiological parameters for the target species. Simpler, physical partitioning type models employ the BCF and are appropriate for use when exchange of the hydrophobic organic chemical between water and fish lipid tissue occurs much faster than internal transit of the pollutant through the fish digestive system.

[3] The Langmuir adsorption isotherm is given by:

$$C_{\text{soil,eq}} = C_{\text{soil,monolayer}} \left(\frac{bC_{w,\text{eq}}}{1 + bC_{w,\text{eq}}} \right)$$

where $C_{\text{soil,monolayer}}$ is the soil concentration with full, monolayer saturation of the chemical on the soil surface adsorption sites, approached as $C_{w,\text{eq}} \rightarrow \infty$, b is a constant. The Langmuir isotherm is nearly linear at very low $C_{w,\text{eq}}$, with $K_d \approx bC_{\text{soil,monolayer}}$.

The Freunlich isotherm is given by $C_{\text{soil,eq}} = K_F C_{w,\text{eq}}^{1/n}$, where K_F and n are constants.

Figure 3.10 Illustration of fish–water partitioning of benzene.

The BCF for an organic pollutant, A, is defined as:

$$\text{BCF} = \frac{C_{f,\text{ss}}^{\text{A}}}{C_{w,\text{ss}}^{\text{A}}} = \frac{\text{Steady-state concentration of A in fish (g of A/kg fish tissue)}}{\text{Steady-state concentration of A in water (g of A/kg of water)}} \tag{3.16}$$

The organic pollutant is primarily exchanged between fish fatty tissue (lipids) and the surrounding water. Since octanol is a good surrogate for lipid tissue, the octanol–water partition coefficient introduced in Section 3.2.2 provides a close approximation of a pollutant's BCF. In much the same manner as the soil–water partition coefficient was estimated from the pollutant's K_{ow} in Section 3.2.4, the pollutant BCF for a variety of fish species may be estimated through statistical correlations with K_{ow} (see Table 3.3). Note the BCF expressions in Table 3.3 and in Eq. (3.16) consider pollutant concentrations per unit mass of the entire fish tissue, that is, not focusing on muscle or lipid layers alone.

In an alternative estimation procedure for BCF developed by Mackay (1982), fish are considered to be composed, on average, of 4.8% lipids, the behavior of which is closely represented by that of octanol. The pollutant BCF for the entire body mass of the fish is therefore obtained as the product of the chemical's affinity for lipids (or octanol), determined by the pollutant K_{ow}, and the lipid content of fish, which is assumed to be 4.8% (or approximately 5%). This simple relationship is expressed in log-linear form and included in Table 3.3. Employing a data set of observed BCF values for a diverse set of chemicals, Devillers et al. (1996) compared the Mackay approximation with the other correlation equations shown in Table 3.3 and found the estimation accuracy of both techniques to be comparable. The reader is referred to Baum (1998) for a more detailed description of this comparative study and for a detailed discussion of BCF estimation techniques. The use of the BCF is illustrated in Example 3.6.

TABLE 3.3 Relationships between Bioconcentration Factors (BCF) in Fish and Chemical K_{ow}[a]

				$\log_{10} \text{BCF} = A + B \log_{10} K_{ow}$			
				n	r^2	Comments	Reference
$A =$	−0.70	$B =$	0.95	55	0.95	Pesticides, PCBs $0 > \log \text{BCF} > 6$ $1 > \log K_{ow} > 7$ fathead minnows	Veith et al. (1979)
	−0.23		0.76	84	0.95	Mixed $0 > \log \text{BCF} > 5$ $1 > \log K_{ow} > 7$ fathead minnows bluegill sunfish	Veith et al. (1980)
	−0.68		0.94	18	0.95	Halogenated chemicals $2.4 > \log \text{BCF} > 4.3$ $3.4 > \log K_{ow} > 5.5$ rainbow trout	Oliver (1984)
	−0.52		0.80	107	0.90	Mixed $0 > \log \text{BCF} > 5$ $1 > \log K_{ow} > 7$ several freshwater fish	Isnard and Lambert (1988)
	−1.32		1.00	50	0.95	Nitrogen and halosubstituted aromatics and aliphatics	MacKay (1982)

[a]A and B are regression constants, r^2 is the correlation coefficient, and n is the number of chemicals used in the regression.
Source: Adapted from Baum (1998).

EXAMPLE 3.6

Fish (fathead minnows) are swimming in waters in which a NAPL spill has occurred. The NAPL is composed of 10% mole fraction benzene, and is assumed to be an ideal NAPL. Estimate the benzene concentration in the fish assuming equilibrium is attained between the NAPL, water, and fish. Use estimates of BCF obtained from the correlations in Table 3.3, including the Mackay model.

Solution 3.6 Considering NAPL–water equilibrium, we apply Eq. (3.13) to compute the equilibrium aqueous benzene concentration to be 0.18 g benzene/L water = 0.18 g benzene/kg of water. The benzene concentration in fish may then be estimated as shown below.

Fish in Water

NAPL with
10% benzene

Using the Mackay model for any fish species:

$$\mathrm{BCF}^{\mathrm{benzene}} = 0.048 \times K_{\mathrm{ow}}^{\mathrm{benzene}} = 0.048 \times 10^{2.13} = 6.48 \frac{\mathrm{kg\ water}}{\mathrm{kg\ fish}}$$

$$C_f^{\mathrm{benzene}} = \mathrm{BCF}^{\mathrm{benzene}} \times C_{w,\mathrm{eq}}^{\mathrm{benzene}} = 6.48 \frac{\mathrm{kg\ water}}{\mathrm{kg\ fish}} \times 0.18 \frac{\mathrm{g\ benzene}}{\mathrm{kg\ water}} = 1.17 \frac{\mathrm{g\ benzene}}{\mathrm{kg\ fish}}$$

Using the specific correlation equation for fathead minnows:

$$\log_{10} \mathrm{BCF}^{\mathrm{benzene}} = -0.23 + 0.76 \left[\log_{10} K_{\mathrm{ow}}^{\mathrm{benzene}}\right] = -0.23 + 0.76 \times 2.13 = 1.389$$

$$\Rightarrow \mathrm{BCF}^{\mathrm{benzene}} = 24.5 \frac{\mathrm{kg\ water}}{\mathrm{kg\ fish}}$$

$$C_f^{\mathrm{benzene}} = \mathrm{BCF}^{\mathrm{benzene}} \times C_{w,\mathrm{eq}}^{\mathrm{benzene}} = 24.5 \frac{\mathrm{kg\ water}}{\mathrm{kg\ fish}} \times 0.18 \frac{\mathrm{g\ benzene}}{\mathrm{kg\ water}} = 4.41 \frac{\mathrm{g\ benzene}}{\mathrm{kg\ fish}}$$

3.4.2 Plant Uptake of Pollutants

Interactions of pollutants with plants can occur in plant roots, shoots, and leaves (see Fig. 3.11). These interactions may be summarized as follows:

1. Uptake and accumulation of pollutants from soil–water into plant roots, where they may be enzymatically transformed
2. Translocation of the pollutants from the roots to the shoots and leaves via plant transpiration, with the potential for accumulation and transformation in aboveground tissues
3. Transfer of certain pollutants, such as trichloroethylene, from the leaf surface to the atmosphere, and vice versa, for example, transfer of PAHs (polyaromatic hydrocarbons) from the atmosphere to the leaf surface.

Thus, plants can serve as important living links between air, water, and soil, while simultaneously having the potential to chemically transform the pollutant as it is transferred between the different media. Interaction between atmospheric pollutants and vegetation occurs on a regional scale, with pollutant concentrations in leaves often being monitored to determine the history and extent of air pollution within geographic boundaries (e.g., PAH analysis of pine needles in Great Britain; Tremolada et al., 1996). Plant uptake of pollutants from soil and water occurs on a more local scale and has been studied intensively over the past decade for application in an

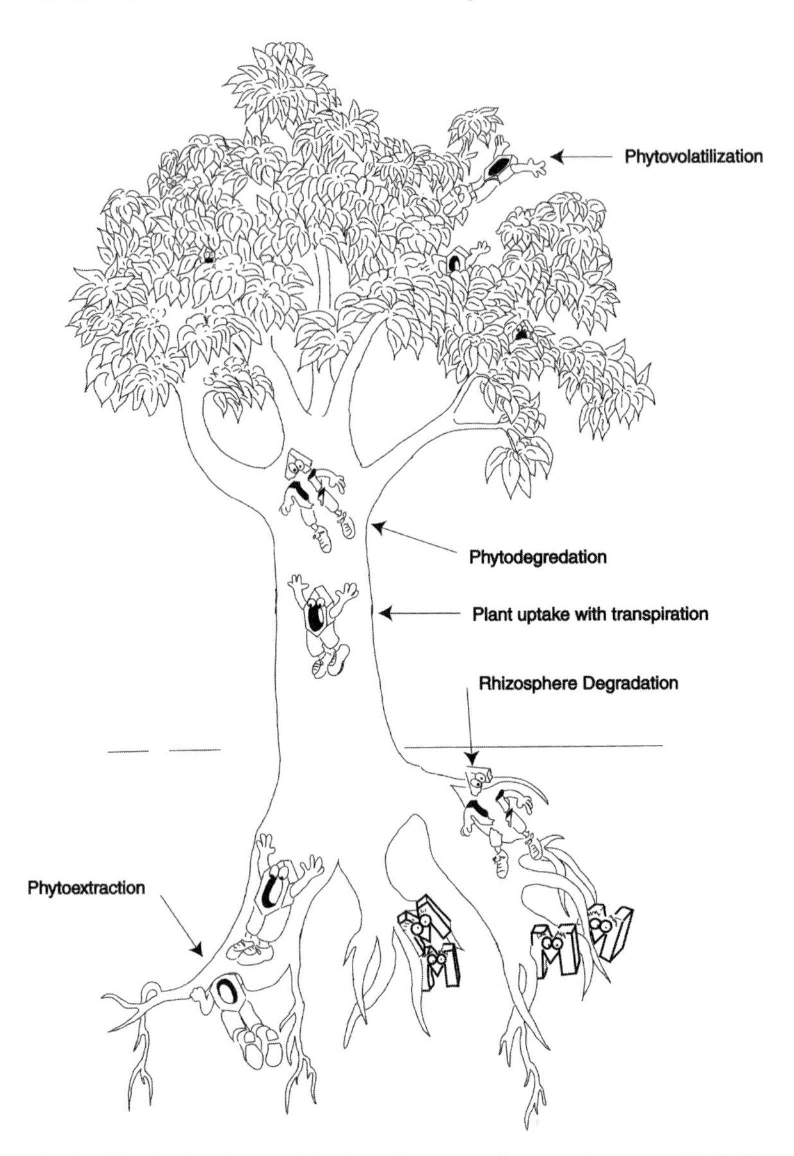

Figure 3.11 Sketch showing partitioning of organics (benzene) and metals (M) in plant systems.

emerging technology termed phytoremediation, that is, plant-assisted remediation at contaminated land sites (Baker and Brooks, 1989; Schnoor et al., 1995; Rubin and Ramaswami, 2001).

Once pollutants are taken up by plants, they may undergo a variety of transformations. Organic pollutants are readily transformed both in the root soil zone (rhizosphere), as well as within the plant, to either carbon dioxide or to intermediate degradation products that may be incorporated inextractably into plant tissue.

Specialized plant enzymes have been discovered that breakdown hazardous organic pollutants such as trichloroethylene and trinitrotoluene (TNT) (Schnoor et al., 1995). In the case of hazardous inorganic pollutants, such as selenium, lead, and zinc, certain specialized plants, termed hyperaccumulators, are able to detoxify and accumulate these metals at extremely high concentrations within plant tissue (Brooks, 1983; Baker and Brooks, 1989). Various partition coefficients have been developed in soil science, geobotany, and environmental engineering, to describe interactions of organic pollutants with plants under equilibrium conditions (Briggs et al., 1982; Brooks, 1983; Burken and Schnoor, 1997). Essentially, these interactions may be described as a series of equilibrium exchanges between the following pairs of compartments:

1. Soil–water and root tissue
2. Soil–water and the plant transpiration stream
3. Pollutant transfer from the plant transpiration stream to the atmosphere
4. Pollutant transfer from plant leaves (foliage) to the atmosphere
5. Pollutant transfer from surrounding water or soil to the entire plant biomass

Each of the above interactions is described briefly below:

1. *Transfer of pollutant from soil–water to plant root* is represented by the root concentration factor (RCF).

$$\text{RCF}^{\text{A}} = \frac{\text{Conc. of pollutant A in wet plant roots (mg/kg)}}{\text{Conc. of pollutant A in soil–water (mg/L)}} \tag{3.17}$$

The RCF is sometimes also represented as a bioconcentration factor for the roots, BCF_{root}. The RCF addresses sorption of organic pollutants from soil to lipophillic root tissues, as well as pollutant uptake by the root xylem–water that flows through the plant during transpiration. However, sorption to the lipophillic root tissues is the major contributor to the RCF, and hence the RCF increases as the pollutant's octanol–water coefficient increases (see Fig. 3.12*a*). Correlations that enable estimation of the RCF from the compound K_{ow} are presented in Table 3.4. Note that the correlations in Table 3.4 were developed for specific plants, for example, barley and poplar trees, and can be used accurately only for those plants that have been experimentally evaluated. In much the same manner as the soil–water partition coefficient, K_d, depends upon the soil organic matter content, plant partition coefficients also depend on the lipid content of the various plants. The correlations shown in Table 3.4 are plant specific and have not been normalized to plant lipid content.

2. *Transfer of pollutant from soil–water to the transpiration stream within the plant* is represented by the transpiration stream concentration coefficient (TSCF):

$$\text{TSCF} = \frac{\text{Conc. of pollutant A in plant transpiration stream (mg/L)}}{\text{Conc. of pollutant A in soil–water (mg/L)}} \tag{3.18}$$

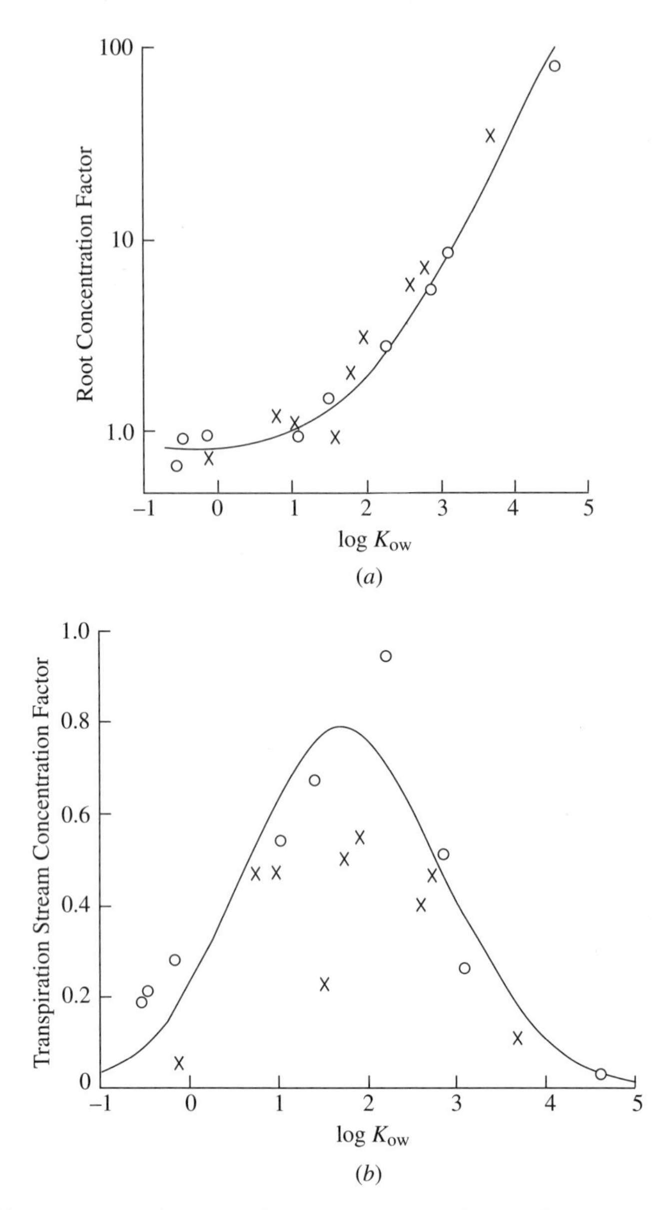

Figure 3.12 (*a*) Impact of compound octanol–water partition coefficient, K_{ow}, on the root concentration factor (RCF): The RCF increases with hydrophobicity of the compound. (*b*) Impact of compound octanol–water partition coefficient, K_{ow}, on the transpiration stream concentration factor (TSCF): TSCF is maximum for compounds with K_{ow} in the range of 10 to 100 (from Briggs et al., 1982).

TABLE 3.4 Correlation Equations for Estimation of Various Partition Coefficients Representing Plant–Pollutant Interactions

Parameter	Equation	Source	Conditions
Root concentration factor (RCF)	$\log_{10}(\text{RCF} - 0.82) = -1.52 + 0.77 \log_{10} K_{ow}$	Briggs et al., 1982	Barley
	$\log_{10}(\text{RCF} - 3) = -1.57 + 0.65 \log_{10} K_{ow}$	Burken and Schnoor, 1997	Poplar
Transpiration stream concentration factor (TSCF)	$\text{TSCF} = 0.784 \exp - \left[\frac{(\log_{10} K_{ow} - 1.78)^2}{2.44} \right]$	Briggs et al., 1982	Barley
	$\text{TSCF} = 0.756 \exp \frac{-(\log_{10} K_{ow} - 2.50)^2}{2.58}$	Burken and Schnoor, 1997	Poplar
Leaf–air coefficient $K_{\text{leaf–air}}$	$\log_{10}(K_{\text{leaf–air}} \cdot K_{aw}) = -1.95 + 1.14 \log_{10} K_{ow}$	Bacci et al., 1990a,b	
	$K_{\text{leaf–air}} = 0.19 + 0.7 K_{wa} + 0.05 K_{oa}$ where $K_{\text{wa(water–air)}} = 1/K_{aw}$; $K_{\text{oa(octanl–air)}} = K_{ow}/K_{aw}$	Paterson et al., 1991	
$\text{BCF}_{\text{plant–soil}}$ Terrestrial plants and soil	$\log_{10} \text{BCF}_{\text{plant–soil}} = 1.54 - 1.18 \log_{10} K_d$	Baes, 1982	
BCF_{PW} Aquatic plants and water	$\log_{10} \text{BCF}_{PW} = -2.24 + 0.98 \log_{10} K_{ow}$	Gobas et al., 1991	

The TSCF is a dimensionless parameter that represents aboveground translocation of the pollutant to plant shoots and leaves. A TSCF of 1 indicates that the plant transpires the pollutant at the same concentration in the xylem sap as in the external soil solution, indicating maximum efficiency of uptake (e.g., Rubin and Ramaswami, 2001), akin to sipping up juice through a drinking straw! In most cases, the TSCF is less than 1 and is found to be maximum for pollutants with log K_{ow} from 1 to 3.5 (Fig. 3.12*b*). Very hydrophobic pollutants (with log $K_{ow} > 3.5$) are sorbed so strongly to the lipophillic root tissues and to soil organic matter that they are not easily translocated above the ground, while hydrophillic compounds (with log $K_{ow} < 1$) are very water soluble and are not actively transported through the plant system. Correlations for estimating the TSCF from the compound K_{ow} are summarized in Table 3.4.

3. *Transfer of volatile pollutants from the plant transpiration stream to air* may be quantified by dividing the pollutant's TSCF by the dimensionless air–water partition coefficient, K_{aw}. This parameter is important because it provides an estimate of the pollutant concentration in air at equilibrium with the water transpired at the surface of plant leaves.

4. *A related parameter is the leaf–atmosphere exchange coefficient* that represents transfer of pollutants from the atmosphere to foliage. This coefficient, represented as $K_{\text{leaf–air}}$, is akin to a leaf bioconcentration factor with respect to air:

$$K^{A}_{\text{leaf–air}} = \frac{\text{Conc. of pollutant A in foliage (mg/L)}}{\text{Conc. of pollutant A in air (mg/L)}} \tag{3.19}$$

Coefficient $K_{\text{leaf–air}}$ includes the interaction of the plant transpiration stream with the atmosphere, as well as the sorption of atmospheric pollutants by lipophillic leaf tissue. Because of these two processes, correlations for estimation of the leaf–air exchange coefficient (Table 3.4) indicate a dependence on K_{oa}, the octanol–air partition coefficient of the chemical that represents its affinity for leaf organic material relative to air, as well as its air–water coefficient, K_{aw}, which quantifies interactions between air and transpiration water.

5. *Exchange between the whole plant and soil or water* compartments is described by the bioaccumulation coefficient (BAC) for metals in soil and by the bioconcentration factor (BCF_{PW}) for whole plants in contact with water contaminated with organic pollutants.

$$\text{BAC}^{A} = \frac{\text{Conc. of metal pollutant A in dry plant biomass (mg/g)}}{\text{Conc. of metal pollutant A in soil (mg/g)}} \tag{3.20}$$

$$\text{BCF}^{A}_{PW} = \frac{\text{Conc. of pollutant A in plant (mg/kg)}}{\text{Conc. of pollutant A in soil–water (mg/L)}} \tag{3.21}$$

Brooks (1983) reviews the BAC of a wide variety of metal hyperaccumulator plants. For example, hyperaccumulators have been identified with BACs as high as 25 for nickel and 130 for selenium (Brooks, 1983, p. 155). The BCF_{PW} for whole plants in aquatic systems contaminated with organic pollutants is found to depend upon the pollutant K_{ow}; correlations depicting this relationship are presented in Table 3.4. An engineering application of the plant-based partitioning coefficients is illustrated in Example 3.7.

EXAMPLE 3.7

Consider a site at which groundwater contaminated with 0.01 g/L of benzene is passing through a screen of poplar trees. Each tree is estimated to have a transpiration rate of 1800 L of water per year. What is the concentration of benzene in the plant transpiration stream, and how much benzene can be removed by each poplar tree over a 1 year period?

Solution 3.7 Since the benzene concentration in the plant transpiration stream is required, the TSCF for benzene must first be estimated for poplar trees using the correlation shown in Table 3.4:

$$\mathrm{TSCF}_{\mathrm{poplar}}^{\mathrm{benzene}} = 0.76 \times \exp\left[\frac{-\left(\log K_{\mathrm{ow}}^{\mathrm{benzene}} - 2.5\right)^2}{2.58}\right] = 0.71$$

The benzene concentration in the plant transpiration stream, TS, may be obtained from Eq. (3.18) as:

$$C_{\mathrm{TS}}^{\mathrm{benzene}} = TSCF_{poplar}^{benzene} \times C_w^{benzene} = 0.71 \times \frac{0.01\ g\ benzene}{L\ water}$$
$$= 0.0071 \frac{\text{g benzene}}{\text{L water}}$$

If the yearly transpiration rate is represented by T, then the mass uptake rate of benzene per tree, represented by U, may be computed as:

$$U = T_{\mathrm{poplar}} \times \mathrm{TSCF}_{\mathrm{poplar}}^{\mathrm{benzene}} \times C_w^{\mathrm{benzene}} = \frac{1800\ \text{L water}}{\text{yr}} \times \frac{0.0071\ \text{g benzene}}{\text{L water}}$$
$$= 12.78 \frac{\text{g benzene}}{\text{yr}}$$

3.5 AVAILABLE COMPUTER PACKAGES: SIMULTANEOUS MULTIMEDIA COMPARTMENT MODELS

With an understanding of how chemicals distribute themselves among different environmental compartments, the reader is now prepared for an introduction to multimedia models for predicting the fate of contaminants. Two fundamentally different approaches are used in multimedia environmental models. In the first approach, models for individual media are linked together with independent or sequential calculation of transport and reactions through each medium. Effective intermedia linkage of individual media models requires an intimate familiarity with the structure, parameters, and workings of each model. While this first approach can allow for detailed spatial representation of pollutant variations within each of the environmental media, special care must be taken to ensure that intermedia mass transfers are properly matched and that mass is balanced. Chapters 5 and 8 to 12 address models for individual media. In the second approach, a single, unified model is developed to predict the partitioning of pollutants among all the available environmental media and compartments. The second approach is usually limited to simple, completely mixed compartments for each medium, which together make up an idealized "unit world." However, simultaneous calculation for all media in the second approach ensures that mass transfers are properly represented and that mass balance is maintained.

This section introduces models based upon the second, simultaneous approach. These models allow for a screening assessment of the broad distribution of pollutants in the environment. They can be utilized when limited observational data are available, and they can answer the question, "Where in the environment should I expect to find a particular chemical once it is released?" As such, the assumed environmental landscape is highly simplified and is often referred to as an "evaluative environment," designed to compare the general fate and partitioning properties of different chemicals.

This second type of *simultaneous* multimedia environmental model was pioneered by Mackay and co-workers (Mackay, 1991). In this section, a formal mathematical introduction to simultaneous multimedia models is provided and level I and level II types of simultaneous multimedia models are described. This is done using example applications of the level I and level II Mackay models now available on the web. The level I model considers equilibrium between multiple environmental compartments within a closed-system framework such as that illustrated in Figure 3.1. Level I models utilize the same types of equilibrium and mass-balance calculations presented in Section 3.2 (see Fig. 3.8). Level II models "open up" the closed-system framework so that flow of the various media in and out of the system is considered as well as simple reactions within a medium. Level II models also employ the equilibrium relationships presented in this chapter but compute concentrations by balancing the rate of contaminant accumulation and loss in the system. The kinetics of intermedia transport are incorporated into successively more sophisticated versions (level III) of the simultaneous multimedia models, which are illustrated in Chapter 4. A more complete presentation of the full capabilities and use of the latest generation of multimedia environmental models, including their integration with human exposure and health effects estimates, is reserved for Chapter 13.

3.5.1 Structure of the Simultaneous Multimedia Model: Level I

Consider an environmental system with n compartments, $i = 1, \ldots, n$. Each compartment is characterized by a volume V_i and the equilibrium concentration of the chemical A within that compartment, denoted by $C_{i,\text{eq}}^{\text{A}}$. The total mass of the chemical in the environment is then given by:

$$M_{\text{total}}^{\text{A}} = \sum_{i=1}^{n} C_{i,\text{eq}}^{\text{A}} \; V_i \tag{3.22}$$

Consider the case where a limited mass of the chemical is released to the environment, as depicted on the extreme right-hand side box in Figure 3.8, so that all chemical concentrations are below saturation values and no pure-phase chemical is present. This is generally a reasonable assumption when the unit environment consists of all the soil, water, air, and biota in a global or even a regional/local environment. The ratios of equilibrium concentrations in each of the media are then given by the respective partition coefficients:

$$K_{i/j} = \frac{C_{i,\text{eq}}^{\text{A}}}{C_{j,\text{eq}}^{\text{A}}} \tag{3.23}$$

For a system with n compartments only $n-1$ partition coefficients need be defined to fully specify the system, as long as each compartment is included in at least one of the defined partition coefficients. Combining Eq. (3.22) with the $n-1$ equations defined by the partition coefficients in Eq. (3.23) yields n equations for the n unknown values of $C_{i,\text{eq}}^{\text{A}}$.

In practice the n equations are readily solved by first defining, or redefining as necessary, the $n-1$ partition coefficients so that they are all expressed as concentration ratios relative to a reference compartment, i^* (usually water). There is thus one partition coefficient for each of the remaining $n-1$ compartments relative to i^*:

$$K_{i/i^*} = \frac{C_{i,\text{eq}}^{\text{A}}}{C_{i^*,\text{eq}}^{\text{A}}} \tag{3.24}$$

The concentration in the reference compartment is first determined as:

$$C_{i^*,\text{eq}}^{\text{A}} = \frac{M_{\text{total}}^{\text{A}}}{V_{i^*} + \sum_{i=1}^{n-1} K_{i/i^*} V_i} \tag{3.25}$$

and the concentrations in the remaining $n-1$ compartments are subsequently determined by simple rearrangement of Eq. (3.24):

$$C_{i,\text{eq}}^{\text{A}} = K_{i/i^*} C_{i^*,\text{eq}}^{\text{A}} \tag{3.26}$$

A quick review of Section 3.2 will reveal that this procedure is simply a generalized version of the method used in Examples 3.2 and 3.4 to solve for equilibrium concentrations in the hypothetical container with fixed volumes of water, air, NAPL, soil, and/or biota. As in those examples, the general method must be modified to account for the fact that concentrations in certain media are commonly reported on a mass per mass basis, rather than as mass per unit volume. For these compartments (in particular, soil, suspended solids, and biota), the compartmental volumes are replaced by compartmental masses (equal to the volume of the compartment, V_i, times the density of the compartment media, ρ_i), and the respective partition coefficients must also have the appropriate units to reflect the mass per mass convention.

To summarize, a model for equilibrium partitioning in a unit environment is defined by the following inputs:

- The total mass of the contaminant, $M_{\text{total}}^{\text{A}}$
- The volume (or mass) of each compartment, V_i (or $M_i = V_i \rho_i$); $i = 1, \ldots, n$
- The partition coefficients relative to a reference compartment, K_{i/i^*}; $i = 1, \ldots, n-1$

Using these inputs, the model computes the concentration in each compartment, $C^{\mathrm{A}}_{i,\mathrm{eq}}$, $i = 1, \ldots, n$, employing a closed-system framework such as that shown in Figure 3.1.

The interested student who follows up on the publications of Mackay and co-workers will note that many of these publications utilize the concept of *fugacity* to motivate and present the equilibrium calculations illustrated above. Fugacity, with units of pressure (e.g., pascals) may be thought of as the tendency (or, if you will, the "desire") of a chemical to leave or escape from a given state or compartment. The concept is similar to that of partial pressure. Differences in fugacity provide the driving force for mass transfer from one medium to another. At equilibrium, the fugacities of the chemical in all compartments are equal.

Once mastered, the concept of fugacity does provide a number of useful insights and a convenient computational basis alternative to the system of equations outlined above. However, students already comfortable with the concept of concentration often prefer to remain in this realm; hence the formulation based on fugacity is not described in this section. In making available their simultaneous multimedia models, Mackay and co-workers at the University of Trent, Ontario, have likewise done so in a manner that they may be utilized without adoption of the fugacity approach. In the following examples, these models, available for downloading from the web pages of the University of Trent Environmental Modelling Centre, are utilized to present the results of equilibrium partitioning calculations. The reader may now wish to spend some time perusing the following web pages before returning to the examples below: http://www.trentu.ca/academic/aminss/envmodel and http://www.trentu.ca/cemc/models/EQC2.html.

EXAMPLE 3.8

Given that 100,000 kg of benzene is released into a six-compartment unit environment, we use the level I model to determine the equilibrium benzene concentrations in the six-compartments with the characteristics listed below:

Properties of Assumed Evaluative Environment

Compartment	V_i (m^3)	ρ_i (kg/m^3)	M_i (kg)
1. Air	1×10^{14}	1.185	11850×10^{10}
2. Water	2×10^{11}	1000	20000×10^{10}
3. Soil	9×10^{9}	2400	2160×10^{10}
4. Sediment	1×10^{8}	2400	24×10^{10}
5. Suspended sediment	1×10^{6}	1500	0.15×10^{10}
6. Fish	2×10^{5}	1000	0.02×10^{10}

In particular, how will benzene mass be distributed among these compartments?

The assumed evaluative environment is representative of a region of a little over 300 × 300 km, with an atmospheric layer 1000 m in depth (a typical mixing height for the surface layer of the atmosphere), surface water covering one-tenth of the land surface to a depth of 20 m, soil covering the remaining nine-tenths of the land surface to a depth of 10 cm, and sediment at the bottom of the surface water to a depth of 1 cm. The suspended solids concentration in the surface water is 7.5 mg/L. Assuming a representative fish size of ~1000 cm^3 (1 L), there are a total of ~200 million fish in the surface water, or one for every 1000 m^3.

Solution 3.8 This unit environment is the default environment assumed when executing the Equilibrium Criterion (EQC) Model (Version 1.01, 16-bit, May 1997) obtained from the Trent Environmental Modelling Centre web pages.[4] The chemical properties and partition coefficients for benzene are available in the model's database and are printed as part of the model output. Note that benzene is specified as a "type 1" chemical for the EQC program. The program requires specification of the chemical as one of the following types:

Type 1: chemicals that partition into all media

Type 2: nonvolatile chemicals (that do not partition into the air)

Type 3: chemicals with zero or near-zero solubility (that do not partition into the water)

In addition, the default temperature of 20°C is changed to 25°C to more closely correspond to the conditions assumed for the examples presented earlier in this chapter. The partition coefficients are all presented relative to water and in dimensionless form. The dimensionless partition coefficients for soil, sediment, and suspended sediment are computed from the organic carbon–water partition coefficient for benzene ($K_{oc} = 55.3$ L/kg), the densities (ρ_i), and the assumed fraction of organic carbon for each compartment ($f_{oc,i}$). As such,

$$K_{soil/water} = K_{oc}\rho_{soil}f_{oc,soil} \Rightarrow 55.3 \text{ L/kg} \times 2.4 \text{ kg/L} \times 0.02 = 2.65$$

$$K_{sed/water} = K_{oc}\rho_{sed}f_{oc,sed} \Rightarrow 55.3 \text{ L/kg} \times 2.4 \text{ kg/L} \times 0.04 = 5.31$$

$$K_{susp.sed/water} = K_{oc}\rho_{susp.sed}f_{oc,susp.sed} \Rightarrow 55.3 \text{ L/kg} \times 1.5 \text{ kg/L} \times 0.20 = 16.6$$

The fish–water partition coefficient is computed employing the Mackay model (Mackay, 1991, p. 81) described in Section 3.4.1 in which fish are assumed to be comprised of approximately 5% lipids, which behave similarly to octanol, so that:

$$K_{fish/water} = 0.05\,K_{ow} \Rightarrow 0.05 \times 135 = 6.75$$

[4] The Trent Environmental Modelling Center has updated EQC to Version 2; however, Version 1 is still available on the web site and can be used. Both versions are similar.

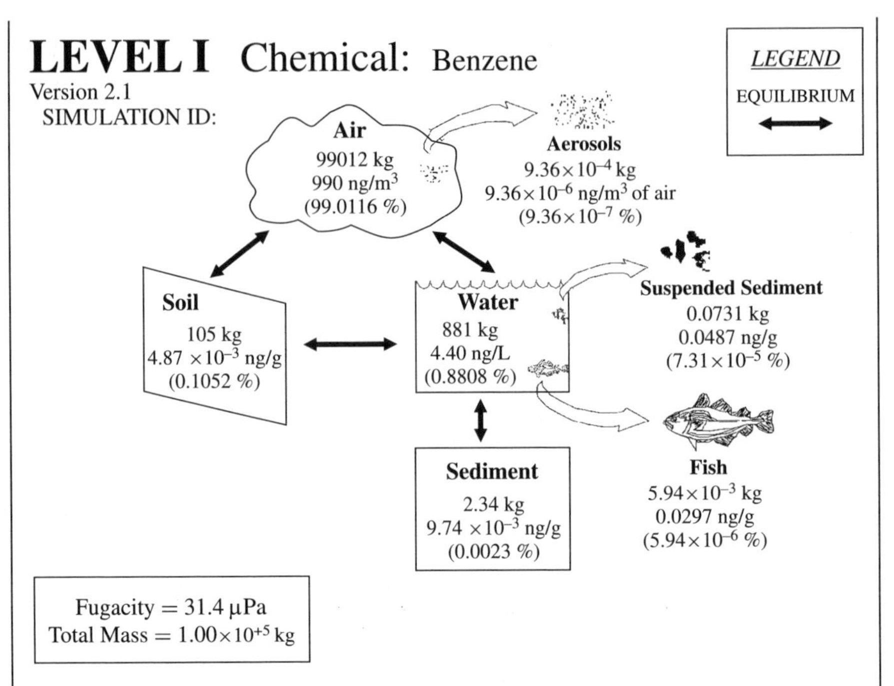

Figure 3.13 Pictorial summary of EQC level I model output (from www.trentu.ca/academic/aminss/envmodel). (*Note:* 1 ng = 1 nanogram = 10^{-9}g; 1 Pa = 1 Pascal = 9.87×10^{-6} atm.)

The results of the level I calculation are summarized in Figure 3.13, as printed by the EQC model. As indicated in Figure 3.13, based solely on equilibrium partitioning and the assumed size of each compartment in the environment, the 100,000 kg of benzene partitions almost completely into the atmosphere (99.0%), with the remaining 1% distributed, in decreasing order, to the surface water (0.88%), soil (0.105%), and the remaining compartments. While the benzene mass is virtually all in the air phase, the concentration in each media, on a mass per volume basis, tells a very different story. As indicated in Figure 3.13, the benzene concentrations in each compartment are ranked in nearly the reverse order, with the highest value in the suspended sediment, followed by fish, sediment, soil, and water, with the lowest concentrations in the air. However, the vast differences in the volumes of these compartments still yield the overall mass distribution described above.

Before moving on to the level II calculation, let us verify that the results of EQC level I model are consistent with Eq. (3.25) for the concentration of benzene in water, the reference compartment:

$$C_{w,\mathrm{eq}}^{\mathrm{benzene}} = \frac{M_{\mathrm{total}}^{\mathrm{benzene}}}{V_w + K_{\mathrm{a/w}}V_a + K_{\mathrm{soil/w}}V_{\mathrm{soil}} + K_{\mathrm{sed/w}}V_{\mathrm{sed}} + K_{\mathrm{susp.sed/w}}V_{\mathrm{susp.sed}} + K_{\mathrm{fish/w}}V_{\mathrm{fish}}}$$

$$= \frac{100{,}000 \text{ kg}}{\left\{2 \times 10^{11} + 0.225 \times 10^{14} + 2.65 \times 9 \times 10^{9} + 5.31 \times 10^{8} + 16.6 \times 10^{6} + 6.75 \times 2 \times 10^{5}\right\} \text{m}^3}$$

$$= 4.41 \times 10^{-9}\,\frac{\text{kg}}{\text{m}^3} = 4.41 \times 10^{-6}\,\frac{\text{g}}{\text{m}^3} = 4.41\,\frac{\text{ng}}{\text{L}}$$

This is the same result as that given by the EQC model. The EQC concentrations for the other compartments can likewise be verified through the application of Eq. (3.26).

3.5.2 Level II Models: Incorporating Reaction and Net Transport

In the level I model described above, a fixed mass of a chemical is introduced to the evaluative environment and is distributed in a closed system among the different compartments. Only equilibrium partitioning processes are operative. To move to the next level of sophistication, consider an environment in which the chemical is removed from different compartments, either by degradation reactions or by advection (flow) to the outside world (i.e., the portion of the environment exogenous to the model). Now assume that the environment is at steady state. In order for steady conditions to be achieved and maintained throughout the unit environment, the reaction and advection removal processes must be balanced by inputs of the chemical at the same total rate. These inputs may be associated with either direct contaminant discharges to one or more of the compartments or with advective inputs. (Note that in order to preclude the accumulation or loss of the compartmental fluid or media, the fluid or media advection rate into a particular compartment must equal its advection rate from that compartment.) At the same time that these mass addition and removal processes are operative, equilibrium among compartments is maintained according to the partition coefficient relationships. This defines the level II model of Mackay and colleagues, illustrated in Figure 3.14.

In addition to the compartment volumes (or masses and densities) and intermedia partition coefficients, the level II model requires:

- Fluid or media advection rates into and out of each compartment, Q_i
- The total mass influx rate of chemical A into the environment, $W^{\mathrm{A}}_{\mathrm{total}}$:

$$W^{\mathrm{A}}_{\mathrm{total}} = \sum_{i=1}^{n} \left[E^{\mathrm{A}}_i + Q_i C^{\mathrm{A}}_{i,\mathrm{inf\ low}}\right]$$

- where
- E^{A}_i = direct mass emission rate of A into compartment i
- $C^{\mathrm{A}}_{i,\mathrm{inflow}}$ = concentration of A in advective inflow to compartment i
- First-order decay rates of A in each compartment, k^{A}_i

The model is solved for the steady-state concentrations: $C^{\mathrm{A}}_{i,\mathrm{ss}}; i = 1, \ldots, n$, which equate the total mass influx rate into the environment with the total loss rate due to advection and reaction:

$$W^{\mathrm{A}}_{\mathrm{total}} = \sum_{i=1}^{n} \left[C^{\mathrm{A}}_{i,\mathrm{ss}} \left(Q_i + V_i k^{\mathrm{A}}_i\right)\right]$$

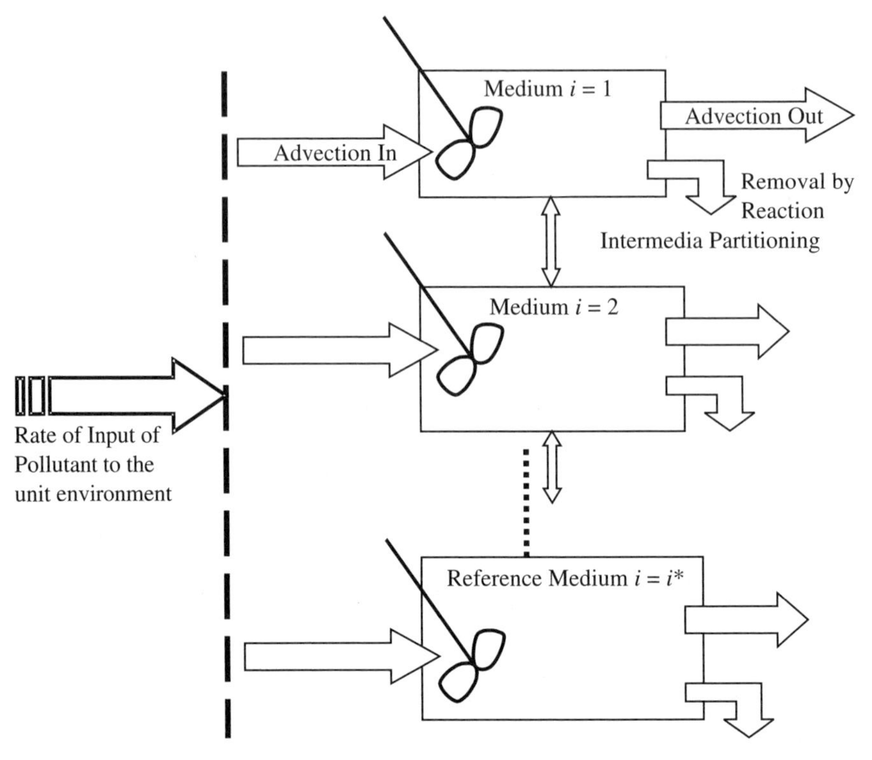

Figure 3.14 Schematic of level II multimedia models illustrating inflow, outflow, and reaction rates occurring in multiple well-mixed media linked by equilibrium partitioning.

With the assumed equilibrium among the concentrations in each compartment, the steady-state concentration in the reference compartment can first be computed as:

$$C_{i^*,\mathrm{ss}}^{\mathrm{A}} = \frac{W_{\mathrm{total}}^{\mathrm{A}}}{\left(Q_{i^*} + V_{i^*}k_{i^*}^{\mathrm{A}}\right) + \sum_{i=1}^{n-1} K_{i/i^*}\left(Q_i + V_i k_i^{\mathrm{A}}\right)} \tag{3.27}$$

The steady-state concentrations in the remaining $n-1$ compartments are subsequently determined from their respective equilibrium relationships in Eq. (3.26). Once again, for those media where chemical concentrations are defined on a mass per mass basis, the compartment volumes are replaced by compartment masses, given by the product of the compartment volume and the associated media density. Comparing Eq. (3.27) to Eq. (3.25) indicates a similar structure for the level I and level II calculations, with the former determining an equilibrium resulting from the addition of a fixed mass with no net loss mechanism from the environment and the latter yielding a steady state in response to a constant mass input rate and constant loss rates.

EXAMPLE 3.9

Computation of steady-state benzene concentrations in a six-compartment unit environment using the level II model.

Solution 3.9 In this example, rather than assuming a fixed mass of benzene partitioned to an equilibrium environment, a constant *loading rate* of benzene to a *steady-state* environment is assumed. The unit environment is the same as that given in the previous example, except that advection of the media material (and accompanying benzene) out of the unit environment occurs for air, water, and sediment, and first-order degradation reactions are operative in the air, water, soil, and sediment compartments. The total benzene loading rate to the environment is given by $W_{\text{total}}^{\text{benzene}} = 1000$ kg/h. The advection and reaction rates for the unit environment are summarized in the first four rows of the EQC level II model output. Again, these correspond to the default assumptions for the version of the model available on the Trent Environmental Modelling Centre web page. The advection flow rates are computed from assumed fluid residence times of 100 h (4.2 days) for the atmosphere, 1000 h (42 days) for the surface water, and 50,000 h (5.7 years) for the sediments. These values are chosen as representative of expected conditions for a 300 × 300 km regional environment, with some mix

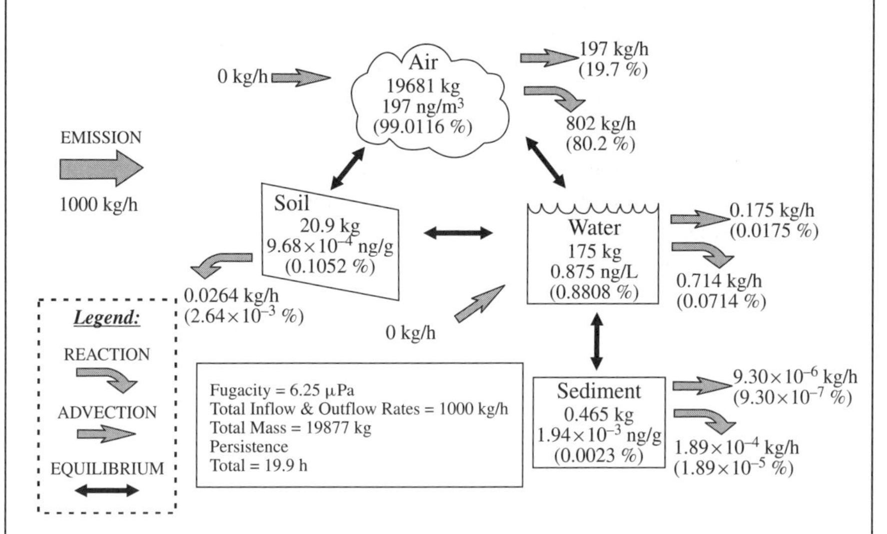

Figure 3.15 Pictorial summary of EQC level II model output (from www.trentu.ca/academic/aminss/envmodel). (*Note:* 1 ng = 1 nanogram = 10^{-9}g; 1 Pa = 1 Pascal = 9.87 × 10^{-6} atm.)

of streams and lakes. Advection is also expected to occur with the suspended sediment and fish associated with the surface water; however, these are negligible in the current problem and have been ignored. The reaction rates are computed from assumed half-lives of benzene ranging from 17 h in the atmosphere to 1700 h (71 days) in the sediments.

The results of the level II calculation are summarized in Figure 3.15. The relative partitioning of benzene between the six environmental compartments is the same as that computed for the level I model in Example 3.9. This is not a coincidence—the advection and reaction terms control the overall steady-state mass of benzene in the unit environment, but the relative distribution of this mass among the compartments remains the same as that computed for level I since this relative amount is controlled by the equilibrium relationships in Eq. (3.23) and the volume (or mass) of each compartment, which is also unchanged. The benzene concentration in all six compartments scales linearly with the benzene mass input loading rate. The advection outflow and reaction loss terms of benzene in the level II model also scale linearly with the input loading rate since each varies linearly with the associated compartment concentration.

Exercise for the Reader Verify the results of the EQC level II model for the water compartment, $C_{\text{water,ss}}^{\text{benzene}} = 0.875$ ng/L, using Eq. (3.27).

In summary, the level I models describe pollutant equilibria between multiple compartments in a closed-system framework, such as that shown in Figure 3.1. Level II models "open up" the system to incorporate flows to and from the outside environment as well as reactions within compartments, while simultaneously solving the basic pollutant equilibria equations (Fig. 3.14). The different environmental compartments are assumed to be well-mixed, and hence the level I and II models provide a measure of the quantities and concentration of pollutants in the different compartments over a large and aggregated spatial scale. Spatial variations within individual compartments are not addressed. The models in Chapters 5 and 8 to 12 consider advection and dispersion within individual media, thereby addressing more local-scale variations in pollutant concentrations in the air, water, and subsurface compartments.

Homework problems for this chapter are provided at www.wiley.com/college/ramaswami.

4 Kinetics of Intermedia Pollutant Transfer

Chapter 3 addressed intermedia pollutant transfers at equilibrium, with emphasis on determining the final concentration and mass distribution of pollutants after a long time period of contact has been established between the various environmental compartments. However, long intermedia contact times are not available in many environmental situations. For example, water flowing in a stream with a linear velocity of u over a NAPL spill zone of length L has a fluid residence time, $\tau (= L/u)$, during which contact between the NAPL and water can occur. If the residence time is less than the time required to achieve equilibrium, NAPL-derived contaminant concentrations in water will show a departure from equilibrium levels. In such situations, the rate at which contaminant mass is transferred between environmental compartments controls their distribution. The focus of this chapter is on quantifying the rate of pollutant mass transfer between pairs of environmental media: air–water, NAPL–water, soil–water, soil–air, NAPL–air, and so forth. Sections 4.1 and 4.2 present molecular diffusion as the principal process resulting in intermedia mass transfer. Intermedia mass-transfer models based on diffusion theory are parameterized in Sections 4.3 to 4.6 for specific environmental situations of interest, for example, the exchange of oxygen between the atmosphere and lakes or streams. A case study is presented in Section 4.7 that employs intermedia mass-transfer models to assess the range and persistence of pesticides in the global multimedia environment composed of air, water, and soil. In contrast with global mixed-box models, spatially resolved models describing pollutant transport in individual media incorporate intermedia mass transfers as localized source or sink terms—this is illustrated in Section 4.8. The topic of pollutant transport within individual media is explored further in Chapter 5 and Chapters 8 to 12.

4.1 MOLECULAR DIFFUSION AND INTERMEDIA MASS TRANSFER

Molecular diffusion is a principal mechanism by which chemical mass is transferred *between* adjacent environmental media. Note that molecular diffusion, along with advection–dispersion due to movement of fluid media, also contributes to chemical transport *within* individual media. For example, if a vial of perfume is left open in a room with no air currents, molecular diffusion will result in movement of the fragrant molecules from the vial. If air currents are present, molecular diffusion will still take place, but the fragrance will spread more quickly due to the advective and dispersive air motions. However, if one considers transfer of the perfume from air to another medium, for example, fabric, molecular diffusion becomes the primary mechanism for the intermedia (air-to-fabric) transport of molecules, diffusion occurring on a microscopic scale both within the air and the fabric.

In some cases, intermedia chemical exchanges can also occur by nondiffusive processes, for example, by the deposition or resuspension of particles that carry chemicals. Deposition of solid particles and liquid droplets can contribute to pollutant transfer from the atmosphere to solid surfaces such as vegetation, the ground surface, buildings, and so forth. Likewise, particle deposition and resuspension play an important role in chemical exchanges between surface waters and underlying sediments. Models for nondiffusive mass transfer processes are described briefly in Section 4.7. A more detailed discusson of nondiffusive mass transfer in surface water and atmospheric systems is reserved for Chapters 10 and 11, respectively. This chapter focuses largely on the direct exchange of chemicals between environmental media, which is governed by molecular diffusion.

Molecular diffusion is the random (Brownian) movement of pollutant molecules from a region of higher chemical concentration to a region of lower chemical concentration within the same medium. According to Fick's law, the diffusive flux is proportional to the chemical concentration gradient within the medium, with the constant of proportionality known as the diffusion coefficient of the chemical within that medium. Consider a medium, j, within which the diffusive mass flux of chemical A is represented by Fick's law as:

$$\text{Diffusive flux of A} = J_j^{\text{A}} = \frac{\text{mass of A transferred}}{\text{area} \cdot \text{time}} = -D_{m,j}^{\text{A}} \frac{\partial C_j^{\text{A}}(x)}{\partial x} \tag{4.1}$$

where $D_{m,j}^{\text{A}}$ represents the molecular diffusion coefficient of chemical A in medium j.

Diffusion coefficients for some chemicals frequently encountered in air and in water are shown in Table 4.1. As is intuitive, molecules diffuse faster in air than in water, the diffusion coefficient being inversely related to the viscosity of the medium in which the diffusion occurs. For a typical organic pollutant, the diffusion coefficient in air, $D_{m,a}^{\text{A}}$, is of the order of 10^{-1} cm^2/s, while that in water, $D_{m,w}^{\text{A}}$, is of the order of 10^{-5} cm^2/s. The diffusion coefficient typically decreases as the size and molecular weight of the chemical increases, and increases with an increase in ambient temperature. The estimation equations presented in Table 4.1 incorporate these influences on the diffusion coefficient. However, the magnitude of the diffusion coefficient does not vary much for a wide spectrum of chemicals (Cussler, 1983). Order-of-magnitude estimates of the molecular diffusion coefficients in air and water (as provided above) are often sufficient for many environmental applications.

Diffusion within soil particles is typically represented as the process of diffusion of the chemical through the water and air entrapped in the porous matrix of soil; solid-phase diffusion, that is, diffusion through the solid matter constituting soil particles, is typically neglected. Solid-phase diffusion is considered only in certain unique environmental situations, for example, diffusion of pollutants through high-density polyethylene (HDPE) fabric that is used to line waste landfills.

TABLE 4.1 Molecular Diffusion Coefficient for Dilute Concentrations of Common Gases in Air and for Liquids or Gases in Water

Gaseous Chemical	Diffusion Coefficient in Air, $D_{m,a}$ (cm^2/s)	Liquid or Gaseous Chemical	Diffusion Coefficient in Water (10^{-5} cm^2/s)
CH_4 (293 K)	0.106	Benzene (293 K)	1.0
CO (293 K)	0.208	Ethanol (298 K)	1.2
CO_2 (293 K)	0.160	Methanol (288 K)	1.3
C_7H_{16} (300 K)	0.075	Toluene (293 K)	0.85
H_2O (293 K)	0.242	O_2 (298 K)	2.4
NO (300 K)	0.180	CO_2 (298 K)	2.0
SO_2 (300 K)	0.126		

Sources: Lide (1997), Mills (1995), and Incropera and Dewitt (1990).

Estimating Diffusion Coefficients in Air (cm^2/s)

$$D_{m,a}^{A} = 9.95 \times 10^{-3}\, T_{\text{env}}^{1.75} \frac{\sqrt{\dfrac{29 + \text{MW}^{A}}{29 \times \text{MW}^{A}}}}{\left[2.7 + \left(v_{\text{mol}}^{A}\right)^{1/3}\right]^{2}}$$

where T_{env} is the environmental temperature in kelvin, MW^{A} is the molecular weight in grams, and v_{mol}^{A} is the molar volume of chemical A in cm^3/g mol. Ambient pressure = 1 atm (Reid et al., 1987).

Estimating Diffusion Coefficients in Water (cm^2/s)

$$D_{m,w}^{A} = 7.4 \times 10^{-8}\, T_{\text{env}} \frac{2.6(\text{MW}_{\text{water}})^{1/2} T_{\text{env}}}{\mu_{\text{water}} \left(v_{\text{mol}}^{A}\right)^{0.6}}$$

where T_{env} is the environmental temperature in kelvin, MW^{A} is the molecular weight in grams, v_{mol}^{A} is the molar volume of chemical A in cm^3/g mol, and μ_{water} is the viscosity of water (cP) (Wilke and Chang, 1955; Bird et al., 1960).

4.2 MODELS OF INTERMEDIA MASS TRANSFER

Diffusion theory is employed to describe intermedia chemical transport in two different conceptual models: (1) the stagnant or thin-film model and (2) the surface renewal model. See Figure 4.1. The stagnant film model was developed by Nernst in 1904 and applied to gas absorption in 1923 (Whitman). The stagnant film model is appropriate when mass transfer between a solid and a fluid, or between two relatively quiescent

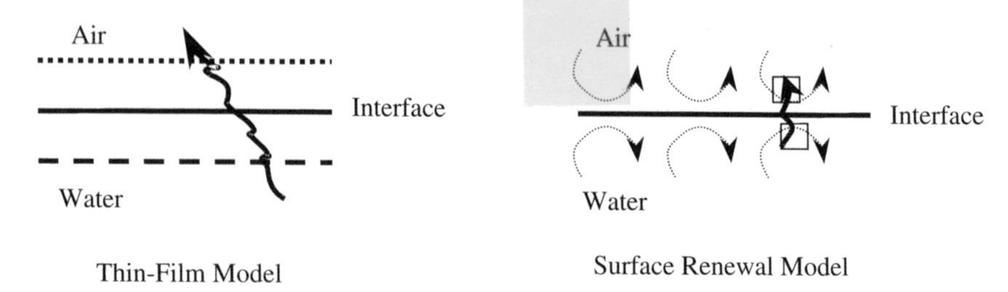

Thin-Film Model | Surface Renewal Model

Figure 4.1 Schematic comparing the stagnant or thin-film model for intermedia mass transfer with the surface renewal model. Molecular diffusion, represented by a bold wavy arrow, occurs across stagnant films in the thin-film model and across moving fluid parcels that meet at the interface in the surface renewal model.

fluids, is being considered. In this model, each fluid medium is subdivided into a bulk, well-mixed compartment and a stagnant (unmixed) film near the intermedia interface. Chemical transfer is modeled to occur by molecular diffusion across the stagnant film in each medium. Applying Fick's first law [Eq. (4.1)], the stagnant film model predicts that the mass flux across a stagnant film is directly proportional to the diffusion coefficient in that medium and the concentration gradient across the film. While chemical concentrations at the edges of the stagnant film can be determined in order to compute the gradient, the thickness of the film itself is difficult to quantify and is an unknown parameter in the stagnant film flux formulation.

When significant mixing and turbulence occur within two fluid media, their interface can no longer be modeled as stagnant. Instead, a surface renewal model (Danckwerts, 1951) is used wherein parcels of the fluid are envisioned as constantly circulating within each medium, periodically renewing the interface and causing pollutant exchange across the two media. The time that each parcel spends at the interface is variable, and diffusive transfer of chemicals occurs across the fluid parcels from each medium during their time period of intersection at the interface. The surface renewal model predicts that the mass flux across an interface is directly proportional to the square root of the molecular diffusion coefficient and the square root of the surface renewal rate.[1] However, the renewal rate of fluid parcels at the interface is difficult to quantify and represents an unknown in the surface renewal model.

Environmental observations indicate that the dependence of intermedia mass-transfer rates on molecular diffusion coefficients ranges from $[D_m]^{0.5}$ (as in surface renewal) to $[D_m]^1$ (as in the stagnant film model), suggesting that both models are merely theoretical constructs used to describe complex physical phenomena. In both the stagnant film model as well as the surface renewal model, the unknown theoretical parameters, namely, film thickness and the surface renewal rate, respectively, cannot be measured directly and must be estimated through correlation with dynamic

[1] $J_j^{\mathrm{A}} \propto \sqrt{D_{m,j}^{\mathrm{A}} \cdot \mathrm{sr}_j}$, where sr_j is the rate of surface renewal of medium j at the interface; J_j^{A} is the mass flux of chemical A across the interface of medium j.

properties of the fluids at the intermedia interface. From a practical point of view, mass-transfer correlations based on the stagnant film model are more widely used, employing an *effective* stagnant layer thickness to represent the complex transition from turbulent flow in the bulk fluid to a stagnant surface condition. Consequently, the main focus of this chapter will be on the stagnant film model for diffusive intermedia mass transfer.

4.2.1 Two-Film Model

Consider two relatively quiescent media, for example, air and water, between which mass transfer of chemical A occurs. A thin film exists in both air and water, across which diffusion of the chemical takes place. The air film lies adjacent to the air–water interface and separates the bulk (well mixed) air from the air–water interface. While the bulk air is well mixed and has a spatially uniform chemical concentration, $C_{\text{air}}^{\text{A}}$ (which may vary with time), concentration gradients exist within the air film as shown in Figure 4.2. Likewise, the water film separates the bulk water from the air–water interface. The bulk water is well mixed and has a spatially uniform chemical concentration, C_w^{A} (which may also be time varying). Spatial concentration gradients exist within the water film. Assuming an absence of interfacial resistances to mass transfer such as from polymer or surfactant films, or biofilms, equilibrium conditions are assumed to be achieved at all times at the air–water interface, such that:

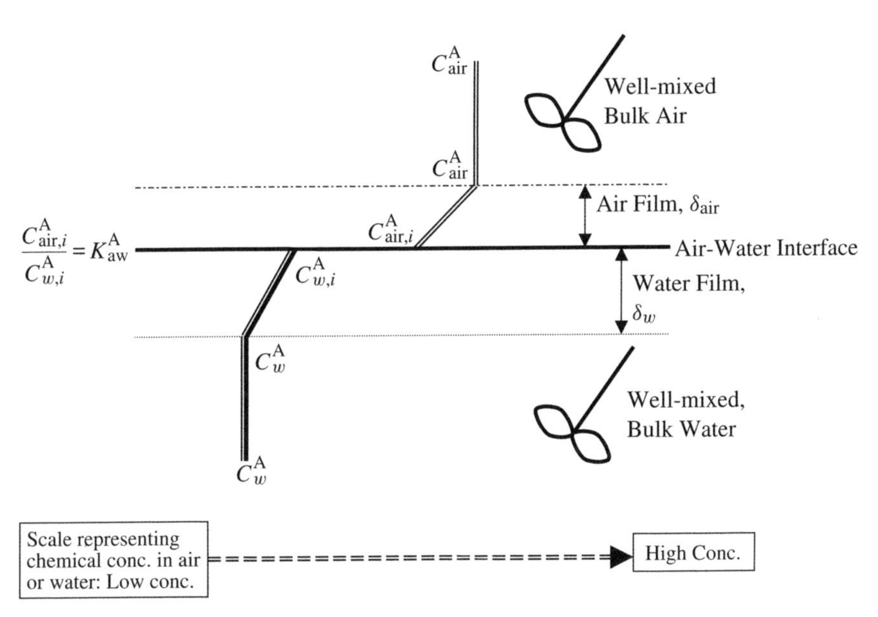

Figure 4.2 Stagnant two-film model depicting the transfer of a chemical A from air to water. Mass transfer occurs from bulk air, across a thin air film to the air–water interface, and thence across a thin water film to bulk water.

$$\frac{C_{\mathrm{air},i}^{\mathrm{A}}}{C_{w,i}^{\mathrm{A}}} = K_{\mathrm{aw}}^{\mathrm{A}} \tag{4.2}$$

where the subscript i refers to interfacial concentrations, and, $K_{\mathrm{aw}}^{\mathrm{A}}$ is the air–water partition coefficient for chemical A, defined previously in Eq. (3.4).

Mass transfer of chemical occurs from zones of higher concentration to lower concentration across the two films in the two media. The direction of mass transfer (air–water or water–air) can be determined by examining the magnitude of the chemical concentrations at the two edges of the films in the individual media. The direction of mass transfer can also be ascertained by comparing the chemical concentrations in bulk water and in bulk air when expressed on the same basis, that is, the chemical concentration in bulk water, C_w, should be compared with the water equivalent of the chemical concentration in bulk air, $(C_{\mathrm{air}}/K_{\mathrm{aw}})$. Example 4.1 illustrates this concept.

EXAMPLE 4.1

The concentration of benzene in the air above a wastewater lagoon at a petroleum refinery is 10 ng/L (of air), while the concentration of benzene in bulk water in the lagoon is 15 ng/L (of water). Determine if benzene is being transferred from the atmosphere to the water or vice versa.

Solution 4.1 If concentrations in bulk air and in bulk water were to be compared as stated, it would appear that benzene mass transfer occurs from water (15 ng/L) to air (10 ng/L).

However, to determine the direction of mass transfer, the bulk air and bulk water benzene concentrations must be compared on the same basis, that is, benzene concentration in air must be expressed on an equivalent aqueous basis. Computing the K_{aw} for benzene from K_H data provided in Table 2.3, the water equivalent of the atmospheric benzene concentration is computed as:

$$\frac{C_{\mathrm{air}}^{\mathrm{benzene}}}{K_{\mathrm{aw}}^{\mathrm{benzene}}} = \frac{C_{\mathrm{air}}^{\mathrm{benzene}}}{K_H^{\mathrm{benzene}}/RT} = \frac{10\ \mathrm{ng/L}}{\dfrac{10^{0.76}\ \mathrm{L\cdot atm/mol}}{0.0821\ \mathrm{L\cdot atm/mol\cdot K} \times 298\ \mathrm{K}}}$$

$$= \frac{10\ \mathrm{ng/L}}{0.225} = 44.4\ \mathrm{ng/L}$$

The air-phase benzene concentration expressed on an equivalent aqueous basis is much larger (44 ng/L) compared to the concentration in bulk water (15 ng/L). Hence, benzene mass transfer actually occurs from air to water.

In Figure 4.2, mass transfer occurs from bulk air to interfacial air. Air–water equilibration takes place at the air–water interface, after which mass transfer occurs in water by diffusion across the water film from the interface to bulk water. If the air compartment is closed and initially contains a fixed mass of the chemical A, the progress of air-to-water mass transfer will result in a decrease in the concentration of A in bulk air, eventually approaching equilibrium with water. Simultaneously, the concentration of chemical A in bulk water will increase, approaching equilibrium conditions with air. In this manner, the equilibrium interfacial concentrations will continuously adjust over time until final equilibrium conditions are achieved in both media.[2] When the bulk air and the bulk water are in equilibrium, no further net mass transfer of the chemical takes place between air and water. This process occurring in a closed system as described above, is illustrated in Figure 4.3*a*. Figure 4.3*b* illustrates what happens when the air compartment is open or contains a large supply of the chemical A as in the case of oxygen being transferred from the open atmosphere to a lake. Because such a large supply of oxygen is available in the atmosphere, the oxygen concentration in bulk air (and hence the interfacial concentrations in both air and water) barely changes with time while the bulk water will (absent any sinks for oxygen in the water column) become saturated with oxygen as a result of transfer from air. Thus, visualizing mass-transfer processes requires analysis of the environmental compartments to assess if these are open or closed or if they contain a large enough supply of the chemical of interest to saturate the other medium.[3] Notice the discontinuity of chemical concentrations at the interface in both cases shown in Figure 4.3, which is associated with the equilibrium partitioning phenomenon [Eq. (4.2)] that is assumed to occur at that location.

4.2.2 Flux Representation

Assuming neither accumulation nor loss (e.g., by degradation) of chemical at the air–water interface shown in Figure 4.2, the rate at which the chemical is transferred across a unit area of the water film must be the same as the rate at which the chemical is transferred per unit area of the air film, which must also equal the rate at which the chemical is transferred per unit area of the combined air-and-water films. In the following equations, net mass transfer of chemical A in Figure 4.2 is assumed to occur from air to water.

Using Fick's law of diffusion, the mass flux (mass transfer rate per unit area) across the water film may be written as:

$$J(\text{water film}) = D_{m,w}^{\mathrm{A}} \frac{C_{w,i}^{\mathrm{A}}(t) - C_{w}^{\mathrm{A}}(t)}{\delta_w} = k_l \left[C_{w,i}^{\mathrm{A}}(t) - C_{w}^{\mathrm{A}}(t)\right] \qquad (4.3a)$$

[2] The final equilibrium chemical concentrations in both media can be determined from concepts detailed in Chapter 3, summarized in Figure 3.8.

[3] In Figure 4.3*b*, if the water column were to include a sink for oxygen (e.g., resulting from the biochemical decay of organic matter), a nonequilibrium *steady-state* condition would result, with a concentration profile similar to that shown for case B at time $t = t_1$. With this profile, there is a continuing flux of oxygen from the air to the water that is balanced by the oxygen uptake in the water column.

where

$$k_l = \frac{D^{\mathrm{A}}_{m,w}}{\delta_w} \tag{4.3b}$$

In Eq. (4.3a) and (4.3b), $D^{\mathrm{A}}_{m,w}$ (L^2/T) is the molecular diffusion coefficient of compound A in water, and δ_w (L) is the thickness of the water film. Since the thickness of the water film cannot readily be measured, it is incorporated into the parameter,

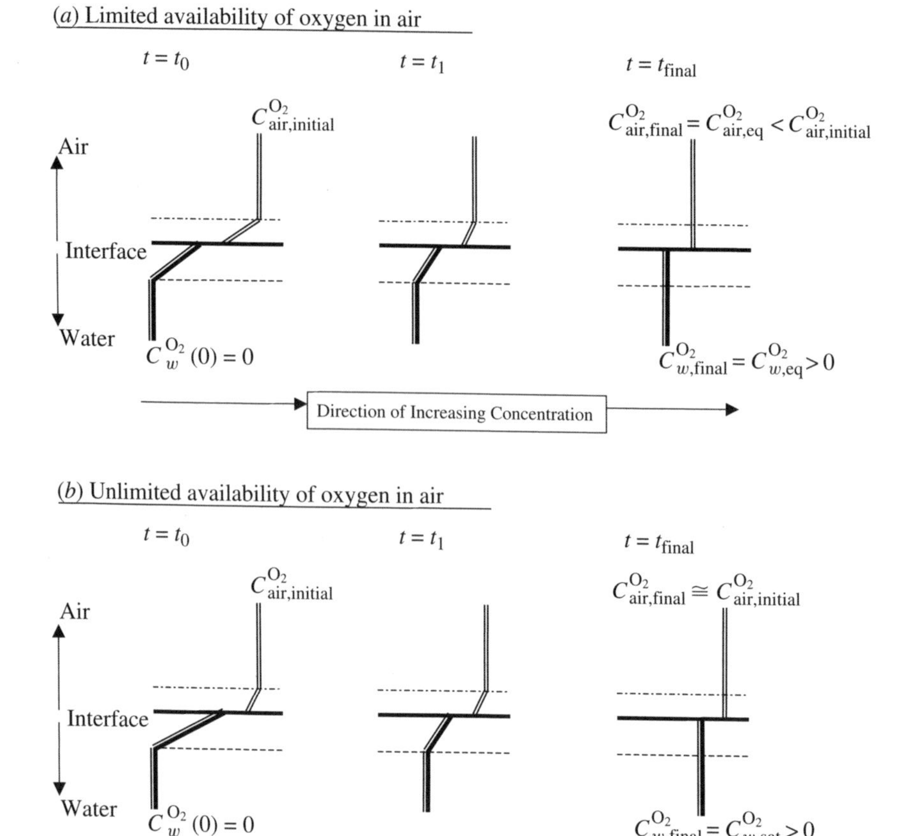

Figure 4.3 Progress of mass transfer of oxygen from air to water. The water is initially completely depleted of oxygen. *Case A*: The air compartment is closed and contains insufficient oxygen to saturate the water. The oxygen concentrations in bulk air decrease and that in bulk water increase until equilibrium is attained. *Case B*: The air compartment is the open atmosphere and contains more than enough oxygen to saturate the small closed water compartment. The oxygen concentration in air remains steady, while the water becomes saturated.

k_l (L/T), called the liquid side mass transfer coefficient, defined as shown in Eq. (4.3b). The mass-transfer coefficient, k_l, has units of (L/T) similar to velocity, and represents the diffusive velocity with which the chemical is transferred across the water film. The mass-transfer coefficient is also referred to as the piston velocity in the medium being considered.

Pollutant mass flux in Eq. (4.3) is expressed in terms of the difference in pollutant concentrations in water at the two edges of the water film. Likewise, chemical mass flux across the air film may be represented employing the difference in chemical concentrations at the two edges of the air film as:

$$J(\text{air film}) = D_{m,a}^{\mathrm{A}} \frac{C_{\text{air}}^{\mathrm{A}}(t) - C_{\text{air},i}^{\mathrm{A}}(t)}{\delta_{\text{air}}} = k_g \left[C_{\text{air}}^{\mathrm{A}}(t) - C_{\text{air},i}^{\mathrm{A}}(t)\right] \tag{4.4}$$

where k_g (L/T) is the gas-phase mass-transfer coefficient, which is defined as: $k_g = D_{m,a}^{\mathrm{A}}/\delta_{\text{air}}$ where $D_{m,a}^{\mathrm{A}}$ (L^2/T) is the molecular diffusion coefficient of compound A in air and δ_{air} (L) is the thickness of the air film.

Equations (4.3) and (4.4) are analogous to flow equations in a variety of situations: electron flow (current) in electrical circuits, water flow in an aquifer, airflow velocities in a pressure field, and so forth. In all cases, the flow rate depends upon a potential gradient and may be expressed as being directly proportional to a potential difference and inversely proportional to a resistance to flow. For example, electric current, which is the rate of flow of electrons in electrical circuits, is directly proportional to the voltage (potential) difference across the circuit, and inversely proportional to the resistance of the wire. In the case of chemical transfer, the mass flux rate in Eq. (4.3) and (4.4) is directly proportional to the concentration (potential) differential across the air and water films, and is inversely proportional to the resistance to mass transfer offered by the gas and liquid films, represented by $[1/k_g]$ and $[1/k_l]$, respectively. As expected, the resistance offered by the gas and liquid films is directly proportional to the thickness of these films. Natural turbulence occurring in the two phases, as well as engineered mixing operations, cause a reduction in the thickness of the individual films. In turn, the film resistance to mass transfer decreases and the magnitude of the mass-transfer coefficient increases, resulting in a corresponding increase in the chemical mass flux.

In general, interfacial chemical concentrations required for use in Eq. (4.3a) and (4.4) are not known. It is more convenient, therefore, to represent chemical mass flux in terms of the difference between bulk air-phase and bulk water-phase pollutant concentrations, when these concentrations are expressed on the same basis (as explained earlier). In this case, the entire film thickness (air film plus water film) contributes a resistance to mass transfer. Carrying the analogy to electrical circuits further, the total resistance to mass transfer may be written as the sum of the individual film resistances, such that the mass flux, denoted in terms of bulk-phase concentration differences expressed on an aqueous-basis, becomes

$$J(\text{across both films}) = k_{\text{LO}} \left[\frac{C_{\text{air}}^{\mathrm{A}}(t)}{K_{\text{aw}}} - C_w^{\mathrm{A}}(t)\right] \tag{4.5a}$$

$$\frac{1}{k_{\text{LO}}} = \left\{\begin{matrix}\text{overall film}\\ \text{resistance}\end{matrix}\right\} = \left\{\begin{matrix}\text{liquid film}\\ \text{resistance}\end{matrix}\right\} + \left\{\begin{matrix}\text{gas film}\\ \text{resistance}\end{matrix}\right\} = \frac{1}{k_l} + \frac{1}{K_{\text{aw}}\, k_g} \tag{4.5b}$$

Equations (4.5a) and (4.5b) were derived by first writing the flux equation across one film [Eq. (4.3a)], then substituting for the interfacial chemical concentration using Eq. (4.2) and the flux equation for the second film [Eq. (4.4)], and finally solving for the flux, J, recognizing that the flux across each of the individual films must equal each other, as well as the flux across both films. In Eq. (4.5a) $[C_a(t)/K_{\text{aw}}]$ represents the aqueous-phase equivalent of the chemical concentration in bulk air at time t, and $C_w(t)$ represents the chemical concentration in bulk water at time t, the difference between which provides the driving force for mass transfer. The term k_{LO} is called the overall liquid-phase mass-transfer coefficient and is utilized with bulk-phase chemical concentrations expressed on an aqueous (liquid) basis. The expression for the flux may also be derived in terms of bulk-phase chemical concentrations expressed on a gas-phase basis, yielding an overall gas-phase mass-transfer coefficient, k_{GO}. Note, $k_{\text{GO}} = k_{\text{LO}}/K_{\text{aw}}$.

In many situations, it may not be necessary to consider the overall mass-transfer coefficient and the combined resistance of the two films, as was done in Eq. (4.5). A simplifying assumption is made that the resistance to mass transfer in one medium is very small and may be neglected. Typically, it is the medium that the chemical "likes" or prefers to be in that offers the least resistance to mass transfer and may be ignored in the kinetic analysis. In this medium, once the film is neglected, interface and bulk concentrations become identical (Fig. 4.4). The medium that the chemical "dislikes" typically offers the dominant resistance to diffusive mass transfer through its stagnant film. Flux relationships for individual media may then be applied using Eq. (4.3) or (4.4), utilizing Eq. (4.2) to compute equilibrium at the interface. The issue of which medium offers the greater resistance to chemical transport can largely be addressed by evaluating the magnitude of the partition coefficient for equilibrium chemical transfer between the two media. For example, examination of Eq. (4.5b) suggests that the gas-phase resistance becomes negligible compared to the water-phase resistance when $K_{\text{aw}} \gg k_l/k_g$. Because diffusion coefficients in air are 4 orders of magnitude larger than in water (see Table 4.1), k_g may be expected to be much larger than k_l, based on

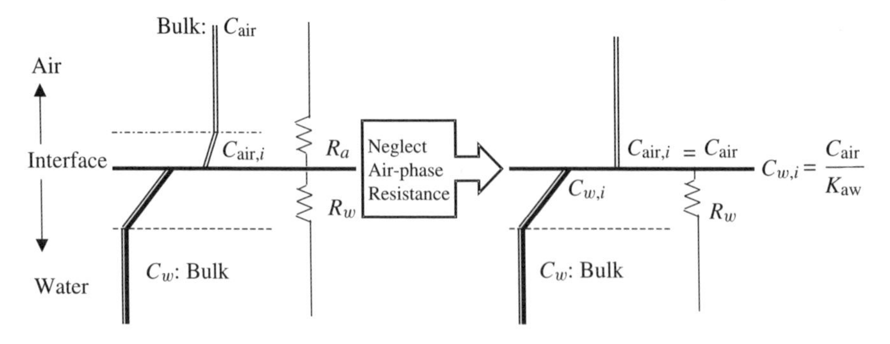

Figure 4.4 Effect of negligibly small air-phase resistance on the two-film configuration.

their definitions in Eq. (4.3b) and (4.4), respectively, assuming the films in air and in water are within an order of magnitude in thickness.[4] As a result, the condition that $K_{\text{aw}} > 0.01$ is often sufficient to ignore the gas-phase resistance (Schwarzenbach et al., 1993; Hemmond and Fechner-Levy, 2000; Mackay, 2001). Thus, in the case of transfer of highly volatile chemicals in air–water systems, film resistance offered by one phase, in this case water, dominates the system, and Eq. (4.3) can be applied with $C_{w,i} = C_{\text{air}}/K_{\text{aw}}$. In the case of highly soluble, nonvolatile chemicals, the air-phase resistance is dominant and Eq. (4.4) can be used. If neither phase resistance can be ignored, for example, for semivolatile chemicals, both resistances must be considered together as was shown in Eq. (4.5).

EXAMPLE 4.2

Indicate which medium offers the greater (dominant) resistance to mass transfer in the following two situations. Explain why. Sketch concentrations at the edges of the film offering the dominant resistance to mass transfer. Assume all mass transfers are occurring at 25°C and ambient pressure of 1 atm.

a. Transfer of oxygen between air and water, with the mole fraction of oxygen in bulk air equal to 21% and the concentration of oxygen in water initially equal to zero.
b. Transfer of naphthalene between a NAPL (e.g., coal tar) and water, with a 10% mole fraction of naphthalene in the NAPL and zero initial concentration of naphthalene in the water.

Also for case (a), compute the chemical mass flux as well as the oxygen mass-transfer rate, assuming an interfacial area of 25 cm^2, as well as a water-side film thickness of 0.001 cm.

Solution 4.2

a. Oxygen is a gas with a large air–water partition coefficient that can be computed from its Henry's law constant (note the inverse definition of K_H for inorganic gases) in Table 2.3.

$$K_{\text{aw}}^{\text{O}_2} = \frac{1}{K_H^{\text{O}_2} RT} = \frac{1}{10^{-2.9}(\text{mol/L} \cdot \text{atm}) \times 0.0821(\text{L} \cdot \text{atm/mol} \cdot \text{K}) \times 298 \text{ K}} = 32.5$$

[4] In natural systems, the effective stagnant film thickness in air is typically of the order of 0.1 to 1 cm, while that in water may range from 0.002 to 0.02 cm, depending on the condition at the interface (Schwarzenbach et al., 1993).

Since the computed $K_{aw} \gg 0.01$, the air-phase film resistance can be neglected such that the oxygen mole fraction in air is uniformly 21%. The water film will offer the dominant resistance to mass transfer. Oxygen concentrations in water at the air–water interface are at equilibrium with its partial pressure in air of 21%, yielding:

$$C_{w,i}^{O_2} = K_H^{O_2} \times p_{air}^{O_2} \times MW^{O_2} = 10^{-2.9}\ \text{mol/L} \cdot \text{atm} \times 0.21\ \text{atm} \times 32\ \text{g/mol}$$
$$= 0.0085\ \text{g/L} = 8.5\ \text{mgO}_2/L_w$$

Oxygen concentrations in water at the two edges of the water film are shown in the following figure:

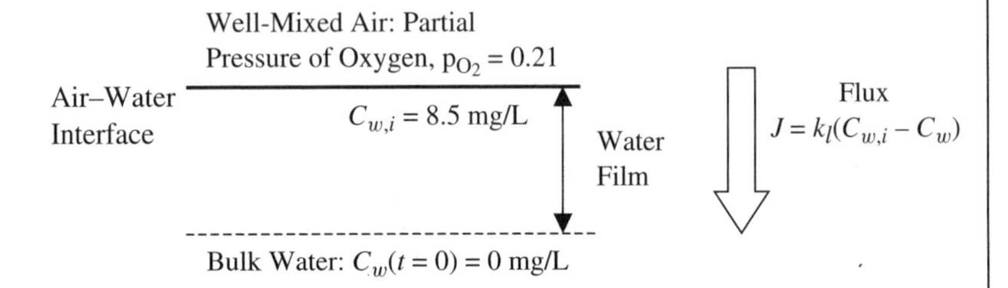

From Table 4.1, $D_{m,w}^{O_2} = 2.4 \times 10^{-5}$ cm²/s, so that the liquid-side mass-transfer coefficient, k_l, is computed as:

$$k_l^{O_2} = \frac{D_{m,w}^{O_2}}{\delta_w} = \frac{2.4 \times 10^{-5}\ \text{cm}^2/\text{s}}{0.001\ \text{cm}} = 0.024\ \text{cm/s}$$

The initial flux,

$$J^{O_2} = k_l^{O_2}\left(C_{w,i}^{O_2} - C_w^{O_2}\right) = 0.024\ \text{cm/s} \times \frac{8.5\ \text{mg O}_2}{L_w} \times \frac{L_w}{1000\ \text{cm}^3}$$
$$= 2 \times 10^{-4} \frac{\text{mg O}_2}{\text{cm}^2 \cdot \text{s}}$$

The rate of oxygen mass transfer across a 25-cm² area is $\dot{m} = J \times$ interfacial area $= 0.005$ mg O_2/s.

b. Provided the NAPL is not highly viscous,[5] diffusion coefficients in NAPL and in water are of similar order of magnitude (see estimation equations for diffusion in liquids in Table 4.1). The film thickness in water and NAPL can also be assumed to be similar in magnitude (Ahn and Lee, 1991). The octanol–water partition coefficient, K_{ow}, provides a good estimate of a

[5] NAPL-phase resistances can become significant for highly viscous, polymer-like NAPLs. See Ortiz et al. (1999).

chemical's affinity for the organic liquid relative to water. Therefore, based on the NAPL–water analog of Eq. (4.5b), the NAPL-phase resistance to diffusive mass transfer can be neglected if the chemical $K_{ow} \gg 1$, which is the case for naphthalene ($K_{ow} = 10^{3.36}$ from Table 2.3). Hence the water film provides the dominant resistance to mass transfer. Naphthalene concentrations in water at the coal tar (NAPL)–water interface can be determined from equilibrium considerations (Chapter 3) employing Raoult's law with the ideal NAPL assumption:

$$C_{w,i}^{\text{naph}} = C_{w,\text{eq}}^{\text{naph}} = X_{\text{NAPL}}^{\text{naph}} \cdot C_{w,\text{sat}}^{\text{naph}(L)} \cdot \text{MW}^{\text{naph}}$$

$$= 0.1 \times 10^{-3.06} \frac{\text{mol naph}}{\text{L water}} \times 128 \frac{\text{g naph}}{\text{mol naph}} = 0.011 \text{ g/L}$$

Naphthalene flux can be computed across the dominant water film, represented as:

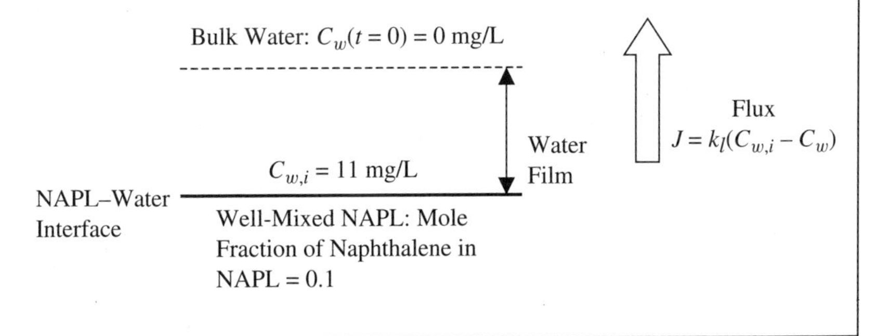

4.2.3 Concentration Representation

While the above analysis enables computation of instantaneous chemical mass fluxes from various environmental compartments, modelers are often interested in determining changes in chemical concentration occurring in the different compartments as a result of intermedia mass transfer. Consider chemical transfer from air to water occurring in a closed system as shown in Figure 4.5. A change in chemical concentration in bulk water occurs because of the chemical mass transferred from air, the rate of which depends on the flux, J, as well as the total interfacial area (IA) available for such mass transfer. Since the *concentration* of chemical A in water is of interest, we must consider the mass of A transferred across the entire air–water interfacial area (IA) per unit volume of water, V_w. With the air–water mass-transfer flux expressed in terms of liquid-side concentrations, the change in aqueous-phase concentration is derived as follows:

$$\frac{\partial C_w^{\text{A}}(t)}{\partial t} = J_{\left[\frac{\text{mass of A}}{(\text{IA})\cdot\text{Time}}\right]} \times a_{\left[\frac{\text{IA}}{V_w}\right]} = k_l a \left[C_{w,i}^{\text{A}}(t) - C_w^{\text{A}}(t)\right] \tag{4.6}$$

where $a(= \mathrm{IA}/V_w)$ is the specific interfacial area $[1/L]$ defined as the interfacial area, IA $[L^2]$ per unit volume of the medium of interest, in this case water, V_w $[L^3]$, since aqueous concentrations are of interest. The mass-transfer coefficient, k_l, and the specific interfacial area, a, are lumped together to yield the liquid-side lumped mass-transfer rate coefficient, $k_l a$ $(= k_l \times a)$. Note that a has dimensions of $[1/L]$, so $k_l a$ has dimensions of $[1/T]$. As in the flux representations in Eq. (4.3) to (4.5), the change in aqueous-phase pollutant concentration can be expressed in terms of the liquid film concentration differences [Eq. (4.6)], gas film concentration differences [Eq. (4.7)], as well as overall bulk-phase concentration difference, expressed on an aqueous basis [Eq. (4.8)]:

$$\frac{dC_w^{\mathrm{A}}(t)}{dt} = k_g a \left[C_{\mathrm{air}}^{\mathrm{A}}(t) - C_{a,i}^{\mathrm{A}}(t)\right] \tag{4.7}$$

$$\frac{dC_w^{\mathrm{A}}(t)}{dt} = k_{\mathrm{LO}} a \left[\frac{C_{\mathrm{air}}^{\mathrm{A}}(t)}{K_{aw}} - C_w^{\mathrm{A}}(t)\right] \tag{4.8}$$

If changes in air-phase pollutant concentrations are of interest, the structure of Eq. (4.6) to (4.8) remains essentially the same; however, the specific interfacial area, a, must now be defined to represent the interfacial area per unit *air* volume in the system.

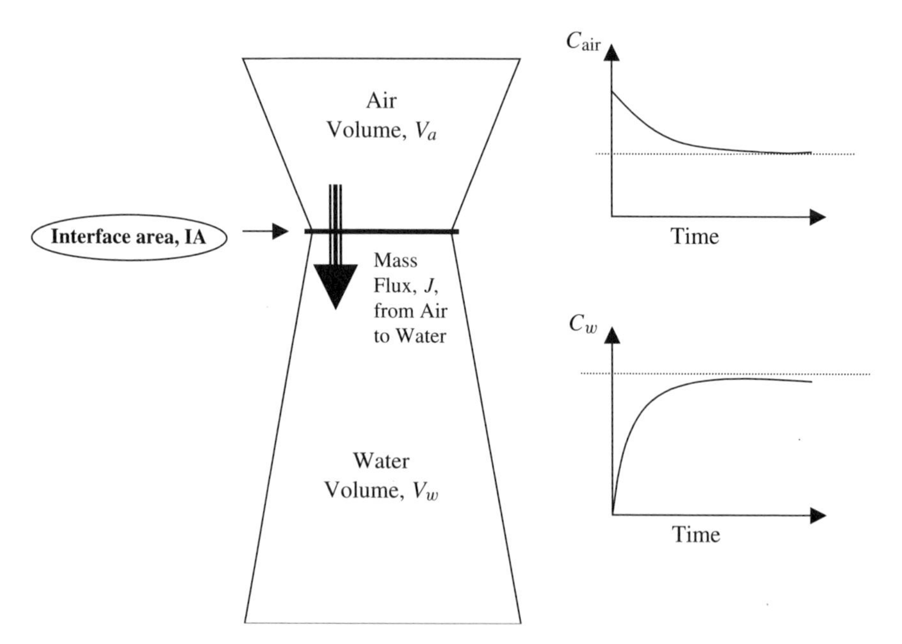

Figure 4.5 Changes in chemical concentration with time in a volume of air, V_a, and a volume of water, V_w, as a result of interphase mass transfer of the chemical from air to water across an interfacial area, IA.

Equations (4.6) to (4.8) are similar to first-order chemical reaction equations with kinetic constant, $k_l a$ (or $k_g a$ or $k_{\mathrm{LO}} a$, depending on the concentrations used). Integrating Eq. (4.6) yields an exponential relationship characteristic of first-order processes. For a system that is gaining mass until the bulk aqueous concentration approaches the interfacial aqueous concentration, $C_{w,i}^{\mathrm{A}}$, which is assumed to be time invariant, for example, when there is a large supply of the chemical in air so that $C_{w,i}^{\mathrm{A}} = C_{w,\mathrm{final}}^{\mathrm{A}} = C_{\mathrm{air}}^{\mathrm{A}} / K_{\mathrm{aw}}^{\mathrm{A}}$, we get

$$C_w^{\mathrm{A}}(t) = C_{w,\mathrm{initial}}^{\mathrm{A}} + \left[C_{w,\mathrm{final}}^{\mathrm{A}} - C_{w,\mathrm{initial}}^{\mathrm{A}}\right] \times \left[1 - e^{-k_l a \cdot t}\right] \tag{4.9}$$

where $C_{w,\mathrm{initial}}^{\mathrm{A}}$ is the initial chemical concentration in the aqueous phase. Likewise, for a system that is losing pollutant mass continuously to another medium, such that the interfacial pollutant concentration, $C_{w,i}^{\mathrm{A}}$, is always zero (for example chemical transfer from water to a large open atmosphere), integration of Eq. (4.6) yields

$$C_{w(t)}^{\mathrm{A}} = C_{w,\mathrm{initial}}^{\mathrm{A}} \left[e^{-k_l a \cdot t}\right] \tag{4.10}$$

When chemical concentrations in any one medium cannot be assumed constant, mass-balance differential equations [e.g. Eq. (4.6)] are written for both media and solved simultaneously using numerical integration methods described in Chapter 6. Analysis of Eq. (4.9) and (4.10) shows that concentrations approach 98% of the final equilibrium condition when the time of interest is larger than $[4/k_l a]$. This is a useful rule of thumb since it enables environmental modelers to distinguish between those situations in which equilibrium assumptions are valid and those in which pollutant mass-transfer kinetics are relevant.

As with flux computations, the determination of which of Eq. (4.6) to (4.8) to use in concentration computations depends on which phase resistances, if any, may be ignored. In general, for mass transfer occurring serially across stagnant films, the medium that offers the dominant resistance to mass transfer is used in flux and concentration computations. Appropriate assumptions of constant interfacial concentrations can be made in gaining systems [Eq. (4.9)] and losing systems [Eq. (4.10)] based on transferring media with high- and near-zero pollutant concentrations, respectively, which facilitate computations. The sequence of steps required for making intermedia kinetic computations is summarized in Figure 4.6.

4.2.4 Parameters Required for Intermedia Kinetic Computations

Figures 4.2 to 4.6 and Eq. (4.2) to (4.10) provide the theoretical framework for the stagnant film model. Partition coefficients (discussed in Chapter 3) along with mass-transfer coefficients for the two stagnant films (k_l, k_g) and the specific interfacial area, a, are needed for flux and concentration computations in air–water systems. The constructs used to describe air–water transport can be employed to describe the behavior of other pairs of media, for example, NAPL–air, soil–air, air–plant and so forth. In particular, the concept of mass-transfer resistance (or conductance) is general and translates well to other media. A larger resistance to molecular diffusion is

Sketch all phases and films:
Draw a schematic diagram showing all the phases and films. Note bulk-phase and interfacial concentrations for all phases.

Determine direction of mass transfer:
Air → water, water → air, NAPL → air, NAPL → water, soil → water, soil → air, etc.

Decide which if any phase resistance may be ignored:
[Qualitative rule: Chemical will resist transfer into the phase it does not have an affinity for; this phase offers the dominant resistance to diffusive mass transfer.]

Delineate concentrations at edges of films:
For the film in the dominant phase, write down pollutant concentrations at the two edges of the film. Note any constant concentration boundaries.
OR
If no single phase dominates transport, consider overall resistance to mass transfer: write down bulk-phase pollutant concentrations on an equivalent basis. Note any constant concentration boundaries in the bulk phases.

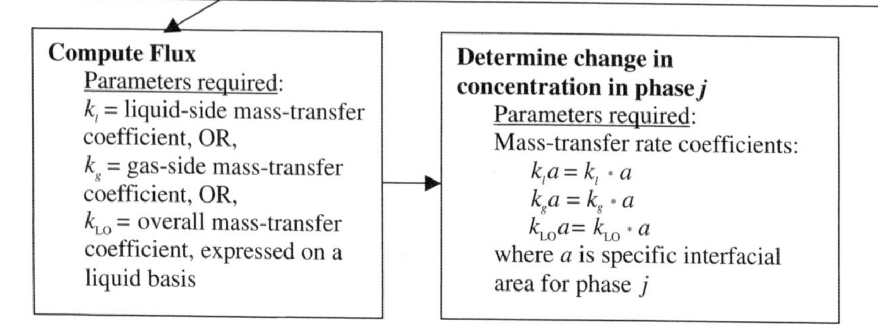

Figure 4.6 Road map for solving intermedia kinetic problems.

expected in media with slow diffusion coefficients and thick stagnant films. The mass-transfer resistance, r_j, offered by a certain medium, j, is thus inversely proportional to its mass-transfer coefficient, that is, $r_j \propto 1/k_j$. The mass-transfer coefficient k_j is analogous to conductance of the medium j, and is defined for the general case [see Eqs. (4.3) and (4.4)] as:

$$k_j = \frac{D_{m,j}}{\delta_j} \tag{4.11}$$

Equation (4.11) *defines* the mass-transfer coefficient of medium j. But since the film thickness (δ_j) is very difficult to measure, Eq. (4.11) is usually not used directly to estimate a numerical value for the parameter k_j. Instead, k_j (or δ_j) is estimated indirectly using one of the following three methods:

1. Experimental measurement of mass-transfer rates
2. Correlation with properties characteristic of specific multimedia configurations
3. Extrapolation from parameter values obtained for a reference chemical

1. *Experimental measurement of intermedia mass transfer* involves either direct measurement of chemical mass flux or measurement of concentration changes caused by intermedia mass transfer, from which the mass-transfer coefficients can be determined by inverting Eqs. (4.3) to (4.10). Often, measurements are made for compounds of widespread environmental significance, for example, oxygen, water vapor, or propane. Example 4.3 illustrates quantification of the mass-transfer coefficient for release of naphthalene from a well-defined pool of coal tar NAPL to water.

EXAMPLE 4.3

Dissolution of naphthalene from a coal tar pool into water was studied in a batch reactor in which 200 mL of high ionic strength water was contacted with a 2.4-cm-diameter circular coal tar pool that was embedded within the reactor (Ramaswami et al., 2001). The aqueous phase of the reactor was mixed well, and naphthalene concentrations in water were monitored over time after contact with coal tar was initiated. The concentration data are shown below. Compute the mass-transfer coefficient and the mass-transfer rate coefficient for naphthalene transfer from the coal tar pool to water. Based on these calculations, estimate the thickness of the mass-transfer boundary layer above the coal tar pool.

Time (min)	Naphthalene Concentration in Water (mg/L)
0	0
20	2.17
60	3.1
120	4.17
240	4.92
1440	4.95

Solution 4.3 Naphthalene concentrations in water are increasing and approach a final aqueous concentration of approximately 4.95 mg/L, which also corresponds

with the naphthalene concentrations in water at equilibrium with this coal tar (after accounting for ionic strength effects). A plot of concentration versus time indicates that concentrations are increasing in a manner consistent with Eq. (4.9). The initial naphthalene concentration is zero, yielding:

$$C_w^{\text{naph}}(t) = 0 \text{ mg/L} + 4.95 \text{ mg/L} \times \left(1 - e^{-k_l a t}\right)$$

Fitting the concentration data to the above equation by transforming it to a log-linear equation in t, and using linear regression to estimate $k_l a$ yields $k_l a = 0.020/\text{min}$.

The specific interfacial area, a, for mass transfer *to water* is computed as the NAPL pool surface area per unit *water* volume: $a = [\pi(1.2)^2 \text{ cm}^2]/200 \text{ cm}^3 = 0.0226/\text{cm}$. Then k_l is computed as $k_l = k_l a/a = 0.020/0.0226 = 0.88$ cm/min. From Eq. (4.11) and our estimate of k_l, the aqueous film thickness above the coal tar pool can then be estimated as:

$$\delta_w = \frac{D_{m,w}^{\text{naph}}}{k_l} \approx \frac{10^{-5} \text{ cm}^2/\text{s}}{(0.88 \text{ cm/min}) \times 1 \text{ min}/60 \text{ s}} = 0.0007 \text{ cm}$$

Accurate knowledge of the interfacial area made the computation of the mass-transfer coefficient, k_l, possible in Example 4.3. The mass-transfer coefficient enabled estimation of the thickness of the aqueous film. A very thin film is indicated in this example because of the mixing generated in the reactor directly above the coal tar pool. Less fluid mixing and/or a smaller specific interfacial area for mass transfer would cause a decrease in the mass-transfer rate coefficient (e.g., Ramaswami and Luthy, 1997). Thus, system geometry and fluid mixing characteristics can strongly influence mass-transfer parameters. These influences are addressed in the second method for estimating mass-transfer coefficients, described next.

2. *Correlation with fluid properties characteristic of specific multimedia configurations* Experimental evaluations of mass-transfer coefficients are expensive and time consuming and cannot be conducted for every chemical across all the multimedia configurations encountered in the environment. Instead, it is more practical to develop generalized correlations of mass-transfer coefficients with media properties characteristic of specific environmental configurations such as lake-to-atmosphere or soil-to-air exchange. Many correlations focus on bulk fluid velocity and mixing characteristics of a medium, increases in which tend to diminish the stagnant film thickness, δ_j, thereby increasing k_j. In this respect, mass-transfer correlations developed for chemical engineering applications may be adapted for use in environmental systems with similar geometry and fluid flow patterns, for example, correlations describing mass transfer from spheres in a packed bed may be used to describe mass transfer from soil particles in the subsurface. Mass-transfer correlations often employ a dimensionless group called the Sherwood

number, $\text{Sh} = dk_j/D_{m,j}$, where d is a length parameter characteristic of the system, k_j is a mass-transfer coefficient defined for a medium j, and $D_{m,j}$ is the molecular diffusion coefficient of the chemical of interest in that medium. The dimensionless Sherwood number is correlated with other dimensionless groups such as the Reynold number and the Schmidt number, which represent dynamic and static properties of the fluid flow field, respectively. *Perry's Chemical Engineering Handbook* (Perry et al., 1997) provides a comprehensive compilation of chemical engineering mass-transfer correlations. However, these correlations were developed in controlled, engineered reactors and need to be evaluated experimentally before being applied to environmental systems. Weber and DiGiano (1996) provide a compilation of mass-transfer correlations that have been tested for environmental systems.

3. *Extrapolation of mass-transfer coefficients obtained for specific reference compounds* Mass-transfer coefficients obtained for reference chemicals, either by direct measurement or using correlations, must be corrected to represent the behavior of other chemicals of interest. The correction factor utilizes the fact that the diffusion coefficient in both air and in water is inversely proportional to the square root of the molecular weight of the diffusing chemical (Table 4.1). Thus, if the stagnant film model is being used, mass-transfer coefficients are proportional to the molecular diffusion coefficient [Eqs. (4.3b), (4.4), and (4.11)], which is inversely proportional to the square root of the molecular weight of the chemical. For a given media geometry and mixing conditions, the film thickness is expected to be the same for all chemicals, so the ratio of liquid (e.g., water) side mass-transfer coefficients for two chemicals, A and B, is

$$\frac{k_l^{\text{A}}}{k_l^{\text{B}}} = \frac{D_{m,w}^{\text{A}}}{D_{m,w}^{\text{B}}} = \left(\frac{\text{MW}^{\text{B}}}{\text{MW}^{\text{A}}}\right)^{\beta} \quad \text{where } \beta = 0.5 \text{ for the stagnant film model} \tag{4.12}$$

Equation (4.12) enables estimation of the mass-transfer coefficient for pollutant A from that known for a reference chemical B, in the same environmental system, assuming a stagnant film model is applicable for the system. Correlations derived from experimental observations indicate exponents slightly different from 0.5 [e.g., $\beta = 0.5$ for water-side films and $\beta = 0.67$ for air-side films, Mackay and Yeun (1979)], reflecting the nature of the films in the systems being considered.

Estimating Mass-Transfer* Rate *Coefficients The three methods outlined above can be used to estimate mass-transfer coefficients for individual media and in turn to compute mass fluxes [Eqs. (4.3) to (4.5)]. However, determining multimedia concentration changes in response to intermedia chemical mass transfers requires use of the mass-transfer *rate* coefficient, $k_j a$ [Eqs. (4.6) to (4.8)], which requires additional knowledge of the specific interfacial area, a. While a can readily be calculated for systems with simple geometry, such as evaporation of a chemical to air from a placid lake or a surface spill of NAPL, estimating the interfacial area in complex systems can be challenging. For example, it is nearly impossible to compute the interfacial area between a subsurface NAPL spill entrapped heterogeneously within soil pores and the surrounding groundwater. Likewise, estimates of contact areas between tree leaves and the atmosphere, or between roots and soil, are inevitably uncertain. Such uncertainty translates to difficulty in modeling many soil and plant mass-transfer rates. However, in less complex systems, the interfacial areas can be derived based on simple geometric assumptions and coupled effectively with mass-transfer coefficients (k_j) determined from either the two-film theory or the surface renewal theory. This approach is illustrated in detail in the following section for some typical air–water exchanges occurring in the environment. Intermedia exchanges involving NAPLs, soils, and vegetation are described in Sections 4.4 to 4.6, respectively.

4.3 AIR–WATER EXCHANGE

Of all the intermedia chemical exchanges occurring in the environment, the transfer of chemicals between air and water has been studied most extensively. Historically, the interest in air–water mass exchange focused on oxygen transfer from air to water bodies, which is essential for sustaining aquatic life. More recently, the exchange of CO_2 between the atmosphere and water bodies has been studied as an essential component of models for surface water acidification and eutrophication and for global climate change. Similarly, the transfer of water vapor from water bodies to the atmosphere has been studied extensively to model evaporation. Both the two-film concept described in Section 4.2 and the surface renewal theory have been used widely to describe air–water mass transfer. Table 4.2 summarizes key relationships and mathematical models that draw upon each of these approaches. Application of these models is illustrated in the cases and examples that follow.

4.3.1 Transfer of Gases and Volatile Pollutants Between Running Streams and the Atmosphere

A complete description of mass transfer in this situation should include a water film and an air film adjacent to bulk water and bulk air, respectively, as shown in Figure 4.2. However, resistance of the air film to chemical transfer may be neglected for compounds with dimensionless air–water partition coefficient, K_{aw}, larger than 10^{-2}, which is the case for the volatile organic compounds (VOCs) and gases being considered here (see Table 2.3 to obtain K_{aw} data). Thus water-side resistances dominate transport of gases and volatile chemicals between running streams and the atmosphere. If the effect of wind waves on the water surface is neglected, then the liquid-side mass-transfer coefficient, k_l, may be derived from surface renewal theory, as given by the O'Connor–Dobbins (1958) equation:

$$k_l = \sqrt{\frac{D^{A}_{m,w}\, v_s}{h_s}} \tag{4.13}$$

where $D_{m,w}$ (L^2/T) is the molecular diffusion coefficient of the chemical in water, v_s (L/T) is the average linear stream velocity, and h_s (L) is the stream depth. In

TABLE 4.2 Water- and Air-Phase Mass-Transfer Correlations for Chemical Mass Transfer in Air–Water Systems.[a] With the exception of the O'Connor–Dobbins equation, specified units must be used for all other empirical correlation equations.

Water-Phase Correlation Equations	Condition	Reference
1. $k_l = \sqrt{\frac{D^{A}_{m,w} v_s}{h_s}}$ $k_l a = \sqrt{\frac{D^{A}_{m,w} v_s}{(h_s)^3}}$ v_s and h_s are stream velocity and depth, respectively.	For moving streams wherein the effect of wind waves on the water surface is neglected. k_l is derived based on surface renewal theory.	O'Connor–Dobbins (1958)
2. $k_l^{O_2}$ (cm/s) $= 4\times10^{-4}+4\times10^{-5}\times u_{10}^2$ u_{10} is wind velocity (m/s) measured 10 m above ground.	For stagnant lakes wherein water-side films are affected by wind velocity. k_l is derived employing the stagnant film model. Oxygen transport is represented.	Schwarzenbach et al. (1993)
3. Average k_l for chemical A used in regional models.		
a. k_l (m/day) $= 0.24$	When v_s (m/s) $< 0.04 \times u_{av}^{0.64}$	CalTox, 1993; Southworth (1979)
b. k_l (m/day) $= 5.64 \left(\frac{v_s^{0.969}}{h_s^{0.673}}\right)\sqrt{\frac{32}{MW^A}}$	For larger v_s and in land units with $u_{av} < 1.9$ m/s.	
c. k_l (m/day) $= 5.64 \left(\frac{v_s^{0.969}}{h_s^{0.673}}\right)\sqrt{\frac{32}{MW^A}}\, e^{0.526(u_{av}-1.9)}$	For land units with $u_{av} > 1.9$ m/s.	
v_s and h_s are stream velocity and stream depth measured in units of m/s and m, respectively; u_{av} represents annual average wind velocity in m/s.	Derived from stagnant film theory employing correlations observed for oxygen transport.	

Air-Phase Correlation Equations	Condition	Reference
1. $k_g^{H_2O}$ (cm/s) $= 0.2\, u_{10} + 0.3$ u_{10} is wind velocity (m/s) measured 10 m above ground.	Water vapor transfer in air. Synthesis of correlations obtained from multiple studies.	Schwarzenbach et al. (1993)
2. $k_g^{H_2O}$ (cm/s) $= 0.27\, u_2 + 0.59$ u_2 is wind velocity (m/s) measured 2 m above ground.	Evaporation of water from pans	Penman (1948, 1956)

(*continued*)

TABLE 4.2 (*Continued*)

Air-Phase Correlation Equations	Condition	Reference
3. $k_g^{H_2O}$ (cm/s) $= 0.065u_{10}(6.1 + 0.63u_{10})^{0.5}$	Evaporation of water in a wind wave tank	Mackay and Yeun (1979)
u_{10} is wind velocity (m/s) measured 10 m above ground.		
4. Average k_g for chemical A used in regional models:	K_H is in the range 0.01 to 1 L · atm/mol	Southworth (1979); CalTox, 1993
a. k_g (m/day) $= 273\,(u_{av} + v_s)\sqrt{\frac{18}{MW^A}}$	$[u_{av} + v_s] > 0.5$ m/s	
b. k_g (m/day) $= 140\sqrt{\frac{18}{MW^A}}$	K_H is in the range 0.01 to 1 L · atm/mol	
v_s is stream velocity in units of m/s.		
u_{av} represents annual average wind velocity in m/s.	$[u_{av} + v_s] < 0.5$ m/s.	

this formula, the rate of surface renewal $(1/T)$ in the stream is assumed equal to v_s/h_s. As is characteristic of the surface renewal model, the mass-transfer coefficient is proportion to $(D_m)^{1/2}$ rather than $(D_m)^1$ as was the case for the film theory model described in Eq. (4.3).

Equation (4.13) can be used to describe the fluxes of gases such as oxygen and VOCs such as chloromethanes across air–water interfaces in a stream environment. The lumped mass-transfer rate coefficient, $k_l a$, for such transfer may be determined by a geometric computation of the specific surface area of the stream. If the stream is modeled as a large rectangular trough with length l, width w, and depth h_s, the interfacial (surface) area per unit volume of water in the stream is given by $a = (l \times w)/(h_s \times l \times w) = 1/h_s$. Hence, the lumped mass-transfer rate coefficient, $k_l a$, is

$$k_l a = k_l \times a = \sqrt{\frac{D_{m,w}^{A}\, v_s}{h_s}} \times \frac{1}{h_s} = \sqrt{\frac{D_{m,w}^{A}\, v_s}{(h_s)^3}} \tag{4.14}$$

The term $k_l a$ in Eq. (4.14) is also called the stream reaeration rate coefficient (k_a) when oxygen transfer is being considered and allows the rate of oxygen reaeration to be determined directly from stream channel geometry and flow estimates, without calibration to in-stream dissolved oxygen concentrations. Other formulas, similar to the O'Connor–Dobbins equation shown above, have been empirically developed and tailored for particular stream settings (Zison et al., 1978; Bowie et al., 1985;

Chapra, 1997). The mass-transfer rate coefficient is used to determine changes in chemical concentration occurring as a result of air–water exchanges, as illustrated in Example 4.4.

EXAMPLE 4.4

An appliance manufacturer accidentally discharges benzene at a concentration of 0.1 g/L into a stream of depth 5 m and velocity 10 m/min. Determine the flux of benzene to the atmosphere and the time required for benzene concentrations in water to be reduced by 90% of their original discharge concentration. To what downstream distance does this correspond? Assume no other loss mechanism for benzene from the stream. Use 1×10^{-5} cm^2/s for the molecular diffusion coefficient of benzene in water.

Solution 4.4 Following the sequence of steps outlined in Figure 4.6, the direction of transfer of benzene is water → air. Because the dimensionless air–water partition coefficient for benzene, $K_{aw}^{benzene} = 0.225$ (Table 2.3) is greater than 0.01, the air film can be neglected and the water film offers the dominant resistance to mass transfer. Since the air-phase film is neglected, the interfacial benzene concentration in air, $C_{air,i}^{benzene}$, is set equal to the bulk-air benzene concentration, $C_{a,bulk}^{benzene} = 0$ g/L. Despite transfer of small amounts of benzene to the atmosphere from the stream, we assume the benzene concentration in bulk air always remains close to 0 g/L because of the large volume of the open atmospheric compartment, which is continuously being refreshed. Using the air–water partition coefficient, the interfacial aqueous-phase benzene concentration, $C_{w,i}^{benzene} = C_{air,i}^{benzene}/K_{aw}^{benzene}$ is also found to be zero. The liquid-side mass-transfer coefficient, k_l, is computed from Eq. (4.13) as 0.00058 cm/s. The benzene flux across the water film is computed from Eq. (4.3) as:

$$J = k_l \times \left[C_{w,\text{bulk}} - C_{w,i}\right] = 5.8 \times 10^{-4}\ \text{cm/s} \cdot \left[0.1\frac{\text{g}}{\text{L}} \cdot \frac{\text{L}}{1000\ \text{cm}^3} - 0\right]$$

$$= 5.8 \times 10^{-8}\ \text{g/s} \cdot \text{cm}^2$$

The rate of change of aqueous-phase benzene concentration is obtained from Eq. (4.6), employing the lumped mass-transfer rate coefficient, $k_l a$, computed from Eq. (4.14):

$$k_l a = k_l \times a = k_l \times \frac{1}{h_s} = 0.00058\,\frac{\text{cm}}{\text{s}} \times \frac{1}{5\ \text{m}\,\dfrac{100\ \text{cm}}{\text{m}}} = 1.15 \times 10^{-6}\,\frac{1}{\text{s}}$$

At time t, the benzene concentration is required to be 10% of the original discharge concentration. Therefore, for the losing system, which is water, solving Eq. (4.10) yields $t = 2 \times 10^6\ s \cong 23$ days. Thus, in a relatively slow-moving stream, with no loss mechanisms other than volatilization to the atmosphere, it takes about a month for 90% of the benzene to be transferred to the atmosphere. In natural systems, biodegradation of benzene in water will significantly increase the removal rate of benzene from water. Example 1.2 illustrated a similar air–water mass transfer process occurring for oxygen dissolving in water. In Example 1.2, the reaeration rate coefficient was used in conjunction with oxygen consumption in the stream due to biodegradation of effluent wastes to describe the net change in aqueous oxygen concentration.

4.3.2 Transfer of Semivolatile Chemicals Between Stagnant Lakes and the Atmosphere

Both air- and water-phase mass-transfer resistances must be considered in modeling the air–water exchange of SVOCs with $K_{aw} < 0.01$. Because we are considering a stagnant lake, a stagnant film theory approach is adopted, and the thickness of both the air and water films is assumed to be controlled by the ambient wind velocity. Water- and air-phase mass-transfer coefficients for SVOCs can be determined from correlations developed for oxygen transport in water and water vapor transport in air, shown in Table 4.2. The empirical relationship between the water-side mass-transfer coefficient for oxygen, $k_l^{O_2}$, and wind velocity, u_{10}, measured 10 m above ground in units of meters/second, is

$$k_l^{O_2}(\text{cm/s}) = 4 \times 10^{-4} + 4 \times 10^{-5} \times [u_{10}\ (\text{m/s})]^2 \tag{4.15a}$$

A similar empirical equation has been developed correlating the air-side mass-transfer coefficient for water vapor, $k_g^{H_2O}$, with wind velocity, u_{10}, measured 10 m above ground and expressed in units of meters/second:

$$k_g^{H_2O}(\text{cm/s}) = 0.2 \times u_{10}\ (\text{m/s}) + 0.3 \tag{4.15b}$$

Since a stagnant film theory approach is employed, the transfer coefficient for any pollutant A may be estimated from the mass-transfer coefficients estimated above for water vapor, or for oxygen, by utilizing the molecular weight correction factors shown in Eq. (4.12). Molecular weight-corrected air- and water-phase mass-transfer coefficients are then combined to evaluate the overall transfer of the SVOC across the air–water interface. This procedure is illustrated in Example 4.5.

EXAMPLE 4.5

How much time is required for aqueous phenanthrene concentrations in a stagnant lake to be reduced by 90%, if wind velocities measured 10 m above the lake surface are 3 m/s? Assume volatilization and subsequent transfer to the atmosphere is the only removal mechanism for phenanthrene. The lake has a volume of 40,000 m^3 and a surface area of 400 m^2. Phenanthrene concentrations in the lake are initially 0.001 g/L.

Solution 4.5 Following the sequence of steps outlined in Figure 4.6, the direction of mass transfer is water → air. Phenanthrene is a semivolatile compound with a $K_{aw} = 0.0015$ (computed from Table 2.3). Since the $K_{aw} < 0.01$, neither the air film nor the water film may be ignored. Hence, the overall mass-transfer coefficient must be computed by combining the resistance offered by both films. First, computing the liquid film coefficient for phenanthrene yields

$$k_l^{O_2}\ (\text{cm/s}) = 4 \times 10^{-4} + [4 \times 10^{-5} \times 3^2] = 7.6 \times 10^{-4}\ \text{cm/s}$$

$$k_l^{\text{phenanthrene}} = k_l^{O_2} \times \sqrt{\frac{\text{MW}^{O_2}}{\text{MW}^{\text{phenanthrene}}}} = 7.6 \times 10^{-4}\sqrt{\frac{32}{178}} = 3.2 \times 10^{-4}\ \text{cm/s}$$

The air film mass-transfer coefficient for phenanthrene is computed next:

$$k_g^{H_2O}\ (\text{cm/s}) = [0.2 \times 3] + 0.3 = 0.9\ \text{cm/s}$$

$$k_g^{\text{phenanthrene}} = k_g^{H_2O} \times \sqrt{\frac{\text{MW}^{H_2O}}{\text{MW}^{\text{phenanthrene}}}} = 0.9\sqrt{\frac{18}{178}} = 0.29\ \text{cm/s}$$

The overall film mass-transfer coefficient, k_{LO}, is computed from Eq. (4.5b) as:

$$\frac{1}{k_{LO}} = \frac{1}{k_l} + \frac{1}{K_{aw}\, k_g} = 3103 + 2329 = 5432\ \text{s/cm} \Rightarrow k_{LO} = 0.00018\ \text{cm/s}$$

Since changes in concentration are required, the lumped mass-transfer rate coefficient, $k_{LO}a$, is computed as $k_{LO}a = k_{LO} \times a$, where a is the surface area per unit volume of the lake, given as 400 m^2/40,000 m^3 = 0.01 m^{-1}.

This yields $k_{LO}a$ = 0.00018 cm/s × (0.0001) cm^{-1} = 1.8 × 10^{-8}/s (which is about 2 orders of magnitude less than that for the more volatile benzene in a stream environment, considered in Example 4.4). At time t, the phenanthrene concentration is required to be 10% of the original concentration. Solving Eq. (4.10), yields $t = 1.3 \times 10^8$ s or 1481 days. Thus, compared with benzene in Example 4.4, much more time is required for the less volatile phenanthrene to be transferred to air from the stagnant lake system.

4.4 MASS TRANSFER FROM NAPL

Single resistance models are often used to describe mass transfer of chemicals from surface NAPL spills to air and to water (see Figure 4.7). As shown in Example

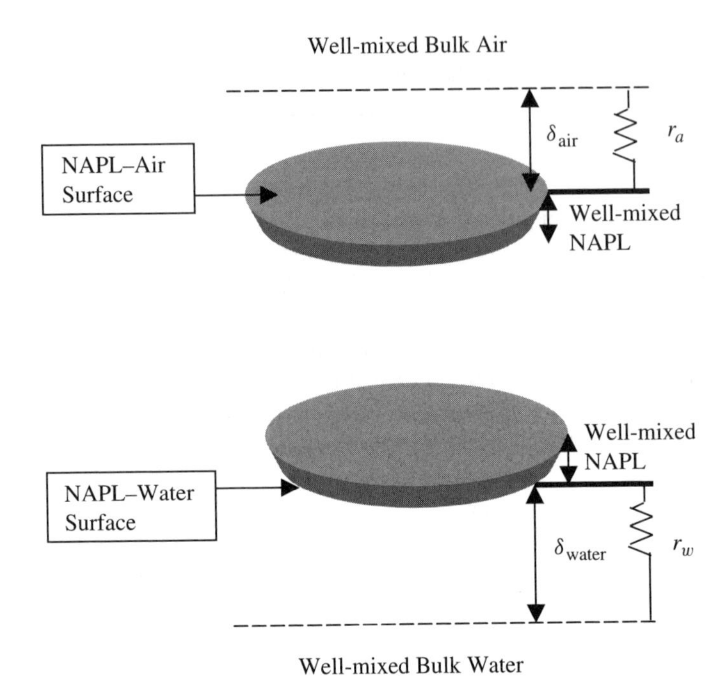

Figure 4.7 Schematic representing mass transfer from surface NAPL spills to air and to water.

4.2b, the NAPL-phase resistance to mass transfer can be assumed to be much less than the resistance to mass transfer offered by water or by air for chemicals with K_{ow} and $K_{oa} \gg 1$ (Ahn and Lee, 1991). Thus, the entire NAPL is assumed to be well mixed, while the film thickness is modeled in air and in water adjacent to the NAPL. Many of the correlations in Table 4.2 that describe resistances in water and air may be applicable for use in conjunction with surface NAPL spills. Hemond and Fechner-Levy (2000) present a simplified empirical relationship to describe the air-phase mass-transfer coefficient for the transfer of volatile organic pollutants to the atmosphere from moderately sized NAPL spills on the ground or on water bodies:

$$k_g^{A}\ (\text{cm/h}) = 1100 \times u_{10}\ (\text{m/s}) \tag{4.16}$$

where u_{10} is the wind velocity measured 10 m above ground in units of meters/second. Estimates of k_g from Eq. (4.16) are similar in magnitude to those obtained from Table 4.2. Eq. (4.16) is employed to describe transfer of pure phase NAPLs to the atmosphere, in which case no concentration gradients exist in the NAPL film, the resistance of which is effectively set to zero. The equation may also be used judiciously for multicomponent NAPLs when it is reasonable to assume that chemical A prefers the NAPL to air. Raoult's law with the ideal NAPL assumption can be utilized to describe equilibrium conditions at the NAPL–air interface, as is illustrated in Example 4.6.

EXAMPLE 4.6

Gasoline is spilled in a layer 2 mm thick on a road. The gasoline is made up of 2% benzene. Make a rough calculation of the time for the benzene to volatilize to air. Assume a wind speed of 2 m/s, 10 m above the ground. Assume all gasoline constituents volatilize at essentially the same rate so that the gasoline composition does not change significantly. The density of gasoline is 0.88 g/mL. Assume standard environmental temperature (25°C) and pressure (1 atm).

Solution 4.6 We apply the sequence of steps outlined in Figure 4.6. The bulk air concentration of benzene is assumed to be zero because the wind is blowing and dispersing the pollutant, $C_{a,\text{bulk}}^{\text{benzene}} = 0$. The direction of mass transfer is thus from NAPL → air. We neglect the NAPL phase resistance because of benzene's affinity for NAPL, quantified by a large K_{ow} (see Table 2.3). We calculate the benzene concentration in air at the gasoline–air interface using Raoult's law [Eq. (3.14)] yielding

$$C_{a,i,\text{eq}}^{\text{benzene}} = 0.008 \, \frac{g^{\text{benzene}}}{L_a} = 8 \, \frac{\text{mg}}{\text{L}}$$

The air-phase mass-transfer coefficient, k_g, is computed from Eq. (4.16) and the benzene flux across the air film above the NAPL is then computed as:

$$k_g^{\text{benzene}} \, \frac{\text{cm}}{\text{h}} = 1100 \times u_{10} \, \frac{\text{m}}{\text{s}} = 2200 \, \frac{\text{cm}}{\text{h}}$$

$$J = k_g \left[C_{a,i,\text{eq}}^{\text{benzene}} - C_{a,\text{bulk}}^{\text{benzene}} \right] = 2200 \, \frac{\text{cm}}{\text{h}} \left[8 \frac{\text{mg}}{\text{L}} - 0 \right] \frac{\text{L}}{1000 \text{ cm}^3}$$

$$= 17.6 \, \frac{\text{mg}_{\text{benzene}}}{\text{cm}^2 \text{ h}} \text{ going to air}$$

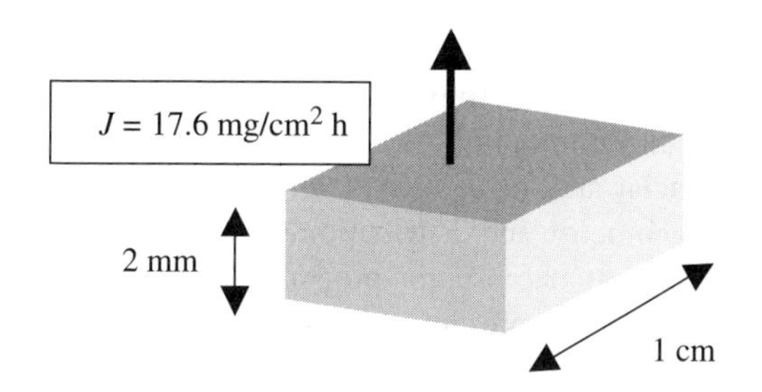

Since the flux yields the mass transfer rate per unit 1 cm^2 area, we can consider a block of NAPL, 1 cm × 1 cm × 2 mm thick. The density of gasoline is given as 0.88 g/cm^3 so the mass of gasoline initially present in this block of volume 0.2 cm^3 is 0.18 g. The mass of benzene in the block is 2% of 0.18 g; i.e., 3.6 mg (0.0036 g). The mass of benzene going to the air from the block per unit time, $\dot{M}$, is computed from the flux, J, as:

$$\dot{M} = J \times \text{surface area} = 17.6 \, \frac{\text{mg}_{\text{benzene}}}{\text{cm}^2\text{h}} \times (1 \text{ cm} \times 1 \text{ cm}) = 17.6 \, \frac{\text{mg}_{\text{benzene}}}{\text{h}}$$

Therefore, the approximate time required to volatilize all the benzene initially present in the block of NAPL is calculated as:

$$\frac{M^{\text{benzene}}}{\dot{M}} = \frac{3.6 \text{ mg}}{17.6 \, \dfrac{\text{mg}}{\text{h}}} = 0.21 \text{ hr}$$

assuming the benzene mole fraction in gasoline remains invariant as volatilization proceeds. In reality, as we saw in Figure 3.7, the equilibrium boundary condition at the interface can decrease as the benzene mole fraction in gasoline decreases, thereby decreasing the concentration driving force and diminishing mass fluxes from the NAPL pool.

The transfer of pollutants from subsurface NAPL bodies to pore water or pore air trapped within a soil matrix is a much more complex problem. Flow bypassing and issues of dimensionality can affect NAPL–water mass-transfer correlations in the subsurface environment. In general terms, the water-side mass-transfer resistance is dominant and affected by the pore water velocity, enabling correlation between the Sherwood, Reynolds, and Schmidt numbers. However, NAPL–water mass-transfer correlations derived experimentally from studies of pure NAPLs in one-dimensional laboratory columns (Miller et al., 1990; Powers et al., 1992; Gellar and Hunt, 1993) are quite different from those obtained using NAPL–water tests conducted in two-dimensional tank systems, suggesting that dimensionality and the ease of water bypassing around the NAPL zone in two-dimensional systems strongly affects the correlations obtained from experiment (Illangasekare et al., 1995). Another large source of uncertainty in NAPL mass-transfer models lies in the estimate of the specific interfacial area of the NAPL globules entrapped in the subsurface, since the exact size, geometry, and distribution of the NAPL globules cannot be not accurately determined from macroscopic properties of the system. Hence, no widely acceptable method exists at the present time for accurately estimating the magnitude of the NAPL–water mass-transfer rate coefficient in large-scale, real-world groundwater systems.

4.5 MASS TRANSFER IN SOIL SYSTEMS

Mass transfer of chemicals from soil to the atmosphere or from sediments to water is often of interest in assessing the fate of pesticides and other agricultural pollutants. Because a well-mixed bulk soil or sediment phase almost never exists in the subsurface, soil–air and sediment–water exchanges are modeled with thin films in air and in water, respectively, coupled with diffusion through the entire depth of the soil matrix, as illustrated in Figure 4.8.

Soils and sediments are multiphase compartments containing solid matter (mineral/soil particles) along with air and water trapped in the porous soil matrix. Molecules are assumed to effectively diffuse through the soil or sediment compartment by diffusing through the air and water present in the porous matrix, that is, diffusion within the solid mineral phases is assumed to be negligible. The rate at which molecules diffuse through the air and water present in the soil matrix must be adjusted for the tortuous paths the molecules must take to go around the soil particles. For this reason, molecular diffusion coefficients in air and in water (Table 4.1) are corrected by a tortuosity factor that accounts for the apparent slower rates of diffusion through structural porous media. The tortuosity factors of Millington and Quirk (Jury et al., 1983) are frequently used to estimate diffusion coefficients in soil–air (sa) and soil–water (sw) as:

$$D^{\mathrm{A}}_{m,\mathrm{sa}} = \frac{[\theta_a]^{10/3}}{n^2} D^{\mathrm{A}}_{m,a} \qquad D^{\mathrm{A}}_{m,\mathrm{sw}} = \frac{[\theta_w]^{10/3}}{n^2} D^{\mathrm{A}}_{m,w} \tag{4.17}$$

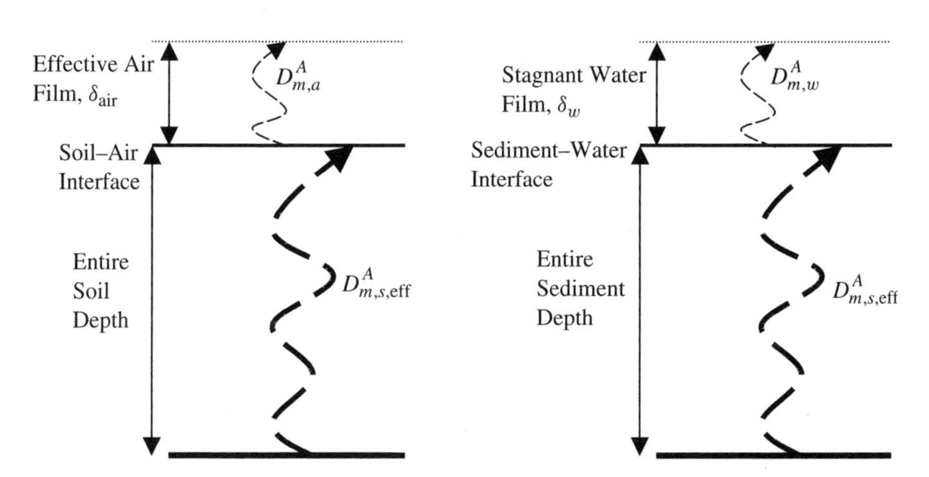

Figure 4.8 Mass transfer from soil to air and from sediment to air is represented by an effective diffusion coefficient in soil, $D_{m,s,\mathrm{eff}}$, which incorporates chemical diffusion through pore water and pore air within the soil or sediment matrix. Diffusion across the entire depth of soil or sediment is coupled with transport across thin films in air and in water, respectively.

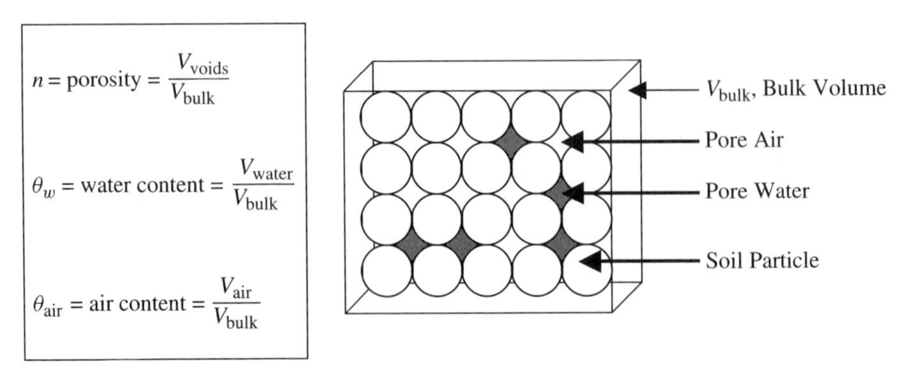

Figure 4.9 Bulk volume of soil, V_{bulk}, contains soil particles and void spaces filled either with water or air.

where θ_a and θ_w are the volumetric air and water content of soil defined as the volume of air or water per unit bulk volume of soil, while n is the soil porosity[6] defined as the void volume per unit bulk volume of soil (see Fig. 4.9).

Diffusive chemical transfer through a porous soil matrix is assumed to occur by a combination of molecular diffusion in soil–air and soil–water such that the total diffusive mass flux is represented as:

$$J = -D_{m,\text{sa}}^{\text{A}} \frac{\partial C_{\text{sa}}^{\text{A}}}{\partial z} - D_{m,\text{sw}}^{\text{A}} \frac{\partial C_{\text{sw}}^{\text{A}}}{\partial z} \tag{4.18}$$

where the subscripts sa and sw represent soil–air and soil–water, respectively and z represents the vertical depth dimension which is often of interest in soil systems. However, chemicals diffusing within water-filled pores can simultaneously partition to soil particles and to pore air. A similar phenomenon can occur in soil–air because of partitioning of the diffusing chemical from pore air to soil particles and to pore water. If partitioning of chemicals between soil particles, pore water and pore air is assumed to occur very rapidly, attaining instantaneous equilibrium between the three subcompartments; then the chemical concentration in total bulk soil (subscripted as s, T) is related to the chemical concentration in soil–water (sw) as:

$$C_{s,T}^{\text{A}} = \frac{\text{Mass of A in soil, pore water, and pore air}}{V_{\text{bulk}}}$$

$$= \frac{M_{\text{soil}} K_d^{\text{A}} C_{\text{sw}}^{\text{A}} + V_w C_{\text{sw}}^{\text{A}} + V_a K_{\text{aw}}^{\text{A}} C_{\text{sw}}^{\text{A}}}{V_{\text{bulk}}} \tag{4.19a}$$

$$\Rightarrow C_{s,T}^{\text{A}} = \left(\rho_b K_d^{\text{A}} + \theta_w + \theta_a K_{\text{aw}}^{\text{A}}\right) \times C_{\text{sw}}^{\text{A}}$$

[6] By these definitions, $n = \theta_a + \theta_w$. In sediments that are fully saturated with water, the air content $\theta_a = 0$.

We define a useful parameter, R, termed the retardation factor in water, as the following:

$$\mathsf{R} = \frac{\rho_b K_d^{\mathrm{A}} + \theta_w + \theta_a K_{\mathrm{aw}}^{\mathrm{A}}}{\theta_{\mathrm{w}}} = \frac{\text{chemical mass in total bulk soil (water, air, soil)}}{\text{chemical mass in pore water only}} \tag{4.19b}$$

Such that

$$C_{s,T}^{\mathrm{A}} = (\mathsf{R}\,\theta_w) \times C_{\mathrm{sw}}^{\mathrm{A}} \tag{4.19c}$$

Likewise, $C_{s,T}$ can also be computed from chemical concentrations in soil–air, C_{sa} as:

$$\begin{aligned} C_{s,T}^{\mathrm{A}} &= \frac{\text{Mass of A in soil, pore water and pore air}}{V_{\mathrm{bulk}}} \\ &= \frac{M_{\mathrm{soil}} K_{\mathrm{sa}}^{\mathrm{A}} C_{\mathrm{sa}}^{\mathrm{A}} + V_w \left(C_{\mathrm{sa}}^{\mathrm{A}}/K_{\mathrm{aw}}^{\mathrm{A}}\right) + V_a C_{\mathrm{sa}}^{\mathrm{A}}}{V_{\mathrm{bulk}}} \end{aligned} \tag{4.19d}$$

$$\Rightarrow C_{s,T}^{\mathrm{A}} = \left[\rho_b \left(\frac{K_d^{\mathrm{A}}}{K_{\mathrm{aw}}^{\mathrm{A}}}\right) + \frac{\theta_w}{K_{\mathrm{aw}}^{\mathrm{A}}} + \theta_a\right] \times C_{\mathrm{sa}}^{\mathrm{A}} = (R'\theta_a) \times C_{\mathrm{sa}}^{\mathrm{A}}$$

where R' is a retardation factor for air, defined as:

$$\mathsf{R}' = \frac{\rho_b K_{\mathrm{sa}}^{\mathrm{A}} + \theta_a + \left(\theta_w/K_{\mathrm{aw}}^{\mathrm{A}}\right)}{\theta_{\mathrm{a}}} = \frac{\text{chemical mass in total bulk soil}}{\text{chemical mass in pore air}} \tag{4.19e}$$

It is useful to recognize that

$$\mathsf{R}' = \frac{R\,\theta_w}{K_{\mathrm{aw}}^{\mathrm{A}}\,\theta_a} \tag{4.19f}$$

In Eqs. (4.19a) to (4.19f), K_{aw}, K_d, and K_{sa} represent air–water, soil–water and soil–air partition coefficients, respectively. V_{bulk}, V_w, and V_a are the volumes of bulk soil, pore water, and pore air, respectively.

Substituting for the pore water and pore air chemical concentrations using relationships provided in Eqs. (4.19c) and (4.19d), the diffusive flux equation (4.18) can now be rewritten in terms of chemical concentrations in the total bulk soil as:

$$J = -D_{m,s,\mathrm{eff}}^{\mathrm{A}} \frac{\partial C_{s,T}^{\mathrm{A}}}{\partial z} \tag{4.20a}$$

where $D_{m,s,\mathrm{eff}}$ is an effective diffusion coefficient for bulk (total) soil defined as:

$$D_{m,s,\mathrm{eff}}^{\mathrm{A}} = \frac{D_{m,\mathrm{sa}}^{\mathrm{A}}}{\theta_a \mathsf{R}'} + \frac{D_{m,\mathrm{sw}}^{\mathrm{A}}}{\theta_w \mathsf{R}} \tag{4.20b}$$

The effective diffusion coefficient incorporates a "slow down" of diffusion in pore air due to partitioning of the chemical from air to the other subcompartments in soil, represented by the air retardation factor, R', as well as a similar slow down of chemical diffusion in water due to partitioning to the other subcompartments represented by the aqueous-phase retardation factor, R. Note that R [Eq. (4.19b)] represents the ratio of the mass of chemical A present in bulk soil (comprising pore air, pore water, and soil particles) to that present in pore water alone.[7] Effectively, $1/\mathsf{R}$ may be viewed as the ratio of the partitioning-retarded contaminant travel velocity in pore water to the contaminant velocity in pore water that would occur in the absence of partitioning (to pore air and soil). Likewise, R' (Eq. (4.19d)] represents the ratio of the contaminant velocity in pore air in the absence of partitioning to the contaminant velocity in pore air retarded by partitioning to soil and pore water; R and R' are always ≥ 1. This phenomenon of retardation caused by chemical partitioning is discussed further in Chapter 9.

Equation (4.20a) indicates that contaminant concentrations in soil vary continuously with depth as a result of diffusive chemical transfers. Closed-form analytical solutions to Eq. (4.20) have been provided by Jury et al. (1990) and Boudreau (1997) for specific soil and sediment contamination scenarios. A simpler approximation for soils has been advocated by Mackay (1991), employing an effective unmixed "stagnant" layer in soil with a depth half that of the entire depth of the soil; the rest of the soil is assumed to be well mixed. McKone (CalTox, 1993; McKone, 1996) modeled three well-mixed layers in soil (grounds surface, root zone, and vadose zone); step changes across the three zones approximated the vertical concentration gradients represented by Eq. (4.20a).

In all cases of soil–air exchange, chemical mass transfer across the soil layer(s) is coupled with mass transfer across an effective stagnant air film that lies above the ground surface (see Fig. 4.2). Flux across the air film is represented in the usual manner as:

$$J_{\text{air}} = -D_{m,a}^{\text{A}} \frac{C_{\text{air},i}^{\text{A}} - C_{\text{air}}^{\text{A}}}{\delta_{\text{air}}} \tag{4.21a}$$

where $C_{\text{air},i}^{\text{A}}$ is the chemical concentration in air at the air–ground interface, which must equal the chemical concentration in soil–air at ground level:

$$C_{\text{air},i}^{\text{A}} = C_{\text{sa}}^{\text{A}}(z=0) = \frac{C_{s,T}^{\text{A}}(z=0)}{(R'\theta_a)}$$

where $C_{\text{air}}^{\text{A}}$ is the concentration of A in bulk air. Estimates of the air film thickness above the ground surface can be obtained from simple rules of thumb that indicate δ_{air} of the order of 0.5 cm [1 cm in still air to 0.1 mm with wind velocities over the surface at 1 m/s (Hanna et al., 1982)]. Jury et al. (1983) present a more rigorous

[7] This can be determined by multiplying the numerator and denominator of Eq. (4.19b) by $\left(V_w C_{w,\text{eq}}^{\text{A}}\right)$.

estimation technique to determine δ_{air} based on measured relative humidity changes above the ground surface after irrigation events.

Chemical transfer between sediments and overlying water is modeled in the same manner as the soil–water exchange described above. Diffusion through the sediment layer is modeled using the Jury et al. (1983) model written in Eqs. (4.19) and (4.20), recognizing that the air content in sediments is zero and the water content therefore equals porosity. As in Eq. (4.21a), mass transfer across an aqueous film is described as:

$$J_{\text{water}} = -D_{m,w}^{\text{A}} \frac{C_{w,i}^{\text{A}} - C_{w}^{\text{A}}}{\delta_w} \tag{4.21b}$$

In this case the aqueous chemical concentration at the sediment–water interface, $C_{w,i}^{\text{A}}$, must equal the chemical concentration in sediment water at that location, that is, $C_{s,w}^{\text{A}}(z = 0)$. An effective aqueous film thickness of 2 cm is often used based on measurement of radon fluxes in water over sediments (Hammond et al., 1975). More sophisticated methods are needed when the characteristics of the sediment layer change with time due to particle deposition, consolidation, and burial processes (Boudreau, 1997).

The approach for describing chemical transfer from soils has been documented in detail by Jury et al. (1983). The modeling framework has been applied to determine the required thickness of landfill covers to minimize the fluxes of VOC released to air (Jury et al., 1990).

4.6 CHEMICAL EXCHANGE WITH PLANTS

Multiple resistance models are used to describe pollutant transfer between the atmosphere and plants. Mass transfer is assumed to occur across an air film and then in parallel, across the stomatal resistance and the cuticle resistance offered by the leaves of the plant. Combining series and parallel resistances, as shown in Figure 4.10, yields the total resistance to mass transfer, r_{total}.

Quantitative methods for estimating the aerodynamic, cuticle, and stomatal resistance are described further in Chapter 11. The aerodynamic resistance over the surface of a single leaf is expected to be inversely proportional to wind speed over the leaf surface since this tends to decrease the air film thickness. However, the effective aerodynamic resistance offered by an entire tree canopy is more complicated since wind speeds typically increase with elevation above ground and also vary within the vegetative canopy. Cuticle resistance is related to the chemical's affinity for cuticular tissue, while stomatal resistances describe the resistance to chemical diffusion offered by stomatal openings typically found on the lower surface of leaves. Cell turgidity controls the opening and closing of stomata. Hence, the conductance through the stomata for any pollutant A, is strongly related to the stomatal conductance for water vapor. Ambient temperature affects water vapor fluxes because of the phase change

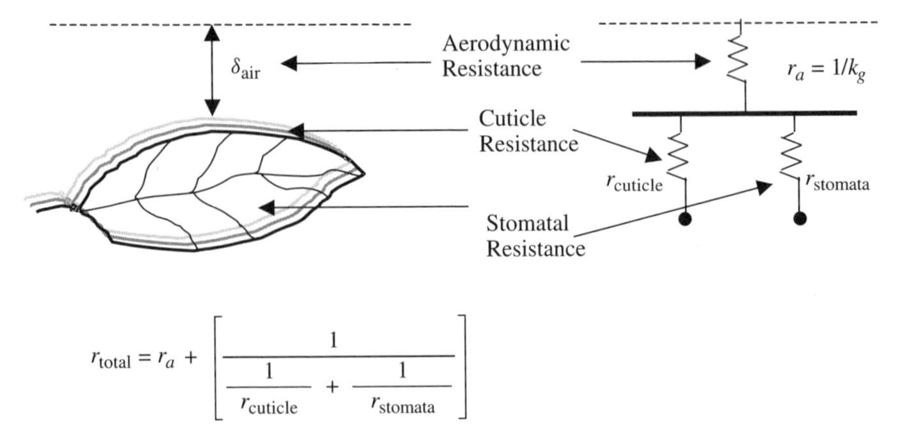

Figure 4.10 Mass transfer from the surface of tree leaves to air is controlled by aerodynamic, cuticle, and stomatal resistances.

of water occurring during transpiration. Heat-transfer constraints were used to model water transpiration from leaves in the pioneering work of Penman (1956), and combined with mass-transfer resistances to yield the combination method, also known as the Penman–Monteith method (Monteith, 1965). Models that compute fluxes from an entire tree or a stand of trees include a term called the leaf area index (LAI), which is a measure of the total leaf surface area offered by a tree to the projected ground area for the tree canopy. The reader is referred to Trapp (1995) for a detailed description of pollutant transfer between plants and the atmosphere. Multiple resistance models have been used widely to describe fluxes of a variety of chemicals, for example, nitrogen and PAHs, between the atmosphere and vegetative covers. Over the past decade, plant transpiration processes have been modeled and measured extensively on a regional scale because of the importance of water vapor in global climate control (e.g., Schmugge and André, 1991).

4.7 INCORPORATING INTERMEDIA CHEMICAL TRANSFERS WITH FLUID FLOW AND REACTION IN MIXED-BOX MODELS

Thus far in this chapter we have discussed models and parameters that describe the rates of diffusive chemical mass transfer from one medium to another. Intermedia mass-transfer processes must be coupled with chemical advection and reaction processes to describe the overall transport of the chemical in the environment. Consider a simple example where a large spill of liquid (NAPL) benzene occurs on the floor of a laboratory. We can assume air is flowing at a volumetric flow rate of Q through the laboratory, and further, for simplicity we assume the room air is well mixed, enabling use of a CSTR model. Benzene is assumed at first to be nonreactive in air, as shown in Figure 4.11*a*. The benzene concentration in air in the room is of interest and can be obtained by writing a mass balance for benzene in air:

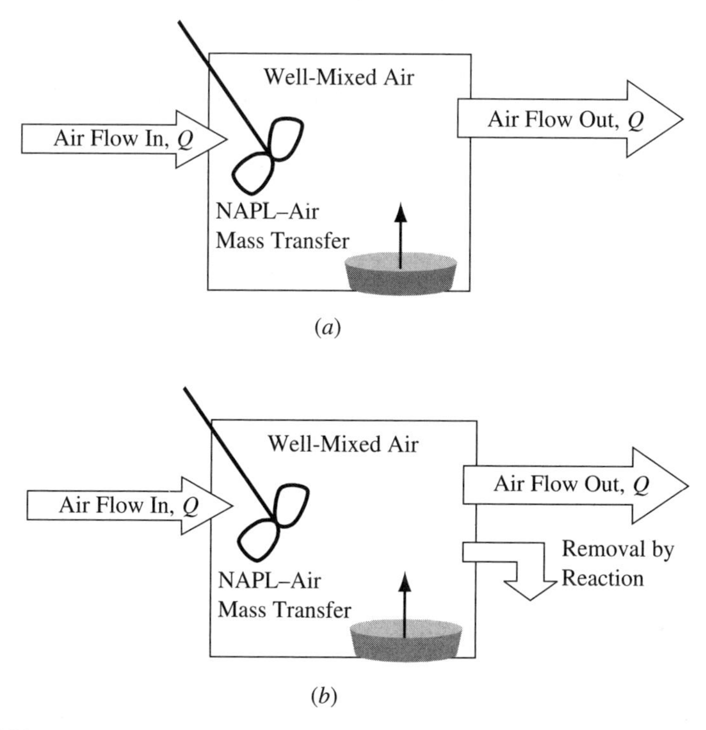

Figure 4.11 Mass transfer from a liquid chemical spill to a well-mixed air compartment. (*a*) Air is flowing in and out of the box at a volumetric flow rate of Q. (*b*) Chemical is also undergoing reaction in the air phase.

$$V_{\mathrm{air}} \frac{dC_{\mathrm{air}}^{\mathrm{A}}(t)}{dt} = V_{\mathrm{air}} k_g a \left[C_{\mathrm{air,eq}}^{\mathrm{A}} - C_{\mathrm{air}}^{\mathrm{A}}(t) \right] - QC_{\mathrm{air}}^{\mathrm{A}}(t) \tag{4.22}$$

where $C_{\mathrm{air,eq}}^{\mathrm{A}}$ represents the benzene concentration in air at the NAPL–air interface, at equilibrium with pure liquid benzene. The lhs of Eq. (4.22) represents the net rate of mass input of benzene to air, which equals the mass-transfer rate of benzene into room air from the NAPL [first term on the right-hand side (rhs)] minus the rate of outflow of benzene from the room due to air flow (second term on rhs). Setting the lhs to zero for steady-state conditions yields

$$\begin{aligned} C_{\mathrm{ss,air}}^{\mathrm{A}} &= \frac{C_{\mathrm{air,eq}}^{\mathrm{A}}}{\left(1 + \dfrac{Q}{Vk_g a}\right)} = \frac{C_{\mathrm{air,eq}}^{\mathrm{A}}}{\left(1 + \dfrac{1}{\tau k_g a}\right)} = \frac{C_{\mathrm{air,eq}}^{\mathrm{A}}}{\left(1 + \dfrac{1}{\mathrm{Da}_{\mathrm{I}}}\right)} \qquad \text{where } \mathrm{Da}_{\mathrm{I}} = \tau k_g a \\ &= C_{\mathrm{air,eq}}^{\mathrm{A}} \qquad \text{when } \mathrm{Da}_{\mathrm{I}} \gg 1 \\ &= \mathrm{Da}_{\mathrm{I}} \times C_{\mathrm{air,eq}}^{\mathrm{A}} \qquad \text{when } \mathrm{Da}_{\mathrm{I}} \ll 1 \end{aligned} \tag{4.23}$$

The dimensionless group, Da_1 is referred to as a Damkohler number (type 1) and may be viewed as the ratio of the fluid residence time (τ) to the mass-transfer time represented by $[1/k_g a]$. Large values of $\mathrm{Da}_1 \gg 1$ (i.e., $k_g a \gg 1/\tau$) indicate that interphase mass transfer occurs quickly, well within the fluid residence time, leading to equilibrium conditions in the system. Conversely, values of Da_1 much less than unity imply slow mass-transfer rates relative to the fluid residence time in the system. In this case, departures from equilibrium are expected, and kinetic analysis is required. Thus, Da_1 indicates the degree to which mass inputs by intermedia transfer are balanced by fluid outflows from the system.

We now incorporate removal of chemical A from air by chemical reactions (Fig. 4.11*b*) with an assumed first-order rate constant of k, to obtain the following mass-balance equation:

$$V_{\mathrm{air}} \frac{dC_{\mathrm{air}}^{\mathrm{A}}(t)}{dt} = V_{\mathrm{air}}\, k_g a \left[C_{\mathrm{air,eq}}^{\mathrm{A}} - C_{\mathrm{air}}^{\mathrm{A}}(t) \right] - Q C_{\mathrm{air}}^{\mathrm{A}}(t) - V_{\mathrm{air}}\, k C_{\mathrm{air}}^{\mathrm{A}}(t) \tag{4.24}$$

Steady-state airborne chemical concentrations are obtained as:

$$C_{\mathrm{ss,air}}^{\mathrm{A}} = \frac{C_{\mathrm{air,eq}}^{\mathrm{A}}}{1 + \dfrac{Q}{Vk_g a} + \dfrac{k}{k_g a}} = \frac{C_{\mathrm{air,eq}}^{\mathrm{A}}}{1 + \dfrac{1}{\mathrm{Da}_1} + \mathrm{Da}_2} \quad \text{where } \mathrm{Da}_2 = \frac{k}{k_g a} \tag{4.25}$$

$$\approx C_{\mathrm{air,eq}}^{\mathrm{A}} \quad \text{when } \mathrm{Da}_2 \ll 1 \text{ (with } \mathrm{Da}_1 \gg 1) \quad \text{Effective reaction rate} = -kC_{\mathrm{air,eq}}^{\mathrm{A}}$$

$$\approx \left(\frac{k_g a}{k} \right) C_{\mathrm{air,eq}}^{\mathrm{A}} \quad \text{when } \mathrm{Da}_2 \gg 1 \text{ (with } \mathrm{Da}_1 \gg 1)$$

For this condition,

$$\text{Effective reaction rate} = -k \times \left(\frac{k_g a}{k} \right) C_{\mathrm{air,eq}}^{\mathrm{A}} \approx -k_g a C_{\mathrm{air,eq}}^{\mathrm{A}}$$

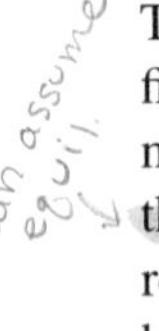

The second type of Damkohler number, Da_2, compares the mass-transfer rate coefficient, $k_g a$, with the reaction rate constant, k, for chemical or biochemical transformation processes occurring in the individual bulk phase. Values of Da_2 much smaller than 1 indicate that interphase mass transfer occurs faster than the transformation reaction in the individual phase, so that steady-state concentrations are unaffected by reaction. In this case, assuming an absence of flow (or at slow flow rates with $\mathrm{Da}_1 \gg 1$), steady-state conditions will approach equilibrium, that is, removal by reaction occurs much slower than the rate of input of the chemical by intermedia mass transfer. The removal rate is controlled by the slower reaction rate constant, which is the rate-limiting phenomenon. Conversely, values of Da_2 much larger than unity imply slow mass-transfer rates relative to the reaction occurring in the system. In this case, departures from equilibrium conditions are expected in air since the chemical undergoes degradation in air much faster than it is released into that medium. The removal rate by reaction is now effectively the same as the mass-transfer rate, that is, slow mass transfer is the rate-limiting step.

Analyses of Damkohler numbers are useful in determining the relative importance of flow, intermedia mass transfer and transformation processes in controlling chemical behavior within the medium of interest (Seagren et al., 1994; Ramaswami et al., 1997). How can this analysis be extended to describe the overall behavior of a chemical within a multimedia environment containing air, water, soil, and vegetation? This issue is addressed in the next section, employing a mixed-box multimedia modeling framework.

4.7.1 Multimedia Mixed-Box Models with Intermedia Mass-Transfer Kinetics

In the scenario illustrated in Figure 4.11 and represented mathematically in Eqs. (4.23) and (4.24), all the "action," in terms of flow and chemical transformation, was occurring in air. No flow or reaction was assumed to occur within the NAPL medium, which, being a pure phase, has no internal concentration gradients and provides a well-defined concentration boundary condition at the air–NAPL interface. However, we can imagine many situations where the second medium is flowing and can support reactive processes. We can also envision more than two media in contact. For example, our "well-mixed" laboratory can now be expanded to represent an air shed where moving air masses exchange chemicals with moving water bodies, vegetation, soils, and sediment. The long-range transport of a chemical in air will now depend not only on reactive processes occurring in air but also on intermedia transfers and subsequent reactions occurring in the other compartments, that is, in water, soil, vegetation, and sediment. This results in a multimedia mixed-box model [referred to by Mackay (2001) as a level III model] with rate-limited mass transfers occurring between all environmental compartments. Chemical reactions can occur within individual compartments, and advection to and from each compartment to the outside environment is also considered. Thus, level III builds upon level II (see Chapter 3) by relaxing the assumption that the different environmental compartments are in equilibrium and, instead, considers the rate of intermedia mass transfers occurring through a combination of diffusive and nondiffusive processes.

4.7.2 Combining Diffusive and Nondiffusive Mass Transfer

Nondiffusive intermedia mass transfer occurs when chemicals move from one environmental compartment to another along with bulk media transfers. For example, chemical transport from the atmosphere to the Earth's surface can occur by air–water partitioning of the chemical into rainfall, termed rain dissolution, as well as by wet and dry deposition of particles (aerosols) that carry the chemical. Likewise, deposition and resuspension of particles in lakes and streams can cause nondiffusive chemical transfers between surface waters and underlying sediments. A mathematical sleight of hand (described next) allows for the convenient addition of diffusive and nondiffusive transport processes, which occur in parallel, and together contribute to the overall removal of a chemical from one compartment to another. The descriptions of nondiffusive mass-transfer processes presented here are abbreviated; more physically detailed descriptions are given in Chapters 10 and 11.

Consider first the diffusive mass transfer of a chemical, A, from air to water. The change in concentration of A in *air* due to diffusion of the chemical to water can be represented using k_{LO}, the overall mass-transfer coefficient expressed on a liquid basis [see Eq. (4.8)], as:

$$V_{air} \left. \frac{dC_{air}^{A}}{dt} \right|_{\substack{\text{diffusive}\\\text{transfer}}} = -k_{LO} \left[\frac{C_{air}^{A}(t)}{K_{aw}^{A}} - C_{w}^{A}(t) \right] \times IA_{w} \tag{4.26a}$$

where IA_{aw} represents the interfacial area between the air and water compartments. However, the overall mass-transfer coefficient on a gas-phase basis, k_{GO}, is related to the overall mass-transfer coefficient expressed on a liquid-phase basis, k_{LO}, as: $k_{GO} = k_{LO}/K_{aw}$. This allows us to rewrite Eq. (4.26a) and interpret it as:

$$V_{air} \left. \frac{dC_{air}^{A}}{dt} \right|_{\substack{\text{diffusive}\\\text{transfer}}} = IA_{aw}\, k_{LO}\, C_{w}^{A}(t) - IA_{aw}\, k_{GO} C_{air}^{A}(t) = \dot{m}^{A}_{\substack{\text{water–air}\\\text{diffusion}}} - \dot{m}^{A}_{\substack{\text{air–water}\\\text{diffusion}}} \tag{4.26b}$$

where the first term on the rhs may be viewed as the rate of mass transfer of chemical A from water to air by diffusion, and the second term as the rate of diffusive mass transfer of chemical A from air to water, the difference between the two representing the net effective diffusive mass-transfer rate that was derived previously in Eq. (4.26a).

The representation in Eq. (4.26b) is convenient because nondiffusive processes that contribute to chemical removal from air to water can now be readily added to the second term in Eq. (4.26b), which represents diffusive removal of the chemical from air to water. Consider dissolution into rainfall as an example of a nondiffusive process contributing to chemical removal from air to water. Assuming equilibrium partitioning between air and rain drops, the rate of change in the air-phase concentration of chemical A due to dissolution into rain occurring at a rate of u_{rain} (L/T), can be written as:

$$\dot{m}^{A}_{\text{rain diss}} = -u_{rain} IA_{air} \left[\frac{C_{air}^{A}(t)}{K_{aw}^{A}} \right] = -k_{\text{rain diss}} C_{air}^{A}(t) \tag{4.27a}$$

where $k_{\text{rain diss}} = u_{rain}/K_{aw}^{A}$. Raindrops falling through air can also intercept and capture suspended particles present in the atmosphere, resulting in wet deposition of chemical A along with the aerosol particles captured by raindrops. If chemical equilibrium is assumed between the atmosphere and aerosol particles, with a dimensionless partition coefficient between aerosols and air defined as

$$K_{qa}^{A} = \frac{\text{g of A per m}^3 \text{ aerosol}}{\text{g of A per m}^3 \text{ air}}$$

the mass loss rate of chemical A from the atmosphere to water by wet deposition is given by:

$$\dot{m}^{\mathrm{A}}_{\text{wet dep}} = \mathrm{IA}_{\mathrm{aw}} u_{\mathrm{rain}} S_r F_q K^{\mathrm{A}}_{qa} C^{\mathrm{A}}_{\mathrm{air}}(t) = \mathrm{IA}_{\mathrm{aw}} k_{\text{wet dep}} C^{\mathrm{A}}_{\mathrm{air}}(t) \tag{4.27b}$$

where $k_{\text{wet dep}} = u_{\mathrm{rain}} S_r F_q K^{\mathrm{A}}_{qa}$, F_q is the volume fraction of air occupied by aerosols, and S_r is the scavenging ratio representing the volume of air swept by raindrops as they descend through the atmosphere divided by the volume of raindrops. In multimedia mixed-box models global average values are used for F_q ($\sim 2 \times 10^{-11}$ $\mathrm{m}^3_{\mathrm{aerosol}}/\mathrm{m}^3_{\mathrm{air}}$), and S_r ($=200{,}000$ $\mathrm{m}^3_{\mathrm{air}}/\mathrm{m}^3_{\mathrm{rain}}$). A typical rainfall rate is assumed to be 1×10^{-4} m/h. The dimensionless aerosol–air partition coefficient, K^{A}_{qa} ($\mathrm{m}^3_{\mathrm{air}}/\mathrm{m}^3_{\mathrm{aerosol}}$) is readily estimated from correlation with the liquid-phase vapor pressure of chemical A (Mackay, 2001) (see Example 4.7).

In addition to wet deposition of aerosol particles, particles can directly undergo settling and removal from air, a process termed dry deposition. Chemicals associated with particles are thence removed from air by dry deposition. Once again, taking a global mass-balance approach, the mass loss rate of chemical A from air due to dry deposition can be written as:

$$\dot{m}^{\mathrm{A}}_{\text{dry dep}} = \mathrm{IA}_{\mathrm{aw}} u_q F_q K^{\mathrm{A}}_{qa} C^{\mathrm{A}}_{\mathrm{air}}(t) = \mathrm{IA}_{\mathrm{aw}} k_{\text{dry dep}} C^{\mathrm{A}}_{\mathrm{air}}(t) \tag{4.27c}$$

where $k_{\text{dry dep}} = u_q F_q K^{\mathrm{A}}_{qa}$, and u_q is the rate at which particles are settled out of the atmosphere—a global average rate of 10.8 m/h is used by Mackay (2001).

Assuming no nondiffusive processes occur that transfer chemicals from water to air, a mass-balance equation for the air compartment can be written by incorporating the processes shown in Eqs. (4.26) and (4.27), as:

$$\begin{aligned}
V_{\mathrm{air}} \frac{dC^{\mathrm{A}}_{\mathrm{air}}}{dt} &= \dot{m}_{\text{water–air,diffusion}} - \left[\dot{m}_{\text{air–water,diffusion}} + \dot{m}_{\text{rain diss}} + \dot{m}_{\text{wet dep}} + m_{\text{dry dep}}\right] \\
&= \begin{bmatrix}\text{net mass-transfer rate} \\ \text{from water to air}\end{bmatrix} - \begin{bmatrix}\text{net mass-transfer rate} \\ \text{from air to water}\end{bmatrix} \qquad (4.28\text{a}) \\
&= \mathrm{IA}_{\mathrm{aw}}\, k_{\mathrm{LO}}\, C^{\mathrm{A}}_{w} - \mathrm{IA}_{\mathrm{aw}} \left[k_{\mathrm{GO}} + k_{\text{rain diss}} + k_{\text{wet dep}} + k_{\text{dry dep}}\right] C^{\mathrm{A}}_{\mathrm{air}} \\
&= \mathrm{IA}_{\mathrm{aw}}\, k_{\mathrm{wa}}\, C^{\mathrm{A}}_{w} - \mathrm{IA}_{\mathrm{aw}}\, k_{\mathrm{aw}}\, C^{\mathrm{A}}_{\mathrm{air}}
\end{aligned}$$

where the net mass-transfer coefficient for chemical transfer from water to air is denoted as k_{wa}, and the net mass-transfer coefficient for chemical transfer from air to water is denoted as k_{aw}:

$$k_{\mathrm{wa}} = k_{\mathrm{LO}} \tag{4.28b}$$

$$k_{\mathrm{aw}} = k_{\mathrm{GO}} + k_{\text{rain diss}} + k_{\text{wet dep}} + k_{\text{dry dep}} \tag{4.28c}$$

Note that diffusive mass-transfer coefficients alone contribute to k_{wa}, while a sum of diffusive and nondiffusive coefficients yields the net air–water mass-transfer coefficient, k_{aw}. Equations (4.28a) to (4.28c) provide an elegant framework in which the mass-transfer coefficient for transfer of chemical from medium i to another medium

TABLE 4.3 Diffusive and Nondiffusive Processes Contributing to Chemical Mass Transfer between Environmental Compartments

Compartments	Processes Contributing to Intermedia Chemical Transfer
Air → water	Diffusion, rain dissolution, wet deposition, and dry deposition
Water → air	Diffusion only
Air → soil (land)	Diffusion, rain dissolution, wet deposition, and dry deposition
Soil → air	Diffusion only
Sediment → water	Diffusion and resuspension
Water → sediment	Diffusion and sedimentation
Water → soil	Soil runoff and water runoff
Soil → water	Diffusion only

j must only be multiplied by concentrations in medium i to obtain the chemical flux from medium i. No interfacial concentrations need be known, and the framework allows for addition of diffusive and nondiffusive processes occurring in parallel as shown in Eqs. (4.28a) to (4.28c). This framework provides the basis for the level III models of Mackay (2001) in which rate-limited mass transfer between multiple media is coupled with transport and reactions occurring within each medium. Processes that contribute to net mass transfer between pairs of environmental compartments are listed in Table 4.3. The reader is referred to Mackay (2001) for a detailed description of these processes.

EXAMPLE 4.7

Compute k_{aw} and k_{wa} for transfer of dieldrin from air to water and from water to air by a combination of diffusive and nondiffusive processes. K_{aw} for dieldrin is $4.53 \times 10^{-4} L_{air}/L_w$ and its liquid-phase vapor pressure, $p_v^{(L)}$ is 0.0188 Pa. Assume a wind velocity of 4 m/s and a stream velocity of 1 m/s, consistent with the multimedia simulations presented in Beyer et al. (2000).

Solution 4.7 From Eqs. (4.28b) and (4.28c), $k_{wa} = k_{LO}$, and $k_{aw} = k_{GO} + k_{rain\,diss} + k_{wet\,dep} + k_{dry\,dep}$; and k_{LO} is computed from Eq. (4.5b), employing values of k_g and k_l computed from Table 4.2, CALTOX regional average correlations. For the wind and stream velocities specified here, k_g is estimated to be 12.4 m/h, k_l is estimated to be 0.027 m/h, and, from Eq. (4.5b), k_{LO} is computed as 0.005 m/h. The water-to-air total transfer coefficient, k_{wa}, incorporates only diffusive processes, hence $k_{wa} = k_{LO} = 4.65 \times 10^{-3}$ m/h.

The air-to-water total transfer coefficient incorporates both diffusive and nondiffusive processes. Recognizing that $k_{GO} = k_{LO}/K_{aw}$, we get $k_{GO} = 10.27$ m/h. We now apply Eqs. (4.27a) to (4.27c) to compute the nondiffusive transfer coefficients. From Eq. (4.27a), $k_{\text{rain diss}} = 0.2$ m/h for a rainfall rate of 10^{-3} m/hr. Employing the correlation equation of Mackay (2001), the aerosol–air partition coefficient, K_{qa}, is estimated as:

$$K_{qa}\left[\frac{\text{m}^3\text{ air}}{\text{m}^3\text{ aerosol}}\right] = \frac{6\times 10^6}{p_v^{\text{dieldrin(L)}}(\text{Pa})} = \frac{6\times 10^6}{0.0188} = 3.2\times 10^8$$

This enables computation of $k_{\text{wet dep}}$ of 0.128 m/h and $k_{\text{dry dep}}$ of 0.069 m/h, applying Eqs. (4.7b) and (4.7c), respectively, using the global average environmental properties recommended for use in these equations. Adding diffusive and nondiffusive transfer coefficients, the total air-to-water transfer coefficient, $k_{aw} = k_{GO} + k_{\text{rain diss}} + k_{\text{wet dep}} + k_{\text{dry dep}} = 10.2 + 0.2 + 0.128 + 0.069 = 10.66$ m/h.

4.7.3 Level III Multimedia Modeling Framework

Consider n environmental compartments as shown in Figure 4.12, with intermedia mass transfer occurring between the various compartments with net mass-transfer rate coefficients k_{ij} representing chemical transfer from compartment i to compartment j. The various compartments may exhibit bulk advective flow at a volumetric flow rate of Q_i. A chemical A may be emitted directly into a compartment i at a source emission rate of S_i, and the chemical may also undergo first-order degradation within the compartment i at a rate constant k_{ri}, representing the reaction rate constant for compartment i. Mass-balance differential equations similar to Eq. (4.28), but with the addition of source terms, advection terms, and reaction terms, are written for each compartment i, with the general form:

$$V_i\frac{dC_i^{\text{A}}}{dt} = S_i^{\text{A}} + Q_i C_{i,\text{upstream}}^{\text{A}} + \text{IA}_i\left(\sum_{j=2}^{n-1} k_{ji}C_j^{\text{A}}\right) - V_i\, k_{r,i}\, C_i^{\text{A}} - Q_i\, C_i^{\text{A}} - \text{IA}_i\left(\sum_{j=2}^{n-1} k_{ij}\right) C_i^{\text{A}} \tag{4.29}$$

where the first three terms on the rhs of Eq. (4.29) represent mass rates of input of chemical A into compartment i from direct source emission into compartment i, inflow with advection, into compartment i, and input from intermedia mass transfers to compartment i from the remaining $(n - 1)$ compartments, respectively. The last three terms represent chemical mass removal from compartment i due to degradation reaction occurring in compartment i, advective outflow from compartment i, and intermedia mass transfers from compartment i to the remaining $(n - 1)$ compartments.

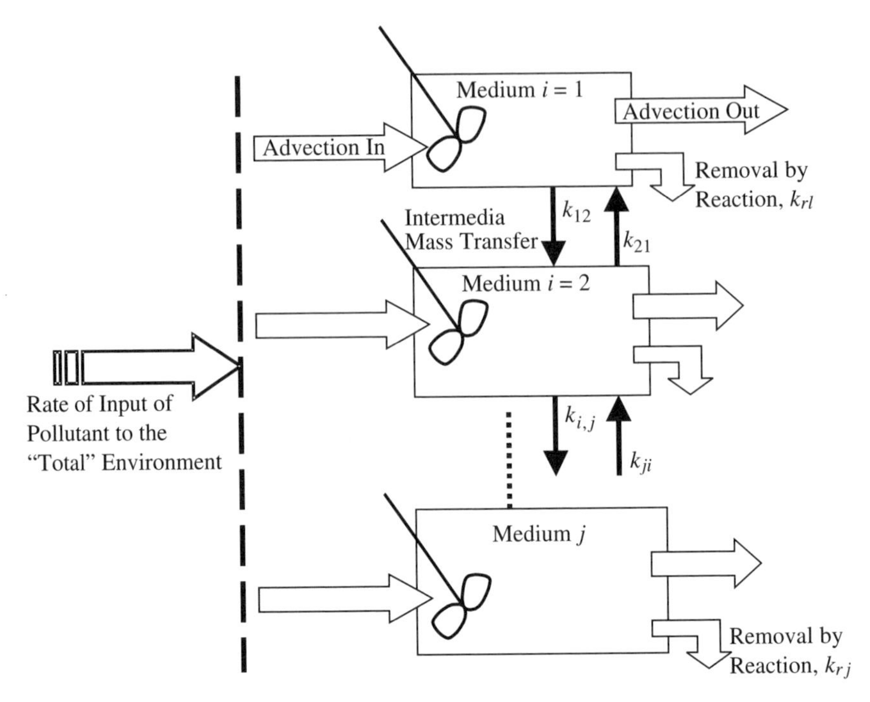

Figure 4.12 Mixed-box multimedia models (level III models) incorporating mass-transfer kinetics with coefficients $k_{i,j}$ and k_{ji} to describe rates of chemical transfer from medium i to medium j and vice versa.

To determine a steady-state concentration in compartment i, in response to constant emission, intermedia transfer, reaction, and advection rates, the left-hand side of this equation is set to zero; n such equations are written for each of the n compartments. The system of n equations is solved simultaneously to determine the steady-state chemical concentration in each of the n compartments. Alternatively, for conditions with variable loads or changing environmental conditions, the n mass-balance equations in Eq. (4.29) can be integrated simultaneously using the methods presented in Chapter 6 (see Section 6.1). This type of calculation could be used, for example, to study the time required for concentrations in different media to be reduced below target threshold levels once the use and emissions of a chemical are phased out.

4.7.4 Case Study: Persistent Organic Pollutants in the Stockholm Accord

Mackay et al. have applied this level III framework to a complex multimedia environment containing more than 10 compartments (Mackay and Paterson, 1991). A similar framework, with far fewer compartments, was used for assessing the global range of pesticides in a coupled soil–air environment (Bennet et al., 1999; Beyer et al., 2000), key results from which are presented here. Beyer et al. and Bennet et al. address the problem of POPs (persistent organic pollutants), focusing on 12 chemicals targeted by the Stockholm accord (see Chapter 2). The use of POPs such as DDT, dieldrin, aldrin, and so forth, has been banned in many developed countries in the northern latitudes. However, continued use of these chemicals in less developed countries, many of which are located in the tropical regions, creates a global fractionation effect where these chemicals volatilize to air after application in the hotter climates and are subsequently transported across continents to northern latitudes. Of interest is how far these chemicals will travel in air and how long they will persist in the environment, given that chemical transport in air is affected by mass transfer from air to soil, air to water, and air to vegetation, as well as by degradation reactions occurring in these compartments. This framework forms the basis of the TaPL3 model (Toxicity and Persistence Level III), which evaluates the potential for long-range chemical transport in one mobile medium into which the chemical is released, when that medium is in contact with other compartments that are assumed stationary (Beyer et al., 2000; Mackay, 2001). Beyer et al. begin with a simple two-compartment problem addressing a moving air mass exchanging POPs with a second immobile compartment (e.g., soil or water). A schematic of the important process parameters representing mass transfer in the two-compartment system is shown in Figure 4.13, with air and soil as the illustrative compartments. The TaPL3 framework (Fig. 4.13) differs from the more generalized level III model shown in Figure 4.12 in that all except the primary compartment into which the chemical is released are assumed to be stationary, with no advective losses occurring in these compartments.

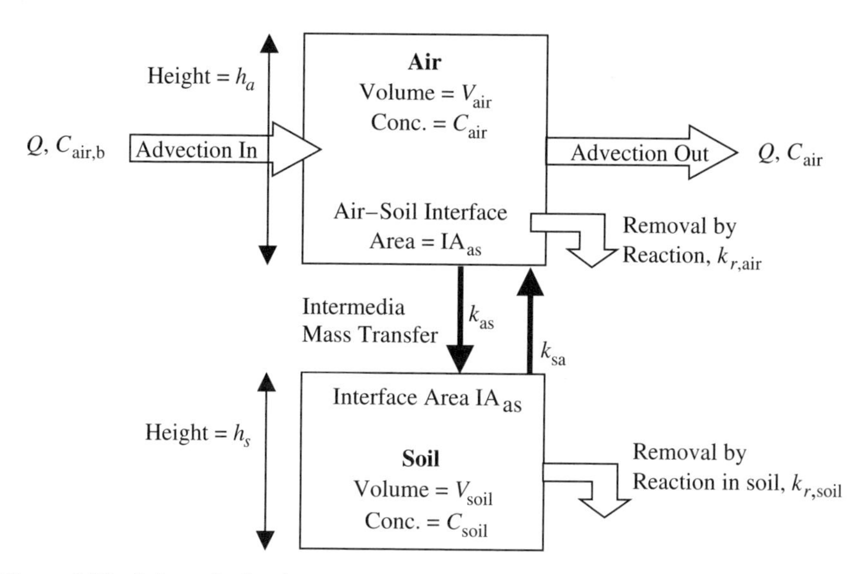

Figure 4.13 Schematic showing processes and parameters controlling movement of a chemical within linked air–soil compartments (after Beyer et al., 2000).

Applying Eq. (4.29) to the soil compartment yields

$$V_{\text{soil}} \frac{dC^{\text{A}}_{\text{soil}}}{dt} = \text{IA}_{\text{as}} k_{\text{as}} C^{\text{A}}_{\text{air}} - \text{IA}_{\text{as}} k_{\text{sa}} C^{\text{A}}_{\text{soil}} - V_{\text{soil}} k_{r,\text{soil}} C^{\text{A}}_{\text{soil}} = 0 \text{ (at steady state, ss)} \tag{4.30a}$$

where k_{as} and k_{sa} represent net mass-transfer coefficients for combined diffusive and nondiffusive mass transfer from air to soil and soil to air, respectively (see previous section, Table 4.3). IA_{as} is the air–soil interfacial area. Dividing by V_{soil} and setting the lhs of Eq. (4.30a) to zero yields the steady-state chemical concentration in soil:

$$C^{\text{A}}_{\text{soil,ss}} = \frac{C^{\text{A}}_{\text{air}} \times k_{\text{as}}/h_s}{k_{r,\text{soil}} + k_{\text{sa}}/h_s} \tag{4.30b}$$

Beyer et al. (2000) then define a stickiness coefficient, F, which is the ratio of the chemical mass consumed by reaction within the soil compartment to the gross chemical mass influx coming into the soil from the air:

$$F = \frac{\text{chemical mass reacted in soil}}{\text{chemical mass input from air}} = \frac{V_{\text{soil}} k_{r,\text{soil}}}{V_{\text{soil}} k_{r,\text{soil}} + k_{\text{sa}} C^{\text{A}}_{\text{soil,ss}} \text{IA}_{\text{as}}} = \frac{k_{r,\text{soil}}}{k_{r,\text{soil}} + (k_{\text{sa}}/h_s)} \tag{4.30c}$$

the stickiness coefficient, F, indicates the fraction of the mass transferred from air to soil that "sticks" to soil because it is consumed by reaction in that compartment.

A steady-state chemical mass balance is written next for the air compartment, incorporating airflow rates in and out of the mixed-box, chemical reactions occurring in air, as well as mass transfer from air to soil and vice versa:

$$\sum \text{mass inputs to air} = \sum \text{mass removal from air} \tag{4.31a}$$

$$QC^{\text{A}}_{b,\text{air}} + k_{\text{sa}} C^{\text{A}}_{\text{soil,ss}} \text{IA}_{\text{as}} = QC^{\text{A}}_{\text{air,ss}} + k_{\text{as}} C^{\text{A}}_{\text{air,ss}} \text{IA}_{\text{as}} + V_{\text{air}} k_{r,\text{air}} C^{\text{A}}_{\text{air,ss}}$$

In Eq. (4.31a) $C_{b,\text{air}}$ represents the background concentration of the chemical in air entering the mixed box. Using Eq. (4.30a), Eq. (4.31a) can be rewritten as:

$$QC^{\text{A}}_{b,\text{air}} = QC^{\text{A}}_{\text{air,ss}} + V_{\text{soil}} k_{r,\text{soil}} C^{\text{A}}_{\text{soil,ss}} + V_{\text{air}} k_{r,\text{air}} C^{\text{A}}_{\text{air,ss}} \tag{4.31b}$$

Substituting for $C_{\text{soil,ss}}$ from Eq. (4.30b) into Eq. (4.31b) and solving for steady-state chemical concentrations in air yields

$$C^{\text{A}}_{\text{air,ss}} = \frac{QC^{\text{A}}_{b,\text{air}}}{Q + (V_{\text{air}} k_{r,\text{air}}) + (\text{IA}_{\text{as}} F\, k_{\text{as}})} = \frac{C^{\text{A}}_{b,\text{air}}}{1 + (\tau_{\text{air}} k_{r,\text{air}}) + (\tau_{\text{air}} F\, k_{\text{as}}/h_a)} \tag{4.31c}$$

where τ_{air} is the residence time of air in the system. Rewriting the above expression for the steady-state airborne POP concentration yields

$$C_{\text{air}}^{\text{A}} = \frac{C_{b,\text{air}}^{\text{A}}}{1 + \tau_{\text{air}}[k_{r,\text{air}} + (Fk_{\text{as}}/h_a)]} \tag{4.31d}$$

Equation (4.31d) indicates that the POP concentration in air exiting the mixed box is attenuated because of reactions occurring in air (given by $k_{r,\text{air}}$) as well as those occurring in soil, represented by the factor (Fk_{as}/h_a), that is, the fraction of the flux from air to soil that undergoes reaction in soil. Notice how Eq. (4.31d) for a two-compartment system relates to Eq. (4.25) for a single compartment. If no reactions occur in soil, $F = 0$ and the last term in the denominator goes to zero. If further, no reactions occur in air, then $k_{r,\text{air}}$ is also zero, so that the exit POP concentrations are equal to the inlet concentrations, that is, no attenuation of POP in air occurs over the region. Thus, an effective reaction rate constant can be defined for the system as:

$$k_{\text{eff,air}} = k_{r,\text{air}} + F\frac{k_{\text{as}}}{h_a} \tag{4.32}$$

Note that k_{as}/h_a has units of $(1/T)$ and is akin to a mass-transfer *rate* coefficient for removal of a chemical from air to soil by a combination of diffusive and nondiffusive processes. For a system with n compartments to which mass transfer from medium i can take place, a more general description of $k_{\text{eff},i}$ for medium i is obtained as:

$$k_{\text{eff},i} = k_{r,i} + \sum_{j=1}^{n-1} F_{ij}\frac{k_i}{h_i} \tag{4.33a}$$

where F_{ij} is a stickiness coefficient describing the fraction of mass flux from compartment i to j that is consumed by reaction in compartment j, and is defined as:

$$F_{ij} = \frac{k_{r,j}}{k_{r,j} + (k_j/h_j)} \tag{4.33b}$$

The total stickiness in the environment is the weighted sum of the stickiness of the $(n-1)$ individual compartments, the weighting factor being the interfacial areas of the various compartments in contact with air.

The effective reaction rate constant [Eqs. (4.33a) and (4.33b)] in a medium i within a multimedia environment can be used to determine two very useful parameters that describe the persistence and potential for long-range transport of chemicals in that medium. The potential for long-range transport (LRT) in air, for example, can be expressed as a characteristic travel distance (CTD) over which the concentration undergoes attenuation by a factor of 2 within the mixed box (i.e., is decreased to 50% of the inlet concentration). The CTD is computed as the ratio of the wind velocity, u, and the effective reaction rate constant:

$$\text{CTD} = \frac{u_{\text{air}}}{k_{\text{eff,air}}} \tag{4.34}$$

Persistence in air is related to the time it takes for the fraction of the chemical mass present in air to be removed by an effective set of reactions parametrized for the medium of air [Eqs. (4.32) and (4.33)]. A measure of how long a chemical will persist in air is given by:

$$T_{\text{persistence,air}} = \frac{M_{\text{air}}^{\text{A}}/M_{\text{Total}}^{\text{A}}}{k_{\text{eff,air}}} \tag{4.35}$$

The concept of persistence in the TaPL3 model can be applied to any of the other mobile compartments in the multimedia environment when all other compartments are assumed to be immobile. Based on the framework described above, Beyer et al. (2000) analyzed the characteristic travel distances, overall "stickiness" to surfaces and persistence for 12 chemicals currently included in the UNEP program on POPs. TaPL3 (http://www.trentu.ca/cemc/models) was used to simulate the behavior of the 12 chemicals in a four-compartment environment consisting of air, soil, surface water, and sediment. Results of their analysis for POP release to air indicate that POPs can broadly be separated into three categories. The first category is characterized by CTDs in excess of 2000 km—these chemicals [e.g., hexachlorohexane (HCH)], have long persistence times and travel large distances in air creating a global reach for pollution. The second category includes chemicals such as dieldrin with CTDs ranging from 700 to 2000 km, that is, intermediate potential for long-range transport. The third category includes chemicals such as benzene with CTD less than 700 km, which are rarely found very far from their sources and hence do not pose a global concern. Thus, the analysis of Beyer et al. (2000) demonstrates how multimedia mixed-box models can provide a rationale for prioritizing POP regulation based on the potential for long-range transport of chemicals. Furthermore, sensitivity analysis (Beyer et al., 2000) indicates that the categorization of chemicals is robust with respect to plausible alternative assumptions regarding the height of the atmospheric compartment, wind speed, and depth of the soil compartment. The reader is directed to the TaPL3 model (http://www.trentu.ca/cemc/models) to further examine, through simulations, chemical behavior in a multimedia mixed-box environment. Example 4.8 illustrates the concept of stickiness and CTD in a simple two-compartment air–water environment.

EXAMPLE 4.8

Compute the stickiness and CTD for dieldrin released into an atmosphere with a mixing height of 1000 m and an average wind velocity of 4 m/s (14.4 km/h). Assume the air is in contact only with water of depth 20 m and stream velocity of 1 m/s. The half lives of dieldrin in air and in water are 55 h and 17,000 h, respectively. Use global average properties of the environmental system as shown in Example 4.7. Confirm your results with a TaPL3 simulation.

Solution 4.8 We use the net air–water and water–air mass-transfer coefficients computed previously in Example 4.7 as $k_{aw} = 10.66$ m/h and $k_{wa} = 0.00465$ m/h. The reaction rate constants for dieldrin in air and in water are computed from the respective half-life data as:

$$k_{r,\text{air}} = 0.693/55\,\text{h} = 0.0126\ h^{-1};\ k_{r,w} = 0.693/17000\,\text{h} = 4 \times 10^{-5}\ h^{-1}$$

With a water compartment depth, h_w, of 20 m, and k_{aw} and $k_{r,w}$ as computed above, we can now apply Eq. (4.30c) to compute the stickiness to water as $F_w = 0.147$. With a height of the air compartment, h_a, of 1000 m, the effective reaction rate constant in air is computed [Eq. (4.32)] as:

$$k_{\text{eff,air}} = \frac{0.0126}{\text{h}} + \frac{0.147 \times 10.66\ \text{m/h}}{1000\ \text{m}} = 0.0142\ \text{h}^{-1}$$

With a wind velocity of 14.4 km/h, the CTD [Eq. (4.34)] is computed to be 1017 km.

These results were confirmed by implementing a two-compartment TaPL3 simulation. In the simulation, the area of the water compartment was set equal to that of the air compartment, such that the effect of the soil compartment becomes negligible. Also, very small compartment heights (< 0.00001 m) and negligible half-lives were chosen for the soil and sediment compartments. The air-side and water-side mass-transfer coefficients (MTC) were set to the values computed in Example 4.7. The resulting two-compartment TaPL3 simulation of dieldrin transport is summarized in Figure 4.14*a*, confirming the theoretical estimates of stickiness and CTD computed for dieldrin in Solution 4.8. Note, the stickiness to water is computed from Figure 4.14*a* as the ratio of the dieldrin mass consumed by reactions occurring in water to that entering water from air [$F = 110/747$ or equivalently, $(747 - 637)/747 = 0.147$]. When soil and sediment compartments are included to create a typical Beyer environment (Beyer et al., 2000), the results look much different, as shown in Figure 4.14*b*. A greater proportion of dieldrin mass now resides in the soil and sediment compartments; less resides in the water. The stickiness from air to the water surface increases to 0.487 [$F = (78 - 40)/78 = 0.487$] because of reaction processes occurring in the sediment compartment. The reader is encouraged to explore the impact of stepwise inclusion of additional compartments into multimedia TaPL3 simulations.

4.8 COUPLING INTERMEDIA MASS EXCHANGE WITH FLOW AND REACTION IN SYSTEMS WITH SPATIAL HETEROGENEITY

Our discussion thus far has focused on chemical concentrations and mass distributions in well-mixed environmental compartments linked by equilibrium (Chapter 3)

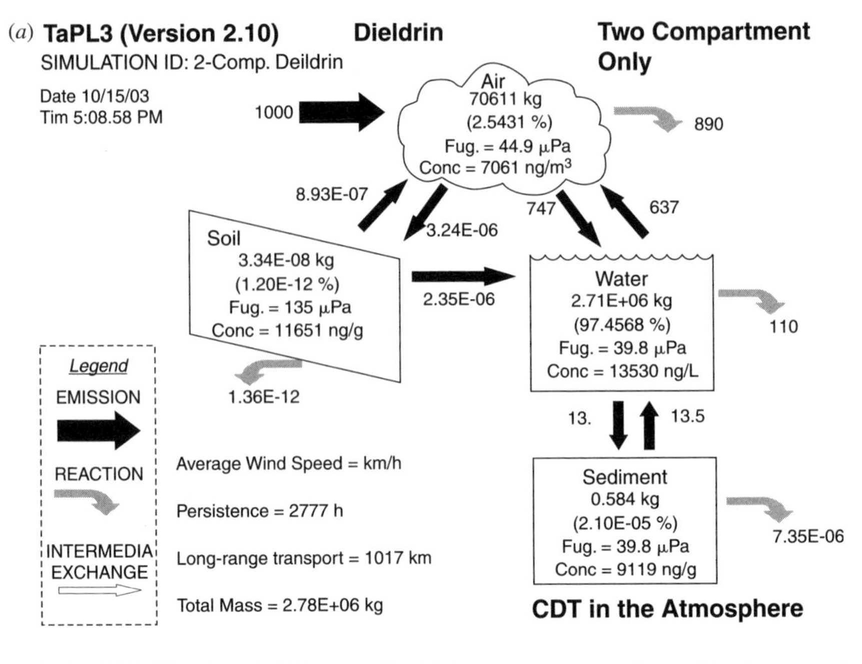

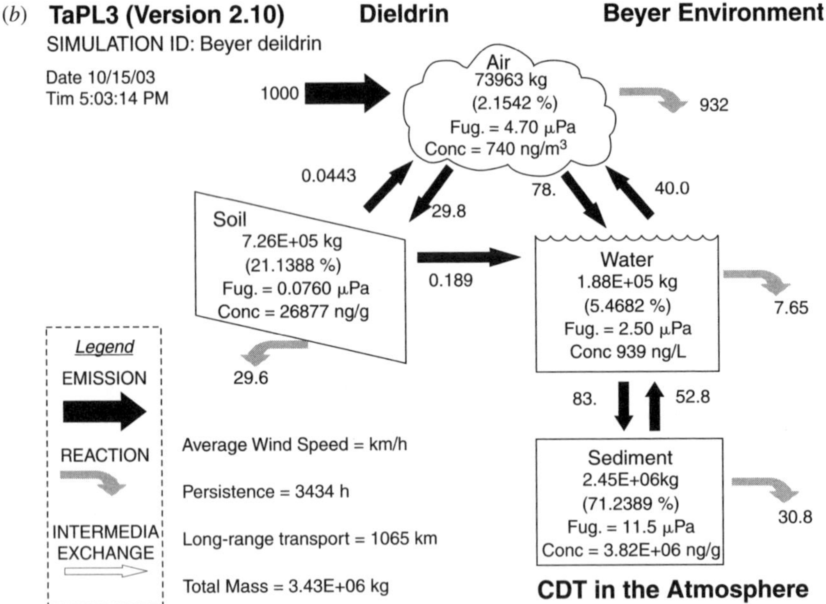

Figure 4.14 TaPL3 simulation (www.trentu.ca/caemc/models) of dieldrin transport and persistence in multimedia environments. Dieldrin is released to an air compartment characterized by a wind velocity of 4 m/s and a compartment height of 1000 m. (*a*) Results for a two-compartment environment: Air is in contact with water only. (*b*) Results for a four-compartment environment with air, soil, water, and sediment (after Beyer et al., 2000). (Note 1 ng = 10^{-9} g; 1 Pa = 1 Pascal = 9.87×10^{-6} atm.)

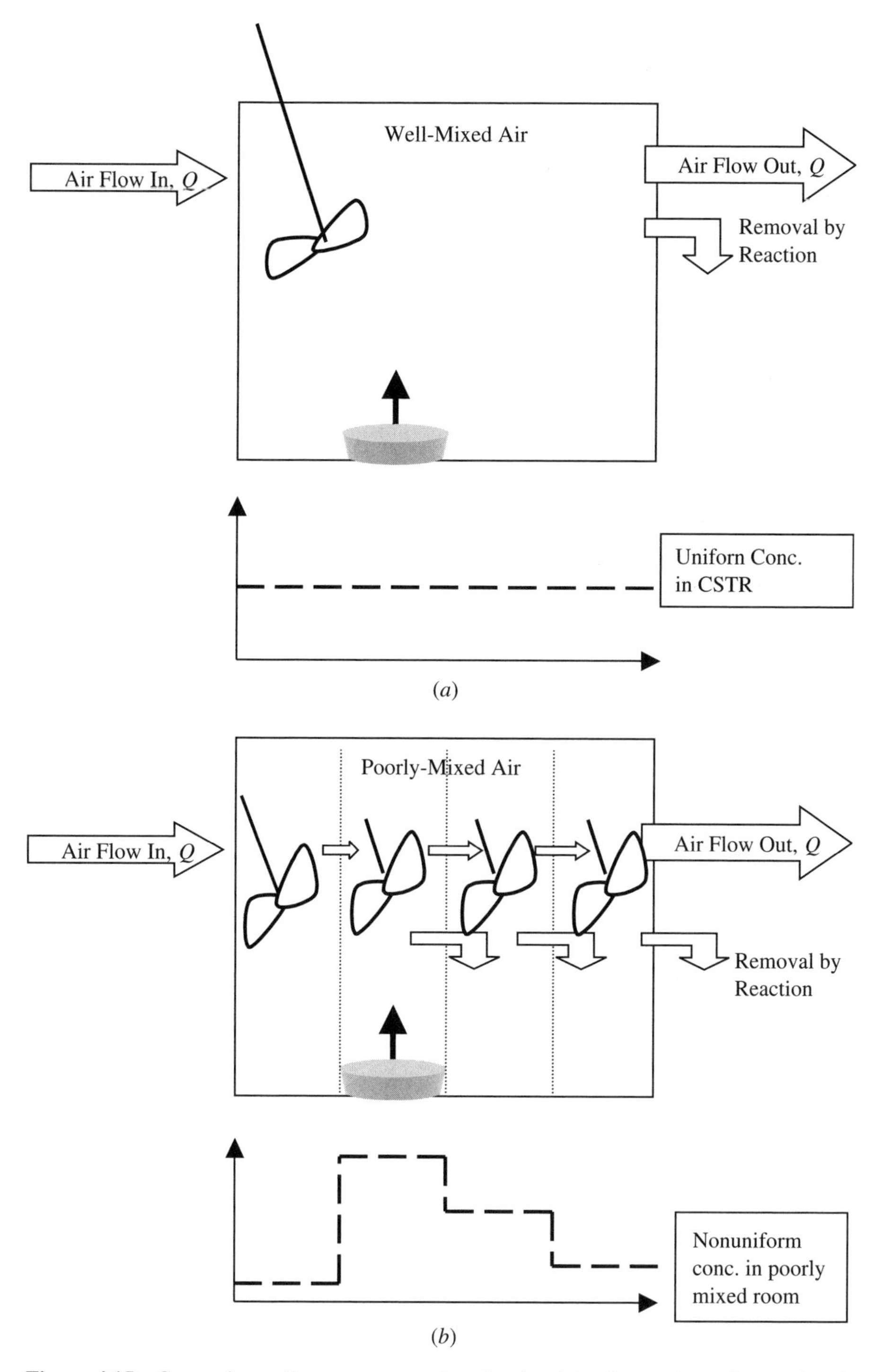

Figure 4.15 Comparing pollutant concentrations in air arising from release from a liquid spill. (*a*) A fully mixed air compartment. (*b*) A poorly mixed air compartment.

or nonequilibrium (Chapter 4) intermedia transport phenomena. Because of the assumption of complete mixing, a single representative chemical concentration was obtained within each bulk medium by normalizing the net chemical mass transferred into the medium by the volume of that medium. If the media are in fact poorly mixed, concentration gradients will appear in the various compartments and the specific locations where intermedia exchanges are occurring become important. For example, consider the liquid chemical spill shown in Figure 4.15*a* that releases pollutant molecules into the well-mixed air in a laboratory. If the room air is poorly mixed, a receptor adjacent to the spill would experience higher pollutant concentrations in the air compared to another receptor situated far away from the spill. A simple technique to account for the incomplete mixing of air in the room is to consider many subcompartments within the room, as shown in Figure 4.15*b*. As a first step, the subcompartment size can be chosen to be the same as the size of the spill, and each subcompartment can be assumed to be well mixed. Thus, concentrations within the subcompartment containing the spill will be high because of mass transfer from the spill. Chemical concentrations in downwind cells are expected to decrease because of the finite rate of advection and dispersion from the source compartment to those adjacent, and subsequently to the other compartments, as well as reactions that remove the chemical from air. Concentrations are thus expected to decrease along the length of the tunnel away from the spill. This can be modeled using a multicompartment level III model with the appropriate intercompartment fluxes incorporated into the mass-transfer formulation. In many cases, however, an explicit representation of the advection and dispersion occurring within key media will be preferred. This need is addressed in subsequent chapters that focus on spatially resolved chemical transport modeling within individual media. Chapter 5 provides an overview of fluid flow and mixing in air, groundwater, and surface water, emphasizing similarities and differences between the three systems. The intermedia mass-transfer phenomena described here are, in subsequent chapters, incorporated as localized pollutant source or pollutant sink terms that interface with fluid flow models to describe pollutant transport in air, groundwater, and surface water systems (Chapters 8 to 10).

Homework problems for this chapter are provided at www.wiley.com/college/ramaswami.

5 Transport Fundamentals

Environmental contaminants can be transported tens of thousands of kilometers from their point of release or never migrate more than a few centimeters. The distance depends on the partitioning and chemical reactivity of the contaminants and on the flow rates of the air and water that carry them. This chapter describes the physical processes that govern transport within a single medium, addressing spatially resolved transport of chemicals from source to receptor locations. The presentation emphasizes the parallel mathematical treatment of transport in air, surface water, and groundwater systems.

Three fundamental processes for moving and mixing contaminants within environmental media were introduced in Chapter 1: advection, diffusion, and dispersion. Advection is the transport of contaminants along with the mean or bulk flow of air or water. Diffusion and dispersion signify the mixing of contaminants due to gradients in contaminant concentration. Molecular diffusion predominates at very fine scales or in quiescent fluids and occurs through random motions employing the kinetic energy of the contaminant molecules. Mixing of contaminants at larger scales occurs due to mixing of the fluid medium carrying the contaminant and is referred to, in the different media, by terms such as turbulent diffusion, shear dispersion, and hydrodynamic dispersion. During turbulent diffusion, contaminants are mixed due to the eddy motion of fluids in the turbulent flow regime. Turbulent diffusion or eddy diffusion is the predominant mixing mechanism in the atmosphere and in large water bodies, such as oceans. Shear dispersion is defined as mixing due to velocity gradients in the fluid and is frequently seen in estuaries and streams. In groundwater applications, the term hydrodynamic dispersion refers to the mixing of a contaminant as the water that carries it encounters channels or other heterogeneities of various scales and undergoes mixing while flowing through the soil pore structure. Both "diffusion" and "dispersion" are sometimes used in a broad sense to mean total mixing, irrespective of whether it occurs by molecular diffusion, turbulent diffusion, channeling, or shear dispersion. As discussed below, each of the environmental modeling disciplines has traditionally used different nomenclature and terminology to describe these phenomena.

The fundamental equations governing fluid flows in the environment are presented in the next section. Section 5.2 discusses advection and gives typical advective transport velocities for each medium. Section 5.3 describes diffusive and dispersive processes in each medium and introduces the random walk model for 1D diffusive transport. Analytical models of advective–diffusive transport are presented in Section 5.4. Sections 5.5 and 5.6 provide further discussion of environmental dispersion, elaborating the point that its treatment depends on the temporal and spatial scales considered in a model. Finally, Section 5.7 introduces the numerical finite cell method for modeling transport in inhomogeneous systems.

5.1 FUNDAMENTAL EQUATIONS OF FLUID DYNAMICS

Modeling contaminant transport by advection, diffusion, or dispersion is relatively straightforward, once the underlying motions of the air or water that carry the contaminant are described. However, the underlying fluid motions are often complex and describing them mathematically can be difficult.

In most environmental modeling applications, both air and water can be treated as Newtonian, incompressible fluids. A Newtonian fluid is one in which viscous or shear stresses are described by the relationship: $\tau = \mu(du/dy)$, where τ is the shear stress ($M\,L^{-2}$), u is a component of fluid velocity ($L\,T^{-1}$), μ is the dynamic viscosity of the fluid ($M\,T\,L^{-2}$), and y is the cross-flow direction (L). An incompressible fluid is one with a constant density. This condition is not strictly true for air, but in the atmospheric boundary layer within about 1 km above Earth's surface, density changes are small enough to be neglected in many applications.

The motion of Newtonian, incompressible fluids in three dimensions is governed by:

$$\frac{\partial u}{\partial x} + \frac{\partial v}{\partial y} + \frac{\partial w}{\partial z} = 0 \tag{5.1a}$$

$$\rho \frac{du}{dt} = \rho g_x - \frac{\partial P}{\partial x} + \mu \left(\frac{\partial^2 u}{\partial x^2} + \frac{\partial^2 u}{\partial y^2} + \frac{\partial^2 u}{\partial z^2} \right) \tag{5.1b}$$

$$\rho \frac{dv}{dt} = \rho g_y - \frac{\partial P}{\partial y} + \mu \left(\frac{\partial^2 v}{\partial x^2} + \frac{\partial^2 v}{\partial y^2} + \frac{\partial^2 v}{\partial z^2} \right) \tag{5.1c}$$

$$\rho \frac{dw}{dt} = \rho g_z - \frac{\partial P}{\partial z} + \mu \left(\frac{\partial^2 w}{\partial x^2} + \frac{\partial^2 w}{\partial y^2} + \frac{\partial^2 w}{\partial z^2} \right) \tag{5.1d}$$

where u, v, and w represent the components of fluid velocity in the x, y, and z directions, respectively; ρ is the fluid density; g_i is the acceleration due to gravity in the ith direction; and P is the fluid pressure. Equation (5.1a) is known as the continuity equation, and Eqs. (5.1b) to (5.1d) as the Navier–Stokes equations. These equations indicate that the rate of change in fluid velocity is determined by three forces acting on the fluid: gravity forces, fluid pressure, and viscous forces. Solutions to the Navier–Stokes equations enable fluid velocity to be characterized over space and time.

The Navier–Stokes equations are coupled, second-order partial differential equations and require specification of boundary and initial conditions to solve. In some cases, for example, if the boundary conditions are simple and the flow is steady and inviscid, the Navier–Stokes equations can be solved analytically. However, randomly fluctuating turbulent flows introduce significant complications into the solution.

Turbulence is induced in surface water primarily by mechanical means such as the flow of water over a rough stream bed or the action of the wind over a lake surface. Atmospheric turbulence is commonly caused by thermal buoyancy as well as by the mechanical effects of variations in terrain height, buildings, or vegetation. The transition from laminar to turbulent flow conditions is marked by a critical value of the Reynolds number, $\mathrm{Re} = \rho u \ell / \mu$, where u is a characteristic velocity and ℓ is a characteristic length scale of the flow. Boundary layer flows with Reynolds numbers above about 4000 are turbulent (Vennard and Street, 1975). In this case, ℓ is defined as the boundary layer thickness and u as the free stream velocity. Similarly, the critical value of Re for flow through an open channel is ~500, with ℓ defined as the channel depth and u as the mean velocity over the depth of the channel. Only approximate numerical solutions of the Navier–Stokes equations are available for turbulent flows in the environment. An introduction to mathematical modeling of turbulent flows is provided in the supplemental information at www.wiley.com/college/ramaswami.

Most fate and transport models rely on a large number of simplifying assumptions in characterizing fluid flows. Often, the flow descriptions are based solely on observed winds, stream flows, or water currents. Interpolation or extrapolation may be applied to produce flow fields from the observations, but no physical modeling of fluid flow is performed. Increasingly, however, hydrodynamic or hydraulic fluid flow modeling is included as part of fate and transport modeling. This is especially common in modeling contaminant transport in groundwater, where flow modeling is usually undertaken as a first step. Moreover, meteorological models are now widely used to provide wind fields for air quality modeling studies conducted at urban to regional scales.

Key characteristics and models of air, surface, and groundwater flows are presented in subsequent chapters that focus on transport in each of the individual media. In these later chapters, brief introductions are also provided to methods for dealing with more complex flow conditions, such as situations where density variations are important and where multiphase or immiscible fluids are present. The remainder of this chapter assumes fully miscible contaminant transport in a single, uniform-density fluid.

5.2 MODELING ADVECTION

In one dimension, the contaminant flux [mass transport rate per unit area ($M\,T^{-1}\,L^{-2}$)] due to advection is $J_{a,x} = uC$, where u is the fluid velocity in the x direction, and C is the contaminant concentration. Performing a mass balance on the control volume shown in Figure 5.1 yields

$$
\begin{aligned}
A\,\Delta x \frac{\partial C}{\partial t} &= A\,[uC(x) - uC(x+\Delta x)] \\
&= A\left\{uC(x) - \left[u\,C(x) + \frac{\partial u\,C(x)}{\partial x}\Delta x\right]\right\} = -A\frac{\partial u\,C(x)}{\partial x}\Delta x
\end{aligned}
\tag{5.2}
$$

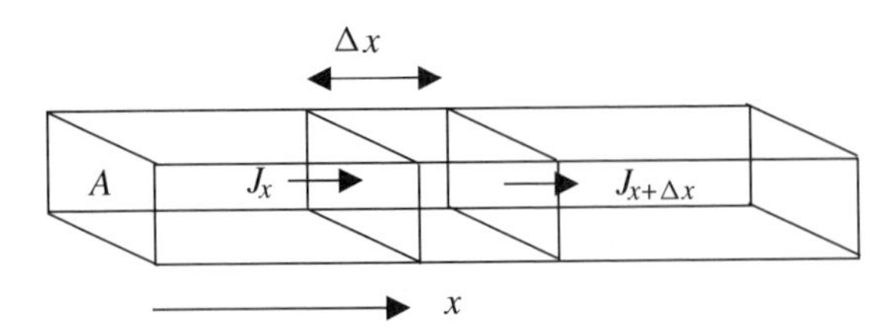

Figure 5.1 Advective flux, J_x, through a control volume with dimensions $A \times \Delta x$.

where A is the cross-sectional area perpendicular to the flow direction. Simplifying Eq. (5.2) and writing it in three dimensions produces the advection equation:

$$\frac{\partial C}{\partial t} = -\frac{\partial (u_j C)}{\partial x_j} \qquad j = 1, 2, 3 \tag{5.3a}$$

In Eq. (5.3a), j indexes the x, y, and z directions, and the repeated index indicates a sum over $j = 1$, 2, and 3. Combining Eq. (5.3a) with the continuity equation (5.1a) results in:

$$\frac{\partial C}{\partial t} = -u_j \frac{\partial C}{\partial x_j} \qquad j = 1, 2, 3 \tag{5.3b}$$

Equation 5.3 indicates that the rate of change in contaminant concentration within a control volume caused by advection depends upon the fluid flow velocity u_j and the concentration gradient around the control volume. Describing advection requires knowledge of fluid velocity and the initial spatial distribution of the contaminant in the environmental media. Simple methods for estimating fluid velocities in air, surface water, and groundwater are presented next. Information on more detailed meteorological and hydrodynamic models is presented in Chapters 8 to 10.

5.2.1 Advection in Air

The atmospheric boundary layer is defined as the portion of Earth's atmosphere in which airflow is affected by surface friction in addition to pressure gradients [see Eq. (5.1)]. For level terrain, the boundary layer typically extends from the surface up to a height of about 1 km. The lowest 50 m or so of the boundary layer is known as the surface layer. Typical wind speeds for advective transport in the atmospheric surface layer are on the order of a few meters per second, measured at the standard height of 10 m above ground level. Summary statistics for wind speed and direction are often presented graphically in the form of a wind rose (Fig. 5.2) in which the segments of the arms correspond to distinct wind speed ranges. In the figure the predominant wind direction is westerly, following the convention that winds are named according to the direction from which they come. The wind rose indicates that winds are westerly with speeds ranging from 2 to 6 m/s about 15% of the time. In general, winds blow from

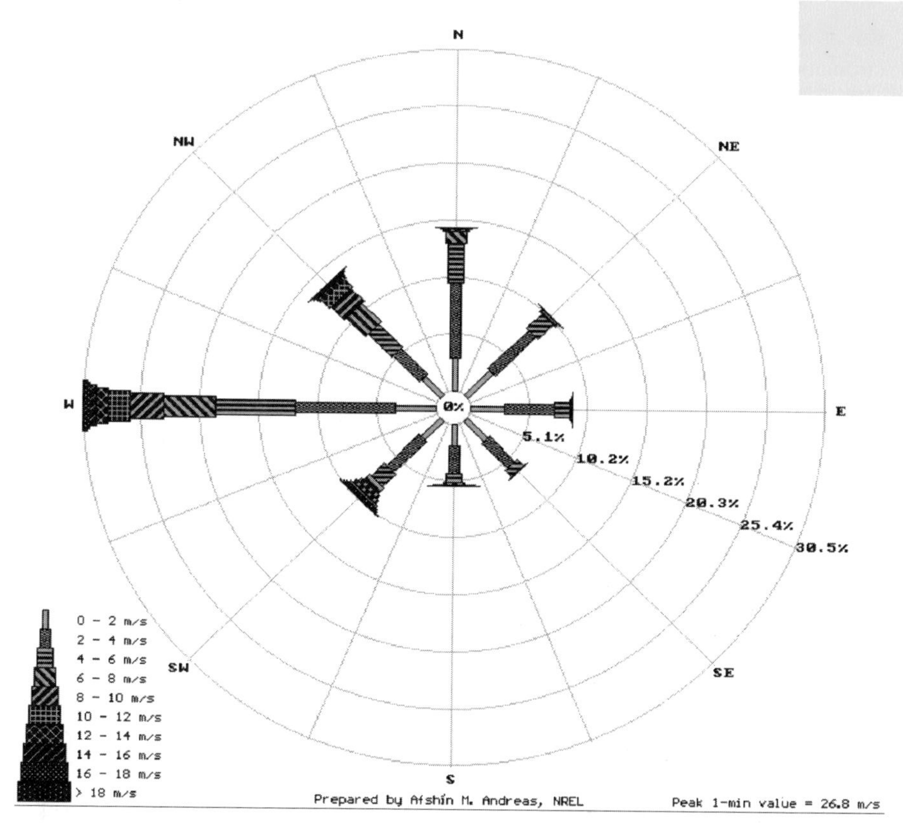

Figure 5.2 A wind rose for winter, 1999, for the Solar Radiation Research Laboratory of the National Renewable Energy Laboratory in Golden, Colorado, summarizing the distribution of 1-min average surface level winds by speed and direction (NREL, 2000).

areas of high pressure to low pressure, being deflected to their right in the Northern Hemisphere (and to their left in the Southern Hemisphere) due to the rotation of Earth. Terrain effects and surface roughness cause variations in wind velocity that are superimposed on these general flow systems.

The increase in wind speed with height in the boundary layer typically follows a power law profile:

$$\frac{u(z)}{u_r} = \left(\frac{z}{z_r}\right)^p \tag{5.4}$$

where the subscript r denotes a reference height and p is a constant exponent. Power law profiles are characteristic of fluid flows over flat surfaces. For the atmosphere, typical values of p range from about 0.15 for flat, open terrain to 0.4 for highly built-up areas. More generally, wind speeds in the atmospheric boundary layer are determined by Earth's rotation, pressure gradient forces, viscous forces, and density variations that arise due to heterogeneous heating and cooling rates. Atmospheric circulations are discussed further in Chapter 8.

5.2.2 Advection in Groundwater

Water flow in the subsurface occurs by a combination of gravity forces and pressure forces; that is, from higher elevation to lower elevation and from high-pressure to lower-pressure regions. The hydraulic head h represents the sum of the elevation head (z) and pressure head (ψ):

$$h = z + \psi \tag{5.5a}$$

where h is the static pressure head of water at a particular location, that is, the height above mean sea level or another reference datum, of water in a test well (see Fig. 5.3) and dh/dL is the hydraulic gradient or change in head in the longitudinal direction of the flow. Groundwater flows in the direction of the head gradient in the subsurface, from areas of high head to low head, perpendicular to equipotential contours. Darcy's law states that the Darcy velocity, q, defined as the volumetric rate of flow of water, Q, per unit gross (total) cross-sectional area (A) of an aquifer, is proportional to the gradient in head in that aquifer. The proportionality constant K is known as the hydraulic conductivity of the medium:

$$q = \frac{\text{volumetric flow rate}}{\text{total cross-sectional area}} = \frac{Q}{A} = K\frac{dh}{\Delta L} \tag{5.5b}$$

The hydraulic conductivity, K, represents the ease with which a porous packed medium such as soil allows water to flow through it. Table 5.1 shows hydraulic conductivity values that are characteristic of different soil types.

The Darcy velocity, q, is also referred to as the specific discharge rate and represents water flow across the total cross section, which is of interest in water supply wells. However, contaminant hydrogeologists are concerned with the effective travel velocity of water in the subsurface, that is, the effective rate of flow of water through the pore cross-sectional area of the medium, termed the average linear groundwater

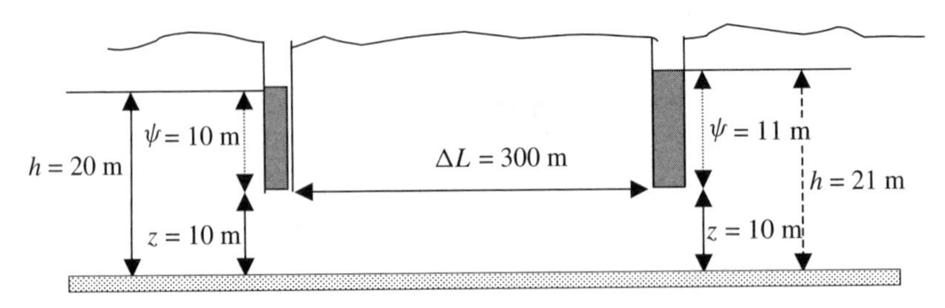

Figure 5.3 Determination of the hydraulic gradient used in Darcy's law to calculate linear groundwater flow velocities in Example 5.1.

TABLE 5.1 Typical Values of Hydraulic Conductivity for Range of Rock and Soil Types

Material	Hydraulic Conductivity (m/day)
Clay	0.0004
Sand	40
Gravel	4000
Sandstone	4
Dense limestone and shale	0.04
Granite	0.0004

Source: Linsley et al. (1975).

velocity.[1] The pore cross-sectional area (CSA) can be determined from the gross cross-sectional area by factoring in the porosity of the medium. Thus, the average linear groundwater velocity, v, may be obtained from the Darcy velocity as:

$$v = \frac{Q}{\text{pore CSA}} = \frac{Q}{nA} = \frac{q}{n} = \frac{K}{n}\frac{\Delta h}{\Delta x} \tag{5.5c}$$

where n is the porosity of the medium, and v is also referred to as the *seepage velocity*.

Linear average groundwater velocities are typically much slower than fluid velocities in air or surface water systems. Values of 1 to 10 m/yr are typical, although values of 100 m/yr or more are also common in porous, sandy aquifers. Groundwater flow rate calculations are discussed in more detail in Chapter 9.

EXAMPLE 5.1

Figure 5.3 shows a one-dimensional example of a situation for which Darcy's law can be applied to calculate a linear flow velocity. Two test wells are sunk into a confined aquifer a distance of 300 m apart. Both wells are screened at the bottom to penetrate the groundwater at a height, z, of 10 m above some reference datum. Water in one well rises up into the well to a height of 10 m, while water in the second well rises up by 11 m, representing the pressure head at the two locations. The total heads, h, are computed as 20 and 21 m, respectively. The difference in static pressure head indicated by the water levels in the two wells is 1 m, so the hydraulic gradient is $dh/dL = 1/300$. If the soil porosity is $n = 0.35$ and its hydraulic conductivity is $K = 4.0$ m/day, Darcy's law indicates that the linear

[1] Effective linear velocity is simply linear distance traveled per unit time, much different from the true travel velocity of water parcels along tortuous paths in the porous matrix.

groundwater velocity in the aquifer would be $v = 0.038$ m/day or 13.9 m/yr. Note that in field applications involving two-dimensional flow, a minimum of three head data points are required to calculate the hydraulic gradient.

5.2.3 Advection in Surface Water

Bulk water flow in surface waters is primarily controlled by gravity forces and viscous or friction forces. Horizontal pressure forces are negligible because surface waters are open to the atmosphere. Thus, elevation gradients within a channel (i.e., slope) and surface roughness and friction forces control water flow rates. Typical flow velocities in streams and rivers range from 0.1 to 1.5 m/s, corresponding to channel slopes of 0.0002 to 0.01 (Fischer et al., 1979). A rough estimate of the flow velocity in an open channel can be obtained by using the Chezy–Manning equation, which applies for uniform flow conditions. Uniform flows of constant velocity and depth occur when gravity and resistance forces are in balance. Under these conditions, flow velocities may be estimated from channel characteristics using:

$$u = \frac{C_0}{m} R_h^{2/3}\,\mathrm{Sl}^{1/2} \tag{5.6a}$$

where C_0 = constant (1.486 for units of feet and seconds, 1.0 for units of meters and seconds)

m = Manning's roughness coefficient

R_h = hydraulic radius of the channel, which is the ratio of the cross-sectional area of water to the wetted perimeter

Sl = channel slope

Typical values of the Manning coefficient m for streams and rivers range from 0.025 for very smooth channels to 0.1 for very rough channels containing an irregular channel surface, significant vegetation, and the like (Chow, 1959, Table 5.6).

It is often desired to estimate the variable wetted channel depth, h, and the corresponding flow velocity u from the volumetric flow rate Q, which is commonly available from stream gage records (see http://water.usgs.gov/ to access real-time flow data for streams and rivers throughout the United States) or the predictions of a hydrologic (rainfall–runoff) model. Assuming a rectangular channel of width w, with (unknown) water depth h, the linear water flow rate u is related to the volumetric flow rate Q as $u = Q/(hw)$. The hydraulic radius is $R_h = wh/(w + 2h)$. When $w \gg h$, $R_h \sim h$. Substituting the above relationships into Eq. (5.6a) and solving for h in terms of Q $(= u \times w \times h)$, yields

$$\frac{Q}{wh} = \frac{C_0}{m} h^{2/3}\,\mathrm{Sl}^{1/2} \quad \text{or} \quad h = \left(\frac{Qm}{wC_0\,\mathrm{Sl}^{1/2}}\right)^{3/5} \tag{5.6b}$$

Correspondingly, the velocity is given in terms of Q as:

for $w >> h$

$$u = \frac{Q}{wh} = \frac{Q}{w}\left(\frac{Qm}{wC_0\,\mathrm{Sl}^{1/2}}\right)^{-3/5} = \left(\frac{Q^{2/5}}{w^{2/5}}\right) \times \left(\frac{m}{C_0\,\mathrm{Sl}^{1/2}}\right)^{-3/5} \tag{5.6c}$$

More generally, the following empirical relationships are often used for velocity and depth: $u = \alpha_1 Q^{\beta_1}$ and $h = \alpha_2 Q^{\beta_2}$, where α_1 and α_2 are constant for a segment of the channel with constant width, slope, and roughness determined from empirical or hydraulic studies of the flow behavior of a specific channel. More accurate methods for physical modeling of open-channel flow are discussed in Chapter 10.

EXAMPLE 5.2

Consider a chemical spill into a river that occurs 10 km upstream from a city water intake. The river is 20 m wide, has a channel slope of $\mathrm{Sl} = 0.0001$ and a flow rate of $Q = 17.5\ \mathrm{m^3/s}$. The value of Manning's coefficient for the river is $m = 0.03$. Assuming that advection is the dominant transport process, how long will it take for the chemical to reach the city's water intake?

Solution 5.2 Combination of Eqs. (5.6a) and (5.6b) yields

$$h = \left(\frac{Qm}{C_0\,\mathrm{Sl}^{1/2} w}\right)^{3/5} = \left(\frac{17.5(0.03)}{1.0 \times 0.0001^{1/2}(20)}\right)^{3/5} = 1.78\ \mathrm{m}$$

With h determined,

$$u = \frac{Q}{wh} = \frac{17.5\ \mathrm{m^3/s}}{20\ \mathrm{m}\ (1.78\ \mathrm{m})} = 0.49\ \mathrm{m/s}$$

Based on this average stream velocity, the chemical will arrive at the water intake in about 5.7 h.

5.3 MODELING DIFFUSION AND DISPERSION

5.3.1 Molecular Diffusion

The fundamental equation describing mass transfer by molecular diffusion follows from the assumption that the rate of contaminant mass transfer across an imaginary interface is proportional to the contaminant mass near the interface. This is illustrated in Figure 5.4.

Assuming the concentration is uniform within each of two differential volume elements of size $\Delta x \times \Delta y \times \Delta z$, the rate of mass transfer in the x direction from

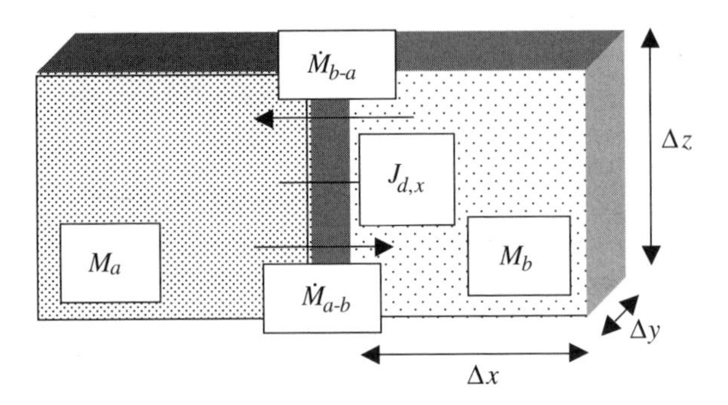

Figure 5.4 Diffusive flux between two adjacent volume elements with different initial contaminant masses, M_a and M_b.

volume a to volume b is $\dot{M}_{a-b} = kM_a$, where k is a rate constant (T^{-1}) and M_a is the contaminant mass in volume a. The rate of mass transfer from volume b to volume a is $\dot{M}_{b-a} = kM_b$, where M_b is the contaminant mass in volume b. The net mass transfer across the interface is thus $\dot{M} = k(M_a - M_b)$. Flux is defined as the mass transfer rate per unit area of the interface, so:

$$J_{d,x} = k\frac{M_a - M_b}{A} \qquad \text{where } A = \Delta y \times \Delta z \tag{5.7a}$$

Assuming by convention that $\Delta x = \Delta y$, the diffusive flux can be expressed in terms of contaminant concentration as:

$$J_{d,x} = k\,\Delta x^2 \frac{C_a - C_b}{\Delta x} \tag{5.7b}$$

Finally, taking the limit as Δx goes to zero and grouping $k\,\Delta x^2$ into a single constant, $D_{m,x}$ gives

$$J_{d,x} = -D_{m,x}\frac{\partial C}{\partial x} \tag{5.7c}$$

and $D_{m,x}$ is referred to as the molecular diffusion coefficient or molecular diffusivity and has units of $L^2\,T^{-1}$. Equation (5.7c) is known as Fick's law, after Adolph Fick, who proposed the hypothesis in 1855 that the mass flux of a solute due to diffusion is proportional to its concentration gradient. Fick's law followed the analogous theory for heat conduction, which was developed by Fourier in 1822.

The equation for the rate of change in contaminant concentration due to diffusion can be derived by applying a mass balance to the control volume shown in Figure 5.4, with the assumption that the molecular diffusion coefficient is constant:

$$A\,\Delta x\frac{\partial C}{\partial t} = -AD_{m,x}\left(\frac{\partial C(x)}{\partial x} - \frac{\partial C(x+\Delta x)}{\partial x}\right) = AD_{m,x}\frac{\partial}{\partial x}\left(\frac{\partial C(x)}{\partial x}\right)\Delta x \tag{5.8a}$$

Simplifying Eq. (5.8a) and extending it to three dimensions produces the diffusion equation:

$$\frac{\partial C}{\partial t} = D_{m,j}\frac{\partial^2 C}{\partial x_j^2} \qquad j = 1, 2, 3 \tag{5.8b}$$

Equation (5.8b) is also known as the heat equation and has applications to numerous physical problems including heat as well as mass transfer. Given the initial and boundary conditions for a particular problem, the second-order partial differential equation can be solved analytically using separation of variables with Fourier series or Fourier integrals. Crank (1979) presents solutions for many cases with simple geometries and boundary conditions. Some of the simplest analytical solutions relevant to environmental applications are those for unbounded domains. This assumption can be invoked if the time scale of interest is short compared to the time required for diffusive transport to the actual system boundaries.

A conceptual picture of a diffusion process is illustrated by the random-walk model, which corresponds to 1D diffusion following instantaneous release of a contaminant spike in an unbounded domain. As shown in Figure 5.5*a*, this model assumes that at $t = 0$ a group of particles is released at the position $x = 0$. In each subsequent time increment, Δt, the particles are assumed to be restricted to moving in steps of either $+\Delta x$ or $-\Delta x$, with an equal probability of moving in either direction. As time passes, the particles spread out but with the highest concentration remaining close to the origin. Most of the particles reverse direction often, with relatively few taking many successive steps in a single direction. In the limit as the number of time steps, n, becomes large and Δx and Δt go to zero, the outcome of this probabilistic behavior is a Gaussian or bell-shaped distribution, as shown in Figure 5.5*b*. The Gaussian distribution results from the central limit theorem, which states that no matter what the distribution of a random variable is [in this case, the distribution is discrete with $p(+\Delta x) = p(-\Delta x) = 0.5$], the distribution of its sum eventually, with many additions, converges to a normal or Gaussian distribution:

$$f(x) = \frac{1}{\sqrt{2\pi}\,\sigma_x}\exp\left[\frac{-(x-\mu_x)^2}{2\sigma_x^2}\right] \tag{5.9}$$

where $\mu_x = 0$ is the mean displacement and σ_x is the standard deviation of the net displacement after n time steps, which is given by $\sigma_x = \sqrt{n}\,\Delta x$ (Chapra and Reckhow, 1983, p. 75).

The mathematical expression for the distribution shown in Figure 5.5*b* can also be deduced from the mass-balance differential equation (5.8b) as follows (Crank, 1979). First, direct differentiation shows that the 1D form of the diffusion equation is satisfied by:

$$C(x,t) = \frac{a}{\sqrt{t}}\exp\left(\frac{-x^2}{4D_{m,x}t}\right) \tag{5.10a}$$

where a is an arbitrary constant. The value of a can be determined by defining $m_a = \int_{-\infty}^{\infty} C\,dx$ as the total mass of particles per unit cross-sectional area. Substituting

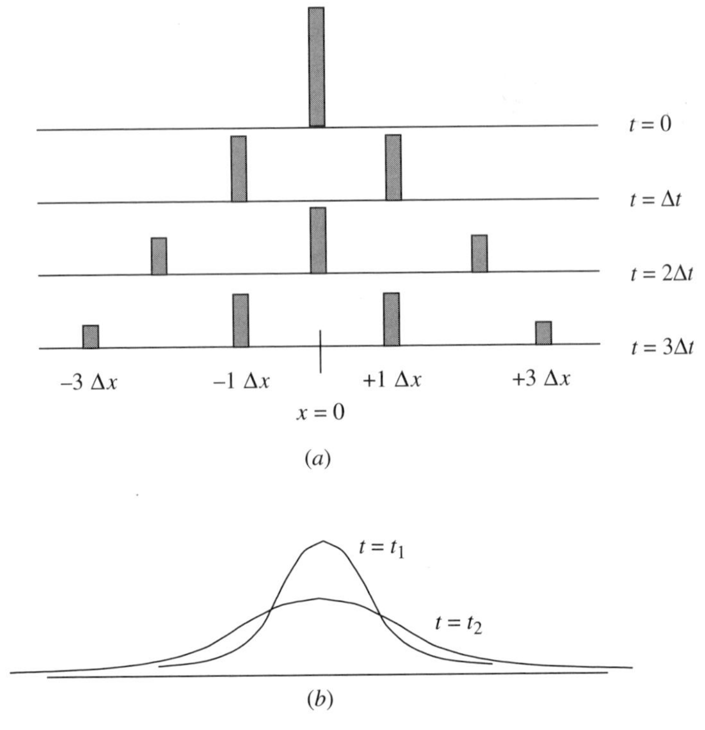

Figure 5.5 Progression of a random walk with particles constrained to move one step $\pm\,\Delta x$ per time step, Δt (after Chapra, 1997). (*a*) Discrete particle jump model. (*b*) Continuous limit.

Eq. (5.10a) into the integral and performing the integration yields $m_a = 2a\sqrt{\pi\, D_{m,x}}$, which can be rearranged to solve for a. The mathematical expression for the random-walk solution is then

$$C(x,t) = \frac{m_a}{2\sqrt{\pi\, D_{m,x} t}} \exp\left(\frac{-x^2}{4\, D_{m,x}\, t}\right) \tag{5.10b}$$

Equation (5.10b) is equivalent to the Gaussian distribution function for the probability that a particle released at time 0 is found at location x at time t [Eq. (5.9)] or, equivalently, for the fraction of the released mass at location x at time t. Comparison of Eqs. (5.9) and (5.10b) shows that the probability distribution function of particle location (given by the variable x in one dimension) is characterized by:

$$\mu_x = 0 \quad \text{and} \quad \sigma_x = \sqrt{2 D_{m,x}\, t} \tag{5.11}$$

Bivariate and trivariate normal distributions that are analogous to Eq. (5.10b) describe molecular diffusion in two and three dimensions, respectively (Table 5.2). For example, in three dimensions, the solution for an instantaneous release of mass

TABLE 5.2 Gaussian Probability Distribution Functions in One, Two, and Three Dimensions for Particle Location and for Chemical Concentrations in Response to an Instantaneous Chemical Release (of Mass M) at the Origin at Time $t = 0$[a]

	Probability Density Function (pdf) of Particle Location = Probability of Occurrence within an Interval Divided by Interval Size	Chemical Concentration Distribution in the Absence of Chemical Reaction
1D	$\mathrm{pdf}(x) = \dfrac{\mathrm{Prob}(x < X \le x + \Delta x)}{\Delta x}$	$C(x, t) = M_a \times \mathrm{pdf}(x)$
	$= \dfrac{1}{\sigma_x \sqrt{2\pi}} \exp\left[-\dfrac{1}{2}\left(\dfrac{x - \overline{x}}{\sigma_x}\right)^2\right]$	$= \dfrac{M_a}{\sigma_x \sqrt{2\pi}} \exp\left[-\dfrac{1}{2}\left(\dfrac{x - \overline{x}}{\sigma_x}\right)^2\right]$
		$M_a = \dfrac{\text{mass of chemical released}}{\text{area}}$
2D	$\mathrm{pdf}(x, y) = \dfrac{\mathrm{Prob}(x < X \le x + \Delta x;\ y < Y \le y + \Delta y)}{\Delta x\, \Delta y}$	$C(x, y, t) = M_z \times \mathrm{pdf}(x, y)$
	$= \dfrac{1}{\sigma_x \sigma_y 2\pi} \exp\left[-\dfrac{1}{2}\left(\dfrac{x - \overline{x}}{\sigma_x}\right)^2\right] \exp\left[-\dfrac{1}{2}\left(\dfrac{y - \overline{y}}{\sigma_y}\right)^2\right]$	$= \dfrac{M_z}{\sigma_x \sigma_y 2\pi} \exp\left[-\dfrac{1}{2}\left(\dfrac{x - \overline{x}}{\sigma_x}\right)^2\right] \exp\left[-\dfrac{1}{2}\left(\dfrac{y - \overline{y}}{\sigma_y}\right)^2\right]$
		$M_z = \dfrac{\text{mass of chemical released}}{\text{depth}}$

(continued)

TABLE 5.2 (*Continued*)

	Probability Density Function (pdf) of Particle Location = Probability of Occurrence within an Interval Divided by Interval Size	Chemical Concentration Distribution in the Absence of Chemical Reaction
3D	$\text{pdf}(x, y, z) = \dfrac{\text{Prob}(x < X \le x + \Delta x;\ y < Y \le y + \Delta y;\ z < Z \le z + \Delta z)}{\Delta x\, \Delta y\, \Delta z}$ $= \dfrac{1}{\sigma_x \sigma_y \sigma_z \left(\sqrt{2\pi}\right)^3} \exp\left[-\dfrac{1}{2}\left(\dfrac{x-\overline{x}}{\sigma_x}\right)^2\right] \exp\left[-\dfrac{1}{2}\left(\dfrac{y-\overline{y}}{\sigma_y}\right)^2\right] \exp\left[-\dfrac{1}{2}\left(\dfrac{z-\overline{z}}{\sigma_z}\right)^2\right]$	$C(x, y, z, t) = M \times \text{pdf}(x, y, z)$ $= \dfrac{M}{\sigma_x \sigma_y \sigma_z \left(\sqrt{2\pi}\right)^3} \exp\left[-\dfrac{1}{2}\left(\dfrac{x-\overline{x}}{\sigma_x}\right)^2\right] \exp\left[-\dfrac{1}{2}\left(\dfrac{y-\overline{y}}{\sigma_y}\right)^2\right] \exp\left[-\dfrac{1}{2}\left(\dfrac{z-\overline{z}}{\sigma_z}\right)^2\right]$ M: Mass of chemical released.

[a] The chemical undergoes molecular diffusion and advection–dispersion within a fluid medium. x, y, z are independent variables representing location; $\overline{x}, \overline{y}, \overline{z}$ are the average locations; and $\sigma_x, \sigma_y, \sigma_z$ are standard deviations in location in x, y, and z. For transport solely by molecular diffusion: $\overline{x} = \overline{y} = \overline{z} = 0$, and $\sigma_x = \sigma_y = \sigma_z = \sqrt{2D_m t}$, where D_m is the molecular diffusion coefficient and t is the time since release of the instantaneous slug. For advection along the x axis with an average fluid velocity of u: $\overline{x} = ut$, $\overline{y} = \overline{z} = 0$, and, $\sigma_x = \sqrt{2D_x t}$, $\sigma_y = \sqrt{2D_y t}$, $\sigma_z = \sqrt{2D_z t}$, where D_x, D_y, and D_z are dispersion coefficients in x, y, and z directions, respectively.

$$M = \int_{-\infty}^{\infty} \int_{-\infty}^{\infty} \int_{-\infty}^{\infty} C \, dx \, dy \, dz$$

is the "three-dimensional puff":

$$C(x, y, z, t) = \frac{M}{8 (\pi t)^{1.5} \sqrt{D_{m,x} D_{m,y} D_{m,z}}} \exp \left\{ \frac{-x^2}{4D_{m,x}t} - \frac{y^2}{4D_{m,y}t} - \frac{z^2}{4D_{m,z}t} \right\} \tag{5.12}$$

As shown later, addition of an advection term simply results in a corresponding displacement of the center of mass. Crank (1979) shows how solutions for finite boundaries, distributed inputs, or continuous sources can also be deduced from the solution for the spike input given in Eq. (5.10a).

The proportionality of the plume spread to $t^{1/2}$, as represented by the particle standard deviation in Eq. (5.11), is characteristic of Fickian diffusion with constant D. However, deviations from this pattern occur in many applications as larger scales of mixing and turbulence are considered. This issue is discussed in Section 5.5 and in the supplemental information at www.wiley.com/college/ramaswami. Similarly, the simple form of the diffusion equation presented above assumes that diffusion coefficients are independent of contaminant concentration. This is a good assumption for common environmental transport applications in which trace quantities of contaminants are diffused through air or water. However, caution is required if aqueous contaminant concentrations exceed a few percent, or if contaminants are present in a concentrated mixture.

Molecular diffusion coefficients for selected compounds in air and water are presented in Chapter 4. Molecular diffusion coefficients in water typically range from $\sim 10^{-6}$ to 10^{-5} cm^2/s. Molecular diffusion coefficients in air are on the order of 10^{-1} cm^2/s. Chapter 4 presents a number of parameter estimation techniques that can be used to estimate diffusion coefficient values at other temperatures and pressures and for compounds with different molecular weights.

5.3.2 Turbulent Mixing in the Atmosphere, Lakes, and Oceans

In many environmental transport situations in air and in water, turbulent "diffusion", that is, contaminant mixing due to mixing of the carrier fluid, occurs much faster than molecular diffusion. For example, in modeling mixing through the height of the atmospheric boundary layer or across a lake, molecular diffusion can be neglected in comparison to turbulent mixing. In contrast, as described in Chapter 4, molecular diffusion is usually the controlling transport mechanism across interfacial layers between soil, vegetation, water, and air and within water droplets. Both processes play a role in other problems. For example, deposition of gas-phase contaminants involves turbulent transport down to the top of a thin laminar layer that coats a liquid or solid surface and then diffusive transport across that laminar layer.

Models of turbulent flows are discussed in the supplement to this chapter (www.wiley.com/college/ramaswami). Here, we discuss how fluid dynamical turbulence affects contaminant transport. Equation (5.3) is equally applicable to laminar or turbulent flows as a description of advective transport. In turbulent flows, however, the velocity components can be viewed as comprised of a mean and a fluctuating term: $u = \overline{u} + u'$. Correspondingly, local contaminant concentrations fluctuate rapidly about a mean value: $C = \overline{C} + C'$. Following the concentration fluctuations is often not feasible or necessary. Instead, transport models track the mean concentration, which is described by applying Reynold's averaging to the advection equation (5.3):

$$\frac{\partial \overline{C}}{\partial t} + \overline{u_j} \frac{\partial (\overline{C})}{\partial x_j} = -\left(\frac{\partial \overline{C' u_j'}}{\partial x_j} \right) \tag{5.13a}$$

The most common procedure for solving Eq. (5.13a) is application of gradient transport theory (also known as first-order closure). In this approach, the terms on the right-hand side of Eq. (5.13a) are treated by analogy with molecular diffusion. For example:

$$\overline{C' u_j'} = -D_{e,x_j} \frac{\partial \overline{C}}{\partial x_j} \tag{5.13b}$$

where D_{e,x_j} is the turbulent or eddy diffusion coefficient in the jth direction and u_j is the jth component of the fluid velocity vector. Substituting Eq. (5.13b) into (5.13a) yields

$$\frac{\partial \overline{C}}{\partial t} + \overline{u_j} \frac{\partial (\overline{C})}{\partial x_j} = \frac{\partial}{\partial x_j} \left(D_{e,j} \frac{\partial \overline{C}}{\partial x_j} \right) \tag{5.13c}$$

Unlike molecular diffusion coefficients, the eddy diffusion coefficients in Eqs. (5.13b) and (5.13c) show directional effects related to velocity fluctuations along the j dimensions, and in addition are not constant, but rather depend on the length scales that characterize mixing by turbulent eddies. For example, $D_{e,z}$ generally increases with height in the atmospheric boundary layer because eddy sizes near the ground are constrained by the surface.

Although widely applied in environmental modeling, gradient transport theory has several important limitations. A fundamental limitation is that the gradient transport hypothesis is not completely generalizable because in buoyantly unstable flows turbulent transport may not actually follow the concentration gradient (Arya, 1999). From a practical point of view, however, the most important limitation is that eddy diffusion coefficients cannot be determined from first principles. Instead, they must be estimated from empirical correlations that are developed from tracer experiments and dimensional considerations. More detailed examination of relationships that are commonly used for estimating turbulent diffusion and dispersion coefficients in air and surface water are presented in Chapters 8 and 10.

Most contemporary models treat turbulent transport as analogous to diffusion and correspondingly use advection only for transport by the mean component of the fluid flow, $\overline{u}$. However, research applications have introduced an alternative approach to modeling turbulent transport, called large-eddy simulation (LES; Moeng, 1984; Sullivan et al., 1996). In LES, the larger turbulent eddies are fully resolved, with transport due to these motions treated as an advective process. In fact, the modeler always has a choice of the temporal and spatial scales over which velocities are averaged. As these scales are increased and more of the temporal and spatial variations in velocity are averaged out, the lost resolution of advection must be replaced by increasing the magnitude of the diffusion or dispersion coefficient. In contrast, finer temporal and spatial disaggregation of the flow field may eliminate the need for some of the modeled dispersion. Dispersion is thus an artificial construct made necessary by our inability to predict the fine scale motions and fluctuations in the velocity field. Deviations from the resolved flow velocities are characterized statistically using relationships shown in Eq. (5.13b), and represented as eddy diffusion or shear dispersion.

5.3.3 Dispersion in Groundwater

The process of dispersion in groundwater transport is physically very different from both turbulent and molecular diffusion, although it is treated mathematically in most models in a manner analogous to molecular diffusion. As water encounters heterogeneities in the subsurface, channeling, entrapment, and bypassing occur (Fig. 5.6), all of which together contribute to what is termed *mechanical dispersion*. As a consequence of mechanical dispersion, some water parcels move faster and some slower than the average groundwater velocity, and parcels carrying various amounts of a contaminant appear to be mixed together.

Mechanical dispersion cannot be separated from effective molecular diffusion in the pore waters (see Chapter 4), and hence the two are combined together to yield a hydrodynamic dispersion coefficient, D_j $\left(L^2 T^{-1}\right)$, where j represents dispersion along the jth dimension:

$$D_j = D_{\text{mechanical},j} + D_{m,\text{eff}} \tag{5.14a}$$

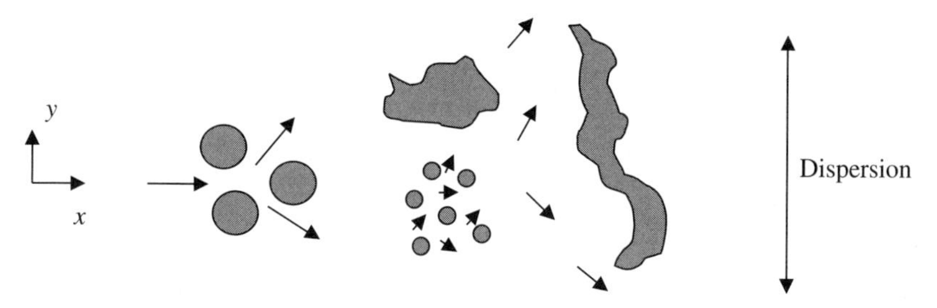

Figure 5.6 Heterogeneities at different scales increase dispersion in the subsurface.

Mechanical dispersion coefficients (like turbulent diffusion coefficients) depend on the characteristics of the flow and in turn on the characteristics of the subsurface. Consequently, values of mechanical (and hence hydrodynamic) dispersion coefficients may differ dramatically between the longitudinal (along the flow direction, D_x or D_L), transverse (horizontal cross flow, D_y or D_T), and vertical directions (D_z), depending on the geologic structure of the soil deposits in the aquifer. Analogous to molecular diffusion, the process of dispersion in groundwater is typically represented by Eq. (5.7), with the molecular diffusion coefficient, D_m, replaced by a direction-specific coefficient, D_j.

Based on field observations, mechanical dispersion coefficients are often modeled as proportional to the average linear groundwater velocity. One can visualize faster flowing waters to be dispersed a greater distance when encountering soils, rocks, and other heterogeneities in the subsurface. Incorporating the relationship between mechanical dispersion and groundwater velocity, the hydrodynamic dispersion coefficient can be rewritten as:

$$D_j = \alpha_j \bar{u} + D_{m,\text{eff}} \tag{5.14b}$$

where the constant of proportionality, $\alpha_j(L)$, is known as the dispersivity of the porous medium and isolates the fixed effects of porous media structure on mechanical dispersion from the variable effects of groundwater velocity. In small-scale porous media systems, such as laboratory columns packed with sand, the longitudinal dispersivity can be assumed to be constant and approximately of the order of the median particle diameter. In field-scale applications, dispersivities are not constant, being a function of the magnitude of subsurface heterogeneities that tend to increase with the scale of the problem as larger scales of heterogeneity impact the transport regime. As such, field-scale dispersion in groundwater is also generally non-Fickian though Fickian (constant α_j) assumptions are often employed. In practice, dispersivity values must normally be determined for a specific site through in situ tracer injection tests. These parameter estimation techniques and methods for accounting for the effects of scale are discussed in detail in Chapter 9.

5.3.4 Dispersion in Streams and Estuaries

In streams and estuaries, contaminant mixing is often controlled by shear dispersion due to velocity gradients rather than turbulent or molecular diffusion. In contrast, turbulent diffusion driven by winds tends to predominate in lakes and bays. Shear dispersion coefficients in the longitudinal direction, $D_{s,x}$ in streams and estuaries may be on the order of 10 to 1000 m^2/s. As illustrated in Figure 5.7, shear dispersion is due to velocity gradients in the direction perpendicular to the flow. In streams and estuaries, strong shear stresses that cause dispersion are developed by rapid mean flow velocities in close proximity to the bed of the channel. Consequently, dispersion coefficients increase with stream flow velocity and decrease with channel depth. This trend with flow can be reversed, however, in very slow moving rivers, with numerous pools and stagnant areas that act more as a sequence of completely stirred tank

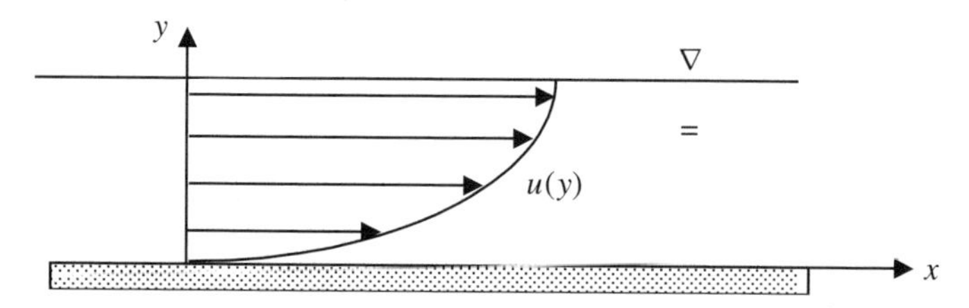

Figure 5.7 Dispersion due to velocity gradients in an open channel.

reactors (CSTRs) (within each of which the dispersion is assumed to be infinite) than the plug flow (or plug flow with dispersion) models customarily assumed.

5.3.5 Comparison of Diffusion and Dispersion Coefficients Across Media

In many environmental modeling applications, the description of transport can be simplified by neglecting all but the dominant mixing process. The mixing processes discussed in this section are all treated as occurring at rates that are directly proportional to the concentration gradient. Therefore, they can be compared strictly on the basis of diffusion or dispersion coefficients, independent of the concentration gradients present in a particular situation. Coefficients for molecular diffusion, turbulent diffusion, and dispersion can differ by several orders of magnitude, as shown in Table 5.3. The values in the table are typical and indicate which mixing processes usually dominate. However, final determinations cannot be made without reference to specific cases because turbulent diffusion and dispersion coefficients depend on the characteristics of the flow.

Throughout the remainder of this book, we use the symbol D for all forms of diffusion and dispersion coefficients. Different subscripts are used for molecular diffusion (D_m), turbulent eddy diffusion (D_e), and shear dispersion (D_s). The symbol

TABLE 5.3 Comparison of Dispersion and Turbulent Diffusion Coefficients in Air and Water

Mixing Process	Diffusion or Dispersion Coefficient (cm^2/s)
Molecular diffusion in water	10^{-6}–10^{-5}
Vertical turbulent diffusion in deep layers of water bodies	10^{-2}–1
Vertical turbulent diffusion in surface layers of water bodies	1–10^{3}
Horizontal turbulent diffusion in surface water bodies	10^{2}–10^{6}
Longitudinal dispersion in streams and estuaries	10^{5}–10^{6}
Field-scale dispersion in porous media	10^{-2}–10
Molecular diffusion in air	10^{-1}
Vertical turbulent diffusion in the atmospheric boundary layer	10^{4}–10^{6}
Horizontal turbulent diffusion in the atmosphere	10^{6}–10^{10}

Sources: Adapted from Chapra (1997), Schnoor (1996), Gelhar et al. (1992), and Stull (1988).

D is used alone and referred to as the *dispersion coefficient* when it is used to represent the combined effects of all sources of mixing in air, surface water, or groundwater, with subscripts used to distinguish directions, such as the longitudinal (D_l or D_x), transverse (D_t or D_y), horizontal (D_h, used when $D_l = D_t$), or vertical dispersion coefficient (D_z).

Although we have opted for consistency in this book, in the environmental modeling literature, different terminology and nomenclature are used by different disciplines. Air pollution scientists and modelers often use the symbol K and the term *diffusivity*, or *eddy diffusivity*, to describe the intensity of mixing that occurs due to turbulent eddies in the atmosphere. Water quality modelers generally use the symbol D for the molecular *diffusion coefficient*, prefer the symbol E for turbulent diffusion coefficients in surface waters, but also refer to E as a *dispersion coefficient* when it combines the effects of turbulent diffusion and shear dispersion. Groundwater modelers use the symbol D for diffusion and dispersion at all scales and refer to it as a dispersion coefficient or *hydrodynamic dispersion coefficient* when applied at the field scale. Groundwater modelers use the term *dispersivity* for the constant α (L) that is multiplied by the average groundwater velocity $\bar{u}$ $\left(L\,T^{-1}\right)$ to yield the dispersion coefficient $\left(L^2\,T^{-1}\right)$.

5.4 ADVECTIVE–DISPERSIVE REACTIVE TRANSPORT

In general, contaminant transport occurs through advection, diffusion, and dispersion. When both advective and dispersive transport processes are significant and reactions and internal sources are present, the applicable mass-balance equation is

$$\frac{\partial C(\mathbf{x}, t)}{\partial t} + \mathbf{u} \cdot \nabla C = \nabla \cdot D\,\nabla C \pm \text{Rxn}(C) + S_v(\mathbf{x}, t) \qquad (5.15)$$

In Eq. (5.15), D represents the tensor of effective dispersion coefficients of the combined mixing processes (e.g., molecular and turbulent diffusion) in multiple dimensions, Rxn is the rate of production or loss of the contaminant through reactions, and S_v is the rate of production of the contaminant due to a source within the control volume. Equation (5.15) is applicable to air, surface water, and groundwater and is called the advection–dispersion–reaction (ADR) equation. Solution of the ADR equation requires specification of the initial and boundary conditions that characterize the nature of chemical contamination in the environment.

Equation (5.15) is written for a generalized system with multidimensional velocity and dispersion components. In practice, the dimensionality of the transport system can often be simplified based on assumptions about the degree of mixing in x, y, and z dimensions, resulting in idealized zero-, one-, two-, and three-dimensional transport systems (for example, see Fig. 1.4). If $\mathbf{u}$, D, Rxn, or S are not constant, Eq. (5.15) must typically be solved using numerical techniques. However, if these parameters are fixed, the advection–dispersion equation can be solved analytically for idealized cases with simple geometries and boundary conditions. Some useful analytical solutions to the ADR equation are presented in the remainder of this section, emphasizing system response to two idealized source conditions: an instantaneous spike input and a continuous input of chemicals.

5.4.1 5.4.1 Comparison of Advection, Dispersion, and First-Order Reaction

Before addressing analytical solutions to the ADR equation, it is useful to understand the relative contributions of advection, dispersion–diffusion, and reaction in contaminant transport. Advection describes the net average forward movement of the contaminant in the fluid flow field, diffusion and dispersion processes cause mixing of the contaminant about its mean position, while reactions transform the contaminant, resulting in removal from the system. The relative importance of advection versus diffusion or dispersion is measured by the Peclet number: $\mathrm{Pe} = \ell u/D$, where ℓ is a characteristic length for the problem at hand. If $\mathrm{Pe} \geq 10$, transport by advection dominates that by diffusion–dispersion, and the system approaches the idealized case of a plug flow reactor (PFR) characterized by zero mixing (see Chapter 1). If $\mathrm{Pe} \leq 0.1$, the system can be treated as completely mixed, that is, as a CSTR. CSTR models were introduced in Chapter 1 and applied in Chapters 3 and 4. In many environmental systems, both advection and diffusion–dispersion are important and a mixed-flow reactor (MFR) ($0.1 < \mathrm{Pe} < 10$) must be considered to fully describe the distribution of chemicals in the environment. Simplification of the advection–dispersion equation may also be possible based on a Damkohler number, which compares chemical reaction rates to transport based on advection:

$$\mathrm{Da}\Big|_{\text{advection}} = \frac{k\ell}{u}$$

or dispersion

$$\mathrm{Da}\Big|_{\text{dispersion}} = \frac{k\ell^2}{D}$$

where k is a rate constant for a first-order reaction. If the Damkohler number is very small, chemical reactions can be neglected compared to transport. Thus, analyses of Peclet and Damkohler numbers enable simplification of the ADR equation, which can then more readily be solved analytically for certain idealized systems, as described next.

5.4.2 Analytical Solutions for Pure Advection and Complete Dispersion: Plug Flow and Completely Stirred Tank Reactor Models

A 1D plug flow reactor model was used in Chapter 1 in the Streeter–Phelps model for biochemical oxygen demand. The model is described by the mass-balance equation:

$$\frac{\partial C}{\partial t} = -u\frac{\partial C}{\partial x} - kC \tag{5.16}$$

which is derived from the 1D form of Eq. (5.15) by neglecting dispersion, assuming the only reaction is a first-order decay process, and assuming there are no internal sources. An instantaneous release of a contaminant spike in a PFR is expected to travel at the average fluid velocity, u, while simultaneously undergoing removal due to reaction. No mixing or dispersion of the contaminant spike occurs in a PFR. When contaminant input occurs continuously at the PFR entrance ($x = 0$), at a fixed chemical concentration, C_0, the constant concentration front moves forward at the average fluid velocity, u. Consistent with the transport dynamics within a PFR, the constant concentration exhibits no dispersion. When chemical transport is balanced by chemical removal through reactions, steady contaminant concentrations are observed in the PFR. Setting the left-hand side (lhs) of Eq. (5.16) to zero, the steady-state solution of the 1D plug flow mass-balance equation yields

$$C(x) = C_0 \exp\left(-k\frac{x}{u}\right) = C_0 \exp(-k\tau) \tag{5.17}$$

where τ is the fluid travel time in the system. The spatial and temporal distributions of contaminants in a PFR in response to instantaneous and continuous inputs are illustrated in Figures 5.8*a* to 5.8*c*. PFR models are frequently used for analysis of chemical transport and fate in drainage channels and wastewater treatment units characterized by well-defined water flow in relatively smooth channels.

In contrast to unmixed PFRs, mixing within a CSTR is assumed to be instantaneous throughout the control volume. Thus, the CSTR may also be considered a zero-dimensional transport system in which concentrations are uniform in all three dimensions. This model was applied for the indoor air pollution example in Chapter 1 and is also commonly used as a simplified model for lakes. Simultaneous multimedia mixed-box models employing the CSTR framework were presented in Chapters 3 and 4. For a single fluid medium, the CSTR mass-balance equation is

$$V\frac{dC}{dt} = QC_{\text{infl}}(t) - QC - kVC + S(t) \tag{5.18}$$

where V is the system volume, Q the volumetric flow rate, and $C_{\text{infl}}(t)$ the influent concentration. No dispersion term appears in Eq. (5.18) because mixing is assumed to be complete and instantaneous. Moreover, there is no concentration gradient ($\partial C/\partial x = 0$), so there is no advection term. Equation (5.18) is a linear, first-order differential equation that can be rearranged to the standard form: $y' + p(t)y = q(t)$. Given the initial condition $[y(t = 0) = y_0]$, the integrating factor solution of this equation is

$$y(t) = y_0 \exp\left[-\int p(t)\,dt\right] + \exp\left[-\int p(t)\,dt\right]\left\{\int_0^{\xi} \exp\left[+\int p(t)\,dt\right] q(t)\,dt\right\} \tag{5.19}$$

In the case of the CSTR with an initial concentration of zero $[C(t = 0) = 0]$, and a constant influent concentration $[C_{\text{infl}}(t) = C_{\text{infl}}]$ and loading $[S(t) = S]$ for $t > 0$, the integrating factor formula leads to the solution:

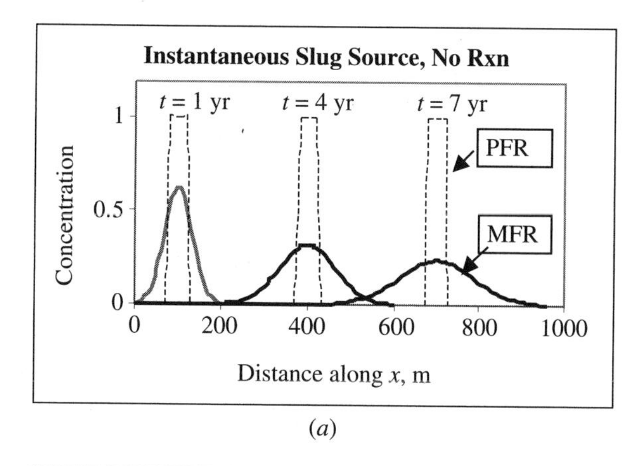

(*a*)

(*b*)

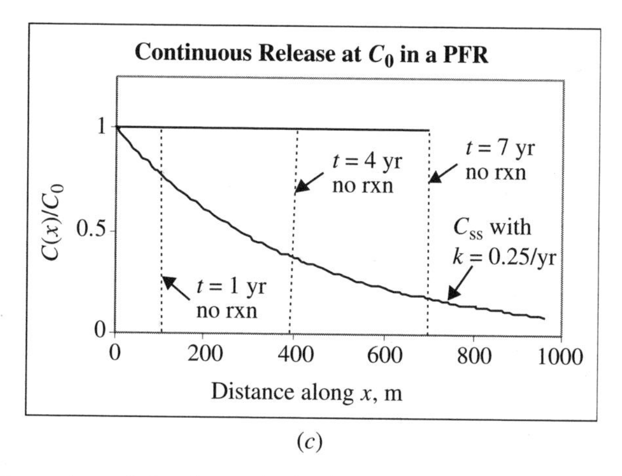

(*c*)

Figure 5.8 Comparison of plug flow and mixed-flow reactor models for transport and decay of a contaminant in a 1D groundwater system. The linear groundwater flow velocity is $v = 100$ m/yr; and the dispersivity is $\alpha = 5$ m. Reactive transport considers first-order decay with $k = 0.25$/yr. (*a*) One-dimensional transport without reaction in response to a short-duration release of 50 g of contaminant. (*b*) One-dimensional transport with reaction in response to a short-duration release of 50 g of contaminant. (*c*) One-dimensional transport with and without reaction in response to a continuous release of contaminant (PFR only).

CSTR
$$C(t) = \frac{QC_{\text{infl}} + S}{Q + kV}\left\{1 - \exp\left[-\left(\frac{Q}{V} + k\right)t\right]\right\} \tag{5.20}$$

The integrating factor formula for the CSTR can be used to model responses for a variety of time-dependent loadings, some of which are explored in the problems posted on the course web site. Models employing single CSTRs and/or a series of CSTRs have been used to describe chemical transport and fate in a variety of environmental systems. A CSTR in series model is used in chapter 10 to simulate watershed runoff and streamflow.

5.4.3 Analytical Solutions for Mixed Advection and Dispersion: Mixed-Flow Reactor Models

A mixed-flow reactor (MFR) (also referred to as a plug flow with dispersion reactor, or PFDR) is intermediate between the plug flow and CSTR cases. It includes both advection and diffusion or dispersion. MFR models are indicated when $0.1 < \text{Pe} < 10$. Many environmental systems of interest, for example, transport in the ambient atmosphere, in rivers, and in groundwater, fall within the domain of an MFR. MFR models can include dispersion in one, two, and three dimensions, depending on the degree of mixing along each dimension.

The 1D mass-balance equation for an MFR with no internal source is

$$\frac{\partial C}{\partial t} = -u\frac{\partial C}{\partial x} + D_x\frac{\partial^2 C}{\partial x^2} - kC \tag{5.21}$$

The remainder of this subsection examines analytical solutions for this equation for idealized cases of an instantaneous (spike) input and a steady continuous input.

Time-Dependent Response of a 1D MFR to a Spike Input In an unbounded domain with a spike input at $x = 0$ and $t = 0$, the solution for the mixed reactor model can be deduced from the solution to the 1D diffusion equation for the same conditions [Eq. (5.7)]. The Gaussian profile [Eq. (5.9)] evolves just as it does in the case of pure diffusion, but the centroid of the distribution is advected down wind or down stream with the flow such that:

$$\mu_x = ut \quad \text{and} \quad \sigma_x = \sqrt{2D_x t} \tag{5.22}$$

where D_x represents the diffusion or dispersion coefficient appropriate for the system of interest. Figure 5.8*a* illustrates this response. Gaussian concentration distributions in one, two, and three dimensions generated by advective–dispersive processes are shown in Table 5.2. If the contaminant simultaneously undergoes first-order decay, the concentration profile for a spike input in a 1D unbounded domain evolves as:

$$C(x, t) = \frac{m_a}{2\sqrt{\pi D_x t}}\exp\left[\frac{-(x - ut)^2}{4D_x t} - kt\right] \tag{5.23}$$

where m_a is the mass per unit cross-sectional area (see Fig. 5.8*b*). The Gaussian "puff" equations in Table 5.2 indicate that dispersion causes the contaminant concentration spike to spread out with a corresponding decrease in peak height, as travel time increases. When removal by reaction is included, a further decrease in contaminant concentration and mass is indicated in Eq. (5.23). The equations in Table 5.2 have been used effectively to describe response to instantaneous releases of chemicals in the atmosphere (3D puff), in groundwater (2D puff), and in streams (1D puff). At large time t, after instantaneous release of the contaminant, contaminant concentrations in the environment are expected to attenuate to zero at all locations because of dilution caused by dispersion and/or removal by reaction. The response to a spike input in a 1D MFR is compared with that in a PFR in Figures 5.8*a* and 5.8*b*.

Steady-State Response to a Continuous Chemical Input Steady-state contaminant concentrations occur in environmental systems with time invariant properties and constant, continuous inputs of chemical mass. We examine the idealized case of 1D advection–dispersion [Eq. (5.21)] with a single continuous source located at $x = 0$. If the source results in a constant concentration condition $C(x = 0) = C_0$, for all $t > 0$, we get a type 1 boundary condition. If, on the other hand, the source results in chemical release at a fixed emission rate (or loading rate), S (mass/time), at $x = 0$, one obtains a constant flux or type 3 boundary condition. In both cases, the steady-state solution is obtained by solving the differential equation written in Eq. (5.21) after setting the lhs to zero and applying appropriate boundary and initial conditions. An illustration of the solution process is presented next.

With $\partial C/\partial t = 0$, the steady-state mass-balance equation for a mixed-flow reactor [Eq. (5.21)] is a homogeneous second-order ordinary differential equation. The equation can be solved by assuming the solution has the form $C(x) = e^{rx}$, and substituting this expression into the original equation:

$$0 = -ure^{rx} + D_x r^2 e^{rx} - ke^{rx} \tag{5.24a}$$

Dividing through by e^{rx} results in the characteristic equation $D_x r^2 - ur - k = 0$, which has the roots:

$$r_1 = \frac{u}{2D_x}\left(1 + \sqrt{1 + \frac{4kD_x}{u^2}}\right) \qquad r_2 = \frac{u}{2D_x}\left(1 - \sqrt{1 + \frac{4kD_x}{u^2}}\right) \tag{5.24b}$$

The complete solution to the equation is

$$C(x) = B_1 \exp(r_1 x) + B_2 \exp(r_2 x) \tag{5.24c}$$

where the values of the constants B_1 and B_2 depend on the boundary conditions specified for the particular problem.

The steady-state 1D mixed-flow model provides an idealized description of transport in an estuary (Chapra, 1997) as depicted in Figure 5.9. For the case of an infinite estuary with a continuous source S, at $x = 0$, the boundary conditions are $C(x = \pm\infty) = 0$ and the concentration must be continuous through the point $x = 0$. These conditions imply that $B_1 = B_2 = C_0$, and

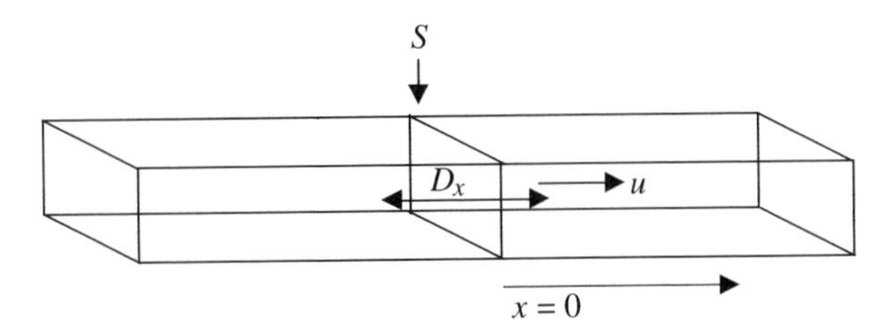

Figure 5.9 Mixed-flow model of an estuary with a continuous source located at $x = 0$.

$$\begin{aligned} C(x \leq 0) &= C_0 \exp(r_1 x) \\ C(x > 0) &= C_0 \exp(r_2 x) \end{aligned} \tag{5.24d}$$

where r_1 and r_2 are the roots determined above, and C_0 is determined by balancing the fluxes through $x = 0$:

$$S + \left(uAC_0 - D_x A \frac{dC}{dx} \right)_- = \left(uAC_0 - D_x A \frac{dC}{dx} \right)_+ \tag{5.24e}$$

The minus sign in Eq. (5.24e) indicates the flux coming into the point from the left, and the plus sign indicates the flux leaving to the right. Substituting the expressions for $C(x)$ into Eq. (5.24e) and rearranging, we get

$$C_0 = \frac{S}{Q} \frac{1}{\sqrt{1 + (4kD_x/u^2)}} \tag{5.24f}$$

EXAMPLE 5.3

Consider an estuary with a cross-sectional area of $A = 1000$ m^2 and a volumetric flow rate of $Q = 60$ m^3/s. A confined hog production operation produces $S = 5000$ kg/day of total ammonia–nitrogen (NH_3–N) that is discharged directly into the estuary, 30 km upstream from the coast. The NH_3–N is removed from the estuary by conversion to NO_2 and algal nitrogen with a rate constant of $k = 0.8$/day.

Solution 5.3 Figure 5.10 shows the resulting total ammonia–nitrogen concentration in the estuary predicted with the mixed-flow model, assuming the discharge from the hog operation is the only source. The results are shown for three different values of the estuary dispersion coefficient, ranging from $D_x = 150$ to 450 m^2/s.

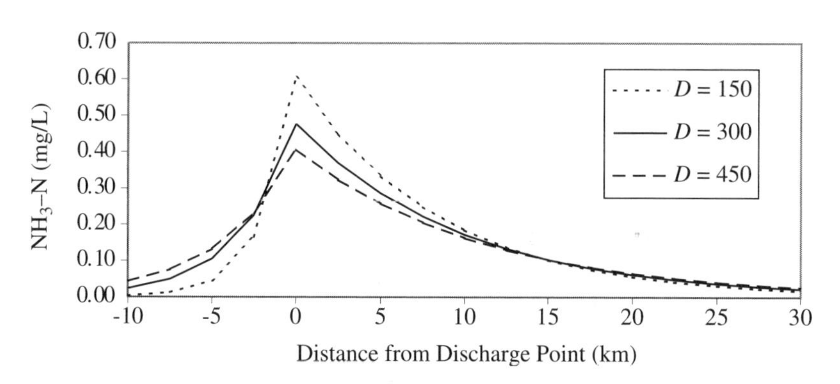

Figure 5.10 Total ammonia–nitrogen predicted with a mixed-flow model as a function of the estuary dispersion coefficient, D_x (m^2/s) and of the distance downstream from the discharge point.

As shown, more dispersion results in a lower concentration at the source discharge point, but relatively more mass is pushed away from the source. This is particularly evident in the higher predicted concentrations upstream of the source when D_x is larger.

Steady-state solutions in MFR systems have been used in a variety of applications, with simplifying assumptions appropriate for the particular setting and physical conditions. O'Connor (1962) and Di Toro (1972) solved a 2D version of the ADR equation for a continuous point-source discharge in a lake or other relatively open water body, assuming isotropic dispersion ($D_l = D_t = D_h$). Hunt (1978), Wilson and Miller (1978), and Ramaswami and Small (1994) applied the equation to steady-state groundwater plumes, allowing for anisotropic dispersion, but assuming a conservative solute ($k = 0$). Applications to long-range atmospheric transport of pollutants are discussed in Section 5.6.

Time-Dependent Response to a Continuous Chemical Input A steady-state solution to the 1D mixed-flow reactor equation was applied in the estuary problem because of the relatively fast rates of advection, dispersion, and reaction observed in many surface waters. In contrast, these processes occur very slowly in groundwater; so in this medium, dynamic solutions that describe how contaminant plumes evolve over time are often of greatest interest.

A long column packed with soil or sand with a contaminant and water flow introduced at one end can be treated as a semi-infinite medium for purposes of analyzing 1D advective–dispersive transport. Assuming instantaneous mixing throughout the cross section of the cylinder, with 1D advection and dispersion occurring in the longitudinal direction, Eq. (5.21) applies. Two boundary conditions and one initial condition are required to solve this partial differential equation. The Ogata–Banks equation (Bear, 1979) provides the solution for the case with no chemical decay ($k = 0$), where a type 1 boundary condition applies at the entrance to the cylinder. The boundary and initial conditions for this case are

$$C(x, 0) = 0 \qquad \text{for all } x \text{ at } t = 0$$

$$C(0, t) = C_0 \qquad \text{for } x = 0 \text{ and } t > 0$$

$$C(\infty, t) = 0 \qquad \text{for } x = \infty \text{ and } t > 0$$

With these boundary and initial conditions, Laplace transforms (Churchill, 1944) can be employed to solve Eq (5.21), yielding:

$$C(x, t) = \frac{C_0}{2}\left[\operatorname{erfc}\left(\frac{x - vt}{2\sqrt{D_L t}}\right) + \exp\left(\frac{vx}{D_L}\right)\operatorname{erfc}\left(\frac{x + vt}{2\sqrt{D_L t}}\right)\right] \tag{5.25}$$

where erfc() is the complementary error function.

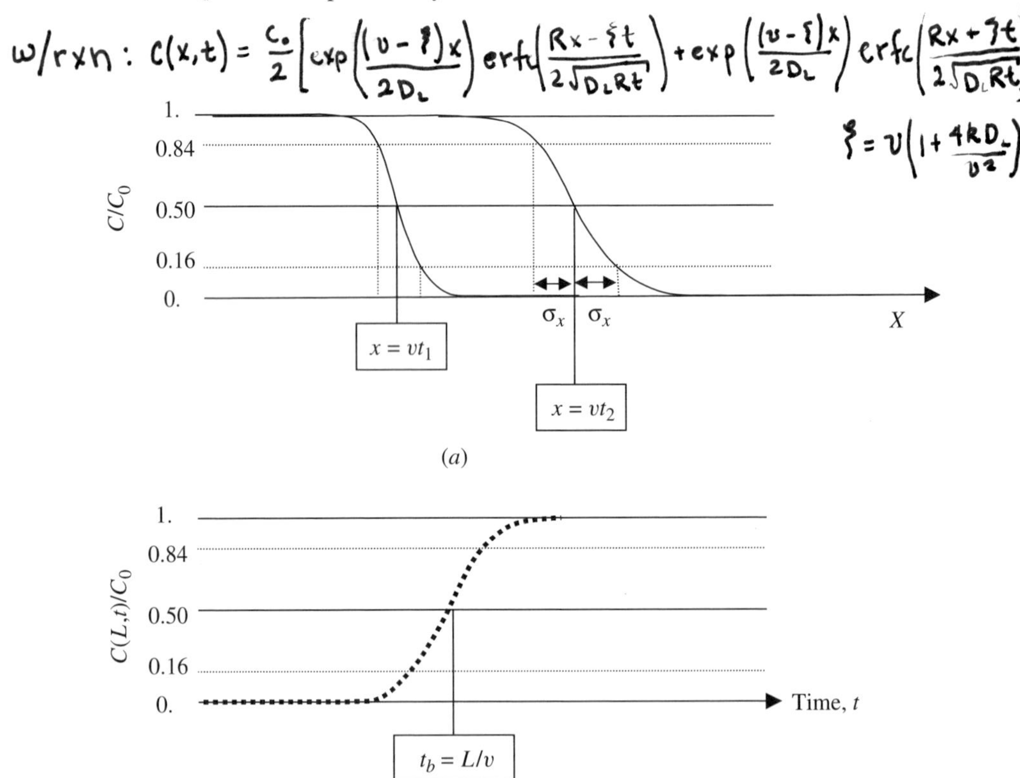

Figure 5.11 Concentration patterns in the subsurface in response to continuous injection of a pollutant at concentration C_0 at $x = 0, t > 0$. (*a*) Tracer concentrations along the x axis at times t_1 and t_2. (*b*) Tracer concentrations as a function of time at an observation well located a distance L from the origin.

The second bracketed term in Eq. (5.25) can be neglected if the Peclet number is larger than 1, which is often the case in groundwater. In the absence of chemical reactions, Eq. (5.25) describes propagation of a concentration "front" in the subsurface, with the degree of dispersion of the front increasing with travel time as shown in Figure 5.11. At any time t, nonreactive chemical (tracer) concentrations in groundwater behind the front are the same as the source concentration, C_0, while the tracer concentration in groundwater far ahead of the front is essentially zero (Fig. 5.11*a*). At steady state ($t = \infty$), $C = C_0$ for all x. Tracer concentrations measured at a fixed location, for example, at an observation well a distance L from the source, are zero initially and then increase as the tracer flows closer to the well (Fig. 5.11*b*), eventually reaching the steady-state concentration of C_0. The breakthrough time at a given location, such as at an observation well, is defined as the time when the tracer concentration at that point is one-half of the source concentration.

EXAMPLE 5.4

A nonreactive tracer is continuously injected at a concentration of 10 mg/L into a cylindrical column that is 20 cm^2 in cross-sectional area and 60 cm long and packed with homogeneous sand. The porosity of the sand is 0.33, its median particle diameter is 0.2 mm, and the steady-state water flow rate through the column is 20 L/h. Determine the breakthrough time for the tracer at the bottom of the column. Determine the tracer concentration in the effluent from the column at $t = 0.2$ h and $t = 1$ h after tracer injection was initiated.

Solution 5.4

$$\text{Pore velocity in the column} = v = \frac{Q}{nA} = \frac{2000\ \text{cm}^3/\text{h}}{0.33 \times 20\ \text{cm}^2} = 300\ \text{cm/h}$$

$$\text{Breakthrough time} = t_b = \frac{x}{v} = \frac{60\ \text{cm}}{300\ \text{cm/h}} = 0.2\ \text{h} = 12\ \text{min}$$

By definition, the effluent concentration will be $0.5C_0 = 5$ mg/L at the breakthrough time, $t_b = 0.2$ h. This same result can be obtained by using Eq (5.25) for $x = 60$ cm and $t = 0.2$ h, since erfc(0) = 1 (see supplemental information for Chapter 9 at www.wiley.com/college/ramaswami).

To apply Eq. (5.25) for $x = 60$ cm and $t = 1$ h, an estimate of the longitudinal dispersion coefficient, D_L, is needed. As discussed in Section 5.3.3, to a first approximation $\alpha_L \sim$ median particle diameter = 0.02 cm, so $D_L = \alpha_L v \sim 0.02$ cm $\times$ 300 cm/h = 6 cm^2/h. The concentration in the column effluent after 1 h is then

$$C(60\ \text{cm}, 1\ \text{h}) = \frac{C_0}{2}\,\text{erfc}\left(\frac{x - vt}{\sqrt{2D_L t}}\right) = \frac{10\ \text{mg/L}}{2}\,\text{erfc}\left(\frac{60 - 300\ \text{cm}}{\sqrt{2 \times 6\ \text{cm}^2/\text{h} \times 1\ \text{h}}}\right)$$

$$= \frac{10\ \text{mg/L}}{2}\,\text{erfc}(-69) \approx 10\ \text{mg/L}$$

The effluent concentration is almost equal to the influent concentration of 10 mg/L at $t = 1$ h.[2]

Although the assumptions of homogeneous systems with spike and continuous chemical inputs that were made in this section represent idealized situations, the analytical solutions available for these cases often provide useful insights that are transferable to real environmental systems. These ideal models are also commonly used to estimate advection and dispersion parameters in tracer studies. This is illustrated in chapter 9 for groundwater and chapter 10 for surface waters. On the other hand, in real environmental systems the assumptions of uniform and constant parameter values for advection, dispersion, and reactions are often inapplicable. Temporal variability and inhomogeneity in source and transport parameters are addressed in subsequent sections of this chapter.

5.5 DISPERSION AS A SCALE-DEPENDENT PROCESS

5.5.1 A Second Look at Dispersion in the Estuary Model

In the example of NH_3–N discharge to an estuary presented in the previous section, a 1D, steady-state advective–dispersive model was assumed. That is, there is a constant flow and velocity from the inland (tidal) river to the sea, with a dispersion coefficient to account for the mixing that occurs in route. However, this is not what actually occurs in a tidal river and estuary. While there is net freshwater flow from the land to the sea, tidal oscillation causes the water to slosh back and forth approximately twice a day with the tidal cycle. The flow is increasingly dominated by the oscillating tidal component as the water moves from an upland river to a tidal river, and eventually through the mouth of the estuary.

The relationship between the oscillating tides in an estuary and alternative approaches for modeling their effects are illustrated in Figure 5.12. One could, as in Figure 5.12*a*, build an *intratidal* model that keeps track of the dynamic flow and velocity at each location in the estuary, transporting the pollutant molecules back and forth until they reach the ocean. Najaran and Harleman (1977) utilized this approach. Chapra (1997, Section 15.2) describes how a sinusoidal flow profile can be used to approximate advection in an estuary during ebb (low) and flood (high) tide. With

[2] $\text{Erfc}(-x) = 2 - \text{erfc}(x)$. Therefore, $\text{erfc}(-69) = 2 - \text{erfc}(69) = 2 - 0 = 2$.

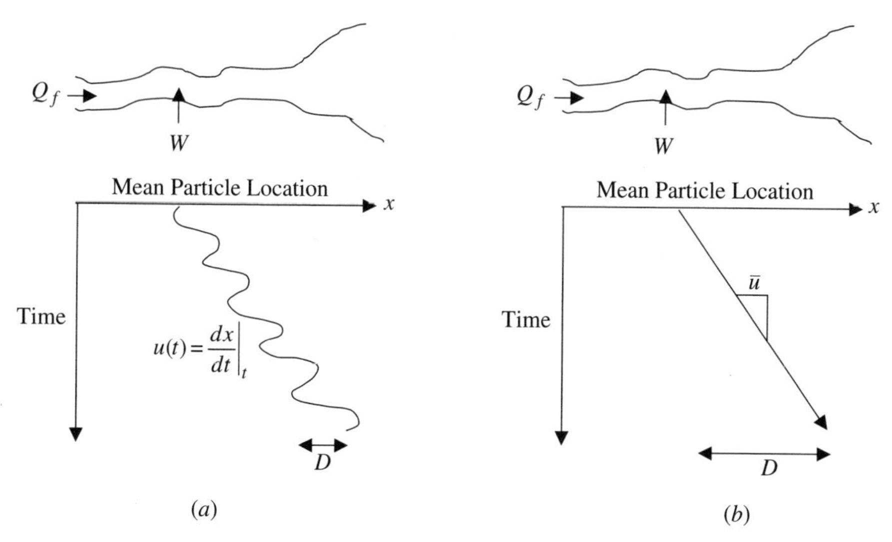

Figure 5.12 Alternative (*a*) intratidal and (*b*) intertidal estuary water quality models and implications for the model dispersion coefficient, D.

this type of model, the concentration profile is tracked throughout the tidal cycle, with higher pollutant concentrations upstream of the discharge point at flood tide and lower concentrations at ebb tide. The dispersion coefficient in this case accounts only for the mixing and agitation that occurs in addition to the oscillating advection profile. Depending on the strength of the tidal currents, channel characteristics, and the magnitude of the vertical mixing and stratification that occurs when the freshwater flow meets the saline ocean, this dispersion can be quite high. However, it is small compared to the dispersion that must be utilized when the analysis is aggregated to the *intertidal* model shown in Figure 5.12*b*.

The steady-state intertidal model in Figure 5.12*b* does not account for variations within a tidal cycle, and only the net advection rate is included. As such, the dispersion coefficient must account for the mixing effects of the oscillating tidal flow, which is now left out of the model. Such models are calibrated and compared to data observed at a single point in the tidal cycle, such as high, mid, or low tide or in some cases data averaged over the tidal cycle. Use of such a model signals that we care first and foremost about temporally averaged results, such as those shown in Figure 5.10. The trade-off for losing the temporal detail provided by the dynamic, intratidal model is the relative simplicity of the steady-state modeling approach.

Typical values for the dispersion coefficient for the oscillating flow model in Figure 5.12*a* are similar to those found in rivers, on the order of 30 m^2/s. Typical values of estuary dispersion for the steady-state model in Figure 5.12*b* are an order of magnitude larger, $\sim$300 m^2/s, as assumed in Example 5.3.

So, what is the correct value of the dispersion coefficient for the estuary in Figure 5.12? There is no single, correct value. The dispersion coefficient depends on the desired scale of temporal aggregation and the associated modeling strategy and structure employed. The more the transport in the model is aggregated over larger temporal and spatial scales, the larger the dispersion coefficient must be to account for lost spatial and temporal resolution in the modeled advection. The remainder of this section provides empirical results, analytical tools and examples to illustrate this important principle, which applies across all media.

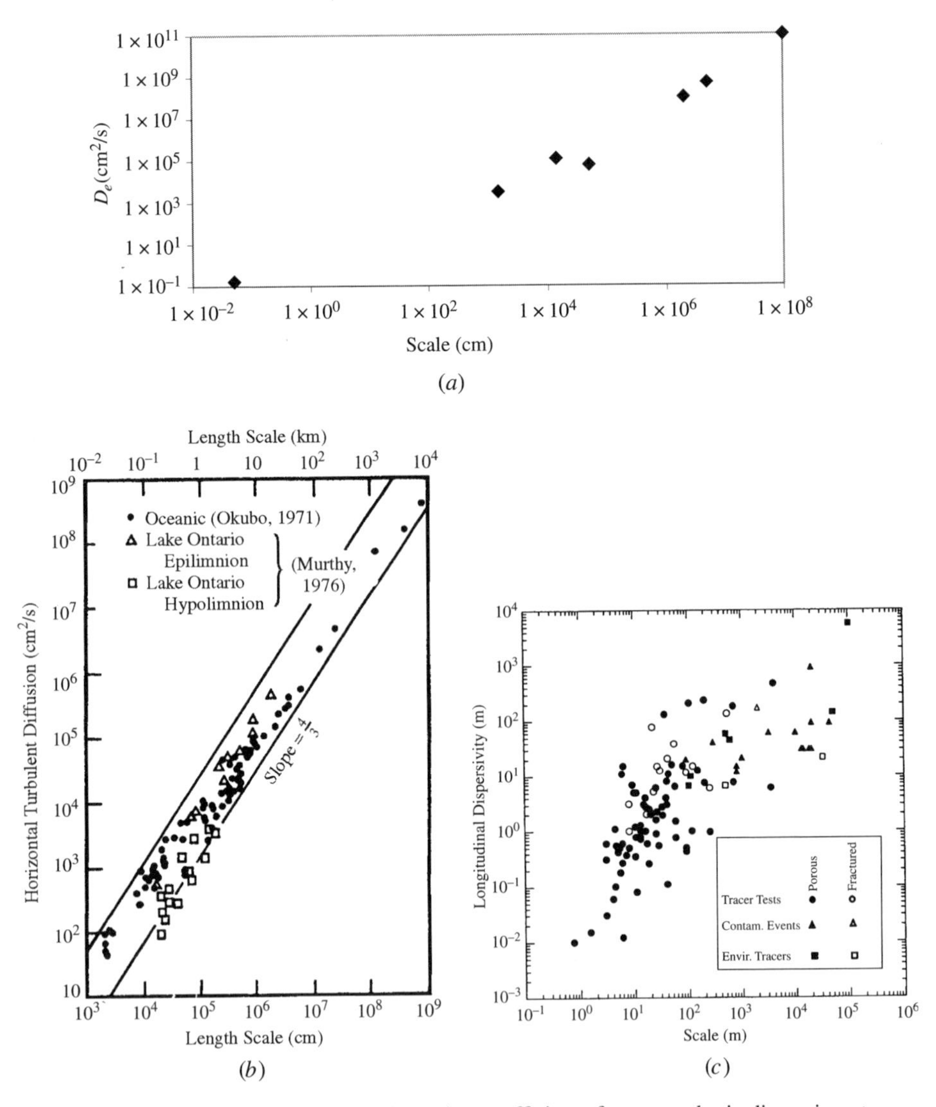

Figure 5.13 (*a*) Values of horizontal dispersion coefficients for atmospheric dispersion at various scales. (Adapted from Richardson, 1926.) (*b*) Relationship between horizontal dispersion coefficients and length scale in surface waters as summarized by Chapra (1997). (*c*) Longitudinal dispersivity versus scale of observation in groundwater systems (Gelhar et al., 1992).

5.5.2 Empirically Observed Dispersion-Scale Relationships

The critical dependence of dispersion on time and space scales is recognized in all environmental transport media. In 1926, Richardson reported that in the atmosphere, horizontal diffusion coefficients determined from the spread of puffs of particles depended upon the size of the puffs involved. In particular, he found that the dispersion coefficient increased according to the length scale of the puffs raised to the 4/3 power. Later, Obukhov (1941) derived a theoretical basis for the 4/3 power law based on the energy transfer of eddies in the inertial range, and this approach has since been widely used to evaluate scale dispersion effects in the atmosphere (e.g., Pasquill, 1962; Gifford, 1968). Studies in oceans and lakes reveal a similar scale effect for dispersion in surface waters (Okubo, 1971; Murthy, 1976; Chapra, 1997).

Dispersion-scale relationships in groundwater have also been studied in great detail during the past few decades (e.g., Gelhar et al., 1992; Dagan, 1986). Here, rather than "different scales of eddies," attention is focused on the spatial structure and distribution of hydraulic conductivity, as determined by subsurface geology and stratigraphy. Statistical methods are presented in Chapter 9 for evaluating dispersion and its dependence on variations in hydraulic conductivity at different scales.

Figure 5.13 summarizes empirical observations of scale effects on dispersion in all three environmental media. Figure 5.13*a* shows the original results of Richardson (1926) for the atmosphere. Figure 5.13*b* shows observations for surface waters summarized by Chapra (1997). Figure 5.13*c* shows longitudinal dispersivities for groundwater systems reported by Gelhar et al. (1992). (The dispersivity values can be multiplied by typical groundwater velocities of 10^{-6} to 10^{-5} m/s to obtain dispersion coefficients with units of m^2/s.) While the scale effects and the dispersion coefficients themselves differ greatly across environmental systems, the need for a higher value of D to model dispersion at larger spatial scales is common to all media. The supplemental information at www.wiley.com/college/ramaswami presents one approach for evaluating scale effects on dispersion, based on a "not-so-random" analogy to the random-walk model presented in Section 5.3.

5.6 TWO-DIMENSIONAL ADVECTIVE–DISPERSIVE SYSTEMS—LONG-RANGE ATMOSPHERIC TRANSPORT AND OTHER APPLICATIONS

The problem of long-range air pollution transport came to world attention in the late 1970s due to recognition of large-scale acid aerosol and acid rain problems. Models were developed for North America and Europe to simulate regional and continental-scale transport of sulfur and nitrogen pollutants and their transformation products (Eliassen, 1980; Johnson, 1983; Fisher, 1983; Hidy, 1984). Because air pollutants usually mix throughout the atmospheric mixing layer within the first 10 to 100 km of transport, vertical gradients in concentration were often ignored; this is implicit in the use of two-dimensional (horizontal) advective–dispersive models for long-range transport.

Figure 5.14 illustrates two of the early approaches taken in long-range transport modeling. The dynamic Lagrangian trajectory and steady-state long-term aggregate

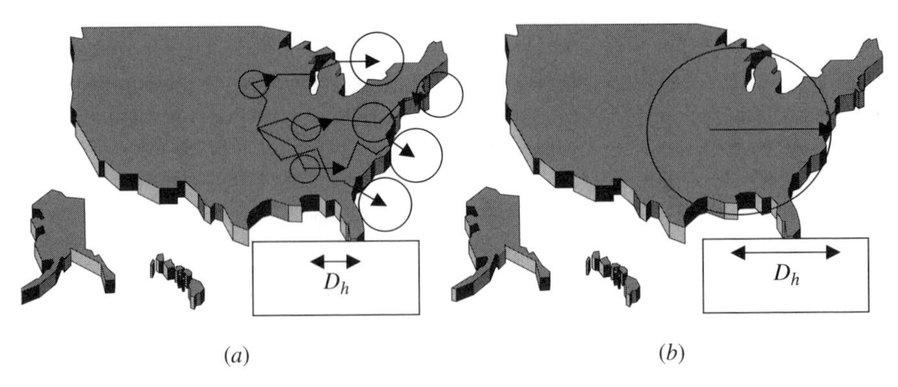

Figure 5.14 (*a*) Individual trajectory and (*b*) long-term aggregate models for long-range atmospheric transport, and implications for the horizontal dispersion coefficient, D_h.

models are applied to time scales of hours versus weeks to months, respectively, and correspondingly use dramatically different dispersion coefficients.

In Figure 5.14*a*, a dynamic, *Lagrangian trajectory* model is depicted. This model, often referred to as a trajectory puff model, simulates the release of a mass of pollutants every few hours (e.g., every 6 h) and tracks the transport and chemical fate of the resulting puffs as they are transported by the mixed-layer winds, or trajectories. For a continuous emission rate S ($M\ T^{-1}$), the simulated mass release M is equal to S multiplied by the time between releases. Dispersion occurs along the trajectories due to wind shear and turbulent eddies, causing the puff to spread with travel time and distance downwind. The 2D version of the dynamic solution for an instantaneous release (see Table 5.2) is used to compute the concentration at any point and time downwind of the source. For pollutant A undergoing first-order transformation and deposition with total removal rate k in an unbounded atmosphere, the puff equation is

$$C^{\mathrm{A}}(x, y, t) = \frac{M}{4\pi z_i \sqrt{D_\ell D_t}\, t} \exp\left\{-\frac{[x - x^*(t)]^2}{4D_\ell t} - \frac{[y - y^*(t)]^2}{4D_t t} - kt\right\} \qquad (5.26)$$

where z_i is the height of the mixed layer, D_ℓ and D_t are the longitudinal and transverse dispersion coefficients, and $x^*(t)$ and $y^*(t)$ denote the location of the trajectory endpoints, that is, the center of mass of the puff, at the corresponding travel times.

Implementation of Eq. (5.26) for long-range transport requires accounting for numerous sources, releases at multiple times, and multiple trajectory endpoints. To determine the average concentration at a receptor location resulting from all of the releases from many sources over an extended time period (e.g., a month or a year), the concentrations computed at that location are summed over all sources and averaged over the period of interest. A linear superposition assumption is thus utilized.

An alternative approach is the steady-state, or average, *net transport* model shown in Figure 5.14*b*. Here all of the individual wind trajectories that leave a source over the period of interest are averaged and an effective steady-state calculation is implemented. The steady-state model uses the average advection computed from many individual trajectories, with dispersion coefficients that incorporate both the dispersion along each trajectory *and* the effective dispersion that results from combining the many divergent trajectories into a single, long-term average wind vector.

The steady-state solution for a continuous release of pollutant A in a two-dimensional advective–dispersive system is given by:

$$C^A(x, y) = \frac{S}{2\pi z_i \sqrt{D_\ell D_t}} \exp\left[\frac{xu}{2D_\ell}\right] K_0\left[\left\{\frac{x^2}{D_\ell}\left(k + \frac{u^2}{4D_\ell}\right) + \frac{y^2}{D_t}\left(k + \frac{u^2}{4D_\ell}\right)\right\}^{1/2}\right] \tag{5.27}$$

where S is the continuous emission rate, z_i is the height of the mixed layer, and the coordinates x and y are aligned with and perpendicular to, respectively, the direction of the average wind velocity u. The expression $K_0[\ ]$ denotes a modified Bessel function of the second type, of order 0. The K_0 function is equal to infinity when the argument is 0 but rapidly approaches zero as the argument increases. Numerical methods for evaluating Bessel functions are provided in Abramowitz and Stegun (1965) and are included in the libraries of many computer math and spreadsheet packages.

What are the appropriate dispersion coefficients for the alternative modeling approaches shown in Figure 5.14? This problem is analogous to the estuary modeling problem posed in the previous section. The value of D used with Eq. (5.27) *must be larger* than that used with Eq. (5.26) since the steady-state model averages out the advection detail of the dynamic model. This is confirmed by the values of D found in the literature for these models, examples of which are summarized in Table 5.4.

TABLE 5.4 Dispersion Coefficients for Alternative Long-Range Atmospheric Transport Models

Model	References	Approach	Dispersion Coefficient (m^2/s)
CAPITA[a]	Patterson et al. (1981)	Equation (5.26) Constant D	$\sim 10^5$
ACID[b]	Small (1982); Samson and Small (1984)	Equation (5.26) $D \propto t$	$\sim 1.5 \times 10^5$ at $t = 3$ days
Fay and Rosenzweig	Fay and Rosenzweig (1980); Fay et al. (1985, 1986)	Equation (5.27) Constant D	2×10^6 to 5×10^6

[a]Center for Air Pollution Impact and Trend Analysis
[b]Atmospheric Contributions to Interregional Deposition

5.7 MODELING TRANSPORT IN INHOMOGENEOUS ENVIRONMENTAL SYSTEMS

At best, the CSTR, PFR, and MFR models provide idealized approximations of environmental systems. Unlike the reactors used in chemical engineering, lakes, streams, watersheds, airsheds, and even the rooms of a building are not homogeneous. The factors that govern transport and mixing in environmental systems are neither steady in time nor spatially uniform. Inhomogeneity in environmental systems is most commonly dealt with in environmental models by segmenting or discretizing the domain into cells and time steps over which conditions may be treated as constant. The size and number of cells and time steps used depends on the variability in the system, as well as the trade-off between model accuracy and requirements for data and computational resources.

In this section, we introduce the finite segment or cell method as one numerical solution technique for discretized models. Finite cell equations treat the environment as a series of interconnected CSTRs, with advection and dispersion between reactors and source emissions and reactions within them. The approach applies most naturally to multicompartment indoor air pollution problems, where each individual room can be treated as a separate control volume or reactor, with air circulation between rooms serving as the source of advection and dispersion. However, the approach can also be applied to surface waters, groundwater, and the ambient atmosphere, where the boundaries between reactor cells are imaginary or "conceptual," created to provide an idealized, discrete representation of the continuous environmental system.

The finite cell approach assumes concentrations are completely mixed within each segment, with step changes in concentration between cells. This leads to a set of numerical equations (and difficulties) that are very similar to those encountered in the finite difference method in which concentrations are tracked at points on a spatial grid. More detailed descriptions of finite difference methods and other numerical techniques are presented in Chapter 6. In the remainder of this section, we illustrate the finite cell approach with applications to one-dimensional steady-state transport problems for surface water quality modeling, as first derived by Thomann (1963).

Figure 5.15 shows a spatially segmented domain for a 1D river or estuary model of uniform cross section, with both advection and dispersion in the longitudinal

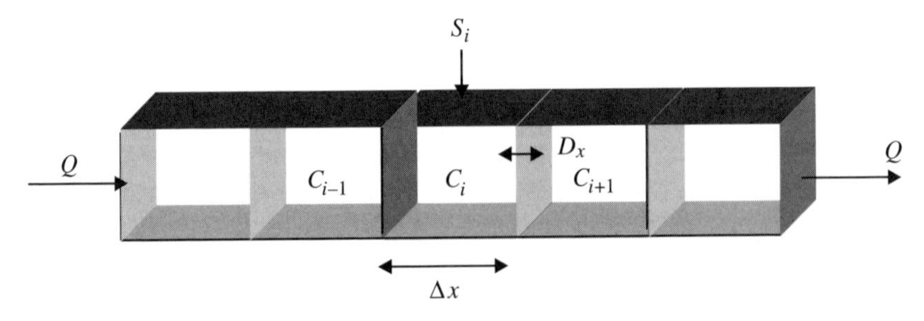

Figure 5.15 Five cells of a spatially segmented domain for an estuary model.

direction. The cells of the domain are assumed to be a series of completely mixed control volumes. Advective transport from one segment to the next is determined from an estimate of the concentration at the interface between the two segments: $C_{i,j} = w_- C_i + w_+ C_j$, where $w_- + w_+ = 1$. The concentration gradient required to calculate dispersion is approximated by the difference between the concentrations in two adjacent segments divided by the distance between the midpoints of the segments: $(C_j - C_i)/\Delta x$. Assuming the cell volumes, dispersion coefficients, and degradation rate constants are time invariant and spatially uniform, the mass balance for segment i is then

$$V_i \frac{dC_i}{dt} = QC_{i-1,i} - QC_{i,i+1} + D_x A \frac{C_{i-1} - C_i}{\Delta x} + D_x A \frac{C_{i+1} - C_i}{\Delta x} - k_i V_i C_i + S_i \tag{5.28}$$

In Eq. (5.28), $V_i = \Delta x\ A$ is the volume of cell i, and S_i is the rate of release of the contaminant from a source located in cell i. The term $D_x A/\Delta x$ has units of flow ($L^3\ T^{-1}$) and is sometimes called the exchange flow because it represents a flow increment between adjacent cells, equal in both directions, that is added to the net flow from one cell to the other. An equation similar to 5.28 is written for each of the N segments in the domain. Because the concentration in each cell depends on those in the two adjacent cells, the equations must be solved simultaneously.

Common choices for the weights required to estimate the interface concentrations are $w_- = 1$, $w_+ = 0$ (backward differencing) and $w_- = w_+ = 0.5$ (central differencing). When the mass-balance equations are solved, backward differencing has the advantage that it is inherently stable (i.e., the results do not oscillate artificially or suddenly diverge from the true solution), whereas central differencing is not inherently stable. However, solutions obtained with backward differencing may exhibit artificially enhanced dispersion. The amount of artificial or "numerical" dispersion introduced by backward differencing depends on the magnitude of advective transport in the system and the length chosen for the control volume: $D_n = (\Delta x/2)u$. As long as $D_n < D_x$, the error introduced by numerical dispersion can be eliminated by using $D'_x = D_x - D_n$ in place of D_x. The problems of numerical dispersion and stability are discussed further in Chapter 6.

At steady state,

$$\frac{dC_i}{dt} = 0 \qquad i = 1, \ldots, N$$

In this case, the mass-balance equations for the N segments form an $N \times N$ system of linear algebraic equations, which can be expressed in matrix form as:

$$\mathbf{AC} = -\mathbf{S}' \tag{5.29}$$

In Eq. (5.29), $\mathbf{C} = \{C_1, C_2, \ldots, C_N\}^T$ and $\mathbf{S}' = \{S'_1, S_2, \ldots, S'_N\}^T$. The terms S'_1 and S'_N include boundary conditions along with the direct source terms. A is a

$N \times N$ matrix comprised of the coefficients a_{ij} multiplying C_i in the jth mass-balance equation. For example, for the ith cell of the ith mass-balance equation (5.28) and using backward differencing, so $C_{i-1,i} = C_{i-1}$ and $C_{i,i+1} = C_i$:

$$a_{ii} = -Q - \frac{2D_x A}{\Delta x} - k_i V_i \tag{5.30}$$

The steady-state solution to the system of Eq. (5.29) is

$$\mathbf{C} = -A^{-1}\mathbf{S}' \tag{5.31}$$

In Eq. (5.31), $-A^{-1}$ is called the unit response matrix because its i,jth element gives the response of the concentration in cell i to a unit loading in cell j.

EXAMPLE 5.5

Consider an estuary that receives heavy-metal loadings from atmospheric deposition and surface runoff. The metal loadings along a 40-km stretch of the estuary have been estimated for 5-km-long segments, as shown in Figure 5.16. The metal is removed from the estuary by sedimentation, with a first-order rate constant of 0.1/day. The estuary has a cross-sectional area of $A = 400$ m^2, a net flow of $Q = 160{,}000$ m^3/day (the net velocity is thus $u = 400$ m/day), and a dispersion coefficient of 2.5×10^6 m^2/day (29 m^2/s). The concentration upstream from the first segment is $C_0 = 12.4$ μg/L. Downstream of the eighth segment, the metal concentration is uniform, that is, $\partial C/\partial x = 0$.

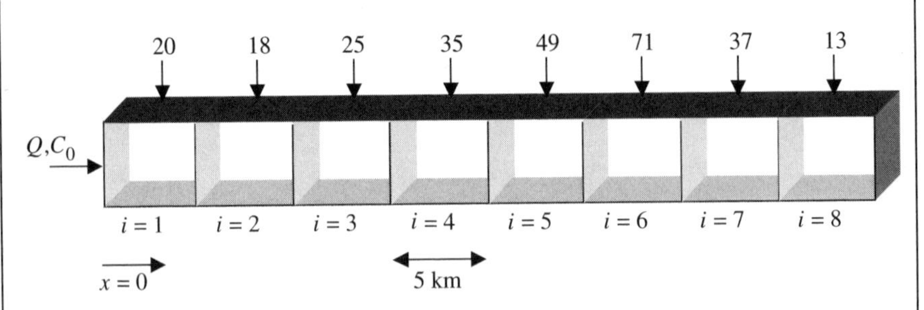

Figure 5.16 Loadings (kg/day) into eight, 5-km-long segments of an estuary.

Solution 5.5 Using backward differencing, the steady-state concentrations in each of the eight segments are given by a set of linear algebraic equations expressing the mass balance for each control volume:

$$i = 1 \qquad QC_0 - QC_1 + D'_x A \frac{C_0 - C_1}{\Delta x} + D'_x A \frac{C_2 - C_1}{\Delta x} - kVC_1 + S_1 = 0$$

$$i = 2, \ldots, 7 \qquad QC_{i-1} - QC_i + D'_x A \frac{C_{i-1} - C_i}{\Delta x} + D'_x A \frac{C_{i+1} - C_i}{\Delta x} - kVC_i + S_i = 0 \tag{5.32}$$

$$i = 8 \qquad QC_{i-1} - QC_i + D'A \frac{C_{i-1} - C_i}{\Delta x} - kVC_i + S_i = 0$$

For the system described above, the numerical dispersion is

$$D_n = \frac{\Delta x}{2} u = \frac{5000 \text{ m}}{2} \times 400 \text{ m/day} = 1 \times 10^6 \text{ m}^2\text{/day}$$

The dispersion input to the model is thus adjusted downward to $D' = D - D_n = 1.5 \times 10^6$ m^2/day. The coefficient matrix A is evaluated by collecting terms for C_{i-1}, C_i, and C_{i+1}, substituting in numerical values for the parameters and rearranging. For example:

$$a_{11} = -Q - \frac{2D'A}{\Delta x} - kV = -160{,}000 - \frac{2 \times 1.5 \times 10^6 \times 400}{5000} - 0.1 \times 400 \times 5000 \text{ m}^3\text{/day} = -600{,}000 \text{ m}^3\text{/day}$$

and

$$a_{12} = \frac{D'A}{\Delta x} = \frac{2 \times 1.5 \times 10^6 \times 400}{5000} \text{m}^3\text{/day} = 120{,}000 \text{ m}^3\text{/day}$$

The complete matrix is

$$A \text{ (m}^3\text{/day)} = \begin{pmatrix} -600000 & 120000 & 0 & 0 & 0 & 0 & 0 \\ 280000 & -600000 & 120000 & 0 & 0 & 0 & 0 \\ 0 & 280000 & -600000 & 120000 & 0 & 0 & 0 \\ 0 & 0 & 280000 & -600000 & 120000 & 0 & 0 \\ 0 & 0 & 0 & 280000 & -600000 & 120000 & 0 \\ 0 & 0 & 0 & 0 & 280000 & -600000 & 0 \\ 0 & 0 & 0 & 0 & 0 & 280000 & 120000 \\ 0 & 0 & 0 & 0 & 0 & 0 & -480000 \end{pmatrix}$$

The elements of the source matrix are taken directly from Figure 5.16, except that the contribution from the upstream boundary condition needs to be included in S'_1:

$$S_1' = QC_0 + \frac{D'AC_0}{\Delta x} + S_1 = 23.47 \text{ kg/day}$$

Thus

$$\mathbf{S}' \text{ (kg/day)} = \{23.47, 18, 25, 35, 49, 71, 37, 13\}^T$$

With A and $\mathbf{S}$ evaluated, matrix inversion was used to solve for the concentrations shown in Figure 5.17.

The use of the unit response matrix, $-\mathsf{A}^{-1}$ is illustrated in Figure 5.17, where the lower concentration response curve is computed assuming that all discharges are eliminated except for the 71 kg/day discharge to segment 6. These results are computed simply as 71 kg/day times the respective entries in the sixth column of the unit response matrix. If this load is reduced by half, the resulting concentrations are halved, and these concentration reductions can likewise be translated to the full model results.

Also shown in Figure 5.17 is the analytical solution for the mixed-flow estuary model, which is given by Eq. (5.24), computed for the single 71-kg/day discharge to segment 6. This solution is available because the example assumes spatially homogeneous flow, dispersion, and channel geometry. Not surprisingly, the results of the numerical and analytical solutions are nearly identical (the predicted concentration at the point of discharge in the analytical solution is slightly higher than that predicted by the numerical cell model, since in the latter the discharge is completely mixed over the 5 km of cell 6). Indeed, one of the first tests for code validity and numerical accuracy of a numerical model is to apply it to a limiting case for which a known analytical solution is available (e.g., where spatial homogeneity is assumed). In most cases in which numerical solutions are utilized,

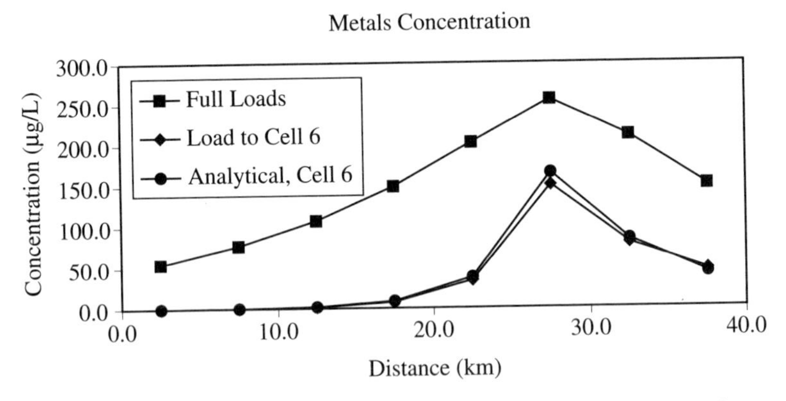

Figure 5.17 Metals concentrations in an estuary, estimated using eight 5-km-long segments with backward differencing and correction for numerical dispersion. The "full-loads" case incorporates all the loading shown in Figure 5.16. The "load-to-cell-6" case is the numerical solution with $C_0 = 0$ and loading only to cell 6 and "analytical, cell 6" the corresponding analytical solution, calculated using Eq. (5.24).

temporal or spatial variations in model inputs and parameters preclude the use of analytical solutions. The tools for addressing these problems are presented in the next chapter. Applications of these methods are illustrated in the chapters that follow for each of the major environmental media.

5.8 SUMMARY OF TRANSPORT FUNDAMENTALS

As shown in this chapter, the basic mathematical descriptions of transport phenomena are similar across the different environmental media. In all cases, the first step in modeling contaminant transport involves characterizing the fluid flow. Often the flow description is based on observed winds, streamflow, or currents, with simple empirical relationships used to relate flow and media characteristics. Increasingly, however, hydrodynamic, hydraulic, or meteorological models based on first principles are being developed as part of the contaminant transport modeling exercise. Chapters 8 to 10 address specific characteristics of fluid flows and contaminant transport in air, groundwater, and surface water, respectively. These chapters build on the fundamental principles of transport modeling that were presented here.

Transport in air, surface water, or groundwater is characterized by advection, with the bulk fluid flow, and diffusion or dispersion from random molecular motions or random or unresolved features of the flow field. The diffusion and dispersion equations, with flux proportional to the concentration gradient, can be applied over a range of physical processes, from molecular diffusion in a stagnant fluid to hydrodynamic dispersion in a porous medium to continental and global-scale atmospheric and oceanic circulation patterns and eddies. The numerical value of the dispersion coefficient can range from 10^{-6} cm^2/s for molecular diffusion in water to 10^{10} cm^2/s for effective atmospheric dispersion over long averaging times. The values of turbulent diffusion, shear, and groundwater dispersion coefficients are scale dependent, increasing as larger averaging times and coarser spatial resolution are considered in a model.

As illustrated here for mixed-flow reactors with instantaneous or continuous releases in unbounded domains, analytical solutions are available for a number of simple advective–dispersive transport problems. For idealized systems with constant flow and dispersion coefficients, the Peclet number, defined as $\text{Pe} = \ell u/D$, is useful for evaluating whether complete mixing, plug flow, or mixed-flow assumptions are applicable. However, when spatially heterogeneous or temporally variable conditions are considered, numerical methods are generally required. The finite segment or finite cell method was introduced in this chapter as one such method. It assumes concentrations are completely mixed within individual cells of a segmented domain, with discrete changes in concentration between cells. The example presented illustrates the utility of the unit response matrix as an algebraic "model of the model" for linear systems. Additional numerical methods for modeling environmental transport are presented in the next chapter.

Homework problems for this chapter are provided at www.wiley.com/college/ramaswami.

6 Overview of Numerical Methods in Environmental Modeling

Up to this point, this book has dealt primarily with idealized models of environmental systems. These models assume uniform geometry, constant system properties, and inputs and boundary conditions that are either constant over time or are described by special time-varying equations such as exponential or sinusoidal functions. The environmental system is represented in many of these models by a single reactor, such as a 0D completely stirred tank reactor, a 1D plug-flow reactor, or a 2D advective–dispersive reactor, with the reaction kinetics usually assumed to be first order. For these idealized conditions, analytical solutions, that is, closed-form equations, can often be derived for the model state variables: concentrations, mass fluxes, or (as introduced in Chapter 13) human exposure and risk. However, when models are developed to provide more detailed and realistic representations of systems exhibiting spatial heterogeneity, temporal variation, and nonlinear reaction kinetics, analytical solutions are generally not available. Numerical solution methods are then needed.

Numerical methods are commonly used to integrate the mass-balance differential equation for contaminant concentration C, with the objective of determining the concentration C at a specific location (x, y, z) or at a specific time t. When the derivative of C appears in the equation with respect to only one variable, usually time (i.e., dC/dt) or a single spatial dimension (such as dC/dx), the equation is referred to as an *ordinary differential equation*, or ODE. A set of coupled equations for multiple contaminants (e.g., describing dC_1/dt, dC_2/dt, etc.) is a *system* of ordinary differential equations. Section 6.1 presents methods for the numerical solution of systems of ODEs. When the concentration derivatives appear in more than one dimension, the equations are referred to as *partial differential equations*, or PDEs. Methods for solving PDEs, which are usually implemented over a spatial system of grid points or cells, are presented in Section 6.2. This section also presents methods for solving systems of linear algebraic equations that arise in numerical techniques for PDEs. Simultaneous nonlinear algebraic equations are often required to describe equilibrium chemistry in environmental models; numerical methods for these systems are described in Section 6.3. The numerical methods described in this chapter are used to develop deterministic models of pollutant transport employing point estimates for various input parameters that appear in the mass-balance equations, for example, fluid flow rates or reaction rate constants. Chapter 7 provides an introduction to random variables and random processes, which are used in stochastic models to simulate both variability in the environment and the uncertainty of model predictions.

This chapter is intended to provide only an overview of the computational methods considered necessary for understanding many of the full-scale models presented in subsequent chapters. Texts devoted to numerical methods (e.g., Chapra and Canale, 1988) provide more detailed discussions. Press et al. (1994) explain and outline computer algorithms for implementing a wide variety of numerical techniques. A more in-depth study of these techniques is recommended for anyone developing models in which they are used or for a full appreciation of the methods used by others.

6.1 ORDINARY DIFFERENTIAL EQUATIONS

Consider the mass-balance equation for the indoor air pollution problem depicted in Chapter 1 (Fig. 1.5), as given by Eq. (1.8):

$$V\frac{dC_i}{dt} = (Q_{\text{in}})\,C_{i,\text{amb}} - (Q_{\text{out}})\,C_i + S - kVC_i \tag{6.1}$$

where C_i represents the pollutant concentration within a well-mixed indoor air compartment, i. Because the indoor air compartment is well-mixed, C_i is independent of location and varies as a function of time only, that is, $C_i = C_i(t)$; $C_{i,\text{amb}}$ represents the influent concentration entering the air compartment from the surroundings. Dividing both sides of the equation by the volume of the room, V, the equation for the concentration derivative is obtained:

$$\frac{dC_i}{dt} = \frac{(Q_{\text{in}})\,C_{i,\text{amb}} + S}{V} - \left(\frac{Q_{\text{out}}}{V} + k\right)C_i \tag{6.2}$$

Equation (6.2) is a first-order ODE. The order of the ODE refers to the order of the derivative. For example, an equation with derivatives up to and including d^2C_i/dt^2 is a second-order differential equation. We limit ourselves in this section to solutions for first-order ODEs. Methods for solving higher-order ODEs are described elsewhere (Hoffman, 1992, p. 296).

For a given set of inputs that define the rhs of Eq. (6.2), we seek a procedure for moving from a known value of C_i at time t to an unknown value of C_i at time $t + \Delta t$.[1] This problem is depicted in Figure 6.1, which shows a true (but unknown to us) value of $C_i(t)$ that we are attempting to reproduce through numerical integration. Figure 6.1*a* shows perhaps the simplest logical solution to this problem (the one you might come up with yourself if you were stuck on a deserted island with no previous knowledge of numerical methods): Use the known slope of the curve at time t to extrapolate the value forward to time $t + \Delta t$:

$$C_i(t + \Delta t) = C_i(t) + dC_i/dt|_t\,\Delta t \tag{6.3}$$

[1] This can be solved for analytically in the case of Eq. (6.2). In the more general case of nonlinear kinetics, multiple contaminants, and the like, it usually cannot.

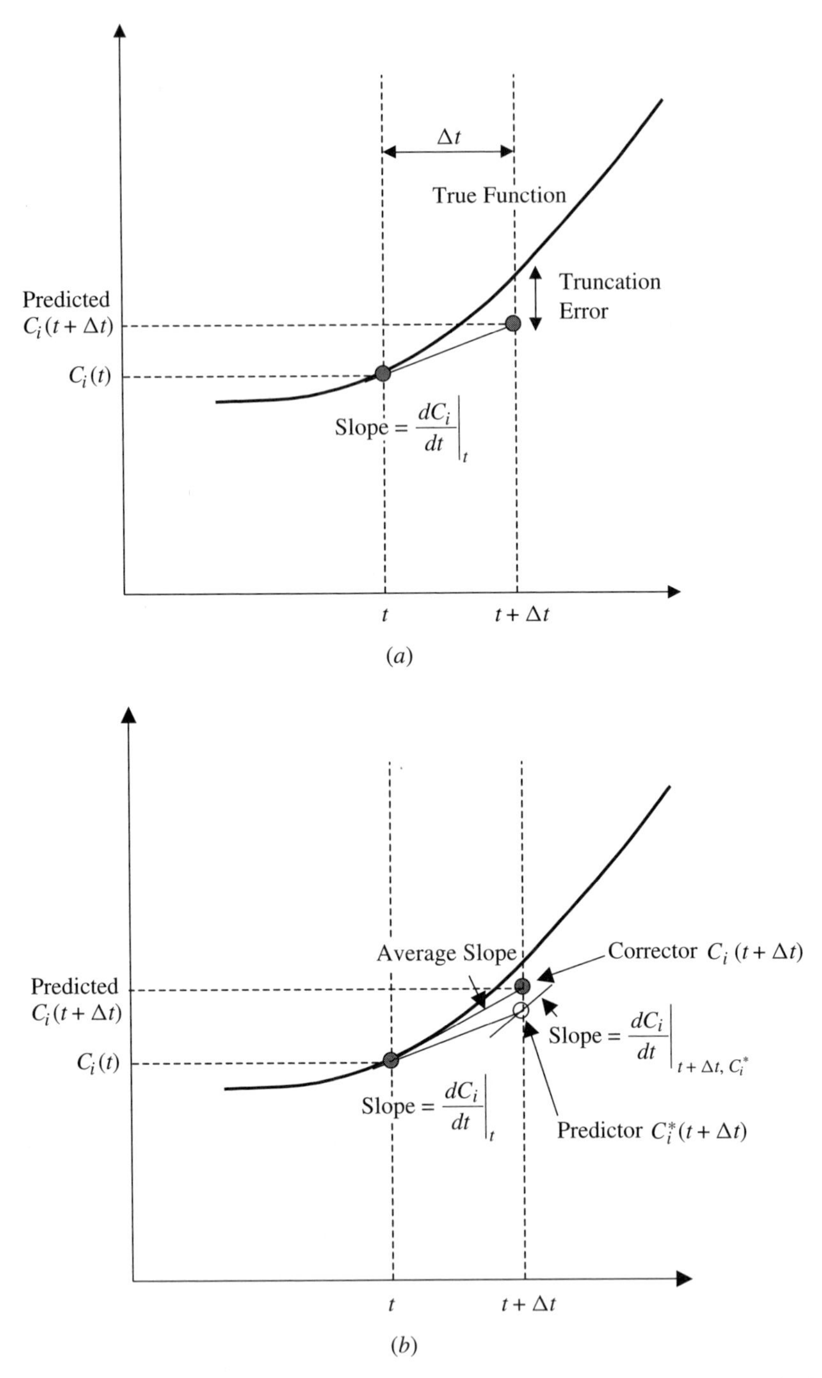

Figure 6.1 Numerical integration of unknown function $C_i(t)$ using (*a*) the Euler method and (*b*) the predictor–corrector method.

Equation (6.3) is known as the Euler or Euler–Cauchy formula. As indicated in Figure 6.1*a*, Eq. (6.3) is likely to introduce some error, especially when the true solution is highly nonlinear and when it is implemented over large time steps, Δt.

How might the error indicated in Figure 6.1*a* be reduced? Figure 6.1*b* suggests an option. The derivative could be recomputed at time $t + \Delta t$, and then the linear extrapolation from time t recomputed using a slope that is the *average* of the initial derivative at time t and that subsequently computed at time $t + \Delta t$. This procedure, known as the predictor–corrector method, is implemented in two steps:

$$\begin{aligned} &\text{Predictor} \qquad C_i^*(t + \Delta t) = C_i(t) + \left.\frac{dC_i}{dt}\right|_t \Delta t \\ &\text{Corrector} \qquad C_i(t + \Delta t) = C_i(t) + \frac{\left.\frac{dC_i}{dt}\right|_t + \left.\frac{dC_i}{dt}\right|_{t+\Delta t, C_i^*}}{2} \Delta t \end{aligned} \tag{6.4}$$

The first predictor step is identical to the Euler method in Eq. (6.3). Note that when multiple model outputs, $C_j(t)$, $j = 1, \ldots, J$, are required either to model J coupled compartments across which a contaminant is distributed or the coupled behavior of J chemical pollutants within a single compartment, a system of J ODEs is obtained. In this case, the predictor step is implemented for each of the J state variables to compute each of the $C_j^*(t)$, before computing the correctors. The second corrector step yields modified estimates, which are expected to be closer to the true values than are the predictors. However, they are still not completely accurate, due to the nonlinearity of the true solutions and the fact that the derivatives at time $(t + \Delta t)$ are computed using only estimates of the $C_j(t + \Delta t)$'s.

A more formal approach for describing the accuracy of numerical integration recognizes that two types of error can occur in the transition from time t to $t+\Delta t$. The first, depicted in Figure 6.1, is known as *truncation error*. This is the inherent error of the method. For the Euler method, the truncation error is proportional to $(\Delta t)^2$, the time step raised to the second power. As such, cutting the time step in half reduces the error by a factor of 4. For the predictor–corrector method, the truncation error is proportional to $(\Delta t)^3$, so that reducing the time step by a factor of 2 reduces the error by a factor of 8. More accurate integration is therefore expected using smaller time steps and the predictor–corrector method, than can be achieved with the Euler method.

If less truncation error occurs with smaller time steps, why not use smaller and smaller values of Δt until the desired accuracy is achieved? A first, practical reason is that the computation time increases as Δt is reduced. (More calculations are needed to integrate over the same time interval.) A second, more fundamental reason involves the second type of error: *round-off error*. The second terms that are added to $C_i(t)$ on the rhs of Eq. (6.3) or (6.4) become smaller and smaller as the time step is reduced. With a limited number of significant digits used for the calculation by the computer (typically 8, or 16 if double precision is used), a greater relative error can occur as round-off eliminates a higher fraction of the addend. While the truncation error is reduced as Δt is made smaller, the round-off error increases. Furthermore, this increased relative error is repeated over more, shorter time steps. The net effect of both truncation error and round-off error over *many* time steps is referred to as the overall *propagation error*.

How can you tell whether a numerical integration is accurate? You do not know the true solution (i.e., from an analytical solution); if you did, you would use it! However, you might be able to simplify your model to a special case, for example, with constant spatial and temporal properties, simplified kinetics, and the like, for which a known analytical solution *is* available. The numerical method should be able to reasonably reproduce the known solution for this special case. Comparison of numerical solutions to known analytical solutions for simplified, idealized cases is thus a common and important first step in testing for accuracy. It does not guarantee that the numerical solution will be accurate for the real, more complex cases that you really care about. However, it does provide some degree of comfort and assurance. If the model cannot reproduce the analytical solution for simplified cases, then something is clearly wrong—either inherently with the method or in its computer implementation.

A second way to diagnose the accuracy of a numerical integration procedure is to evaluate the model with varying time steps. Initially, the time step is chosen based on the time scales of variation in model inputs and responses. For example, a model with variations in emissions, transport terms, reaction rates, and resulting concentrations over time scales of minutes and hours will typically require time steps of seconds for numerical integration; models with variations over weeks, months, and years typically require time steps of days or fractions of a day. A high estimate of the time step is first selected, the model executed, and the results recorded. A second, smaller time step (e.g., one-half of the value of Δt used for the first test run) is used and the results compared to those from the first case. If the initial time step was indeed too large, the results should be different. Successively smaller time steps are tested and the differences between runs should diminish, until reducing the time step further no longer yields a change in the results. This indicates that a sufficiently small time step has been selected and that accurate numerical integration has most likely been achieved. Eventually, reducing the time step further should once again yield changes in model predictions, as round-off error comes into play. The assumption is that initially reductions in Δt act to reduce the truncation error and that round-off error does not become significant until the time step is reduced to a very small value. This is usually the case, especially if the initial time step is chosen to be "conservatively large" for the problem under consideration. Other methods are available for diagnosing model accuracy, such as checking the model for mass balance. These and other quality control procedures for model evaluation are illustrated in the examples presented in this chapter and discussed further in Chapter 14.

The Euler and the predictor–corrector methods are among the simplest of the available procedures for numerical integration of ODEs but are not especially accurate. For some problems, they can yield unstable results. Instability occurs when the results deviate so far from the correct values that wildly diverging or oscillating predictions are made. Instability is not easy to define, but you know it when you see it. More sophisticated numerical methods may be able to maintain stability and achieve significantly greater accuracy. Some of these methods utilize model results prior to t (e.g., at time $t - \Delta t$ and $t - 2\,\Delta t$) in making the transition from t to $t + \Delta t$. Use of these multiple values allows the higher-order shape and associated derivatives of the function to be taken into consideration. While such multistep methods are more accurate, they are not self-starting, since at time $t = 0$, values at $t - \Delta t$ and $t - 2\,\Delta t$ are not available. A self-starting method is thus needed for the initial calculations over the first few time steps. Some of the stability and accuracy of multistep methods can be achieved by self-starting methods if each time step is broken up into partial steps. Among the most widely used of these partial step procedures is the Runge–Kutta method.

The Runge–Kutta procedure is actually a family of methods, each with different "order" depending on how many partial steps are utilized within each time step. The fourth-order Runge–Kutta method is especially popular, due to its very high accuracy and stability, yet relative simplicity. The accuracy of the fourth-order Runge–Kutta method is related to that of Simpson's method for numerical integration, illustrated in Figure 6.2. An accurate estimate of $A = \int f(x)\,dx$ (i.e., the shaded area under the curve in Fig. 6.2) is computed using Simpson's rule as follows:

$$A = (\Delta x/6)\,[f(a) + 4f(b) + f(c)] \tag{6.5}$$

Where a, b, and c are evenly-spaced points in the interval $(x, x + \Delta x)$. In numerical solution of chemical mass-balance ODEs, the function $f(x)$ corresponds with the time derivative dC_i/dt that is integrated over a time step, Δt, to determine the unknown function $C_i(t)$.

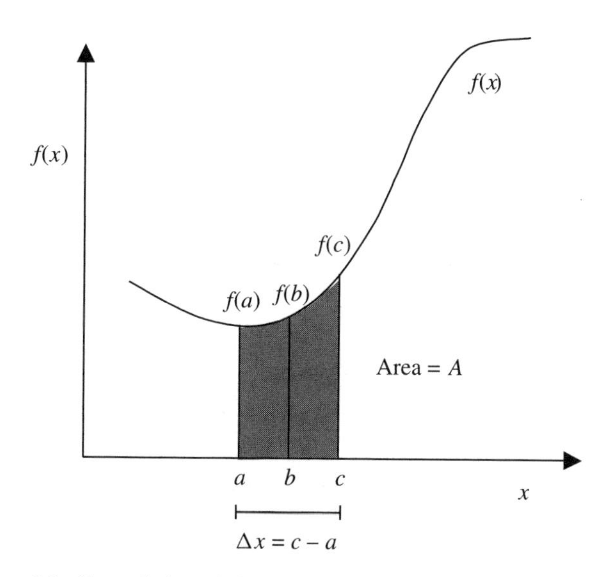

Figure 6.2 Formulation of Simpson's method for integration [Eq. (6.5)].

As in Simpson's method, the fourth-order Runge–Kutta method similarly divides the time step in half and computes values of $C(t)$ at the next time step from the derivatives, F_1, F_2, F_3, and F_4, as:

$$C_i(t + \Delta t) = C_i(t) + \tfrac{1}{6}\left[F_1^i + 2F_2^i + 2F_3^i + F_4^i\right] \tag{6.6}$$

where

$$F_1^i = \left[\left. \frac{dC_i}{dt} \right|_{t, C_i = C_i(t)} \right] \Delta t$$

$$F_2^i = \left[\left. \frac{dC_i}{dt} \right|_{t+(\Delta t/2), C_i = C_i(t) + 0.5\, F_1^i} \right] \Delta t$$

$$F_3^i = \left[\left. \frac{dC_i}{dt} \right|_{t+(\Delta t/2), C_i = C_i(t) + 0.5\, F_2^i} \right] \Delta t$$

$$F_4^i = \left[\left. \frac{dC_i}{dt} \right|_{t+\Delta t, C_i = C_i(t) + F_3^i} \right] \Delta t$$

As with the predictor–corrector method, when the Runge–Kutta method is applied to a system of ODEs for a suite of constituents, the F_1^j's must be computed for each constituent $j = 1, \ldots, J$, before moving on to calculate each of the F_2^j's, and so on. The derivative of C_i used to calculate each of the F's is evaluated with model inputs set at the indicated times ($t, t + \Delta t/2, t + \Delta t/2$ and $t + \Delta t$, for F_1, F_2, F_3 and F_4, respectively) and with "predictor" values of C_i computed as shown. The fourth-order Runge–Kutta method has an error proportional to $(\Delta t)^5$, so very high accuracy can be achieved as the time step is reduced.

EXAMPLE 6.1 NUMERICAL INTEGRATION OF A SIMPLE FOOD CHAIN MODEL

To illustrate procedures for numerical integration and the sensitivity of numerical results to different methods and time steps, the idealized system for nutrient uptake and growth of phytoplankton and zooplankton shown in Figure 6.3 is considered. [This example is based on Section 14.1.3 of Chapra and Reckhow (1983).] The model simulates the cycling of phosphorus between three species: inorganic phosphorus, p_1; phytoplankton, p_2; and zooplankton, p_3. The phytoplankton grow via uptake of inorganic phosphorus and are subsequently consumed by the zooplankton. The zooplankton grow as a result of this consumption, but die and degrade

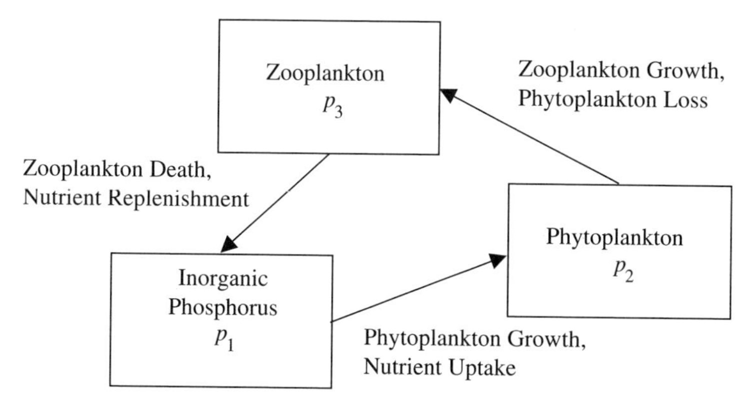

Figure 6.3 Simple phosphorus food chain cycle for Example 6.1.

back into inorganic P. Lotka–Volterra predator–prey relationships are defined for the phytoplankton and zooplankton, with nutrient uptake and growth dependent on both the "prey" (inorganic P for the phytoplankton; phytoplankton for the zooplankton) and predator concentrations. The concentrations of each of the three species are represented in terms of their phosphorus content $[M(\mathrm{P})\,L^{-3}]$.

The model assumes nutrient (inorganic phosphorus)-limited uptake and growth of the phytoplankton, described by Michaelis–Menten kinetics (see Chapter 12):

$$\text{Phytoplankton growth rate} = k_m \left(\frac{p_1}{K_s + p_1} \right) p_2$$

where k_m (T^{-1}) is the maximum growth rate and K_s $[M(\mathrm{P})\ L^{-3}]$ is the half saturation constant, equal to the nutrient concentration at which the growth rate of phytoplankton is half of its maximum value. The phytoplankton are consumed by zooplankton grazing:

$$\text{Zooplankton growth rate} = k_{23}\, p_2\, p_3$$

where k_{23} $[\{M(\mathrm{P})\ L^{-3}\}^{-1}\ T^{-1}]$ is a second-order rate constant, referred to as the zooplankton feeding rate. The rate of zooplankton death and consequent nutrient replenishment is given by:

$$\text{Nutrient replenishment rate} = k_z\, p_3$$

where k_z (T^{-1}) is the first-order zooplankton death rate.

The phosphorus–phytoplankton–zooplankton food chain is simulated for a batch reactor, considering only kinetic processes with assumed constant rate coefficients. In real aquatic systems these kinetic processes are supplemented by a seasonal pattern of loadings and discharge from the water body, with temperature- and light-driven variations in the rate constants. Examination of the growth patterns and nutrient cycling predicted to occur in a closed system is nonetheless useful to begin to understand the dynamics and cyclical nature of the food chain. This idealized system also provides a good illustration of the behavior of numerical solutions for ODEs.

The three simultaneous, nonlinear ordinary differential equations for the system are

$$\frac{dp_1}{dt} = k_z\, p_3 - k_m \frac{p_1}{K_s + p_1} p_2$$

$$\frac{dp_2}{dt} = k_m \frac{p_1}{K_s + p_1} p_2 - k_{23}\, p_2\, p_3 \tag{6.7}$$

$$\frac{dp_3}{dt} = k_{23}\, p_2\, p_3 - k_z\, p_3$$

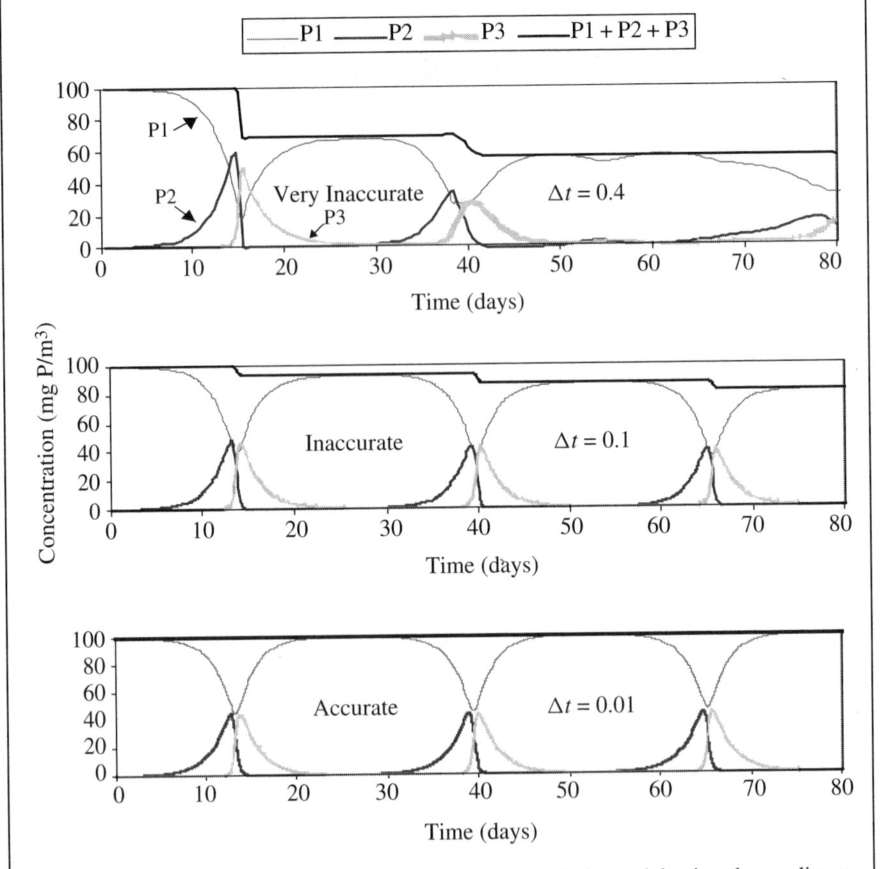

Figure 6.4 Numerical solution of the phosphorus food chain model using the predictor–corrector method.

Following Chapra and Reckhow (1983), the following coefficients are assumed for the model: $k_m = 0.5\ \text{day}^{-1}$; $K_s = 2\ \text{mg}\ P/\text{m}^{-3}$; $k_{23} = 0.1\ (\text{mg}\ P/\text{m}^{-3})^{-1}\ \text{day}^{-1}$; and $k_z = 0.5\ \text{day}^{-1}$; with initial conditions: $p_1(0) = 99.8$ and $p_2(0) = p_3(0) = 0.1\ \text{mg}\ P/\text{m}^{-3}$.

Numerical simulation results for this model are calculated using the predictor–corrector method and the Runge–Kutta method, with results shown in Figures 6.4 and 6.5, respectively. For each method, time steps of $\Delta t = 0.4, 0.1,$ and 0.01 days are utilized. Figures 6.4 and 6.5 also show the computed total P concentration: $p_T(t) = p_1(t) + p_2(t) + p_3(t)$; which, because the system is closed (and constant volume), should remain equal to the initial value of 100 mgP/m^{-3}, as long as an accurate mass balance is maintained for the system.

As indicated in the figures, the results are sensitive to the time step chosen. With too large a time step, inaccurate results are obtained, especially with the less

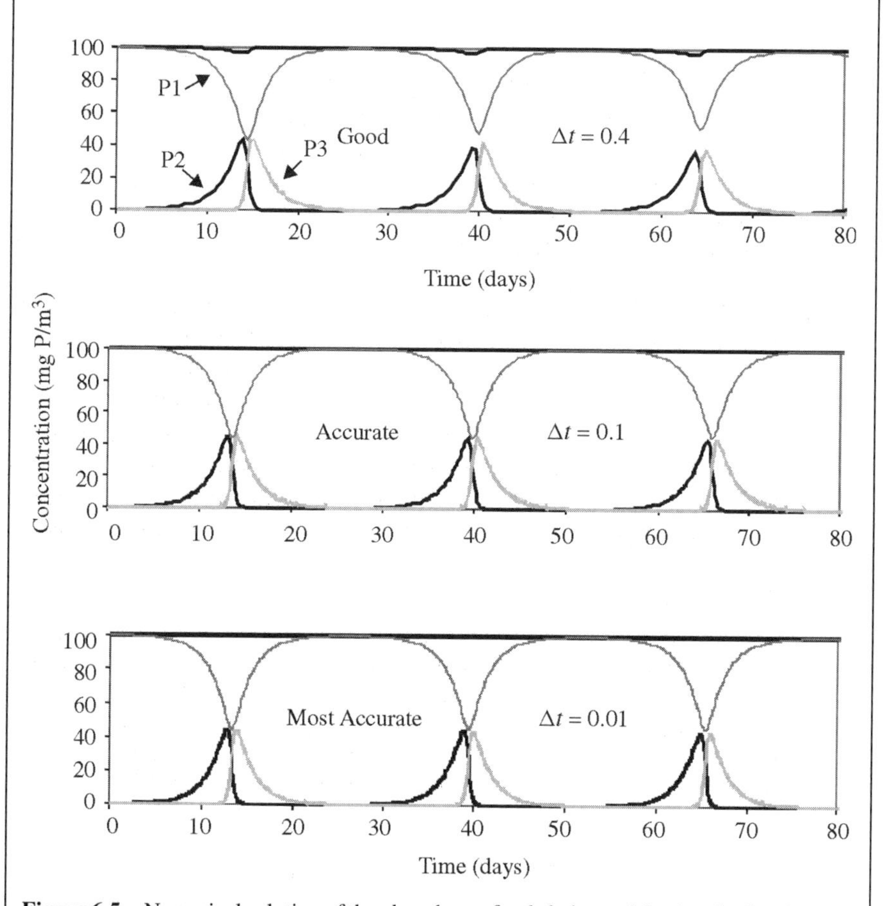

Figure 6.5 Numerical solution of the phosphorus food chain model using the fourth-order Runge–Kutta method.

accurate predictor–corrector method.[2] If too large a time step is chosen, irregular results can be generated, with significant deterioration of the mass balance and (though not evident in these simulations) even predictions of negative concentrations. When state variables that must be nonnegative violate this physical requirement, this is a clear indication of error and usually a precursor to instability. It is tempting in this situation to consider simple corrections to the algorithm, such as setting all predicted negative concentrations equal to zero or to a very small number. However, such ad hoc corrections generally introduce further mass-balance errors into the solution. Rather, the preferred approach is to try smaller time steps or utilize an alternative, more accurate solution method.[3] These steps and the results in Figures 6.4 and 6.5 illustrate the type of trial-and-error testing that typically must be done when developing and implementing a numerical solution.

Finding the right time step to use in solving systems of ordinary differential equations is particularly difficult if the system is *stiff*, meaning that changes in the magnitude of some of the state variables occur orders of magnitude more quickly than for other variables. The mathematical consequence of stiffness is that the solution to a system of ordinary differential equations requires inverting a matrix that is nearly singular. Chapter 11 addresses solution techniques for stiff systems of ODEs because the rate equations for ozone formation in the atmosphere are a prime example of a stiff system.

Over the past decade, a number of convenient and powerful mathematics software packages have become widely available. Mathcad (MathSoft, 1997), Matlab (MathWorks, 1995), and Mathematica (Wolfram, 1991) all include functions that will numerically integrate systems of ODEs. In addition to Runge–Kutta schemes, Mathcad and Mathematica provide specialized functions with adaptive time steps that can handle stiff systems. The primary limitation of these software packages is that they offer relatively little flexibility for formatting model inputs and outputs. However, when such flexibility is not required, they offer a useful shortcut for numerical analysis.

[2] The "correct" results are understood to be those achieved with an accurate method at small time steps (though not *too* small, due to round-off error), exhibiting stable and repeatable behavior, and maintaining overall mass balance. See also Figure 14.5d of Chapra and Reckrow (1983, p. 262).

[3] Another trick that can be used when negative concentrations are simulated is to transform the problem to one described in terms of variable(s) that ensure that the positivity requirement is met. For example, dividing both sides of Eq. (6.2) by C_i yields an equation in terms of the transformed variable $Y_i = \ln(C_i)$, since

$$\frac{dC_i}{C_i\,dt} = \frac{d\ln C_i}{dt} = \frac{dY_i}{dt} = \frac{(Q_{\text{in}})C_{\text{amb}} + S}{V\exp(Y_i)} - \left(\frac{Q_{\text{out}}}{V} + k\right)$$

This model may be solved in terms of Y_i and subsequently transformed back to the targeted variable, $C_i(t) = \exp[Y_i(t)]$. No matter what the value of $Y_i(t)$, $C_i(t)$ remains nonnegative.

6.2 PARTIAL DIFFERENTIAL EQUATIONS

When mass-balance equations are specified with derivatives in more than one dimension, such as time, and one or more spatial dimensions for dynamic models or more than one spatial dimension for steady-state models, the system is then described by *partial* differential equations. Consider the equation for two-dimensional advective–dispersive transport with first-order decay:

$$\frac{\partial C}{\partial t} = -u\frac{\partial C}{\partial x} + D_x\frac{\partial^2 C}{\partial x^2} + D_y\frac{\partial^2 C}{\partial y^2} - kC \tag{6.8}$$

In this application concentration variations are considered over time and in the x and y directions, that is, $C = C(x, y, t)$. While analytical solutions to this equation are available when u, D_x, D_y, and k are constant over time and space (as presented in Chapter 5, dependent on the assumed boundary and initial conditions), numerical methods are required when these parameters vary temporally and/or spatially. The methods involve discretized calculations over both the temporal and spatial dimensions. Two of the most common methods for implementing this type of solution, finite difference and finite element methods, are described in this section.

6.2.1 Finite Difference Method

The finite difference method solves the mass-balance equation(s) by forcing them to be satisfied at a set of discrete points in space. Figure 6.6 illustrates a two-dimensional grid that might be used to solve Eq. (6.8). Nodes in the longitudinal (x) direction are indexed by i, while nodes in the transverse (y) direction are indexed by j, with $C(i, j, t)$ indicating the concentration at grid point i, j at time t. The key step in the finite difference method is to express the derivatives for concentration at each point i, j as appropriate differences between concentrations at adjacent nodes. For example, for the first (advection) term on the rhs of Eq. (6.8), the first derivative of $C(i, j, t)$ with respect to x is needed and could be expressed using either:

$$\frac{\partial C(i, j, t)}{\partial x} = \frac{C(i+1, j, t) - C(i, j, t)}{\Delta x} \tag{6.9}$$

or

$$\frac{\partial C(i, j, t)}{\partial x} = \frac{C(i, j, t) - C(i-1, j, t)}{\Delta x} \tag{6.10}$$

where Δx is the distance between nodes in the x direction. Equation (6.9) uses *forward differencing* for the advection term. It approximates the derivative at point i, j using the difference between the concentration one node *downstream* of the point (in the direction of advection) and $C(i, j, t)$. Alternatively, Eq. (6.10) uses *backward differencing* since the first derivative is computed by taking the difference between $C(i, j, t)$ and the concentration at the node one step *upstream*. Neither approach is expected to provide an especially accurate estimate of the concentration derivative

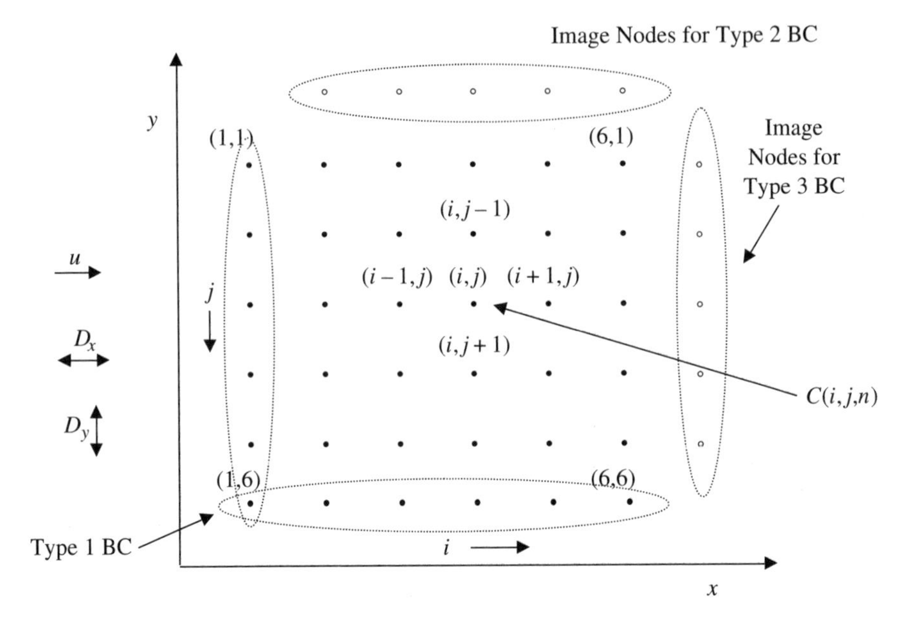

Figure 6.6 Finite difference grid with 6 × 6 (solid) node system. Image nodes (open) added for subsequent implementation of boundary conditions.

at point i, j, especially if its value is changing rapidly as a function of x. A third alternative is achieved by taking the average of the forward and backward differencing equations:

$$\frac{\partial C(i,j,t)}{\partial x} = \frac{\text{forward} + \text{backward}}{2} = \frac{C(i+1,j,t) - C(i-1,j,t)}{2\,\Delta x} \tag{6.11}$$

This approach, referred to as *central differencing* for the advection term, does not even use the value of the concentration at point i, j; rather, it calculates the derivative by drawing a straight line between the concentrations immediately up- and downstream of the target location.

Referring back to the finite *cell* method introduced in Section 5.7, the same terminology—backward, forward, or central differencing—was used to describe the weighting of concentrations passing with the advective flow across the interface of two cells. There is a clear parallel between the finite cell method, where concentrations are averaged over volumetric compartments and mass transport occurs between them, and the finite difference method in which mass-balance equations are satisfied and concentrations computed at points in space. In the latter, the points are still thought to be representative of the spatial domain around them, that is, the degree of spatial representation is still limited by the coarseness of the grid. Likewise, many of the same issues that arise with respect to accuracy, stability, and numerical dispersion in the finite cell method apply to the finite difference method, especially with regard to the use of backward, forward, or central differencing. These issues are addressed in more detail below. First, equations for the remaining derivatives in Eq. (6.8) are developed.

The dispersion terms in Eq. (6.8) require a differencing expression for second derivatives. Recognizing that the second derivative describes the rate of change in the first derivative with distance, it can be computed by taking the difference between the forward difference estimate of the first derivative (which best describes the value of the first derivative midway between node i, j and node $i+1, j$) and the backward difference estimate (for the point midway between node i, j and node $i-1, j$), and dividing by the distance between these points, Δx:

$$\frac{\partial^2 C(i,j,t)}{\partial x^2} = \frac{\partial\left(\frac{\partial C}{\partial x}\right)}{\partial x} = \frac{\left[\frac{C(i+1,j,t)-C(i,j,t)}{\Delta x}\right]-\left[\frac{C(i,j,t)-C(i-1,j,t)}{\Delta x}\right]}{\Delta x}$$

$$= \frac{C(i+1,j,t)-2C(i,j,t)+C(i-1,j,t)}{(\Delta x)^2} \tag{6.12}$$

A similar expression describes the second derivative for the dispersion term in the y direction:

$$\frac{\partial^2 C(i,j,t)}{\partial y^2} = \frac{C(i,j+1,t)-2C(i,j,t)+C(i,j-1,t)}{(\Delta y)^2} \tag{6.13}$$

The second derivative in each case is estimated by adding the concentrations at the adjacent nodes and subtracting twice the concentration at the target node, and then dividing by the square of the internode distance.

The final derivative that must be specified for dynamic solution of Eq. (6.8) is the time derivative. Indexing discrete points in time by $t = 1, 2, \ldots n, n+1, \ldots,$ where n is the current time in the simulation, the time derivative is first expressed using a simple Euler expression (forward differencing in time):

$$\frac{\partial C(i,j,n)}{\partial t} = \frac{C(i,j,n+1)-C(i,j,n)}{\Delta t} \tag{6.14}$$

Rearranging to solve for $C(i, j, n+1)$:

Taylor Series expansion

$$C(i,j,n+1) = C(i,j,n) + \frac{\partial C(i,j,n)}{\partial t}\Delta t \tag{6.15}$$

Finally, substituting for the terms on the rhs of Eq. (6.8), including the central difference expression for the advection term, the following equation is obtained:

$$C(i,j,n+1) = C(i,j,n)$$

$$+\,\Delta t\left\{\begin{array}{l} -u\left[\dfrac{C(i+1,j,n)-C(i-1,j,n)}{2\Delta x}\right] \\ +D_x\left[\dfrac{C(i+1,j,n)-2C(i,j,n)+C(i-1,j,n)}{(\Delta x)^2}\right] \\ +D_y\left[\dfrac{C(i,j+1,n)-2C(i,j,n)+C(i,j-1,n)}{(\Delta y)^2}\right]-kC(i,j,n) \end{array}\right\} \tag{6.16}$$

Equation (6.16) is *explicit*, that is, $C(i, j, n+1)$ at node i, j at time step $n+1$ is computed solely from values of C available at the current time n, at node i, j and the four surrounding nodes. While relatively easy to implement, explicit solutions may exhibit problems with stability. These problems are discussed in the following section, followed by presentation of an implicit solution method that addresses them.

Stability, Numerical Dispersion, and Implicit Solution Methods As with the finite cell method, finite difference solutions using central differencing are prone to instability when applied to highly advective, low-dispersion systems. Central differencing exhibits inherent "static" instability whenever the grid size is too large relative to the ratio of the dispersion to velocity. Stability is maintained when

$$\Delta x \leq \frac{2D}{u} \tag{6.17}$$

This is equivalent to requiring that the Peclet number be less than or equal to 2, where the Peclet number is defined as $\text{Pe} = \Delta x\, u/D$. Highly advective (high Peclet number) domains, such as the riverine portion of a coastal water system, thus require finer grid spacing. This requirement applies when implementing either a steady-state or a dynamic solution. When a dynamic solution such as Eq. (6.16) is implemented, a further requirement is placed on the time step, Δt. To avoid dynamic instability, the time step must be chosen so that (for a one-dimensional problem):

$$\Delta t \leq \frac{(\Delta x)^2}{2D} \tag{6.18}$$

For a two-dimensional problem [such as Eq. (6.16)], the restriction is greater still, with the two in the denominator of Eq. (6.18) replaced by 4.

One approach for avoiding the restrictions on Δx imposed by the static stability requirement [Eq. (6.17)] is to use backward, instead of central, differencing. However, as with cell models, this introduces numerical dispersion. Backward differencing computes the advective transport from a value of the derivative further upstream. The resulting error has the same effect as increasing the dispersion coefficient, moving additional mass from nodes with high concentration to nodes with low concentration. In the dynamic solution the magnitude of the numerical dispersion is given by:

$$D_n = \frac{1}{2} u\, \Delta x \left(1 - \frac{u\, \Delta t}{\Delta x}\right) \tag{6.19}$$

When backward differencing is employed to address stability problems, the extra numerical dispersion must be recognized. It may be necessary to reduce the input values of the dispersion coefficient(s), that is, reduce assigned values of D (D_x and/or D_y) so that the assigned values *plus* the numerical dispersion introduced through the numerical method equals the level of dispersion targeted for the problem. If the numerical dispersion exceeds that targeted for the application, this clearly will not work (negative dispersion coefficients cannot be assigned). Furthermore, in problems

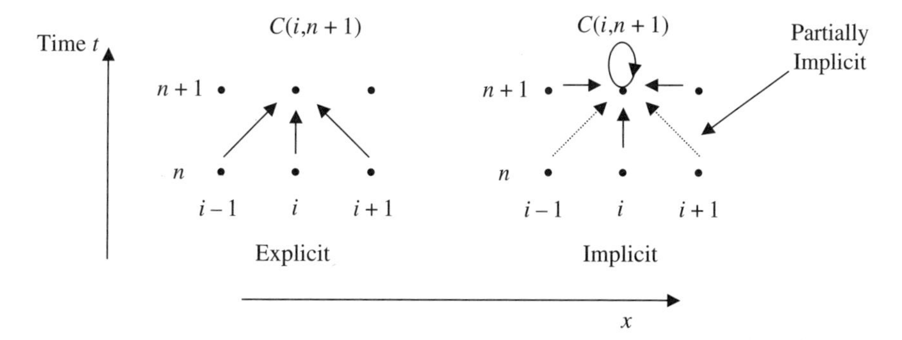

Figure 6.7 Explicit versus implicit solution of a finite difference problem over time for a spatially one-dimensional system. The explicit method uses only information available at time n to compute derivatives and move forward to time $n+1$. (Arrows show flow of information.) Implicit methods compute derivatives from the model state at time $n+1$ (and at time n if partially implicit) so that solution of simultaneous equations is required to move forward to time $n+1$.

with variable time steps and complex grid systems, it may be difficult to ascertain the magnitude of the numerical dispersion. The use of backward differencing is thus an imperfect solution to stability problems, but one that in some cases cannot be avoided, especially when solving steady-state models.

For dynamic models, an alternative approach is available for maintaining stability. The explicit approach depicted in Eq. (6.16) may be replaced by an *implicit* method. The key difference between explicit and implicit time integration is illustrated in Figure 6.7. The explicit method uses available node concentrations from time n to compute all concentrations at time $n+1$. The implicit method expresses the equations for concentration at each node at time $n+1$ in terms of concentrations at other nodes at time $n+1$. The equations are implicit because the information needed to move forward from time n to time $n+1$ is not all available at time n; rather the equations for the concentrations at the different nodes at time $n+1$ must be solved simultaneously.

A *fully* implicit solution defines all terms in the time derivative, $\partial C/\partial t$, using concentrations (and model inputs and parameters) at time $n+1$. The fully implicit equivalent of Eq. (6.16) is given by:

$$C(i,j,n+1) = C(i,j,n) + \Delta t \left\{ \begin{array}{l} -u\left[\dfrac{C(i+1,j,n+1) - C(i-1,j,n+1)}{2\,\Delta x}\right] \\ +D_x\left[\dfrac{C(i+1,j,n+1) - 2C(i,j,n+1) + C(i-1,j,n+1)}{(\Delta x)^2}\right] \\ +D_y\left[\dfrac{C(i,j+1,n+1) - 2C(i,j,n+1) + C(i,j-1,n+1)}{(\Delta y)^2}\right] - kC(i,j,n+1) \end{array} \right\} \tag{6.20}$$

If model inputs u, D_x, D_y, and k vary with time, their values at time $n+1$ would also be used in Eq. (6.20).

A *partially* implicit solution defines the time derivative as a weighted average of values computed using concentrations (and model inputs) at times n and $n+1$:

$$C(i, j, n+1) = C(i, j, n) + \Delta t \left\{ \alpha \left. \frac{\partial C}{\partial t} \right|_{n+1} + (1-\alpha) \left. \frac{\partial C}{\partial t} \right|_{n} \right\} \tag{6.21}$$

The case where $\alpha = 0$ corresponds to the explicit method, $\alpha = 1$ to the fully implicit method, and $0 < \alpha < 1$ to a partially implicit solution. The special case in which $\alpha = 0.5$ is known as the Crank–Nicolson method. The partially implicit solution corresponding to the explicit Eq. (6.16) and the fully implicit Eq. (6.20) is

$$\begin{aligned} C(i, j, n+1) &= C(i, j, n) \\ &+ \Delta t\, \alpha \left\{ \begin{aligned} &-u \left[\frac{C(i+1, j, n+1) - C(i-1, j, n+1)}{2\,\Delta x} \right] \\ &+D_x \left[\frac{C(i+1, j, n+1) - 2C(i, j, n+1) + C(i-1, j, n+1)}{(\Delta x)^2} \right] \\ &+D_y \left[\frac{C(i, j+1, n+1) - 2C(i, j, n+1) + C(i, j-1, n+1)}{(\Delta y)^2} \right] - kC(i, j, n+1) \end{aligned} \right\} \\ &+ \Delta t(1-\alpha) \left\{ \begin{aligned} &-u \left[\frac{C(i+1, j, n) - C(i-1, j, n)}{2\,\Delta x} \right] \\ &+D_x \left[\frac{C(i+1, j, n) - 2C(i, j, n) + C(i-1, j, n)}{(\Delta x)^2} \right] \\ &+D_y \left[\frac{C(i, j+1, n) - 2C(i, j, n) + C(i, j-1, n)}{(\Delta y)^2} \right] - kC(i, j, n) \end{aligned} \right\} \end{aligned} \tag{6.22}$$

Implicit or partially implicit solution methods are very effective at maintaining stability. However, they increase the computational burden significantly. For a system with N total nodes, a set of N simultaneous equations must be solved *at each time step*. Efficient methods for implementing this type of solution are discussed below. First though, the remaining building blocks of the solution, the initial and boundary conditions, must be specified.

Specification of Initial and Boundary Conditions With any of the finite difference solution methods thus far described, initial and boundary conditions are needed. For dynamic simulations, initial conditions are needed for the model state variables (e.g., contaminant concentrations) at *all nodes* at time $t = 0$ in order to initiate the computations. In some problems these conditions are unknown or highly uncertain, especially when the model is used for long-term simulations involving historic reconstruction. Even when the primary interest is in recent, current, or future values predicted by the model, historic conditions, perhaps beginning many years earlier, may be needed to initiate the calculations. In some cases accurate selection of these initial conditions may not be necessary—the model equations may have a short enough "memory" that the initial conditions do not influence the solution. This is fortunate, especially if you have little information for specifying the initial conditions. For other problems, the initial conditions do matter, and they must be determined as part of the calibration or parameter estimation process. The only way to tell is to try solving the equations with different initial conditions and see how the solutions differ.

Boundary conditions are required to specify the system state variables and/or their derivatives (e.g., concentrations and mass flux conditions) at all system boundaries, whether a dynamic or a steady-state solution is implemented. Consider the set of boundary nodes identified in Figure 6.6. Along the left-hand side of the domain (where $i = 1$), values of $C(i-1, j,)$ are off the grid and unavailable. How then can the backward or central differencing expressions for the first derivative [Eqs. (6.10) and (6.11), respectively] or the expression for the second derivative [Eq. (6.12)] be included in the mass-balance equation since these both include $C(i-1, j, t)$? Similar problems arise with the identification of $C(i, j-1, t)$ for nodes along the top row, $C(i+1, j, t)$ for nodes in the last (right-hand) column, and $C(i, j+1, t)$ for nodes along the bottom row. A creative solution is needed for this dilemma.

Three types of boundary conditions may apply:

Type 1, or *Dirichlet* boundary conditions: Here the model state variables are specified, for example, the concentrations along the boundary are known (at all times);

Type 2, or *Neumann* boundary conditions: The derivatives of the model state variables are known. In the case of advective–dispersive transport, this involves specification of the spatial derivative of the concentration normal to the boundary, either $\partial C/\partial x$, $\partial C/\partial y$, or $\partial C/\partial z$.

Type 3, *Cauchy,* or *mixed* boundary conditions: Linear combinations of the state variables and their derivatives are specified.

When the concentrations along a boundary are known, a Dirichlet boundary condition is used. Given the fluid flow rate, $Q = uA$, this is equivalent to specifying the advective flux across the boundary, $= uAC$, where A is the cross-sectional area associated with the grid point (normal to the boundary). A Neumann boundary condition is used to specify the dispersive flux across a boundary $[-DA(\partial C/\partial x)]$, since it involves fixing the value of the derivative. The total (advective + dispersive) flux across a boundary, $uAC - DA(\partial C/\partial x)$, is specified using a mixed boundary condition.[4]

[4] Confusion may arise in recognizing that advection at the boundary is determined by the concentration at the boundary and not the concentration derivative since the term representing advection in the mass-balance equation includes the derivative. However, in the derivation of the mass-balance equation, this term actually expressed the *change* in advection that occurs at the point, leading to mass accumulation or loss, and not the advection across the point itself. Similarly, specifications related to dispersion are implemented by fixing the first derivative, even though the second derivative appears in the dispersion term of the mass-balance equation.

To implement a Dirichlet boundary condition, the concentrations along the appropriate boundary row or column are set to their known values throughout the calculation. To implement a Neumann or a mixed boundary condition, image nodes are added as an extra row or column beyond the grid boundary. Concentrations at these grid points are set equal to the values necessary to maintain the known flux (total or dispersive) across the boundary. The exact equation used depends upon the type of differencing used to represent the derivatives. The steps taken to specify boundary conditions and associated node calculations are illustrated in the following example.

EXAMPLE 6.2 BOUNDARY CONDITION SPECIFICATION FOR A FINITE DIFFERENCE MODEL

To illustrate different types of boundary conditions and how they might be implemented in a finite difference solution, consider once again the system depicted in Figure 6.6. The following assumptions are made for each boundary, with the indicated implications for numerical computation.

1. Concentrations are known at all times along the lhs of the domain (where $i = 1$) and along the bottom boundary ($j = 6$). The mass-balance equation [Eq. (6.22)] is not needed for these nodes. However, the known values at these boundary points are used when implementing Eq. (6.22) for the nodes in the second column ($i = 2$) and for the nodes in the next-to-the-last row ($j = 5$), since these known concentrations constitute the respective values of $C(i-1, j, n)$ (for calculations in the second column) and $C(i, j+1, n)$ (for calculations in the next-to-the-last row). As a result, given the 6×6 grid in Figure 6.6, the number of equations that must be solved is reduced by $6 + 5 = 11$. The use of type 1 boundary nodes thus reduces the number of equations that must be solved. Note that the total fluxes across the upstream and bottom boundaries are *not* specified since the derivatives (and with them, the dispersive fluxes) across these boundaries are not input, but are computed as part of the model solution.

2. Assume that zero fluxes occur across the top boundary and the downstream (right-hand) boundary. Since there is no flow across the top boundary and the advective flux is therefore zero (by definition), this is equivalent to stating that the dispersive flux across the top boundary is zero. At the downstream boundary, where fluid flow leaves the system, the *total* flux is set equal to zero.[5] In both cases a set of fictitious image nodes is needed along the boundary, at a

[5] Like the semi-infinite column boundary condition used for one-dimensional analytical solutions ($\partial C/\partial x = 0$ at $x = \infty$), this boundary condition only makes sense when the downstream boundary is "far downstream," so that the concentrations computed there are not really of concern and have little effect on the computed concentrations that are of interest, further upstream. A bigger grid system, extending further downstream in the x direction, may be necessary to accomplish this, and this assumption should be tested by varying the length of extension to ensure that the downstream boundary does indeed not affect the concentrations computed at the upstream points of interest.

distance Δy above the top row and Δx to the right of the last column. Assuming that the central difference expression is used to define the first derivative [as it is in Eqs. (6.16), (6.20), and (6.22)], the concentrations at the image nodes are computed as follows:

For the top boundary: Beginning with the equation for the dispersive flux,

$$- D_y A \frac{\partial C}{\partial y} = 0$$

and substituting the appropriate node concentrations into the central difference expression gives

$$- D_y A \left[\frac{C(i, 2, n) - C(i, 0, n)}{2\,\Delta y} \right] = 0$$

where $j = 0$ refers to the image nodes in the row above the top boundary.

This equation can be solved for the image node concentrations at each time step. In particular, the zero-dispersive flux requirement can only be satisfied when the image node concentrations are set equal to those of their "partner" nodes in the second row:

$$C(i, 0, n) = C(i, 2, n)$$

For the downstream boundary: Beginning with the *total* flux equation

$$uAC - D_x A \frac{\partial C}{\partial x} = 0$$

and substituting the appropriate node concentrations:

$$uAC(6, j, n) - D_x A \left[\frac{C(7, j, n) - C(5, j, n)}{2\,\Delta x} \right] = 0$$

yields the desired expression for the image node concentrations $[C(7, j, n)]$ at each time step:

$$C(7, j, n) = \left[\frac{2\,\Delta x\, u}{D_x} \right] C(6, j, n) - C(5, j, n)$$

Setting the top and downstream image nodes as indicated above implements their respective no-flux boundary conditions but does nothing to reduce (or increase) the number of equations that must be solved. Thus, for the 6×6 grid system in Figure 6.6, a total of $36 - 11 = 25$ equations must be solved at each time step. With the explicit method, each of the 25 equations is solved individually at each time step, as in Eq. (6.16). With the implicit or partially implicit method, the 25 equations must be solved simultaneously at each time step.

This example does not exhaust the different types of boundary conditions that might arise in environmental models. Variations can occur in different physical systems and at different media interfaces (in problems involving mass transport across media). A further example of boundary condition selection and implementation is found in Chapter 9, where a finite difference model is used to simulate groundwater flow and contaminant transport. In each case, the boundary condition and its implementation in the model must be deduced from the underlying mass or momentum conservation and transport conditions assumed at the boundary. When poorly understood or highly uncertain boundary conditions have a significant impact on the overall mass balance of the system and predicted concentrations at points of interest, it may be necessary to determine at least some of the boundary conditions through a model calibration or parameter estimation effort. Methods for parameter estimation of unknown or uncertain model coefficients and inputs are presented in Chapter 14.

Steady-State Solution Consider now the steady-state two-dimensional advective–dispersive transport problem. The concentration time derivative in Eq. (6.8) is set equal to zero:

$$0 = -u\frac{\partial C}{\partial x} + D_x\frac{\partial^2 C}{\partial x^2} + D_y\frac{\partial^2 C}{\partial y^2} - kC \tag{6.23}$$

This equation must be satisfied at all nodes. The same differencing expressions are used for the spatial derivatives and the boundary conditions as implemented above. Using central differencing for the advection term, Eq. (6.23) becomes

$$0 = -u\left[\frac{C(i+1,j) - C(i-1,j)}{2\,\Delta x}\right] + D_x\left[\frac{C(i+1,j) - 2C(i,j) + C(i-1,j)}{(\Delta x)^2}\right]$$
$$+ D_y\left[\frac{C(i,j+1) - 2C(i,j) + C(i,j-1)}{(\Delta y)^2}\right] - kC(i,j) \tag{6.24}$$

Note that the time dimension (n) is no longer included. The equation is rewritten to isolate $C(i,j)$ on the lhs:

$$C(i,j) = \left\langle [-u]\left[\frac{C(i+1,j) - C(i-1,j)}{2\,\Delta x}\right] + D_x\left[\frac{C(i+1,j) + C(i-1,j)}{(\Delta x)^2}\right]\right.$$
$$\left. + D_y\left[\frac{C(i,j+1) + C(i,j-1)}{(\Delta y)^2}\right]\right\rangle \Bigg/ \left\{\frac{2D_x}{(\Delta x)^2} + \frac{2D_y}{(\Delta y)^2} + k\right\} \tag{6.25}$$

For the example system above, the boundary conditions again determine known concentrations for the first column and the bottom row. Similarly, image nodes are added above the top row and beyond the last column with appropriately fixed concentrations to implement the no-flux assumptions. This leaves a set of 25(= 36 − 11) simultaneous linear algebraic equations that must be solved *once*. In general, steady-state problems using finite difference or finite cell methods require solution of a set of simultaneous equations once. Dynamic models with explicit solution methods require (the much simpler) solution of a sequence of independent equations at each time step, while dynamic models with implicit methods require solution of a set of simultaneous equations at each time step. We next illustrate a particular set of iterative methods that are especially useful when solving "sparse" systems of equations of the type considered thus far.

Iterative Solution Methods for Simultaneous Linear Algebraic Equations In the example above, a set of 25 simultaneous equations must be solved at each time step to implement a (partially) implicit solution, or once to determine the steady-state solution. While matrix inversion or related procedures can be used to accomplish this, this problem is especially suited to solution using iterative techniques. These techniques are relatively easy to implement when the problem matrix is sparse, as occurs when the mass-balance equation for a node includes concentrations at only a few other system nodes: typically the upstream and downstream nodes for a one-dimensional problem, the four surrounding nodes for a two-dimensional problem, or the six surrounding nodes for a three-dimensional problem.

The first step in an iterative solution is to isolate the targeted unknown state variables on the lhs of each equation, with one equation for each of the unknowns. Thus, for the dynamic model with implicit solution described above, Eq. (6.22) is rewritten for each of the 25 nodes by moving all terms involving $C(i, j, n+1)$ to the lhs of the equation:

$$C(i, j, n+1)\left\{1 + \Delta t\,\alpha\left[\frac{2D_x}{(\Delta x)^2} + \frac{2D_y}{(\Delta y)^2} + k\right]\right\} = C(i, j, n)$$

$$+ \Delta t\,\alpha\left\{\begin{aligned} &-u\left[\frac{C(i+1, j, n+1) - C(i-1, j, n+1)}{2\,\Delta x}\right] \\ &+D_x\left[\frac{C(i+1, j, n+1) + C(i-1, j, n+1)}{(\Delta x)^2}\right] \\ &+D_y\left[\frac{C(i, j+1, n+1) + C(i, j-1, n+1)}{(\Delta y)^2}\right] \end{aligned}\right\}$$

$$+ \Delta t(1-\alpha)\left\{\begin{aligned} &-u\left[\frac{C(i+1, j, n) - C(i-1, j, n)}{2\,\Delta x}\right] \\ &+D_x\left[\frac{C(i+1, j, n) - 2C(i, j, n) + C(i-1, j, n)}{(\Delta x)^2}\right] \\ &+D_y\left[\frac{C(i, j+1, n) - 2C(i, j, n) + C(i, j-1, n)}{(\Delta y)^2}\right] - kC(i, j, n) \end{aligned}\right\}$$

Dividing both sides of the equation by

$$\left\{1 + \Delta t\,\alpha\left[\frac{2D_x}{(\Delta x)^2} + \frac{2Dy}{(\Delta y)^2} + k\right]\right\}$$

yields the targeted expression, with $C(i, j, n+1)$ only found on the lhs:

$$C(i, j, n+1) = \left[\frac{1}{1 + \Delta t\,\alpha\left(\dfrac{2D_x}{(\Delta x)^2} + \dfrac{2D_y}{(\Delta y)^2} + k\right)}\right]$$

$$\times \left((C(i, j, n) + \Delta t\,\alpha \left\{ \begin{array}{l} -u\left[\dfrac{C(i+1, j, n+1) - C(i-1, j, n+1)}{2\,\Delta x}\right] \\ +D_x\left[\dfrac{C(i+1, j, n+1) + C(i-1, j, n+1)}{(\Delta x)^2}\right] \\ +D_y\left[\dfrac{C(i, j+1, n+1) + C(i, j-1, n+1)}{(\Delta y)^2}\right] \end{array} \right\} \right.$$

$$\left. + \Delta t(1-\alpha) \left\{ \begin{array}{l} -u\left[\dfrac{C(i+1, j, n) - C(i-1, j, n)}{2\,\Delta x}\right] \\ +D_x\left[\dfrac{C(i+1, j, n) - 2C(i, j, n) + C(i-1, j, n)}{(\Delta x)^2}\right] \\ +D_y\left[\dfrac{C(i, j+1, n) - 2C(i, j, n) + C(i, j-1, n)}{(\Delta y)^2}\right] - kC(i, j, n) \end{array} \right\}) \right) \tag{6.26}$$

Equation (6.26) looks daunting but is easy to implement in the iterative solution framework. To do this, the full set of 25 equations corresponding to each unknown value of $C(i, j, n+1)$ is written, an initial guess for each of the $C(i, j, n+1)$ is made, and the equations are solved repeatedly over many iterations until convergence is achieved. That is, beginning with an initial guess, $C(i, j, n+1)^0$ for each of the 25 nodes where solution is required, values of $C(i, j, n+1)^1$ are computed, values of $C(i, j, n+1)^2$ computed from these, and eventually $C(i, j, n+1)^{m+1}$ computed from the $C(i, j, n+1)^m$ until:

$$C(i, j, n+1)^{m+1} - C(i, j, n+1)^m < \varepsilon \tag{6.27}$$

where ε is a very small value chosen to test for convergence. When the convergence criterion is satisfied at all nodes, the values of the target concentrations at iteration number m (or $m+1$) are accepted and used to move forward to the next time step.

Within each iteration, the calculation proceeds across the grid, and Eq. (6.26) is evaluated for all of the target nodes. At iteration m, the unknown concentration values at time $n+1$ on the rhs of Eq. (6.26) $[C(i-1, j, n+1), C(i+1, j, n+1), C(i, j-1, n+1)$ and $C(i, j+1, n+1)]$ can all be evaluated using their values from iteration $m-1$:

$$C(i,j,n+1)^m = \left[\frac{1}{1+\Delta t\,\alpha\left(\dfrac{2D_x}{(\Delta x)^2}+\dfrac{2D_y}{(\Delta y)^2}+k\right)}\right]$$

$$\times\left(\begin{array}{l} (C(i,j,n) + \Delta t\,\alpha \left\{\begin{array}{l} -u\left[\dfrac{C(i+1,j,n+1)^{m-1} - C(i-1,j,n+1)^{m-1}}{2\,\Delta x}\right] \\ +D_x\left[\dfrac{C(i+1,j,n+1)^{m-1} + C(i-1,j,n+1)^{m-1}}{(\Delta x)^2}\right] \\ +D_y\left[\dfrac{C(i,j+1,n+1)^{m-1} + C(i,j-1,n+1)^{m-1}}{(\Delta y)^2}\right] \end{array}\right\} \\ \\ +\Delta t(1-\alpha)\times\left\{\begin{array}{l} -u\left[\dfrac{C(i+1,j,n) - C(i-1,j,n)}{2\,\Delta x}\right] \\ +D_x\left[\dfrac{C(i+1,j,n) - 2C(i,j,n) + C(i-1,j,n)}{(\Delta x)^2}\right] \\ +D_y\left[\dfrac{C(i,j+1,n) - 2C(i,j,n) + C(i,j-1,n)}{(\Delta y)^2}\right] - kC(i,j,n) \end{array}\right\}) \end{array}\right) \tag{6.28}$$

This calculation, known as *Jacobi iteration*, can proceed in any order through the grid system since all the unknown concentrations at iteration $m-1$ are computed, stored, and used in the calculation for iteration m. An alternative approach, the *Gauss–Seidel* iteration, uses unknown concentrations that have already been calculated in iteration m for the calculation of subsequent unknown concentrations during the same iteration. To envision this, imagine sequentially sweeping across each row from left to right, beginning each iteration at the upper left-hand corner (at $i = 1, j = 1$), moving across the first row until it is completed, then moving to the beginning of the second row ($i = 1, j = 2$), continuing to the right, and so on, until the full set of unknown nodes is evaluated. With this order of calculation, values of $C(i-1, j, n+1)^m$ (from the column to the left) and $C(i, j-1, n+1)^m$ (from the row above) will already be available when it is time to compute $C(i, j, n+1)^m$. Why not use them in Eq. (6.26) instead of the old values of $C(i-1, j, n+1)^{m-1}$ and $C(i, j-1, n+1)^{m-1}$ from the previous iteration? If (as we hope!) each iteration brings the concentration estimates closer to their correct values, the latter values should be more accurate and allow for more rapid convergence to the correct values. Gauss–Seidel iteration does indeed allow for more rapid convergence. It has the further advantage that it is *easier* to program since values of the unknown concentrations from the previous iteration no longer need to be stored once $C(i-1, j, n+1)^m$ and $C(i, j-1, n+1)^m$ are computed and their tests for convergence are implemented. The resulting Gauss–Seidel iterative equation is given by:

$$
\begin{aligned}
C(i, j, n+1)^m &= \left[\frac{1}{1+\Delta t\,\alpha\left(\dfrac{2D_x}{(\Delta x)^2}+\dfrac{2D_y}{(\Delta y)^2}+k\right)}\right] \\
&\times \left((C(i,j,n) + \Delta t\,\alpha \left\{ \begin{array}{l} -u\left[\dfrac{C(i+1,j,n+1)^{m-1} - C(i-1,j,n+1)^m}{2\,\Delta x}\right] \\ +D_x\left[\dfrac{C(i+1,j,n+1)^{m-1} + C(i-1,j,n+1)^m}{(\Delta x)^2}\right] \\ +D_y\left[\dfrac{C(i,j+1,n+1)^{m-1} + C(i,j-1,n+1)^m}{(\Delta y)^2}\right] \end{array} \right\} \right. \\
&\left. + \Delta t(1-\alpha) \left\{ \begin{array}{l} -u\left[\dfrac{C(i+1,j,n) - C(i-1,j,n)}{2\,\Delta x}\right] \\ +D_x\left[\dfrac{C(i+1,j,n) - 2C(i,j,n) + C(i-1,j,n)}{(\Delta x)^2}\right] \\ +D_y\left[\dfrac{C(i,j+1,n) - 2C(i,j,n) + C(i,j-1,n)}{(\Delta y)^2}\right] - kC(i,j,n) \end{array} \right\}) \right)
\end{aligned}
\tag{6.29}
$$

This same approach can be used to solve the steady-state model equations. To do this, note that the working equation has already been modified to isolate $C(i, j)$ on the lhs [see Eq. 6.25)]. Equation (6.25) is then written for Gauss–Seidel iteration as follows:

$$
\begin{aligned}
C(i, j)^m = \Bigg\langle &[-u]\left[\frac{C(i+1,j)^{m-1} - C(i-1,j)^m}{2\,\Delta x}\right] \\
&+ D_x\left[\frac{C(i+1,j)^{m-1} + C(i-1,j)^m}{(\Delta x)^2}\right] \\
&+ D_y\left[\frac{C(i,j+1)^{m-1} + C(i,j-1)^m}{(\Delta y)^2}\right]\Bigg\rangle \Bigg/ \left\{\frac{2D_x}{(\Delta x)^2} + \frac{2D_y}{(\Delta y)^2} + k\right\}
\end{aligned}
\tag{6.30}
$$

With any of the iterative solution methods, initial guesses are required for each of the unknowns. For the steady-state model, these may be estimated by interpolating between boundary conditions, if they are known. For the dynamic model with implicit solution, the initial guesses for each time step may be set to the computed values from the previous time step, or to estimates for the new time step computed using the explicit solution [e.g., Eq. 6.16)].

Other techniques, such as *successive over relaxation (SOR)*, are available to further speed the rate of conversion of an iterative solution (Hoffman, 1992, p. 56; Wang and Anderson, 1982, p. 27). With SOR, the change that occurs with each iteration is extended by first noting the change, multiplying it by a factor ω (generally between 1.0 and 2.0), and adding the product to the previous iteration. A two-step procedure (similar to a predictor–corrector method) thus results. For example, for modification of the Gauss–Seidel solution of the steady-state model [Eq. (6.30)], SOR is implemented by first computing a "predictor" value of $C(i, j)^{m*}$ using Eq. (6.30), then computing the "corrector" as:

$$C(i, j)^m = C(i, j)^{m-1} + \omega\left[C(i, j)^{m*} - C(i, j)^{m-1}\right] \tag{6.31}$$

Though convergence is generally faster with SOR, problems with overshooting the correct answer can occur, causing oscillation around the correct answer. A method that utilizes decreasing values of ω (decreasing toward 1.0) as m increases can be used if this problem occurs. In most situations it is safest to stick with the Gauss–Seidel method, unless the computation time resulting from the iterative solution is excessively large (as may be the case when, as with the implicit dynamic solution, the iteration is required at each time step).

6.2.2 Finite Element Methods

A second numerical approach for solving partial differential equations is the use of finite element methods. Finite element methods originated in the field of solid mechanics and are the standard methods for solving structural analysis problems. They are less widely used than finite difference methods for fluid mechanics and heat and mass transfer but offer significant advantages for some applications. Finite element methods are especially useful for problems with irregular geometry, inhomogeneous properties, and complicated loadings. It is relatively easy to vary the size of the elements over the model domain, whereas grid cell size variations can introduce significant complications with finite differences. Finite element methods are also more accurate and stable than finite difference methods for comparable element and grid cell sizes.

As discussed in the previous section, finite difference methods are developed by replacing spatial or temporal derivatives in a partial differential equation with differences defined between nodes on a regular rectangular space–time grid. Finite element methods also start with a discretized domain and produce algebraic equations from differential equations. However, with finite elements, the value of the state variable of interest across a small element of the domain is approximated by an interpolating function, which is usually a polynomial. Elements are regular polygons in one, two, or three dimensions. Values at the vertices or nodes of each element are determined to optimally satisfy the partial differential equations that apply to the system. The solution is assembled over the full domain by requiring that values of the state variables along the edges of adjacent elements match and that boundary conditions be satisfied.

A description of the main steps required in finite element analysis, with an illustration for the 1D steady-state advection–diffusion equation, is found in the supplemental information for this chapter at www.wiley.com/college/ramaswami. More in-depth presentations of finite element methods are provided in textbooks dedicated to the subject (e.g., Desai, 1979; Wait and Mitchell, 1985; and Zienkiewicz and Taylor, 2000). Bear (1979), Wang and Anderson (1982), and Zheng and Bennett (1995) provide accessible introductions to finite elements in the context of groundwater flow and contaminant transport modeling; more in-depth presentations are found in (Guyman (1970), Guyman et al. (1970), Pinder and Gray (1977), Huyakorn and Pinder (1983), and Carey (1995).

6.3 SOLUTIONS TO NONLINEAR SYSTEMS OF EQUATIONS

Section 6.2.1 described the Jacobi and Gauss–Siedel methods for solving systems of linear algebraic equations, as employed for numerical solution of the finite difference representation of advective–dispersive transport phenomena. Indeed, these and other matrix inversion techniques are the basic numerical methods for simultaneously solving any general system of linear equations. However, linear relationships are often not adequate to represent real-world systems, particularly when chemical reactions are involved. Consideration of chemical equilibria, dependence of equilibrium constants and rate constants on temperature (often an exponential relationship), and inclusion of chemical kinetics of orders greater than 1, introduce a high degree of nonlinearity into the relationships between different variables in physicochemical systems. This section describes numerical methods used to solve the systems of nonlinear algebraic equations that may apply to these problems. As always, for any solution technique to work, the number of equations must be equal to the number of variables in the system. Typically, the variables in a system may include the concentrations of various chemical species involved in reactions, ambient temperature, pressure, and so forth. Kinetic and equilibrium relationships and charge and mass balances provide the framework for the nonlinear equations relating these variables to each other. At a single time step, the multispecies nonlinear system of equations is solved using the techniques described below.

Before addressing multispecies systems of equations, it is useful to first consider the Newton–Raphson method for solving a nonlinear equation involving just a single variable. Consider a polynomial equation, $f(x)$, of order greater than 1. Solving this equation involves finding values of x at which $f(x)$ will go to 0, that is, $f(x) = 0$. We do not know the roots of this equation a priori, and use an iterative scheme to approach the roots from an initial guess, x_0. If x were the real root of the equation, $f(x)$ would be exactly 0. In addition, if the initial guess, x_0, was sufficiently close to the real root, x, $f(x)$ could be computed from the first derivative $f'(x)$ evaluated at $x_0[=f'(x_0)]$ as shown below, and set to 0:

$$f(x) = f(x_0) + f'(x_0) \times [x - x_0] = 0 \tag{6.32}$$

which, upon rearrangement, yields

$$x = x_0 - \frac{f(x_0)}{f'(x_0)} \tag{6.33}$$

In reality, since we have no a priori knowledge of the value of the roots of the equation, Eq. (6.33) is used iteratively to converge on the roots of the equation $f(x) = 0$, starting with an initial guess, x_0. If we specify each iteration number as m:

$$x_{m+1} = x_m - \frac{f(x_m)}{f'(x_m)} \tag{6.34}$$

Equation (6.34) is called Newton's method or the Newton–Raphson method for finding the roots of nonlinear equations. The method rapidly converges to the solution but is computationally expensive since the exact derivative $[f'(x)]$ needs to be first determined analytically and then evaluated at each iteration. Note that rearrangement of Eq. (6.34) yields the discrete definition of the slope $f'(x)$ when $f(x) = 0$, that is, $f'(x_m) = [0 - f(x_m)]/[x_{m+1} - x_m]$. The Newton–Raphson method is closely related to other slope-type iteration methods such as the bisection method and the method of Regula Falsi, as illustrated in Figure 6.8.

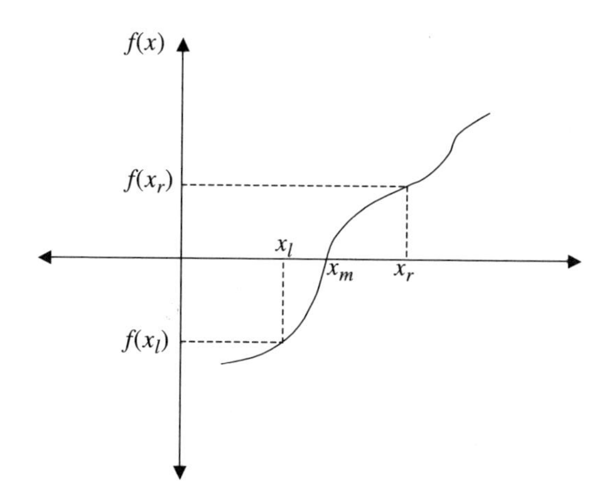

Bisection Method: Slope = $\left[\frac{0 - f(x_l)}{x_m - x_l}\right] = \left[\frac{f(x_r) - f(x_l)}{x_r - x_l}\right]$

Newton's Method: Slope = $f'(x) = \left[\frac{0 - f(x)}{x_m - x}\right]$ for any x near the root

Figure 6.8 Using the concept of slope in the bisection method where the root, x_m, is converged on from a point x_r to the right and a point x_l to the left, and in the Newton–Raphson method where the derivative function $f'(x)$ is used to converge toward the real root $x_0 = x_m$.

Problems that can arise with the Newton–Raphson method include the presence of an inflection point in the function near the location of the root, which can cause divergence from the real solution. The solution may oscillate around a local minimum or maximum in the function if one is encountered on the way to the root. The presence of multiple roots in the interval within which iterations are being commenced can also cause masking of one root by another, that is, depending upon the choice of the initial guess x_0, the solution may converge consistently toward one root, missing the other solution.

Next consider a set of n equations, F_1, F_2, F_3, ... F_n, involving n variables, x_1, $x_2, x_3, \ldots, x_n$. The set of equations may be written as:

$$
\begin{aligned}
F_1(x_1, x_2, x_3, \ldots, x_n) &= 0 \\
F_2(x_1, x_2, x_3, \ldots, x_n) &= 0 \\
F_3(x_1, x_2, x_3, \ldots, x_n) &= 0 \\
&\vdots \\
F_n(x_1, x_2, x_3, \ldots, x_n) &= 0
\end{aligned}
\tag{6.35}
$$

This set of equations is solved when the values of $(x_1, x_2, x_3, \ldots, x_n)$ simultaneously satisfy $F_n = 0$, for all n shown in Eq. (6.35). If our initial guesses of the solution set are $(x_{1,0}, x_{2,0}, x_{3,0}, \ldots, x_{n,0})$, an expression analogous to Eq. (6.32) for the multivariate problem specified in Eq. (6.35) can be written as:

$$
\begin{aligned}
F_1(x_1, x_2, x_3, \ldots, x_n) &= F_1(x_{1,0}, x_{2,0}, x_{3,0}, \ldots, x_{n,0}) + \frac{\partial F_1}{\partial x_1}\delta x_1 + \frac{\partial F_1}{\partial x_2}\delta x_2 + \ldots \frac{\partial F_1}{\partial x_n}\delta x_n = 0 \\
F_2(x_1, x_2, x_3, \ldots, x_n) &= F_2(x_{1,0}, x_{2,0}, x_{3,0}, \ldots, x_{n,0}) + \frac{\partial F_2}{\partial x_1}\delta x_1 + \frac{\partial F_2}{\partial x_2}\delta x_2 + \ldots \frac{\partial F_2}{\partial x_n}\delta x_n = 0 \\
F_3(x_1, x_2, x_3, \ldots, x_n) &= F_3(x_{1,0}, x_{2,0}, x_{3,0}, \ldots, x_{n,0}) + \frac{\partial F_3}{\partial x_1}\delta x_1 + \frac{\partial F_3}{\partial x_2}\delta x_2 + \ldots \frac{\partial F_3}{\partial x_n}\delta x_n = 0 \\
&\vdots \\
F_n(x_1, x_2, x_3, \ldots, x_n) &= F_n(x_{1,0}, x_{2,0}, x_{3,0}, \ldots, x_{n,0}) + \frac{\partial F_n}{\partial x_1}\delta x_1 + \frac{\partial F_n}{\partial x_2}\delta x_2 + \ldots \frac{\partial F_n}{\partial x_n}\delta x_n = 0
\end{aligned}
\tag{6.36}
$$

In Eq. (6.36), $\delta x_i = (x_i - x_{i,0})$. Note the similarity between the system of n equations in Eqs. (6.36) and (6.32), which was written for a single variable. As above, our first guess may not be very good, so multiple iterations may be required. In matrix notation, the expansion equation for iteration $m + 1$ may be written as follows:

$$
\mathbf{F}(\mathbf{x}_{m+1}) = \mathbf{F}(\mathbf{x} + \boldsymbol{\delta}\mathbf{x}) = \mathbf{F}(\mathbf{x}_m) + J(\mathbf{x}_m)\boldsymbol{\delta}\mathbf{x} = \mathbf{0} \tag{6.37}
$$

where $\mathbf{x} = \{x_1, x_2, \ldots, x_n\}^{\mathrm{T}}$, $\mathbf{F} = \{F_1, F_2, \ldots, F_n\}^{\mathrm{T}}$, and $\boldsymbol{\delta}\mathbf{x} = \{\delta x_1, \delta x_2, \ldots, \delta x_n\}^{\mathrm{T}}$, with $\delta x_i = (x_{m+1} - x_{i,m})$. The matrix J in Eq. (6.37) is known as the Jacobian matrix, and is defined as:

$$J(x_m) = \begin{bmatrix} \left[\frac{\partial F_1}{\partial x_1}\right] & \left[\frac{\partial F_1}{\partial x_2}\right] & \cdots & \left[\frac{\partial F_1}{\partial x_n}\right] \\ \left[\frac{\partial F_2}{\partial x_1}\right] & \left[\frac{\partial F_2}{\partial x_2}\right] & \cdots & \left[\frac{\partial F_2}{\partial x_n}\right] \\ \cdots & \cdots & \cdots & \cdots \\ \left[\frac{\partial F_n}{\partial x_1}\right] & \left[\frac{\partial F_n}{\partial x_2}\right] & \cdots & \left[\frac{\partial F_n}{\partial x_n}\right] \end{bmatrix} \tag{6.38}$$

The partial derivatives in this matrix are evaluated using values from the mth iteration, x_m. With $\mathbf{F}(\mathbf{x}_{m+1}) = \mathbf{0}$, Eq. (6.37) can be rearranged into the standard form for a system of linear algebraic equations:

$$J(x_m)[\delta\mathbf{x}_m] = -\mathbf{F}(\mathbf{x}_m) \tag{6.39}$$

Rearrangement of Eq. (6.57) yields

$$\mathbf{x}_{m+1} = \mathbf{x}_m - [J(\mathbf{x}_m)]^{-1}\,\mathbf{F}(\mathbf{x}_m) \tag{6.40}$$

Once again, note the similarities between the iteration algorithm for Newton's method for a single variable shown in Eq. (6.34) and that represented in matrix form in Eq. (6.40) for the multivariate problem. While the matrix representation for multivariate nonlinear systems of equations provides convergence when certain matrix constraints are met, the evaluation and inversion of the Jacobian matrix can be difficult. Instead, a modified Newton–Raphson method is used for faster computations. In this method, the partial derivative matrix is simplified and approximated by using only the diagonal terms of the matrix shown in Eq. (6.38). Thus, the iteration algorithm becomes

$$\begin{aligned} x_{1,m+1} &= x_{1,m} - \frac{F_1(x_1, x_2, \ldots, x_n)_m}{\left[\frac{\partial F_1}{\partial x_1}\right]_{(X_1, X_2, \ldots, X_n)_m}} \\ &\vdots \\ x_{n,m+1} &= x_{n,m} - \frac{F_n(x_1, x_2, \ldots, x_n)_m}{\left[\frac{\partial F_n}{\partial x_n}\right]_{(X_1, X_2, \ldots, X_n)_m}} \end{aligned} \tag{6.41}$$

Example 6.3 illustrates the use of the modified Newton–Raphson method for a two-variable problem, mimicking (for a relatively simple case) some of the chemical equilibrium equations that are presented in Chapters 2 and 12.

EXAMPLE 6.3

Consider a reversible reaction in which a chemical species A1 is transformed into another species A2, with equilibrium constant $K = 2$. By stoichiometry, it is given that 2 [A1] ⇔ [A2]. The total molar concentration of A introduced into the system is 1 mol/L, initially introduced as A1. Hence the two relevant equations linking the two variables A1 and A2 are

$$K = \frac{[\text{A2}]}{[\text{A1}]^2} = 2 \qquad \text{Equilibrium relationship}$$

$$[\text{A1}] + 2\,[\text{A2}] = 1 \qquad \text{Mass-balance relationship}$$

Substituting the first equation ($[\text{A2}] = 2[\text{A1}]^2$) into the second one, and solving the resulting quadratic equation, we get the exact solution (to the sixth decimal place!) as: A1 = 0.390388 and A2 = 0.304806. This indicates that, of 1 mol of A1 introduced into a 1-L reactor, 0.61 mol are converted to A2, yielding 0.305 mol of A2 (due to stoichiometry) with 0.39 mol of A1 remaining in solution.

Now, we shall see if the exact solution can be found by the modified Newton–Raphson iteration method. To be consistent with the notation used in the discussion of theory, we see that:

$$F_1(\text{A1}, \text{A2}) = -2[\text{A1}]^2 + [\text{A2}] = 0 \qquad \partial F_1/\partial \text{A1} = -4[\text{A1}]$$

$$F_2(\text{A1}, \text{A2}) = [\text{A1}] + 2[\text{A2}] - 1 = 0 \qquad \partial F_2/\partial \text{A2} = 2$$

We can begin with an initial guess that A1 = 1 and A2 = 0, that is, the situation at the point when the reaction had not yet commenced. The first 27 iterations are shown below:

Iteration	A1	A2	F_1	F_2	$\partial F_1/\partial A1$	$\partial F_2/\partial A2$
1	1	0	−2	0	−4	2
2	0.5	0	−0.5	−0.5	−2	2
3	0.25	0.25	0.125	−0.25	−1	2
4	0.375	0.375	0.09375	0.125	−1.5	2
5	0.4375	0.3125	−0.07031	0.0625	−1.75	2
6	0.397321	0.28125	−0.03448	−0.04018	−1.58929	2
7	0.375627	0.301339	0.019148	−0.02169	−1.50251	2
8	0.388371	0.312186	0.010522	0.012744	−1.55348	2
9	0.395144	0.305814	−0.00646	0.006773	−1.58058	2
10	0.391055	0.302428	−0.00342	−0.00409	−1.56422	2
11	0.388868	0.304473	0.002035	−0.00219	−1.55547	2

(continued)

Iteration	A1	A2	F_1	F_2	$\partial F_1/\partial A1$	$\partial F_2/\partial A2$
12	0.390177	0.305566	0.00109	0.001308	−1.56071	2
13	0.390875	0.304912	−0.00066	0.000698	−1.5635	2
14	0.390456	0.304562	−0.00035	−0.00042	−1.56182	2
15	0.390232	0.304772	0.000209	−0.00022	−1.56093	2
16	0.390366	0.304884	0.000112	0.000134	−1.56147	2
17	0.390438	0.304817	−6.7E–05	7.16E-05	−1.56175	2
18	0.390395	0.304781	−3.6E-05	−4.3E-05	−1.56158	2
19	0.390372	0.304802	2.15E-05	−2.3E-05	−1.56149	2
20	0.390386	0.304814	1.15E-05	1.38E-05	−1.56154	2
21	0.390393	0.304807	−6.9E-06	7.34E-06	−1.56157	2
22	0.390389	0.304803	−3.7E-06	−4.4E-06	−1.56156	2
23	0.390387	0.304806	2.2E-06	−2.4E-06	−1.56155	2
24	0.390388	0.304807	1.18E-06	1.41E-06	−1.56155	2
25	0.390389	0.304806	−7.1E-07	7.53E-07	−1.56155	2
26	0.390388	0.304806	−3.8E-07	−4.5E-07	−1.56155	2
27	0.390388	0.304806	2.26E-07	−2.4E-07	−1.56155	2

Notice that the solution converges to the exact solution within the third decimal place ($A1 = 0.390 \pm 0.001$) by the 12th iteration, and to the sixth decimal place by the 26nd iteration. Also note how the iterative solutions oscillate above and below the true solution in consecutive iterations. This is a feature of the slope technique, in which we are approaching the true solution from the right, then from the left, and so on, as also occurs in the bisection method (Fig. 6.8) for a single variable nonlinear equation. The solutions obtained with the full Jacobian matrix would converge with fewer iterations, but with more intensive matrix inversions that would become increasingly demanding as the number of equations is increased.

Chapter 12 presents applications of nonlinear solution techniques to determine chemical equilibria involving multiple chemical species participating in acid–base dissociation, dissolution–precipitation, oxidation–reduction, and surface complexation reactions. Computer packages such as MICROQL (Westall, 1986) and MINTEQ (Allison et al., 1991) implement techniques such as the Newton–Raphson method and are readily available for solution of a wide array of nonlinear equations involving multispecies chemical equilibria in aqueous systems.

6.4 SUMMARY

Methods for numerically solving four classes of mathematical models have been described in this chapter. Numerical integration techniques, such as the Euler–Cauchy method, the predictor–corrector method, and the Runge–Kutta method are described in Section 6.1 and can be used to solve chemical mass-balances ODEs with concentration represented as a function of a single variable, typically time. Initial concentrations in the system must be known to initiate these integration techniques.

Methods for solving partial differential equations, applicable in solving mass-balance differential equations over time and multidimensional space, were presented in Section 6.2. Section 6.2.1 focused on finite difference methods for solving PDEs, formulating forward differencing, backward differencing, and central differencing schemes for the spatial derivative, and implicit and explicit methods to step forward in time. Finite element methods for solving PDEs were introduced in Section 6.2.2. Both techniques for solving PDEs require the statement of initial (temporal) and (spatial) boundary conditions to initiate the iterations. Techniques for simultaneously solving a set of linear algebraic equations are presented in Section 6.2.1 as tools for implementing finite difference schemes. Solution techniques that address systems of nonlinear algebraic equations are presented in Section 6.3 and are typically used to model complex chemical reactions with multispecies equilibria or higher-order kinetics.

The numerical methods briefly presented here form the foundation of many computer codes and simulation packages designed to model contaminant transport and fate in the environment. In the applications presented in this chapter, all the input parameters to the models are assumed to be "single valued," that is, they have fixed, or "point-estimate" values assumed to be known a priori by the user. Single-valued, point-estimate input parameter values, while easy to incorporate into models, are not often representative of our understanding of the real world. In particular, modelers must also consider systematic (seasonal or spatial) variability in model parameter values, random fluctuations in these values with no known systematic underlying cause, as well as uncertainty in model parameters associated with a lack of a priori knowledge of the system. To address this need, Chapter 7 presents probabilistic techniques for the incorporation of random variables and random processes into environmental models.

7 Overview of Probabilistic Methods and Tools for Modeling

Environmental systems are highly variable in their properties and response to inputs. Furthermore, there is a great deal of uncertainty about these properties, future inputs, and responses. In this chapter we introduce the basic tools of probability used to model variable and uncertain environmental systems.

Variability refers to the inherent differences in environmental properties that occur over space and time and from one sample to another (e.g., the differences in exposure, susceptibility, and risk that occur between one individual and another in a target population). *Uncertainty* reflects a lack of knowledge of environmental processes and properties. While many of the same tools of probability and statistics can be applied to characterize variability and uncertainty, the need to carefully distinguish between them is widely recognized (e.g., Bogen and Spear, 1987; Burmaster and Wilson, 1996; Cullen and Frey, 1999), and we are careful to do so in the applications that follow.

In Section 1.5.1 we characterized deterministic models as those that calculate a single value for each model output, in contrast to stochastic models that produce a distribution of values for each prediction. While probabilistic methods provide the basic building blocks for stochastic models, the distinction between deterministic and stochastic models is not always clear-cut. For example, the random motions of fluid elements described in Sections 5.3 and 5.5 that lead to Fickian and non-Fickian dispersion are typically aggregated over many fluid elements and treated as deterministic processes at the continuum scale. Similarly, chemical transformations that are stochastic at the scale of individual particles and molecules are usually aggregated, using kinetic models to provide deterministic representations of bulk reaction processes. Individual chemical and fluid elements can be statistically simulated to yield the same results as deterministic process models for transport and reaction; this is the basis for the *population balance* method that tracks discrete pollutant particles through the environment (Patterson et al., 1981; Koch and Prickett, 1993; Visser, 1997).

Probability models describe the likelihood, or probability, of different outcomes or events. A *random variable* is a quantity that can take on different values for a variety of reasons, but with no specific mechanism or underlying cause that allows an outcome to be predicted with certainty. Instead, the variable is described by a *probability distribution function*. When the quantity is *ordered* in space and/or time, the probability model must also describe the nature of this ordering. Such quantities are referred to as *random processes*. We begin this chapter by describing models for random variables, followed by formulations for random processes often encountered in environmental systems.

The tools presented in the following sections include both probability models and statistical methods. Probability models are mathematical in nature and are used to describe the theoretical properties of an idealized *population*. Statistical methods are used to estimate probability models based on an observed *sample* from the population. Although the terms "probabilistic" and "statistical' are often used interchangeably, it is important to keep their different meanings in mind. We identify four main objectives in applying these methods:

1. To estimate the statistical properties of a population based upon physical models and constraints as well as actual observations of a (usually small) sample
2. To make predictions using environmental process models that employ the fitted probability models
3. To update the statistical properties of model input variables with new information and to assess the impact of this updated knowledge on model predictions and management decisions
4. To design monitoring programs that allow the collection of environmental data to accomplish objectives 1 to 3 in an effective and efficient manner

The primary purpose of this chapter is to provide tools for addressing the first two objectives. Objectives 3 and 4 are addressed in Chapter 14

7.1 MODELS FOR RANDOM VARIABLES

Random variables can be either *discrete* or *continuous*. Discrete random variables can only be assigned particular values (usually integers). Examples include the number of trees in a forest stand or the number of storm events or wet days in a year. Continuous random variables can be assigned all (integer and noninteger) real values, though their range may be limited to a subset of the real numbers, often between zero and a physical or practical upper bound. Examples include temperatures, stream flow rates, wind speeds (water and air *velocities* are also continuous but can be either positive or negative), soil densities and porosities, and pollutant concentrations.

The probability distribution for a discrete random variable, X, is described by its *probability mass function* (pmf):

$$p_X(x) = \text{Prob}[X = x] \tag{7.1}$$

Note that (uppercase) X is the name of the random variable, while (lowercase) x is the particular value of X. The pmf is dimensionless. Correspondingly, the probability distribution for a continuous random variable is described by its *probability density function* (pdf):

$$f_X(x) = \lim_{\Delta x \to 0} \left[\frac{\text{Prob}[x < X < x + \Delta x]}{\Delta x} \right] \tag{7.2}$$

The pdf has units of $[X^{-1}]$, representing the density of probability per unit interval of X. The pdf must be integrated over a given range to find the probability of occurrence over that range:

$$\text{Prob}[x_1 < X < x_2] = \int_{x_1}^{x_2} f_X(x)\,dx \tag{7.3}$$

The probability that a continuous random variable is equal to any *exact* value is thus zero.

Mixed random variables have discrete and continuous parts, with corresponding probability mass and density. Examples include daily rainfall depth and annual oil spill volumes in a harbor, each with probability mass at zero (when no events occur) and probability density at values greater than zero (when there is one or more event). Mixed variable representations may also be necessary for continuous random variables that are reported at discrete values (due, e.g., to roundoff), such as the age or height of individuals in a target population.

A fully equivalent representation of the probability distribution function for either a discrete, continuous, or mixed random variable is provided by the *cumulative distribution function* (cdf):

$$\begin{aligned} F_X(x) &= \text{Prob}[X \leq x] \\ &= \sum_{u \leq x} p_X(u) \qquad \text{for a discrete random variable} \\ &= \int_{u-\infty}^{x} f_X(u)\,du \qquad \text{for a continuous random variable} \end{aligned} \tag{7.4}$$

The cdf is the sum of the pmf values up to and including the value of interest, x, for a discrete random variable or the area under the pdf curve to the left of x for a continuous random variable.

The cdf is a monotonic function, increasing from zero to one over the range of X. The quantiles of a random variable are the values of X corresponding to particular percentiles of the cdf. For example, the 99th percentile value (quantile) of X is given by $X_{99} = F_X^{-1}(0.99)$, where $F_X^{-1}(\,)$ denotes the inverse function of the cdf.

Random variables are also characterized by their *moments*. The kth moment of a random variable is the expected value of X^k, given by:

$$\begin{aligned} E[X^k] &= \sum_{all\ x} x^k p_X(x) \qquad \text{for a discrete random variable} \\ &= \int_{x=-\infty}^{\infty} x^k f_X(x)\,dx \qquad \text{for a continuous random variable} \end{aligned} \tag{7.5}$$

The first moment, obtained by setting $k = 1$, is the mean of the random variable. The symbol μ_x ($= E[X]$) is often used for the mean.

Central moments are also used to described random variables. The kth central moment is the expected value of $\{X - E[X]\}^k$, given by:

$$E\left[\{X - E[X]\}^k\right] = \sum_{\text{all } x} (x - \mu_X)^k \, p_X(x) \quad \text{for a discrete random variable}$$

$$= \int_{x=-\infty}^{\infty} (x - \mu_X)^k \, f_X(x) \, dx \quad \text{for a continuous random variable} \tag{7.6}$$

The second central moment ($k = 2$) is known as the variance of the random variable. It may also be calculated directly from the first two (regular) moments as:

$$\begin{aligned} \text{Var}[X] &= E\left[\{X - \mu_X\}^2\right] \\ &= E[X^2] - \mu_X^2 \end{aligned} \tag{7.7}$$

The symbol σ_x^2 is often used for the variance, with units of $[X^2]$. The standard deviation, σ_x, is the square root of the variance, with units of $[X]$.

The mean characterizes the central tendency of a random variable, while the variance characterizes its spread or dispersion. Other measures of central tendency include the median:

$$\text{Median} = X_{50} = F^{-1}(0.5)$$

and the mode, which is the point of maximum probability density or mass. Alternative measures of dispersion include the mean absolute deviation:

$$\text{Mean absolute deviation} = E\left[|X - \mu_X|\right]$$

and various quantile intervals, such as the 90th percentile range:

$$I_{90} = F^{-1}(0.95) - F^{-1}(0.05)$$

the 95th percentile range:

$$I_{95} = F^{-1}(0.975) - F^{-1}(0.025)$$

and the interquartile range:

$$I_{50} = F^{-1}(0.75) - F^{-1}(0.25) \tag{7.8}$$

Further properties of the shape of a distribution can be inferred from higher-order central moments, including the skewness:

$$\text{Skewness coefficient} = \gamma_1 = \frac{E[(X - \mu_X)^3]}{\sigma_X^3} \tag{7.9}$$

and the kurtosis:

$$\text{Kurtosis coefficient} = \gamma_2 = \frac{E[(X - \mu_X)^4]}{\sigma_X^4} \tag{7.10}$$

The skewness coefficient is a measure of the asymmetry of the distribution around its mean. A perfectly symmetric distribution has a skewness coefficient of 0, while distributions with a long right tail are positively skewed and distributions with a long left tail are negatively skewed. Many random variables in environmental systems are positively skewed, with many small values (e.g., near zero) and some high values. The kurtosis of a random variable indicates the degree of flatness in its distribution. This is a subtle quantity not often used in applied problems; a normal distribution has a kurtosis of 3, and some texts modify Eq. (7.10) by subtracting 3 so that the kurtosis of the normal distribution is 0.

A random variable is fully characterized by its probability distribution function (pmf, pdf, or cdf). However, when the underlying distribution is not known, all moments ($k = 1, \infty$) are needed for an equivalent characterization. The exception occurs when a *parametric* form of a probability distribution function is used for a random variable, using a closed-form equation with a fixed number of parameters (usually one, two, or three). Once the parameters are known, the full distribution is characterized and all moments are determined. Moreover, j moments can always be used to compute j parameters, thereby determining the distribution and all of its other moments (this is one reason why the kurtosis is rarely considered, since we rarely use parametric probability functions with more than three parameters).

7.1.1 Parametric Probability Distribution Functions

Common parametric forms for discrete random variables include the binomial, geometric, negative binomial, and Poisson distributions. These are often used to describe the number of occurrences of discrete events such as storms, floods, spills, or upset conditions at sources such as incinerators, and are discussed further in Section 7.5 in the context of random processes. Common parametric forms for continuous random variables are shown in Table 7.1, including the uniform, exponential, gamma, normal, and lognormal distributions. Continuous random variables are often used to describe the variation in inputs and process parameters employed in mathematical models.

A uniform distribution can arise when describing the unknown location of a leak or discharge along a pipe or stream segment. Exponential distributions commonly arise for the interarrival times of purely random (Poisson) events. Gamma distributions are positively skewed and are often used to describe rainfall volumes and fluid residence times in environmental systems. Normal distributions are often used to represent small measurement errors. The normal distribution is generally inappropriate for variables that are constrained to be nonnegative. However, it can provide a reasonable approximation for nonnegative variables when the standard deviation is small compared to the mean. Typically, the coefficient of variation, $\nu_x = \sigma_x/\mu_x$, must be less than $\sim$0.3 for a normal approximation to be reasonable for a nonnegative random variable. By the central limit theorem, normal distributions arise from summing many random variables.

Lognormal random variables, like variables that have a gamma distribution, are nonnegative and positively skewed. However, the lognormal distribution exhibits a

TABLE 7.1 Continuous Probability Distributions Frequently Used in Environmental Applications[a]

Distribution: Sketch of the pdf: $f_X(x)$ versus x	Parametric Equation for the pdf (and cdf if available)	Moments of X	Parameter Estimation
Uniform 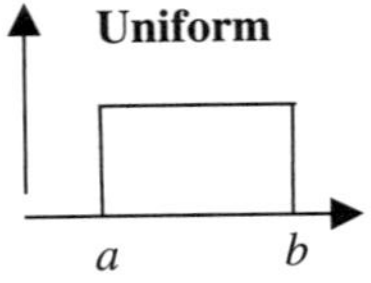	$f_X(x) = \frac{1}{b-a}$; $F_X(x) = \frac{x-a}{b-a}$ for $a \le x \le b$	$\mu_x = \frac{a+b}{2}$ $\sigma_x^2 = \frac{(b-a)^2}{12}$ $\gamma_1 = 0$	Method of moments: $a = \bar{x} - \sqrt{3}\, s_x$; $b = \bar{x} + \sqrt{3}\, s_x$ Maximum likelihood: $a = x_{\min}$; $b = x_{\max}$
Exponential 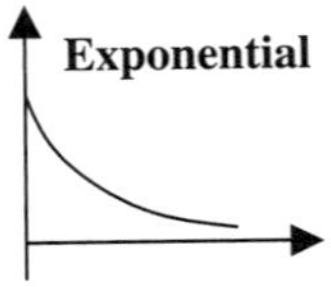	$f_X(x) = \lambda e^{-\lambda x}$; $F_X(x) = 1 - e^{-\lambda x}$ for $0 \le x$, $\lambda > 0$	$\mu_x = \frac{1}{\lambda}$ $\sigma_x^2 = \frac{1}{\lambda^2}$ $\gamma_1 = 2$	Method of moments: $\lambda = 1/\bar{x}$ Maximum likelihood: Same as method of moments
Gamma 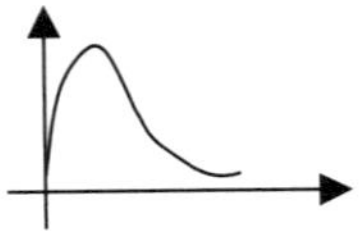	$f_X(x) = \frac{1}{\Gamma(a) b^a} x^{a-1} e^{-x/b}$ for $0 \le x$, $a, b > 0$ (no closed-form equation for cdf)	$\mu_x = ab$ $\sigma_x^2 = ab^2$ $\gamma_1 = 2a^{-1/2}$	Method of moments: $a = \left(\frac{\bar{x}}{s_x}\right)^2$; $b = \frac{s_x^2}{\bar{x}}$

Normal 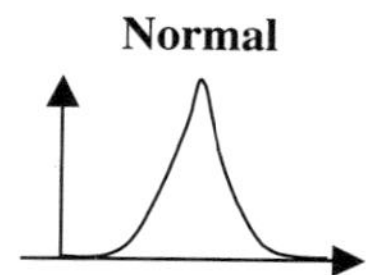	$f_X(x) = \frac{1}{b\sqrt{2\pi}} \exp\left[\frac{-(x-a)^2}{2b^2}\right]$ for $-\infty < x < \infty, \quad b > 0$ (no closed-form equation for cdf)	$\mu_x = a$ $\sigma_x^2 = b^2$ $\gamma_1 = 0$	Method of moments: $a = \bar{x}; \quad b = s_x$ Maximum likelihood: Same as method of moments
Lognormal	$f_X(x) = \frac{1}{bx\sqrt{2\pi}} \exp\left[\frac{-(\ln x - a)^2}{2b^2}\right]$ for $0 \le x, \quad b > 0$ (no closed-form equation for cdf)	$\mu_x = \exp\left[a + 0.5b^2\right]$ $\sigma_x^2 = \omega(\omega - 1)\left[e^{2a}\right]$ $\gamma_1 = (\omega - 1)^{1/2}(\omega + 2)$ where $\omega = \exp\left(b^2\right)$	Method of moments: $b^2 = \ln\left[\left(\frac{s_x}{\bar{x}}\right)^2 + 1\right]; \quad a = \ln[\bar{x}] - 0.5b^2$ Maximum likelihood: $a = \bar{y}; \quad b = s_y$ where $y = \ln x$

[a]Population mean (μ_x), variance (σ_x^2), and skewness coefficient (γ_1) are defined in Eq. (7.5), (7.7), and (7.9), respectively. Sample mean ($\bar{x}$) and standard deviation (s_x) are calculated as indicated in Eq. (7.11).

heavier "right tail" than the gamma distribution, yielding higher probabilities for extremely high values. When X is lognormally distributed with parameters a and b, $Y = \ln X$ is normally distributed with parameters a and b. For the normal distribution, a and b are the mean and standard deviation of the variable, while for the lognormal distribution, a and b are the mean and standard deviation of the natural logarithm of the variable. Lognormal distributions arise from multiplicative processes, which can be seen by applying the central limit theorem to the sum of the logarithms. Multiple random environmental dilutions (equivalent to applying many random multiplications to an initial concentration) yield distributions that are approximately lognormal (Ott, 1995), and the lognormal distribution provides a good representation for many environmental variables (Ahrens, 1954; Stedinger, 1980; Georgopoulos and Seinfeld, 1982; Di Toro, 1984; Small et al., 1995; Lockwood et al., 2001).

Other common distributions include the Weibull, often used in reliability studies for the time to system failure, but also found to provide a good empirical model for wind speeds (Troen and Petersen, 1989; Zaphiropoulos et al., 1999); the beta distribution, used to describe uncertainty in a parameter constrained between zero and one (such as a fraction or a probability); and the Gumbel and log-Pearson type III distributions, often used for extreme stream flows. The Weibull distribution, like the gamma distribution, is a more general form of the exponential, while the beta distribution serves as a more general form for the uniform distribution.

7.1.2 Sample Statistics and Parameter Estimation

The first three columns of Table 7.1 describe the population characteristics of uniform, exponential, gamma, normal, and lognormal random variables—when their respective parameters are *known*. When parameters must first be *estimated* from an observed data set, the methods of statistics are needed. A "statistic" is formally defined as some function of an observed sample. Consider a sample of the random variable X: $\vec{x} = x_i$; $i = 1, \ldots, n$. If the sample is *random*, then all values of x_i are independent and identically distributed (iid) according to $f_X(x)$ [or $p_X(x)$ if X is discrete]. Properties of the sample include the minimum, $x_{\min}$ = the smallest of the n sample values; the maximum, $x_{\max}$ = the largest of the n sample values; the sample range, $R = x_{\max} - x_{\min}$; and the sample moments:

$$
\begin{aligned}
&\text{Sample mean} = \hat{\mu}_x = \overline{x} = \frac{1}{n}\sum_{i=1}^{n} x_i \\
&\text{Sample variance} = \hat{\sigma}_x^2 = s_x^2 = \frac{1}{n-1}\sum_{i=1}^{n}(x_i - \bar{x})^2 \\
&\text{Sample skewness} = \hat{\gamma}_1 = g_1 = \frac{n}{(n-1)(n-2)}\sum_{i=1}^{n}(x_i - \bar{x})^3 \Big/ s_x^3 \\
&\text{Sample kurtosis} = \hat{\gamma}_2 = g_2 = \frac{n(n+1)}{(n-1)(n-2)(n-3)}\sum_{i=1}^{n}(x_i - \bar{x})^4 \Big/ s_x^4
\end{aligned}
\qquad (7.11)
$$

The "hats" over the population moments indicate that these are estimates derived from a sample. These estimates follow in a relatively straightforward manner from the definitions of the population moments in Eqs. (7.5) to (7.10), except for the terms in front of the summation signs for the variance, skewness, and kurtosis. These terms replace the anticipated $1/n$ with functions of $n-1$, $n-2$, and $n-3$, to ensure that the estimates are *unbiased*, that is, that the expected values of the sample estimates are equal to the population moments when a random sample (of size n) is drawn many times from a (known) population. Note that the correction terms before the summation are larger than $1/n$ in each case but approach $1/n$ as n becomes large.

Sample estimates of the probability distribution function (pmf or pdf) and the cdf are also of interest, and these are sometimes used to provide an empirical estimate of the population distribution, or to check how well a parametric distribution (fitted using the sample moments or using other statistical procedures as described below) matches the observed data. Sample probability distribution functions are estimated using a sample frequency distribution or histogram, which counts the number of observations over consecutive intervals of the random variable and plots these against the variable, on the X axis. The sample frequency distribution provides an empirical (nonparametric) approximation to the pmf or the pdf. Because different interval selections and smoothing functions can vary the shape of the empirical frequency distribution, different intervals, and kernel estimation techniques are possible and should be compared (Newton, 1988).

The empirical cdf is somewhat easier to compute than a pmf or pdf but still requires a "plotting position formula" to ensure that $x_{\max}$ (which was only estimated from a sampled subset of the true population) is not assigned a cdf value of 1.0 [as might be inferred from the definition of $F_X(x)$]. This is necessary in order to allow for the possibility that values larger than $x_{\max}$ can occur in the population, even though they are not found in a particular sample. Plotting position formulas are implemented by ordering the sample and assigning a rank, $\mathrm{Rank}(x_i)$ to each observation, from $\mathrm{Rank}(x_{\min}) = 1$ to $\mathrm{Rank}(x_{\max}) = n$. Common empirical cdf formulas include:

$$\left.\begin{aligned}
&\text{The Weibull plotting position formula:} && \hat{F}_X(x_i) = \frac{\mathrm{Rank}(x_i)}{n+1} \\
&\text{The Hazen plotting position formula:} && \hat{F}_X(x_i) = \frac{\mathrm{Rank}(x_i) - 0.5}{n} \\
&\text{The Blom plotting position formula:} && \hat{F}_X(x_i) = \frac{\mathrm{Rank}(x_i) - 0.375}{n+0.25}
\end{aligned}\right\} \quad (7.12)$$

The Weibull formula is among the simplest and most commonly used, although the Blom formula is often preferred for constructing normal and lognormal probability plots and is used in the examples that follow.

Normal probability plots are another way of depicting the empirical cdf, transformed so that the curve is plotted as a straight line. Normal probability plots are a special case of a q–q (quantile-quantile) plot, which graphs the value of x_i against the value of $x_{i,\mathrm{hypothesized}}$, the value that would be theoretically expected if X exactly followed the hypothesized distribution, $F_{X,\mathrm{hypothesized}}(x)$. To construct a q–q plot:

$$\text{Graph } x_i \quad \text{versus} \quad x_{i,\text{hypothesized}} = F^{-1}_{X,\text{hypothesized}}\left[\hat{F}_X(x_i)\right] \tag{7.13}$$

where the term inside the brackets [ ] is computed from sample data using a plotting position formula. In a normal probability plot the value of x_i is plotted against the corresponding standardized normal variate, z_i. To construct a normal probability plot:

$$\text{Graph } x_i \quad \text{versus} \quad z_i = F_Z^{-1}\left[\hat{F}_X(x_i)\right] \tag{7.14}$$

where Z follows the standardized normal distribution with mean zero and variance one ($Z \sim N[0, 1]$). As noted above, $\hat{F}_X(x_i)$ for this application is often computed using the Blom formula. A lognormal probability plot is constructed by graphing $\ln(x_i)$ [or $\log_{10}(x_i)$ for ease of interpretation] against the expected Z variate. The z_i axis also serves as a surrogate for the equivalent fraction less than or equal to the given value of x_i or $\ln(x_i)$; that fraction $= F_Z(z_i)$.

7.1.3 Parameter Estimation Methods

Consider a random variable X with hypothesized distribution $f_X(x|\boldsymbol{\theta})$, where $\boldsymbol{\theta}$ is the vector of parameters for the distribution. Assume that a random sample $\mathbf{x} = x_i$; $i = 1, \ldots, n$ is available. Principal techniques for estimating the parameters include: (a) the method of moments (MOM), (b) the maximum-likelihood estimation (MLE) method, (c) cdf regression, and (d) Bayesian methods. With MOM, the parameters, $\boldsymbol{\theta}$, are chosen so that the moments of $f_X(x|\boldsymbol{\theta})$ match those of the sample. The MLE technique chooses $\boldsymbol{\theta}$ to maximize the likelihood function, $L(\boldsymbol{\theta}|\mathbf{x})$, which, for a discrete random variable is the probability (or, for a continuous random variable, the equivalent probability density) of obtaining the sample $\mathbf{x}$, given that the "true" probability distribution is $f_X(x|\boldsymbol{\theta})$. For a random sample of a continuous random variable, the MLE method is implemented by:

$$\text{Maximize } L(\boldsymbol{\theta}|\mathbf{x}) = \prod_{i=1}^{n} f_X(x_i|\boldsymbol{\theta}) \tag{7.15}$$

This maximization must often be carried out using a numerical search or optimization routine. However, for many common distribution functions (including all but the gamma distribution in Table 7.1), closed-form statistical formulas for estimating $\boldsymbol{\theta}_{\text{MLE}}$ have already been derived. When such formulas are available, the MLE method is usually preferred to MOM estimates. As indicated in Table 7.1, the MLE and MOM estimates are the same for some important distributions like the exponential and normal distributions.

When numerical determination of the MLE is necessary, the maximization is usually carried out using the log-likelihood function, $\ln[L(\boldsymbol{\theta}]$, instead of directly using $L(\boldsymbol{\theta})$. The value of $\boldsymbol{\theta}$ that maximizes $\ln[L(\boldsymbol{\theta})]$ is the same as that which maximizes $L(\boldsymbol{\theta})$. Equation (7.15) can yield values of $L(\boldsymbol{\theta})$ that are very close to zero, especially when n is large, which can cause underflows in numerical computation. This problem is solved by using the log-likelihood function. The log-likelihood calculation is applied to a sum rather than a product:

$$\text{Maximize } \ln [L(\boldsymbol{\theta})] = \sum_{i=1}^{n} \ln [f_x(x_i|\boldsymbol{\theta})] \tag{7.16}$$

The log-likelihood is usually a moderate to large negative number but is still easily computed.

In the following example, MOM and MLE methods for fitting a probability distribution function are illustrated for a positively skewed sample typical of those found in environmental applications.

EXAMPLE 7.1

Determine an appropriate probability distribution function and the values of its parameters for the following observations of hydraulic conductivity, K (cm/day), for the soil at a contaminated waste disposal site. Consider lognormal and gamma distributions as alternatives, fitting the former using both the MOM and MLE methods and the latter using the MOM procedure from Table 7.1. Compare the fitted distributions and indicate which appears to provide the best fit to the data.

Obs.	K (cm/day)	ln[K]	Obs.	K (cm/day)	ln[K]
1	7.77	2.0503	26	9.98	2.3006
2	4.04	1.3962	27	4	1.3863
3	13.85	2.6283	28	8.66	2.1587
4	37.3	3.6190	29	9.54	2.2555
5	20.74	3.0321	30	3.75	1.3218
6	89.99	4.4997	31	7.44	2.0069
7	19.75	2.9832	32	8.8	2.1748
8	4.28	1.4540	33	101.9	4.6240
9	2.69	0.9895	34	35.93	3.5816
10	23.48	3.1561	35	28.22	3.3400
11	15.91	2.7669	36	14.09	2.6455
12	54.08	3.9905	37	28.04	3.3336
13	4.2	1.4351	38	12.36	2.5145
14	9.29	2.2289	39	10.21	2.3234
15	12.76	2.5463	40	1.8	0.5878
16	1.39	0.3293	41	40	3.6889
17	21.36	3.0615	42	3.52	1.2585
18	11.63	2.4536	43	10.81	2.3805
19	3.52	1.2585	44	19.66	2.9786

(*continued*)

Obs.	K (cm/day)	ln[K]	Obs.	K (cm/day)	ln[K]
20	35.29	3.5636	45	16.77	2.8196
21	13.55	2.6064	46	8.11	2.0931
22	10.79	2.3786	47	54.24	3.9934
23	2.86	1.0508	48	13.68	2.6159
24	6.09	1.8066	49	3.73	1.3164
25	27.35	3.3087	50	5.06	1.6214
			Average =	18.285	2.4383
			St. dev =	20.505	0.9810

Solution 7.1 As a first, exploratory step, we examine the qualitative features of the data distribution. The coefficient of variation, $\nu_K = (20.505/18.285) = 1.12$, is much greater than 0.3, indicating that a normal distribution is inappropriate since K is a nonnegative random variable. Inspection of the 50 observations of K indicates a predominance of values less than 15, and the less frequent occurrence of much larger values, for example, 54 and 101. Using the data analysis tool in Excel, histograms can be constructed for both K and $\ln(K)$ as shown in Figure 7.1. The empirical histogram for K is positively skewed with a heavy right tail, while the histogram for $\ln[K]$ shows a more symmetric distribution. These histograms for K and $\ln[K]$ are consistent with the qualitative features of positively skewed distributions, such as the lognormal and the gamma, and either might provide an appropriate model for the data.

For the lognormal (LN) distribution, the parameters are estimated by the method of moments and by maximum-likelihood estimation. The MOM estimates are determined from the mean ($\bar{K}$) and standard deviation (s_K) of the 50 observations of K:

$$b^2 = \ln\left[\left(\frac{s_k}{\bar{K}}\right)^2 + 1\right] = \ln\left(\left[\frac{20.50513}{18.2852}\right]^2 + 1\right) = 0.814281 \Rightarrow b_{\text{LN,MOM}} = 0.9024$$

$$a_{\text{LN,MOM}} = \ln[\bar{K}] - 0.5b^2 = \ln[18.2852] - 0.5(0.814281) = 2.4990$$

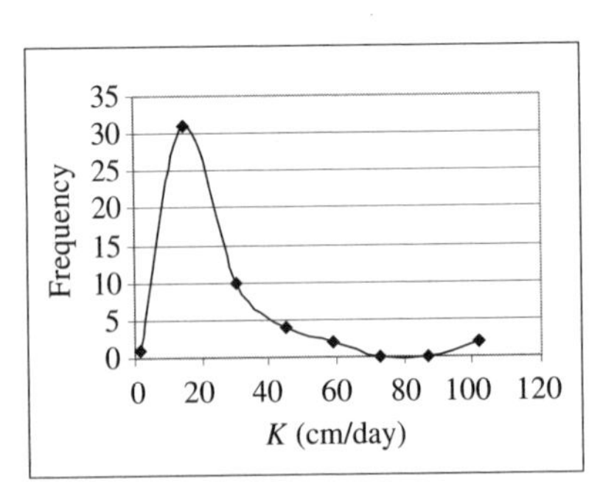

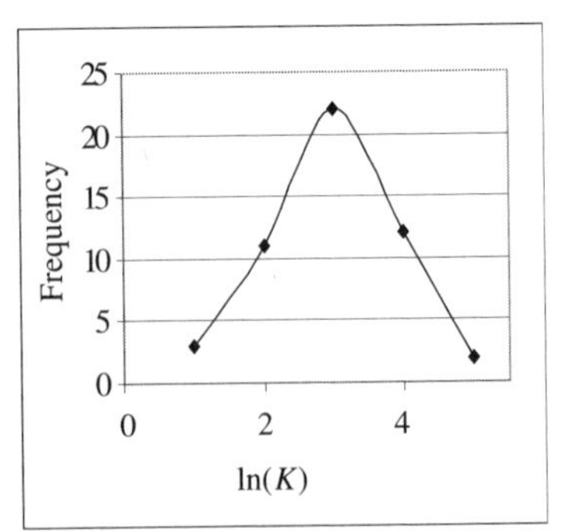

Figure 7.1 Empirical frequency distribution for hydraulic conductivity data and its logarithm.

For the MLE method, the parameters, a and b, of the lognormal distribution are estimated from the mean and standard deviation of the log-transformed data, $Y = \ln K$.

$$a_{\text{LN,MLE}} = \bar{Y} = 2.4383 \quad \text{and} \quad b_{\text{LN,MLE}} = s_Y = 0.9810$$

The estimates of a and b obtained from the MOM and MLE methods are similar in magnitude, but somewhat different.

The parameters of the gamma distribution are estimated using the method of moments equations in Table 7.1:

$$a_{\gamma,\text{MOM}} = \left(\frac{\bar{K}}{s_K}\right)^2 = \left(\frac{18.2852}{20.50513}\right)^2 = 0.7952$$

$$b_{\gamma,\text{MOM}} = \frac{s_K^2}{\bar{K}} = \frac{20.50513^2}{18.2852} = 23.0$$

The three fitted distributions (two lognormal and one gamma) are compared to each other and to the observed distribution in Figure 7.2. Figure 7.2*a* is a standard cdf plot of $F_K(K)$ vs. K; Figure 7.2*b* plots the cdf as a function of $\ln K$; while Figure 7.2*c* graphs $\ln K$ as a normal probability plot. The observed cdf points are computed using the Blom plotting position formula. The fitted distributions are difficult to distinguish on the standard cdf plot in Figure 7.2*a*, though the gamma distribution appears to deviate somewhat from the observed data. This lack of fit is more apparent in Figure 7.2*b*, where the gamma distribution is indicated to be particularly inaccurate at low values of $\ln K$. This plot also indicates a better fit at low $\ln K$ for the lognormal distribution using the MLE parameters, as compared to the MOM fit. Similar differences are apparent in the lognormal probability plot in Figure 7.2*c*, where the observed data plot as a straight line (suggesting immediately that the lognormal distribution is appropriate for this data set), with the MLE parameters providing the best fit over the full range of the data.

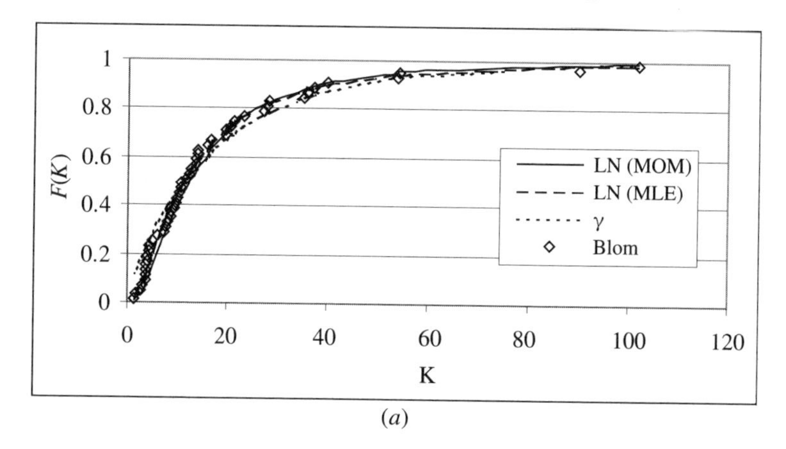

(*a*)

(*continued*)

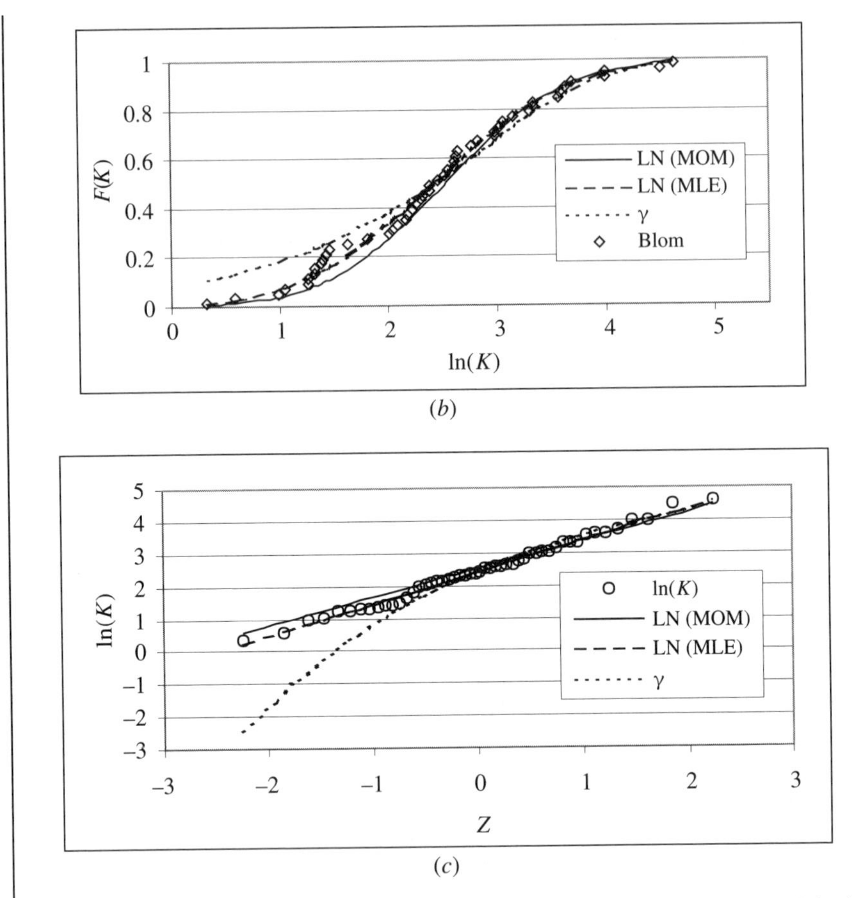

Figure 7.2 Comparison of fitted and observed distributions of hydraulic conductivity in Example 7.1: (*a*) standard cdf plot, (*b*) ln K cdf plot, (*c*) lognormal probability plot.

To confirm qualitative inferences such as those drawn from Figure 7.2, various quantitative goodness-of-fit measures can be computed and compared across distributions. These goodness-of-fit measures include:

1. The root-mean-square error of the fitted cdf (RMSE_F), in the probability dimension (this is the square root of the average square vertical distance between the observed and fitted cdf for each data point):

$$\text{RMSE}_\text{F} = \left\{ \frac{1}{n} \sum_{i=1}^{n} \left[\hat{F}_X(x_i) - F_X(x_i) \right]^2 \right\}^{1/2} \tag{7.17}$$

2. The log-likelihood as computed by Eq. (7.16).

TABLE 7.2 Goodness-of-Fit Comparison for Alternative Fitted Distributions for *K*

Distribution	Parameters		Goodness-of-Fit Measure	
			$RMSE_F$	ln(Likelihood)
LN(MOM)	$a = 2.4990$	$b = 0.9024$	0.0366	−69.9
LN(MLE)	$a = 2.4383$	$b = 0.9810$	0.0245	−69.5
Gamma(MOM)	$a = 0.7952$	$b = 23.0$	0.0602	−197.2

These metrics are compared for the fitted hydraulic conductivity distributions in Table 7.2.

The $RMSE_F$, commonly used in statistical applications, indicates the goodness of fit in the order inferred from the plots: The best fit is achieved by the LN(MLE), followed by the LN(MOM), with the worst fit for the Gamma(MOM). The log-likelihood values indicate a similar ordering but with less apparent preference for the LN(MLE) model as compared to the LN(MOM), and both being greatly preferred to the Gamma(MOM) model (–69.5 is somewhat greater, i.e., less negative, than –69.9, but both are substantially greater than –197.2). Of course, it is not surprising that the computed likelihood for the LN(MLE) model is greater than that for the LN(MOM) model since $a_{LN,MLE}$ and $b_{LN,MLE}$ are explicitly chosen to *maximize the likelihood*.[1]

The difference in log-likelihood provides a widely recognized mechanism for comparing the fits of alternative models to observed data and is used to compute the *likelihood ratio*, $LR_{1,2}$, between two models (1 and 2): $LR_{1,2} = L_1/L_2 = \exp[\ln(L_1) - \ln(L_2)]$. In this case the likelihood ratio between the LN(MLE) model and the LN(MOM) model is $LR_{LN(MLE),\,LN(MOM)} = \exp[(-69.5) - (-69.9)] = e^{0.4} = 1.49$. The likelihood ratio between these models and the Gamma(MOM) is much larger (greater than 10^{55} !). The use of likelihood ratios and related measures is revisited in Chapter 14, when broader issues of model uncertainty are considered.

A third method of parameter estimation, cdf regression, involves regressing the parametric form for $F_X(x|\boldsymbol{\theta})$ vs. $\hat{F}_x(x_i)$. This may, in general, be accomplished using nonlinear regression. Alternatively, linearized forms of the cdf can be evaluated using linear regression, if these are available. For example, the parameters of a normal (or lognormal) distribution can be estimated from the normal probability plot by regressing the values of x_i (or $\ln[x_i]$) against the computed values of z_i; the fitted intercept is the mean, a, while the fitted slope is the standard deviation, b. Probability plot regression methods can also be used when some of the data are censored, due, for example, to reported below-detection-limit data in the sample, by including these data

[1] The MLE for the normal and lognormal distributions are *precisely* computed using the *biased* estimator for the variance of the variable (or its ln, for the lognormal model):

$$S^2_{x,\text{biased}} = \frac{1}{n}\sum_{i=1}^{n}(x_i - \bar{x})^2$$

so that slightly different values of a and b actually maximize the likelihood in this application. However, for simplicity and consistency in defining and computing variances and standard deviations, we assume that use of the unbiased value of s_x in the MLE equations for a and b is acceptable.

in the calculation of the ranks and $\hat{F}_x(x_i)$, but using only the above-detection-limit data (e.g., the above-detection limit $x_i - z_i$ pairs from the normal probability plot) in the regression calculation for a and b (El-Shaarawi, 1989; Helsel and Hirsch, 1992; Piegorsch et al., 1998). The reader may wish to verify that the (log)normal probability plot regression method yields parameter estimates $a_{\mathrm{LN}} = 2.4383$; $b_{\mathrm{LN}} = 0.9949$ for the preceding example. These results are very similar to the MLE estimate for this case. (This is not surprising since the MLE line in Figure 7.2*c* fits the observed points on the lognormal probability plot quite closely.)

As indicated by Example 7.1, given a sample, different parameter estimates for a probability distribution function are possible, and alternative forms for the probability distribution function can even be considered (although the gamma distribution considered in this case does not appear to provide an especially accurate alternative). The distribution of hydraulic conductivity is intended to characterize the variability of K at a site. However, with a limited sample and an idealized model form, there is uncertainty in this characterization. Explicit methods for estimating the uncertainty in fitted distribution parameters (in addition to best, point estimates) are available, including methods based on the shape of the likelihood function around the point estimate, estimated standard errors and covariances for fitted regression parameters, and methods based on bootstrap-data resampling techniques (Brattin et al., 1996; Burmaster and Thompson, 1998; Frey and Burmaster, 1999). Like the MOM, MLE, and cdf regression methods for obtaining point estimates of distribution parameters, these techniques for assessing uncertainty consider *only* the information in the sample $\mathbf{x}$, and are referred to as *classical* statistical methods. When further information is available outside of the data set, for example, from the scientific literature or data sets collected at other sites, Bayesian methods for parameter estimation can be utilized.

7.1.4 Bayesian Estimation

The Bayesian approach is often referred to as nonclassical, or subjectivist, since it allows for subjective differences in probability assessments that arise when different experts and analysts consider and incorporate *extra*-sample information in their analyses. While this feature has in the past made the use of Bayesian methods somewhat controversial, their use has gained growing acceptance in recent years (e.g., Massman et al., 1991; Smith and French. 1993; Abbaspour et al., 1996; Ellison, 1996; Small, 1997; Stow and Qian, 1998; Bergin and Milford, 2000), especially when differences among experts are explicitly recognized, and the effects of these differences are considered and analyzed as part of the assessment (Wolfson et al., 1996; Stiber et al., 1999).

The Bayesian method is particularly well-suited for evaluating the uncertainty in fitted probability distribution functions since (unlike the classical methods considered above) it explicitly allows for a distribution function, $f(\boldsymbol{\theta})$, for the uncertain parameter values of the distribution of X. The method considers exogenous information and expert judgment in the formulation of a *prior* distribution for the parameters, $f^{\mathrm{o}}(\boldsymbol{\theta})$, and combines this with the information of the observed sample $\mathbf{x}$, to obtain a *posterior* distribution $f(\boldsymbol{\theta}|\mathbf{x})$. The posterior distribution is computed from the Bayes rule, using the likelihood function:

$$f(\boldsymbol{\theta}^*|\mathbf{x}) = \frac{L(\boldsymbol{\theta}^*|\mathbf{x})\, f^{\text{o}}(\boldsymbol{\theta}^*)}{\int_{\text{all}\,\boldsymbol{\theta}} L(\boldsymbol{\theta}|\mathbf{x})\; f^{\text{o}}(\boldsymbol{\theta})\, d\boldsymbol{\theta}} \tag{7.18}$$

where $\boldsymbol{\theta}^*$ denotes a particular set of parameter values. Implicit in this formulation is a nested, or hierarchical, set of distributions for the original variable X *and* the parameters of the distribution that characterizes X:

$$\begin{aligned} X &\sim f_x(\mathbf{x}|\boldsymbol{\theta}) \\ \boldsymbol{\theta} &\sim f(\boldsymbol{\theta}|\boldsymbol{\gamma}) \end{aligned} \tag{7.19}$$

As described in Chapters 13 and 14, this can provide an effective framework for characterizing the variability in X [as represented by $f_x(\mathbf{x}|\boldsymbol{\theta})$] and the uncertainty in this characterization [as represented by $f(\boldsymbol{\theta}|\boldsymbol{\gamma})$]. In this context the parameters, $\boldsymbol{\gamma}$, of the uncertainty distribution are referred to as *hyperparameters*.

If very broad, "noninformative" prior distributions are chosen for the parameters in Bayesian analysis, then the posterior parameter distributions will be dominated by the information contained in the data. The results are then similar to those obtained using classical estimation methods. As tighter, more precise prior distributions are selected based on more knowledge and/or exogenous data, the influence of the priors increases. Similarly, when only small or imprecise observed data sets are available, the posterior parameter distributions remain similar to the prior distributions.

The mechanics of implementing the Bayesian calculation are described in a number of texts (e.g., Press, 1989; Gelman et al., 1995; DeGroot and Schervish, 2002). Bayesian updating is relatively "simple" when a closed-form *conjugate* distribution is used to describe the prior and posterior parameter distributions. When a conjugate is used, the prior and posterior parameter uncertainties are described by the same analytical functions. Only the hyperparameters in $f(\boldsymbol{\theta}|\boldsymbol{\gamma})$ change as data are collected and incorporated with the prior estimates. While conjugate models are desirable bccause of their relative ease of implementation, the necessary assumptions and conditions are overly restrictive for many applications. In these cases, numerical integration is necessary to compute posterior distributions. Among the most popular recent advances for numerical solution involves use of Markov Chain Monte Carlo (MCMC) sampling (Gelman et al., 1995; Monahan, 2001).[2]

[2] In the MCMC method, the posterior sample space of the model parameters is numerically generated using a sequence of dependent sample points. Thus, each sample in the simulated distribution is related to the previously simulated point by a Markov chain. (In contrast, in traditional Monte Carlo sampling, the sample points are mutually independent.) Popular MCMC methods include Gibbs sampling, where each parameter of the model is sampled sequentially from its conditional distribution, given the current value of all other model parameters, and the Metropolis–Hastings algorithm, where a new set of parameter values can be probabilistically accepted or rejected based on its posterior probability relative to the current parameter vector (if the new set is rejected, the current set is repeated). Both Monte Carlo simulation methods and Markov chains are discussed further in later sections of this chapter.

The Bayesian approach can also be used for model discrimination, computing relative posterior probabilities of alternative model *structures*, that is, different candidate parametric forms for $f_x(\mathbf{x}|\boldsymbol{\theta})$ (e.g., Wood and Rodriguez-Iturbe, 1975; Iman and Hora, 1989; Draper, 1995; Gaganis and Smith, 2001). A simple demonstration of Bayesian model discrimination can be made for the three alternative distributions fitted to the hydraulic conductivity data in Example 7.1. If alternative probability models, $M_j;\ j = 1, \ldots N_M$, are assigned relative prior probabilities $p^{\text{o}}(M_j)$, then their relative posterior probabilities can be computed using the discrete form of Bayes rule:

$$p(M_j) = \frac{L(M_j|\mathbf{x})p^{\text{o}}(M_j)}{\sum_{u=1}^{N_M} L(M_u|\mathbf{x})p^{\text{o}}(M_u)} \tag{7.20}$$

Assuming equal prior probabilities of one-third for the three fitted distributions of K, the relative posterior probabilities of the candidate distributions are computed as:

$$p[\text{LN(MOM)}] = \frac{e^{-69.9}(1/3)}{e^{-69.9}(1/3) + e^{-69.5}(1/3) + e^{-197.2}(1/3)} \approx 0.401$$

$$p[\text{LN(MLE)}] = \frac{e^{-69.5}(1/3)}{e^{-69.9}(1/3) + e^{-69.5}(1/3) + e^{-197.2}(1/3)} \approx 0.599$$

$$p[\Gamma\text{(MOM)}] = \frac{e^{-197.2}(1/3)}{e^{-69.9}(1/3) + e^{-69.5}(1/3) + e^{-197.2}(1/3)} \approx 0$$

The LN(MLE) model has the greatest posterior probability (~0.6), followed by the LN(MOM) model (with posterior probability of ~0.4). The Gamma(MOM) model has a posterior probability of ~0. The Bayesian approach for model discrimination is discussed further in Chapter 14.

TABLE 7.3 Rules for Determining the Expectation and Variance for Z^a

Estimation of the Mean Value of Z, $E[Z]$	
Linear function: $Z = aX + bY + c$	$E[Z] = aE[X] + bE[Y] + c$
Quadratic function: $Z = aX^2$	$E[Z] = aE[X^2]$
Multiplicative function: $Z = X * Y$	$E[Z] = E[X]^*E[Y] + \text{Cov}[X, Y]$
Quotient function: $Z = X/Y$	$E[Z] = E[X]^*E[1/Y] + \text{Cov}[X, 1/Y]$
Estimation of the Variance of Z, $Var[Z]$	
Linear function: $Z = aX + bY + c$	$\text{Var}[Z] = a^2\,\text{Var}[X] + b^2\,\text{Var}[Y] + 2ab\,\text{Cov}[X, Y]$

[a] Z is a function of random variables X and Y, with a, b, and c being constants. Cov[X,Y] is the covariance, Eq. (7.25).

7.2 DERIVING OUTPUT DISTRIBUTIONS FOR MECHANISTIC MODELS

Once the distribution type and its parameter values are determined for a random variable, these can be included in a mechanistic chemical transport and fate model using either of two methods: one based on probability theory, and the other using computer simulations of random variables. Where applicable, probability theory allows analytical derivation of the distribution function of model output induced by input variability (or uncertainty) with a known distribution. In particular, the pdf of input X can be used to derive the distribution and statistical properties of model output C, if C is a monotonic mathematical function of X. If C (e.g., chemical concentration) and X are related by a contaminant transport mass-balance equation, then the concentration pdf, $f_C(c)$, can be determined from the pdf of X, using the following rule for derived distributions:

$$f_c(c) = f_x[x(c)] \times \left| \frac{dX}{dC} \right| = f_x[x(c)] \times \left| \frac{dC}{dX} \right|^{-1} \tag{7.21}$$

Equation (7.21) applies because there is a one-to-one correspondence between C and X; hence the probability of finding an observation of X within a certain range, Δx, should be the same as finding an observation of C in a corresponding range, ΔC. To apply Eq. (7.21), the environmental model must be "invertible," to allow the calculation of the input from the output, that is, $x(c)$. Rules for the expectation and variance for mathematical functions of a random variable X are also readily derived from probability theory and are summarized in Table 7.3. Example 7.2 illustrates the use of some of these probability rules for a situation where C is a function of a single random variable X.

EXAMPLE 7.2

Consider the change in aqueous chemical concentration with time in a batch reactor when the chemical is undergoing first-order decay, where the decay constant k is now a random variable that is uniformly distributed between 0.1 and 0.5 per day. Determine the pdf, cdf, expected value, and standard deviation of the aqueous chemical concentration in the reactor at $t = 10$ days, assuming an initial concentration of 1 mg/L.

Solution 7.2 For a species that undergoes first-order decay starting at an initial concentration of 1 mg/L, the chemical concentration at $t = 10$ is given by $C = 1 \times e^{-10k}$ where k is a random variable with

$$f_k(k) = \begin{cases} \dfrac{1}{0.5 - 0.1} = \dfrac{1}{0.4} & 0.1 \leq k \leq 0.5 \\ 0 & \text{otherwise} \end{cases}$$

$dC/dk = -10 \times e^{-10k}$ from which the pdf of C can be computed as:

$$f_c(c) = f_k[k(c)] \times \left|\frac{dC}{dk}\right|^{-1} = \begin{cases} \dfrac{e^{10k}}{4} = \dfrac{1}{4C} & e^{-5} \leq k \leq e^{-1} \\ 0 & \text{otherwise} \end{cases}$$

The cdf,

$$F_c(c) = \int_{u=e^{-5}}^{c} f_c(u)\,du = \int_{u=e^{-5}}^{c} \frac{1}{4u}\,du = \frac{\ln(c)}{4} + \frac{5}{4}$$

with C varying over the range e^{-5} to e^{-1}. The mean,

$$E[C] = \int Cf_c(c)\,dC = \int_{0.0067}^{0.3679} \frac{C}{4C}\,dc = \left.\frac{C}{4}\right|_{0.0067}^{0.3679} = \frac{1}{4}[0.3679 - 0.0067]$$

$$= 0.090 \text{ mg/L}$$

The variance,

$$\text{Var}[C] = E[C^2] - E^2[C]$$

with

$$E[C^2] = \int_{0.0067}^{0.3679} \frac{C^2}{4C}\,dc = \left.\frac{C^2}{8}\right|_{0.0067}^{0.3679} = \frac{1}{8}[0.3679^2 - 0.0067^2] = 0.0169 \text{ [mg/L]}^2$$

So,

$$\text{Var}[C] = 0.0169 - 0.090^2 = 0.00876[\text{mg/L}]^2$$

$$\rightarrow \sigma_C = \sqrt{\text{Var}[C]} = \sqrt{0.00876} = 0.094 \text{ mg/L}$$

The results for the rth moment of C could also be derived directly from $f_k(k)$, without first deriving $f_C(c)$, as:

$$E\left[[C(k)]^r\right] = \int [C(k)]^r f_k(k)\,dk \tag{7.22}$$

The reader may wish to verify that the same results for $E[C]$ and σ_c are obtained using this approach.

7.3 JOINT RANDOM VARIABLES

The method illustrated in Example 7.2 and related analytical approaches for deriving the probability distribution of model output from the distribution of inputs have been used in a number of interesting applications in the literature.[3] However, analytical probabilistic solutions for model output are usually only possible under fairly restrictive assumptions and when there are only a few variable (or uncertain) model inputs. To address the effects of variations in multiple model inputs, the joint distributions of these inputs must be characterized. We consider methods for treating multiple random variables by starting with two random variables and then generalizing to more than two random variables as needed.

The joint probability distribution is defined analogously to that for a single random variable, as in Eqs. (7.1) and (7.2). For continuous variables X and Y, the joint pdf, $f_{X,Y}(x, y)$ is the limit of the probability of finding an observation of X within a specified interval Δx *and* an observation of Y within a specified interval Δy, divided by the size of the interval as it converges to zero:

$$f_{X,Y}(x, y) = \lim_{\Delta x,\, \Delta y \to 0} \left[\frac{P(x < X < x + \Delta x) \cap P(y < Y < y + \Delta y)}{\Delta x\, \Delta y} \right] \tag{7.23}$$

When X and Y are independent, the joint probability distribution is the product of the individual distributions of X and Y:

$$\begin{aligned} p_{X,Y}(x, y) &= p_X(x) \times p_Y(y) && \text{for discrete } X, Y \\ f_{X,Y}(x, y) &= f_X(x) \times f_Y(y) && \text{for continuous } X, Y \\ F_{X,Y}(x, y) &= F_X(x) \times F_Y(y) && \text{for both discrete and continuous } X, Y \end{aligned} \tag{7.24}$$

Thus, when dispersion is modeled as a Gaussian process occurring independently in the X, Y, and Z dimensions, the joint probability distribution describing particle location in 3D space is the product of the individual probability distributions in the X, Y, and Z dimensions. When X and Y are not fully independent, that is, they are correlated in some way, their degree of dependency can be characterized by their joint second moment, the *covariance*:

$$\begin{aligned} \text{Cov}[X, Y] &= E\left[\{X - E[X]\} \times \{Y - E[Y]\}\right] \\ &= E[XY] - \mu_X \mu_Y \end{aligned} \tag{7.25}$$

[3] Examples include the derivation of the statistical properties of soil moisture and runoff (Eagleson, 1978; Cordova and Bras, 1981; Milly, 2001); non-point-source loadings, capture, treatment, and water quality impacts (Di Toro and Small, 1979; Hydroscience, 1979; Di Toro, 1984; Adams and Papa, 2000); air pollution concentrations (Gifford, 1959; Hanna, 1984; Brown, 1987; Munro et al., 2001); and the regional distribution of lake chemistry impacted by acid deposition (Small and Sutton, 1986; Small et al., 1988).

This association is often represented by the dimensionless correlation coefficient, $\rho_{x,y}$, computed from the covariance as:

$$\rho_{X,Y} = \frac{\mathrm{Cov}[X, Y]}{\sigma_X \, \sigma_Y} \tag{7.26}$$

where, in all cases, $-1 \leq \rho_{X,Y} \leq 1$. The correlation coefficient is estimated from a joint sample of X, Y as:

$$r = \hat{\rho}_{X,Y} = \frac{\sum_{i=1}^{n}(x_i - \bar{x})(y_i - \bar{y})}{\sqrt{\sum_{i=1}^{n}(x_i - \bar{x})^2 \sum_{i=1}^{n}(y_i - \bar{y})^2}} \tag{7.27}$$

When X and Y are independent, $\mathrm{Cov}[X, Y] = \rho_{x,y} = 0$. However, zero covariance does not necessarily imply independence, unless X and Y are jointly normal.

Probability rules for determining the expectation of a variable Z, which is a simple function of two variables X and Y, are included in Table 7.3. As indicated in Table 7.3, the covariance plays an important role in determining the means of various functions. Rules for determining the variance of Z can be written out explicitly only if it is a linear combination of the variables X and Y. Thus, deriving a complete description of Z, including characterization of its distribution, can become very difficult as functional relationships between Z, X, and Y become more complex. For most environmental models, the distribution of Z cannot be derived analytically, and numerical computer simulations are needed to assess the impact of variable parameters on model outcomes.

7.4 NUMERICAL SIMULATION METHODS FOR DERIVING DISTRIBUTIONS

The standard numerical method used to calculate model output distributions is referred to as *stochastic* or *Monte Carlo simulation*.[4] With this method, random variable generators are used to generate many (N) samples of X, which are then employed in N computations of the model output C. A numerical distribution of C is thus obtained, and the statistical properties of C can be estimated as if the simulation results are a sample from the population of C, albeit one associated with the "model world," rather than the real one.

Most personal computers are equipped with versatile random number generators that can be used to implement Monte Carlo analysis. For example, the data analysis tool in Excel includes a random number generator for most standard parametric

[4] The term Monte Carlo simulation is used by many to refer to any numerical sampling technique that generates multiple realizations of model inputs and outputs. Others limit the term Monte Carlo simulation to refer to methods that generate a random (i.e., iid) sample. Other methods, such as importance sampling and Latin hypercube sampling, yield a stratified sample of inputs and outputs and can be very effective at characterizing distributions with a smaller sample size. However, it is then more difficult to apply the standard tools of statistics to analyze the properties of the simulation output.

distributions. Furthermore, specialized stochastic simulation packages, such as @Risk and Crystal Ball, can be linked to Excel and provide a number of additional tools for joint distributions, stratified samples, and graphical and postsimulation analysis of model results. However, for some models, interfacing with these programs is difficult and stochastic simulation routines must be written and incorporated directly into the model code. Knowledge of the basic methods for simulating random variables (and for testing the adequacy of prepackaged random number generators) is thus important.

Computer programs written for random number generation typically begin by sampling a random variable that is uniformly distributed between 0 and 1, termed $U(0,1)$. The $U(0,1)$ random variable can be transformed to any distribution by applying the inverse cdf transform:

$$X = F_X^{-1}(U):$$

$$\begin{aligned} &1.\ \text{Generate } U_i \\ &2.\ X_i \leftarrow F_X^{-1}(U_i) \end{aligned} \tag{7.28}$$

Implementation of the inverse cdf method requires a cdf that is analytically or numerically invertible (preferably the former). For example, for $X\sim$ exponentially distributed, the inverse cdf is readily derived:

$$\begin{aligned} F_X(x) &= 1 - e^{-\lambda x} \\ x &= \frac{-\ln[1 - F_X(x)]}{\lambda} \end{aligned} \tag{7.29}$$

Simulation of X can thus be implemented with the following algorithm:

$$\begin{aligned} &1.\ \text{Generate } U_i \\ &2.\ X_i \leftarrow \frac{-\ln[1 - U_i]}{\lambda} \end{aligned} \tag{7.30}$$

In practice, the $[1 - U_i]$ term in step 2 is replaced by U_i since the distribution of $1 - U$ is the same as U (both are $U[0, 1]$) and the subtraction is thus unnecessary and inefficient. Other techniques can be used to simulate different distributions beginning with $U(0, 1)$ variates, using methods such as acceptance–rejection sampling or special properties of the random variable. These techniques are described in detail in textbooks on probability, statistics, and simulation (e.g., Brately et al., 1987; Ross, 1989; DeGroot and Schervish, 2002).

Typically, the number of samples and associated model replications, N, is increased in a Monte Carlo simulation until the results for the output variable consistently converge toward a certain statistical distribution with stable parameter values. Ross (1989) recommends at least $N = 30$ for an exploratory simulation run, which will provide the modeler with a "flavor" of the output generated by the model. The number of simulations may then be increased until the desired precision in the predicted output distribution is obtained. Alternatively, the central limit theorem can be used to estimate the number of runs required to obtain a desired degree of confidence in the expected value of the output variable (e.g., $E[C]$), since the estimate of the expected value should converge to a normal distribution as N becomes large. The sample mean $\bar{C} = E[C]$ is an unbiased estimator of the population mean, μ_c, while an estimate of the variance of μ_c is given by $[s_c]^2/N$, where s_c is the sample standard deviation for C obtained from the N simulations. The number of simulations can be increased to obtain the desired confidence interval around μ_c, computed as a function of $s_{\mu c} = \{[s_c]^2/N\}^{1/2}$.

EXAMPLE 7.3

Implement the problem posed in Example 7.2 using Monte Carlo computer simulation. Determine if the output variable, C, behaves as predicted by the probability rules used in Example 7.2.

Solution 7.3 We initiate the simulation by first generating 50 realizations of k drawn from a uniform distribution between 0.1 and 0.5, that is, $U(0.1, 0.5)$. The random number generator in Excel (found under "tools" in data analysis) was used to generate the 50 values of k, as shown in column 1 of Table 7.4. The sample mean of k is found to be 0.3081, close to the population mean for $k \sim U(0.1, 0.5)$ of 0.3, while the sample standard deviation of 0.1212 is reasonably close to the population value of $\{[0.5-0.1]^2/12\}^{1/2} = 0.1155$. The corresponding values of C are computed as shown in column 2, where C is the concentration in the reactor at $t = 10$ days, which is computed as $C = 1 \times e^{-10k}$. The average of the 50 ($N = 50$) simulated observations of C is 0.0898 mg/L, which is quite close to the population mean predicted in Example 7.2 (0.090 mg/L); similarly the simulated standard deviation (0.1033 mg/L) is reasonably close to the derived population value of 0.094 mg/L. The overall agreement between the derived and simulated distribution of C is assessed in Figure 7.3 by comparing the empirical cdf of the simulated concentrations to the analytically derived cdf from Example 7.2:

$$F_C(C) = \tfrac{1}{4}\ln(C) + \tfrac{5}{4} \qquad 0.00674 \le C \le 0.368$$

The comparison suggests that the 50 simulations are able to provide a reasonable approximation of the population distribution for C. Further sets of realizations with sample size $N = 50$ (generated with a different initial seed number) are suggested for the reader to see whether this is true in all (or even most) cases. Simulations with a larger sample size (or equivalently, concatenation of multiple simulations of $N = 50$ samples) would be expected to yield closer agreement with the (known, in this case) population distribution.

TABLE 7.4 Example Simulation of Model Input k and Model Output $C(t = 10)$ with $N = 50$

k	$C = \exp(-10k)$	k	$C = \exp(-10k)$
0.136586	0.255162	0.421519	0.01477
0.266985	0.069263	0.115931	0.313704
0.329316	0.037136	0.34134	0.032929
0.259661	0.074526	0.336592	0.03453
0.282855	0.059099	0.344087	0.032037
0.47196	0.008919	0.119019	0.304163
0.224491	0.105937	0.270513	0.066862
0.119703	0.302091	0.181497	0.162843
0.43867	0.012442	0.369137	0.024938
0.47871	0.008337	0.393454	0.019555
0.13567	0.257509	0.111817	0.326878
0.130164	0.272084	0.32392	0.039195
0.160402	0.201086	0.362313	0.026699
0.298688	0.050445	0.494006	0.007154
0.186477	0.154931	0.345686	0.031529
0.44619	0.01154	0.488488	0.00756
0.404672	0.01748	0.252263	0.080248
0.195926	0.140963	0.327143	0.037952
0.383554	0.02159	0.375546	0.02339
0.451952	0.010894	0.496265	0.006994
0.206632	0.126651	0.281451	0.059934
0.101819	0.361249	0.248527	0.083303
0.421885	0.014716	0.4854	0.007797
0.394675	0.019317	0.442344	0.011993
0.227934	0.102352	0.323347	0.039421
		Mean = 0.3081 Std. Dev. = 0.1212	Mean = 0.0898 Std. Dev. = 0.1033

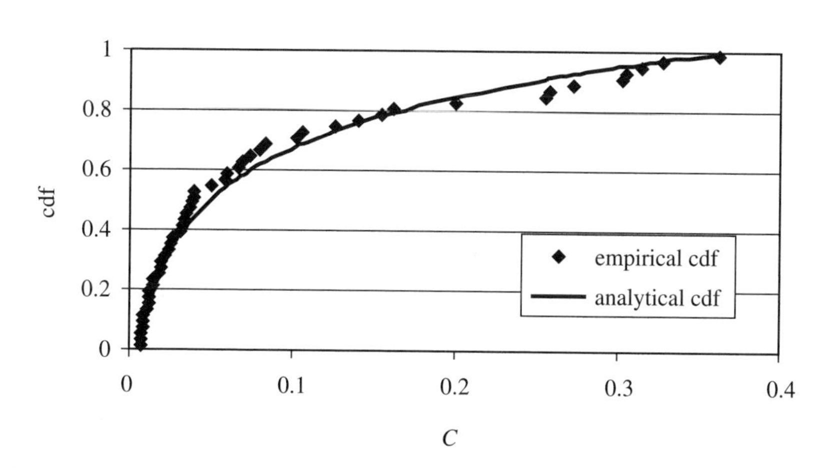

Figure 7.3 Comparison of analytical cdf of concentration, C (mg/L) from Example 7.2 with a simulated (empirical) distribution from Example 7.3.

7.5 RANDOM PROCESSES

Thus far we have considered our parameters to be random variables, which are not ordered in space or time. A random process is a collection of random variables that are mapped to certain state variables, typically spatial coordinates or a timeline. With random processes, the order or location of the random variables is important. For example, 50 random samples of soil conductivity taken at 50 unordered locations at a contaminated site would represent a sample of 50 unordered observations of the random variable, K. When the 50 samples are arranged sequentially, say spatially along the X axis, in the direction of groundwater flow, we have a representation of $K(x)$ as a random process. Random processes are also called stochastic processes.

We first consider a continuous-value, continuous-state random process, that is, a random process in which the value of some variable, say P, is continuous and mapped to a state variable, t, which varies continuously as well. Such a random process $P(t)$ is shown in Figure 7.4. The process $P(t)$ could represent hydraulic conductivity varying over space, or wind speed or pollutant concentration (at a single location) varying over time. Random processes can also vary over multiple dimensions of space and time—and in fact, usually do. We can consider such a process to be the same as having several random variables, for example, $P(t_1)$, $P(t_2)$, $P(t_3)$, ..., all of which are jointly distributed. Looking at the wavelike nature of Figure 7.4, the random process $P(t)$ can also be considered as a summation of waves of different wavelengths (or different frequency)—such a representation yields a frequency domain, or spectral representation of the random proccss.

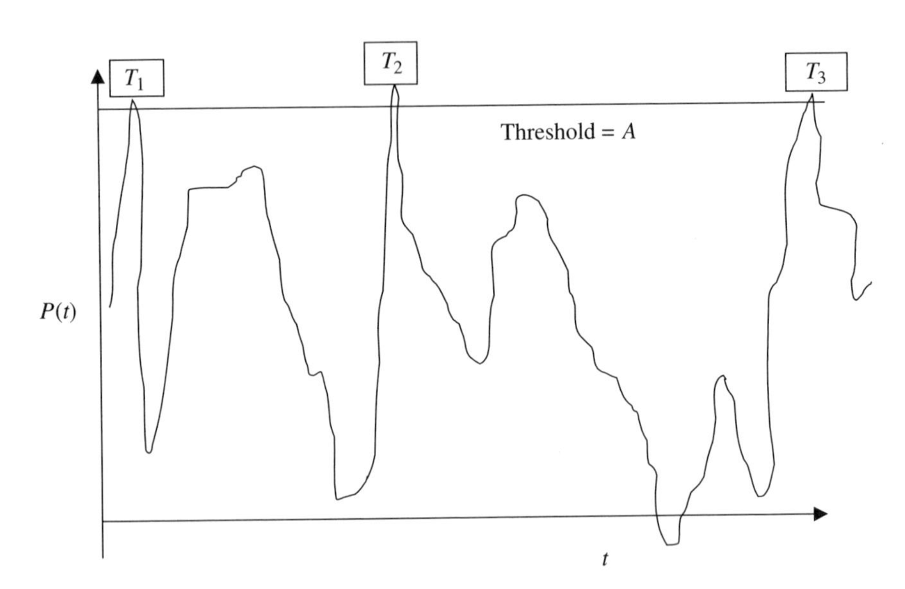

Figure 7.4 Sketch of a continuous random process $P(t)$. Note the wavelike form of $P(t)$ and the occurrences of maxima (at $t = T_1$, T_2, and T_3) that exceed the threshold, A.

7.5.1 Stationarity

Stationarity implies that the statistical properties of a random process do not change with time or space. The stochastic process $P(t)$ is stationary and homogeneous if the following condition is met for the multivariate distribution $P(t)$, $P(t-\Delta t)$, $P(t-2\Delta t), \ldots,$:

$$
\begin{aligned}
&F_{P(t),P(t-\Delta t),P(t-2\Delta t),\ldots,}[P(t), P(t-\Delta t), P(t-2\,\Delta t), \ldots,] \\
&\quad = F_{P(t+\tau),P(t+\tau-\Delta t),P(t+\tau-2\Delta t),\ldots,}[P(t+\tau), P(t+\tau-\Delta t), P(t+\tau-2\,\Delta t), \ldots,] \\
&\quad = \text{Constant} \qquad \text{for all } \tau
\end{aligned}
\tag{7.31}
$$

That is, the joint distribution of $P(t_1)$ and $P(t_2)$ is invariant, depending only on $\Delta t = t_2 - t_1$. A stochastic process is stationary, but nonhomogeneous, if the joint distribution varies in some periodic manner that is recurrent. This is common for random processes that exhibit diurnal and/or seasonal patterns of variation. A random process is wide sense stationary (WSS) if its first two moments are constant with time. Here the first moments include the mean, standard deviation and the autocovariance, or autocorrelation function of the process.

Figure 7.5 depicts different types of nonstationarity that might be present, including (a) a long-term trend in the mean, (b) changes in the variance with time, or (c) changes in the spectral (autocorrelation) properties of the process. By definition, nonstationary random processes are nonhomogeneous. More in-depth discussion of autocorrelation and wide sense stationarity is presented later in the discussion of regionalized variables and geostatistical methods.

7.5.2 Types of Random Processes

When $P(t)$ is normally distributed, the process is referred to as a Gaussian random process. A Gaussian random process is fully characterized by its mean, variance, and autocorrelation function, so that WSS implies stationarity in all respects. A special type of *purely random* stochastic process has zero autocorrelation for all values of $\tau > 0$ and is referred to as "white noise":

$$
\begin{aligned}
\rho(\tau) &= 1 \quad \text{for} \quad \tau = 0 \\
&= 0 \quad \text{for all other } \tau
\end{aligned}
\tag{7.32}
$$

As noted above, Figures 7.4 and 7.5 depict continuous-value, continuous-parameter random processes. Random processes can also be discrete in the variable and/or the indexing parameter(s). Continuous parameter random processes are often only measured—and simulated in environmental models—at discrete points in time or space. Models for *time-series analysis*, such as the *autoregressive integrated moving average* (ARIMA) model, are often used to represent this type of random processes (Box et al., 1994; Chatfield, 1996; Bras and Rodriguez-Iturbe, 1994; Robeson and Steyn, 1990; Berthouex and Brown, 1994, Chapter 41).

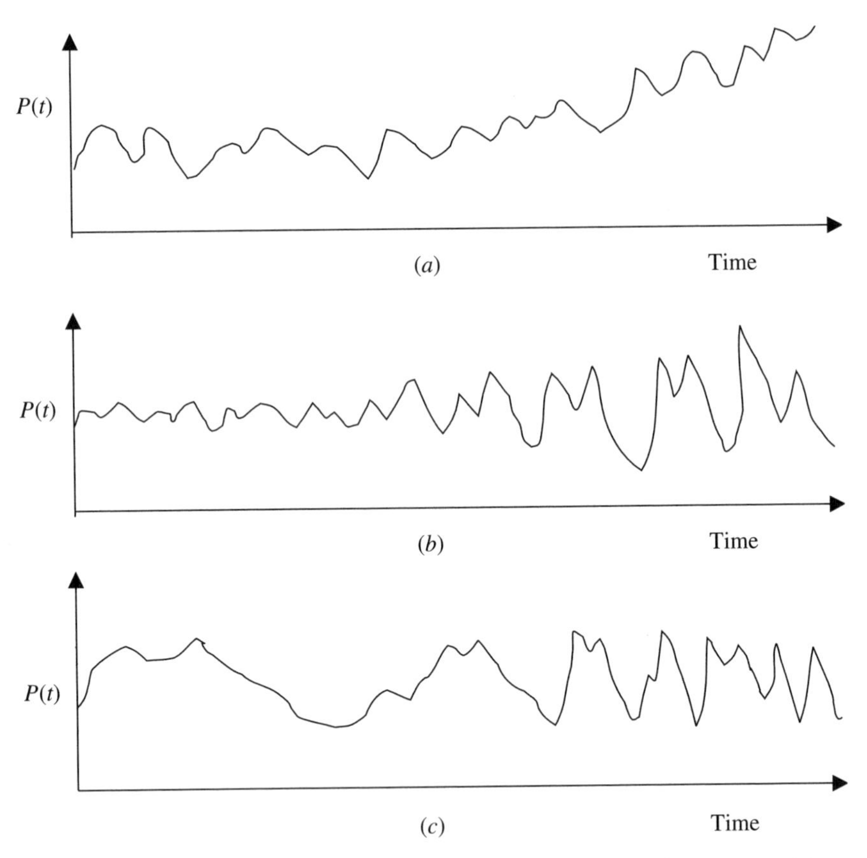

Figure 7.5 Common types of nonstationarity that can occur in a random process: (*a*) nonstationary in the mean, (*b*) nonstationary in the variance, and (*c*) nonstationary in the autocovariance.

Discrete variable random processes commonly arise when discrete categories are used to describe environmental conditions. Examples include:

- Soil type in an aquifer, for example, sand, silt, and clay, or permeable versus impermeable
- Dry versus wet rainfall conditions or days
- The physical-chemical state of a lake, such as stratified versus well-mixed (typically varying over seasons) or oligatrophic, mesotrophic, or eutrophic (typically varying over decades)

Discrete state Markov chain models are often used to describe this type of process (Parzen, 1962; Davis, 1986; Ross, 1989; Higgins and Keller-McNulty, 1995). A random process is Markov if the conditional probability for a future state depends only on its present value but not on its past:

$$F_{P(t+\Delta t),P(t),P(t-\Delta t),P(t-2\Delta t),\ldots,}[P(t+\Delta t), P(t), P(t-\Delta t), P(t-2\,\Delta t)\ldots,]$$
$$= F_{P(t+\Delta t),P(t)}[P(t+\Delta t), P(t)] \qquad [\text{independent of } P(t-\Delta t), P(t-2\,\Delta t),\ldots,] \tag{7.33}$$

A first-order Markov chain model has this property and is characterized (for n possible states) by an $n \times n$ transition probability matrix, with elements p_{ij} = probability of a transition from state i to state j over the interval from t to $t + \Delta t$. Inclusion of all possible states for the system implies that:

$$\sum_{j=1}^{n} p_{ij} = 1 \qquad \text{for all rows } i \tag{7.34}$$

First-order Markov chains have been used in environmental applications for meteorological variables (Gabrial and Neuman, 1962; Richardson, 1981; Small et al., 1989; Johnson et al., 1996) and to describe chemical kinetics and transport in population balance models (Patterson et al., 1981; Chou, et al., 1988; van Kampen, 1992).

When the Markov property does not apply, that is, the probability of the state at time $t + \Delta t$ depends upon the state at times prior to t as well as t; then, a higher-order Markov chain model is required, with transition probabilities to the state at time $t + \Delta t$ depending on the current and previous states (e.g., Green, 1964; Chin, 1977; Small and Morgan, 1986). Discrete value, continuous-time Markov models apply when there is a constant rate at which a transition from one state to another could occur as a function of time (or space) (Parzen, 1962; Rodhe and Grandell, 1972; Ripley, 1992; Gross and Small, 1998).

Another type of random process of particular interest in environmental systems is a point process, denoting the occurrence of discrete events in time or space (see Fig. 7.6*a*). A purely random point process is referred to as a Poisson process, in which the rate of event occurrence, $\lambda(t)$ [events/T], is constant, independent of the occurrence of any other events (or, e.g., the time since the previous event).[5] The events in Figure 7.6*a* may appear to exhibit some degree of clustering; this is due to the exponential distribution of interevent times associated with the Poisson process. Non-Poisson ("less random") point process models, with events that occur in a more regular manner (Fig. 7.6*b*) or in a more clustered manner (Fig. 7.6*c*), are often needed to represent observed environmental processes, such as storm events in certain locations (Waymire and Gupta, 1981; Rodriguez-Iturbe et al., 1988; Cowpertwaite, 1991) and the location of individuals or communities in different ecosystems or habitats (Diggle, 1983; Cressie, 1991; Plotkin et al., 2000). Since the Poisson and related processes are found in a number of different environmental applications, further discussion of their genesis and properties is provided in the following section.

[5] The condition $\lambda(t)$ = constant actually defines a homogeneous Poisson process. If $\lambda(t)$ varies in some periodic but recurrent manner, the Poisson process is still stationary, though nonhomogeneous. A nonstationary Poisson process occurs when there is a long-term trend in $\lambda(t)$.

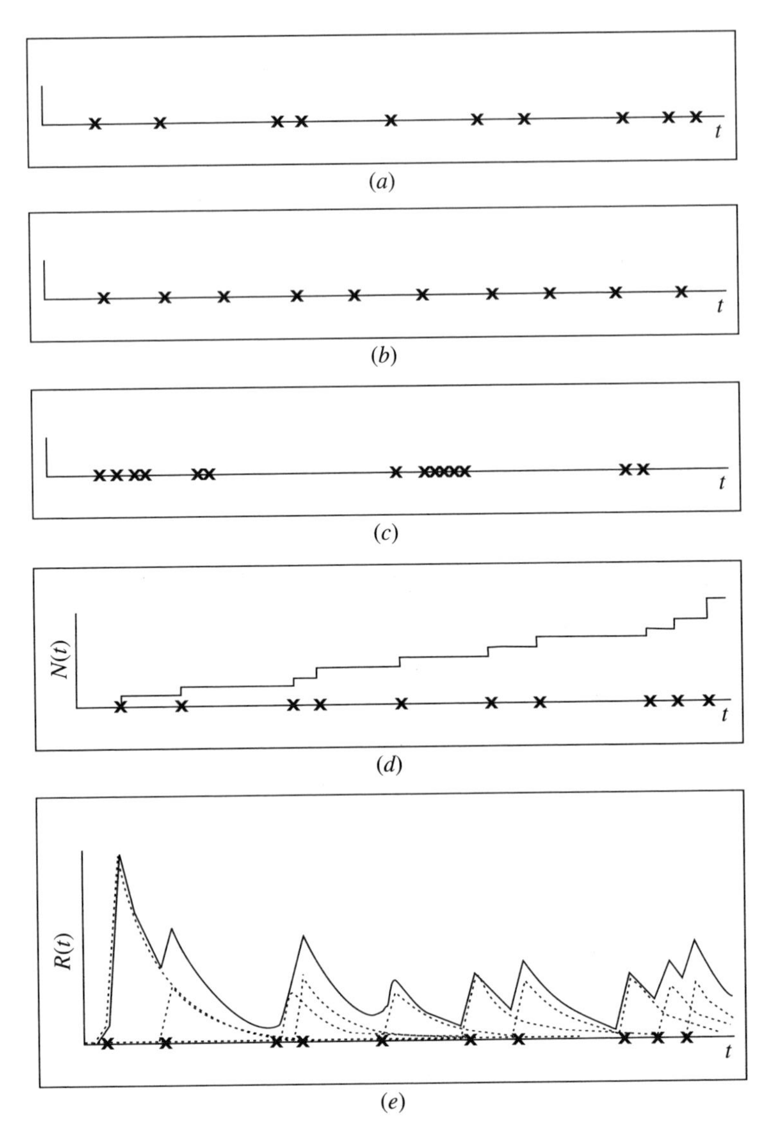

Figure 7.6 Poisson and related processes: (*a*) Poisson point process, (*b*) more regular point process, (*c*) highly clustered point process, (*d*) counting function for Poisson process in (*a*), and (*e*) filtered Poisson process.

7.5.3 Poisson Processes

One mechanism by which Poisson processes arise is through the occurrence of upward excursions or maxima above a certain high threshold level of a random process (see Fig. 7.4). The mathematical analysis of such random processes is known as level-crossing theory (Rice, 1964; Leadbetter et al., 1982; Vanmarcke, 1983; Resnick, 1987). Based on level-crossing theory, for a purely stationary, homogeneous random process, the number of occurrences of maxima above a high threshold, A, approximates a Poisson point process with rate λ, and the time interval between these occurrences is an exponentially distributed random variable, with mean $= E(\tau) = 1/\lambda$. This result ideally applies when the underlying random process is white noise. Then successive occurrences are related to the return period of the extreme excursion:

$$\text{Return period} = E[\tau] = \lambda = \frac{1}{1 - F_{P(t)}[A]} \tag{7.35}$$

However, when there is positive near-field (short-term) autocorrelation in $P(t)$, extreme excursions can occur in clusters, as in Figure 7.6*c*. Negative near-field correlation results in nonrandom regularity, as in Figure 7.6*b*. These deviations from Poisson behavior are reduced for larger threshold values. These results from crossing theory have been applied in water resources modeling for extreme flood events (Rodriguez-Iturbe, 1986; Desmond and Guy, 1991), in the analysis of air pollution exceedances (Smith and Shively, 1995), and for trace element deposits and their effects on groundwater quality (Ramaswami and Small, 1994).

A Poisson process is often formally characterized by the counting function, $N(t)$, for the number of events that have occurred up to and including time t (see Fig. 7.6*d*). Other related processes often used in environmental applications include the compound Poisson process and the filtered Poisson (or "shot noise") process. A compound Poisson process, $X(t)$, tallies the cumulative sum of random variables, Y_i, associated with each event i (and assumed to be independent from one event to the next):

$$X(t) = \sum_{i=1}^{N(t)} Y_i \tag{7.36}$$

Examples include the total amount of rainfall resulting from a Poisson storm event process or the total amount of a contaminant released due to random spill/emission events. The counting function $N(t)$ is actually a special case of a compound Poisson process with $Y_i \equiv 1$, for all events. The moments of $X(t)$ can be computed from the moments of the individual event amounts ($E[Y]$ and $\text{Var}[Y]$) and the rate of the Poisson event process (λ) as:

$$\begin{aligned} E[X(t)] &= E[N(t)]E[Y] = \lambda t E[Y] \\ \text{Var}[X(t)] &= E[N(t)]\text{Var}[Y] + \text{Var}[N(t)]E^2[Y] = \lambda t E[Y^2] \end{aligned} \tag{7.37}$$

Note that the coefficient of variation of $X(t)$

$$\begin{aligned} \nu_{X(t)} &= \frac{\{\text{Var}[X(t)]\}^{1/2}}{E[X(t)]} \\ &= (\lambda t)^{-1/2} \frac{E^{1/2}[Y^2]}{E[Y]} \end{aligned} \tag{7.38}$$

decreases with time t, so that the cumulative sum can be predicted with greater relative precision over long time periods as compared to short ones. (This "law of large numbers" is relied on by insurance companies to ensure that the cumulative claims that they receive over the long term are fairly stable and predictable, even if highly variable for short periods of time.) The probability distribution function of $X(t)$ is a mixed random variable because there is a finite probability of no events (i.e., $\text{Prob}[N(t) = 0] > 0$). For longer time intervals the distribution function of $X(t)$ can be approximated by that for a continuous random variable, for example, by a gamma distribution or a normal distribution (especially the latter as the coefficient of variation becomes small for large t), both of which can be specified by the moments in Eq. (7.37).

Filtered Poisson Process The filtered Poisson process tracks the sum of system responses (due to an impulse–response function) caused by a sequence of random events (see Fig. 7.6*e*). Examples include the stream flow induced by a Poisson-distributed storm event runoff process or water pollution concentrations resulting from Poisson-distributed oil spills to a harbor. The total response is computed from the superposition of the response to each event, so that the effects of each event are assumed independent and additive (as is the case when a linear environmental model is used). With the unit impulse–response function given by $g(t - t_i)$ [i.e., $g(t - t_i)$ is the computed response when $Y_i = 1$, so that the response to any event i is given by $r_i(t) = Y_i g(t - t_i)$], the filtered Poisson process is given by $R(t)$:

$$\begin{aligned} R(t) &= \sum_{i=-\infty}^{+\infty} r_i(t - t_i) \\ &= \sum_{i=-\infty}^{+\infty} Y_i g(t - t_i) \end{aligned} \tag{7.39}$$

When time is the indexing parameter, physical systems are affected only by events that have occurred up to and including time t [i.e., events cannot affect the past, so $g(\tau) = 0$ for $\tau < 0$], and the summation in Eq. (7.39) is from $i = 1, \ldots, N(t)$. However, this is often not the case for spatial filtered Poisson processes, that is, $R[x]$, since in some cases (especially when dispersion is present) downstream events can affect the upstream response.

The moments of $R(t)$ are derived from statistical theory (Parzen, 1962) using Campbell's theorem:

$$\begin{aligned} E[R(t)] &= \lambda E[Y] \int_{\tau=-\infty}^{\infty} g(\tau)\, d\tau \\ \text{Covariance}\,[R(t), R(t+s)] &= \lambda E[Y^2] \int_{\tau=-\infty}^{\infty} g(\tau) g(\tau + s)\, d\tau \end{aligned} \tag{7.40}$$

The variance, $\mathrm{Var}[R(t)]$ is computed from the covariance by setting $s = 0$. The filtered and related Poisson process models have been used to characterize the response of stream flow and water quality to random precipitation events (Weiss, 1977; Di Toro, 1980; Cowpertwait and O'Connell, 1993), groundwater pollutant recharge in response to stochastic water inputs (Small and Mular, 1987), indoor air quality from random cooking events (Borrazzo et al., 1992), and the concentration response in an aquifer due to the occurrence of mineral-enriched deposits modeled to occur as a spatial Poisson process (Ramaswami and Small, 1994).

Simulation of Poisson and related processes is achieved by first generating a sequence of random events and then computing the quantities or effects associated with each. Poisson events can be simulated as a sequence of occurrence times, t_i, with exponential interarrival times:

$$\tau_i = t_i - t_{i-1} \sim \exp(\lambda) \tag{7.41}$$

Alternatively, the number of events $N(T)$ over a time interval of length T can first be generated from a Poisson distribution with parameter λT, with each of the $N(T)$ events then distributed uniformly over the time period. This technique is especially useful when simulating the location of random Poisson events in a two- (or three-) dimensional spatial domain with area (or volume) S. Here the parameter λ of the Poisson process has units of L^{-2} (for a two-dimensional system) or L^{-3} (for a three-dimensional system), with the parameter of the Poisson distribution for $N(S)$ equal to λS. Each of the $N(S)$ events simulated for the domain is subsequently distributed uniformly over its spatial dimensions.

EXAMPLE 7.4 POISSON PROCESS MODEL FOR INDOOR AIR POLLUTION FROM CIGARETTE SMOKING

Consider the indoor air pollution problem described in Section 1.3.2, where the effects of cigarette smoke emissions are modeled for a single room, assumed to behave as a CSTR. As before, each cigarette results in an emission of $M_c = 125$ mg of CO, however, instead of assuming that each of the four card players smokes four cigarettes per hour, resulting in a total (steady) emission rate of $4 \times 4 \times 125 = 2000$ mg CO/h, assume that these cigarettes are smoked in a random fashion, each over a period of $\tau_s = 6$ min. The occurrence of cigarette smoking events is thus assumed to follow a Poisson process with rate $\lambda = 16$/h, so that the expected number of cigarettes smoked in an hour is $E[N(1\text{h})] = 16$, while the standard deviation of this number is $\{\mathrm{Var}[N(1\text{h})]\}^{1/2} = 4$.

Solution 7.4 During each 6-min smoking event, a steady emission rate of $M_c/\tau_s = 125$ mg/0.1 h $= 1250$ mg/h occurs. The concentration response function in the room for each cigarette is given by:

$$\begin{aligned} g(\tau) &= \frac{M_c/\tau_s}{\alpha V}\left(1 - e^{-\alpha\tau}\right) \qquad 0 < \tau \le \tau_s \\ &= \frac{M_c/\tau_s}{\alpha V}\left(1 - e^{-\alpha\tau_s}\right)e^{-\alpha(\tau-\tau_s)} \qquad \tau > \tau_s \end{aligned} \tag{7.42}$$

The mean and variance of the filtered Poisson process concentration response are derived by substituting the event-response function into Eq. (7.40) (note that since each emission event is the same, and its magnitude has already been incorporated into $g(\tau)$, $E[Y] = E[Y^2] = 1$ in this case) to yield:

$$\begin{aligned} E[C] &= \frac{\lambda M_c}{\alpha V} \\ \text{Var}[C] &= \frac{\lambda M_c^2}{\alpha^3 \tau_s^2 V^2}\left[\alpha\tau_s - 1 + e^{-\alpha\tau_s}\right] \end{aligned} \tag{7.43}$$

The result for the expected value is not at all surprising; the mean concentration is simply the average long-term emission rate, λM_c, divided by αV, the same as the steady-state concentration associated with a steady emission rate of $W = \lambda M_c$. However, the variance is more interesting. Further insight is gained by solving for the standard deviation ($= \{\text{Var}[C]\}^{1/2}$):

$$\text{Std.Dev. } [C] = \frac{\lambda^{1/2} M_c}{\alpha^{3/2}\tau_s V}\left[\alpha\tau_s - 1 + e^{-\alpha\tau_s}\right]^{1/2} \tag{7.44}$$

and the coefficient of variation (COV) ($=$ Std.Dev. $[C]/E[C]$):

$$v_c = \text{COV}[C] = \frac{[\alpha\tau_s - 1 + e^{-\alpha\tau_s}]^{1/2}}{\alpha^{1/2}\tau_s\lambda^{1/2}} \tag{7.45}$$

The coefficient of variation increases as λ is reduced (the emission events are more intermittent), as τ_s decreases (the emission events occur over shorter, more intense time intervals), and as α increases (the concentration response is more rapid and "spikier").

Substituting the following problem characteristics:

$$\lambda = 16/\text{h} \qquad M_c = 125 \text{ mg} \qquad \tau_s = 0.1 \text{ h} \qquad V = 80 \text{ m}^3 \qquad Q = 40 \text{ m}^3/\text{h}$$

$$k = 0.1/\text{h} \qquad \alpha = k + Q/V = 0.1 + 40/80 = 0.6/\text{h}$$

into the solutions for the moments yields the following results:

$$E[C] = \frac{16 \times 125}{0.6 \times 80} = 41.67 \text{ mg/m}^3$$

$$v_c = \frac{[0.6 \times 0.1 - 1 + e^{-0.6\times 0.1}]^{1/2}}{0.6^{1/2} \times 0.1 \times 16^{1/2}} = 0.136$$

Again, the mean is the same as the steady-state result computed in Section 1.6.1 (see Fig. 1.6), but now the variability in the concentration response is also computed.

A 100-h realization of the stochastic process was simulated and is shown in Figure 7.7*a* (this would indeed be a long card game, but we wish to simulate a long-enough time period to allow the calculation of stable statistical properties). The mean and standard deviation of the simulated concentration profile were calculated (ignoring the initial ~10-h ramp-up period needed to achieve stationary

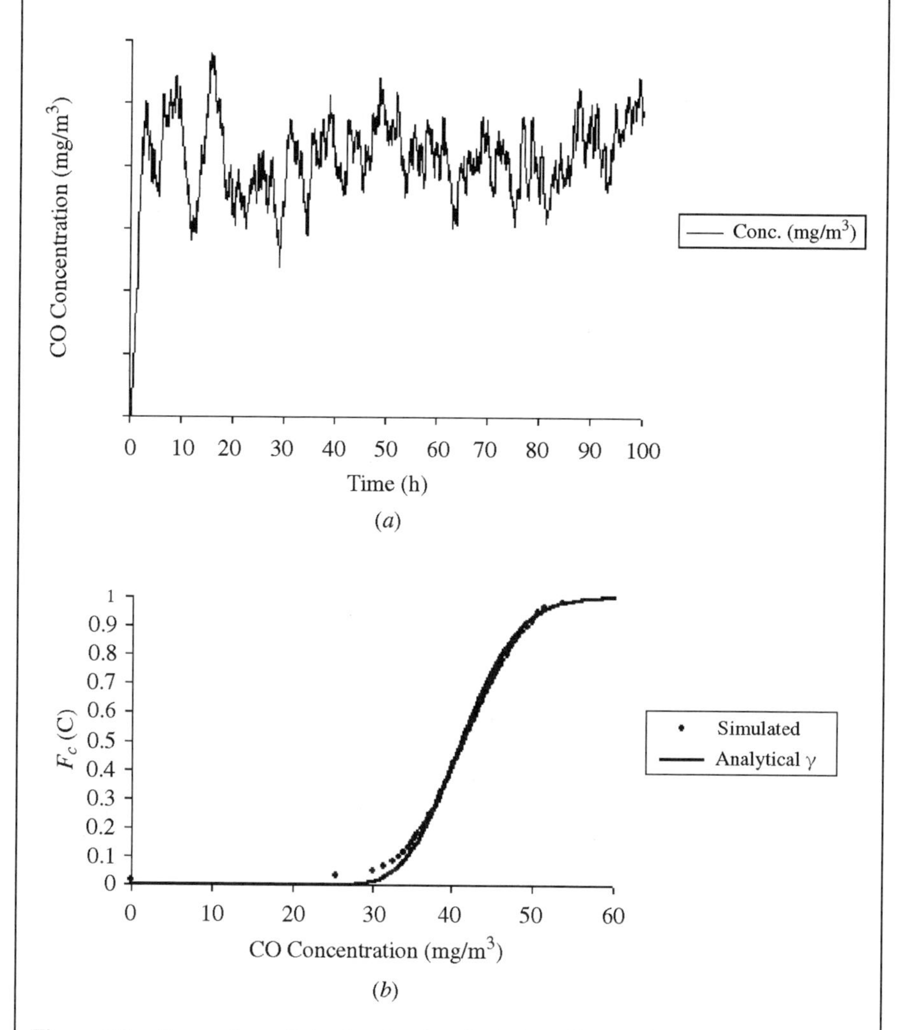

Figure 7.7 Comparison of simulated and analytical predictions for filtered Poisson process indoor air pollution problem given in Example 7.4: (*a*) simulated indoor CO concentration and (*b*) comparison of simulated distribution with approximate analytical gamma distribution.

conditions) to be $E[C] = 41.37$ and Std.Dev. $[C] = 5.56$ ($\nu_c = 0.134$), very similar to the analytical results calculated with the filtered Poisson process model. The cumulative distribution function of the simulated concentrations is compared in Figure 7.7*b* to a gamma distribution with the analytical moments, $E[C] = 41.67$; Std.Dev. $[C] = 5.65$, used to derive the gamma-distribution parameters:

$$a_\gamma = \left(\frac{E[C]}{\text{Std.Dev}[C]}\right)^2 = \left(\frac{41.67}{5.65}\right)^2 = 54.5$$

$$b_\gamma = \frac{\text{Std.Dev}^2[C]}{E[C]} = \frac{5.65^2}{41.67} = 0.766$$

The analytically derived gamma distribution, while only approximate, provides a very close match to the simulated concentration distribution.

This filtered Poisson process model for indoor air pollution resulting from varying cigarette smoke emissions can be compared to one developed by Ott et al. (1992), based on a times-series model formulation.

7.5.4 Averages of Random Processes

Averaging is important in many environmental applications. Many environmental measurements are averages over time or space. For example, particulate air pollution measurements are typically made by drawing air through a filter at a known flow rate for a given period of time (the targeted compounds are subsequently extracted from the filter and analyzed). Inputs to environmental models (e.g., wind speeds or soil properties) are often averaged over time or space to match the temporal and spatial dimensions of the model. Subsequent averaging of model output is often needed to compute average environmental exposures and health effects. Environmental standards for ambient concentrations often include explicit consideration of the averaging time of the sample. For example, see the different averaging times for the U.S. EPA's ambient ozone standard (Fact Sheet for EPA's Revised Ozone Standard, July 17, 1997, at http://www.epa.gov/ttn/oarpg/naaqsfin/o3fact.html, accessed September 20, 2004) and the more recent adoption of similar approaches for water quality standards (*Federal Register*, July 13, 2000, Vol. 65, No. 135, at http://www.epa.gov/fedrgstr/EPA-WATER/2000/July/Day-13/w17831.htm, accessed September 20, 2004).

Averaging is a smoothing process with effects that can be derived directly from the statistical properties of a random process. Consider the average value of the random process, $P(t)$ over an averaging time T_A:

$$P_A(t) = \frac{1}{T_A}\int_{t=0}^{T_A} P(t)\,dt \tag{7.46}$$

For a stationary random process, the expected value of $P_A(t)$ is the same as $P(t)$:

$$E[P_A(t)] = E[P(t)] \tag{7.47}$$

The variance, however, is reduced by the smoothing of the averaging process and can be computed from the autocorrelation function, $\rho(s)$, as:

$$\frac{\mathrm{Var}[P_A(t)]}{\mathrm{Var}[P(t)]} = \frac{2}{T_A}\int_0^{T_A}\left(1 - \frac{s}{T_A}\right)\rho(s)\,ds \tag{7.48}$$

To illustrate the application of this equation, consider an exponential autocorrelation function:

$$\rho(s) = \exp(-\phi s) \tag{7.49}$$

where $1/\phi$ is the characteristic time (or length) of correlation. This function is often appropriate in environmental applications and yields a Markov random process [equivalent to an AR(1) model]. The exponential autocorrelation function is assumed for wind speeds in Appendix 5A, and is often used to described the spatial variation of soil properties and observed concentration fields. Substituting Eq. (7.49) into (7.48) yields:

$$\frac{\mathrm{Var}[P_A(t)]}{\mathrm{Var}[P(t)]} = \frac{2}{(\phi T_A)^2}\left[\phi T_A - 1 + \exp(-\phi T_A)\right] \tag{7.50}$$

Figure 7.8 demonstrates how the variance reduction factor in Eq. (7.50) decreases as the averaging time increases relative to the characteristic time of correlation.

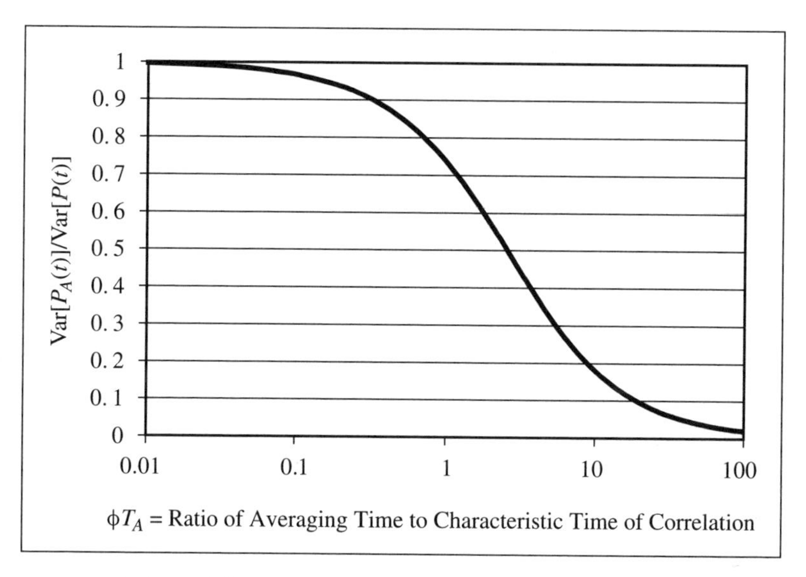

Figure 7.8 Plot of Eq. (7.50): variance reduction achieved by time-averaging a Markov continuous random process with exponential autocorrelation function (variance reduction shown as a function of ϕT_A).

The distribution of $P_A(t)$ is similar to that of $P(t)$ when T_A is small and approaches a normal distribution as T_A becomes large. Time-averaged Gaussian processes are thus themselves always Gaussian, irrespective of T_A. An application of the variance reduction calculation is demonstrated in Chapter 10 for stream flow rates associated with storm runoff events.

In dealing with spatial random processes, the autocorrelation functions necessary for averaging or other applications such as interpolation are often computed using a special set of techniques, referred to as geostatistical methods or "kriging." The following section provides an introduction to these methods.

7.5.5 Stochastic Processes and Regionalized Variables

Environmental data that exhibit a high degree of spatial correlation are often modeled using the techniques of random fields and processes. Such data may be treated as a sample of a *regionalized variable*, which is assumed to behave as a wide sense stationary random process. In the geostatistics literature, this hypothesis is also called the intrinsic hypothesis (Journel and Huijbregts, 1981; Conradsen et al., 1992). A wide sense stationary random process, $P(t)$ has the following properties:

Stationary in the mean: $E[P(t)] = E[P(t+\tau)] = \mu$ (constant population mean)
Stationary in variance: $\sigma^2(t) = \sigma^2(t+\tau) = \sigma^2$ (constant population variance)
Covariance between $P(t)$ and $P(t+\tau)$ depends only on the separation, τ:

$$\mathrm{Cov}(\tau) = E\{[P(t) - \mu(t)] \times [P(t+\tau) - \mu(t+\tau)]\} = h(\tau)$$

The covariance depends only upon the separation τ, and is some function $h(\tau)$. In the literature on regionalized variables, the covariance function, $h(\tau)$, is related to another property of the regionalized variable, called the variogram function, $\gamma(t)$, which is easier to compute from raw spatially ordered data. The variogram function, also referred to as the semivariance, is defined as follows:

$$\gamma(\tau) = \tfrac{1}{2} \times E[P(t) - P(t+\tau)]^2 \tag{7.51}$$

which can be expanded to yield

$$\gamma(\tau) = \tfrac{1}{2} \times \left\{E[P^2(t)] - 2E[P(t) \times P(t+\tau)] + E[P^2(t+\tau)]\right\} \tag{7.52}$$

Recognizing that $\mathrm{Cov}(\tau) = E[P(t) \times P(t+\tau)] - \mu^2$, the semivariance can be rewritten as:

$$\gamma(\tau) = \mathrm{Cov}(\tau = 0) - \mathrm{Cov}(\tau) = \sigma^2 - \mathrm{Cov}(\tau) \tag{7.53}$$

The semivariogram $\gamma(\tau)$ is easily computed from raw data using Eq. (7.51). The covariance function $\mathrm{Cov}(\tau)$ can then be obtained using Eq. (7.53). This allows specification of a regionalized autocorrelation function $[\rho(\tau)]$, which can then be used in spectral density analysis.

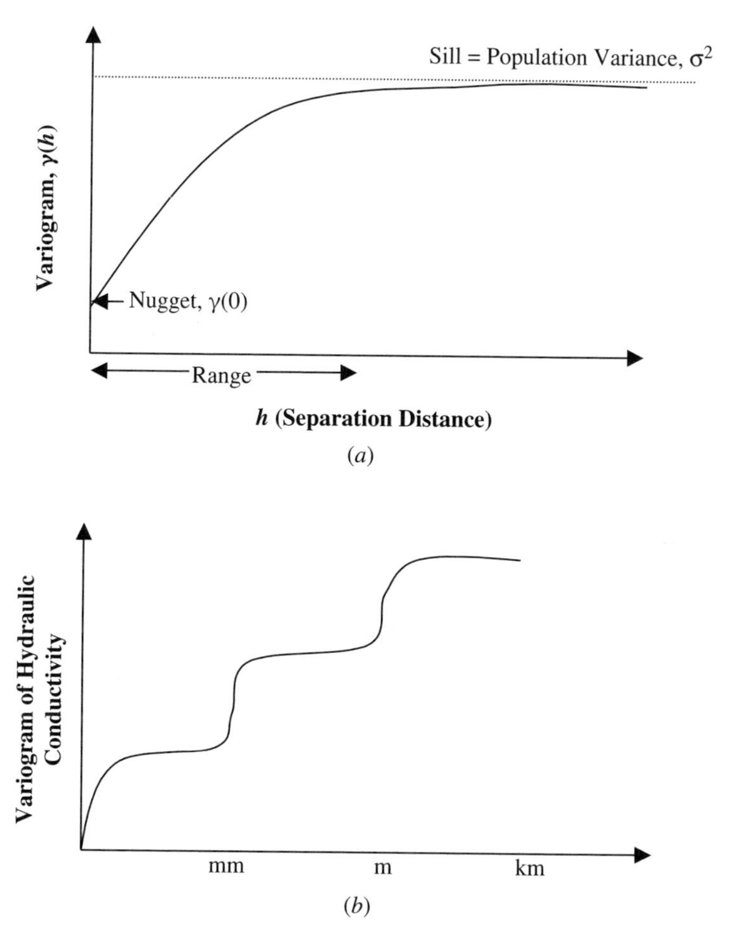

Figure 7.9 (*a*) Classic features of a spherical variogram function with sill and nugget and (*b*) nested variograms with sills representing different scales of spatial correlation.

Several standardized variogram models are used in geostatistics to fit the empirical function computed from Eq. (7.51). The most common structure is called a spherical model (Conradsen et al., 1992) in which the semivariance increases from a noise value, $\gamma(0)$, called the nugget effect, to the population variance σ^2 specified by the sill of the semivariogram (Fig. 7.9*a*). The distance over which correlation between random variables is observed is called the range of the variogram—within the range, the semivariance is less than the population variance due to proximity correlation effects. Random variables may be correlated over different spatial scales, for example, the hydraulic conductivity of porous media can be spatially correlated at the grain scale, at the site scale, and at the regional scale, giving rise to a nested variogram structure as shown in Figure 7.9*b* (Gelhar, 1986).

Knowledge of the variogram structure (and hence the spatial covariance structure) enables more efficient interpolation between sparse observations of the random process $P(t)$. A minimum least squares error technique that employs the variogram function is called kriging. The kriged estimate of a stationary random process at interpolation point I is determined as a weighted sum of the measured values at observation points $j = 1, \ldots, n$, as follows:

$$\hat{P}_I = \sum_{j=1}^{n} w_j P_{\text{obs},j} \tag{7.54}$$

The weights w_j must sum to one for the estimate to be unbiased. The ordinary kriging estimates of the w_j are chosen to minimize the error variance of Eq. (7.54) [based on the Lagrange multiplier method—see Davis (1986) for a derivation] and can be determined as the solution $\mathbf{W}$ to

$$[A] \cdot \mathbf{W} = \mathbf{B} \tag{7.55}$$

where

$$\mathbf{W} = \begin{bmatrix} w_1 \\ w_2 \\ \vdots \\ w_j \\ \vdots \\ w_n \\ \lambda \end{bmatrix} \qquad \mathbf{B} = \begin{bmatrix} \gamma(\tau_{1,I}) \\ \gamma(\tau_{2,I}) \\ \vdots \\ \gamma(\tau_{j,I}) \\ \vdots \\ \gamma(\tau_{n,I}) \\ 1 \end{bmatrix}$$

$$[A] = \begin{bmatrix} \gamma(\tau_{1,1}) & \gamma(\tau_{1,2}) & & \gamma(\tau_{1,j}) & & \gamma(\tau_{1,n}) & 1 \\ \gamma(\tau_{2,1}) & \gamma(\tau_{2,2}) & & \gamma(\tau_{2,j}) & & \gamma(\tau_{2,n}) & 1 \\ \cdot & \cdot & & \cdot & & \vdots & \vdots \\ \gamma(\tau_{j,1}) & \gamma(\tau_{j,2}) & \vdots & \gamma(\tau_{j,j}) & \vdots & \gamma(\tau_{j,n}) & 1 \\ \cdot & \cdot & & \cdot & & \vdots & \vdots \\ \gamma(\tau_{n,1}) & \gamma(\tau_{n,2}) & & \gamma(\tau_{n,j}) & & \gamma(\tau_{n,n}) & 1 \\ 1 & 1 & & 1 & & 1 & 0 \end{bmatrix}$$

Furthermore, the estimation variance is given by:

$$s_\epsilon^2 = \text{Var}[\hat{P}_I - P_I] = \sum_{j=1}^{n} w_j \gamma(\tau_{j,I}) + \lambda \tag{7.56}$$

In these equations, note that the semivariance between any two points j and k is simply a function of the distance $\tau_{j,k}$ between the points, so that $\gamma(\tau_{j,k}) = \gamma(\tau_{k,j})$. Also, λ is the Lagrange multiplier of the equation set.

Kriging has been used extensively in spatial interpolation schemes to map the extent and concentrations of ore bodies, spatial hydraulic conductivity fields, soil porosity fields, and so forth. The EPA has produced public domain software called GEOEAS for fitting standard variogram models to raw data and subsequently using the variogram structure in kriging to produce interpolated contour maps of various environmental properties of interest (http://www.epa.gov/ada/csmos/models/geoeas.html, accessed on April 7, 2002). Gelhar (1986) and Dagan (1986) present excellent reviews of the theory and application of variogram functions in stochastic modeling of subsurface hydrology. The spatial variation (and correlation) in hydraulic conductivity is the principal process that causes random variations in head as well as macrodispersivity in aquifers. A brief description of stochastic models used in the subsurface is provided in Chapter 9. Geostatistical methods and kriging have been applied in other domains as well, including interpolation of meteorological data (Barancourt et al., 1992; Holdaway, 1996; Biau et al., 1999) and air pollutant concentrations and deposition (Tang et al., 1986; Haas, 1990; Schaug et al., 1993; Casada et al., 1994).

8 Models of Transport in Air

8.1 INTRODUCTION

This chapter introduces models of contaminant transport and mixing in the atmosphere. Techniques for modeling atmospheric chemistry and pollutant removal are covered in Chapter 11. Building on the transport fundamentals presented in Chapter 5, we first review the composition and structure of the atmosphere and the nature of the flows that transport air pollutants. Analytical dispersion models applicable over distances of tens of kilometers are presented next. We then discuss numerical models of air pollutant transport and dispersion, which have commonly been applied for urban and regional-scale (i.e., long-range transport) problems.

The descriptions of atmospheric mixing and transport that are widely used today draw on fundamental studies of turbulence and boundary layer flows conducted in the late nineteenth and early twentieth centuries (e.g., Boussinesq, 1877; Reynolds, 1883, 1894; Prandtl, 1904, 1925; Von Karman, 1936) and on empirical and theoretical studies of atmospheric flows conducted in the 1920s (e.g., Richardson, 1920). Many of the semiempirical dispersion parameterizations adopted in current models were developed in the 1950s and 1960s (e.g., Monin and Obukhov, 1954; Sutton, 1953; Gifford, 1961). Development of numerical models that combine transport with the sunlight-driven chemistry of smog formation began in the early 1970s, with much of the initial effort focused on the notorious problem in the Los Angeles area. During the following decade, attention also turned to regional-scale models of acid deposition in Europe and the United States. Current research frontiers in air quality modeling include coupled models of gaseous and aerosol chemistry and transport (discussed in Chapter 11), development and incorporation of new approaches for modeling turbulent dispersion at increasingly fine scales, and improved modeling of chemistry and transport on the global scale.

Contaminant transport in the atmosphere is governed by the advection–diffusion equation (5.15), with mixing driven primarily by atmospheric turbulence. The winds and turbulent eddies that carry contaminants and control their rates of mixing are driven at the largest scales by latitudinal variations in solar heating and by the rotation of Earth. At finer scales, land–sea differences in heating and terrain effects are influential. Surface friction effects come into play at still finer scales.

Atmospheric motions and contaminant transport are frequently classified according to horizontal distance scales. In air pollution meteorology, microscale phenomena are those that occur at horizontal scales of $\sim$1 km or less. Microscale motions extend down to the size of the smallest turbulent eddies that can exist in the atmosphere, on the order of 10^{-3} m. Buoyant plume rise and plume dispersion in the immediate vicinity of a source are microscale processes. Phenomena with horizontal scales on the order of 100 km and corresponding time scales of about one day are classified as mesoscale phenomena. They include migratory high- and low-pressure systems, the flow and dispersion of urban smog plumes, and the transport of urban air pollution on- and offshore with diurnal sea breeze cycles. At the upper end of the mesoscale category, synoptic scale motions have a horizontal scale of about 1000 km. The transport and conversion of sulfur dioxide emissions from Ohio Valley power plants to acid precipitation that falls over New England is an example of a synoptic-scale phenomenon. Macro- or global-scale phenomena are characterized by distances ranging from 1000 km up to about 40,000 km, the circumference of Earth. As an example, sulfate particles lofted into the stratosphere by volcanoes are transported over the global scale.

8.2 ATMOSPHERIC COMPOSITION, STRUCTURE, AND MOTIONS

8.2.1 Atmospheric Composition

Table 8.1 lists selected chemical constituents of the atmosphere in order of decreasing concentration. Concentrations of N_2, O_2, and argon and the other noble gases are essentially constant over time and uniform across locations. Concentrations of the other gases listed in Table 8.1 vary on time scales shorter than human lifetimes and are strongly affected by human activity. Once airborne, a chemical's lifetime in the global atmosphere is determined by its removal rate due to reaction (e.g., photodissociation or thermal degradation) and deposition to land surfaces, vegetation, and the oceans (Junge, 1975; Scheringer, 1997; Bennett et al., 1999).

TABLE 8.1 Atmospheric Constituents

Gas	Typical Surface Concentration (ppm)	Lifetime	Source
N_2	780,840		Biological cycle
O_2	209,460		Biological cycle
H_2O	Variable		Hydrologic cycle
Ar	9340	Inert	
CO_2	355	7 y	Fossil fuel combustion, deforestation
CH_4	1.72	7 y	Wetlands, natural gas use, agriculture
N_2O	0.31	120 y	Oceans, soils, combustion
CO	0.10	1 month	Methane oxidation, combustion
O_3	0.04		Photochemistry
NH_3	10^{-3}	10 d	Livestock, fertilizer, wildlife, oceans
NO/NO_2	10^{-3}	1 d	High-temperature combustion
SO_2	10^{-3}	1 d	Coal combustion
$CFCl_3$	2.7×10^{-4}	50 y	Refrigerant, solvent

Source: Adapted from Seinfeld and Pandis (1998).

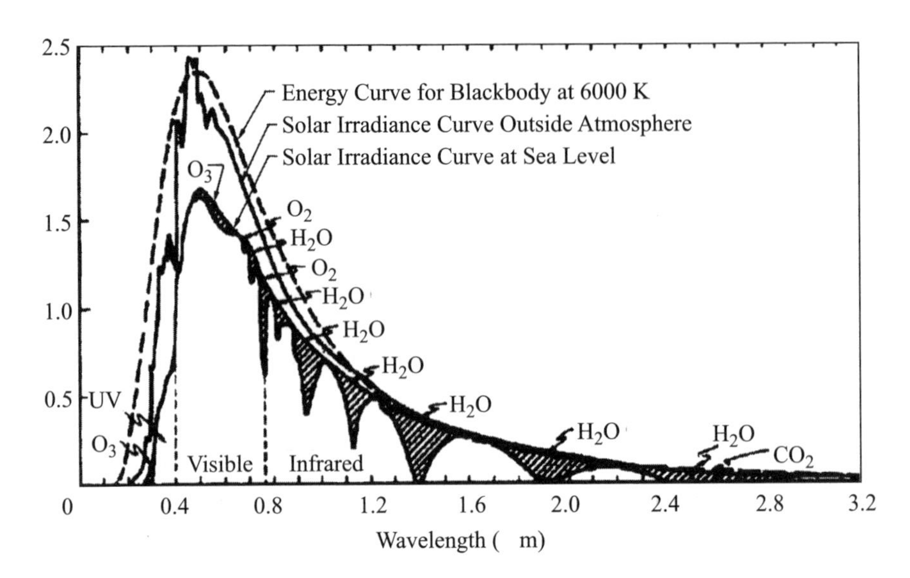

Figure 8.1 Absorption spectra for atmospheric gases. (From *Atmospheric Chemistry and Physics: From Air Pollution to Climate Change*, J. H. Seinfeld and S. N. Pandis, © 1998. Reprinted by permission of John Wiley & Sons, Inc.)

The thermal structure and consequent circulation of the atmosphere are affected by the subset of its constituent gases that interact with solar or terrestrial radiation. Figure 8.1 shows absorption spectra for O_2, O_3, water vapor, and CO_2 over the wavelength range from 0.1 to 3 μm. Solar radiation closely approximates that emitted by a blackbody at about 5800 K, peaking in intensity at approximately 0.5 μm. About 40% of the energy in the sun's radiation falls in the visible wavelength band between 0.4 and 0.8 μm. Molecular oxygen and ozone absorb almost all of the solar radiation with wavelengths in the ultraviolet range below 0.29 μm. In contrast, little atmospheric absorption occurs in the visible range. Earth emits radiation approximately as a blackbody at a temperature of 300 K, with a maximum intensity at wavelengths just above 10 μm and very little energy below 1 μm. Both solar and terrestrial radiation are absorbed strongly by water vapor and CO_2 in several bands spanning the infrared region above 1 μm.

Ozone is formed in the stratosphere from the photodissociation of O_2, and reaches a maximum concentration of more than 1 ppm at an altitude of about 30 km. In combination, O_2 and O_3 efficiently absorb enough ultraviolet (UV) radiation that temperature increases with altitude in the stratosphere (Fig. 8.2). Little light in this wavelength region penetrates into the atmosphere below an altitude of 10 km. The trend for temperature to decrease with altitude in the troposphere is primarily determined by radiant heating and convection at Earth's surface. Because of the temperature inversion that starts at the tropopause, the stratosphere is highly stable and has little turbulence or mixing. In contrast, turbulence is a common feature of the troposphere, the lowest 10 to 15 km of the atmosphere.

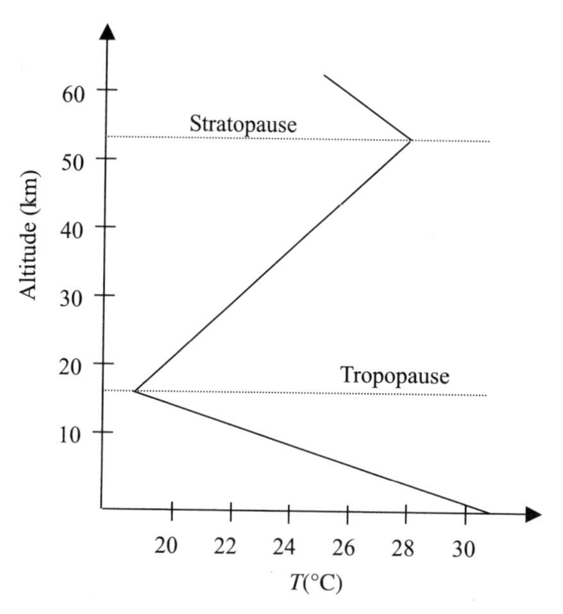

Figure 8.2 Temperature structure of Earth's atmosphere. Profile shown is approximately that for tropical latitudes. Tropopause height at the poles is approximately 5 km lower.

Figure 8.3 shows that the troposphere can be further subdivided into layers based on transport characteristics. The planetary boundary layer (PBL) is defined as the part of the atmosphere where transport is affected by surface features. Over relatively flat terrain, the midday PBL height, z_i, also called the *mixing height*, is typically about 1 to 2 km. In the free troposphere, turbulence can arise from buoyant convection but

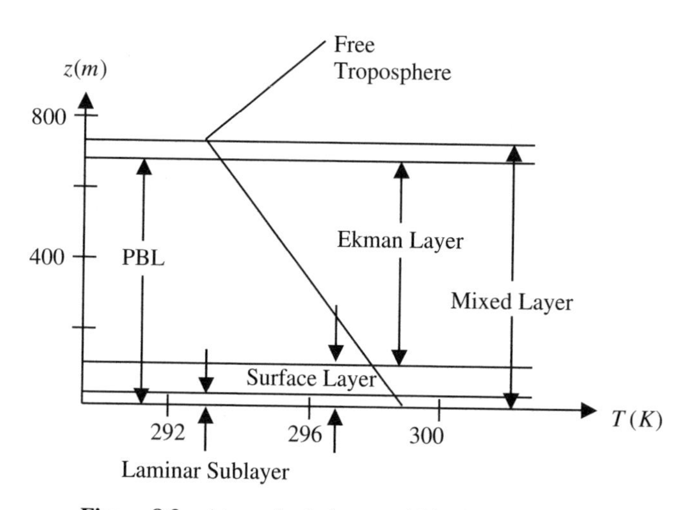

Figure 8.3 Atmospheric layers within the troposphere.

not from mechanical effects of surface roughness. Winds in the free troposphere are determined only by pressure gradient forces and the Coriolis force.

Within the PBL, the surface layer extends from the surface up to a height of about $0.1z_i$. Fluxes of heat, momentum, and moisture in the surface layer are approximately constant with height. The Ekman layer, extending from z_s to z_i, is affected by the Coriolis force and by surface shear and pressure gradients. This combination of forces turns the winds through this layer in the Ekman spiral. The laminar sublayer is the layer of air adjacent to the surface in which turbulence is not fully developed. The laminar sublayer is not shown to scale in Figure 8.3 but is actually less than a centimeter thick. Nevertheless, the absence of turbulent mixing in this layer presents a significant resistance to vertical transport, which must be factored into air pollutant deposition calculations (see Chapter 11).

8.2.2 Atmospheric Circulation

On the global scale, atmospheric circulation is driven by a combination of thermal circulation and the Coriolis force imparted by Earth's rotation. The difference in net radiation at the poles versus the equator is dramatically illustrated by the 50°C difference in the average surface temperature at the two locations. Figure 8.4 shows the classic three-cell model that qualitatively describes the major features of global atmospheric circulation by assuming Earth's surface is uniformly covered by water. Focusing on the Northern Hemisphere (the three-cell model is symmetric), the "Hadley cell" circulation consists of warm air rising at the equator and sinking at about 30°N. In the polar cell, cold air sinks at the pole and rises at about 60°N. The intermediate "Ferrel" cell balances these two flows. Because of Earth's rotation,

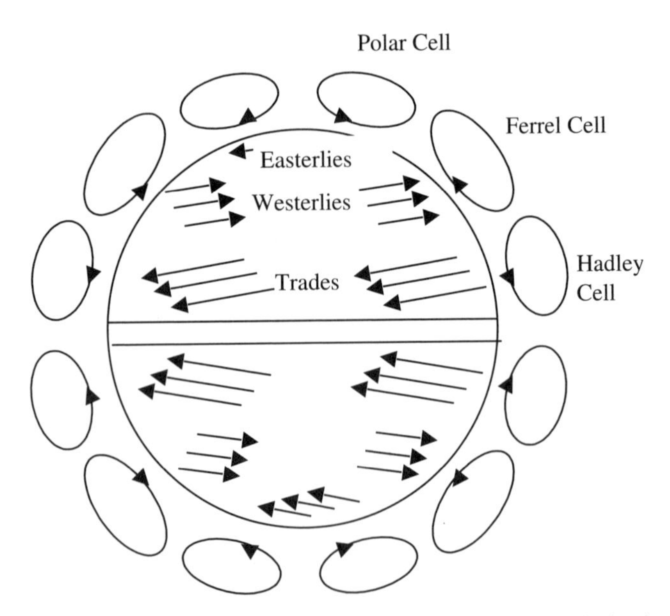

Figure 8.4 Schematic illustration of the three-cell model of atmospheric circulation.

winds are deflected to their right in the Northern Hemisphere and to their left in the Southern Hemisphere. Thus, an observer at low latitudes experiences the north–south surface flow of the Hadley cell as a northeasterly wind. (Winds are named according to the direction from which they are blowing.) The low-latitude northeasterly wind is the consistent feature of atmospheric circulation known as the trade winds. The area of intense convective activity where the north- and southeasterly trade winds meet at the equator is known as the Intertropical Convergence Zone (ITCZ). The prevailing westerlies in the temperate latitudes are also explained by the combined effects of Earth's rotation and the thermal circulation, which in this case flows from south to north at the surface.

The winds in the free troposphere that result from the balance of pressure gradient forces and the Coriolis force are referred to as geostrophic winds. The derivation of the relationship between the geostrophic wind speed and the pressure gradient illustrates how the equations of fluid motion can be applied to describe atmospheric dynamics. The Coriolis force is a "virtual" force that describes motion relative to a rotating frame of reference such as Earth. The Coriolis acceleration is given by $-2(\mathbf{\Omega} \times \mathbf{u})$, where $\mathbf{\Omega}$ is an angular rotation vector and $\mathbf{u} = \{u, v, w\}$ is a velocity vector. Assuming that the geostrophic winds in the free troposphere are two dimensional ($w = 0$) and inviscid and that air has constant density, the applicable governing equations of continuity and motion are

$$\frac{\partial u}{\partial x} + \frac{\partial v}{\partial y} = 0 \tag{8.1a}$$

$$\rho\left(\frac{\partial u}{\partial t} + u\frac{\partial u}{\partial x} + v\frac{\partial u}{\partial y}\right) = -\frac{\partial P}{\partial x} + \rho 2\Omega v \sin\beta \tag{8.1b}$$

$$\rho\left(\frac{\partial v}{\partial t} + u\frac{\partial v}{\partial x} + v\frac{\partial v}{\partial y}\right) = -\frac{\partial P}{\partial y} - \rho 2\Omega u \sin\beta \tag{8.1c}$$

where Ω is the angular velocity of Earth's rotation (2π/day) and β is the latitude. The last terms in Eqs. (8.1b) and (8.1c) are the x and y components of the Coriolis force, per unit volume. Further assuming that the flow is steady ($\partial/\partial t = 0$) and oriented in the x direction ($v = 0$) and solving Eqs. (8.1a) to (8.1c) for u, gives

$$u = \frac{-\dfrac{1}{\rho}\dfrac{\partial P}{\partial y}}{2\Omega \sin\beta} \tag{8.2}$$

EXAMPLE 8.1

To illustrate the magnitude of the geostrophic wind, consider a case with a pressure gradient of 1 mb per 100 km at a latitude of 30°N, altitude of 500 mb, and temperature of 260 K. Compute the geostrophic wind speed for these conditions.

Solution 8.1 The density of air under these conditions is

$$\rho = \frac{P}{RT} = \frac{50{,}000\dfrac{\text{N}}{\text{m}^2}}{286.8\dfrac{\text{N m}}{\text{kg K}}(260\text{ K})} = 0.67\frac{\text{kg}}{\text{m}^3}$$

The geostrophic wind speed is thus

$$|u| = \frac{\dfrac{1}{0.67\text{ kg/m}}\dfrac{100\text{ N/m}^2}{100{,}000\text{ m}}\left(\dfrac{1\text{ kg m}}{\text{s}^2\text{ N}}\right)}{2(2\pi/\text{day}^{-1})\left(\dfrac{1\text{ day}}{24\times 3600\text{ s}}\right)(0.5)} = 20.5\ \frac{\text{m}}{\text{s}}$$

Beyond the broad conceptual picture provided by the three-cell model, the other major features of the general circulation are strongly affected by the distribution and topography of the continents. These relatively constant features of global circulation include semipermanent high- and low-pressure systems that steer the winds in both hemispheres and the westerly subtropical and polar jet streams that flow through the upper troposphere and lower stratosphere.

Synoptic-scale circulation is generally characterized by large air masses with horizontal dimensions on the order of 1000 km, which form over large homogeneous areas under stationary high-pressure systems. High-pressure systems that bring fair weather to the midwestern region of the United States, for example, are often associated with dry continental polar air masses that sweep down from Canada. In contrast, winter weather in the southwestern United States and weather year-round in the Southeast is influenced primarily by maritime tropical air masses. Weather fronts are the transition zones between air masses with different densities, which usually occur along troughs of low pressure. Fronts are characterized as warm or cold based on the temperature of the surface air mass moving into an area.

Under certain circumstances, large-scale atmospheric disturbances can fold fronts into cyclonic eddies with diameters of hundreds of kilometers, which slowly rotate around low-pressure centers. Cyclones rotate counterclockwise in the Northern Hemisphere. Similar sized anticyclones rotate in the opposite direction around high-pressure centers. Slow-moving anticyclones are associated with high levels of air pollution more than any other weather system. Because the surface air flows outward from a high-pressure center, the upper air slowly subsides in these systems, forming temperature inversions that trap pollutant emissions relatively close to the ground. In addition, warm temperatures and sunny skies associated with high-pressure systems promote the photochemistry that leads to the formation of ozone and other secondary air pollutants.

The mesoscale features of circulation that affect flows over distances on the order of 100 km are primarily associated with local topography or land–water interfaces.

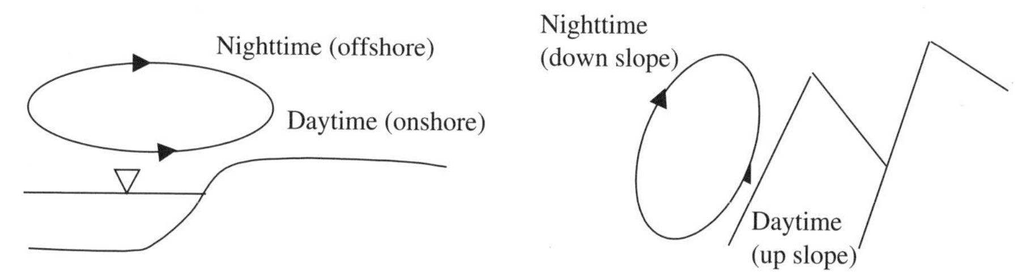

Figure 8.5 Mesoscale land–sea and mountain–valley circulations due to differential heating.

Such phenomena include mountain–valley flows and land–sea breezes (Fig. 8.5). In mountainous areas, upslope flows occur at the surface during the day as low-lying areas are heated by solar radiation. At night, the upper mountain slopes cool most rapidly, with the result that surface winds tend to flow down the mountain. Both mountain and valley flows are shallow, however, so the breezes felt at the surface may not be representative of the winds that most strongly affect contaminant transport. The land and sea breeze cycle is onshore at the surface and offshore aloft during hot sunny days and reverses at night. Over a diurnal cycle, the sea surface temperature remains relatively constant compared to the temperature of adjacent coastal areas. As the land surface warms during the day, heated air above it tends to rise, drawing in relatively cool air from over the ocean to fill the gap. At night, relatively cool air over land moves out over the ocean at the surface, with onshore flow occurring aloft.

8.2.3 Atmospheric Stability

In the planetary boundary layer, vertical transport of momentum and mixing of contaminants are driven by turbulence created from mechanical shear and buoyancy forces. The degree of static atmospheric stability characterizes the tendency of buoyancy forces to accelerate or damp air parcel motions. In unstable conditions (Fig. 8.6) temperature drops off sharply as altitude increases. Buoyancy forces cause relatively warm air at lower altitudes to rise, and cooler air at higher altitudes to sink, resulting in vigorous mixing. In stable conditions, the temperature drops off less steeply or even increases with altitude. The latter condition is called a temperature *inversion*. Buoyancy forces oppose air parcel motion under stable conditions. In the intermediate neutral case, buoyancy forces have no effect on air parcel motions. Unstable atmospheric conditions are prevalent on sunny days when the ground is heated by solar radiation and convection warms the layers of air next to the surface. In contrast, temperature inversions often form at night due to rapid radiative heat loss from the ground. Subsidence inversions occur when air masses of different temperatures collide, with the warmer air mass sliding over the cooler one.

The atmospheric *lapse rate* is defined as $\Lambda = -dT/dz$, where T is temperature and z is height above some reference level. The average lapse rate of the troposphere is about 6.5°C/km, which is slightly stable. Actual atmospheric lapse rates are commonly compared to a reference value, the *dry adiabatic lapse rate*, which

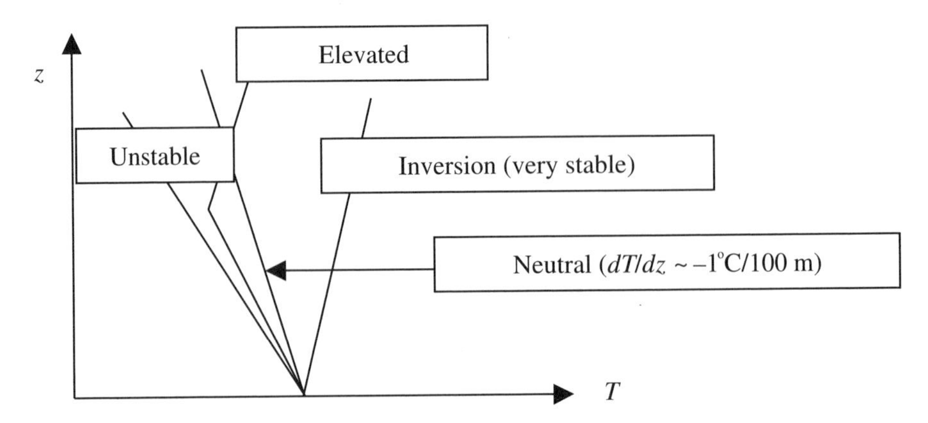

Figure 8.6 Temperature profiles in the lower atmosphere.

characterizes neutral stability in dry air. The value of the dry adiabatic lapse rate can be derived by assuming the atmosphere is at hydrostatic equilibrium and that air parcels move adiabatically, that is, without exchanging heat with the surroundings (Seinfeld and Pandis, 1998).

Let $\Gamma = -dT/dz$ represent the lapse rate of a dry, adiabatic atmosphere. The magnitude of Γ can be found using the chain rule:

$$\frac{dT}{dz} = \frac{dT}{dP}\frac{dP}{dz}$$

where P is pressure. Pressure, temperature, and density ρ are related by the ideal gas law:

$$P = \frac{\rho RT}{\text{MW}} \tag{8.3}$$

where R is the universal gas constant (8.134 J/K/mol) and MW is the molecular weight of air (28.97 g/mol). The hydrostatic equation expresses how air pressure changes with height under conditions of hydrostatic equilibrium:

$$\frac{dP}{dz} = -g\rho \tag{8.4}$$

where g is the acceleration due to gravity (9.8 m/s^{-2}). The hydrostatic equation is strictly valid only for still air. It is a good approximation for descriptions of large-scale flows, but at smaller scales, the effect of vertical accelerations cannot always be neglected. Combining Eqs. (8.3) and (8.4) gives

$$\frac{dP}{dz} = \frac{-g\,\text{MW}\,P}{RT} \tag{8.5}$$

The first law of thermodynamics relates pressure and volume changes to changes in the internal energy, U, or temperature of an ideal gas:

$$dU = c_v\, dT = dQ_{\mathrm{Th}} + dW \tag{8.6}$$

In Eq. (8.6), c_v is the heat capacity at constant volume, dQ_{Th} is the heat input to the system, and $dW = -P\,\mathrm{d}V = -d(PV) + V dP$ is the work done on the air parcel by the surroundings. For an adiabatic process, $dQ_{\mathrm{Th}} = 0$. Thus Eq. (8.6) can be written: $c_v\, dT = -d(PV) + V\, dP$. Eliminating volume from this equation and rearranging it gives an expression for how temperature changes with pressure:

$$\frac{dT}{dP} = \frac{RT}{\mathrm{MW}\; P\left(\hat{c}_v + \dfrac{R}{\mathrm{MW}}\right)} = \frac{RT}{\mathrm{MW}\; P\, \hat{c}_p} \tag{8.7}$$

where $\hat{c}_v$ is the heat capacity per unit mass or specific heat at constant volume, and $\hat{c}_p$ is the specific heat at constant pressure. Finally, combining Eqs. (8.6) and (8.7) by the chain rule:

$$\frac{dT}{dz} = -\frac{g}{\hat{c}_p} = -\frac{9.8\ \mathrm{m/s^2}}{1003\ \mathrm{J/K\,kg}} = -0.0098\ \mathrm{K/m} \quad \text{or} \quad \Gamma = -\frac{dT}{dz} = 0.0098\ \mathrm{K/m} \tag{8.8}$$

The adiabatic lapse rate for conditions when water vapor is present but the air is unsaturated can be found from Eq. (8.8) by correcting the specific heat, $\hat{c}_p$, to account for the water.

For an air parcel moving with acceleration a through surrounding air that is at hydrostatic equilibrium, a balance of forces gives

$$-\frac{dP'}{dz} - g\rho = a\rho \tag{8.9}$$

where the prime denotes the properties of the surrounding air. The hydrostatic equation holds for the air surrounding the parcel:

$$\frac{dP'}{dz} = -g\rho'$$

Combining these equations, the force balance for the parcel reduces to

$$a = g\left(\frac{\rho' - \rho}{\rho}\right) = g\left(\frac{T - T'}{T'}\right) \tag{8.10}$$

Assuming that the lapse rate of the surrounding atmosphere is constant, $T'(z) = -\Lambda z + T_0'$, where $T'(z = 0) = T_0'$. For the air parcel, $T(z) = -\Gamma z + T_0$. For the case in which $T_0 = T_0'$, the relationships for temperature as a function of height can be substituted into Eq. (8.10) to give

$$a = g\left(\frac{\Lambda - \Gamma}{T'}\right)z \tag{8.11}$$

Thus, for $\Lambda > \Gamma$, buoyancy forces produce a positive acceleration and conditions are unstable. For $\Lambda < \Gamma$, the acceleration is negative and conditions are stable.

The departure of the atmosphere from an adiabatic profile is clearly seen in plots of potential temperature, θ, versus altitude. Potential temperature is defined as the temperature that an air parcel originally at height z would have if brought adiabatically to the surface or to a reference pressure of 1000 mb. Based on this definition, for neutral (i.e., adiabatic) atmospheric conditions, $d\theta/dz = 0$. The relationship between potential temperature and actual temperature at a given height is

$$\theta = T\left(\frac{1000}{P}\right)^{R/\hat{c}_p} \tag{8.12}$$

Potential temperature profiles are approximately related to actual temperature profiles by the relationship:

$$\frac{d\theta}{dz} \cong \frac{dT}{dz} + \Gamma \tag{8.13}$$

Atmospheric stability is also characterized using the Richardson number (Ri), a dimensionless parameter defined as the ratio between the rate of consumption of turbulent kinetic energy by buoyancy to its rate of production by shear stresses:

$$\text{Ri} = \frac{g}{T_0}\frac{(d\theta/dz)}{(du/dz)^2} = \frac{g}{T_0}\frac{(\gamma - \Gamma)}{(du/dz)^2} \tag{8.14}$$

Flows with negative Richardson numbers are statically (and dynamically) unstable. Flows with Richardson numbers greater than 0 are statically stable. However, the criterion for dynamic instability is a Richardson number value less than 0.25. For Ri values between 0 and 0.25, turbulence is maintained by wind shear. Larger values of Ri indicate very weak or dissipating turbulence.

8.2.4 Meteorological Models

Descriptions of meteorological conditions are important inputs to air quality models. However, the specific requirements vary considerably depending on the type of model being applied. The simpler models discussed below require only a single representative wind velocity and information on cloud cover from which the atmospheric stability can be inferred. In contrast, three-dimensional, urban-scale photochemical models typically require hourly, spatially resolved wind, temperature, humidity, mixing depth, and solar radiation fields. Meteorological inputs to air quality models may be determined from observations, in some cases using objective analysis techniques to interpolate them over the required domain (e.g., Harley et al., 1993). Increasingly, however, air quality modeling applications are utilizing prognostic meteorological models that calculate the necessary meteorological variables from the fundamental mass, momentum, and energy conservation equations (e.g., Lu et al., 1997; Byun and Ching, 1999).

Objective analysis or diagnostic techniques interpolate observations from a relatively sparse network of meteorological monitoring stations to calculate two- or three-dimensional fields of meteorological variables covering the modeling domain. For wind fields, it is critical that these techniques ensure mass consistency. Interpolation barriers may be used to account for terrain features that decouple surface flows (Goodin et al., 1979). Three-dimensional wind fields can be generated if upper level wind velocity and mixing depth data are available, together with surface observations of wind speed and direction.

Prognostic meteorological models solve the Navier–Stokes equations for mass and momentum conservation in the dynamic atmosphere, together with the ideal gas law and energy conservation equation. Turbulence in air motions must be treated in these models, as discussed in the supplemental information for Chapter 5 at www.wiley.com/college/ramaswami. Meteorological models also track the atmospheric transport and phase transformations of water. Meteorologists have gained extensive experience with these models through weather forecasting applications. This experience has lead to appreciation of the sensitivity of the system to initial conditions. Any errors in the initial or boundary conditions, or in the numerical solution techniques, can be rapidly amplified. The growth of these errors severely limits the period of time over which meteorological forecasts are accurate. One method that has been developed to reduce the growth of errors in modeling historical conditions is to "nudge" the solution toward observed values of key variables. In this approach, called four-dimensional data assimilation (Stauffer and Seaman, 1990), the predicted solution from the dynamic model is continually nudged toward the observed meteorological fields by averaging the model results and observations using various weightings. Two of the most popular meteorological models that are designed for use with air quality modeling systems are the RAMS model developed at Colorado State University (Pielke et al., 1992) and MM5, which was developed in a collaboration between the Pennsylvania State University and the National Center for Atmospheric Research (Grell et al., 1994).

8.3 GAUSSIAN PLUME MODELS

The most widely used models for atmospheric dispersion over distances out to a few tens of kilometers and averaging times of up to an hour are based on the steady-state advection–diffusion equation:

$$u\frac{\partial C}{\partial x} = D_y\frac{\partial^2 C}{\partial y^2} + D_z\frac{\partial^2 C}{\partial z^2} \tag{8.15}$$

which can be solved analytically for (among other configurations) a continuous point source in an unbounded atmosphere with the assumptions that u, D_y, and D_z are constant. Implicit in the form of the equation are the assumptions that steady-state conditions apply and that advection occurs only in the x direction (i.e., the x axis is oriented directly downwind from the source) and diffusion only in the y (cross-wind) and z (vertical) directions. As in the steady-state model for water quality in a stream or river, longitudinal dispersion (in the x direction) though present, makes a negligible contribution to mass transport along the primary direction of fluid flow, compared to advection.

For release height h and constant emissions rate S ($M\ T^{-1}$) the solution to Eq. (8.15) with the stated assumptions is

$$C(x, y, z, t) = \frac{S/u}{4\pi\, t\left(D_y\, D_z\right)^{1/2}} \exp\left\{\left(\frac{-1}{4t}\right)\left[\frac{y^2}{D_y} + \frac{(z-h)^2}{D_z}\right]\right\} \tag{8.16}$$

The *Gaussian plume equation* (GPE) for an unbounded atmosphere follows from Eq. (8.16) by letting:

$$\sigma_y = \sqrt{\frac{2D_y x}{u}} \qquad \sigma_z = \sqrt{\frac{2D_z x}{u}} \qquad t = \frac{x}{u}$$

Then

$$C(x, y, z) = \frac{S}{2\pi\sigma_y\sigma_z u} \exp\left\{\frac{-1}{2}\left(\frac{y}{\sigma_y}\right)^2 + \frac{-1}{2}\left(\frac{h-z}{\sigma_z}\right)^2\right\} \tag{8.17}$$

Figure 8.7 illustrates the model configuration and downwind concentration profiles given by Eq. (8.17).

w/rxn multiply by:
exp (-tk)

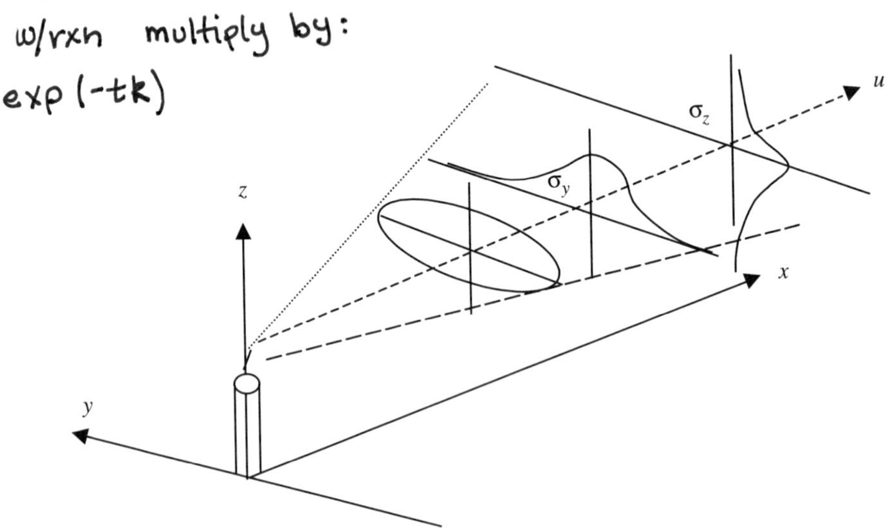

Figure 8.7 Gaussian profile of time-averaged air pollutant concentrations downwind of a point source.

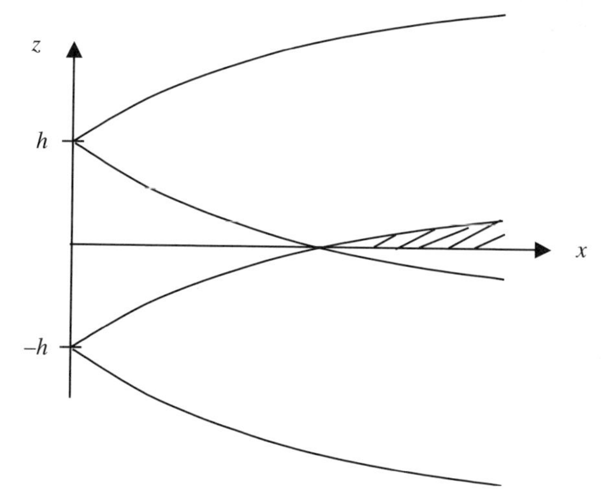

Figure 8.8 Concept of a virtual source for treating perfect reflection at the ground.

In most applications, Eq. (8.17) is modified to account for the fact that the atmosphere is bounded. The lower boundary is commonly treated by assuming that contaminants contacting the ground are reflected back into the atmosphere. This is modeled by introducing a virtual source at a height $z = -h$, as shown in Figure 8.8. The resulting form of the Gaussian plume equation is

$$C(x, y, z) = \frac{S}{2\pi\sigma_y\sigma_z u} \exp\left[\frac{-1}{2}\left(\frac{y}{\sigma_y}\right)^2\right]\left\{\exp\left[\frac{-1}{2}\left(\frac{z-h}{\sigma_z}\right)^2\right] + \exp\left[\frac{-1}{2}\left(\frac{z+h}{\sigma_z}\right)^2\right]\right\} \tag{8.18}$$

EXAMPLE 8.2

Consider a chemical plant that releases 60 g/s of xylene from an effective stack height of 40 m. The windspeed is 4 m/s. At a distance 5 km downwind from the plant, the dispersion coefficients are $\sigma_y = 438$ m and $\sigma_z = 264$ m (which correspond to slightly unstable conditions, as discussed below). Estimate the xylene concentration along the plume centerline at ground level, 5 km downwind from the plant. Assume decay of xylene is negligible.

Solution 8.2 Using Eq. (8.18) and assuming perfect reflection at the ground,

$$C(5000\,\mathrm{m}, 0, 0) = \frac{60\mathrm{g/s}}{2\pi(438\ \mathrm{m})(264\ \mathrm{m})(4\ \mathrm{m/s})}\exp(0)\left\{2\exp\left[-\frac{1}{2}\left(\frac{40\ \mathrm{m}}{264\ \mathrm{m}}\right)^2\right]\right\}$$

$$= 4.1\times10^{-5}\ \mathrm{g/m^{-3}} = 41\ \mu\mathrm{g/m^{-3}}$$

Sometimes, a strong inversion above the emissions release height is assumed to cap the dispersion of the plume, forming an upper boundary. In this case, the mathematical solution corresponds to the case of an infinite number of reflections from the elevated inversion height, z_i:

$$C(x, y, z) = \frac{S}{2\pi\sigma_y\sigma_z u}\exp\left[\frac{-1}{2}\left(\frac{y}{\sigma_y}\right)^2\right]$$

$$\sum_{j=0,\pm1,\pm2,\ldots} \times\left\{\exp\left[\frac{-1}{2}\left(\frac{z+2jz_i+h}{\sigma_z}\right)^2\right] + \exp\left[\frac{-1}{2}\left(\frac{z+2jz_i-h}{\sigma_z}\right)^2\right]\right\} \tag{8.19}$$

Equation (8.19) converges slowly in some cases, but the infinite series can be truncated after $j = 0, \pm1$ if $\sigma_z/z_i \leq 0.63$ (Zannetti, 1990).

EXAMPLE 8.3

Estimate the centerline, ground-level xylene concentration 5 km downwind of the plant described in Example 8.2 if the conditions are the same as in that example except that a strong temperature inversion exists at a height of 300 m.

Solution 8.3 In this case Eq. (8.19) applies, with $z_i = 300$ m. An adequate approximation to the infinite series can be obtained by summing over $j = 0, \pm1$. The solution with $S = 60$ g/s, $\sigma_y = 438$ m, $\sigma_z = 264$ m, and $h = 40$ m is 47 μg/m^3. Including the terms for $j = \pm2$ changes the solution by less than 0.1%.

At distances beyond the point at which the vertical dispersion coefficient exceeds the inversion height, complete mixing can be assumed over the mixed layer and Eq. (8.19) approaches the *fumigation* solution (Zannetti, 1990):

$$C(x, y) = \frac{S}{\sqrt{2\pi}\,\sigma_y u\, z_i}\exp\left[-\frac{1}{2}\left(\frac{y}{\sigma_y}\right)^2\right] \tag{8.20}$$

The basic form of the GPE applies to continuous emissions from point sources. Because the governing advection–diffusion equation is linear, the effect of multiple point sources on concentrations at a given location can be determined by superposition of the concentrations resulting from each individual source. Likewise, equations for idealized line, area, and volume sources can be obtained from the point-source equation by integrating over these geometrical shapes. For example, the effects of a line source oriented perpendicular to the wind can be determined by integrating with respect to differential line segments that individually can be treated as point sources:

$$C(x, y, z) = \int_{y_1}^{y_2} \frac{s}{2\pi u \sigma_y \sigma_z} \exp\left(\frac{-y^2}{2\sigma_y^2}\right) \left\{ \exp\left[\frac{-(z-h)^2}{2\sigma_z^2}\right] + \exp\left[\frac{-(z+h)^2}{2\sigma_z^2}\right] \right\} dy \tag{8.21}$$

In Eq. (8.21), y_1 and y_2 are the line source endpoints and s is the emissions rate per unit length. The x axis is assumed to be aligned with the wind and to pass through the receptor, that is, the receptor is located at $y = 0$. Substituting $\xi = y/\sigma_y$ into Eq. (8.21) and changing the limits of integration, the concentration due to a line source can be calculated from:

$$C(x, y, z) = \frac{s}{\sqrt{2\pi} u \sigma_z} \left\{ \exp\left[\frac{-(z-h)^2}{2\sigma_z^2}\right] + \exp\left[\frac{-(z+h)^2}{2\sigma_z^2}\right] \right\} \times \int_{\xi_1}^{\xi_2} \left(\frac{1}{\sqrt{2\pi}}\right) \exp\left(\frac{-\xi^2}{2}\right) d\xi \tag{8.22}$$

The term inside the integral in Eq. (8.22) corresponds to the Gaussian or normal distribution function for which definite integrals are widely tabulated in statistics books.

EXAMPLE 8.4

A 500-m-long segment of roadway oriented north–south carries heavy traffic that backs up frequently. When traffic backs up, carbon monoxide is estimated to be emitted in vehicle exhaust at a rate of 20 g/km s. Estimate the CO concentration at ground level, 1 km east of the roadway and 150 m south of its northern end if the wind is coming from the west at a speed of 4 m/s. Consider a slightly stable situation with $\sigma_y(1000 \text{ m}) = 110$ m and $\sigma_z(1000 \text{ m}) = 62$ m. Assume the CO emissions are released at an effective height of 1 m.

Solution 8.4 Treating the roadway segment as a line source with $y = 0$ corresponding to the receptor location, the endpoints of the roadway are at $y_1 = -350$ m and $y_2 = 150$ m. The limits of integration are then $\xi_1 = -3.18$ and $\xi_2 = 1.36$. The integral can be evaluated from cumulative distribution functions for the normal distribution as:

$$\int_{\xi_1}^{\xi_2} \left(\frac{1}{\sqrt{2\pi}}\right) \exp\left(\frac{-\xi^2}{2}\right) d\xi = \Phi(\xi_2) - \Phi(\xi_1) = 0.913 - 0.001 = 0.912$$

The CO concentration 1 km downwind of the roadway segment is

$$C(1\text{ km}, 0, 0) = \frac{20 \times 10^{-3}\text{ g/m s}}{\sqrt{2\pi}\, 4\text{ m/s}\ 110\text{ m}\ 62\text{ m}} \left\{2\ \exp\left(-\frac{1}{2}\left(\frac{1\text{ m}}{62\text{ m}}\right)^2\right)\right\} (0.912)$$

$$= 5.3 \times 10^{-7}\text{ g/m}^3 = 0.53\ \mu\text{g/m}^3$$

The Gaussian plume equation applies to steady-state conditions, including steady winds, dispersion, and emissions rates. For instantaneous or rapidly varying emissions and/or varying meteorological conditions, the Gaussian puff equation (Zannetti, 1990) is commonly used:

$$\Delta C(x, y, z, t) = \frac{\Delta M}{(2\pi)^{3/2}\sigma_x\sigma_y\sigma_z} \times \exp\left\{\frac{-1}{2}\left(\frac{x_p - x}{\sigma_x}\right)^2\right\} \exp\left\{\frac{-1}{2}\left(\frac{y_p - y}{\sigma_y}\right)^2\right\} \exp\left\{\frac{-1}{2}\left(\frac{z_p - z}{\sigma_z}\right)^2\right\} \tag{8.23}$$

Equation (8.23) gives the contribution of a puff released at time $t = 0$ to the concentration at location (x, y, z) at time t. In this equation, ΔM represents the mass released over the time interval represented by the puff, and (x_p, y_p, z_p) is the position of the puff center at time t. Note that wind speed, direction, and time do not appear explicitly in Eq. (8.23) but affect the solution by determining the location of the puff center and the dispersion coefficients.

The puff approach is used in the CALPUFF (California Puff) modeling system (Scire et al., 2000) that EPA recommends for modeling pollutant dispersion and transport over distances of 50 to 200 km. The CALPUFF model and user's guide can be downloaded from a link accessible through EPA's web site at http://www.epa.gov/ttn/scram/. Non-steady-state puff models are potentially more accurate than steady-state plume models at these distances because puff models can take into account shifts in wind direction that occur downwind of the release point. Additional meteorological data are required to take advantage of this feature. Accordingly, CALPUFF is commonly run using meteorological fields from the MM5 prognostic meteorological model, along with observed meteorological data.

A common modification to the Gaussian plume and puff equations is incorporation of first-order decay or loss processes. Because the governing equation is linear, the effect of such processes is readily modeled by multiplying the right-hand side of the relevant equation [(8.18) to (8.23)] by $\exp(-t/\tau)$, where $t = x/u$ is the travel time and $\tau = k^{-1}$ is the lifetime of the contaminant with respect to a loss process with decay rate k.

8.3.1 Parameter Estimation for Gaussian Plume Models

The Gaussian plume equation includes several parameters that must be estimated for particular atmospheric conditions: u, σ_y, σ_z, h, and z_i. Because the form of the model is highly simplified, the parameterizations used in practice are rough approximations and some modeling experience is required to assess their accuracy.

Wind Speed The GPE assumes that the wind is oriented strictly in the x direction and is constant with height. In reality, both wind speed and direction are likely to vary with height in the planetary boundary layer. A best approximation might be to set u equal to the mean value of the wind speed, averaged over the vertical extent of the plume (e.g., from $h - 2\sigma_z$ to $h + 2\sigma_z$). However, many applications only have access to surface wind data measured at the standard height of 10 m. A power law profile, $u(z)/u_r = (z/z_r)^p$, is commonly used to correct the wind speed from a reference height z_r to the height of the plume centerline. Values of p depend upon atmospheric stability and surface roughness. More unstable atmospheric conditions even out the velocity profile, that is, p decreases, while stable atmospheric conditions accentuate the change in velocity with height. Typical values of p range from about 0.1 to 0.7 (Huang, 1979).

Dispersion Coefficients The effect of atmospheric stability on the rate of spreading of the plume is introduced through the dispersion coefficients $\sigma_y(x)$ and $\sigma_z(x)$. Empirical relationships such as those depicted in Figure 8.9 are used to estimate $\sigma_y(x)$ and $\sigma_z(x)$. The curves shown in Figure 8.9 were estimated from data collected in Project Prairie Grass (Barad, 1958), which released tracers in open, flat terrain, from a height of about 0.5 m. Three to 10-min average concentrations were measured out to a distance of about 1 km away from the release point. In the Pasquill–Gifford scheme, dispersion coefficient estimates were developed from the tracer concentration data for six atmospheric stability classes (A to F), as shown in Table 8.2. The stability classes can be assigned using data on surface (10 m) wind speeds and the intensity of surface heating as estimated from the sun's angle and cloud cover. Expressions for $\sigma_y(x)$ and $\sigma_z(x)$ that match the curves shown in Figure 8.9 are given by:

$$\sigma_y(x) = ax^b$$
$$\sigma_z(x) = cx^d + f \tag{8.24}$$

σ in meters, x in km

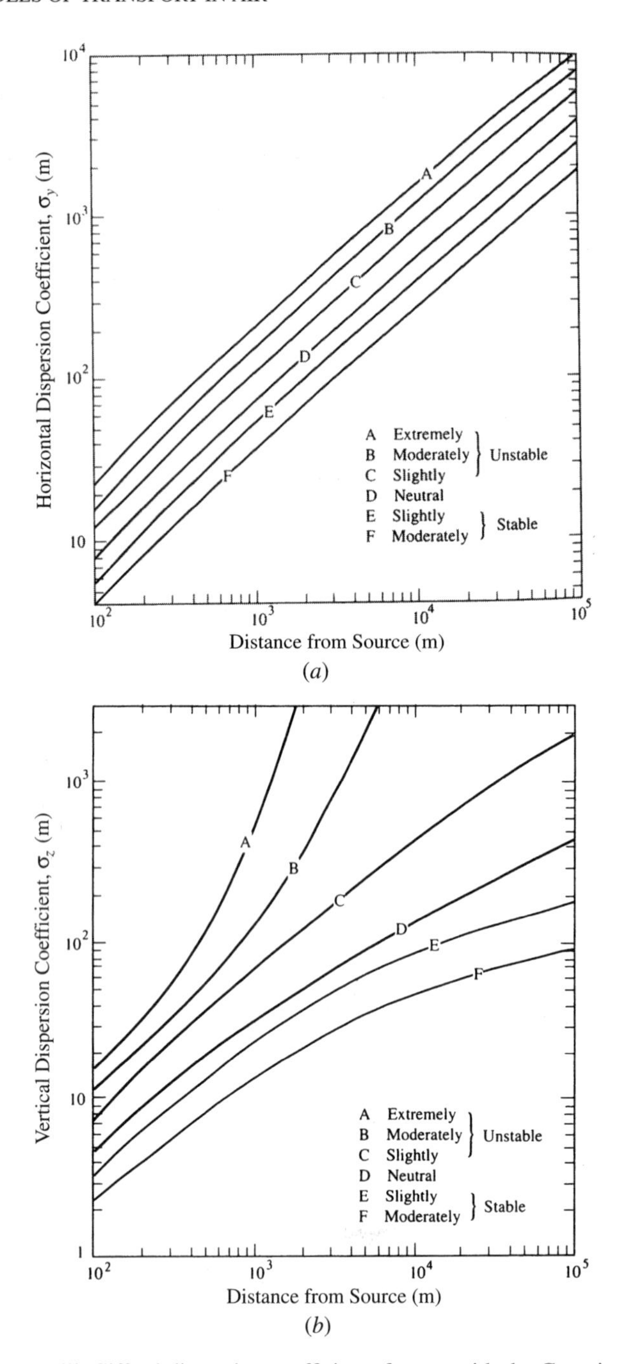

Figure 8.9 Pasquill–Gifford dispersion coefficients for use with the Gaussian plume equation. (From *Atmospheric Chemistry and Physics: From Air Pollution to Climate Change*, J. H. Seinfeld and S. N. Pandis, © 1998. Reprinted by permission of John Wiley & Sons, Inc.) (*a*) Cross-wind dispersion coefficient (σ_y). (*b*) Vertical dispersion coefficient (σ_z).

TABLE 8.2 Pasquill Stability Classes[a]

Wind Speed (m/s)	Daytime Solar Radiation			Nighttime Cloud Cover	
	Strong	Moderate	Slight	$> \frac{1}{2}$	$< \frac{3}{8}$
< 2	A	A–B	B		
2–3	A–B	B	C	E	F
3–5	B	B–C	C	D	E
5–6	C	C–D	D	D	D
> 6	C	D	D	D	D

[a] A = extremely unstable, B = moderately unstable, C = slightly unstable, D = neutral, E = slightly stable, and F = moderately stable.

Source: Turner (1969).

TABLE 8.3 Coefficients for the Pasquill–Gifford Expressions for σ_y and σ_z [Eq. (8.24)]

Stability	a	b	$x < 1$ km c	d	f	$x > 1$ km c	D	F
A	213	0.894	440.8	1.941	9.27	459.7	2.094	−9.6
B	156	0.894	106.6	1.149	3.3	108.2	1.098	2.0
C	104	0.894	61.0	0.911	0	61.0	0.911	0
D	68	0.894	33.2	0.725	−1.7	44.5	0.516	−13.0
E	50.5	0.894	22.8	0.678	−1.3	55.4	0.305	−34.0
F	34	0.894	14.35	0.740	−0.35	62.6	0.180	−48.6

Source: Martin (1976).

with x in kilometers and the coefficients and exponents given in Table 8.3. For flat terrain, the resulting parameter values are viewed as being accurate to within a factor of 2 for distances out to a few hundred meters under all stability conditions, distances out to a few kilometers under neutral or moderately unstable conditions, and distances up to about 10 km for unstable conditions beneath a marked inversion (Zannetti, 1990).

Briggs (1973) utilized the Pasquill–Gifford scheme but added data from experiments conducted on rougher terrain and with elevated tracer releases to derive more broadly applicable expressions for σ_y and σ_z (Table 8.4). His "open-country" expressions match the Pasquill–Gifford estimates for short distances downwind from the source.

TABLE 8.4 Briggs Formulas for σ_y and σ_z, Using x in m

Stability Class	σ_y (m)	σ_z (m)
	Urban Parameters (100 m < x < 10,000 m)	
A–B	$0.32x(1+0.0004x)^{-0.5}$	$0.24x(1+0.001x)^{0.5}$
C	$0.22x(1+0.0004x)^{-0.5}$	$0.20x$
	Urban Parameters, continued	
D	$0.16x(1+0.0004x)^{-0.5}$	$0.14x\,(1+0.0003x)^{-0.5}$
E–F	$0.11x(1+0.0004x)^{-0.5}$	$0.08x\,(1+0.00015x)^{-0.5}$
	Rural Parameters (100 m < x < 10,000 m)	
A	$0.22x(1+0.0001x)^{-0.5}$	$0.20x$
B	$0.16x(1+0.0001x)^{-0.5}$	$0.12x$
C	$0.11x(1+0.0001x)^{-0.5}$	$0.08x(1+0.0002x)^{-0.5}$
D	$0.08x(1+0.0001x)^{-0.5}$	$0.06x(1+0.0015x)^{-0.5}$
E	$0.06x(1+0.0001x)^{-0.5}$	$0.03x(1+0.0003x)^{-1}$
F	$0.04x(1+0.0001x)^{-0.5}$	$0.016x(1+0.0003x)^{-1}$

Sources: From Zannetti (1990); originally from Panofsky and Dutton (1984).

EXAMPLE 8.5

Compare the Pasquill–Gifford and Briggs urban formulas for dispersion coefficients by calculating ground-level, centerline concentrations for distances from 200 to 5000 m, for the case described in Examples 8.2 and 8.3, assuming that at 7 a.m. on a partly sunny spring day the solar radiation is slight. Include the effects of a temperature inversion at 300 m.

Solution 8.5 The conditions described suggest stability class C. (For $x = 5000$ m, the Pasquill–Gifford sigmas for stability class C are $\sigma_y = 438$ m and $\sigma_z = 264$ m, which are the values used in the previous examples.) Concentrations calculated using Pasquill–Gifford expressions with values from Table 8.4 and Briggs urban expressions from Table 8.3 in Eq. (8.19), summing over $j = 0, \ldots, \pm 3$ are shown in Figure 8.10. Concentrations obtained with Eq. (8.20) are also shown for comparison. As seen in the figure, the Briggs urban dispersion formulas lead to more rapid dispersion, which produces higher ground-level concentrations within a few hundred meters from the source and lower concentrations further downwind. Equation (8.20) provides a good approximation to Eq. (8.19) beyond about 2000 m with the Briggs urban dispersion coefficients and beyond about 4000 m with the Pasquill–Gifford coefficients.

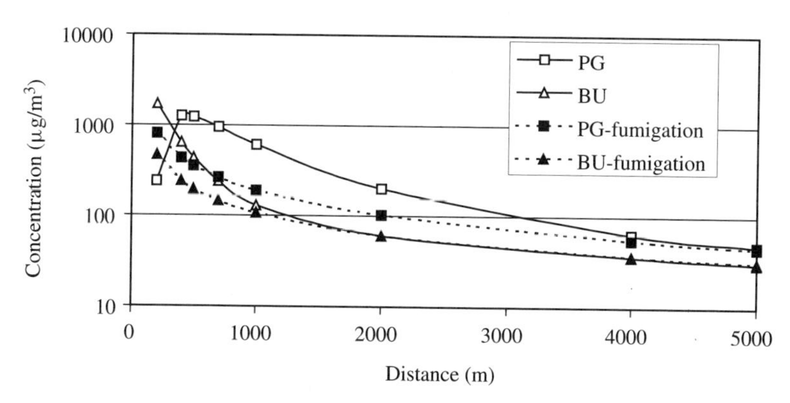

Figure 8.10 Solution to Example 8.5 comparing concentrations estimated with Pasquill–Gifford and Briggs urban formulas for dispersion coefficients. Results obtained using Eq. (8.19) for reflection from an elevated inversion and using the fumigation equation [Eq. (8.20)] are also compared.

The Pasquill–Gifford and Briggs formulas have the advantage of requiring minimal data. However, better estimates of the dispersion coefficients can be made if more information is available. For example, lapse rates or wind direction fluctuations can be used to determine the applicable stability class more objectively. Zannetti (1990) and Seinfeld and Pandis (1998) review several schemes for doing so.

The coefficients of x in Eq. (8.24) for $\sigma_y(x)$ and $\sigma_z(x)$ in the Pasquill–Gifford method and their functional forms in the Briggs formulae are of special interest to the question of whether dispersion in the Gaussian plume model is Fickian, as discussed in Chapter 5. Because with constant wind velocity u the travel time is directly proportional to the downwind location (i.e., $t = x/u$), the coefficients of x in these expressions determine whether the constant-dispersion Fickian model is assumed or whether a non-Fickian dispersion process is represented. When the standard deviation increases as $t^{0.5}$ (or equivalently as $x^{0.5}$), dispersion is Fickian. Values of the coefficient greater than 0.5 indicate that the dispersion is non-Fickian, with the dispersion coefficient increasing with travel time and distance. As shown in Table 8.3, the coefficient b for the expression $\sigma_y(x) = ax^b$ in the Pasquill–Gifford scheme is assigned the value 0.894 for all stability classes, while for $x < 1$ km the coefficient d for $\sigma_z(x) = cx^d + f$ ranges from 0.678 to 1.941, depending on the stability class. These values all portray a non-Fickian dispersion process with the effective dispersion coefficient increasing with travel time and distance. For $x > 1$ km, d exceeds 0.5 for all but the nighttime stability classes, E and F.

In the Briggs scheme, $\sigma_y(x)$ and $\sigma_z(x)$ are expressed as follows for all but a few stability classes:

$$\sigma_y(x) \quad \text{or} \quad \sigma_z(x) = \frac{\alpha x}{(1 + \beta x)^{0.5}} \tag{8.25}$$

Because in all cases, $\beta \ll 1$, the standard deviations increase with x for small x, increase as $x^{0.5}$ for large x, and undergo a transition from the power of 1.0 to 0.5 in the intermediate range. As such, the transport undergoes a transition from an initial non-Fickian regime where D increases with x at short travel distances to an asymptotic Fickian regime at longer travel times.

Plume Rise The next parameter of the GPE that must be estimated is the effective stack height h. The effective stack height is defined as $h = h_s + \Delta h$, where h_s is the actual stack height and Δh is the "plume rise" that the emissions undergo due to their initial forced momentum and buoyancy. Eventually, the plume is knocked over by the wind; the extent of rise that occurs before this happens depends on the temperature and stability of the surrounding atmosphere. Numerous semiempirical equations have been developed for the purpose of estimating plume rise. These equations usually incorporate the exit velocity of the emissions as they leave the stack, the exit temperature of the emissions, the stack diameter, and ambient wind speed, temperature, and atmospheric stability. The formulas typically incorporate the momentum flux parameter, F_m, or the buoyancy flux parameter, F_b, depending on whether the initial vertical velocity is dominated by the buoyancy or momentum of the stack gases:

$$F_m = \frac{\rho_s d_s^2 w_s^2}{4\rho}$$
$$F_b = \frac{g\, d_s^2 w_s \,(T_s - T)}{4T_s} \qquad (8.26)$$

In Eq. (8.26), ρ_s, w_s, and T_s are the density, velocity, and temperature, respectively, of the stack gases at the exit, ρ and T are the corresponding density and temperature of the ambient air, and d_s is the stack diameter at the exit. The plume rise formulas developed by Briggs (1969) are among the best known. For example, for unstable and neutral conditions, the Briggs plume rise formula for buoyancy dominated plumes is

$$\Delta h = \frac{1.6 F_b^{1/3} x^{2/3}}{\bar{u}} \qquad (8.27)$$

For momentum-dominated plumes under neutral and unstable conditions, the corresponding formula is (Arya, 1999)

$$\Delta h = \frac{2.0 F_m^{1/3} x^{1/3}}{\bar{u}^{2/3}} \qquad (8.28)$$

Inversion Height The final parameter that may need to be estimated to apply the GPE is the height of an elevated inversion, if present. This height, z_i, can be estimated from temperature sounding data if they are available. Relationships for estimating z_i if no sounding data are available are discussed below in Section 8.4.2.

8.3.2 Accounting for Variation of Wind Speed and Dispersion Coefficients with Height

Though widely used, the Gaussian plume equation gives a flawed description of pollutant dispersion in the atmosphere, even for the temporal and spatial scales and flat terrain for which it was designed. One problem, as discussed in Section 8.5, is that in convective (unstable) conditions vertical plume dispersion is not symmetric because downdrafts tend to be weaker and more frequent than updrafts. Another problem is that winds and dispersion are not constant with height as assumed for the GPE solution. Huang (1979) gives a steady-state solution for the advection–diffusion equation:

$$u\frac{\partial C}{\partial x} = \frac{\partial}{\partial y}D_y\frac{\partial C}{\partial y} + \frac{\partial}{\partial x}D_z\frac{\partial C}{\partial z} + S, \tag{8.29}$$

for the case in which the wind speed and vertical eddy diffusion coefficient vary with height according to:

$$\begin{aligned} u(z) &= az^p \\ D_z(z) &= bz^n \end{aligned} \tag{8.30}$$

with D_y assumed to be constant. As in the Gaussian plume formulation, advection is assumed to occur only in the downwind (x) direction and dispersion only in the vertical and cross-wind directions. For a continuous, constant point source located at $(x = 0, y_s, z_s)$:

$$S = \Sigma\, \delta(x)\delta(y - y_s)\delta(z - z_s) \tag{8.31}$$

where Σ is the source strength and δ is the Dirac delta function. The boundary conditions are $C \to 0$ as $x \to \infty$, $y \to \pm\infty$, and $z \to \pm\infty$, with perfect reflection at $z = 0$.

For an elevated point source, with D_y defined by:

$$D_y = \frac{u}{2}\frac{d\sigma_y^2}{dx}$$

the solution given by Huang (1979) is

$$C = \frac{Q}{\sqrt{2\pi}\sigma_y}\exp\left[-\frac{(y-y_s)^2}{2\sigma_y^2}\right]\frac{(zz_s)^{(1-n)/2}}{b\alpha x}\exp\left[\frac{-a\left(z^\alpha + z_s^\alpha\right)}{b\alpha^2 x}\right]I_{-\nu}\left[\frac{2a(zz_s)^{\alpha/2}}{b\alpha^2 x}\right] \tag{8.32}$$

where $\alpha = 2 + p - n$, $\nu = (1 - n)/\alpha$, and $I_{-\nu}$ is the hyperbolic Bessel function of the first kind, of order $-\nu$, which is tabulated in mathematical handbooks. In general, a, b, p, and n depend on atmospheric stability and surface roughness. When $p = n = 0$ and $\nu = 0.5$, the form of Eq. (8.32) is that of the Gaussian plume equation.

8.4 NUMERICAL MODELS

Modeling the transport and transformation of secondary air pollutants formed in the atmosphere through complex systems of chemical reactions or phase transformations requires numerical solution techniques. Similarly, numerical models are required to account for complex terrain or detailed treatment of spatial and temporal variations in atmospheric flow fields. Arguably the most generally applicable approach available today to modeling atmospheric dispersion is the numerical technique known as large eddy simulation (LES), in which the large-scale turbulent motions containing most of the kinetic energy are resolved (Arya, 1999) with transport associated with these motions treated as advection. However, LES is computationally expensive; applications to date have been limited to domains of a kilometer or less. Here we focus on numerical treatment of advection and dispersion in urban- and regional-scale models that include both chemistry and transport. The reader is referred to reviews by Mason (1994) and Arya (1999) for more information on large eddy simulation techniques for modeling microscale dispersion.

Numerical models for air pollutant advection, dispersion, and reaction on urban and regional scales have been developed using both Lagrangian and Eulerian frameworks. The Eulerian framework employs a coordinate system that is fixed at Earth's surface. In this framework, a succession of air parcels is viewed as being carried by the wind past a stationary observer. In contrast, the Lagrangian reference frame moves with the flow of air, in effect keeping the observer in contact with the same air parcel over an extended period of time. The Lagrangian reference frame forms the basis of Lagrangian trajectory models and particle models, which were introduced in Chapter 5.

Particle models track the center of mass of a certain number of computational parcels of emissions (i.e., particles) as they travel with the mean wind. These models simulate diffusion about the center of mass by adding a random translation for members of the parcel cloud. Zannetti (1990) provides an overview of this computational technique and reviews a number of applications.

Lagrangian trajectory models can be viewed as following a column of air as it is advected across an air basin at the local wind velocity (Fig. 8.11). Simultaneously, they treat vertical dispersion and entrainment of pollutants, deposition, emissions into the air parcel, and chemical reactions. Trajectory models solve a simplified form of the advection–diffusion equation for the concentration of species i, in which advection is implicit in the location of the air column at a particular time:

$$\frac{\partial C}{\partial t} = \frac{\partial}{\partial z} D_z \frac{\partial C}{\partial z} + S(t) + \text{Rxn} \tag{8.33}$$

Initial conditions are specified as $C(t = 0) = C_0$. Typical boundary conditions are

$$D_z \frac{\partial C}{\partial z}\bigg|_{z=0} = v_g C - E$$

at the surface and

$$D_z \frac{\partial C}{\partial z}\bigg|_{z=z_i} = 0$$

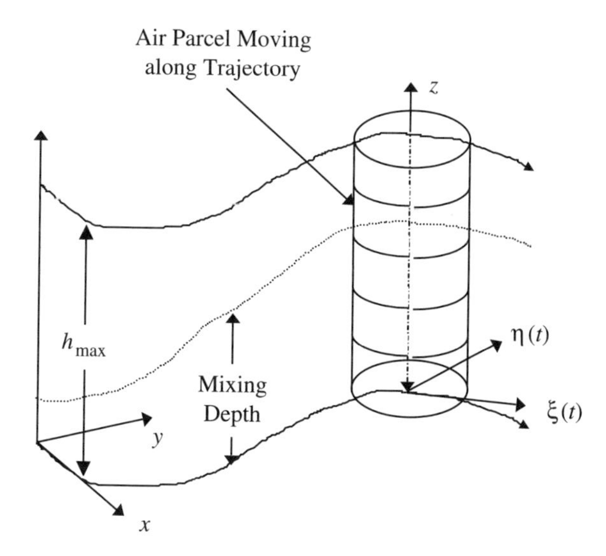

Figure 8.11 Schematic diagram of a Lagrangian trajectory model. $h_{max}(t)$ represents the height of the air parcel. Whereas the coordinate system in an Eulerian framework remains fixed with respect to Earth's surface, the horizontal coordinates in a Lagrangian model move with the air parcel.

at the top of the mixed layer, where E represents the surface emissions flux of the contaminant. Lagrangian trajectory models require horizontally and temporally resolved wind fields from which trajectory paths can be estimated, and correspondingly resolved estimates of emissions, meteorological, and deposition parameters. Depending on the application, trajectory paths can either be projected forward from a given source or backward from a specified receptor location.

The primary limitation of Lagrangian trajectory models is their neglect of vertical wind shear and horizontal dispersion of pollutants (Liu and Seinfeld, 1975). Their primary advantage compared to multidimensional Eulerian models is that their lower computational requirements facilitate incorporation of more detailed descriptions of atmospheric chemistry as well as exploration of system behavior through sensitivity and uncertainty analysis (Pandis et al., 1992; Bergin et al., 1999). Incorporation of chemical reactions into Lagrangian trajectory models is discussed in Chapter 11.

The most widely used framework for numerical air pollution modeling is the Eulerian framework. In Eulerian "grid" models, spatial variations in model inputs or parameters such as emissions or vertical dispersion coefficients are discretized by dividing the modeling domain into a large number of interacting cells (Fig. 8.12). Concentrations are uniform within each cell, but spatial variations are represented by concentration differences from one cell to the next. Numerically solved Eulerian grid models are capable of treating all of the processes represented in Eq. (5.15), including coupled nonlinear chemistry. However, their accuracy is limited in practice by the resolution of the computational grid and associated inputs.

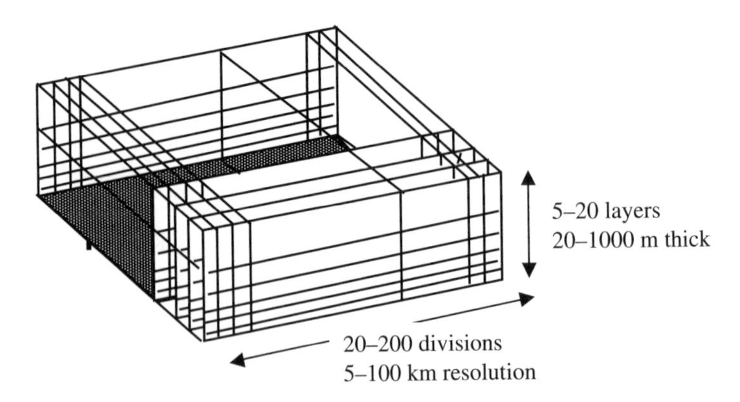

Figure 8.12 Schematic of a discretized Eulerian model domain. A typical application would include about 60 × 100 horizontal cells and 5 to 20 vertical layers. Vertical cell spacing is usually nonuniform, with the finest resolution (tens of meters) near the ground.

Eulerian grid models are most widely used for applications involving coupled, nonlinear chemistry. The level of grid resolution used in most models results from a trade-off between capturing spatial variations in pollutant concentrations and the computational demands of modeling transport and chemical reaction kinetics of a large number of interacting species. Typical fixed grid sizes have been 4 to 5 km on a side for urban-scale models and 20 to 80 km on a side for regional-scale models. Alternatively, nested grid schemes allow a fine grid covering the portion of the model domain where sharp gradients are expected to be embedded within a coarse grid that covers a larger domain (Russell and Dennis, 2000). In one-way nesting, outputs from a model run completed using the coarse domain are used as boundary conditions for a subsequent simulation in the fine domain. In two-way nesting, air and contaminants flow from coarse- to fine-grid regions and back out in a single simulation. Together with continuing increases in computing power, the nested grid approach should facilitate increased horizontal grid resolution in Eulerian models down to about 1 km by 1 km (Seaman, 2000). Numerical modeling with much finer resolution than this enters the regime of large eddy simulation, requiring revised parameterization of dispersion to avoid double counting transport by the largest turbulent eddies.

Uneven terrain is a complication encountered in applying air quality models in many domains. In fact, hilly or mountainous terrain contributes to air pollution problems in many areas by hindering ventilation. To simplify numerical solution of transport equations over complex terrain, the vertical coordinate can be replaced with the terrain following coordinate:

$$\zeta = \frac{z - z_t(x, y)}{h_{\max} - z_t(x, y)} \tag{8.34}$$

where $z_t(x, y)$ represents the local terrain height and $h_{\max}$ is the height of the top of the model domain relative to a fixed reference height. McRae et al. (1982) show

how the components of the wind field are expressed in terms of the terrain following coordinates. Similarly, regional-scale models in which the vertical domain extends from the surface through the whole troposphere commonly use normalized pressure coordinates, known as *sigma* coordinates:

$$\sigma = \frac{P - P(h_{\max})}{P(z_t) - P(h_{\max})} \tag{8.35}$$

Historically, transport in Eulerian grid models has most often been calculated using finite difference techniques. Finite element and spectral methods are also used. Numerical solution of the advection–diffusion equation usually starts by applying operator splitting. This is especially important when chemical reactions are included, due to the wide ranges of time scales involved. A typical scheme would separate out horizontal transport, vertical transport, and physical or chemical production and loss terms. Operator splitting treats these processes as occurring independently over time. The solution sequence can be designed so longer time steps are used to advance slower processes, with shorter time steps reserved for relatively fast processes. Moreover, completely different solution techniques can be applied to most effectively describe each process.

Although finite difference methods are widely applied, their use in modeling advection causes artificial dispersion that can significantly degrade the accuracy of the solution if not corrected. As discussed in Chapter 6, this problem occurs with backward differencing in all media. A variety of techniques have been proposed and used to address this error in atmospheric transport models (e.g., Boris and Book, 1973; Smolarkiewicz, 1983; Odman and Russell, 1993). Chock and co-workers (Chock and Dunker, 1983; Chock, 1985, 1991) have reviewed and compared many of these techniques. Compared to advection, diffusion is a relatively easy operation to treat numerically. Because diffusion smooths out gradients, the process does not amplify or preserve errors as advection does. Implicit finite difference schemes (e.g., Crank–Nicolson) or finite element methods are often used for numerical treatment of diffusion, because the time steps required to ensure stability of explicit finite difference methods are prohibitively small.

8.4.1 Transport Parameter Estimates for Numerical Models

When numerical solution techniques are used to solve the advection–diffusion equation, the simple power law relationships used by Huang (1979) for variation of wind speed and vertical dispersion coefficients with height can be replaced with more complicated forms developed from similarity theory. This approach hypothesizes the set of variables that are important for describing a particular phenomenon, and then uses dimensional analysis to group these variables into generally applicable relationships.

The application of similarity theory to atmospheric flows can be illustrated by the relationship of mean wind speed with height in the adiabatic surface layer over uniform flat terrain (Arya, 1999). In this case, turbulence is purely mechanical, and the change in wind speed with height is hypothesized to depend only on the height, z, and the kinematic surface shear, τ_0/ρ:

$$\frac{\partial \bar{u}}{\partial z} = f\left(z, \frac{\tau_0}{\rho}\right) \tag{8.36}$$

Defining the friction velocity $u_* = \sqrt{\tau_0/\rho}$, dimensional analysis shows that $(\partial \bar{u}/\partial z)(z/u_*) = \kappa$, a constant. Integration of this relationship from z_0 to z gives

$$\frac{\bar{u}(z)}{u_*} = \frac{1}{\kappa} \ln\left(\frac{z}{z_0}\right) \tag{8.37}$$

where z_0 is a surface roughness parameter defined so that the profile goes through $u = 0$ at $z = z_0$. The empirical constant, κ, is known as Von Karman's constant and has a value of about 0.4. Note that the friction velocity cannot be measured directly but can be determined from wind speed measurements at two different heights.

For nonadiabatic conditions, wind profiles are affected by buoyancy as well as mechanical shear, which either dampens or accelerates the transfer of momentum. The Monin and Obukhov (1954) similarity theory is widely used for these cases. The Monin–Obukhov hypothesis argues that the mean wind profile and characteristics of atmospheric turbulence in the surface layer can be described by four scaling parameters (Arya, 1999) including z and u_*. The other scaling parameters introduced are the friction temperature:

$$\theta_* = \frac{-H_0}{\rho \hat{c}_p u_*}$$

where H_0 is the surface heat flux, and a buoyancy length scale referred to as the Monin–Obukhov length:

$$L_{\text{MO}} = \frac{-\rho \hat{c}_p T_0 u_*^3}{\kappa g H_0} = \frac{u_*^2}{\kappa \left(\frac{g}{T_0}\right) \theta_*} \tag{8.38}$$

where L_{MO} can be interpreted as the height above the ground at which mechanical and buoyancy forces contribute equally to the production of turbulence. For $L_{\text{MO}} > 0$, the atmosphere is stable; for $L_{\text{MO}} < 0$ unstable; and for $L_{\text{MO}} = \infty$ neutral. Dimensional analysis shows that:

$$\begin{aligned} \frac{\partial \bar{u}}{\partial z} \frac{\kappa z}{u_*} &= \Phi_m\left(\frac{z}{L_{\text{MO}}}\right) \\ \frac{\partial \bar{\theta}}{\partial z} \frac{\kappa z}{\theta_*} &= \Phi_h\left(\frac{z}{L_{\text{MO}}}\right) \end{aligned} \tag{8.39}$$

Equation (8.39) may be integrated to find the mean wind and potential temperature as a function of height. The expressions for Φ_m given by Businger et al. (1971) are widely used:

$$\Phi_m = 1 + 4.7\left(\frac{z}{L_{\text{MO}}}\right) \quad \text{for} \quad \left(\frac{z}{L_{\text{MO}}}\right) > 0 \text{ (stable)}$$
$$= 1 \quad \text{for} \quad \left(\frac{z}{L_{\text{MO}}}\right) = 0 \text{ (neutral)} \tag{8.40}$$
$$= \left[1 - 15\left(\frac{z}{L_{\text{MO}}}\right)\right]^{-1/4} \quad \text{for} \quad \left(\frac{z}{L_{\text{MO}}}\right) < 0 \text{ (unstable)}$$

Above the surface layer, buoyant convection, not surface shear, is expected to dominate production of turbulence. Deardorff (1970) hypothesized that turbulent momentum and heat transfer in the "convective" boundary layer require consideration of a length scale given by the height of the boundary layer, z_i, a convective velocity scale w_*:

$$w_* = \left(\frac{g H_0 z_i}{T_0 \rho \hat{c}_p}\right)^{1/3} \tag{8.41}$$

and a temperature scale:

$$T_* = \frac{H_0}{w^* \rho \hat{c}_p} \tag{8.42}$$

Along with temperature and mean wind speeds, the parameter that varies most dramatically with height in the atmosphere is the vertical eddy diffusion coefficient. (The vertical eddy diffusion coefficient is referred to here as D_z, for consistency with nomenclature in other chapters. In the air pollution literature, the symbol K_z is standard.) Although Huang's power law expression for D_z is convenient because it permits an analytical solution, other semiempirical expressions provide a better fit to observed mixing rates. For example, the California/Carnegie Institute of Technology (CIT) model uses (McRae et al., 1982; Harley et al., 1993)

$$D_z = \frac{\kappa u_* z}{0.74 + 4.7\,(z/L_{\text{MO}})} \quad \text{for} \quad z < L_{\text{MO}}$$
$$= D_L \quad \text{for} \quad L_{\text{MO}} \le z \le z_i \qquad \text{for the stable atmosphere} \tag{8.43a}$$
$$= 0.05 D_L \quad \text{for} \quad z \ge z_i$$

$$D_z = \kappa u_* z \exp\left(\frac{-8 z f_{\text{cor}}}{u_*}\right) \qquad \text{for the neutral atmosphere} \tag{8.43b}$$

$$D_z = 2.5 w_* z_i \left(k\frac{z}{z_i}\right)^{4/3}\left[1 - 15\left(\frac{z}{L_{\text{MO}}}\right)\right]^{1/4} \qquad \text{for the unstable surface layer} \tag{8.43c}$$

and

$$D_z = w_* z_i \phi\left(\frac{z}{z_i}\right) \qquad \text{for the unstable PBL above the surface layer} \tag{8.43d}$$

In Eqs. (8.43a) to (8.43d), D_L is the vertical diffusivity evaluated at $z = L_{MO}$ using the first expression given for the stable case; z_i is the height of the base of the lowest strong inversion layer; f_{cor} is the Coriolis parameter; and $\phi(z/z_i)$ is an empirical function that has different expressions depending on z. If no sounding data are available, z_i can be estimated for neutral conditions from $z_i \cong 0.2(u_*/f_{cor})$. For stable conditions, $z_i \cong 2.4 \times 10^3 u_*^{3/2}$ (Venkatram, 1980). In unstable conditions, z_i can be estimated from the surface heat budget (Panofsky and Dutton, 1984).

Accounting for horizontal dispersion in grid models is somewhat problematic due to the coarse horizontal grid resolution used in most numerical models and the possibility of artificial dispersion from the numerical treatment of advection. As mentioned above, most urban-scale grid models utilize cell dimensions of 4 km by 4 km or larger. Unless a special "plume-in-grid" treatment is included, cross-wind spreading of air pollutant concentrations near to their source is overestimated due to the assumption of complete mixing throughout a grid cell of this size. For urban-scale models, horizontal dispersion is typically included only for unstable conditions. The CIT model uses $D_H \cong 0.1 w_* z_i$ for this case (McRae et al., 1982).

8.5 MODELING PACKAGES

This section introduces two dispersion modeling packages, the Industrial Source Complex Model, version 3 (ISC3), and the AERMOD system, which are designed to predict concentrations of inert air pollutants or pollutants undergoing first-order decay, within distances of tens of kilometers from emissions sources. Models that include complex chemistry will be discussed in Chapter 11. Both the ISC and AERMOD packages are intended for use in regulatory permitting applications, in which proposed sources or modifications of existing sources are being evaluated to determine what emissions controls might be required. The ISC model has been the foremost model used in air pollution permitting for three decades. The AERMOD system is a new package that has been proposed as a replacement for the short-term mode of the ISC model. Both are available on the U.S. EPA's web site at http://www.epa.gov/ttn/scram/.

ISC3 applies the Gaussian plume equation to industrial facilities with multiple emissions locations. The model can calculate both airborne concentrations and deposition; however, the deposition treatment does not conserve mass because no material is removed from the plume as deposition occurs. The model can handle emissions from point, line, area, or volume sources. Algorithms for estimating plume rise as well as the downwash effects of building wakes are included in the package. Dispersion coefficients can be estimated using urban, rural, or complex terrain parameterizations.

Inputs required for the ISC model include source location and configuration (e.g., stack height, stack diameter, emissions rate, release temperature), receptor locations, and meteorological data. The model can be run in either short-term (ISCST3) or long-term (ISCLT3) mode. The short-term model uses sequential meteorological data with wind speeds, wind directions, stability conditions, and mixing heights supplied for any number of times. The model is run sequentially for each time and the user can select from a variety of options for averaging or reporting maximum values of the output concentrations. In contrast, the long-term model uses joint frequency distributions of wind speed, wind direction, and stability category compiled over seasons or years. This mode predicts long-term averages and maximum concentrations relatively efficiently, but any association of peak concentrations to a specific date or time is lost. Historical meteorological data for many locations across the United States are available from the EPA web site for use with the ISC models. The PCRAMMET processor available on the web site calculates the required hourly mixing height, stability class, and wind speed and direction inputs from twice-daily soundings and hourly surface observations reported by the National Weather Service. For the long-term model, the STAR (Stability Array Program) processor, which is also on EPA's web site, can be used to generate frequency distributions from sequential meteorological data.

EXAMPLE 8.6

Hourly meteorological data from Sioux City, Iowa, were used with the ISCST3 model to calculate 24-h average PM10 (particulate matter less than 10 μm in diameter) concentrations from three hypothetical sources. Figure 8.13 shows the fifth highest concentrations of the year, calculated using 1991 meteorological data. The

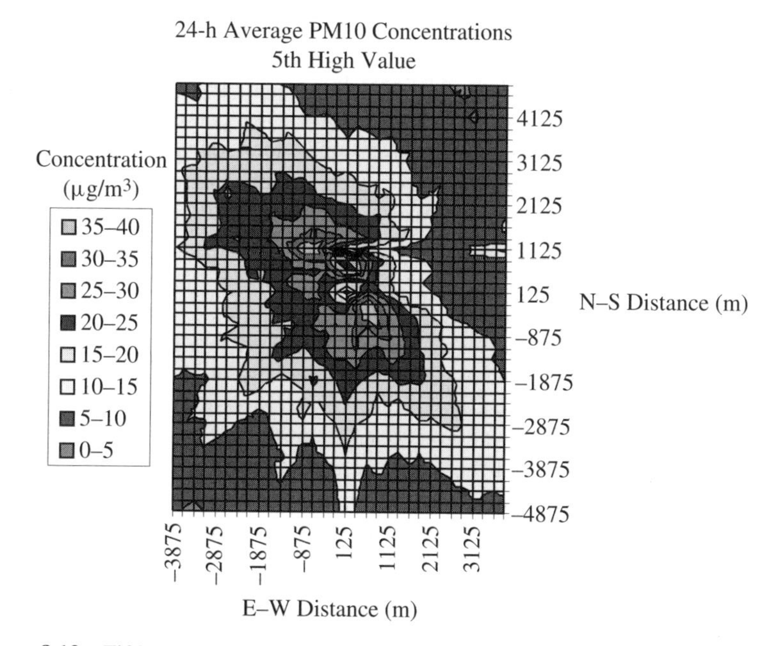

Figure 8.13 Fifth highest 24-h average PM10 concentrations calculated in the vicinity of three elevated release points, as shown in Table 8.5.

TABLE 8.5 Emissions Estimates and Stack Parameters Used in Example 8.6

Source	Stack 1	Stack 2	Stack 3
PM10 emissions (g/s)	20	10	5
Stack height (m)	75.0	25	25
Stack diameter (m)	0.50	1.4	1.4
Exit velocity (m/s)	10.0	12.5	15.0
Exit temperature (K)	305.0	320.0	330.0
East–west location (m)	0	250	0
North–south location (m)	0	0	−250

concentration patterns reflect the predominant wind directions at Sioux City, which are northwesterly and southeasterly. Emissions rates and stack variables used in the calculation are shown in Table 8.5. Rural dispersion parameters were used. Particle deposition and building downwash were neglected. With only three sources and no building downwash interactions, this is a relatively simple case. In some ISC applications, emissions from a hundred or more release points are modeled simultaneously.

During the mid-1990s, the U.S. EPA and the American Meteorological Society collaborated on a new steady-state plume model for short-range dispersion of air pollutants emitted from industrial source complexes. The AERMOD dispersion model (Cimorelli et al., 1998) and associated meteorological and terrain preprocessors (AERMET and AERMAP, respectively) incorporate advances in understanding of dispersion in the planetary boundary layer that have been made since the 1970s. Compared to the ISCST3 model, the AERMOD system includes improved algorithms for dispersion in both convective and stable boundary layers; treatment of plume lofting and plume penetration into elevated inversions under convective conditions; estimation and use of vertical profiles of winds, temperature, and turbulence parameters; and treatment of receptors in complex terrain (Cimorelli et al., 1998). The source and receptor information and formats used in AERMOD match those used in the ISC model. The AERMOD system has been submitted to EPA for approval for regulatory applications.

AERMOD assumes Gaussian concentration distributions in the vertical and lateral directions for the stable boundary layer (SBL) and in the lateral direction for the convective boundary layer (CBL). The vertical distribution in the CBL is bi-Gaussian, reflecting the prevalence of downdrafts versus updrafts under convective conditions. The resulting concentration distribution is sketched in Figure 8.14.

In AERMOD, concentration distributions in the stable boundary layer, $C_s(x_r, y_r, z_r)$, are calculated as a function of distance downwind of the source from:

$$C_s = \frac{Q}{\sqrt{2\pi}\, u \sigma_z} F_y \sum_{m=-\infty}^{\infty} \left\{ \exp\left[-\frac{(z_r - h_e - 2mz_i)^2}{2\sigma_z^2}\right] + \exp\left[-\frac{(z_r + h_e + 2mz_i)^2}{2\sigma_z^2}\right] \right\} \tag{8.44}$$

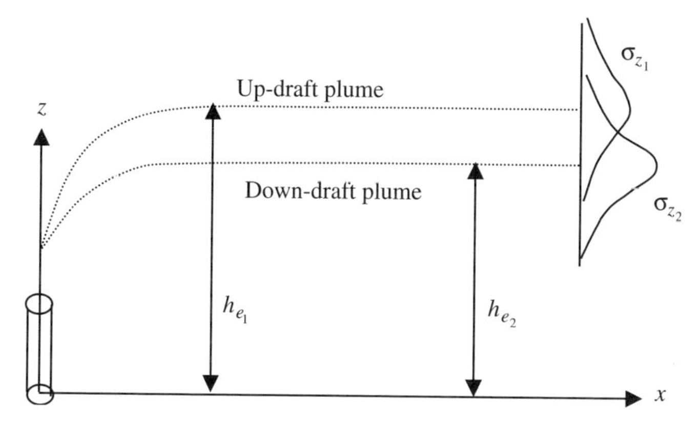

Figure 8.14 Illustration of the bi-Gaussian concentration distribution used in AERMOD for the convective boundary layer.

The lateral distribution function, F_y, is a weighted combination of contributions from dispersion associated with unresolved turbulence and from meander of the plume location due to random shifts in wind direction over the averaging period.

As mentioned, for the convective boundary layer a bi-Gaussian distribution is used to represent vertical dispersion, resulting in the following form for the concentration distribution $C_c(x_r, y_r, z_r)$:

$$C_c = \frac{Q}{\sqrt{2\pi}u\sigma_y} \exp\left(-\frac{y_r^2}{2\sigma_y}\right) \sum_{j=1}^{2} \sum_{m=0}^{\infty} \frac{\lambda_j}{\sigma_{zj}} \times \left\{ \exp\left[-\frac{(z_r - h_{ej} - 2mz_i)^2}{2\sigma_{zj}^2}\right] + \exp\left[-\frac{(z_r + h_{ej} + 2mz_i)^2}{2\sigma_{zj}^2}\right] \right\} \quad (8.45)$$

In Eq. (8.45), λ_j, $j = 1, 2$ weights the contributions of updrafts and downdrafts, respectively, to the vertical spread in time-averaged concentrations.

In AERMOD, the wind speed and dispersion parameters in Eqs. (8.44) and (8.45) represent averages across the vertical layer of interest, the top and bottom of which are determined by the heights of the plume and the receptor at a given downwind distance. These averages are calculated within the model from vertical profiles that are derived from similarity theory of scaling in the surface and mixed layers of the planetary boundary layer. The scaling parameters required by AERMOD are estimated within the system's meteorological preprocessor, AERMET, from available meteorological data and surface characteristics. Required meteorological data include wind speed and direction at the standard 10-m height, surface temperature, cloud cover, and the temperature profile from the morning vertical sounding. Estimates of surface roughness, albedo, and the ratio of sensible to latent heat fluxes (Bowen ratio) are also required. The model has been found to be very sensitive to surface roughness (Sakulyanontvittaya, 2003).

During its development, AERMOD was evaluated in a multistep process that included peer review, performance evaluation, and comparison with ISCST3 and other dispersion models (Paine et al., 1998). The evaluation included comparison of AERMOD concentrations with observations for 4 year-long data sets of hourly SO_2 concentrations in the vicinity of isolated sources, with no "tuning" of AERMOD algorithms or parameters. The source locations included both flat and hilly terrain. The ratios of predicted to observed "robust peak" concentrations ranged across these four data sets from 1.06 to 1.31 for 3-h averaging times and from 0.72 to 1.72 for 24-h averaging times. AERMOD's performance was better than that for ISCST3 in all cases for the peak 24-h average and in all but one case for the peak 3-h average. Based on these and other evaluations conducted to date (Hanna et al., 2000; Sakulyanontvittaya, 2003), AERMOD thus appears to perform better than ISCST3, in addition to being viewed as the more scientifically defensible model. At present, however, experience with AERMOD is relatively limited.

Homework problems for this chapter are provided at www.wiley.com/college/ramaswami.

9 Models of Transport in Individual Media: Soil and Groundwater

9.1 OVERVIEW OF SOIL AND GROUNDWATER SYSTEM

The subsurface soil–water system consists of the unsaturated zone and the saturated zone. The unsaturated zone occurs in the upper, near-surface layers of soil that are not fully saturated with water, while the saturated zone typically represents deeper soils wherein all the void spaces between the soil particles are filled up with water. Saturation occurs as water from precipitation events percolates through the upper layers of the soil and seeps down, recharging the deeper layers, which then form the saturated zone. The water table is the dividing "line" between the saturated and unsaturated zone. The water table is not a thin, crisp line, but rather a fuzzy zone called the capillary fringe. The nature of the geologic formations that occur below the water table governs the ability of these formations to carry water. An aquifer refers to a fertile water-bearing geologic unit in the saturated zone, while aquitards and aquicludes refer to formations that are largely impervious to water and exclude its passage through the subsurface. Aquifers are tapped for water supply purposes by means of bore wells (in deep aquifers) or near-surface pit-type wells (for shallow aquifers). Aquifers may be bounded above and below by aquitards—such aquifers are termed confined aquifers. An aquifer that is bounded by an aquitard below, but is open to the atmosphere via the water table and the unsaturated zone, is termed an unconfined or phreatic aquifer (see Fig. 9.1).

The unsaturated zone is also referred to as the vadose zone. The unsaturated zone is packed with soil particles. The interparticle void spaces contain air as well as water, so that the soil porosity is the sum of the air content and the moisture content in the system (see Fig. 4.9). Porosity, moisture content, and air content are also defined in the inset in Figure 9.1. As one proceeds deeper into the vadose zone, the moisture content increases until it equals the porosity, that is, all the void spaces are filled with water and the air content is zero. At this depth, the water table has been reached. The fluid pressure at the water table is equal to atmospheric pressure, while the pressure below the water table, in the saturated zone, is greater than the atmospheric pressure, increasing with depth because of the increasing weight of the layers of water above. The fluid pressure above the water table, i.e., in the vadose zone, is less than atmospheric pressure due to the suction created by the capillary action of the intraparticle void spaces. The suction pressure is also termed matric potential and is represented as a negative pressure, that is, pressure less than atmospheric. The

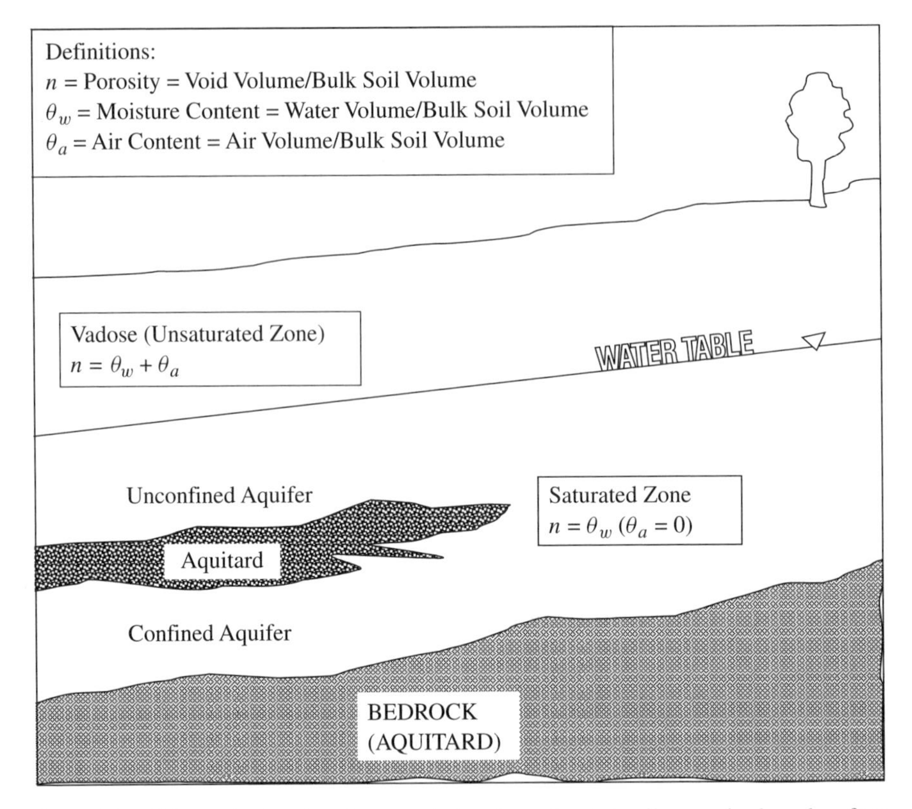

Figure 9.1 Schematic showing the saturated and unsaturated soil zones in the subsurface. Geologic units in the saturated zone include water-bearing units called aquifers and water-excluding formations called aquitards. Confined aquifers are bounded above and below by aquitard materials, while unconfined (phreatic) aquifers are open to the atmosphere.

matric potential increases as the soil moisture content decreases—an effect easily illustrated by inverting a test tube filled with water over a dry pot of soil (unsaturated). The drier the soil is, the more water will be sucked out of the tube, demonstrating greater suction, and hence a larger matric potential. At the capillary fringe, the soil pores are saturated due to capillary action although the fluid pressure is less than atmospheric, and hence the capillary fringe is also called the tension-saturated zone. Figure 9.2 illustrates the vertical variation in soil moisture content and fluid pressure in the unsaturated and saturated zones.

The objective of this chapter is to describe the movement of water and the associated transport of chemical pollutants in both saturated and unsaturated zones of the subsurface. Contamination of the saturated zone has been of utmost concern because the waters in these layers are typically tapped for drinking water. Saturated zone processes are therefore the primary focus in the first part of this chapter (Sections 9.2 and 9.3). The unsaturated zone is described in Section 9.4, with transport parameters defined and described in terms of analogous parameters applicable to the saturated

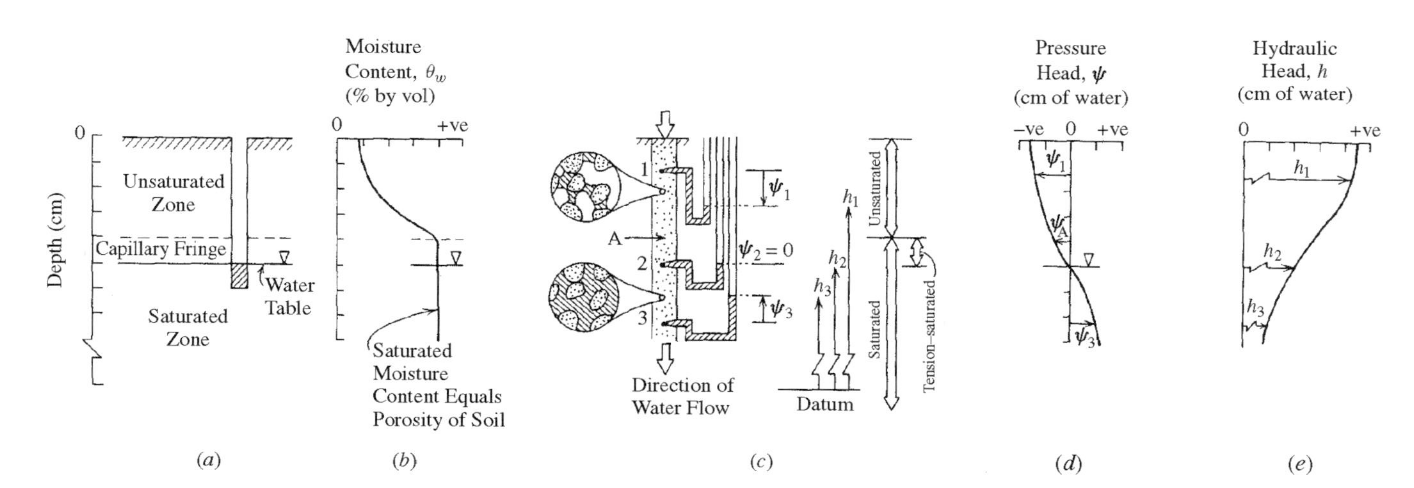

Figure 9.2 Vertical profile showing fluid pressure and moisture content in the saturated and unsaturated zones of an aquifer. (From *Groundwater* by A. R. Freeze and J. A. Cherry, © 1979. Reprinted by permission of Pearson Education, Inc., Upper Saddle River, NJ.

zone. Parameter estimation techniques for saturated and unsaturated zone pollutant transport models are presented in Section 9.5. The primary focus in Sections 9.2 to 9.5 is on chemical pollutants that are dissolved in groundwater, their movement in water being impacted by partitioning to air and soil present in the subsurface. Section 9.6 addresses pollutants that exist as a separate phase in the subsurface, either as separate organic-phase liquids (NAPLs), or as colloidal particles as in the case of viruses and bacteria. Section 9.7 presents a review of available computer packages for modeling subsurface contamination. Two case studies are included in Section 9.7 to illustrate computer-aided modeling of subsurface contaminant transport in real-world systems with uncertainties related to contaminant source and transport parameters.

9.2 ADVECTION–DISPERSION OF TRACERS IN THE SATURATED ZONE

This section describes transport in the saturated zone of a tracer chemical that neither sorbs to soil nor undergoes degradation, so that its mass in groundwater is conserved at all times. Advection and diffusion–dispersion processes that determine tracer transport in the saturated zone are described in more detail here, building upon the introduction presented in Chapter 5.

9.2.1 Advection in Saturated Groundwater: Darcy's Law

As described in Chapter 5, the effective average linear velocity of water in the saturated zone is derived from the Darcy velocity q, which is defined as the volumetric rate of flow of water, Q, per unit gross (total) cross-sectional area (A) of an aquifer perpendicular to the flow direction. According to Darcy's law (Hubbert, 1940), the Darcy velocity q is proportional to the gradient of hydraulic head in the aquifer, with a proportionality constant denoted by the hydraulic conductivity K of the aquifer medium:

$$q = \frac{\text{Volumetric water flow rate}}{\text{Total cross-sectional area}} = \frac{Q}{A} = K\frac{\Delta h}{\Delta x} \tag{9.1a}$$

The effective average linear groundwater velocity v is derived from the Darcy velocity q as:

$$v = \frac{\text{Volumetric water flow rate}}{\text{Effective pore cross-section area}} = \frac{Q}{nA} = \frac{q}{n} = \frac{K}{n}\frac{\Delta h}{\Delta x} \tag{9.1b}$$

The hydraulic head h in the above equations represents the sum of the pressure head, ψ, and the elevation head, z, of groundwater at a certain location, the velocity head being neglected due to the very slow velocities of water in the subsurface:

$$h = z + \psi \tag{9.1c}$$

The head h at a certain location is measured as the height above some datum reference point to which water rises in a well screened at that location (see Fig. 5.3). Hydraulic head typically varies with spatial location, the gradient in head $(\Delta h/\Delta x)$ providing the driving force for groundwater flow. For a given head gradient, the conductivity K determines the ease with which water flows through the aquifer medium. Hydraulic conductivity K depends upon the properties of the porous medium, for example, particle size and packing, as well as properties of the fluid (in this case water) such as fluid density ρ_l and viscosity μ. The dual dependence of the conductivity K upon media as well as fluid properties is quantitatively expressed as:

$$K = \mathbf{k}\left[\frac{\rho_l g}{\mu}\right] \tag{9.1d}$$

where **k** is termed the intrinsic permeability of the medium—the ability of a medium to allow passage of any fluid through its porous packed structure, a property that depends upon particle size and packing. Empirically, $\mathbf{k} = Cd^2$ where C is a shape factor related to packing, and d is the median particle diameter. Thus, well-agglomerated, evenly sized, fine-grained particles of small diameter, such as clay, will exhibit low intrinsic permeability. On the other hand, unevenly sized, loosely packed, large-grained particles, such as unconsolidated gravels, will exhibit high intrinsic permeability. In addition to the intrinsic permeability of the medium, the ease of fluid flow depends on fluid properties such as density and viscosity, lumped together in the parenthetical term in Eq. (9.1d). Higher density promotes gravity and momentum driven flows, while higher fluid friction (viscosity) restricts fluid flow. The intrinsic permeabilty and hydraulic conductivity of various geologic media are summarized in Table 9.1.

TABLE 9.1 Range of Intrinsic Permeability and Hydraulic Conductivity of Various Aquifer Materials[a]

Degree of Consolidation	Particle Size	Aquifer Material	Intrinsic Permeability, k (cm^2)	Hydraulic Conductivity, K (cm/s)
Unconsolidated	Coarse	Gravel	10^{-6}–10^{-3}	10^{-1}–10^{2}
	↓	Clean sand	10^{-9}–10^{-5}	10^{-4}–1
		Silty sand	10^{-10}–10^{-6}	10^{-5}–10^{-1}
		Silt, loess	10^{-12}–10^{-8}	10^{-7}–10^{-3}
		Glacial till	10^{-15}–10^{-9}	10^{-10}–10^{-4}
	Fine	Clay	10^{-15}–10^{-12}	10^{-10}–10^{-7}
Highly consolidated rocks	Coarse/porous	Karst limestone	10^{-9}–10^{-5}	10^{-4}–1
	↓	Fractured i/m rock	10^{-11}–10^{-7}	10^{-6}–10^{-2}
		Limestone and	10^{-12}–10^{-9}	10^{-7}–10^{-4}
		dolomite	10^{-13}–10^{-9}	10^{-8}–10^{-4}
		Sandstone	10^{-16}–10^{-12}	10^{-11}–10^{-7}
	Fine/well-sorted	Shale	10^{-17}–10^{-13}	10^{-12}–10^{-8}
		Unfractured i/m rocks		

[a] Note: i = igneous rocks, e.g., granite; m = metamorphic rocks. Limestone, dolomite, and sandstone belong to the class of sedimentary rocks.

Source: Adapted from Freeze and Cherry (1979).

Equation (9.1d) is useful because it implies that knowledge of the intrinsic permeability of a material can be combined with knowledge of fluid properties to predict the conductivity K of any fluid in that medium. In later sections, this relationship is used to quantify the conductivity of soil for fluids other than water, for example, air and NAPL, essential for the assessment of airflow in the unsaturated zone, and of NAPL flow rates in the subsurface.

The hydraulic head gradient, $\Delta h/\Delta x$, is computed from head data gathered at steady-state conditions from several locations in the aquifer. Head contours, that is, equal head lines, are plotted as shown in Figure 9.3*a* and the gradient determined from these plots. Groundwater flow occurs from a region of high head to low head, with the flow lines being perpendicular to the equipotential contours. When limited head data are available such that contour plotting is not possible, the hydraulic gradient may be

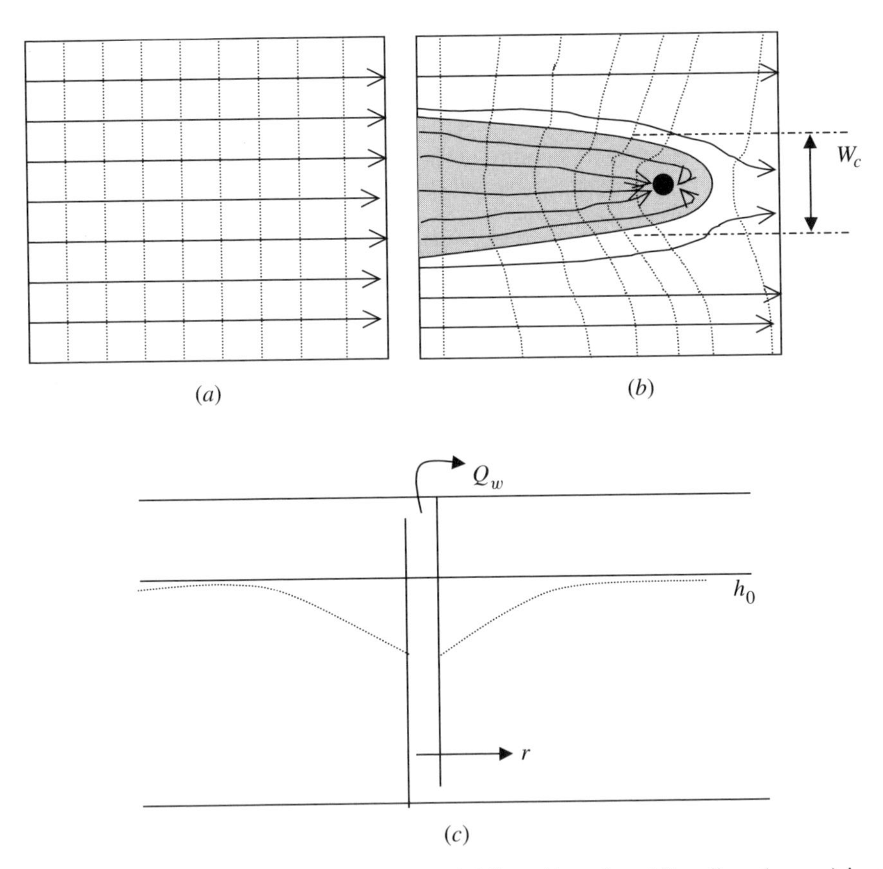

Figure 9.3 (*a*) Plan view showing equipotential lines (dotted) and flow lines (arrows) in a homogeneous aquifer of uniform thickness under natural gradient conditions (no pumping well). (*b*) Plan view showing equipotential lines (dotted) and flow lines (arrows) in response to a pumping well (solid circle). All flow lines within the capture zone (gray tone) of width W_c enter the pumping well. (*c*) Elevation view showing the drawdown (dotted line) created by a pumping well that is withdrawing water at a volumetric flow rate of Q_w.

computed using a minimum of three data points, employing a technique called the three-point method. The three-point method is illustrated in Example 9.1, and requires hydraulic head measured at three locations, with the highest, lowest, and intermediate head locations being associated with points A, B, and C, respectively, on a map drawn to scale. The three-point method assumes a linear change in hydraulic head over distance and interpolates between the highest and lowest head locations (A and B) to construct an equipotential line corresponding to the intermediate head at location C. The flow direction is from the region of high head to low head, perpendicular to the constructed equipotential line. The three-point method is a useful tool to visualize groundwater flow direction when limited head data are available. The constructed flow field must be confirmed and reconciled with regional geologic observations in the area.

EXAMPLE 9.1

Groundwater heads were measured at three existing wells on a site as shown below. Porosity of the soil is 0.32 and hydraulic conductivity is 1×10^{-6} m/s. Calculate the hydraulic gradient, Darcy velocity, and average groundwater velocity.

1. Draw the well layout to scale. Draw a line between the two wells with the highest head (well A) and the lowest head (well B).

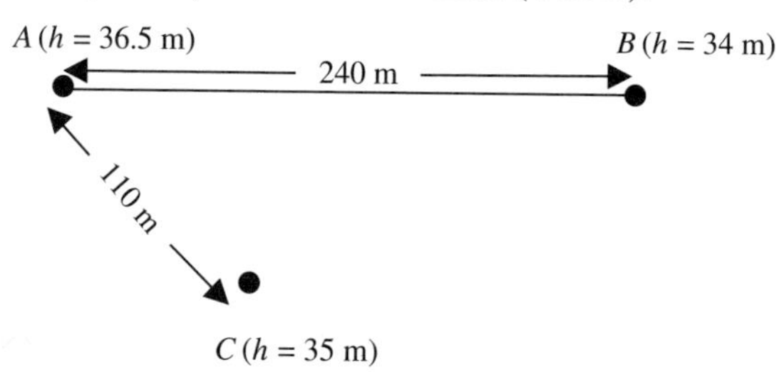

2. Find, by interpolation, point D on line AB with head equal to the intermediate head (well C). CD (dashed line) is an equipotential line. Water flows perpendicular to this line from high to low head.

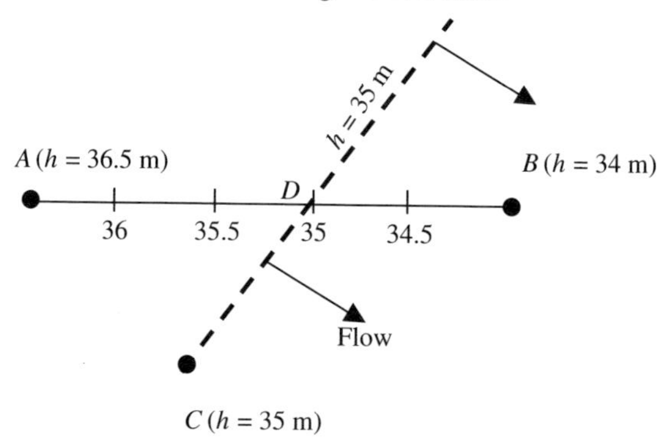

3. Draw a line perpendicular to the equipotential line through the well with the lowest head (or through the well with the highest head) and measure the length of that line, to scale. See line BE.

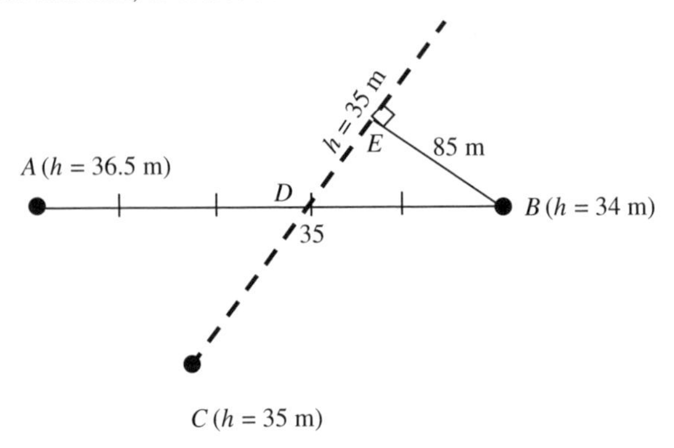

4. The hydraulic gradient,

$$\frac{\Delta h}{\Delta L} = \frac{35.0\text{ m} - 34.0\text{ m}}{85\text{ m}} = 0.012$$

$$\text{Darcy velocity, } q = K\frac{\Delta h}{\Delta L} = 1 \times 10^{-6}\frac{\text{m}}{\text{s}}(0.012)\left(\frac{3600\text{ s}}{\text{h}}\frac{24\text{ h}}{\text{day}}\right) = 0.00104\frac{\text{m}}{\text{day}}$$

$$\text{Average groundwater pore velocity, } v = \frac{q}{n} = \frac{0.00104\dfrac{\text{m}}{\text{day}}}{0.32} = 0.003\frac{\text{m}}{\text{day}}$$

Under unsteady-state conditions, that is, when the hydraulic heads are changing with time, for example, while a well is pumping, the continuity equation is used with appropriate initial and boundary conditions, to solve for the time-variant head distribution in the aquifer. Assuming uniform and invariant hydraulic conductivity in the aquifer medium, a water balance across a control volume yields the transient flow equation for one-dimensional flow along the x axis:

Volumetric water flow rate in − Volumetric water flow rate out
= Rate of change in storage of water in unit aquifer volume

$$-(\Delta y\,\Delta z) \times K\left.\frac{\partial h}{\partial x}\right|_{x} - \left[-(\Delta y\,\Delta z) \times K\left.\frac{\partial h}{\partial x}\right|_{x+\Delta x}\right] = S_s \times \frac{\partial h}{\partial t} \times (\Delta x\,\Delta y\,\Delta z) \tag{9.2a}$$

where S_s is the specific storage of the aquifer, defined as the volume of water released from a unit volume of the aquifer for a unit drop in head. Dividing by the control volume ($\Delta x\ \Delta y\ \Delta z$), the left-hand side reduces to the second spatial derivative of the hydraulic head when Δx is infinitesimally small, such that the transient flow equation can be written as:

$$\frac{\partial^2 h}{\partial x^2} = \frac{S_s}{K}\frac{\partial h}{\partial t} \tag{9.2b}$$

Equation (9.2b) can also be written in equivalent form by employing aquifer transmissivity T ($T = Kb$) and storativity S ($S = S_s b$), where b is the thickness of the aquifer. The storativity S represents the volume of water released per unit area of the aquifer for a unit drop in head.

Solving the unsteady (transient) flow equation with appropriate boundary and initial conditions yields time-varying head distributions in the aquifer, resulting in dynamic groundwater velocity fields. The transient flow equation is written in radial coordinates to determine the dynamic change in head caused by the radial flow of water into a pumping well. The action of a pumping well is to lower the head at the well location ($r = r_w$); this creates a drawdown in the water table, promoting radial inflow of water toward the well. The drawdown is maximum at the pumping well and decreases with increasing radial distance away from the well, approaching the initial water table level (h_0) at a large distance away from the well. In an ideal, homogeneous, infinite aquifer of uniform thickness, b, the drawdown created by a well that is pumping water at a constant volumetric flow rate of Q_w is given by (Theis, 1935):

$$h_0 - h(r,t) = \frac{Q}{4\pi T} \times W(u) \quad \text{where } u = \frac{r^2 S}{4Tt} \quad \text{and} \quad W(u) \quad \text{is the well function} \tag{9.3}$$

The Theis solution employs the well function,[1] $W(u)$, to determine the drawdown, $[h_0 - h(r,t)]$, at any location r at time t after pumping commences. The ideal drawdown curve indicated by the equation above is sketched in Figure 9.3*c* (elevation view). Modifications to Eq. (9.3) for nonideal, leaky aquifers and bounded aquifers with impermeable boundaries and recharge zones are discussed further in classic texts on groundwater (e.g., Freeze and Cherry, 1979; Fetter, 1993).

In an ideal aquifer, a pumping well lowers the head at the well radius and also impacts the hydraulic head gradients around each pumping well in a manner described by Eq. (9.3). Long-term pumping of water from a well will result in a new steady-state groundwater flow field illustrated in Figure 9.3*b* (plan view). The well, pumping water at a volumetric flow rate, Q_w, creates a capture zone of width W_c; flow lines within the capture zone are diverted into the well and withdrawn. Knowledge of the width of the capture zone is important when planning site remediation employing pumping wells. As expected, W_c increases with the well water-pumping rate, Q_w, and W_c decreases with an increase in the natural gradient Darcy velocity q in the

[1] For small values of u (i.e., at short distances from the well or at long times after pumping has commenced), $W(u)$ can be approximated as $[-0.5772 - \ln u]$, which yields

$$h_0 - h(r,t) = \frac{Q}{4\pi T} \times (-0.5772 - \ln u) \quad \text{for} \quad u < 0.1$$

aquifer (prior to pumping), and with an increase in the thickness b of the aquifer. For a homogeneous isotropic aquifer of uniform thickness, the width of the capture zone in the *near vicinity* of a single pumping well may be approximated as (Javendel and Tsang, 1986): $W_c = Q_w/2bq$. Thus, the pumping well creates a distortion in the steady-state groundwater flow field that is manifested within the capture zone around each well. Outside the capture zone when one takes a larger view of the aquifer, the groundwater velocity is once again described by v, which is computed from the water flow equation [Eq. (9.2b)] with appropriate initial and boundary head conditions.

The discussion thus far has been in the context of a homogeneous, isotropic aquifer with uniform and invariant properties. Very simple boundary conditions have been considered—essentially an aquifer of infinite extent. Real-world subsurface systems are rarely homogeneous, being composed of a variety of geologic materials. Often aquifers contain impermeable materials (rocks) and low-permeability (clay) zones and lenses interspersed within the matrix. These impermeable and low-permeability units cause flow by-passing in the subsurface. In addition, aquifer thickness can vary regionally and layering effects are also commonly observed. Some aquifers exhibit preferential conductivity in a certain flow direction, a phenomenon called anisotropy, for example, the hydraulic conductivity in the x direction may be much different from that for flow in the y direction. Complex boundary conditions may exist at the site, for example, a stream may run through a site creating a constant head boundary; plants and trees may be transpiring water via evapotranspiration creating local sinks within unconfined aquifers, concrete structures may create a zero flow boundary, and so forth. A conceptual model of an aquifer must first be established to include all layering phenomena, boundary conditions, and key recharge and discharge zones in an aquifer. Numerical methods, such as the finite difference and finite element methods presented in Chapter 6, are then employed to solve the water flow equation, using a sequence of steps outlined in Figure 9.4. Simulation packages and numerical models for water flow modeling are reviewed briefly in Section 9.7. In Sections 9.2 and 9.3, however, we develop subsurface contaminant transport principles employing the notion of an ideal, relatively homogeneous, isotropic aquifer of uniform thickness, in which an effective average linear groundwater velocity, represented by v, describes the advection of dissolved chemicals in saturated groundwater.

Exercise for the Reader Simulate regional water flow in an ideal homogenous unconfined aquifer, 1 km long and 500 m wide, located between two streams at elevations of 101 m and 95 m above sea level, respectively. Verify that the regional flow field is consistent with Figure 9.3*a*. Add a pumping well at the center point of the aquifer and verify the flow field with Figures 9.3*b*,*c*. Finally, incorporate nonideal factors such as heterogeneities and layering. Modeling techniques and results using MODFLOW are summarized and cached at: www.wiley.com/college/ramaswami.

9.2.2 Dispersion in Ideal, Saturated Groundwater Systems

Dispersion of chemicals in groundwater occurs due to channeling and mixing of water as it moves through the porous medium (mechanical dispersion), as well as molecular

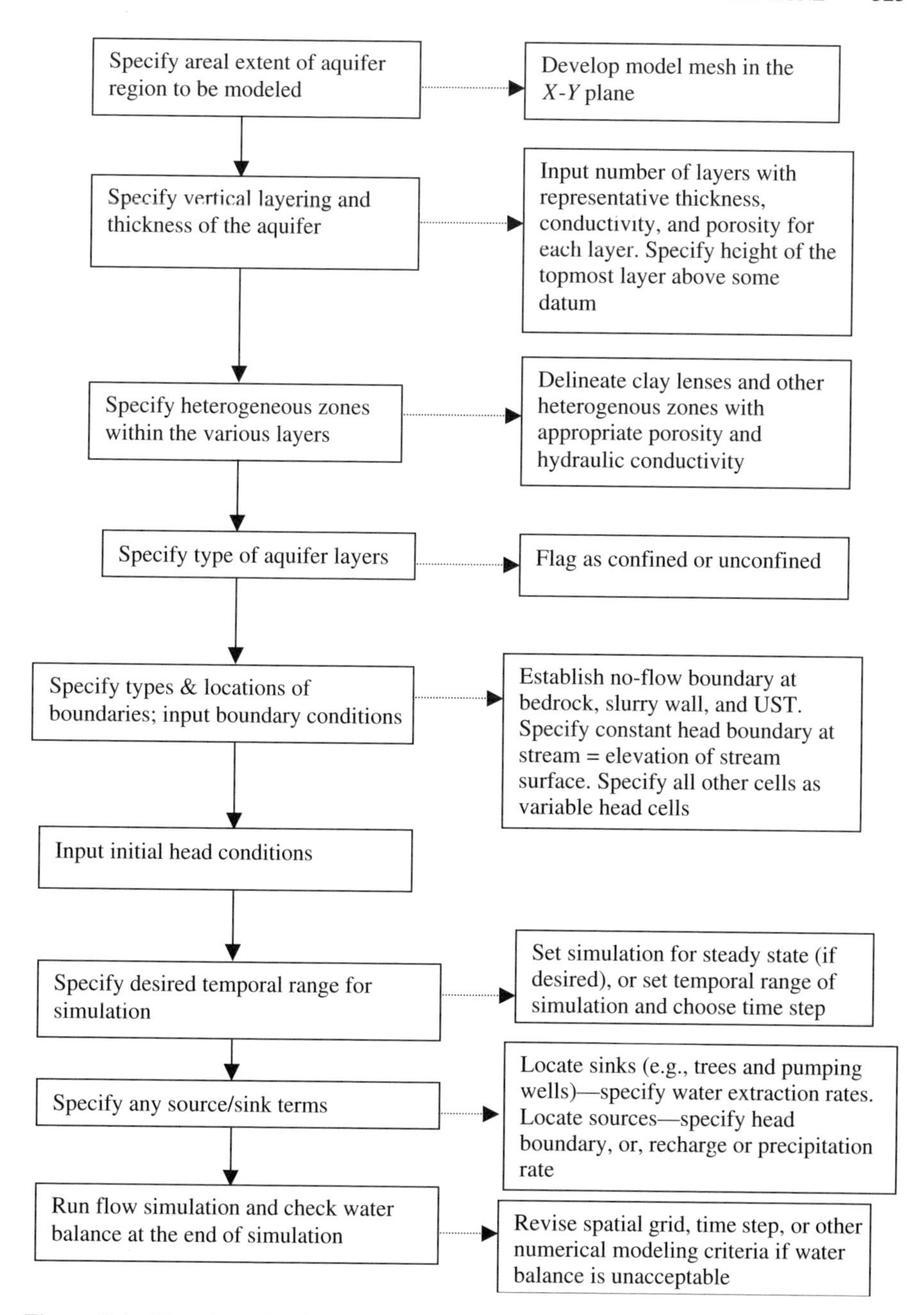

Figure 9.4 Flowchart showing sequence of operations in flow simulation in a complex aquifer employing numerical models that solve the water flow equation. (*Note:* UST = Underground Storage Tank)

diffusion of the chemical in groundwater. In Chapter 5 we saw that the two processes are combined and represented by hydrodynamic dispersion coefficients, D_j (L^2/T), defined in three dimensions as:

$$\begin{aligned} D_L &= \alpha_L\, v + D_{m,sw} \\ D_T &= \alpha_T\, v + D_{m,sw} \\ D_z &= \alpha_z\, v + D_{m,sw} \end{aligned} \tag{9.4}$$

where the subscript L represents the longitudinal direction aligned with the primary direction of groundwater flow (typically the x axis); and the subscripts T and Z represent directions perpendicular to flow in the horizontal and vertical plane, that is, the y and z axes, respectively. The first term on the rhs of Eq. (9.4) represents mechanical dispersion that results when groundwater flowing at an average velocity, $v(L/T)$, encounters porous media with channeling properties that yield a dispersivity, $\alpha_j(L)$. The second term on the rhs represents the effective molecular diffusion coefficient for the chemical of interest in soil water, $D_{m,sw}$ (L^2/T), which incorporates the tortuosity effects in soil discussed previously in Chapter 4 (see Section 4.5). Molecular diffusion in groundwater typically occurs very slowly and can often be neglected in comparison with hydrodynamic dispersion. Hence, the molecular diffusion terms on the rhs of Eq. (9.4) are often dropped such that $D_L \approx \alpha_L v$, $D_T \approx \alpha_T v$, and $D_Z = \alpha_Z v$.

When Fickian dispersion is assumed to occur, that is, the dispersivities (α) and hence the dispersion coefficients are assumed to be constant, the variance (σ^2) in the location of chemical parcels relative to their mean position increases with travel time t as:

$$\begin{aligned} \sigma_x^2 &= 2D_L\, t \\ \sigma_y^2 &= 2D_T\, t \\ \sigma_z^2 &= 2D_z\, t \end{aligned} \tag{9.5}$$

The dispersion coefficients, D_L, D_T, and D_Z, as determined from Eq. (9.5), are used directly in groundwater models, instead of the standard deviation parameters (σ_x, σ_y, σ_z) used in the Gaussian models for air pollution (Chapter 8).

The assessment of groundwater pollution would be greatly simplified if, indeed, the dispersivites in Eqs. (9.4) and (9.5) were constant. However, experimentalists noticed that the dispersivity increased with the scale of transport, that is, dispersivity in a longer sand core was found to be greater than the dispersivity observed in a shorter core, even when both were packed with the same porous medium. This effect is called the "scale effect of dispersivity" and is even more dramatic when field dispersivities computed at different spatial scales are compared for similar aquifer materials. Figure 5.13*c* shows how field-scale dispersivities increase with the distance scale. The spatial scale effect is caused by the tendency to encounter increasing heterogeneity in soil size, texture, and hydraulic conductivity over larger spatial scales, resulting in an observed apparent increase in deviations from average groundwater flow velocities.

If one were to be able to fully characterize media heterogeneities and the resulting changes in groundwater velocity in the subsurface, the need for spatial scaling of dispersivity would disappear. Indeed, a probabilistic representation of the variability in hydraulic conductivity of the medium is used by stochastic modelers to characterize dispersivity in aquifers (Gelhar et al., 1979; Dagan, 1982; Gelhar and Axness, 1983; Neuman et al., 1987; Neuman, 1990; Gelhar et al., 1992). However, for the most part, the state of practice in contaminant hydrogeology is to employ a scale-dependent dispersivity that is estimated either empirically or by means of on-site tracer tests in which a best-fit value of the aquifer dispersivity is obtained from the observed transport of tracers in the aquifer.

Tracer tests to estimate ideal aquifer dispersion parameters are described in the next section, and, some approaches for stochastic modeling of aquifer hydraulic conductivity and associated dispersion are demonstrated in the simulations presented in Section 9.7. At this point, the reader is only made aware that dispersivity is a parameter that is strongly influenced by the spatial scale of the problem, and its values must be estimated with attention to this detail. We proceed with the assumption of an idealized aquifer with a unique and uniform dispersivity (albeit scale-dependent). Empirically, one can use a simplified approximation that the dispersivity in the longitudinal dimension is one-tenth of the distance scale, that is, $\alpha_L = 0.1x$ (see Fig. 5.13*c*). Other rules of thumb are that the transverse and vertical dispersivities are one-tenth of the longitudinal dispersivity. In small packed soil columns used in laboratories, longitudinal dispersivity is typically of the order of the median particle diameter. The above guidelines provide rough estimates of dispersivities. Field tracer tests, described in the next section, are typically conducted to determine α_L, α_T, and α_z for modeling purposes.

9.2.3 Advective–Dispersive Transport of Tracers in Ideal Aquifers

In an ideal aquifer, with steady water flow and uniform dispersivity, the advection and dispersion processes described in the previous sections can be coupled together to yield a contaminant mass-balance equation employing the control volume approach described in detail in Chapter 5. For a tracer chemical that undergoes no reaction or sorption to soil, the mass-balance equation in three dimensions is written as:

$$\begin{aligned}\frac{\partial C(x,y,z,t)}{\partial t} &= -v\frac{\partial C(x,y,z,t)}{\partial x} + D_L\frac{\partial^2 C(x,y,z,t)}{\partial x^2} \\ &\quad + D_T\frac{\partial^2 C(x,y,z,t)}{\partial y^2} + D_Z\frac{\partial^2 C(x,y,z,t)}{\partial z^2}\end{aligned} \tag{9.6}$$

As per the discussion in Chapter 1, the modeler may choose to simplify the dimensionality of the problem, as appropriate. Thus, advection in a large, deep aquifer may be modeled in three dimensions, whereas shallow aquifers are often assumed to be well-mixed in the vertical dimension, reducing the problem to that of two-dimensional transport (x-y only). Subsurface transport in a narrow stream bed may be further reduced to that of one-dimensional transport (x only), when complete mixing across the cross section of the stream bed is a reasonable assumption. See Figure 1.4.

Equation (9.6) assumes constant advection and dispersion parameters in the aquifer that are invariant with time and space, at the chosen dimensionality and spatial scale of interest. Analytical solutions to the above equation may be determined if initial pollutant concentrations in the aquifer are known, and appropriate and simple boundary conditions are provided for specific pollution problems (Wilson and Miller, 1978; Hunt, 1978; Bear, 1979; Bear and Verruijt, 1987). Analytical solutions for advective–dispersive tracer transport in saturated groundwater occurring under two idealized source conditions—an instantaneous spike release of mass M of a chemical to groundwater and a continuous point input of a chemical at a concentration C_0—were presented in Chapter 5 and are summarized in Tables 9.2 and 9.3, respectively. These idealized conditions help visualize main features of contaminant transport in the subsurface, which are sketched alongside the analytical equations. Note the similarity between the specific Gaussian puff equations for groundwater shown in

Compare to mixed reactor & air sols

TABLE 9.2 Gaussian Distributions in Saturated Groundwater in Three-, Two-, and One-Dimensions for Instantaneous Release of Mass M, of a Tracer from Point Source Located at Origin

Dimensions	Analytical Solution
Three-Dimension M = mass (Hunt, 1978)	$C(x,y,z,t) = \dfrac{M}{8n(\pi t)^{3/2}\sqrt{D_L D_T D_Z}} \times \exp -\left[\dfrac{(x-vt)^2}{4D_L t} + \dfrac{y^2}{4D_T t} + \dfrac{z^2}{4D_Z t}\right]$
Two-Dimension M_z = mass/depth (Wilson and Miller, 1978)	$C(x,y,t) = \dfrac{M_z}{4\pi n t\sqrt{D_L D_T}} \exp -\left[\dfrac{(x-vt)^2}{4D_L t} + \dfrac{y^2}{4D_T t}\right]$ $t = 0$, $t = t_1$, $t = t_2$, Y, X, $v \rightarrow$
One-Dimension M_a = mass/area (Hunt, 1978)	$C(x,t) = \dfrac{M_a}{2n\sqrt{\pi t D_L}} \exp -\left[\dfrac{(x-vt)^2}{4D_L t}\right]$ $t = 0$, $t = t_1$, $t = t_2$, X, $v \rightarrow$

Table 9.2 and the general Gaussian puff equations listed in Table 5.2, when the relationships shown in Eq. (9.5) are incorporated. Example 9.2 illustrates the use of the Gaussian puff equations to describe transport of a nonreactive tracer in saturated groundwater.

EXAMPLE 9.2

Benzene (1000 g) was instantaneously released one year ago into a saturated aquifer, which has a thickness of 10 m, a hydraulic gradient of 0.1, a hydraulic conductivity of 0.1 m/day, and a soil porosity of 0.33. Assume no sorption, volatilization, or reaction/degradation of benzene occurs in this confined aquifer.

1. Determine present-day benzene concentrations in a residential well located 100 m down-gradient from the spill.
2. A consultant tells the homeowners that this is the maximum benzene concentration that they will ever see. Make calculations and determine if you agree.

Solution 9.2 Choose the simplest and most appropriate model to describe benzene transport in the aquifer. Since the aquifer is fairly shallow (10 m), we can assume that the water is well-mixed in the z direction and use the two-dimensional instantaneous release puff equation from Table 9.2.

M_z, the total benzene mass released to water per unit aquifer thickness is given as 1000 g/10 m, that is, 100 g/m. The average groundwater velocity, v, is computed as:

$$v = \frac{K}{n}\frac{\Delta h}{\Delta L} = \frac{0.1\ \text{m/day}}{0.33}(0.1)(365\ \text{day/yr}) = 11.1\ \text{m/yr}$$

With no site-specific information, the longitudinal dispersion coefficient, α_L, is estimated as:

$$\alpha_L = 0.10(x) = 0.10(100\ \text{m}) = 10\ \text{m}$$

So,

$$D_L = \alpha_L \times v = 10\ \text{m} \times 11.1\ \text{m/yr} = 111\ \text{m}^2/\text{yr}$$

Another rule-of-thumb is

$$\alpha_T = 0.10(\alpha_L) = 0.10(10\ \text{m}) = 1\ \text{m}$$

from which

$$D_T = \alpha_T \cdot v = 1\ \text{m}(11.1\ \text{m/yr}) = 11.1\ \text{m}^2/\text{yr}$$

The concentration of benzene in the groundwater at the well one year after release and 100 m down-gradient of release is calculated as:

$$C(100\text{ m}, 0\text{ m}, 1\text{ yr}) = \frac{1000\text{ g}/(0.33 \times 10\text{ m})}{4 \times \pi \times 0.33 \times 1\text{ yr}\sqrt{111\text{ m}^2/\text{yr} \times 11.1\text{ m}^2/\text{yr}}}$$

$$\times \exp\left[-\frac{(100\text{ m} - 11.1\text{ m}^2/\text{yr} \times 1\text{ yr})^2}{4(111\text{ m}^2/\text{yr} \times 1\text{ yr})} - \frac{0^2}{4(11.1\text{ m}^2/\text{yr} \times 1\text{ yr})}\right]$$

$$C(100\text{ m}, 0\text{ m}, 1\text{ yr}) = 1.28 \times 10^{-8}\frac{\text{g}}{\text{m}^3} = 1.28 \times 10^{-8}\text{ mg/L}$$

When will C_{max} occur at the well 100 m down-gradient? The center of the benzene puff will be at the 100-m down-gradient well at a travel time t, computed as:

$$t = \frac{x}{v} = \frac{100\text{ m}}{11.1\text{ m/yr}} = 9.0\text{ yr}$$

This is approximately[2] when the maximum benzene concentration at the house will be seen. See also Homework Problem 9.3. So, the concentration of benzene arriving at the well after *one year* will not be the *maximum* concentration of benzene that will ever be seen at that well. In fact, the maximum concentration will be seen close to $t = 9$ years, and can be estimated to be approximately:

$$C_{(max)} \cong C(100\text{ m}, 0\text{ m}, 9\text{ yr}) = \frac{M/(n \times b)}{4\left(\sqrt{\pi t}\right)^2\sqrt{D_L D_T}} = 0.076\text{ mg/L}$$

assuming no sorption, reaction, or degradation of the benzene occurs over the 9-year period. Plotting the 2-d puff equation, conc. versus time at location (100 m, 0 m), will yield the exact time and concentration at the maxima.

While "puffs" generated by instantaneous chemical releases eventually dissipate due to dispersion, long-lasting groundwater plumes are produced in the subsurface in response to a continuous release of a chemical. When advective transport dominates dispersive transport such that the Peclet number, Pe, is greater than 200, the response in saturated groundwater to a continuous tracer injection at $x = 0$ starting at time $t = 0$, at a concentration C_0, is shown in Table 9.3. Tracer transport in response to continuous injection at a constant concentration into a one-dimensional flow field is characterized by the Ogata–Banks equation (Ogata and Banks, 1961) (see Section 5.4.3 and Fig. 5.11*a* and 5.11*b*). Tracer transport in two dimensions with similar injection conditions is illustrated in Figure 9.5. Analytical equations for tracer transport

[2] The reader should plot the change in benzene concentration with time at the well and note that the maximum concentration at the well occurs a little before the center of the puff reaches the well.

TABLE 9.3 Analytical Solutions for Continuous Release of a Tracer at $x = 0, t = 0$, in Three-, Two-, and One-Dimensional Transport Systems in Saturated Groundwater[a]. (For conditions when Pe > 100; Pe is the Peclet number.[b])

Dimensions	Analytical Solution
Three-Dimension $\dot{M} = \left(\frac{\text{Mass}}{\text{Time}}\right)$ (Hunt, 1978)	$C(x, y, z, t) = \frac{\dot{M}}{8\pi n r\sqrt{D_T D_Z}} \exp\left[\frac{(x-r)v}{2D_L}\right] \operatorname{erfc}\left(\frac{r - vt}{2\sqrt{D_L t}}\right)$ $r = \sqrt{\left(x^2 + y^2\frac{D_L}{D_T} + z^2\frac{D_L}{D_Z}\right)}$
Two-Dimension $\dot{M}_z = \frac{\text{Mass}}{\text{Depth} \times \text{Time}}$ (Wilson and Miller, 1978)	$C(x, y, t) = \frac{\dot{M}_z}{4n\sqrt{\pi r v D_T}} \exp\left[\frac{(x-r)v}{2D_L}\right] \operatorname{erfc}\left(\frac{r - vt}{2\sqrt{D_L t}}\right)$ $r = \sqrt{\left(x^2 + y^2\frac{D_L}{D_T}\right)}$
One-Dimension $\dot{M}_a = \left[\frac{\text{Mass}}{\text{Area} \times \text{Time}}\right]$ $C_0 = \frac{\dot{M}}{nv}$ (Ogata-Banks, 1961, Hunt, 1978)	$C(x, t) = \frac{\dot{M}_a}{2nv} \operatorname{erfc}\left(\frac{x - vt}{2\sqrt{D_L t}}\right) = \frac{C_0}{2} \operatorname{erfc}\left(\frac{x - vt}{2\sqrt{D_L t}}\right)$

[a] Groundwater flows at an average velocity v in the x direction.

[b] $\text{Pe} = \frac{\sqrt{2L}}{D_L}$

in one- and two-dimensional systems are often calibrated with experimental observations of tracer transport in laboratory columns and controlled field tests, yielding best-fit estimates of groundwater advection and dispersion parameters, as illustrated in the examples that follow.

EXAMPLE 9.3

A 50-mg/L bromide (tracer) salt solution was injected continuously starting from time $t = 0$ into a porous packed laboratory column of length 22.78 cm with a a square cross section of 5.4 cm × 5.4 cm. The volumetric flow rate of the aqueous solution through the column was measured at 1.10 mL/min. The effluent tracer concentrations were measured periodically and a breakthrough curve was plotted as shown below. First, visually estimate groundwater velocity v and dispersivity α_L. Next, use an optimization package (e.g., Excel's Solver) to obtain estimates of v and α_L that would generate the best fit between the Ogata–Banks analytical equation (see Table 9.3) and the experimental observations.

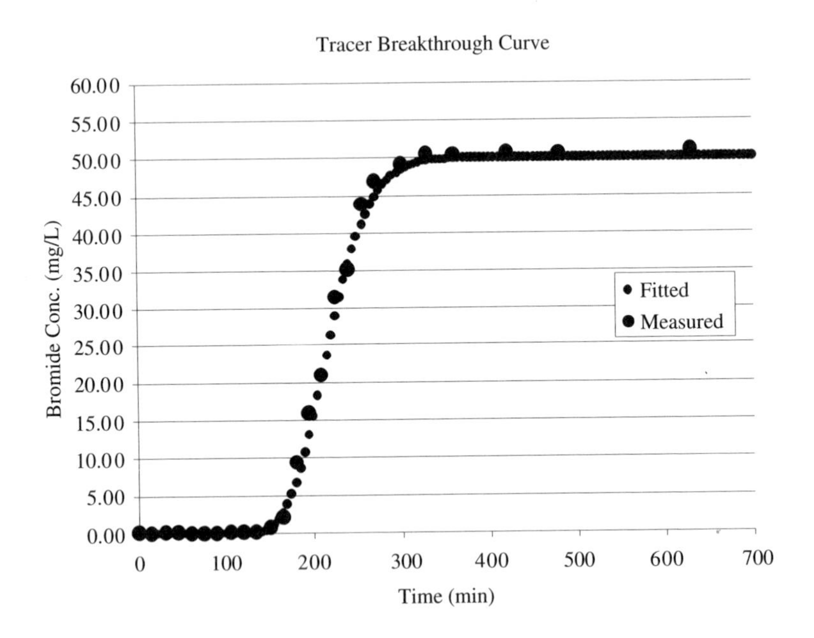

Time (min)	Measured Effluent Tracer Concentration	Time (min)	Measured Effluent Tracer Concentration
0	0.12	180	9.48
15	0.00	195	16.03
30	0.13	210	20.99
0	0.12	225	31.40
45	0.16	240	35.20
60	0.00	255	43.79
75	0.00	270	46.87
90	0.00	300	49.10
105	0.13	330	50.50
120	0.22	360	50.36
135	0.17	420	50.73
150	0.78	480	50.63
165	2.14	630	50.98

Solution 9.3 The tracer concentration is one-half of the source concentration (50th percentile or 25 mg/L) at the breakthrough time t_b, which can be visually estimated or obtained from interpolation of the data in the above table as $t_b =$ 215.8 min. The average pore water velocity is estimated as L/t_b, where L is the length of the column (= 22.78 cm), yielding $v = 0.105$ cm/min. The times at which one observes 16th and 84th percentile concentrations, that is, 8 and 42 mg/L, are approximately 177 and 248 min, their difference representing two standard deviation measures about the breakthrough time, that is, $2\sigma_{\text{time}} = 71$ min. The standard deviation in units of distance along the length of the column is $\sigma_x = v\sigma_{\text{time}} = 3.73$ cm. Applying Eq. (9.5) to estimate the dispersion coefficient yields $D_L = (\sigma_x^2/2\,t_b) = 0.032$ cm^2/min, from which the dispersivity can be computed. Using our estimate of v, the dispersivity α_L is estimated as $\alpha_L = D_L/v = 0.31$ cm. All the above relationships can be coupled together to yield

$$\alpha_L = \frac{L \times (t_{84} - t_{16})^2}{8t_b^2} = 0.31 \text{ cm} \tag{9.7}$$

It is important to note that the estimate of v enables us to also estimate the porosity of the packed bed from Eq. (9.1b), since the volumetric flow rate and pore cross-sectional area are known from measurement. The estimated porosity is $n = 0.36$.

Parameter estimation by visual inspection/interpolation of the breakthrough curve has been demonstrated above. For greater accuracy, the Ogata–Banks equation can be fitted to the data using the Excel Solver function to obtain best-fit parameter valued for velocity v (or porosity n) and dispersivity α_L, which minimize the error between the fitted results and the measurements. The best-fit parameter values were found to be $v = 0.1$ cm/min (or porosity, $n = 0.36$) and dispersivity $\alpha_L = 0.34$.

For a continuous tracer injection in a two-dimensional groundwater flow field, a single observation well is installed at location A along the centerline of the tracer plume (see Fig. 9.5*a*). Water velocity and longitudinal dispersivity are determined from the tracer breakthrough curve observed at A (see Fig. 9.5*b*), in a manner similar to that shown in Example 9.3. Additionally, a row of wells is installed along a transect perpendicular to the water flow direction at A (see Fig. 9.5*a*). From Table 9.3, for the two-dimensional case, we see that at any time t, the concentration along the Y transect will be Gaussian with respect to the centerline concentration at well A, which is the peak concentration along the Y dimension. See Figure 9.5*c*. Application[3] of the

[3] We seek a location along the y axis, y_b, such that the tracer concentration at y_b is one-half that at the central well A (at $y = 0$). Recognize from symmetry considerations that Γ is double the distance represented by $[y_b - 0]$. Solve for y_b in terms of σ_y in the Gaussian equation:

$$C^{\text{tracer}}(y_b, t) = C^{\text{tracer}}_{\text{well}A,t} \times \exp\left(\frac{-1}{2}\left(\frac{y_b - 0}{\sigma_y}\right)^2\right) = 0.5 \times C^{\text{tracer}}_{\text{well}A,t}$$

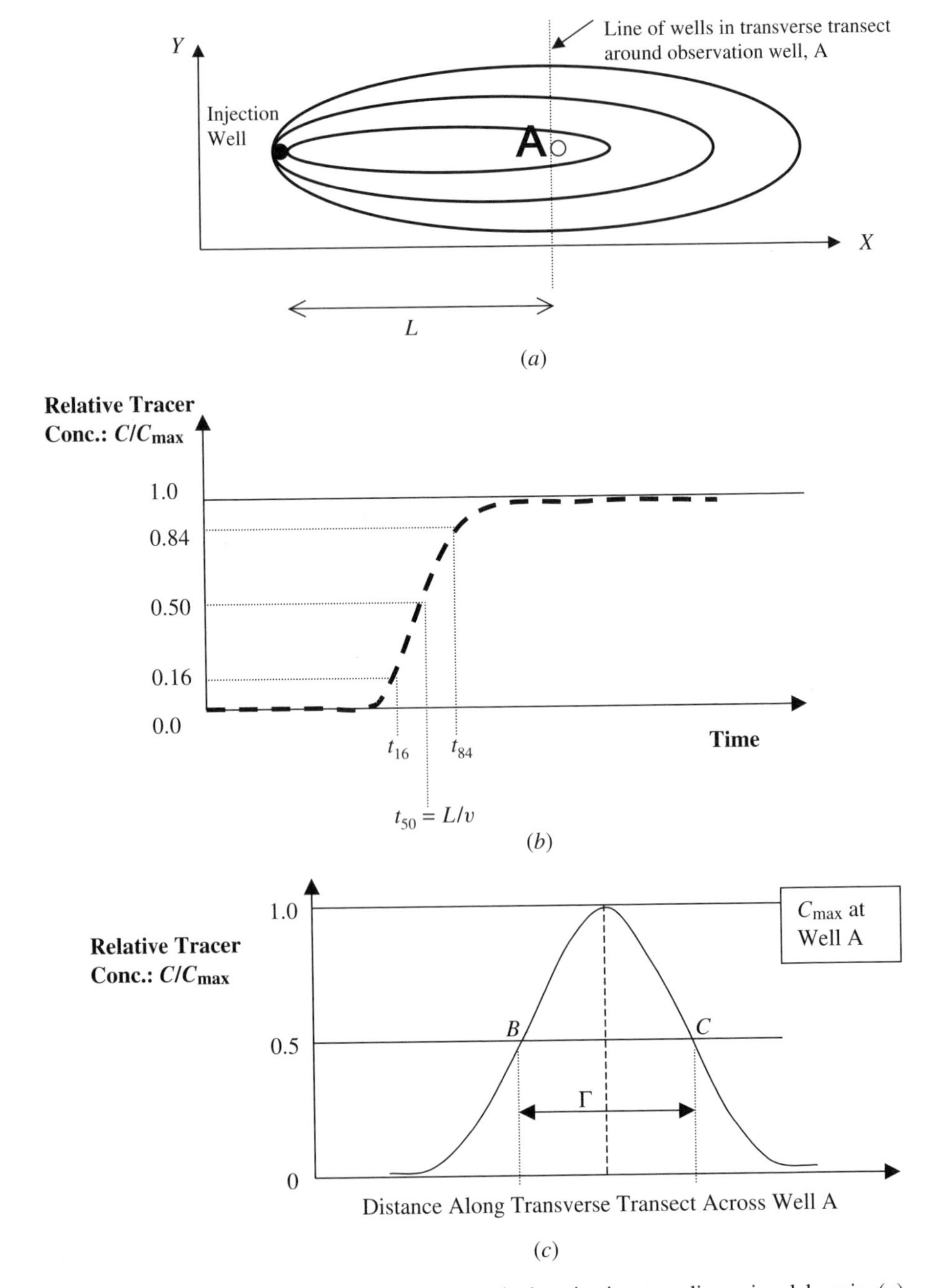

Figure 9.5 Continuous injection of tracer at a point location in a two-dimensional domain. (*a*) Two-dimensional plume. (*b*) Breakthrough curve (versus time) at observation well A. (*c*) Tracer concentrations along transverse transect.

Gaussian dispersion equation in the Y dimension in Figure 9.5c yields the following estimate for the standard deviation, σ_y, associated with transverse dispersion:

$$\sigma_y = \frac{\Gamma}{2.34} \tag{9.8}$$

where Γ is the distance between the two symmetrical locations (B and C) in Figure 9.5c where the observed tracer concentration in well water is half that observed in the centerline well, A. Then D_T is computed from the estimated σ_y by applying Eqs. (9.5), knowing the time t, since continuous tracer injection commenced in the aquifer. The above examples demonstrate the utility of tracer tests in determining transport parameters pertaining to advection and dispersion in saturated groundwater.

9.3 SORPTIVE–REACTIVE CONTAMINANT TRANSPORT IN THE SATURATED ZONE

9.3.1 Effect of Sorption to Soil

The analytical solutions presented in Section 9.2 describe transport of a nonsorbing and nonreactive tracer chemical in a homogeneous saturated groundwater system. For the case of pollutants that sorb to the aquifer matrix, the soil acts as an additional compartment in the groundwater system. This affects the control volume mass balance in the following way.

Assuming instantaneous equilibrium between chemical concentration in water represented on a mass/volume bases, $C(x, t)$, and the chemical concentration in the adjacent soil, $C_s(x, t)$, represented on a mass/mass basis, the total mass of chemical, M_{total}, present within a bulk volume, V_{bulk}, of an aquifer is given as:

$$M_{\text{total}}(x, t) = nV_{\text{bulk}}C(x, t) + \rho_b V_{\text{bulk}} K_d C(x, t) = V_{\text{bulk}}(n + \rho_b K_d)C(x, t) \tag{9.9a}$$

where K_d is the soil–water partition coefficient $[K_d = C_s(x, t)/C(x, t)]$, assuming such a linear sorption isotherm as defined in Eq. (3.6) is valid; and ρ_b is the bulk density of the aquifer matrix, that is, the mass of the soil particles divided by the total (bulk) volume of the matrix, including the pore space. The rate of change of total pollutant mass in the control volume is thus given by:

$$\frac{dM_{\text{total}}(x, t)}{dt} = V_{\text{bulk}}(n + \rho_b K_d)\frac{dC(x, t)}{dt} \tag{9.9b}$$

Recognizing that the advective and dispersive transport processes for pollutant mass to and from the control volume occur in the liquid volume nV_{bulk}, the mass-balance equation in a one-dimensional system is given by:

$$\frac{\partial M_{\text{total}}(x, t)}{\partial t} = V_{\text{bulk}}(n + \rho_b K_d)\frac{\partial C(x, t)}{\partial t} = nV_{\text{bulk}}\left(\frac{-v\,\partial C(x, t)}{\partial x}\right) + nV_{\text{bulk}} D_L \frac{\partial^2 C(x, t)}{\partial x^2} \tag{9.10a}$$

Dividing by nV_{bulk} yields the following equation:

$$\left(1+\frac{K_d\rho_b}{n}\right)\frac{\partial C(x,t)}{\partial t}=-v\frac{\partial C(x,t)}{\partial x}+D_L\frac{\partial^2 C(x,t)}{\partial x^2} \tag{9.10b}$$

The terms within the parentheses on the left-hand side of Eq. (9.10), together represent the retardation factor, R, defined as:

$$\begin{aligned}\mathsf{R}=1+\frac{K_d\rho_b}{n}&=\frac{VnC(x,t)+K_d\rho_bVnC(x,t)}{VnC(x,t)}\\&=\frac{\text{Pollutant mass in soil and water}}{\text{Pollutant mass in water alone}}\end{aligned} \tag{9.11}$$

Equation (9.10b) can therefore be rewritten as:

$$\mathsf{R}\frac{\partial C(x,t)}{\partial t}=-v\frac{\partial C(x,t)}{\partial x}+D_L\frac{\partial^2 C(x,t)}{\partial x^2} \tag{9.12}$$

or, further, after dividing by R, as:

$$\frac{\partial C(x,t)}{\partial t}=-v_s\frac{\partial C(x,t)}{\partial x}+\alpha_L v_s\frac{\partial^2 C(x,t)}{\partial x^2} \tag{9.13a}$$

where v_s represents the sorbed contaminant velocity, obtained as:

$$v_s=\frac{\text{Groundwater velocity}}{\text{Retardation factor}}=\frac{v}{\mathsf{R}} \tag{9.13b}$$

The contaminant velocity v_s represents the apparent reduction in the advective velocity of the pollutant, which is reduced by a factor of R, due to sorption to the soil, which effectively "retards" the contaminant migration. Both advective transport and the dispersive spread[4] of the contaminant appear to be retarded. Thus, water flows at a faster rate than the contaminant, which appears to be slowed down but, in fact, is merely retained on the soil. It is useful to note that the advection–dispersion–retardation equation above [Eqs. (9.13a) and (9.13b)] is very similar to the advective–dispersive equation for a tracer shown in Eq. (9.6), with v now being replaced with v_s for the sorbed contaminant. Therefore, assuming that a instantaneous linear soil–water equilibrium partition coefficient is valid, sorbed contaminant transport may be described with the same models that describe nonsorbing tracers (Tables 9.2 and 9.3), except that the water velocity v is now replaced with the sorbed contaminant velocity v_s. Correspondingly, the dispersion coefficient for the sorbed contaminant, s, in the longitudinal, transverse, and vertical dimension [Eq. (9.5)] effectively becomes $D_{L,s}\approx\alpha_L v_s$; $D_{T,s}\approx\alpha_T v_s$; and $D_{Z,s}=\alpha_Z v_s$. The (tracer) mass released term in the

[4] The impact of sorption to soil on the rate of molecular diffusion in groundwater is described in Section 4.5 and employs the concept of a retardation factor, R, identical to that defined here.

numerator of the equations in Table 9.2 is equivalent to the pollutant mass released to the aqueous phase alone, that is

$$M = M_{aq} = \frac{M_{total}}{\mathsf{R}}$$

where M_{total} is the total mass of chemical released to the aquifer (prior to partitioning to the soil). For continuous sources releasing pollutants at an aqueous concentration of C_0, the constant concentration term is not significantly impacted by retardation because an infinite supply of pollutants is assumed to be present such that the soil is assumed to contain pollutant at equilibrium with C_0, and "breakthrough" occurs in the system. Example 9.4 illustrates the changes that must be incorporated to pollutant distribution models to account for sorption-retardation.

EXAMPLE 9.4

Start with Example 9.2 but now include the effects of retardation. Assume $K_d = 0.5\ \mathrm{L_w/kg_{soil}}$ and the bulk density of the soil is 2 g/cm^3.

Solution 9.4 Determine present-day benzene concentrations in a residential well located 100 m down-gradient from the spill. We now calculate the travel velocity of the benzene incorporating the effect of sorption: $v_s = v/\mathsf{R}$ where the retardation factor, $\mathsf{R} = 1 + (\rho_b K_d)/n = 4$ (based on data given).

Hence,

$$v_s = \frac{11.1\ \text{m/yr}}{4.0} = 2.8\ \text{m/yr}$$

The benzene mass released into water is influenced by sorption and is determined as:

$$M = M_{aq} = \frac{M_{total}}{\mathsf{R}} = \frac{1000\ \text{g}}{4.0} = 250\ \text{g}$$

Using the same values for dispersivities as in Example 9.2, the dispersion coefficients for the sorbed contaminant are computed as $D_{L,s} = \alpha_L v_s = 28\ \text{m}^2/\text{yr}$ and $D_{T,s} = \alpha_T v_s = 2.8\ \text{m}^2/\text{yr}$. The concentration of benzene in the groundwater at the 100-m down-gradient well one year after release is calculated as:

$$C(100\ \text{m}, 0\ \text{m}, 1\ \text{yr}) = 1.6 \times 10^{-37}\text{mg/L} \approx 0$$

Approximately when will C_{max} occur at the well 100 m down-gradient?

$$t = \frac{x}{v_s} = \frac{100\ \text{m}}{2.8\ \text{m/yr}} = 35.7\ \text{yr}$$

At which time the concentration will be

$$C_{(\max)} \cong \frac{M_{\text{aq}}/(n \times b)}{4\left(\sqrt{\pi t}\right)^2 \sqrt{D_{L,s} D_{T,s}}} = 0.02\ \text{mg/L}\ (= \text{g/m}^3)$$

The above computation assumes benzene sorbs instantaneously and reversibly to soil, but no reaction or degradation of benzene occurs in groundwater over 35 years. Thus, of the total 1000 g of benzene released into the subsurface, 250 g travels in the aqueous phase (M_{aq}), while 750 g are sorbed under instantaneous equilibrium conditions to the soil in contact with the moving groundwater "puff." Benzene concentrations 100 m down-gradient are very small (much below any existing detection method) just 1 year after the spill, but are predicted to peak to approximately 0.02 mg/L at around 35 years, assuming no reaction or degradation of benzene takes place in groundwater.

The impact of sorption–retardation on contaminant transport is frequently seen in the field and produces, on a large scale, separation phenomena for various pollutants as is seen in bench-scale chromatography columns in chemistry laboratories. Thus, one can expect tracers and water to travel fastest, at the average velocity, v, followed by various contaminants, each traveling at a contaminant velocity given by v/R. For more hydrophobic contaminants, the soil–water partition coefficient is larger, causing a corresponding increase in R and a resultant decrease in v_s. When a mixture of contaminants is released to water, the lighter, less hydrophobic chemicals travel faster while the more hydrophobic contaminants move much slower. The separation of contaminants due to sorption has been documented at many sites; Figure 9.6 shows observations from a controlled field test conducted in the sandy Borden Aquifer in Canada (Roberts et al., 1986).

The inverse of the retardation factor is a useful parameter that represents the fraction of pollutant released into the aquifer that is present in water and will be flushed under pump-and-treat conditions, assuming soil–water equilibrium is always maintained. See the definition of R in Eq. (9.11). Calculations employing the retardation factor provide a "first-cut" estimate of the degree of pollutant removal that occurs with water removal by pumping, under optimistic conditions of rapid soil–water partitioning equilibrium. For example, let M_{total} represent the total mass of benzene initially present in an aquifer, including the benzene mass present in water and that sorbed on soil. After the first pore volume of water is extracted, the mass of benzene remaining in the aquifer is given as $M_{\text{total}}(1 - \mathsf{R}^{-1})$, of which the fraction $1/\mathsf{R}$ is flushed out with the second pore volume. The benzene mass remaining in the aquifer after the second pore volume is flushed $= M_{\text{total}}\left(1 - \mathsf{R}^{-1}\right)^2$. By induction, the mass

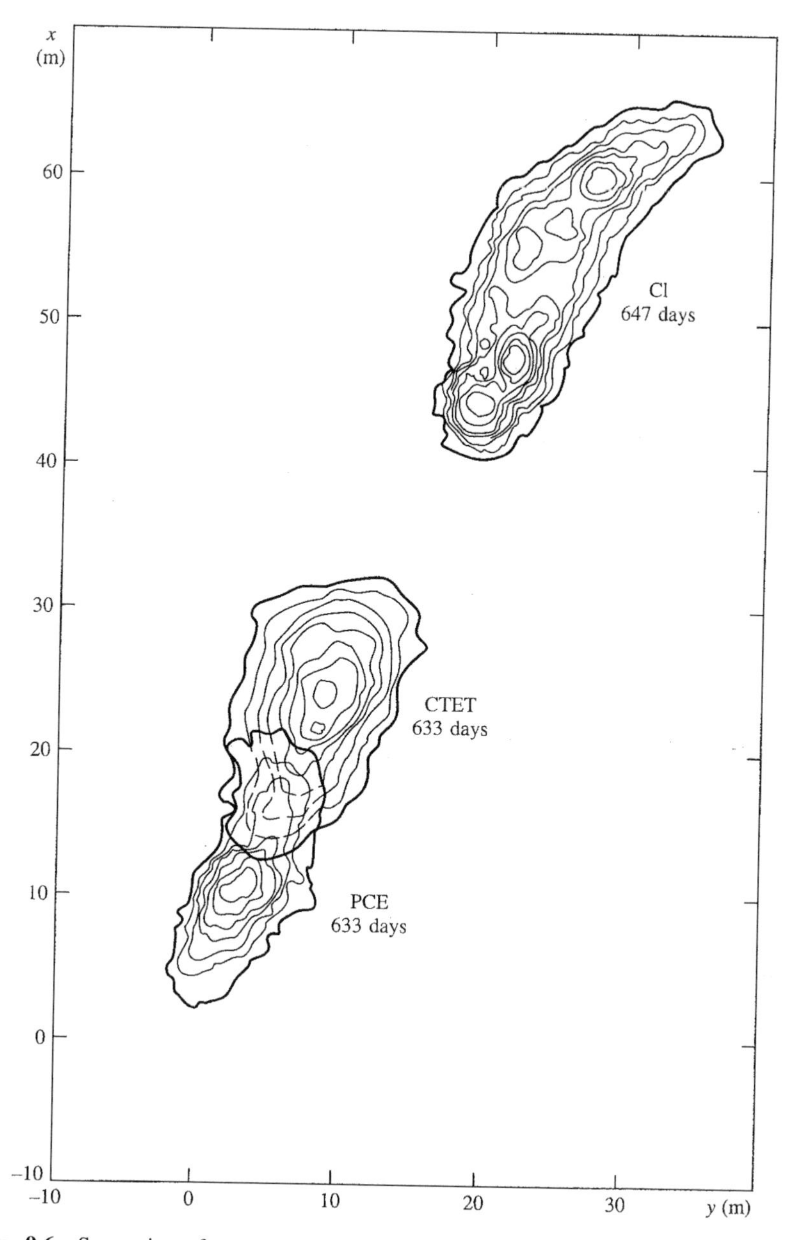

Figure 9.6 Separation of tracer and chemical "puffs" in groundwater after simultaneous release at the Borden Aquifer, Canada (from Roberts et al., 1986).

of pollutant remaining in the aquifer after the jth pore volume has been flushed is $M(j) = M_{\text{total}}(1 - \mathsf{R}^{-1})^j$. The cumulative fraction of initial mass, M_{total}, flushed after j pore volumes:

$$F_{(j)} = \left[\frac{[M_{\text{total}} - M(j)]}{M_{\text{total}}}\right] = 1 - \left(1 - \mathsf{R}^{-1}\right)^j \tag{9.14}$$

The projected removal for benzene with an assumed retardation factor of 4 is shown in Figure 9.7. The graph shows the law of diminishing returns in pump-and-treat operations at contaminated sites. The maximum pollutant mass removal occurs when the first pore volume of water is flushed, after which the mass removed keeps decreasing because of decreasing total contaminant mass in the aquifer and hence a corresponding decrease in equilibrium concentrations in the water. Groundwater pump-and-treat becomes increasingly impractical as a remedy for extracting hydrophobic contaminants with large retardation factors from the subsurface.

Equation (9.14) enables a rough calculation of the number of pore volumes required, under the best of conditions, to achieve a certain desired cumulative fraction of removal (say 90% removal) of contaminant from the subsurface. However, as noted earlier, an important assumption is that soil–water sorption–desorption equilibrium occurs instantaneously. When pollutant desorption from soil occurs much slower than the flow rate of water through the contaminated zone, groundwater pollutant concentrations will be lower than the predicted equilibrium, resulting in even lower pollutant mass fractions removed. Such a phenomenon has been observed at many field sites where pump-and-treat operations were discontinued when aqueous pollutant concentrations fell below the target treatment limit, suggesting that the groundwater had been remediated. However, when pumping was stopped, aqueous pollutant concentrations were found to slowly increase, regaining equilibrium with

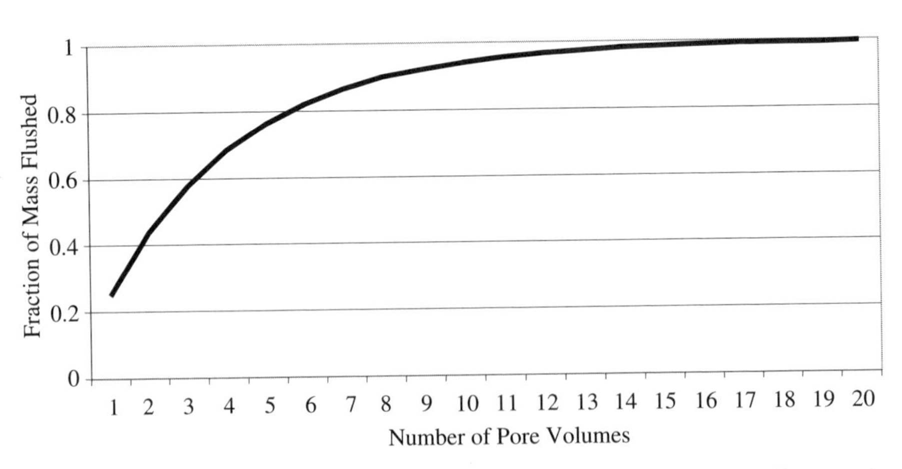

Figure 9.7 Fraction of benzene mass initially present in an aquifer that is removed by groundwater extraction. The retardation factor for benzene in this aquifer is 4. Of the benzene mass 25% is removed after 1 pore volume of water is extracted. However, more than 16 pore volumes are needed to remove 99% of the benzene mass initially present in the aquifer.

soil-phase contamination after the passage of time. In some field studies, even for natural flow conditions, that is, with no rapid water flow due to pumping wells, desorption disequilibrium (rate-limited desorption) was found to result in the slow release of contaminant to groundwater (Ball and Roberts, 1989). In such situations, the pollutant concentration distributions in groundwater show a heavy "tail," that is, instead of nearly symmetric Gaussian puff distributions over space, aqueous concentrations were found to be higher behind the center of the puff due to slow re-release of contaminants sorbed to soil. For the case of a continuous release of pollutant, the observed breakthrough curve is also found to be asymmetric, with a heavy tail. These patterns (such as those illustrated in Fig. 9.8) are indicators of slow sorption–desorption occurring in the aquifer, which would need to be included in the modeling effort.

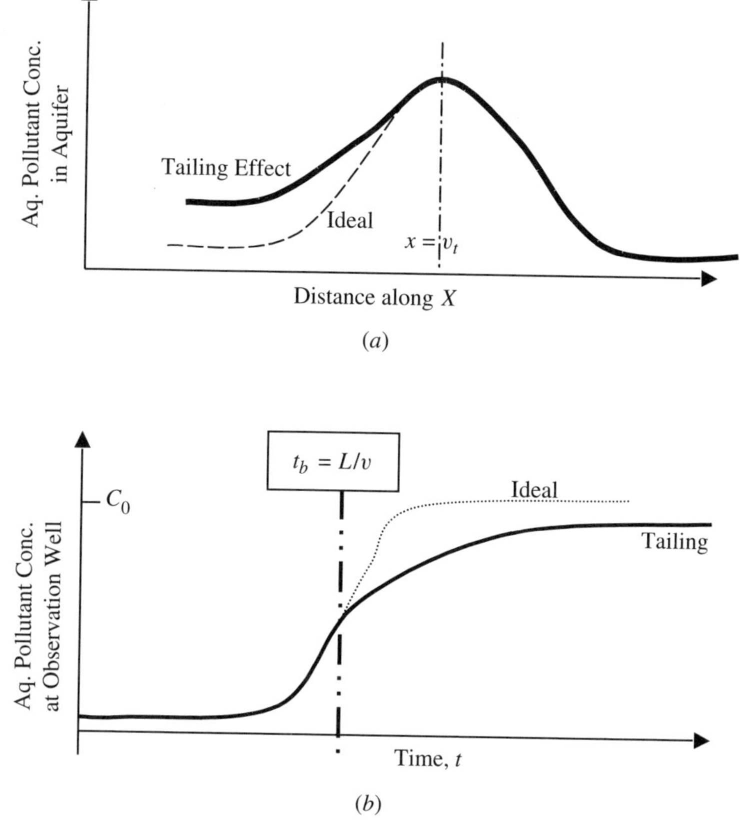

Figure 9.8 Tailing in patterns of pollutant concentration caused by rate-limited desorption of the pollutant from soil. (*a*) Asymmetry in spatial Gaussian concentration patterns in response to an instantaneous release. (*b*) Asymmetric breakthrough of a pollutant at an observation well in response to a continuous release. (Note that such tailing when observed for a tracer would indicate heterogeneities that result in spatial variation in aquifer dispersivity.)

Two approaches have been taken to address rate-limited sorption. The first examines microscale phenomena underlying sorption. Sorption (or desorption) of a pollutant to soil is viewed to occur in two steps: first, rapid transport via diffusion into a more accessible compartment of organic matter occurs, followed by slow diffusion into a second compartment with "glassy" or heavy polymeric qualities that sequesters the pollutant (Webber and Huang, 1996). The two-compartment model accounts for a rapid phase in sorption–desorption kinetics, followed by a slow phase as has been observed in laboratory and field tests. Efforts have also been made to characterize the glassy organic matter in soil—for some sediments, this compartment has been found to be associated with very fine particles containing organic carbon from soot (Ghosh et al., 2000). While microscale approaches such as those referenced above are useful in understanding phenomena leading to slow sorption–desorption of pollutants in soil–water systems, these microscale parameters have not yet been scaled-up to enable field-scale modeling efforts. A lumped parameter approach has been espoused by Brusseau et al., in which a two-compartment model for sorption is used with fitted parameters derived from field data to represent the rate-limited sorption phenomena (Brusseau, 1991a,b). Since these parameters must necessarily be site-specific, having been fitted to field data, they cannot be translated to other sites. It is expected that ongoing research on sorption processes, occurring on many fronts, will result in convergence between the micro- and macroscale approaches summarized very briefly above.

9.3.2 Effect of First-Order Removal Processes

First-order removal processes (e.g., radioactive decay or pseudo-first-order biodegradation kinetics) are incorporated into the advective–dispersive–sorptive equation (9.12) as:

$$\mathsf{R}\frac{\partial C(x,t)}{\partial t} = -v\frac{\partial C(x,t)}{\partial x} + \alpha_L v\frac{\partial^2 C(x,t)}{\partial x^2} - k_{\text{eff}}C(x,t) \tag{9.15a}$$

where

$$k_{\text{eff}} = \left(k_w + \frac{\rho_b K_d}{n} k_{\text{soil}}\right) \tag{9.15b}$$

where k_w represents the degradation rate constant for the aqueous-phase contaminant and k_{soil} that of the sorbed contaminant. Thus, k_{eff} represents the effective contaminant degradation rate constant in the combined soil–water system in the most general case where the rate constants for pollutant decay are not the same in water and in soil, for example, when organic chemicals biodegrade rapidly in water but not when sorbed to soil. When chemical degradation rates in both media can be described by the same first-order rate constant, k, as in the case of radioactive decay, then $k_w = k_{\text{soil}} = k$ and $k_{\text{eff}} = \mathsf{R}k$ [from Eq. (9.15b)]. Analytical solutions to Eq. (9.15) for instantaneous and continuous release of reactive chemicals that undergo degradation with $k_{\text{eff}} = \mathsf{R}k$ are detailed in Bear (1979) and Van Genuchten (1981).

Effectively, for an instantaneous release of a pollutant into groundwater, the chemical concentration of the reactive species at a certain location and time, $C^{w/\text{rxn}}(x, y, z, t)$, is obtained by factoring exponential first-order decay into the sorption-corrected Gaussian puff concentrations computed at that same location and time, in the absence of reaction:

$$C^{w/\text{rxn}}(x, t) = C^{w/o\ \text{rxn}}(x, t) \times e^{-kt} \tag{9.16}$$

where t is the travel time in the aquifer, that is, time since release of the chemical. Thus, as shown in Figure 9.9*a*, the contaminant center of mass travels at the sorbed velocity v_s, but the concentrations are decreased by a factor of e^{-kt} to account for first-order reaction that occurs during travel.

For a continuous constant concentration input into a one-dimensional system, first-order reaction is incorporated into the Ogata–Banks equation (Table 9.3) to yield the following analytical solution that addresses advection, dispersion, sorption, and reaction:

$$C^{w/\text{rxn}}(x, t) = \frac{C_0}{2} \exp\left[\frac{(v - u)x}{2D_L}\right] \operatorname{erfc}\left(\frac{\mathsf{R}x - ut}{2\sqrt{\mathsf{R}D_L t}}\right) \tag{9.17a}$$

where u is quantified as:

$$\frac{u}{v} = \sqrt{1 + \frac{4\alpha_L k_{\text{eff}}}{v}} \tag{9.17b}$$

The relative importance of reaction to advection is represented by the ratio $4\alpha_L k_{\text{eff}}/v$. When $4\alpha_L k_{\text{eff}} \ll v$, that is, when reaction does not significantly impact the effective contaminant velocity in the aquifer, u is approximately equal to v and[5] $v - u \sim -2k_{\text{eff}}\alpha_L$. With $k_{\text{eff}} = \mathsf{R}k$, $D_L = \alpha_L v$, and $v_s = v/\mathsf{R}$, Eq. (9.17a) simplifies to:

$$C^{w/\text{rxn}}(x, t) = \frac{C_0}{2} \exp\left(\frac{-kx}{v_s}\right) \operatorname{erfc}\left(\frac{x - v_s t}{2\sqrt{\alpha_L v_s t}}\right) \quad \text{for} \quad k_{\text{eff}} \ll \frac{v}{4\alpha_L} \tag{9.17c}$$

Equation (9.17c) is identical to the simplified representation in Eq. (9.16), wherein first-order reaction is incorporated by multiplying the analytical solution to the advection–dispersion equation [Eq. (9.13)] by a correction factor of e^{-kt}, where t is the travel time for the contaminant in the aquifer.[6] Steady-state conditions are attained in one-dimensional transport systems when continuous release of a chemical into groundwater is effectively countered by removal by reaction/degradation. For the case of first-order degradation, the steady-state concentration patterns are obtained by simplifying Eq. (9.17) for large times, t, yielding:

[5] The series expansion formula, $\sqrt{1 + (4\alpha_L k_{\text{eff}}/v)} \cong 1 + 0.5(4\alpha_L k_{\text{eff}}/v)$, yields the expression for $v - u$.
[6] When reaction rates do not significantly impact the effective contaminant velocity, the travel time t for a continuously released chemical is equivalent to x/v_s.

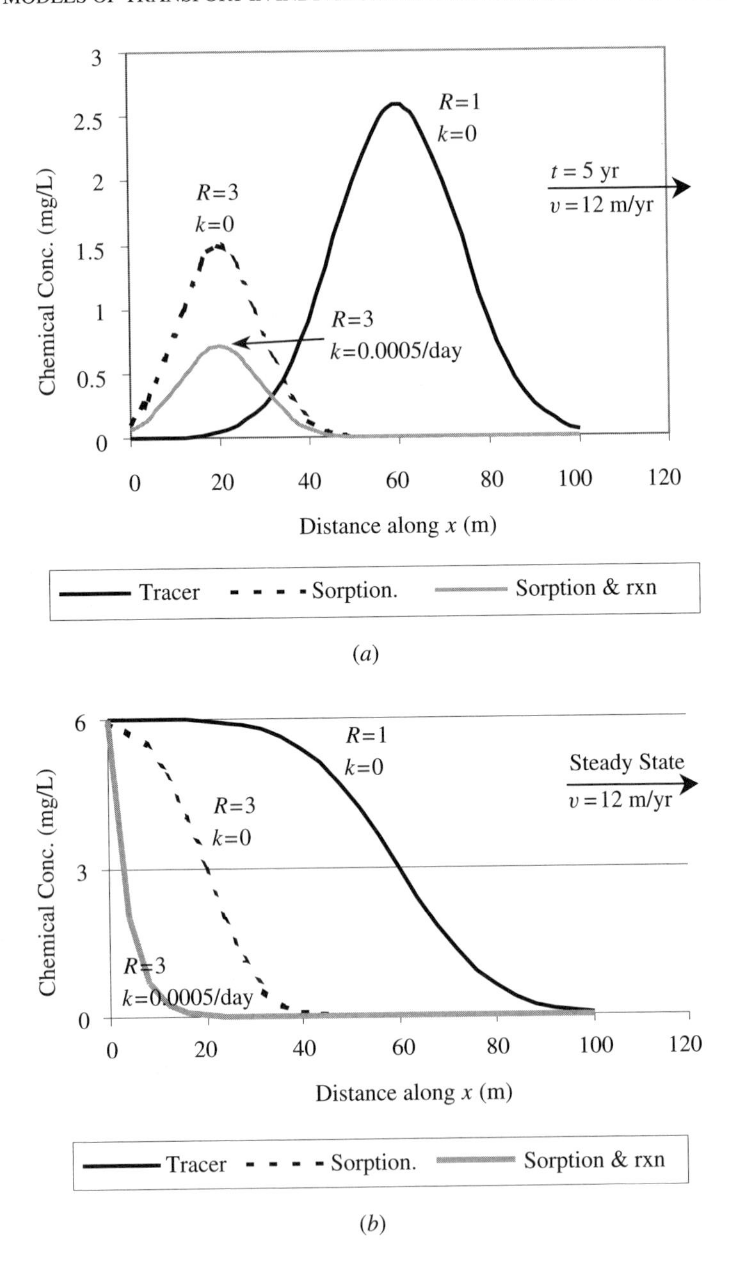

Figure 9.9 Impact of soprtion-retardation (R) and first-order reaction (k, day^{-1}) on pollutant concentration simulated in a model aquifer with $v = 12$ m/yr, $\alpha_L = 2$ m. (*a*) Groundwater concentrations at t = 5 yr in response to an instantaneous release of a fixed mass (100 mg) of a chemical. (*b*) Steady state groundwater concentrations in response to a continuous release of a chemical at a concentration, $C_0 = 6$ mg/L at the origin.

$$C_{ss}^{w/\mathrm{rxn}} = C_0 \exp\left[\frac{(v-u)x}{2D_L}\right] \tag{9.18a}$$

$$\cong C_0 \exp\left(\frac{-kx}{v_s}\right) \quad \text{when } k_{\mathrm{eff}} \ll \frac{v}{4\alpha_L} \tag{9.18b}$$

Note the similarity between Eq. (9.18b) and the concentration profile within a plug flow reactor with first-order degradation [Eq. (5.17)]. Figure 9.9*b* illustrates the combined effect of advection, dispersion, sorption, and first-order reaction on chemical concentrations in an aquifer in response to a continuous release of reactive pollutants in a one-dimensional system. Contaminant fate and transport models incorporating reactions with more complex kinetic or equilibrium representations are described in Chapter 12.

EXAMPLE 9.5

Consider that benzene in Example 9.4 now undergoes microbial degradation. Assume a half-life of 2 days in this aquifer and compute benzene concentrations at the receptor location when the center of the puff reaches the receptor. Compare your answer with that in Example 9.4.

Solution 9.5 Assuming benzene undergoes degradation at the same rate in soil and water, the assumed first-order rate constant $k = 0.693/t_{1/2} = 0.346/\text{day}$. The maximum benzene concentration with no degradation was predicted (in Example 9.4) to occur at the receptor location at $t = 36$ years at approximately 0.02 mg/L. Incorporating reaction yields:

$$C(x = 100\text{ m},\ t = 36\text{ yr}) = 0.02\text{ mg/L}$$
$$\times \exp(-[0.346/\text{day}] \times [365\text{ day/yr}] \times 36\text{ yr}) = 0\text{ mg/L}$$

After 36 years (with a half-life of 2 days) the benzene is essentially all degraded, unlike the results in Example 9.4.

A comparison of Examples 9.2, 9.4, and 9.5 indicates large differences in predicted benzene concentrations at the receptor, depending upon the processes considered in the model. A suitable model for contaminant transport at a particular site must incorporate all relevant processes operational at that site. Simple analytical equations such as those presented in this section enable screening level comparisons of model predictions with the broad trends observed in the field. Numerical models can then be implemented to incorporate important site-specific features related to media heterogeneities, or temporal changes in hydraulic head, chemical reaction rates, and so forth, as well as links between surface water systems and saturated and unsaturated zones in an aquifer. Fundamental processes governing contaminant transport in the unsaturated zone are discussed next, after which parameter estimation techniques for both saturated and unsaturated zone models are reviewed briefly.

9.4 ADVECTION–DISPERSION IN THE VADOSE ZONE

Pollutant transport in the vadose zone is best understood by analogy with processes occurring in the saturated zone. As shown in Figures 9.1 and 9.2, the main difference between the saturated and unsaturated zone is the presence of air-filled void spaces in the latter. As a result, the moisture content in the unsaturated zone is less than the porosity, the difference being the air content of soil. Changes in moisture content, θ_w, affect many physical properties within the vadose zone. As the moisture content of soil decreases, the suction pressure or matric potential increases, as illustrated in Figure 9.10*a*.[7] When the soil is wetted so that θ_w increases, the suction pressure decreases, that is, the pressure head ψ becomes less negative, becoming zero at the water table. It is important to note that the wetting and drying cycles of the pressure versus moisture content curve are not identical but exhibit hysteresis. Hysteresis is caused primarily by the constricted pore neck structure of void spaces in soil (also called the ink bottle effect or the dumb-bell effect), due to which the pressure difference needed to drain a pore is different from that needed to fill one. Figure 9.10*a* also indicates the presence of an irreducible water content in soil, θ_r, which represents the residual water trapped in dead-end pores that cannot be extracted by suction pressure. When the irreducible water saturation is reached, an infinite increase in suction pressure causes no further decrease in moisture content.

Much as the pressure head decreases as a soil is drained, the conductivity of the soil also decreases as the soil moisture is reduced (Fig. 9.10*b*). This effect is due to the presence of fewer interconnected pores filled with water that facilitate water flow through the porous medium. When the soil is saturated, the conductivity approaches the saturated zone conductivity, K, now referred to as K_s, which was discussed in detail in Section 9.2 and can be estimated from Eq. (9.1d). The hydraulic conductivity in the unsaturated zone is not constant but varies as a function of soil moisture and therefore also the pressure head, ψ, and is represented as $K(\theta_w)$, that is, K is now a function of θ_w, or equivalently as $K(\psi)$, a function of the matric potential, ψ. Several empirical relationships have been developed to correlate $K(\theta_w)$ with moisture content, θ_w, the simplest being the Campbell equation shown in Table 9.4.

The primary movement of water in the vadose zone occurs vertically, that is, from the top surface of soil toward the water table after precipitation and infiltration events. Although the pressure head at the top layers of soil is negative (when dry), and the

[7] The impact of soil moisture content on fluid pressure is easily illustrated by inverting a tube filled with water over a pot of dry, unsaturated soil. Drier soil will "suck up" more water from the tube compared to moist soils, creating increased suction pressures as moisture content decreases.

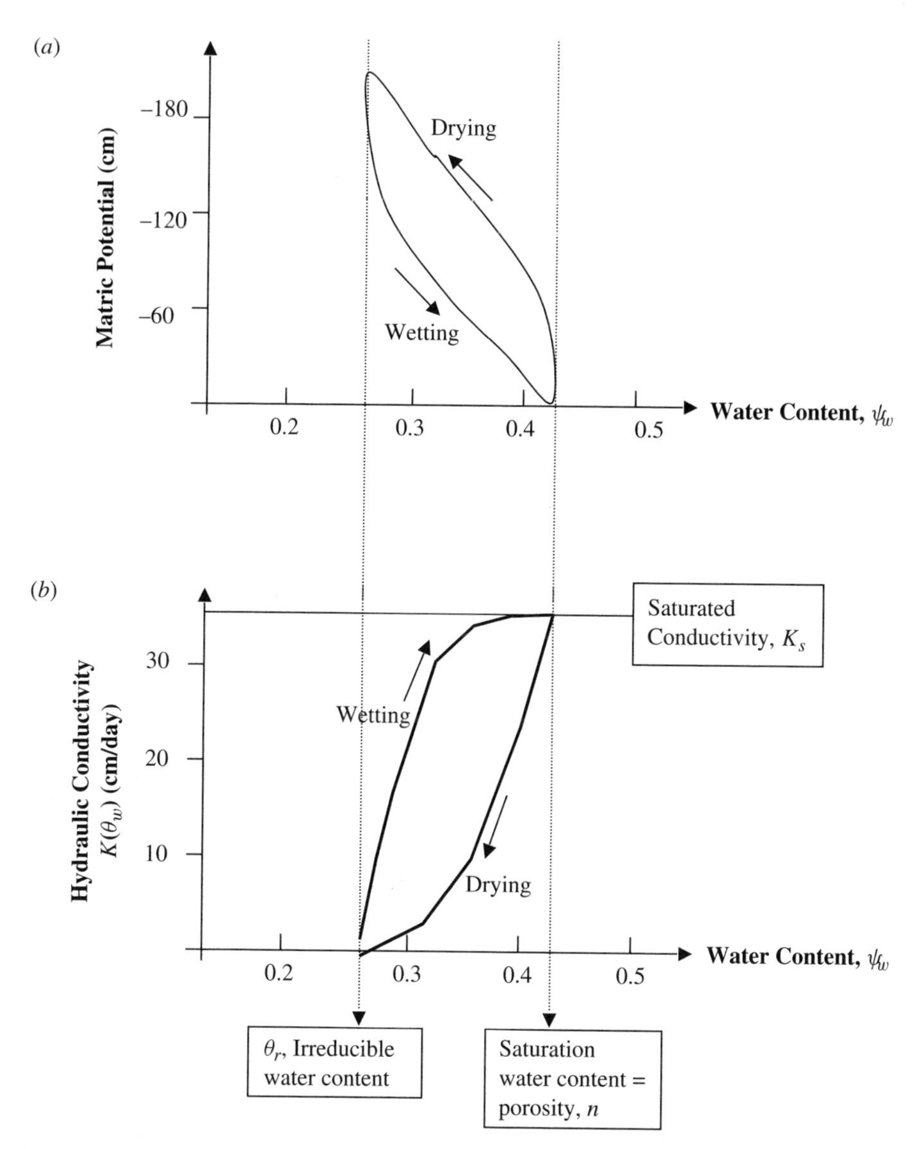

Figure 9.10 Impact of volumetric moisture content on (*a*) matric potential and (*b*) hydraulic conductivity in the vadose zone.

pressure head at the water table is zero, the total head at the top surface of soil is positive and greater than that at the water table due to the addition of elevation z. Thus, $h = z + \psi(z)$, with $\psi(z)$ becoming increasingly negative as one moves upward from the water table (see Fig. 9.2). As a result, water flows typically from the top surface of soil, vertically downward toward the water table, although plant transpiration processes can reverse this flow direction. A relationship equivalent to Darcy's law, called the Richards equation, can be written to describe the vertical

TABLE 9.4 Soil–Water Retention and Hydraulic Conductivity Relationships

Brooks–Corey Relationships (1966)

λ = pore size distribution index

ψ_b = bubbling pressure (cm)

Both λ and ψ_b can be estimated from soil texture and content (% sand and % clay)

Soil moisture retention	Unsaturated hydraulic conductivity, $K(\theta_w)$
$\frac{\theta_w - \theta_r}{n - \theta_r} = \left(\frac{\psi_b}{\psi}\right)^{\lambda}$	$\frac{K(\theta_w)}{K_s} = \left(\frac{\theta_w - \theta_r}{n - \theta_r}\right)^{(3+2/\lambda)}$

Campbell's Equations (1974)

b = constant related to inverse slope of ψ vs. θ_w curve

Soil moisture retention	Unsaturated hydraulic conductivity, $K(\theta_w)$
$\frac{\theta_w}{n} = \left(\frac{h_b}{h}\right)^{1/b}$	$\frac{K(\theta_w)}{K_s} = \left(\frac{\theta_w}{n}\right)^{(3+2b)}$

Van Genuchten Estimation Equations (1980)

a, c, and d are constants

Soil moisture retention	Unsaturated hydraulic conductivity, $K(\theta_w)$
$\frac{\theta_w - \theta_r}{n - \theta_r} = \left[\frac{1}{1 + (\alpha\psi)^d}\right]^c$	$\frac{K(\theta_w)}{K_s} = \left(\frac{\theta_w - \theta_r}{n - \theta_r}\right)^{1/2} \times \left\{1 - \left[1 - \left(\frac{\theta_w - \theta_r}{n - \theta_r}\right)^{1/c}\right]^c\right\}^2$

Definition of Parameters

θ_w = moisture content (variable)

θ_r = irreducible moisture content

n = porosity

ψ = capillary suction head (cm) (variable)

K_s = saturated hydraulic conductivity (cm/h)

$K(\theta_w)$ = unsaturated hydraulic conductivity varying as a function of moisture content

Source: From Morel-Seytoux (1989).

gradient-driven flow of water in the vadose zone. Under steady flow conditions, assuming uniform moisture (and hence conductivity) in the vadose zone, the Richards equation is written as:

$$q = -K(\theta_w)\frac{\partial h(z)}{\partial z} = -K(\theta_w)\frac{\partial[z + \psi(z)]}{\partial z} = -K(\theta_w)\left[1 + \frac{\partial\psi(z)}{\partial z}\right] \tag{9.19}$$

For uniform moisture conditions in the soil, that is, when ψ does not vary significantly with depth, Eq. (9.19) yields an approximation of the specific discharge rate of water:

$$q \cong -K(\theta_w) \tag{9.20}$$

where q is also referred to as the infiltration rate in the vadose zone. Steady-state infiltration of water under uniform soil moisture conditions, as given by Eq. (9.20), is an idealized condition representing the passage of infiltrating precipitation water through the vadose zone to the water table, without significantly changing the moisture content of the unsaturated soils. This may occur under conditions of slow and steady irrigation, but, more often one observes *unsteady* infiltration after precipitation events, yielding a rise of the water table and finally resulting in ponding when the entire vadose zone is completely saturated and the subsurface cannot hold any more moisture. These events occur under conditions of heavy precipitation and subsequent ponding followed by flooding. Several models have been developed to describe the unsteady response of unsaturated soils to precipitation events; in these models the unsteady form of the Richards equation[8] is solved numerically (Celia et al., 1990; Morel-Seytoux, 1989; Nielsen et al., 1986; Pan and Wierenga, 1995).

For the purpose of discussion in this chapter, we proceed with the assumption that steady infiltration is occurring in the vadose zone at a rate, q, determined from Eq. (9.20). The infiltration rate is equivalent to the Darcy velocity in the saturated zone and needs to be corrected for the true cross-sectional area available for water flow, to obtain the unsaturated zone pore water velocity v. Noting that only moisture-filled pores participate in the water flow process, the effective average linear pore water velocity in the unsaturated zone is computed as:

$$v \cong \frac{q}{\theta_w} \tag{9.21}$$

Thus, the moisture content, θ_w, determines the advective velocity in the vadose zone. For the saturated zone, the porosity, n, is used since porosity is equal to the moisture content in the saturated zone.

Dispersion in the vadose zone is defined as in the saturated zone [Eqs. (9.4) and (9.5)]; however, the longitudinal direction of flow is in the vertical (z) direction. Since vadose zone transport occurs over much smaller distance scales, for example, a few meters to the water table from the ground surface, vadose zone transport is often modeled as a one-dimensional transport system, with dispersion occurring primarily in the longitudinal direction (corresponding to the vertical z axis). The magnitude of the dispersivity can be assumed to be similar to the grain size of soil; thus, α_L is typically of the order of a few millimeters in the vadose zone.

9.4.1 Contaminant Transport in the Unsaturated Zone: Analogy with Saturated Zone

With advection and dispersion in the unsaturated zone characterized above, the two processes can be coupled using mass-balance differential equations with numerical

[8] For one-dimensional flow in the z direction:

$$\frac{\partial \theta_w}{\partial t} = -\frac{\partial q}{\partial z} = \frac{\partial}{\partial z}\left[K(\theta_w) \times \frac{\partial h}{\partial z}\right] \approx \frac{\partial}{\partial z}\left[K(\theta_w) \times \frac{\partial \psi}{\partial z}\right]$$

or analytical solutions similar to those outlined for the saturated zone transport of pollutants in groundwater. A one-dimensional transport framework is used for the vadose zone. Thus the one-dimensional transport equations outlined in Section 9.2 (with analytical solutions specified in Tables 9.2 and 9.3) are suitable for use in the vadose zone as well as the saturated zone, after orienting the flow direction to the vertical axis. In the absence of reaction, the center of mass of pollutant will travel through the vadose zone at an average velocity of v, determined from Eq. (9.21); v is strongly controlled by soil moisture content. A small amount of dispersion is expected to occur, with α_L of the order of the grain size.

The concept of retardation in the vadose zone groundwater is similar to that in the saturated zone, except that the pollutant can now partition both to air as well as to soil. Both partitioning processes, together, retard pollutant velocity in pore water. Similar to the definition of the retardation factor for the saturated zone in Eq. (9.11), we obtain

$$\mathsf{R} = \frac{\text{Total pollutant mass in air, water, and soil}}{\text{Pollutant mass in water alone}} = \frac{\theta_w + \rho_b K_d + \theta_a K_{aw}}{\theta_w} \tag{9.22}$$

The treatment of retardation and reaction is the same as in the case of the saturated zone. Thus, by learning the analogous parameters representing contaminant transport in the saturated and unsaturated zones, presented in Table 9.5, one can develop a consistent framework for solving subsurface pollutant transport problems. Analytical solutions for various vadose zone contamination problems are reviewed by Van Genuchten (1981). Example 9.6 illustrates pollutant transport in the vadose zone.

EXAMPLE 9.6

Compute the retardation factor in water, R, for benzene in the vadose zone. The moisture content of the soil is uniformly 0.2 and the air content, 0.4. Assume $K_d = 0.5\ \mathrm{L_w/kg_{soil}}$ and the bulk density of the soil is 2 g/cm^3. What is the average travel velocity of benzene through this soil if the water infiltration rate, q, is 1 m/d, and is assumed steady.

Solution 9.6 From Table 2.3,

$$\log K_H^{\text{benzene}} = 0.74 \frac{\mathrm{L \cdot atm}}{\mathrm{mol}} \quad \text{so} \quad K_H^{\text{benzene}} = 10^{0.74} = 5.5 \frac{\mathrm{L \cdot atm}}{\mathrm{mol}}$$

$$K_{aw}^{\text{benzene}} = \frac{K_H^{\text{benzene}}}{RT} = \frac{5.5 \dfrac{\mathrm{L \cdot atm}}{\mathrm{mol}}}{\left(0.0821 \dfrac{\mathrm{L \cdot atm}}{\mathrm{mol \cdot K}}\right)(273 + 25)K} = 0.22$$

$$\mathsf{R} = 1 + \left[\frac{\rho_b K_d + \theta_a K_{a/w}}{\theta_w}\right] = 1 + \left[\frac{2 \dfrac{\mathrm{g}}{\mathrm{cm^3}}\left(\dfrac{\mathrm{kg}}{1000\ \mathrm{g}}\right) 0.5 \dfrac{\mathrm{L}}{\mathrm{kg}}\left(\dfrac{1000\ \mathrm{cm^3}}{\mathrm{L}}\right) + 0.4(0.22)}{0.2}\right] = 6.4$$

$$v_s^{\text{benzene}} = \frac{v}{\mathsf{R}} = \frac{q}{\theta_w \mathsf{R}} = \frac{1\ \mathrm{m/d}}{0.4 \times 6.4} \cong 0.39\ \mathrm{m/d}$$

TABLE 9.5 Comparison of Transport Parameters in Groundwater Across Saturated and Unsaturated Zone

Parameter	Saturated Zone	Unsaturated Zone
Porosity, n Moisture content, θ_w Air content, θ_a	$n = \theta_w = \text{constant}$ $\theta_a = 0$	$n = \theta_w + \theta_a$ $\theta_w = \theta_w(z)$ Moisture content varies with depth and is a function of pressure head, ψ.
Fluid pressure, P	$P = \rho g \psi \qquad P > P_{\text{atm}}$	$P < P_{\text{atm}}$
Pressure head, ψ	$\psi > 0$	$\psi < 0$ suction head
Hydraulic head, h	Measure with a piezometer (open pipe with a well point at base) $h = z + \psi$	Measure with a tensiometer (tube with a gage at top and a porous cup at base) that measures pressure head, ψ $h = z + \psi$
Hydraulic conductivity, K	$K = K_s = \dfrac{k\rho g}{\mu} = \text{constant}$	$K = K(\theta_w)$ K is variable, a function of moisture content
Darcy velocity, q	$q = -K\dfrac{dh}{dL}$	$q = -K(\theta_w) \times \left[1 + \dfrac{d\psi}{dz}\right]$
Average groundwater pore velocity, v	$v = \dfrac{q}{n}$	$v = \dfrac{q}{\theta_w}$
Retardation, R, of pollutant in water.	$\mathsf{R} = \dfrac{n + \rho_b K_d}{n}$	$\mathsf{R} = \dfrac{\theta_w + \rho_b K_d + \theta_a K_{\text{aw}}}{\theta_w}$
Retardation, R′, of pollutant in air.	$C_{\text{water}} = \dfrac{M_{\text{total}}/\mathsf{R}}{V_w}$	$\mathsf{R}' = \dfrac{\theta_a + \dfrac{\theta_w}{K_{\text{aw}}} + \dfrac{\rho_b K_d}{K_{\text{aw}}}}{\theta_a}$
Contaminant flow rate in water $= v_s = \dfrac{v}{\mathsf{R}}$		$C_{\text{water}} = \dfrac{M_{\text{total}}/\mathsf{R}}{V_w}$ $C_{\text{air}} = \dfrac{M_{\text{total}}/\mathsf{R}'}{V_a}$

The analysis above has considered pollutant transport in water only, whether it be in the saturated or the unsaturated zones. For very volatile pollutants, however, volatilization through air in the vadose zone can contribute as well to contaminant transport. Jury et al. (1990) present an analysis of pollutant transport in the vadose zone incorporating movement of pollutants in air via diffusion, coupled with water-phase transport by advection–dispersion–retardation. When the air phase is in motion, such as in soil vapor extraction remedial operations at contaminated sites, advection in air in the vadose zone also needs to be considered. Airflow rates in soil are modeled

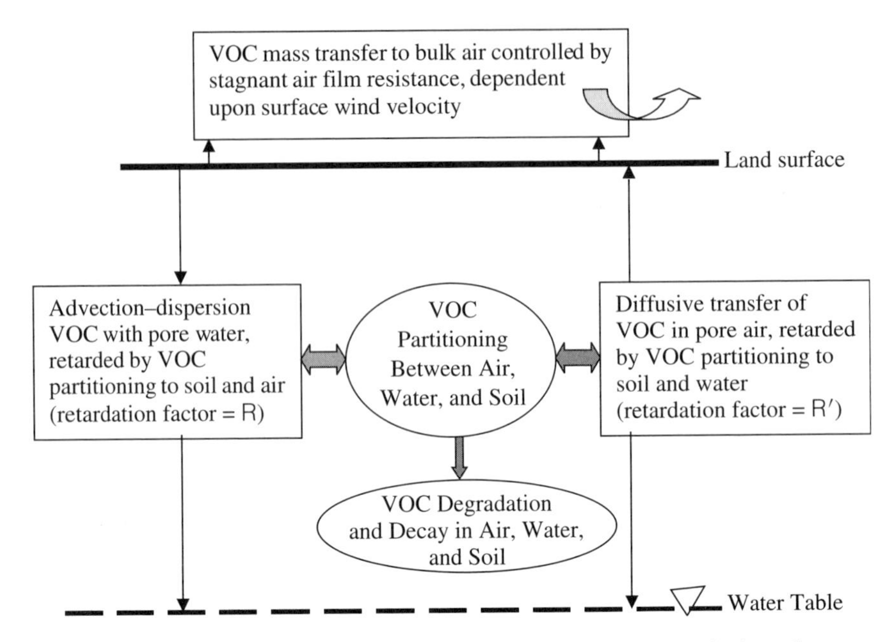

Figure 9.11 Depiction of important processes that govern VOC transport in the vadose zone.

using Darcy's law, driven by air pressure gradients. The media conductivity for air is determined from Eq. (9.1d), employing the fluid properties of air. The movement of contaminants in air, whether by diffusion or advection, will appear to be retarded by chemical partitioning to water and to soil, represented by the retardation factor for air, R'. Section 4.5 presents a detailed description of effective diffusion rates in soil–water and soil–air, retarded by the retardation factors, R and R', respectively. Advection due to water flow in vadose zone is coupled with the diffusion processes described in Chapter 4, as shown in Figure 9.11, to describe the overall movement of pollutants in the vadose zone (Jury et al., 1990). A soil behavior assessment model that considers all of these processes has been developed for spreadsheet implementation on Excel (Dawson, 1995) and can be used for screening the behavior of chemicals in soil.

EXAMPLE 9.7

Write a mass-balance differential equation for volatile organic compound (VOC) transport in the vadose zone incorporating all the processes shown in Figure 9.11.

Solution 9.7 We assume (as shown in Fig. 9.11) that VOC advection occurs due to vertical movement of groundwater in the unsaturated zone. Dispersion in pore water is assumed to be slow compared to molecular diffusion in pore air. Molecular diffusion in pore air and pore water is together described employing the effective diffusion coefficient for soil, $D_{m,s,\text{eff}}$, as defined in Section 4.5. First-order degradation of contaminant from the total soil matrix volume is assumed to occur with a rate constant, k. Mass-balance differential equations are written in terms of the contaminant concentration in the total soil matrix, C_{sT}, including soil particle, pore air, and pore water. Consistent with the definition used in Chapter 4, C_{sT} can be computed from contaminant concentrations in pore water as:

$$C_{s,T} = \frac{\text{Mass of chemical in soil, pore water, and pore air}}{V_{\text{bulk}}}$$

$$= \frac{M_{\text{soil}} K_d C_{\text{sw}} + V_w C_{\text{sw}} + V_a K_{\text{aw}} C_{\text{sw}}}{V_{\text{bulk}}}$$

$$\Rightarrow C_{s,T} = (\rho_b K_d + \theta_w + \theta_a K_{\text{aw}}) \times C_{\text{sw}} = (\mathsf{R}\theta_w) \times C_{\text{sw}}$$

The rate of change in total chemical mass within a bulk volume of soil equals net advective input due to pore water movement plus net diffusive input due to effective molecular diffusion through bulk soil [including pore air and pore water; see Eqs. (4.18) and (4.19)] minus the removal by reaction in bulk soil.

$$(\Delta x\ \Delta y\ \Delta z)\frac{\partial C_{s,T}(z,t)}{\partial t} = [v \times C_{\text{sw}}(z,t) - v \times C_{\text{sw}}(z+\Delta z,t)] \times (\theta_w\ \Delta y\ \Delta x)$$
$$+ \left[D_{m,s,\text{eff}} \times \left(\left.\frac{-\partial C_{s,T}(z,t)}{\partial z}\right|_z - \left.\frac{-\partial C_{s,T}(z+\Delta z,t)}{\partial z}\right|_{z+\Delta z}\right)\right] \times (\Delta y\ \Delta x)$$
$$- (\Delta x\ \Delta y\ \Delta x) \times k \times C_{s,T}(z,t)$$

Advection occurs in pore water alone and hence the advection term employs the VOC concentration in pore water, C_{sw}, along with the water velocity v, and the water-filled pore cross-sectional area given by $\theta_w\ \Delta y\ \Delta x$. Reaction and diffusion terms are based on total chemical concentration in bulk soil, and hence use the bulk volume $(\Delta x\ \Delta y\ \Delta z)$ and total cross-sectional area $(\Delta x\ \Delta y)$, respectively. Substituting the relationship between $C_{s,T}$ and C_{sw}, and dividing both the lhs and rhs by the bulk volume of soil $(\Delta x\ \Delta y\ \Delta z)$, yields

$$\frac{\partial C_{s,T}(z,t)}{\partial t} = -\frac{v}{\mathsf{R}}\frac{\partial C_{s,T}(z,t)}{\partial z} + D_{m,s,\text{eff}}\frac{\partial^2 C_{s,T}(z,t)}{\partial z^2} - kC_{s,T}(z,t)$$

At steady state:

$$\frac{\partial C_{s,T}(z,t)}{\partial t} = 0 \Rightarrow \frac{v}{\mathsf{R}}\frac{\partial C_{s,T}(z,t)}{\partial z} + kC_{s,T}(z,t) = D_{m,s,\text{eff}}\frac{\partial^2 C_{s,T}(z,t)}{\partial z^2}$$

Examples 9.2 to 9.6 demonstrated the use of analytical models in describing contaminant transport in the subsurface. Input parameter values in these examples, such as the groundwater flow rate, soil conductivity or dispersivity, are specified in the problem statement. When developing models for application at field sites, whether analytical models or numerical models, many of the input parameter values are not known a priori and must be estimated from laboratory and/or field tests. Parameter estimation techniques for subsurface contaminant transport models are described in Section 9.5.

9.5 PARAMETER ESTIMATION

Groundwater contaminant transport parameters can be estimated in several ways:

- From literature reports, empirical approximations, and rules of thumb
- From laboratory tests in bench-scale systems
- From field tracer tests at a single point (well)—single-well injection tests
- From field tracer recovery tests—multiple-well injection and recovery systems

Many of the parameter estimation techniques are based on the theoretical models for groundwater and contaminant transport presented in Sections 9.2 to 9.4. Each of these methods is briefly described below.

Literature Reports, Empirical Approximations Soil moisture retention and conductivity curves, soil porosity, and permeabilities for different soil types and textures are provided in the literature (Gupta and Larson, 1979; Cosby et al., 1984; Saxton et al., 1986). These parameters can be combined with measured hydraulic head for the site of interest to estimate v, employing Darcy's law and Richards equation, for the saturated and unsaturated zones, respectively. The average groundwater velocity v provides a first estimate of tracer advection parameters in the subsurface. Rules of thumb for tracer dispersion are as follows. In small-scale, homogeneous systems, that is, in laboratory columns or for short travel distances across the vadose zone, the longitudinal dispersivity, α_L, is approximately of the order of the median soil grain size. For larger scale problems, that is, regional-scale transport in the saturated zone, a first approximation of α_L is of the order of one-tenth the distance scale of the problem; α_T and α_Z are assumed to be one-tenth of α_L. Sorption parameters can be estimated from estimates of K_{ow} and therefore K_d for the pollutant of interest, as described in detail in Chapter 3. First-order degradation constants can be estimated from reported half-life data (Howard et al., 1991) as shown in Chapter 2.

Laboratory Tests Several laboratory tests are available for the determination of the hydraulic conductivity, porosity, and moisture retention curves for soils. Soil cores from the field are studied in flow-through columns in the lab with a constant pressure drop applied across the column (constant head permeator) or with variable head across the column as water levels drop at the head of the vertical column (falling head permeator). Freeze and Cherry provide a comprehensive description of methods for obtaining point estimates of hydraulic conductivity from laboratory tests on soil cores (Freeze and Cherry, 1979, pp. 336). Soil porosity is measured in the lab by saturating a known volume of soil with water, then drying it in an oven at 105°C to evaporate all the water, and measuring the weight change that can be then correlated to the volume of the void spaces. Along with measurements of porosity and conductivity, site-specific head gradient data are needed to estimate v. The three-point method for simple estimation of head gradients in the saturated zone was discussed in Example 9.1. Advection in the vadose zone is more difficult to measure since moisture content affects both the pressure head as well as conductivity. Typically, soil moisture retention–permeability curves are constructed from laboratory data, fitted to models such as those summarized in Table 9.4, and these models are then used to compute advection rates in the field based on infiltration rates and the presumed texture and properties of the soil. Saturated zone dispersivity in laboratory tests is quantified through column studies with continuous injection of tracer from one end of the column and analysis of temporal changes in effluent tracer concentration, as was illustrated in Example 9.3.

Field Tests All the above laboratory methods with soil cores provide microscale point estimates of aquifer conductivity, porosity, water flow rates, and dispersivity that are strictly valid only for the small cores that were tested. Single-well or multiple-well field tests are often conducted to obtain macroscopic parameter values representative of a larger area surrounding these wells. Prior to conducting the field well tests, geophysical data are collected from uncased bore holes dug into the ground at the site. The geophysical logs measure naturally occurring electric potential differences between vertical layers at the site, as well as the resistivity change versus depth. The logs are interpreted to yield information on the geologic strata present at the site. The geophysical data are used to identify primary aquifer units, within which specific well water pumping tests and/or tracer tests are conducted as summarized in Table 9.6.

The tracar test methods summarized in Table 9.6 enable the user to look at pollutant patterns in space and time and quickly estimate likely aquifer transport, sorption, and degradation parameters. Due to the noise, variability, and measurement errors associated with field plume measurements, more accurate techniques for parameter estimation are needed. Field-scale (or even lab-scale) break-through curves look much different from the smooth curves shown in Figures 9.5 and 9.9. Not only the noise in field data but also the sparseness of data can make visual comparisons with the theoretical curves inaccurate. For rigorous field-scale modeling, parameter estimation occurs by inverse modeling of the tracer release experiments, that is, the observed tracer data are fitted to the appropriate model, whether analytical or numerical, with parameter values optimized by minimizing the square error between field observations and model predictions. MODFLOWP is a package that solves the inverse problem for groundwater flow problems, while UCODE is designed for contaminant transport problems (Poeter and Hill, 1998). These packages provide a powerful tool for estimating field and site-appropriate parameters from tracer test data. Thus far, we have described

TABLE 9.6 Summary of Some Important Field Tests Conducted to Measure Aquifer Transport Parameters

Description	Parameter Estimation
	Single-Well Tests
Slug Test or Bail Test Observe head recovery in a point piezometer of total length L, intake length L_i, radius r_w, and intake radius r_i, when a slug of water is rapidly added to the well (slug) or rapidly removed (bail). Head recovery is modeled as a first-order process, depending upon the unrecovered head at any time: H: head prior to test H_0: head at start of test, after bail $h(t)$: head in well at time t	Saturated zone hydraulic conductivity K, estimated as: $K = \frac{r_w^2 \ln(L_i/R_i)}{2LT_0}$ T_0 is the basic time lag obtained from analysis of the head recovery: $\frac{H - h(t)}{H - H_0} = e^{-t/T_0}$ (Estimation is valid when $L_i > 8R_i$, for a well in a homogeneous aquifer of uniform thickness) (Hvorslev, 1951) Use Cooper's graphical method for a well open over the entire thickness of the aquifer. (Cooper et al., 1967)
Borehole Dilution Test A concentrated salt solution is introduced into a well of inner diameter d, screened over a length L. The salt concentration $C(t)$ in the well decreases exponentially over time due to dilution caused by water flow into the well.	Darcy velocity in saturated groundwater $q = -\frac{\pi d^2 L}{4At} \ln\left[\frac{C(t)}{C_0}\right]$ A is the vertical midwell cross-section area. C_0 is salt concentration in well at $t = 0$ (Halevy et al., 1967; Drost et al., 1968)
	Multiple-Well Tests
Tracer Injection–Recovery Test A known amount of tracer chemical is injected over a period of time into an aquifer, followed by pumping from that same well to recover the tracer	Since the tracer disperses in the aquifer, the recovery time will be expanded compared to the injection time, analysis of which yields an estimate of aquifer dispersivity (Fetter, 1993; pp. 69–70)

Water Pump Tests Water is pumped out of a well, and head drawdown is monitored over time at observation well(s) located at a radial distance r, out from the pumping well (see Fig. 9.3). Saturated zone conductivity K and storativity S are determined from analysis of the head drawdown equation	Inverse estimation of K and S using Eqs. (9.8) and (9.9): The Theis method and the Jacob method (Freeze and Cherry, 1979, pp. 343–350) Aquifer characteristics such as recharge and discharge boundaries, impermeable boundaries, and vertical "leakiness" must be considered for accurate determination of conductivity and storativity from pump test data.
Continuous Tracer Injection and Observation Tests A tracer is injected continuously into a two-dimensional flow field (Fig. 9.5). Tracer breakthrough is observed in a down-gradient well, located at a distance L in line with the injection well [See Fig. 9.5*b*; Eq. (9.7)]	Groundwater velocity and longitudinal dispersion coefficients are estimated from the tracer breakthrough curve observed at the downgradient well: Advection: $v = \frac{L}{t_{50}}$, where t_{50} is the breakthrough time Dispersion: $\sigma_x = \frac{(t_{84} - t_{16}) \times v}{2}$ $\quad D_L = \frac{L^2(t_{84} - t_{16})^2}{8(t_{50})^3}$
Tracer concentrations in a transverse transect of wells are also measured. [See Fig. 9.5*c*; Eq. (9.8)]	Transverse Dispersion: $\sigma_y = \frac{\Gamma}{2.34} = \sqrt{2D_T t_{50}}$
Simultaneous Instantaneous Injection of Tracer and Chemical A spike mass of a tracer chemical along with the pollutant of interest are instantaneously injected into a well and concentrations of both are observed in a series of down-gradient wells. [See Eqs. (9.13b) and (9.16)]	$v = \frac{\text{Location of center of mass of tracer}}{\text{Travel time } t}$ $\mathsf{R} = \frac{\text{Location of center of mass of tracer at time } t}{\text{Location of center of mass of pollutant at } t}$ $e^{-kt} = \frac{\text{Pollutant mass recovered in water at time } t}{\text{Initial pollutant mass released to groundwater}}$ $= \frac{M_{aq}(t)}{M_{total}/\mathsf{R}} \Rightarrow k = \frac{-1}{t} \ln\left(\frac{M_{aq}(t)}{M_{total}/\mathsf{R}}\right)$

parameter estimation at the site scale. When surface water and groundwater linkages are needed on a dynamic basis, for example, groundwater recharge in response to precipitation events, a larger, regional view of the groundwater system is needed, with appropriate linkages to surface water processes, which are described in the next chapter.

9.6 MOVEMENT OF SEPARATE PHASE POLLUTANTS IN THE SUBSURFACE

Sections 9.2 to 9.5 addressed transport of chemical contaminants that are dissolved in groundwater, their migration in the subsurface being impacted by partitioning to air and soil, as well as by chemical/biochemical reactions occurring in the subsurface. At many sites, contaminants can be present as a separate phase from water and migrate in this form. In this section we discuss two such scenarios: (1) behavior of NAPLs in the subsurface and (2) transport of colloidal particles in the subsurface.

9.6.1 Behavior of NAPLS in the subsurface

When NAPL spills occur at soil surfaces, a multiphase transport problem results with three phases—air, water, and NAPL—migrating in the subsurface. Pollutants can dissolve from the NAPL into water or volatilize to air, while also moving in bulk as the NAPL migrates. Multiphase models describing NAPL migration are described in Abriola and Pinder (1985). A discussion of NAPL migration models is beyond the scope of this book. However, a qualitative discussion on the behavior of NAPLs in the subsurface is presented in this section. NAPLs are subdivided into two categories—LNAPLs, that is, *light* NAPLs that are lighter than water such as gasoline, diesel fuel, aviation fuel, and so forth, and DNAPLs, that is, *dense* NAPLS that are heavier than water, for example, coal tar, transformer oils, chlorinated solvents (chloroform, TCE, DCE, etc.). The behavior of NAPLs in the subsurface can be understood to some extent by analogy with air–water behavior in the vadose zone. A wetting fluid is one whose contact angle with a solid surface is less than 90°, that is, the liquid does not "bead" on the soil surface like mercury does on a watch-glass or like water does on a waxed car! Instead, the wetting fluid "coats" the surface and will preferentially enter any micropores within the solid. Nonwetting fluids preferentially occupy the macropore spaces between solid particles. Wettability depends upon the surface tension between the fluid and the surface in question.

Entry of a fluid into soil pores always involves displacement of the other fluid (air or water) that was previously residing in the pores. When two fluids are in contact with a solid surface, whether they are air–water in contact with soil or NAPL–water in contact with soil, it is water that functions as the wettable fluid. The relationship between capillary pressure and the percent saturation of wetting and nonwetting fluids in the void space in soil is shown in Figure 9.12*a*. Note that the curves for water removal (drying) and water addition (wetting) are not identical and show hysteresis, much like the capillary pressure–moisture content curves discussed in Figure 9.10*a* for the vadose zone. It is also important to note that for a nonwetting fluid, that is, NAPL, to begin to displace water from soil, a nonzero capillary pressure must be

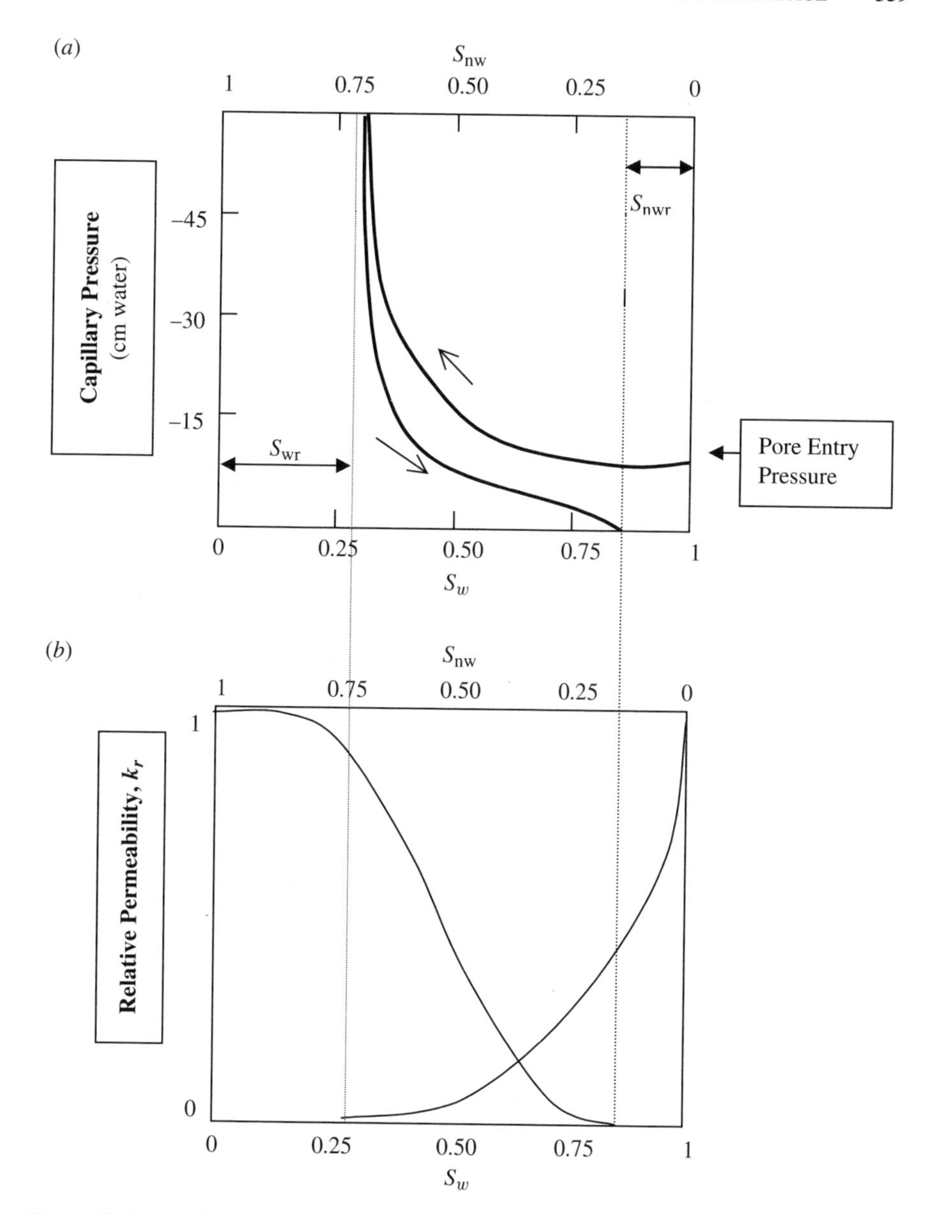

Figure 9.12 (a) Capillary pressure versus saturation curves for wetting (w) and nonwetting fluids (nw) in porous media. S_{wn} and S_{nwr} represent residual saturation of the wetting (w) and non-wetting (nw) fluids, respectively. (b) Relative permeability versus saturation of wetting and nonwetting fluids in porous media. Saturation S = Volume of fluid (w or nw)/Volume of voids.

overcome, which is termed the threshold pressure, "air" entry pressure, or imbibation pressure. The capillary pressure is inversely proportional to the throat radius of the soil pores and determines the thickness of NAPL ponded on the land surface that can create sufficient pressure to exceed the entry pressure and thence enter the soil. For very fine-grained media, the capillary pressure to be overcome can be very large (due to the small interparticle spaces), creating thick NAPL pools above fine-grained lenses in heterogeneous soil systems.

Figure 9.12*a* also indicates that, as in the vadose zone (Fig. 9.10*a*), no fluid can be completely displaced from soil by another fluid. There exists a point of irreducible saturation beyond which additional pressure gradients do not produce any flow of that fluid—the fluid is considered to be entrapped in soil at residual saturation. Thus, as NAPL migrates through the subsurface, it leaves behind areas of residual NAPL saturation, that is, zones where water has displaced NAPL "completely" to the point of irreducible NAPL saturation. Likewise in areas of NAPL accumulation or "pooling," the NAPL has displaced water "completely" from the pore structure, though water at its residual (irreducible) saturation still remains. Consequently, over the long-term, NAPL migration through the subsurface results in three saturation regimes. The first zone is one of residual NAPL saturation that typically occurs at the original entry location of the NAPL into the subsurface where water has now fully displaced migrating NAPL. The residual (irreducible NAPL saturation) NAPL is held immobile, but water flows can occur freely in this zone. A second zone of mixed NAPL and water flow occurs in the deeper subsurface where NAPL blobs and ganglia may be interconnected to facilitate NAPL flow. The maximum length and thickness of interconnected NAPL ganglions can be estimated by writing force balance equations in the horizontal and vertical axes, respectively (Hunt et al., 1988), as shown in Table 9.7. The vertical and horizontal extent of the NAPL ganglion zone is useful in estimating the extent of the NAPL source zone. Note that NAPL mobilization with water flow occurs when the

TABLE 9.7 Maximum Size of Ganglion Zone in Areas of Residual NAPL Saturation

Dimension	Force Balance Equation	Maximum Size of Interconnected Ganglia Sustained in Aquifer
Vertical	Gravity force/Area = Capillary pressure $g\,(\rho_N - \rho_w) = 2\sigma_{\text{NW}}\left(\frac{1}{r_t}\right)$	Maximum thickness of ganglion zone $L_{v,\max} \approx \frac{2\sigma_{\text{NW}}}{r_t\, g\, \lvert\rho_N - \rho_w\rvert}$
Horizontal	Water pressure drop (due to flow) = Capillary pressure $\rho_w g\, \Delta h = \frac{q\mu_w L_h}{k} = 2\sigma_{\text{NW}}\left(\frac{1}{r_t}\right)$	Maximum length of ganglion zone $L_{h,\max} = \frac{\sigma_{\text{NW}}}{q\mu_w}\frac{2k}{r_t} = \frac{1}{Ca}\frac{2k}{r_t}$ where Ca = Capillary no. = $\frac{q\mu_w}{\sigma_{\text{NW}}}$

Source: Adapted from Hunt et al. (1988).

[a]Densities of NAPL and water are given by ρ_N and ρ_w, respectively, μ_w is the viscosity of water, σ_{NW} is the NAPL–water interfacial tension, q is the Darcy velocity, k the intrinsic permeability of the aquifer material, and, r_t is the throat radius, i.e., the smallest pore throat radius in the soil.

capillary number (Table 9.7) is much larger than 5×10^{-3}, achieved primarily by lowering the NAPL–water interfacial tension by steam or surfactant injection. Finally, a NAPL accumulation zone occurs in which the NAPL has displaced all the water in the soil pore structure and forms "pools" of pure NAPL that exhibit gradient-driven flows (though some water must remain immobile at irreducible water saturation). Figure 9.13 illustrates the different NAPL saturation regimes for both L- and DNAPLs.

As seen in Figure 9.13, LNAPLs are typically entrapped in the unsaturated zone in a smear zone created at the fluctuating water table, where they are more readily accessible for removal and remediation actions. Residual LNAPL ganglia are created in both the saturated and the unsaturated zone as the combined LNAPL–water table moves up and down. Due to the "light" nature of LNAPL components, this type of pollution can be treated by air sparging techniques and bioventing processes that capitalize on the ready movement of air through the unsaturated zone.

In contrast, DNAPLs are gravity driven to migrate through the saturated zone down to bedrock, forming large pools that flow to conform with the slope of the bedrock.

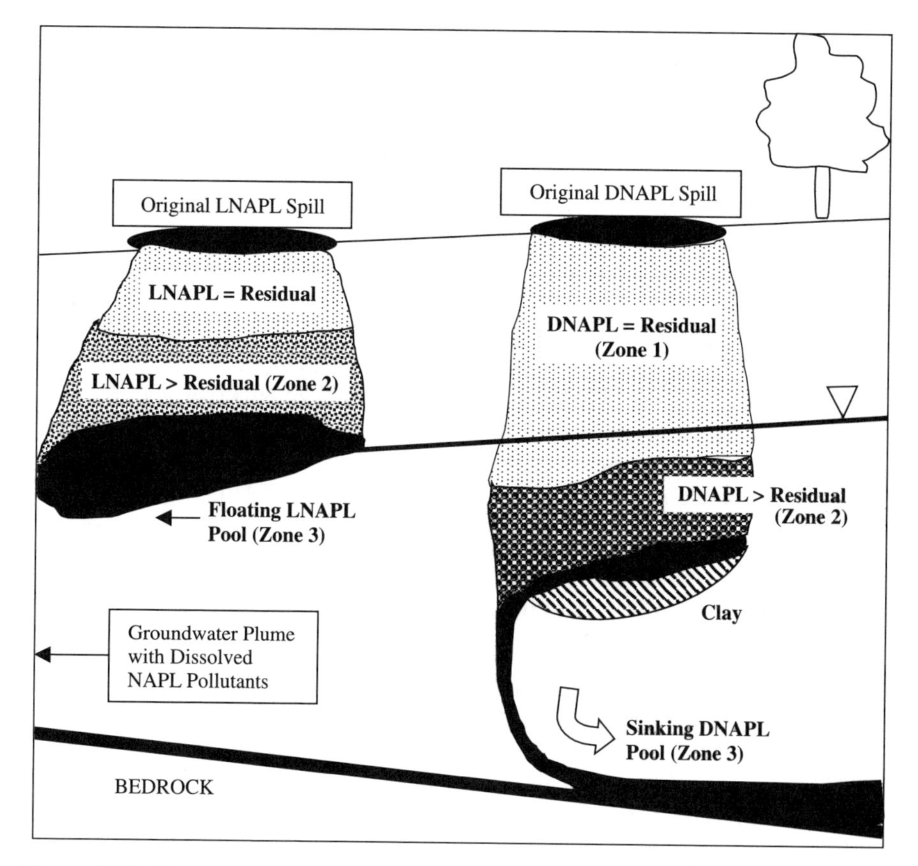

Figure 9.13 LNAPL and DNAPL migration patterns in the subsurface showing different regimes or zones of NAPL saturation.

DNAPL pools can also be formed at grain heterogeneities, as shown in Figure 9.13, where fine-grained lenses create trapped DNAPL pools (Schwille, 1988; Mercer and Cohen, 1990). Because of the uncertainty about heterogeneities in the subsurface and the slope of bedrock, we are often "blind" about the exact location and direction of migrating DNAPL pools. DNAPL pools are difficult to access or pinpoint, making remediation at DNAPL-contaminated sites very challenging (USEPA, 1994; Luthy et al., 1994). The utmost care must be exercised in placing wells and interpreting well data at NAPL-contaminated sites. Improperly installed wells can create a short circuit, promoting downward movement of DNAPL in layered aquifer systems. Wells designed to capture LNAPLs must be screened above the water table to allow mobile LNAPL to enter the well; likewise DNAPL wells should be screened carefully just above lower confining layers. The thickness of free NAPL product observed in the wells must be corrected for density effects to obtain the true thickness of mobile NAPL in the subsurface (Farr et al., 1990).

The simultaneous flow of NAPL and water in the subsurface is modeled by gradient equations, much like Darcy's law in Eqs. (9.1a) to (9.1d), with one difference. Since flow rates for each fluid would depend on the availability of interconnected pores filled with that same fluid, the intrinsic permeability k is replaced with a relative intrinsic permeability k_r for each fluid (Fig. 9.12*b*); k_r is a function of fluid saturation in soil and represents the proportion of the "total" intrinsic permeability that can be apportioned to partial saturation by one fluid. Theoretical formulation of multiphase fluid flow is described in detail in Abriola and Pinder (1985), based upon which numerical modeling of NAPL flow and migration has also been accomplished (Abrioloa and Pinder, 1985; UTCHEM, 1999). Understanding NAPL migration is important because the NAPL-contaminated zone functions as the source region from which contaminated groundwater plumes originate. In slow-moving groundwater, NAPL–water partitioning equilibrium (see Chapter 3) determines the source zone pollutant concentration, C_0. However, for NAPL source zones with insufficient contact time or contact area between the NAPL and water, the kinetics of NAPL–water dissolution (Chapter 4) can be slow. As a result, the source zone aqueous pollutant concentration is less than that expected from NAPL–water equilibria. Because NAPL components are not very soluble in water, and because the kinetics of NAPL–water dissolution can often be very slow, NAPLs can persist in the subsurface for several decades.

9.6.2 Transport of Colloids in the Subsurface

Submicron-sized particles suspended in water are called colloids. Colloids in groundwater include small organic or mineral particles generated by attrition and abrasion in the subsurface, as well as microorganisms that function as biocolloids, or living colloids. Subsurface transport of colloids has received much attention in recent years, not only because some living colloids, that is, some bacteria and viruses, are pathogens, but also because colloids can carry along harmful chemical pollutants sorbed on their surfaces. Because colloids can travel faster than the average groundwater velocity, pollutants strongly sorbed to colloids can travel longer distances than otherwise expected (e.g., Penrose et al., 1990; observed transport of plutonium at rates of 500 m/yr in the presence of colloids compared with 4 cm/yr based on sorption–retardation in groundwater). The enhanced transport of pollutants with colloids is called colloid-facilitated transport (McCarthy and Zachara, 1989; Corapcioglu and Jiang, 1993), models for which build upon fundamental colloid transport phenomena described briefly here.

The movement of colloids in groundwatcr is similar to that of dissolved solutes with some notable differences. First, because colloids are larger in size compared with 10 to 100-Å-sized solute molecules dissolved in water, colloidal particles are size excluded from traveling in very small pores in soil. This size exclusion phenomenon forces colloids to travel through larger pores in the subsurface, as a result of which average linear colloid transport velocities in groundwater can be larger than the average groundwater flow rate. Assuming spherically shaped particles of radius r_p, the average velocity of these particles in groundwater, v_p, can be estimated as (Johnson et al., 1996; Grolimund et al., 1998):

$$v_p = v \times \left[2 - \left(1 - \frac{r_p}{r_t}\right)\right]^2 \tag{9.23}$$

where v is the average groundwater velocity and r_t is the throat radius of soil pores. Johnson and Elimelech (1995) present methods for computing the average throat radius for a porous packed medium containing well-sorted soils. Equation (9.23) indicates that when particle size is much smaller than the soil pore size, the particles have full access to all the soil pores and hence travel at the same average velocity as groundwater. However, when particle size is comparable or larger than soil pore throat radii, the particle velocity v_p is greater than the average groundwater velocity v, as shown in Eq. (9.23).

As particles travel through the subsurface, they undergo dispersion, much as solutes do. However, diffusion coefficients for colloids are much smaller than those estimated for dissolved solutes. The Stokes–Einstein equation[9] is used to determine the diffusion coefficients for colloids (Johnson et al., 1996). The third primary difference between solute transport and colloid particle transport is that colloids can be captured on immobile soil surfaces. Colloids do not sorb per se to soil as do solutes but are removed from aqueous solution because of deposition on soil particles encountered in the subsurface. Assuming a first-order deposition removal rate constant represented as $k_{r,p}$, the advective–dispersive–reactive equation [Eq. (9.15a)] applied to colloids becomes

$$\frac{\partial C_p(x,t)}{\partial t} = -v_p \frac{\partial C_p(x,t)}{\partial x} + D_L \frac{\partial^2 C_p(x,t)}{\partial x^2} - k_{r,p} C_p(x,t) \tag{9.24}$$

where C_p is the concentration of colloids in groundwater. The primary challenge in characterizing the transport of colloids in groundwater is to measure and model the removal rate of colloids from water due to deposition on soil particles. Most often, colloids are assumed to be irreversibly captured on soil barriers. Filtration theory has

[9] $D_{m,w,p} = K_B(273 + T_{\text{env}})/6\pi\mu r_p$ where KB is Boltzman's constant $= 1.38 \times 10^{-23}$ J/K.

been found to be effective in describing colloid deposition during the early stages of colloid transport in a porous packed medium, during which time the first-order removal model for deposition is valid. According to filtration theory, the overall efficiency η of capture of particles on a single spherical grain of diameter, d_g, is given by the product of a mechanical collection efficiency η_0 and a collision efficiency α (Ryan and Elimelech, 1996):

$$\eta = \alpha\eta_0 \tag{9.25a}$$

The mechanical collector efficiency η_0 quantifies the fraction of particles in the path of a collector (soil grain) that are intercepted by it. The collision efficiency, α, (also known as stickiness coefficient) represents the fraction of particles intercepted by the collector that actually stick to it. Mechanical collection of particles can occur by a combination of three processes (see also Fig. 11.12): diffusion of particles onto a barrier, direct capture of particles due to interception by the barrier, and inertial impaction that occurs when the particle is unable to change its trajectory adequately to avoid encountering the barrier and hence gets "collected" on the barrier. Larger particles and/or particles with larger mass (density) will have too much momentum to bypass the barrier and hence will collect or impact with high efficiency onto the barrier. In contrast, larger barriers or slower fluid approach velocities will allow sufficient time for particle trajectory correction and barrier avoidance and will decrease the collection efficiency. These competing processes, as well as diffusional capture of particles on barriers, are represented in the form of dimensionless groups that reflect a combination of particle and barrier properties. Such dimensionless groups have been used in both groundwater and atmospheric systems to describe mechanical particle collection by barriers. In groundwater systems, semiempirical equations (Ryan and Elimelech, 1996; Rajagopalan and Tein, 1976; Spielman and Friedlander, 1974) are often used to determine the mechanical collection efficiency of a single soil grain, η_0. The overall collection efficiency η is also affected by the collision efficiency, α, which is a chemical interaction parameter that quantifies attraction or repulsion between colloids and soil surfaces. Soil surfaces typically carry a negative charge and can repel colloids, which also carry a negative charge. The repulsive effects between the two solid surfaces are minimized when the ionic strength of water is high and amplified at low aqueous ionic strengths. pH changes in aqueous solution can reverse the charge on soil surfaces, causing attractive forces that yield a higher collision efficiency. Likewise, soil zones coated with iron, aluminum, or manganese oxides can also carry a positive charge promoting collision efficiency (Ryan and Elimelech, 1996). Surface chemistry equilibrium models enable more accurate prediction of the effects of pH and ionic strength on collision efficiency, α.

In porous packed media, we are not concerned with the collection and collision efficiency of a single isolated spherical soil grain, but rather the total removal achieved by all the grains encountered by pore water within a unit bulk volume of the aquifer. Based on aquifer porosity and particle geometry considerations, the rate constant $k_{r,p}$ for particle removal from pore water by porous packed media composed of soil grains of size d_g is given as:

$$k_{r,p} = \frac{3}{2}\left(\frac{1-n}{d_g}\right) \times \eta \times v_p \tag{9.25b}$$

where n is the porosity of the packed matrix.

The overall particle collection efficiency η is constant only during the initial period of particle deposition. When attractive surface forces are in effect, favorable conditions exist for particle capture. Captured particles provide additional surfaces for further incoming particles to be captured, creating a process called "ripening of the filter," during which the deposition rate increases with time. In contrast, if colloid deposition occurs with repulsive surface forces in effect (e.g., at low ionic strength), particles are deposited only until monolayer surface coverage is attained, after which all further incoming particles are repelled from the soil surface. This phenomenon is called blocking and causes a decrease in deposition rates over time. Blocking functions and numerical integration techniques are used to track the available surfaces and the number of particles deposited to model the long-term deposition of particles, incorporating blocking (e.g., Johnson et al., 1996; Ryan and Elimelech, 1996).

The advective–dispersive equation with average particle velocity described by Eq. (9.23), appropriate diffusion computations, and first-order removal as shown in Eq. (9.24) can be used to describe the early period of colloid deposition. Particle release tests conducted in the field or in laboratory columns can also be analyzed to determine the average particle velocity, dispersivity, and removal rate constants for particles in porous media. More details on parameter estimation for colloidal transport can be found in Grolimund et al. (1998), Schijven and Hassanizadeh (2000), and Ryan et al. (1999).

When bacteria are transported in the subsurface and simultaneously feed upon organic substrates, their numbers can increase dramatically resulting in the formation of biofilms coating soil particles (Rittman, 1993). In some cases, biofilm growth has been shown to cause changes in hydrodynamic properties of porous media, including aquifer porosity and dispersivity (e.g., Taylor and Jaffe, 1990). Biofilm formation is controlled by microbial growth and subsequent bacterial transport in groundwater that is impacted by attachment to and detachment from soil surfaces. Many of the important physical processes governing bacterial transport and attachment in the subsurface have been presented in this section. These physical processes can be coupled with biogrowth processes described in Chapter 12 to describe biofilm formation in the subsurface.

9.7 NUMERICAL MODELS AND SUBSURFACE HETEROGENEITY

Previous sections have described important physical–chemical phenomena that control the movement of water and contaminants through the subsurface, with focus on four distinct subsystems:

- Chemical transport in the saturated zone: Analytical solutions were presented that describe advection, dispersion, equilibrium sorption, and first-order degradation of pollutants in idealized, homogeneous aquifers enabling broad behavior assessment of pollutants in the saturated zone. However, it was noted that heterogeneous aquifers abound in which channeling and layering significantly impact chemical transport and cause the emergence of observed "macrodispersion" in aquifers.

- Chemical transport in the vadose zone was also discussed for idealized situations, employing analogies with transport in the saturated zone. Heterogeneity and macropore effects caused by rooting plants are noted to significantly affect chemical transport in the vadose zone, along with other parameters such as evapotranspiration, unsteady-state infiltration, ponding, and other linkages with the regional surface water system.
- NAPL transport in the subsurface was discussed qualitatively in the context of multiphase fluid flow. Simultaneous transport of water, air, and NAPL must be addressed to predict migration of NAPL-derived pollutants in the subsurface.
- Transport of colloidal particles in the subsurface was described employing process models similar to those used to describe the transport of solutes in water, with emphasis on lowered particle diffusion coefficients and faster, size-excluded particle velocities in groundwater, coupled with deposition of mobile particles on immobile soil grains encountered in the subsurface.

Numerical models have been developed to describe the transport of dissolved and separate-phase pollutants in heterogeneous, multiphase subsurface environments that are closely linked with the regional surface water system. No model to date has integrated all the above subsystems together, in their entire complexity. However, several models address, separately and to varying degree, the four pollutant transport subsystems encountered in the subsurface: solute transport in the saturated zone, solute transport in the unsaturated zone, NAPL systems, and colloid transport systems. A summary classification of some widely used numerical modeling packages that address these applications is provided in the supplementary materials for Chapter 9 at www.wiley.com/college/ramaswami.

Two case studies are presented to demonstrate the use of computer models in understanding contaminant behavior in the subsurface environment. No computer model can truly and uniquely represent all physical, chemical, and biological processes occurring at a site, particularly when there are many unknowns pertaining to site-specific source release rates, macrodispersion, and degradation rates. However, a site conceptual model that includes the most relevant processes and is calibrated with current field observations presents a useful tool for evaluating the effectiveness of future risk management actions taken at the site. The primary objective of this section is to give the reader an overview of the science and the art involved in developing a site conceptual model, implementing the model numerically at the appropriate scales and dimensions of interest, and calibrating the model based on field observations. The presentation of the two case studies in this chapter is limited to conceptual model development and water and solute/contaminant transport simulations; the addition of complex reactive processes to contaminant transport models is reserved for discussion in Chapter 12.

Case Study 1

The first case study examines polyaromatic hydrocarbons (PAHs) and pentachlorophenol (PCP) contamination of groundwater at an abandoned creosoting site in Conroe, Texas. The description is excerpted from Borden et al. (1986). Unknown quantities of wastes from wood-preserving operations were disposed from 1946 to 1972 into two unlined ponds on site property as shown by the hatched regions in Figure 9.14*a*. In addition to organic contaminants such as naphthalene, anthracene, and PCP, inorganic compounds that originated from the raw wood as well as chloride ion are found in the groundwater plume generated from the waste ponds. The chloride is believed to be a contaminant or breakdown product of PCP. Chloride ion, naphthalene, and PCP were chosen as the principal representative chemicals within the contaminated plume. Field measurements showed concentrations of all organic chemicals decreasing along the water flow direction on the site. Naphthalene concentrations were consistently found to be very low in areas with high dissolved oxygen in water, suggesting that the availability of oxygen limits the rates of degradation of naphthalene at the site. Microbial populations at the site were well-characterized with about 10^6 to 10^7 cells per gram of aquifer material. Laboratory studies of microbes obtained from the vicinity of the waste plume indicated that they were acclimated to the range of organic compounds found in groundwater, being able to degrade them fairly rapidly (to less than 20 ppb within 20 days) in aerated laboratory microcosms. Systems maintained under nitrogen showed greatly lowered degradation rates suggesting that oxygen supply in the subsurface may be the primary rate-limiting step controlling in situ biodegradation of these chemicals. The objective of the modeling exercise was to characterize the transport and attenuation, attributed to biodegradation, of the organic chemicals found in groundwater at the site.

Water Flow Parameters The water flow parameters for the site were determined from geologic logs and well pumping tests. Geologic logs showed a surficial layer of sandy clay, beneath which is a shallow unconfined water-bearing sand aquifer. The surface sand–clay layer reduced the rate of infiltration of wastewater from the ponds into the underlying unconfined sand aquifer. Beneath the unconfined sand aquifer lies a semicontinuous, impermeable clay that overlies deeper sand units. Much of the pollutant transport occurs in the unconfined sand layer. The water flow in this zone is to the south as shown in Figure 9.14*a*, with a gradient of 0.6%. Soil porosity was measured to be approximately 29%. Soil conductivity was determined from slug tests conducted in selected wells at the site. An average hydraulic conductivity of $K = 0.74$ m/day was used for much of the site, except a small low-permeability zone near SW-2. Determination of aquifer conductivity, porosity, and head gradient provided the input parameters for modeling water flow at the site with constant head boundaries at the northern and southern edges of the site. Simulated groundwater heads matched well with observations, consistent with the movement of chloride onsite at an average rate of about 5 m/yr.

Reasonable source zone loading rates for the two ponds were assumed to simulate the releases of chloride and organic contaminants from 1946 to 1972 and thereafter

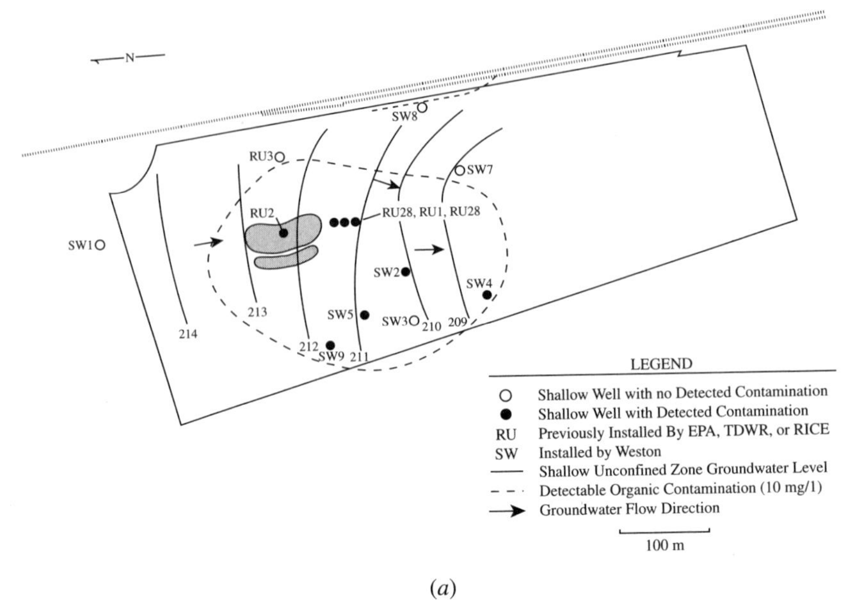

(*a*)

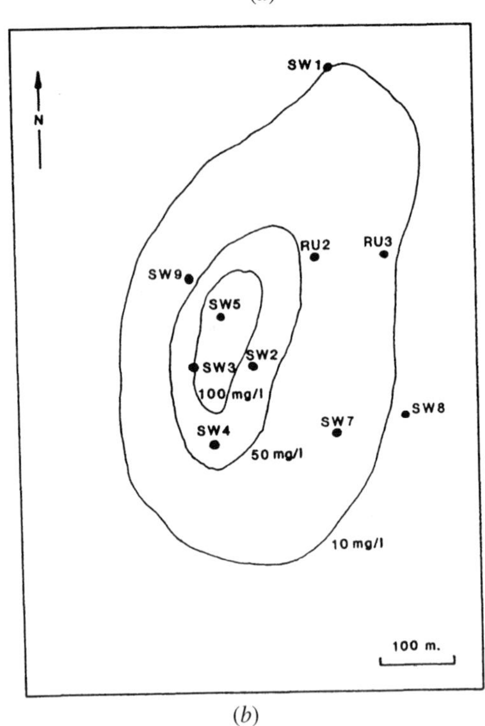

(*b*)

(*continued*)

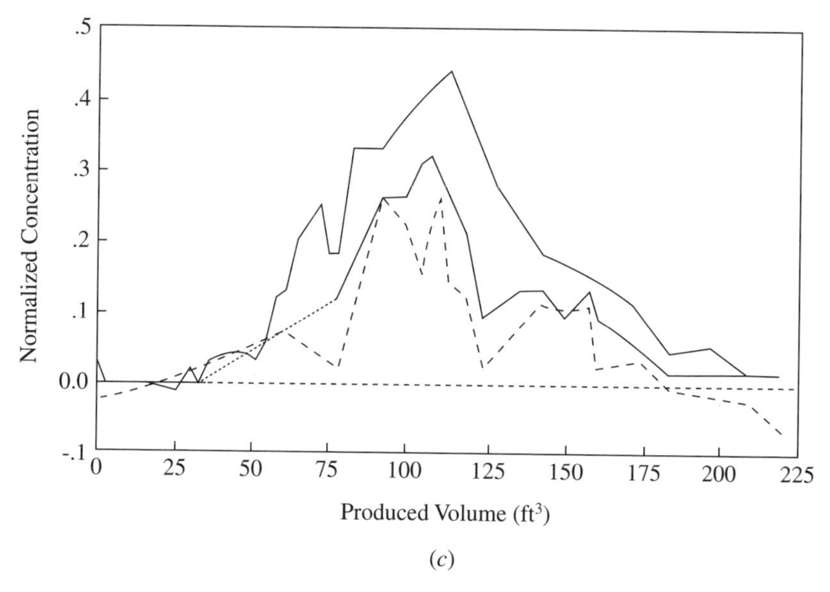

Figure 9.14 (*a*) United Creosoting Co., Inc. site map showing groundwater flow and contamination zone (from Borden et al., 1986), (*b*) simulated chloride plume in the contamination zone (from Borden et al., 1986), (*c*) comparison of normalized concentration distributions for chloride (——), pDCB(· · · · ·), and naphthalene (- - - -), in RU 28 (from Borden and Bedient, 1987).

(Table 9.8a). Both ponds were together assumed to provide 5 m^3/day recharge water to the shallow aquifer. The concentrations of chloride ion and organic pollutants were specified in the assumed discharges from the two ponds. Higher concentrations were assumed in the 1946 to 1972 period when the waste ponds were in active use with lower concentrations for the post-1972 period. Because this site had a chloride source and resulting plume, tracer transport properties could be estimated by assessment of the chloride plume already present on site. With the assumed chloride ion source loading specified in Table 9.8a and with the average groundwater flow characterized as described above, aquifer dispersion parameters were estimated by fitting appropriate longitudinal and transverse dispersivity values to match the chloride plume already present at the site. Fitted values of $\alpha_L = 9.1$ m and $\alpha_T = 1.8$ m were obtained yielding a simulated chloride plume shown in Figure 9.14*b*. Simulated and measured chloride concentrations matched fairly well as shown in Table 9.8b. These estimates of dispersivity were subsequently used in modeling the transport of other contaminants in the aquifer.

Retardation and degradation factors must be coupled with advection–dispersion to describe the overall transport of naphthalene and PCP in the aquifer. Retardation and degradation parameters were determined by conducting a controlled pulse injection and recovery test at an uncontaminated area of the site—water containing known amounts of chloride ion, naphthalene, and dichlorobenzne was injected at one

TABLE 9.8 Simulated and Observed Data from the Conroe Site

(a) Source Zone Characterization

Source Parameters (1947–1972)	Small Pond	Large Pond
Water injection rate, m^3/day	0.9	4.1
Chloride conc. (mg/L)	270	45
Oxygen conc. (mg/L)	0.0	0.0
Hydrocarbon conc. (mg/L)	45	45
Source Parameters (1972–1977)	**Small Pond**	**Large Pond**
Water injection rate, m^3/day	0.09	0.41
Chloride conc. (mg/L)	10	10
Oxygen conc. (mg/L)	0.0	0.0
Hydrocarbon conc. (mg/L)	45	45
Source Parameters (1977–1986)	**Small Pond**	**Large Pond**
Water injection rate, m^3/day	0.01	0.04
Chloride conc. (mg/L)	10	10
Oxygen conc. (mg/L)	0.0	0.0
Hydrocarbon Conc. (mg/L)	45	45

(b) Chloride Concentration (mg/L)

Well Location	Observed	Simulated	Error (mg/L)
SW-1	15	10	−5
SW-2	77	83	6
SW-3	84	96	12
SW-4	71	66	−5
SW-5	152	152	0
SW-7	19	30	11
SW-8	26	10	−16
SW-9	41	44	3
RU-2	59	49	−10
RU-3	8	17	9
			(rms error = 9)

Source: Adapted from Borden et al. (1986).

well, and the recovery of these chemicals was monitored at an adjacent recovery well (Borden and Bedient, 1987). Normalized recovery for all three chemicals is shown in Figure 9.14*c*. The peaks of all three curves appear at approximately the same time, suggesting that retardation is not significant in this aquifer, confirmed by the very low f_{oc} (<0.1%) of the aquifer materials yielding low K_d values for naphthalene and PCP. However, the percent mass recovery for naphthalene (at 35%) and PCP (at 52%) is much less than the recovery for chloride ion, which was close to 100%. A combination of laboratory microcosm tests, along with the observed degradation of these chemicals in the field, suggests that significant microbial degradation of these chemicals could be occurring in the aquifer. The authors hypothesize that oxygen supply controls the rate of biodegradation of organic chemicals at the site. An early version of the BIOPLUME (USEPA-CSMOS) model was used to simulate the biodegradation of naphthalene and PCP at the site. The simulation results were then used to evaluate the impact of two alternative remedial options at the site. The BIOPLUME simulations for the Conroe site are described in Chapter 12, focusing on the reaction mechanisms operational at the site.

Case Study 2

In addition to many of the challenges in source zone characterization highlighted in Case Study #1, subsurface media heterogeneities can affect the patterns of contamination predicted in any simulation, as well as the results of parameter estimation from tracer test data. If detailed data on the spatial distribution of heterogeneous media properties are available, these can directly be input into the model by specifying unique and different parameter values within each cell of the model mesh. When a rich spatial database is lacking at the level of detail needed for the simulation, the probability distribution and spatial autocorrelation structure of soil properties can be determined from sampling data and employed in stochastic modeling of the aquifer. This approach is described in this second case study conducted at a large Superfund site in California. The discussion is excerpted from Sohn (1998) and Sohn et al. (2000).

The site shown in Figure 9.15*a* includes several industrial and federal facilities that have contributed to a groundwater plume that extends over an area ~3.7 km × 0.9 km. The source areas are denoted as SA-1, SA-2, and SA-3, with chlorinated organic pollutants (TCE and DCE) being detected widely, along with less widespread contaminants such as PCE, fuel constituents, and metals. The chlorinated compounds were detected at levels close to solubility, suggesting the occurrence of DNAPLs at the site. The site contains extensive paved areas that limited subsurface sampling. Site geology indicates that a two-layer model may be sufficient to describe contaminant transport—an upper unconfined aquifer and a lower confined aquifer (Fig. 9.15*b*). Regional water flow is from south to north, toward the bay. A highway cuts across the site in the east–west direction, north of which are airport runways. A 50 m × 50 m mesh was used to model both aquifers, using MODFLOW (McDonald and Harbaugh, 1988) for water flow simulation and MT3D (Zheng, 1992) for contaminant transport modeling. Since hydraulic conductivity data were not available for all the mesh points, the spatial correlation of the hydraulic conductivity field was first determined from analysis of spatial conductivity data gathered from 800 and 600 monitoring wells located in the upper and lower aquifers, respectively. In Section 7.5.5, we discussed variogram structures that can be used to describe the spatial correlation of soil properties, in this case hydraulic conductivity. Note, the variogram function [Eq. (7.51)] is the complement of the spatial autocovariance function [Eq. (7.53)]. Using the field measurement of hydraulic conductivity at 1400 locations, variograms [Eq. (7.51)] were constructed first assuming that the site was isotropic with respect to spatial correlation (Figs. 9.16*a* and 9.16*b*), that is, the horizontal and vertical hydraulic conductivity was correlated to the same degree in all directions (x, y, and z). Directional variograms were then constructed exploring correlation along the

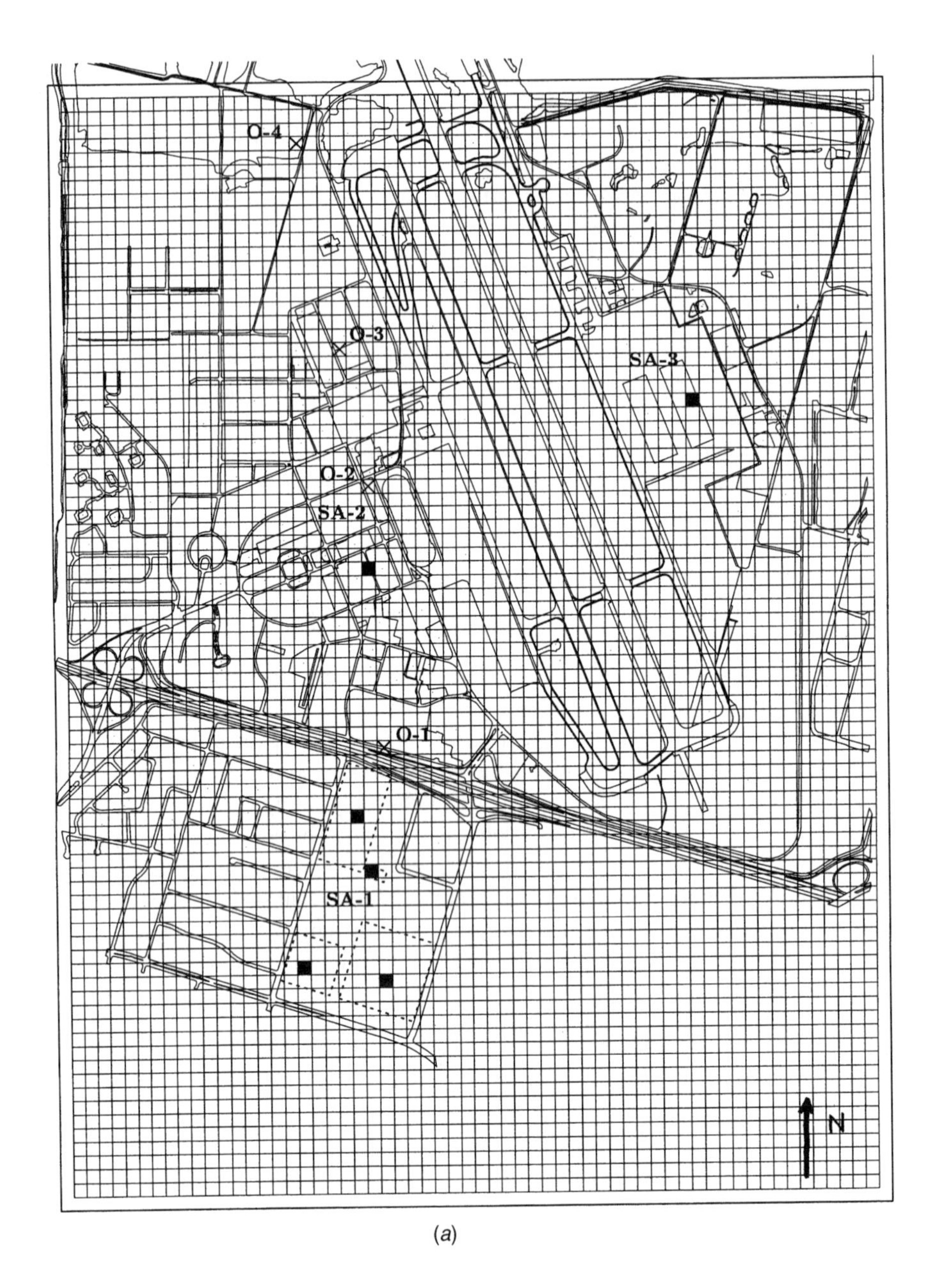

(*a*)

(*continues*)

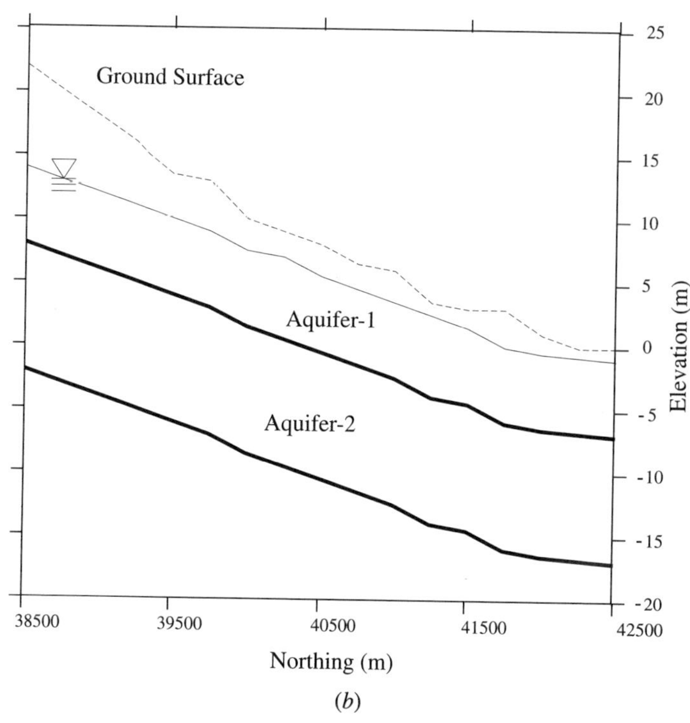

(*b*)

Figure 9.15 Site map of the Superfund Site in California (from Sohn, 1998): (*a*) Plan view showing the study area oriented in the N-S direction, with a 50 m × 50 m grid. Source areas of contamination are delineated as SA1, SA2, and SA3. A highway cuts across the site in the east–west direction, north of which is an airport with runways. Groundwater flows toward the north. (*b*) Elevation view showing a two-layer aquifer system, with an upper unconfined aquifer (aquifer 1) and a lower confined aquifer (aquifer 2).

longitudinal, transverse, and vertical dimensions. Figures 9.16*c* and 9.16*d* show correlation of horizontal and vertical conductivity along the longitudinal (south–north) direction. A comparison of the range of the variograms shows a distinct anisotropy in the spatial correlation structure, that is, there is a greater scale of correlation in hydraulic conductivities in the longitudinal (south–north) direction. Exponential variogram models were fitted to the data points on the variograms, yielding parametrized variogram functions that were then used in a sampling procedure called kriging (see Section 7.5.5) to generate many (1000) realizations of the hydraulic conductivity field at the site. The mean and standard deviation of the hydraulic conductivity at each mesh point was thus obtained from the 1000 simulations. Mean horizontal and vertical conductivity as shown in Figures 9.17 and 9.18, developed from the anisotropic variograms shown in Figures 9.16*c* and 9.16*d*, respectively. Note the distinct channels that emerge in these figures, particularly for vertical conductivity, that are indicative of preferential flow paths in the aquifer and demonstrate the importance of considering the anisotropic spatial correlation of hydraulic conductivity at this site.

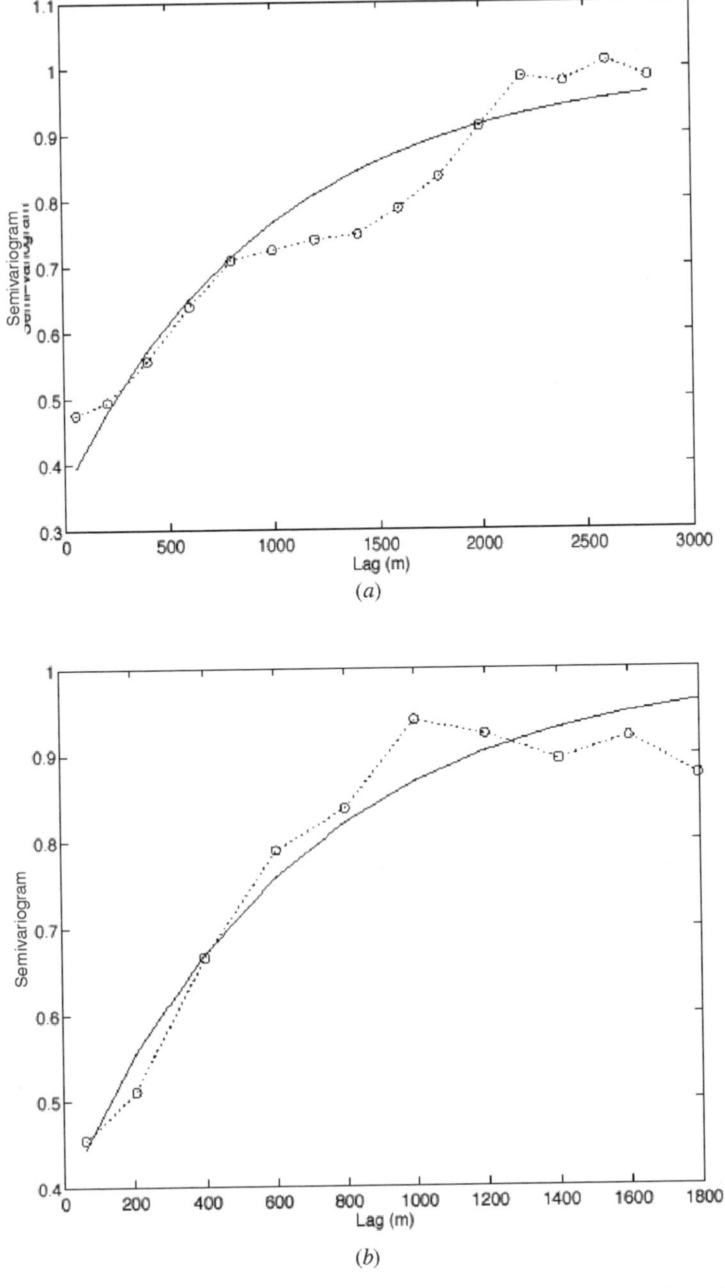

Figure 9.16 Variograms showing the spatial correlation of hydraulic conductivity at the California Superfund site (from Sohn, 1998). (*a*) Isotropic variogram for horizontal hydraulic conductivity [$\ln(K_x)$; m/day] with a fitted range of 3000 m. (*b*) Isotropic variogram for vertical hydraulic conductivity [$\ln(K_z)$; m/day] with a fitted range of 1950 m. *(continued)*

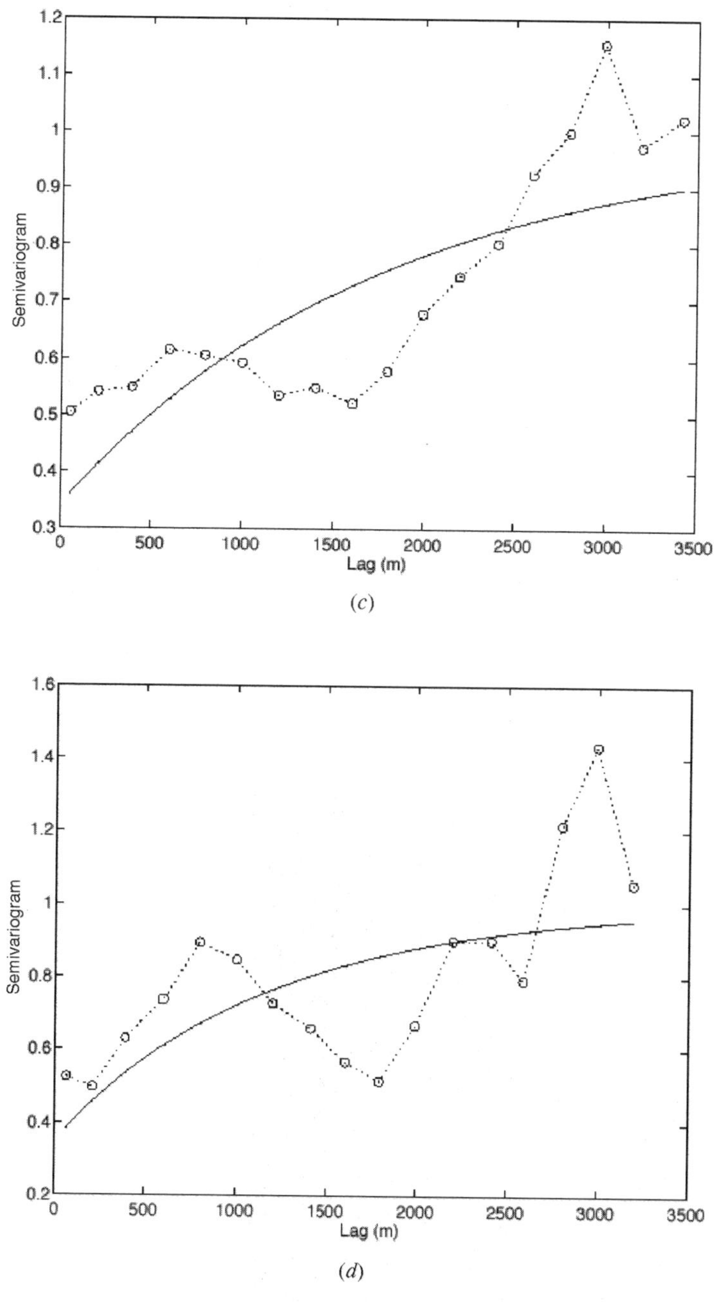

Figure 9.16 (*Continued*) (*c*) Anisotropic variogram for horizontal hydraulic conductivity [ln(K_x); m/day] in the longitudinal (N–S) direction with a fitted range of 5400 m. (*d*) Anisotropic variogram of vertical hydraulic conductivity [ln(K_z); m/day] in the longitudinal (N–S) direction with a fitted range of 3600 m.

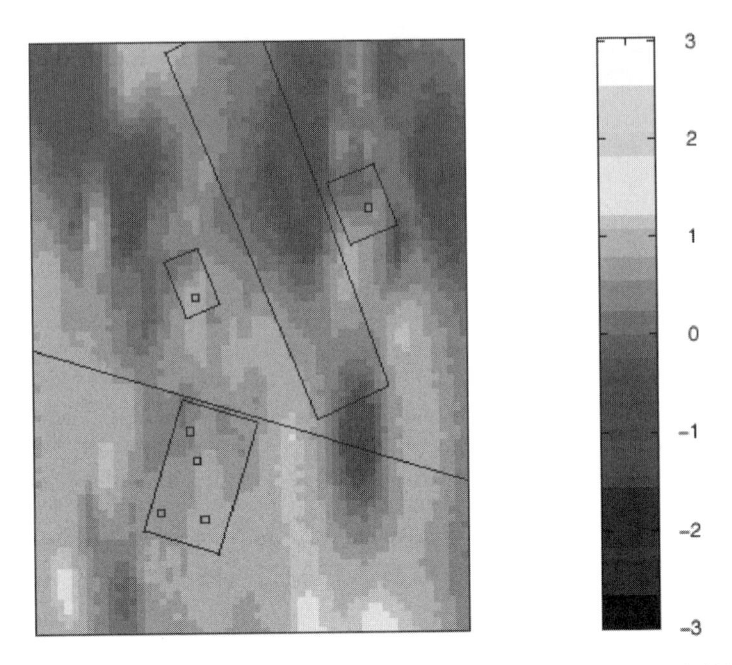

Figure 9.17 Mean of 1000 realizations of the horizontal hydraulic conductivity field [ln(K_x); m/day] at the California Superfund site generated from anisotropic variograms (From Sohn, 1998). Light areas represent regions of high hydraulic conductivity through which flow and pollutant transport tends to be channelled.

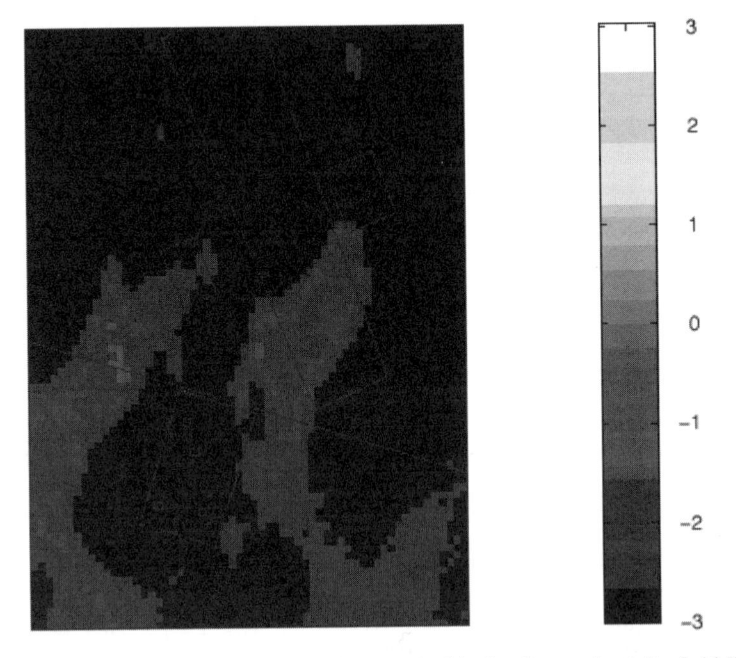

Figure 9.18 Mean of 1000 realizations of the vertical hydraulic conductivity field [ln(K_z); m/day] at the California Superfund site generated from anisotropic variograms (from Sohn, 1998). In this figure *darker* areas represent regions of higher hydraulic conductivity through which vertical aquifer connectivity occurs.

Case study 2 thus illustrates the use of available modeling packages such as GEOEAS (USEPA-CSMOS) and GSLIB (Deutsch and Journel, 1992) to capture and simulate spatially correlated hydraulic conductivity fields.

When the spatial covariance function of hydraulic conductivity is described mathematically from analysis of field data, Fourier transformations can also be applied to determine the spectral density function of the soil property (in this case, K or $\ln K$). Water flow and contaminant mass-balance differential equations are written in stochastic form, analysis of which yields quantitative assessment of macrodispersion in the aquifer (e.g., Gelhar, 1986). In an alternative approach, a "turning bands" method is used to generate an ensemble of aquifers with spatially correlated media properties (Tompson et al., 1989) that mimic the observed correlation in the field. Water flow and contaminant transport is then simulated numerically in Monte Carlo simulations (e.g., MacQuarrie and Sudicky, 1990). Although much work has been conducted in the theory and simulation of stochastic processes, the primary challenge is that of collecting sufficient field data to adequately describe the spatial covariance of soil properties. General methods for characterizing model uncertainty are presented in Chapter 14, including those that make use of observed field data to reduce the uncertainty of model inputs and predictions.

Homework problems for this chapter are provided at www.wiley.com/college/ramaswami.

10 Models of Transport in Surface Water

10.1 INTRODUCTION

Like the models of pollutant transport in the atmosphere and groundwater described in Chapters 8 and 9, models for surface water quality have evolved from simple analytical equations based on idealized reactors to sophisticated numerical codes for complex, multidimensional systems. Reactor analogs employed in simple models include the one-dimensional plug flow reactor used to describe steady-state transport in a stream or river (Section 1.6.2), the one-dimensional plug-flow-with-dispersion reactor used to model an intertidal estuary (Sections 5.4.3 and 5.7), and the CSTR reactor used to represent a well-mixed pond or lake (as illustrated in Section 10.3). Since the introduction of the classic Streeter–Phelps model for steady-state BOD and dissolved oxygen profiles in a stream in the 1920s, water quality models have been developed to characterize and help manage a wide range of water quality problems. Examples include microbial pathogens from municipal wastewater, soil erosion and related non-point-source pollution from agricultural and urban land uses, toxic chemicals from industrial plants and contaminated hazardous waste sites, and metals pollution and acidification caused by mining activities and atmospheric deposition.

Water quality models include two principal steps and associated inputs: (i) a description of the flow and mixing process of the water that constitute the surface water, which results in contaminant transport, and (ii) a characterization of the chemical and biological transformations of pollutants in the water column and associated sediments or biota. This chapter focuses on the former, describing the climatologic, hydrologic, hydraulic, and hydrodynamic processes that determine water flow rates and mixing. Models for both non-point-source pollution generation and instream concentrations are introduced. While simple, first-order reaction mechanisms are included, a more advanced treatment of chemical and biological transformation processes in surface (and ground) waters is reserved until Chapter 12.

The chapter begins with a review of the different types of surface water systems that occur in the environment and the key factors that distinguish them (Section 10.2). Section 10.3 provides a framework for determining the implications of these factors for choosing the dimensionality, time scales, and processes that are included in surface water quality models. In Section 10.4, hydrologic models are described for determining the quantity and rate of runoff and stream flow in a watershed, including a statistical model that builds on the principles of probabilistic modeling presented in

Chapter 7. Hydraulic modeling and flow routing methods for a stream are presented in Section 10.5. Models for water quality in river systems are introduced in Section 10.6, including new approaches based upon consideration of transient storage zones in streams. The chapter concludes with a summary of available computer models for estimating non-point-source loadings and receiving water quality, including the newest generation of sophisticated models that link hydrodynamic calculations for water flow with water quality predictions for complex surface water systems.

The chapter perforce provides only a brief and selective introduction to the field of water quality modeling. More complete presentations can be found in a number of excellent, dedicated texts, including Thomann and Mueller (1987), Chapra (1997), and Martin and McCutcheon (1999). A web appendix on techniques for modeling sediment erosion, deposition, and resuspension in surface waters is also provided.

10.2 SURFACE WATER SYSTEMS

Major surface water bodies include streams, rivers, lakes, impoundments, tidal rivers, estuaries, harbors, bays, inlets, and the open ocean. Figure 10.1 summarizes some of the key characteristics of these systems that affect water quality and the formulation of an appropriate water quality model. The transition from headwater streams to rivers, impoundments, and lakes is characterized by a change from advection-dominated to dispersion-dominated transport. A similar process occurs as streams and rivers flow into tidal rivers, estuaries, harbors, and the ocean, and this transition is also accompanied by a progression from fresh to saline waters.

Headwater, lower-order streams[1] are characterized by rapid hydrologic response times, shallow flows with very little variation in the vertical direction, high suspended sediment transport, little opportunity for benthic (bottom) sediments to accumulate, low anthropogenic influence, and low biological activity. Biochemical processes on rock surfaces in headwater streams can be of some significance and can have significant impact on water quality due to the shallowness of the water column.

Higher-order, main river channels exhibit intermediate response times and moderate levels of vertical variation. As flow rates vary, major rivers often experience alternating periods of sediment deposition, resuspension, and transport in different portions of the stream channel. As a result, they exhibit diverse and variable benthic sediments and biota. Benthic sediments become more prevalent as rivers become bigger, deeper, and slower, though the increased depth of these systems implies less net benthic influence on the quality of the overlying water column. Major rivers typically exhibit an intermediate level of biological activity and are often subject to high anthropogenic influence, as their channels are modified for navigation and flood control and their waters used for agriculture, industry, municipal water supply, and wastewater assimilation.

[1] The order of a stream identifies the number of branches that are upstream in the channel network. Zero-order streams have no identifiable tributary channels; these are the headwater streams where channel flows originate. Zero-order channels combine to yield first-order streams, and so on. Main river channels are typically of 5th to 9th order.

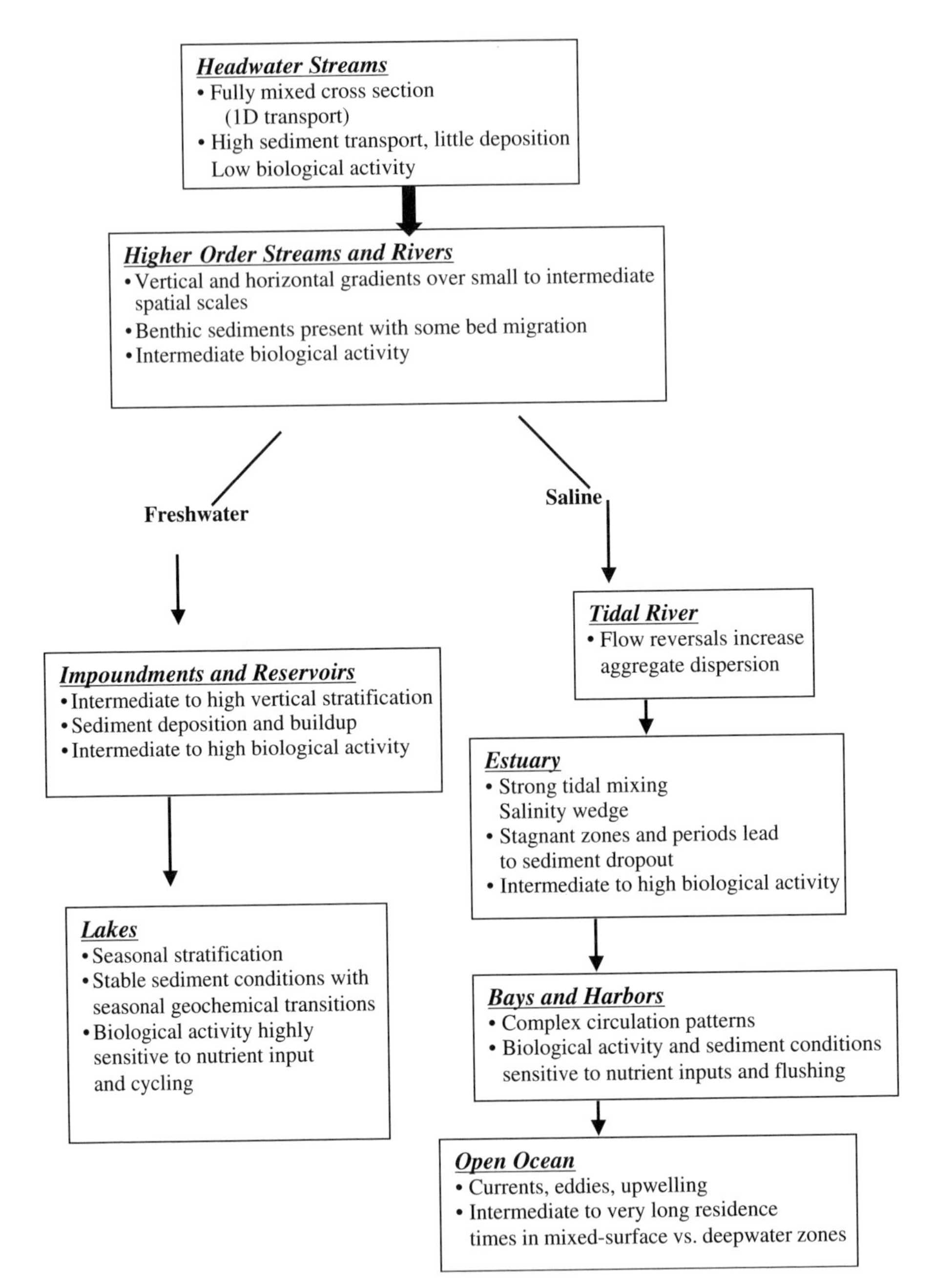

Figure 10.1 Surface water systems and key processes for water quality.

The transition from rivers to impoundments and lakes results in increasingly longer hydrologic response times and higher degrees of vertical stratification. Impoundments, often constructed to provide flood protection, improved navigation, more reliable water supply, recreation, and/or power can cover the range from slow-moving rivers to large, lakelike systems, depending on their depth, flow-through rate, and resulting retention time. Thermal stratification occurs due to density differences associated with the vertical temperature profile induced by surface heat input, but the development of stratification can be overcome by the momentum of the flowing water. Reservoirs deeper than 10 m with residence times greater than 20 days are usually stratified during warm weather periods (Ford and Johnson, 1986; Martin and McCutcheon, 1999). A quantitative indication of stratification potential in a flow-through impoundment is provided by the densiometric Froude number (Martin and McCutcheon, 1999, p. 348):

$$F_d = \sqrt{\frac{1}{g d_{\text{ng}}} \frac{LQ}{HV}} \tag{10.1}$$

where g is the acceleration due to gravity, d_{ng} is a representative, normalized density gradient (change in density relative to an average density of the water body per unit depth, typically $\sim 10^{-6}/m$), L is the length of the reservoir (m), Q is the flow (m^3/s), H is the average depth of the reservoir (m), and V is the volume of the reservoir (m^3). The reservoir will remain well mixed when $F_d \gg 1/\pi$, will stratify when $F_d \ll 1/\pi$, and will be weakly or only intermittently stratified for values of $F_d \sim 1/\pi$.

Most natural lakes are of the *drainage* variety, fed by one or more streams and emptied by one outflow channel. Flow exchanges with surrounding groundwater aquifers can occur in either direction and often vary over time (Cheng and Anderson, 1994; Kim et al., 2000). However, *headwater* lakes can also occur, especially in areas impacted by intense glacial advance and retreat, leaving flat, elevated land features. Kettle lakes are formed by glacial remains and receive all of their water input by direct precipitation, with no channel inflow. Since the incoming waters have little or no contact with soils in the watershed, there is no exposure to the cation exchange and weathering processes that impart alkalinity and neutralize acidity in the coupled soil–stream watersheds that more commonly occur in drainage lakes. Kettle lakes are thus highly sensitive to the effects of acid deposition. Seepage lakes have no outflow stream; all water loss is due to evaporation or seepage to the underlying groundwater. Impoundments and lakes serve as collectors of sediments carried to them by the faster moving portions of upstream rivers. The benthic sediments that form can play an important role in the long-term storage and release of nutrients and toxic chemicals.

The varying vertical temperature distribution and seasonal stratification that occurs in many large lakes and impoundments is critical to their annual cycles of chemical flux and plays an important role in regulating biological populations. Figure 10.2*a* illustrates a typical seasonal pattern of temperature variation with depth for a dimictic lake (one that mixes twice a year), resulting in well-mixed conditions during fall and spring overturn periods but a high degree of thermal stratification in the summer. During this period, a steep temperature gradient (or thermocline) separates the

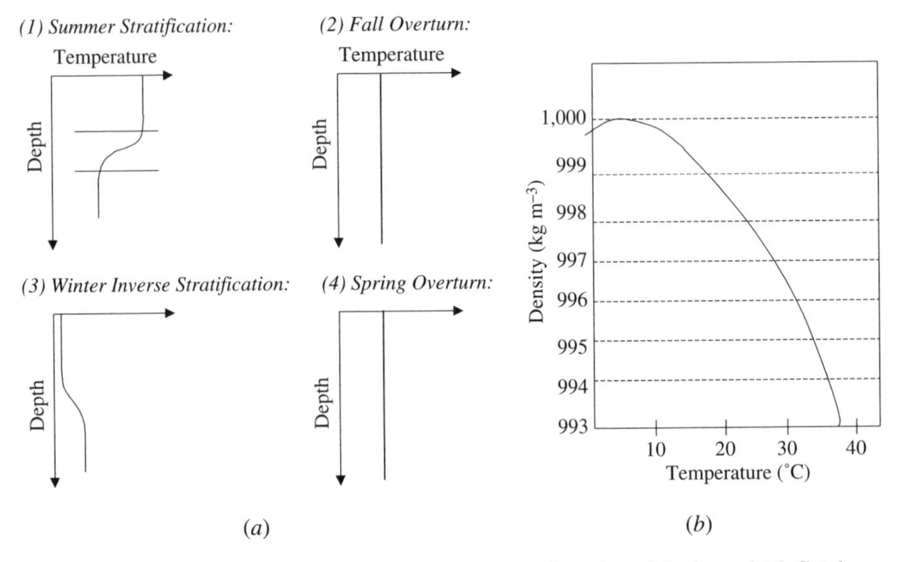

Figure 10.2 Thermal stratification in surface waters (based on Martin and McCutcheon, 1999). (*a*) Seasonal pattern of stratification in a dimictic lake. (*b*) Temperature–density relationship for water.

well-mixed, warmer, and oxygenated surface layer (the epilimnion) from the deeper, colder, and anoxic hypoliminion.

The stratification that occurs during the summer at the thermocline (metaliminion) is similar to the stability that occurs in the atmosphere in an inversion layer (see Section 8.2.3): Water that is displaced upward is cooler and more dense than the warmer water it encounters, so it sinks back to its place; water displaced downward is less dense than its cooler surroundings, so it returns upward. Since the density–temperature relationship for water is reversed below 4°C (see Fig. 10.2*b*), a reverse stratification develops in the winter, with frozen or nearly frozen water near the surface and denser water at 4°C in the bottom layer.[2]

Lakes and impoundments usually exhibit intermediate to high levels of biological activity, with intermediate levels usually preferred. Anthropogenic impacts can artificially reduce biological activity through acidification or increase it through an oversupply of nutrients and resulting eutrophication. Eutrophication is a natural geo-biochemical process resulting from the long-term capture and accumulation of nutrients from the atmosphere and the land surface. However, this process can be greatly accelerated (and made effectively irreversible) by anthropogenic loadings (Horne and Goldman, 1994; Arrow et al., 2000).

The transition on the right-hand side of Figure 10.1, from rivers to estuaries, harbors, and the ocean, also brings with it a lengthening of system response times.

[2] The reversal of the density–temperature relationship in Figure 10.3*b* is essential to the maintenance of aquatic life in lakes, streams, and the polar oceans during the winter and to the sport of ice skating.

Estuaries and other coastal areas can be locally impacted in the short-term (e.g., by stormwater runoff or spills) and exhibit regular variations over the (~twice daily) tidal cycle. However, significant changes in water quality in coastal waters generally occur only over time scales of months, years, and decades, depending on the pollutant. Tidal rivers are influenced by tidal motion but are upstream of the near-coastal zone that is affected by salinity intrusion. As such, they exhibit only minor to intermediate levels of vertical stratification. In contrast, estuaries are often influenced by a significant tidal wedge, where the denser saltwater from the ocean intrudes in a lower vertical layer, while the freshwater from the land flows seaward near the surface (see Fig. 10.3). This often necessitates the use of a two-layer model for estuaries, with diffusive mixing and flux between the layers.

To account for changes in water density that occur with both changing water temperature and salinity, an equation of state is needed (Neumann and Pierson, 1966; Thomann and Mueller, 1987). A simple approximation can be obtained from the relationship of Crowley (1968) for conditions of atmospheric pressure (pressure corrections for density at depth are usually minor, except in the deep oceans):

$$\rho_w \left(\text{g/cm}^{-3}\right) = 1 + \{10^{-3}[(28.14 - 0.0735T - 0.00469T^2) + (0.802 - 0.002T)(S - 35)]\} \tag{10.2}$$

where T is the temperature in degrees Celsius and S is the salinity in parts per thousand (ppt). The density of the near-shore ocean waters at a salinity of $S = 30$ ppt and a temperature of $T = 15°\text{C}$ is $\rho_w \sim 1.022\text{g/cm}^{-3}$. Salinity concentrations at this level also shift the temperature at which the maximum density of water occurs downward, from 4°C to about –2.4°C (Thomman and Mueller, 1987).

The wider, slower moving reaches of estuaries and bays promote sediment deposition, creating active and important benthic communities (and replenishing beaches). Estuaries, harbors, inlets, and bays exhibit intermediate to high biological activity, important for maintaining healthy ecosystems, but again making them susceptible to overfertilization and eutrophication from the accumulated nutrients of land runoff.

The world's oceans are probably the most difficult surface water systems to classify into simple categories of behavior and response and certainly the most difficult

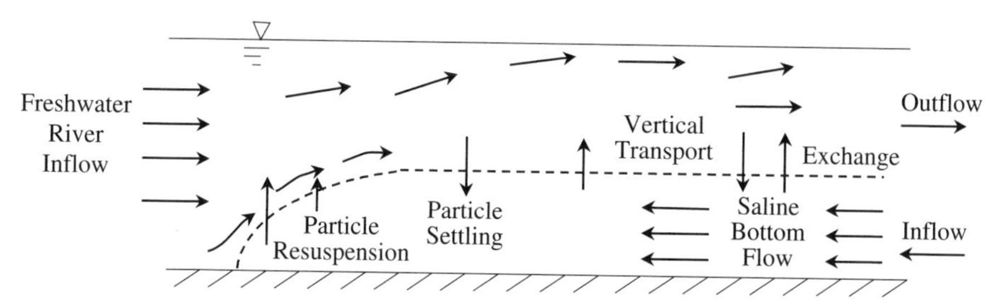

Figure 10.3 Schematic of salinity wedge in vertically stratified estuary flow (based on Thomann and Mueller, 1987).

to model. A complex set of eddies and currents, which both depend upon and greatly affect oceanic heat fluxes and temperatures, create hydrodynamic patterns extending from local to global scales (see Section 8.2.3 for a discussion of the atmospheric components of this circulation). Highly aggregated ocean water quality models are a component of the global ocean–atmospheric CO_2 heat-flux models used to predict future climate change and sea level rise. In these models, vertical segmentation often includes a mixed (isothermal) surface layer and an intermediate diffusion-only (or diffusion with upwelling) layer, with the deep ocean assumed to be at constant temperature (Oeschger et al., 1975; Harvey and Schneider, 1985; Patwardhan and Small, 1992; Harvey et al., 1997). Other water quality problems in the oceans require more localized assessment and finer spatial resolution but can be affected by large-scale circulation patterns as well. Examples include the development and growth of offshore hypoxic (low dissolved oxygen) regions, thought to be stimulated by major discharges of nutrients from streams and estuaries (Goolsby et al., 1999; National Research Council, 2000), and impacts at former ocean disposal sites (Takada et al., 1994; Bothner et al., 1998).

10.3 SELECTING SURFACE WATER QUALITY MODEL FEATURES AND COMPONENTS

Key decisions in developing or selecting a surface water quality model for a given application include:

1. Determining whether a zero-, one-, two-, or three-dimensional (3D) spatial model is appropriate
2. Determining whether transport is captured purely by advection terms or whether dispersion is also to be included
3. Determining whether a separate hydraulic or hydrodynamic submodel is used to compute water flow profiles, depth, and velocity or whether the advection field is instead entered into the model as exogenous input
4. Determining whether separate transport calculations must be made for non-aqueous phases, including particles and immiscible fluids
5. Selection of system boundaries and boundary condition types
6. Determining whether a dynamic or steady-state calculation is needed

Shanahan et al. (2001) provide a framework for addressing a number of these issues in the selection of river water quality models. A number of these decisions dictate further choices that must be made in model specification. For example, if in decision 3 a hydraulic or hydrodynamic submodel is used to compute flow fields, the modeler must determine whether these calculations are affected by computed water quality concentrations (e.g., of salinity or particles) requiring that the flow and water quality calculations be implemented simultaneously, or whether they can be implemented separately and sequentially over the selected time and spatial steps of the model. If a dynamic model is selected in decision 6, the modeler must then determine how much temporal detail is needed in the specification of model inputs—flow rates, temperatures, and loadings. Are annual or monthly averages sufficient or do daily, hourly, or even shorter variations matter? The following example illustrates insights that can be gained into the scale of temporal disaggregation needed for input to a water quality (or any environmental) system model, based on the system's characteristic time of response.

EXAMPLE 10.1 CSTR RESPONSE TO A SINUSOIDAL LOAD (ADAPTED FROM CHAPRA AND RECKHOW, 1983)

Consider a completely mixed lake of volume V, modeled as a CSTR receiving a discharge of a reactive pollutant that degrades with first-order rate k, as shown in Figure 10.4. As indicated, the discharge rate W varies sinusoidally, with a period

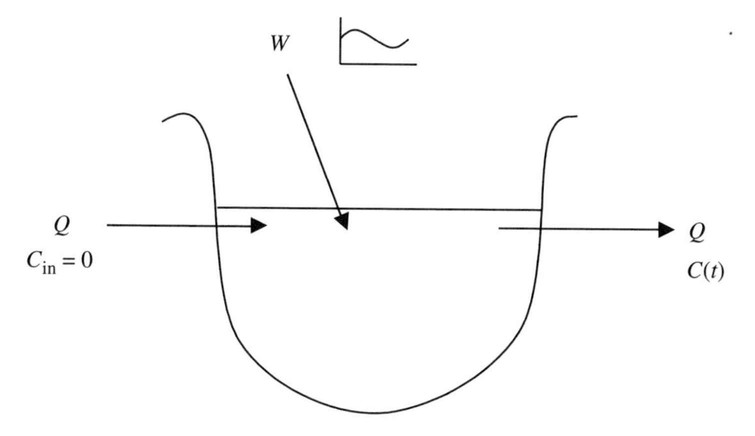

Parameters

$V = 10^6\ \text{m}^3$

$Q = 10^6\ \text{m}^3/\text{yr}$

$\tau_F = \dfrac{V}{Q} = 1$ year

$k = 1.5\ \text{year}^{-1}$

$\alpha = \dfrac{1}{\tau_F} + k = 2.5\ \text{years}^{-1}$

$W(t) = \overline{W} + A_W \sin(\omega t - \theta)$

$\overline{W} = 100$ kg/yr

$A_W = 25$ kg/yr

$\theta = 0$

Scenarios

Case 1: Intermediate Frequency

Annual period of variation: $T = 1$ year; $\omega = 2\pi/T = 6.283\ \text{year}^{-1}$

$\omega/\alpha \sim 1$

Case 2: High Frequency

Monthly period of variation: $T = 0.0833$ year; $\omega = 2\pi/T = 75.40\ \text{year}^{-1}$

$\omega/\alpha >> 1$ highly damped

Case 3: Low Frequency

100-year period of variation: $T = 100$ year; $\omega = 2\pi/T = 0.06283\ \text{year}^{-1}$

$\omega/\alpha << 1$ no damping

Figure 10.4 Completely mixed lake with sinusoidal loading input.

of oscillation T and amplitude A_W. The mass-balance differential equation for the pollutant concentration, $C(t)$, is given by (see Sections 1.2.3 and 6.1):

$$\frac{dC}{dt} + \alpha C = \frac{W(t)}{V} \tag{10.3}$$

where

$$\alpha = \frac{Q}{V} + k$$

The loading term is given by:

$$W(t) = \overline{W} + A_W \sin(\omega t - \theta)$$

Here the system flow (Q), volume (V), and first-order reaction rate coefficient (k) are assumed constant (so that α is constant), and only the loading rate varies—in the indicated sinusoidal manner with mean $\overline{W}$, half-amplitude A_W, frequency ω, and phase shift θ. The solution can be determined by analytical integration, employing the integrating factor to yield [assuming $C(t = 0) = 0$]:

$$C(t) = \frac{\overline{W}}{\alpha V}(1 - e^{-\alpha t}) + \frac{A_W}{V\sqrt{\alpha^2 + \omega^2}} \left\{ \sin\left(\omega t - \theta - \arctan\frac{\omega}{\alpha}\right) - e^{-\alpha t} \sin\left(-\theta - \arctan\frac{\omega}{\alpha}\right) \right\} \tag{10.4}$$

In the example shown, the mean loading rate is 100 kg/yr, and the amplitude of the sinusoidal variation of the loading rate is one-fourth of this amount, 25 kg/yr. A plot of the concentration response for this system computed by Eq. (10.4) is shown in Figure 10.5*a*. The first term in Eq. (10.4) begins with the concentration at the initial condition value $C(0) = 0$ and exponentially approaches the eventual mean of the resulting sinusoid. This mean value is equal to the steady-state concentration associated with constant discharge at the mean loading rate, $\overline{W}$:

$$\overline{C(t)} = C_{SS} = \frac{\overline{W}}{\alpha V} \tag{10.5}$$

In the example system $C_{SS} = 40\ \mu$g/L. The second sinusoidal term is also transient (since it is multiplied by $e^{-\alpha t}$), allowing a gradual transition to the eventual lag (phase-shift) angle, given by $\arctan(\omega/\alpha)$. The concentration profile thus eventually settles into the (lagged and dampened) sinusoidal response:

$$C(t) = \overline{C(t)} + \frac{A_W}{V\sqrt{\alpha^2 + \omega^2}} \left\{ \sin\left(\omega t - \theta - \arctan\frac{\omega}{\alpha}\right) \right\} \tag{10.6}$$

To gain further insight into the system response, consider the amplitude of the resulting concentration sinusoid *relative* to the amplitude of the forcing load term:

$$\text{Relative amplitude} = \frac{A_C/\overline{C(t)}}{A_W/\overline{W}} = \frac{A_W/V\sqrt{\alpha + \omega^2}/\overline{W}/\alpha V}{A_W/\overline{W}}$$

$$= \left[1 + \left(\frac{\omega}{\alpha}\right)^2\right]^{-1/2} \tag{10.7}$$

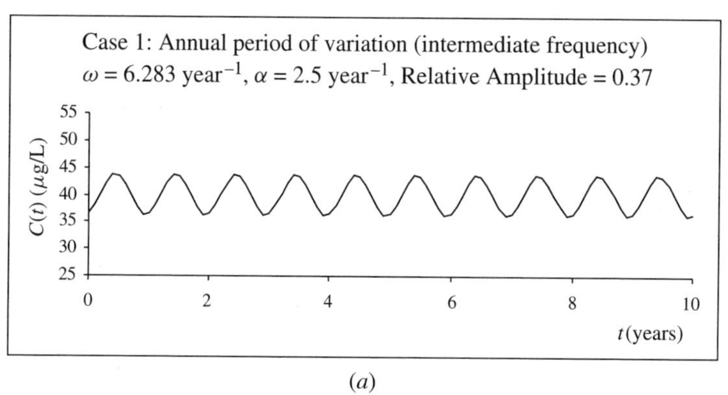

(*a*)

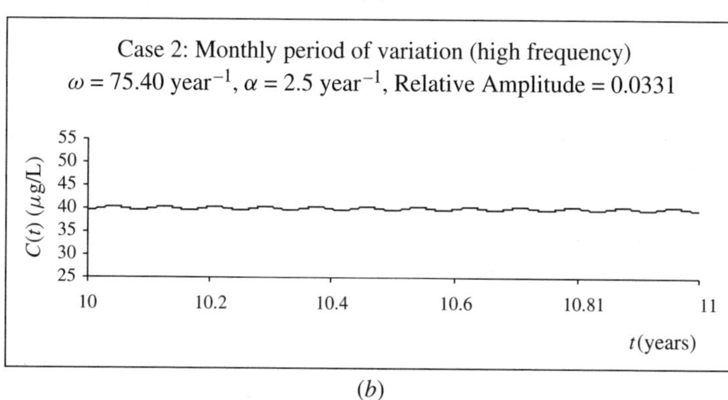

(*b*)

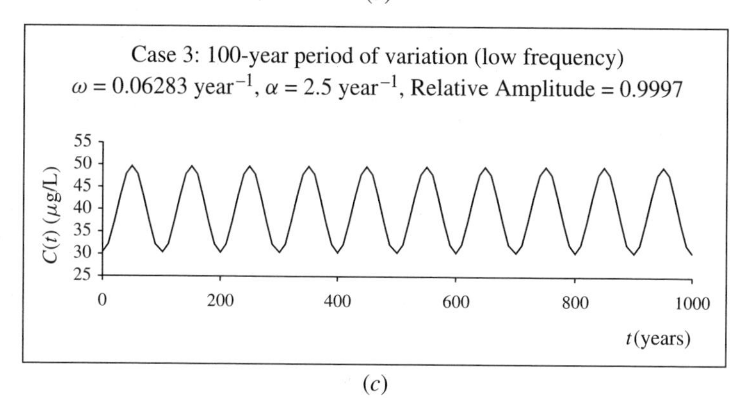

(*c*)

Figure 10.5 Concentration response to sinusoidal input.

Three distinct domains can be identified for the solution, depending on the time scale of the forcing function (ω) relative to the time scale of the water body (α), with the results illustrated in Figures 10.5*a*, 10.5*b*, and 10.5*c* (note the different time scales used for each plot):

- *Case 1: Intermediate Frequency* When the time scale of the forcing function is similar to that of the water body, that is, $\omega/\alpha \sim 1$ (in this case, $\sim$2.5), then the concentration response is partially dampened by a factor between 0 and 1 (Fig. 10.5*a*).
- *Case 2: High Frequency* When the time scale of the forcing function is very short (i.e., with high frequency, large ω) relative to the time scale of the water body, that is, $\omega/\alpha \gg 1$ (in this case, $\sim$30), then the relative amplitude approaches zero and the sinusoidal variation of the loading is completely dampened in the concentration response [Fig. 10.5*b*—note that in this plot, $C(t)$ is plotted only after the initial phase-in period has been completed, beginning at $t = 10$ years, and then over only a 1-year period to reflect the short, monthly period of variation].
- *Case 3: Low Frequency* When the time scale of the forcing function is very long (i.e., with low frequency, small ω) relative to the time scale of the water body, that is, $\omega/\alpha \ll 1$ (in this case, $\sim$0.025), then the relative amplitude approaches one and the sinusoidal variation of the loading, with a relative amplitude equal to 25% of its mean, is carried through completely, undampened, in the concentration response (Fig. 10.5*c*—plotted over a period of 1000 years).

Clear insights emerge from this analysis as to the necessary degree of temporal disaggregation and averaging appropriate for the loading input to a water quality model.[3] When the time scale of the variation is much shorter than the response time of the system, these variations may be ignored and only the average loading rate ($\overline{W}$) is needed to compute the (nearly constant) response, $\overline{C(t)}$. Indeed, if there are no other temporal variations in the parameters or inputs to the model that occur on *long enough* time scales to be translated through to the concentration response, then an equivalent steady-state model[4] can be used, in this case given by Eq. (10.5), $C_{SS} = \overline{C(t)} = \overline{W}/\alpha V$. In large lakes, such as the larger of the Great Lakes (where hydraulic retention times are on the order of decades to

[3] Equivalent insights and results are obtained from spectral and frequency-domain analysis of the input–output response of a water quality system. See, for example, Thomann (1972) and Duffy et al. (1984).

[4] Since the concentration response is linear with the loading W, the average value, $\overline{W}$, can be used in the equivalent steady-state model. For parameter variations that do not yield linear, proportional changes in the concentration response (e.g., for the case where first-order reaction rates vary with time), the average (steady-state) response is not equal to the steady-state model evaluated with the average value of the parameter [e.g., $\overline{k(t)}$]. Instead, the equivalent value of k used in the steady-state model must account for this nonlinearity. Problem 10.2 in the web appendix for this chapter addresses this situation. See www.wiley.com/college/ramaswami

hundreds of years), hourly, daily, and seasonal variations in loading rates can be safely ignored—with annual-average values used in the model—especially for conservative, nonreactive substances (such as chlorides from road salt), since α is very small and ω/α is large, as in Figure 10.5*b*. For highly reactive substances, however, where high values of k in Eq. (10.3) can increase α and shorten the response time of the system, consideration of short term variations in loadings and other model inputs may still be necessary. Interestingly, as ω/α decreases, moving from the case in Figure 10.5*b* to that in 10.5*a*, a dynamic model must be employed to capture the temporal variations in the concentration response, but when ω/α is further reduced to the case in Figure 10.5*c*, a short-term assessment of the system at a particular time t^* can once again be carried out using a steady-state model. This is implemented using the loading rate at the time of interest, $W(t^*)$, recognizing that the concentration is in fact slowly changing with time (i.e, the assumed steady-state conditions are slowly evolving) in direct response to the changing value of $W(t)$.[5]

10.4 SURFACE RUNOFF AND STREAM FLOW

Rainfall–runoff modeling is often the first step in an integrated water quality assessment, addressing the generation of land surface runoff and its routing through natural and engineered channels and conveyance systems. The following sections address the hydrologic and hydraulic principles upon which these models are based, leading to consideration of alternative approaches for predicting water flow and its effect on pollutant transport. Particular emphasis is placed on the development of statistical models derived from the input–output response of hydrologic systems and the statistical properties of meteorological inputs.

10.4.1 Hydrologic Modeling

The fundamental objective of hydrologic modeling is to account for the balance of water inputs and flows through a watershed. Hydrologic models apply the continuity equation for a water balance introduced in Chapter 5. Models that apply both the continuity and the momentum equations for water flow are referred to as hydraulic or hydrodynamic models and are addressed in Section 10.5.

The elements of the water balance equation for the land surface include

P = precipitation $(L^3\,T^{-1})$

I = infiltration to groundwater $(L^3\,T^{-1})$

[5] This transition from steady-state to dynamic and back to steady-state models corresponds to the three scales of temporal averaging and variation considered by Shanahan et al. (2001).

R = return flow from groundwater[6] $(L^3\,T^{-1})$

Q = flow through surface channels $(L^3\,T^{-1})$

ET = evapotranspiration $(L^3\,T^{-1})$

S = storage on the land surface and in water channels (L^3)

The water fluxes, P, I, R, Q, and ET, and the storage, S, can be divided by the drainage area of the basin, A_B, to yield the equivalent quantities expressed in terms of water depth over the watershed, that is, $P_A = P/A_B\ (L\,T^{-1})$, $I_A = I/A_B\ (L\,T^{-1})$, and so on, and $S_A = S/A_B\ (L)$. The surface water flow rate can be expressed using either convention:

$$Q = P - I + R - \mathrm{ET} - \frac{dS}{dt} \qquad (L^3\,T^{-1})$$
$$Q_A = P_A - I_A + R_A - \mathrm{ET}_A - \frac{dS_A}{dt} \qquad (L\,T^{-1}) \tag{10.8}$$

Long-term average water balance models, applied for a period of many years over which *dS/dt* is assumed to be zero, focus on the fraction of incoming precipitation that is partitioned to net infiltration ($I - R$) and evapotranspiration (ET), with the remainder flowing through the basin channel network (Q). Short-term hydrologic models keep running tabs of these flows and reservoirs over time steps of minutes, hours, days, or months (for an example of the latter, see Hay and McCabe, 2002), depending on the targeted degree of temporal (and spatial) aggregation.

Precipitation can occur as rainfall, snow, or hail, with snow and hail contributing directly to storage in the watershed and leading to delayed runoff during periods of melt. Snow accumulation and snowmelt are components of complete watershed hydrologic models; the importance of these processes obviously depends on the location and season to which the model is applied.

Precipitation is measured at individual gages, while the required input to a hydrologic model is the precipitation rate averaged over the entire watershed or over individual modeled subcatchments. Spatial interpolation and averaging schemes, such as the Theissen polygon method (e.g., Bedient and Huber, 1992, pp. 26–29) and various approaches based on spatial correlation, geostatistics, and other features of the rainfall field can be applied to compute areal precipitation rates from those measured at individual gages (Bastin et al., 1984; Bras and Rodriguez-Iturbe, 1985; Tabios and Salas, 1985; Barancourt et al., 1992).

[6] A number of recent watershed hydrologic models include explicit treatment of groundwater flows, with coupled surface water/groundwater calculations (Amin and Campana, 1996; Person et al., 1996; Swain and Wexler, 1996; Yu and Schwartz, 1998). Similarly, coupled atmospheric/land-surface models consider the effects of surface moisture and evapotranspiration on atmospheric circulation, water vapor and precipitation (Eagleson, 1978; Wood et al., 1992; Entekhabi et al., 1999). Here we focus solely on surface water flows, with the groundwater and atmosphere treated as exogenous sources of water input or extraction. A review of available models and the state of science of hydrologic modeling is provided by Singh and Woolhiser (2002).

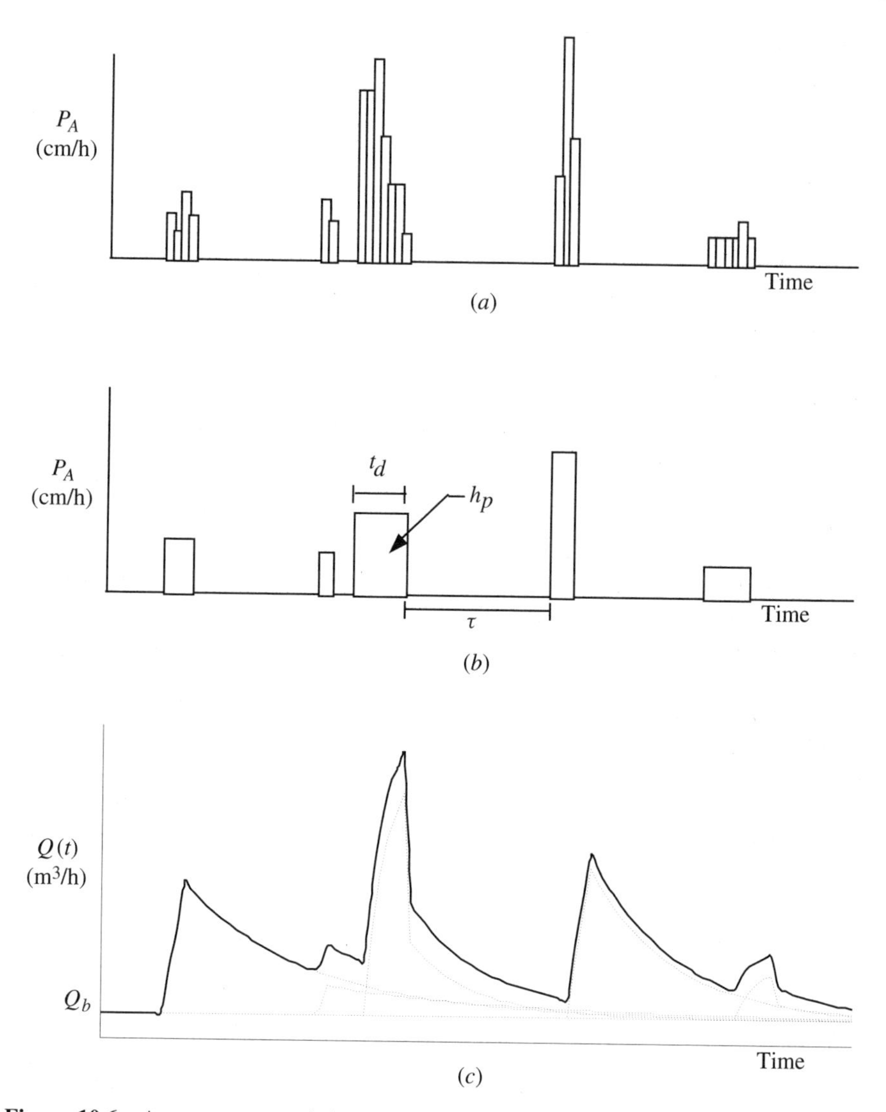

Figure 10.6 An aggregate statistical approach for storm precipitation and runoff. (*a*) Precipitation variation within events. (*b*) Precipitation variation between events. (*c*) Event runoff hydrographs and resulting stream flow.

Precipitation rates are typically recorded as average values over short-term intervals ranging from a few minutes to an hour. Models for short-term hydrologic response (usually for smaller catchments) use these inputs directly. More aggregate models may use precipitation features averaged over a storm (see Figs. 10.6*a* and 10.6*b*), including the storm duration (t_d), the total depth of precipitation (h_p), and the average precipitation rate or intensity during the storm ($p = h_p/t_d$). The overall precipitation process (including dry and wet periods) is characterized by including the time between storms (τ). Models to describe the statistical properties of these storm variables in different locations are presented in Howard (1976), Eagleson (1978), Hydroscience (1979), Di Toro and Small (1979), Small and Morgan (1986), and Adams and Papa (2000).

The aggregation of storm precipitation over monthly, seasonal, and annual intervals has been studied from a number of perspectives (e.g., Murphy and Katz, 1985; Small and Morgan, 1986; Small and Mular, 1987; Small et al., 1989). As discussed in Chapter 7, longer time periods of aggregation yield statistical distributions that are smoother, with relatively less variability. Similarly, the spatial averaging of precipitation from individual gages to a watershed results in a smoothing of rainfall statistics, and methods are available for considering the spatial and temporal variations of precipitation in a unified manner (Bras and Rodriguez-Iturbe, 1985, Chapter 6).

Evaporation and transpiration are important processes in any long-term analysis of water flow. Approximately 70% of the mean annual precipitation in the United States is extracted from the hydrologic cycle through these mechanisms (Bedient and Huber, 1992, p. 30). However, rates of evapotranspiration are highly variable over time and location, depending on soil, land surface, and climatic conditions. Direct evaporation from a water surface can be estimated using a mass-transfer calculation driven by turbulence in the overlying air (e.g., wind speed) and the difference between the water vapor pressure in the thin layer just above the water surface and that in the bulk air, since water evaporation is an air-side limited mass-transfer process (see Sections 4.2 and 4.3). Alternatively, an energy balance method can be used to compute the energy available for evaporation. Examples of methods for computing evaporation rates from the land surface and surface waters that consider both air-side limited turbulent mass transfer and heat energy inputs are found in Penman (1948), Priestley and Taylor (1972), Bedient and Huber (1992), Singh (1992), and Dingman (1993).

Infiltration and Excess Precipitation Since evapotranspiration is most important in the times between storms, it is often ignored in short-term rainfall–runoff simulations. Evapotranspiration does, however, determine the antecedent soil moisture and watershed storage conditions that are important for initiating the subsequent storm rainfall–runoff calculations. At the beginning of a storm an initial portion of precipitation is captured by surface depression storage on vegetation, ponding on building surfaces, and so forth. Following this, infiltration begins to occur through the vadose zone of the soil. As described in Section 9.4, this can be modeled numerically using Richards equation [Eq. (9.19)], given the required relationships between soil moisture content, conductivity, and pressure head. Since solution of the Richards equation is difficult, a number of rainfall–runoff models use approximate methods, such as the Green–Ampt (1911) equations (Mein and Larson, 1973; Rawls et al., 1983). These equations involve separate calculations for (Bedient and Huber, 1992):

- Time periods when the precipitation rate, $p(t)$, is less than or equal to the saturated hydraulic conductivity of the soil [i.e., $p(t) \leq K_s$] and early in the storm when surface saturation has not yet occurred—in these cases the infiltration rate $i(t) = p(t)$.

- Later time periods in the storm after surface saturation has been achieved and $p(t) > K_s$. For this condition the infiltration rate can exceed K_s but does so by decreasing amounts as the available void space in the vadose zone is filled.

The quantity of precipitation that is not captured by initial depression storage and not infiltrated into the ground surface is referred to as *effective* or *excess* precipitation and is available for surface runoff. A hydrograph translates the excess precipitation, $p_e(t)$ into a profile of subsequent stream flow, $Q_r(t)$, in the downstream channel network.

10.4.2 Linear System Response and the Unit Hydrograph

Total hydrographs for a storm are often modeled by assuming a linear, proportional response to each unit of excess precipitation, and applying unit hydrograph theory. A unit hydrograph, $Q_u(t)$, describes the flow resulting from a unit of excess precipitation over the time step of the hydrologic model. These time steps are typically 1 h for larger basins, with shorter time periods required for smaller catchments. The unit hydrograph equation is written in terms of a transfer function, $q_u(t)$ (T^{-1}), for the temporal distribution of the flow resulting from each unit volume of excess precipitation during time interval i:

$$
\begin{aligned}
Q_u(t) &= \quad 1 \quad \times A_B q_u(t) \\
Q_{R,i}(t) &= p_{eA,i}(t-\delta)\,\Delta t \times A_B q_u(t) \\
&= p_{eA,i}(t-\delta)\,\Delta t \times Q_u(t)/1 \qquad (10.9) \\
& \qquad\quad \vdots \\
& \qquad \longleftrightarrow \\
(L^3\,T^{-1}) & \qquad\quad (L) \qquad (L^2)(T^{-1})
\end{aligned}
$$

where $Q_u(t)$ is the runoff flow resulting from a unit depth of excess precipitation, that is, $p_{eA,i} \times \Delta t = 1$, over a drainage area A_B; and $Q_{R,i}(t)$ is thus the flow resulting from the excess precipitation volume $A_B \times (p_{eA,i} \times \Delta t)$.

The total runoff from a storm is computed as the linear convolution of the individual contributions from each time interval:

$$Q_r(t) = \sum_{i}^{n} p_{eA,i}(t-\delta_i) Q_u(\delta_i) \qquad (10.10)$$

where δ_i is the time between the excess precipitation and the unit hydrograph response, and n is the number of time intervals necessary to capture the earliest excess precipitation still contributing to flow. As indicated in Figure 10.6c, a base flow, Q_b, is added to the storm runoff to yield the total streamflow:

$$Q(t) = Q_r(t) + Q_b \tag{10.11}$$

Idealized conceptual models for water flow through a watershed yield useful theoretical forms for the unit hydrograph (Singh, 1988). Among these are *instantaneous* unit hydrographs (IUH) that represent transfer functions to predict the runoff produced by an instantaneous pulse of excess rainfall. The IUH, $q_{\mathrm{IUH}}(t)$ (T^{-1}), is multiplied by the excess precipitation rate, $p_{eA}(t)$ $(L\ T^{-1})$ and drainage area A_B (L^2), and integrated over time to compute the resulting runoff flow rates $(L^3\ T^{-1})$. The total runoff flow is thus computed through a convolution integral (rather than through summation of discrete excess rainfall volumes) as:

$$Q_r(t) = A_B \int_0^t p_{eA}(t-\tau) q_{\mathrm{IUH}}(\tau)\, d\tau \tag{10.12}$$

One the most widely used instantaneous unit hydrograph models is that which derives from a cascade of linear reservoirs, shown in Figure 10.7. This model, developed by Nash (1958, 1959; see also Aron and White, 1982; Boufadel, 1998; Weiler et al., 2003), yields an instantaneous unit hydrograph transfer function that corresponds to a gamma distribution for the probability that an element of excess precipitation generated at time $t - \tau$ appears at the discharge point in the stream at time t:

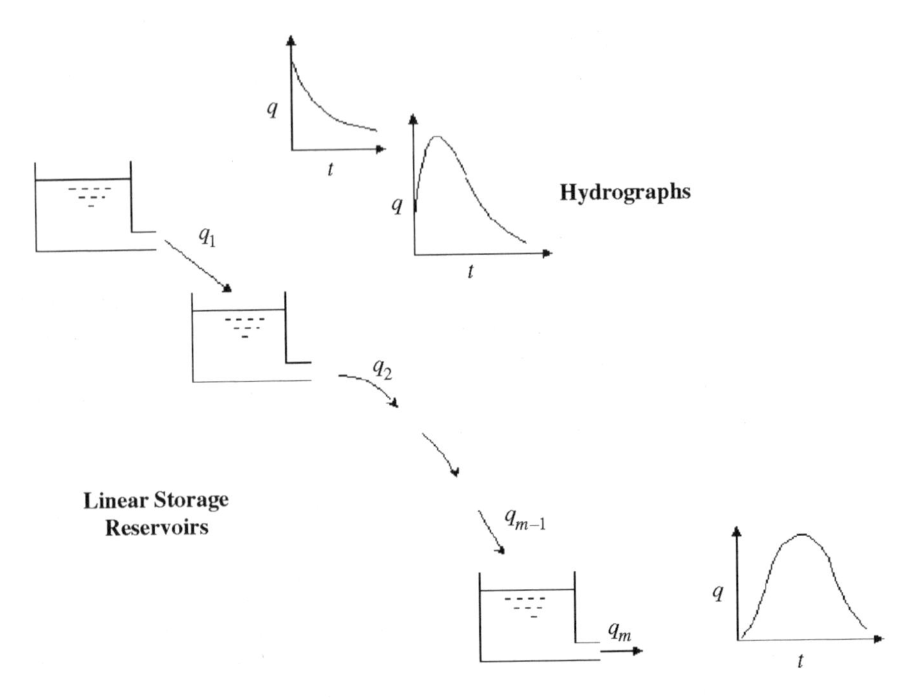

Figure 10.7 Cascade of equal linear reservoirs or the Nash model (from Singh, 1992).

$$q_{\text{IUH}}(\tau) = \frac{1}{\Gamma(m)\tau_F^m}\tau^{m-1}\exp\left(\frac{-\tau}{\tau_F}\right) \tag{10.13}$$

where m is the number of linear reservoirs and τ_F is the fluid residence time in each reservoir (τ_F is equivalent to V_i/Q, where V_i is the volume of each reservoir, assumed the same for each reservoir, and Q is the runoff flow). Comparing Eq. (10.13) to the gamma pdf in Table 7.1, it is seen that m is equivalent to the shape parameter a, and τ_F equivalent to the scale parameter b.[7]

The temporal moments of the gamma IUH are thus given by:

$$\begin{aligned} \tau_M &= E[\tau] = m\tau_F \\ \text{Std.Dev.}[\tau] &= m^{1/2}\tau_F \end{aligned} \tag{10.14}$$

and the temporal moments of the runoff resulting from brief storm events can be used to estimate the gamma IUH parameters.[8] The value of τ_M corresponds to the mean time required for a drop of water impacting the watershed as precipitation to reach the point of interest in the discharge stream. The IUH parameters can also be derived to provide an equivalent (or approximate) linear system representation for a more complex hydrologic/non-point-source pollution model, by running the model with a brief storm event and computing and fitting the flow response to the gamma moments in Eq. (10.14). In this way, the gamma (or other IUH linear system model) may provide a simplified representation of the more complex model.[9]

10.4.3 Hydrologic Flow Routing

Flow rates at different points along a stream channel can be related using methods based on the continuity equation to route flows from upstream locations to points downstream. For a particular section of the channel, the change in the water storage in the channel, $S_c(L^3)$, is related to the inflow Q_{in} and outflow Q_{out} as:

$$\frac{dS_c}{dt} = Q_{\text{in}} - Q_{\text{out}} \qquad (L^3\,T^{-1}) \tag{10.15}$$

[7] If a linear (plug flow) channel is also added to allow for a possible further displacement time, T_{PF}, needed for the flow to reach the point of interest due to channel translation, a three-parameter gamma distribution results with location parameter T_{PF} (and $\tau - T_{\text{PF}}$ follows the standard two-parameter gamma distribution with $a = m$ and $b = \tau_F$). This model is used by Dooge (1973).

[8] See Section 10.6 for further details on how temporal moments are calculated. There, in Eqs. (10.37) to (10.40) and (10.44) to (10.47), the concentration-weighted temporal moments are presented and used to fit alternative advective–dispersive transport models. Here, the flow-weighted temporal moments would be computed from the storm hydrograph and used to fit the instantaneous unit hydrograph model.

[9] If the model includes nonlinear runoff processes, the accuracy of the linear approximation should be tested by checking whether changing the excess precipitation depth of a storm does in fact result in a proportional change in the runoff response over the same time period.

Note that multiple inflows or losses can occur along a stream segment, including the upstream flow, groundwater inflow or outflow, evaporation, and runoff flow from the intervening drainage area.

The model in Eq. (10.15) is implemented by relating inflows and outflows to channel storage. In the Muskingum method (Bedient and Huber, 1992; Martin and McCutcheon, 1999) depth-flow rating curves applicable to periods of increasing and decreasing flow are linearized to allow the storage in a section of the channel to be expressed as the product of a characteristic travel time for the section and a weighted average of the inflow and outflow rates.

$$S_c = g_\tau \left[g_x Q_{\text{in}} + (1 - g_x) Q_{\text{out}} \right] \tag{10.16}$$

where g_τ is a travel time constant for the reach (T) and g_x is a dimensionless weighting coefficient. Typical values of g_x range from 0.1 to 0.3, indicating the greater association between outflow and storage than between inflow and storage. The change in storage over a discrete time interval Δt is then given by:

$$\begin{aligned} S_c(t + \Delta t) - S_c(t) = g_\tau \big\{ & g_x [Q_{\text{in}}(t + \Delta t) - Q_{\text{in}}(t)] \\ & + (1 - g_x) [Q_{\text{out}}(t + \Delta t) - Q_{\text{out}}(t)] \big\} \end{aligned} \tag{10.17}$$

This yields the routing equation for the outflow at time $t + \Delta t$ as a weighted average of the inflows at time t and $t + \Delta t$ and the outflow at time t:

$$Q_{\text{out}}(t + \Delta t) = G_1 Q_{\text{in}}(t + \Delta t) + G_2 Q_{\text{in}}(t) + G_3 Q_{\text{out}}(t) \tag{10.18}$$

where

$$G_1 = \frac{0.5\Delta t - g_\tau g_x}{g_\tau(1 - g_x) + 0.5\Delta t}$$

$$G_2 = \frac{0.5\Delta t + g_\tau g_x}{g_\tau(1 - g_x) + 0.5\Delta t}$$

$$G_3 = \frac{g_\tau(1 - g_x) - 0.5\Delta t}{g_\tau(1 - g_x) + 0.5\Delta t}$$

(note that $G_1+G_2+G_3 = 1$). The travel time and inflow–outflow weighting constants for the routing equation, g_τ and g_x, are fit to the flow–storage relationship for the channel determined from historical inflow and outflow hydrographs. In particular, g_x is selected so that the weighted flow–storage relationship is as linear as possible, and g_τ is determined as the inverse of the slope of the relationship between the weighted flow and storage. Methods for estimating the values of g_x and g_τ are presented in Wu et al. (1985). Once these constants are determined for each reach, the flow routing can be implemented iteratively and sequentially along the stream channel using Eq. (10.18).

Whether the hydrograph at a point in the stream is derived by routing the hydrographs from several upstream subcatchments or estimated directly assuming a single, aggregated upstream area, the temporal moments of the rainfall–runoff relationship can be fit to obtain a unit hydrograph for the overall basin. To demonstrate how unit hydrograph theory can lead to a simple conceptual model of stream flow, a statistical model for the instream response to a random sequence of storm events is developed and illustrated in the following section.

10.4.4 A Filtered Poisson Process Model for Storm Runoff and Stream Flow

Consider a sequence of storm events that occur as a Poisson process, each with total excess precipitation depth, P_{EA} (equal to the integral of the excess precipitation rate over the duration of the storm). We demonstrate here how the filtered Poisson process model introduced in Chapter 7 can be used to derive a simple but insightful model for storm runoff and stream flow. With the gamma IUH model, the resulting stream flow is a filtered Poisson process with moments that can be derived from Eq. (7.40) as:[10]

$$E[Q_r] = \lambda E[P_{\mathrm{EA}}]A_B \int_{\tau=0}^{\infty} \frac{1}{\Gamma(m)\tau_f^m}\tau^{m-1}\exp\left[\frac{-\tau}{\tau_F}\right]d\tau$$

$$= \lambda E[P_{\mathrm{EA}}]A_B$$

Covariance$[Q_r(t), Q_r(t+s)]$

$$= \lambda E[P_{\mathrm{EA}}^2]A_B^2 \int_{\tau=0}^{\infty} \frac{1}{\Gamma^2(m)\tau_F^{2m}}[\tau(\tau+s)]^{m-1}\exp\left[\frac{-(2\tau+s)}{\tau_F}\right]d\tau$$

$$= \frac{\lambda E[P_{\mathrm{EA}}^2]A_B^2 s^{m-1/2}}{2^{m-1/2}\sqrt{\pi}\,\Gamma(m)\tau_F^{m+1/2}}K_{1/2-m}\left[\frac{s}{\tau_F}\right] \tag{10.19}$$

where $K_{1/2-m}[\;]$ denotes the modified Bessel function of the second kind of (fractional) order $1/2 - m$. (Note, when m is an integer, $K_{1/2-m}[\;] = K_{m-1/2}[\;]$.) The

[10] The solutions that follow utilize the following definite integrals from Gradshteyn and Ryzhik (1980):

$$\int_{x=0}^{\infty} x^{v-1}e^{-\mu x}\,dx = \frac{\Gamma(v)}{\mu^v} \qquad \text{(Eq. 3.3814, page 317)}$$

$$\int_{x=0}^{\infty} x^{v-1}(x+\beta)^{v-1}e^{-\mu x}\,dx = \frac{\Gamma(v)}{\sqrt{\mu}}\left(\frac{\beta}{\mu}\right)^{v-1/2} e^{\beta\mu/2}K_{1/2-v}[\beta\mu/2] \qquad \text{(Eq. 3.3838, page 319)}$$

They also utilize the relationships for fractional order Bessel functions in Abramowitz and Stegun (1964), in particular, Eqs. 10.2.16 and 10.2.17 on page 444.

TABLE 10.1 Statistical Moments of Stream Runoff Flow Resulting from Filtered Poisson Process Model with Cascade of Linear Reservoirs (Gamma) IUH

Number of Reservoirs = m	Mean = $E[Q_r(t)]$	Variance = $\text{Var}[Q_r(t)]$	Autocorrelation function = $\rho(s)$
1	$\lambda E[P_{\text{EA}}]A_B$	$\dfrac{\lambda E[P_{\text{EA}}^2]A_B^2}{2\tau_M}$	$\exp\left(-\dfrac{s}{\tau_M}\right)$
2	$\lambda E[P_{\text{EA}}]A_B$	$\dfrac{\lambda E[P_{\text{EA}}^2]A_B^2}{2\tau_M}$	$\left[1+\left(\dfrac{2s}{\tau_M}\right)\right]\exp\left(-\dfrac{2s}{\tau_M}\right)$
3	$\lambda E[P_{\text{EA}}]A_B$	$\dfrac{9\lambda E[P_{\text{EA}}^2]A_B^2}{16\tau_M}$	$\left[1+3\left(\dfrac{s}{\tau_M}\right)+3\left(\dfrac{s}{\tau_M}\right)^2\right]\exp\left(-\dfrac{3s}{\tau_M}\right)$

covariance implies the following expressions for the variance and the autocorrelation function:

$$\text{Variance}[Q_r(t)] = \frac{\lambda E[P_{\text{EA}}^2]A_B^2\Gamma(2m-1)}{2^{2m-1}\tau_F\Gamma^2(m)}$$

$$\rho(s) = \frac{2^{m-1/2}\Gamma(m)}{\sqrt{\pi}\Gamma(2m-1)}\left(\frac{s}{\tau_F}\right)^{m-1/2} K_{m-1/2}\left[\frac{s}{\tau_F}\right] \tag{10.20}$$

Simplifications for the variance and autocorrelation function for cases where $m = 1, 2, 3$ are presented in Table 10.1, with τ_F replaced by τ_M/m, so that models with different m can be directly compared for the same total mean residence time τ_M. As indicated, increasing the number of linear reservoirs does not affect the mean flow rate but does increase the variance of the stream flow and results in a lower extent of temporal autocorrelation.

Assuming that $Q_r(t)$ in the filtered Poisson process model is well approximated by a gamma distribution (Weiss, 1977, indicates that the gamma distribution is exact for $m = 1$), the moments in Table 10.1 can be used to characterize the distribution of stream flow and to derive some interesting statistical features that are important in the application of a water quality model. This is illustrated in the following example.

EXAMPLE 10.2 FILTERED POISSON PROCESS MODEL FOR STREAM FLOW AND LOW-FLOW STATISTICS

The flow at a location in a stream is modeled using a cascade of linear reservoirs IUH. Storm events occur as a Poisson process with an average time between storms of $1/\lambda = 3$ days ($\lambda = 0.333$ day^{-1}), a mean storm excess precipitation depth of $E[P_{\text{EA}}] = 0.3$ cm, and a coefficient of variation of excess precipitation depth of $\nu_{P_{\mathrm{EA}}} = 1.5$ [as noted in Di Toro and Small (1979) and Hydroscience (1979), this coefficient of variation is typical of storm event precipitation and runoff depths]. The area of the upstream drainage basin is $A_B = 100\ \mathrm{km}^2$. Assume that the behavior of the observed runoff hydrograph indicates a gamma IUH with $m = 2$ and $\tau_F = 1$ day, so that the mean residence time of runoff reaching the discharge point in the stream is 2 days. A base flow of $Q_b = 0.25$ m^3/s is assumed for the stream.

The $E[P_{\mathrm{EA}}^2]$ must first be computed as:

$$\text{Std.Dev.}[P_{\mathrm{EA}}] = \nu_{P_{\mathrm{EA}}} \times E[P_{\mathrm{EA}}] = 1.5 \times 0.3 = 0.45\ \mathrm{cm}$$

$$E\left[P_{\mathrm{EA}}^2\right] = \mathrm{Var}[P_{\mathrm{EA}}] + E^2[P_{\mathrm{EA}}] = 0.45^2 + 0.3^2 = 0.2925\ \mathrm{cm}^2$$

The mean and standard deviation of the runoff flow are then computed from the solutions in Eqs. (10.19) and (10.20) (see Table 10.1 for $m = 2$):

$$\begin{aligned} E[Q_r] &= \lambda E[P_{\mathrm{EA}}] A_B \\ &= \frac{0.333}{\mathrm{day}} \times 0.3\ \mathrm{cm} \times \left[\frac{1\ \mathrm{m}}{100\ \mathrm{cm}}\right] \times 100\ \mathrm{km}^2 \times \left[\frac{10^6\ \mathrm{m}^2}{\mathrm{km}^2}\right] \times \left[\frac{\mathrm{day}}{86400\ \mathrm{s}}\right] \\ &= 1.156\ \mathrm{m}^3/\mathrm{s} \end{aligned}$$

$$\text{Std.Dev.}[Q_r] = \left\{\frac{\lambda E\left[P_{\mathrm{EA}}^2\right] A_B^2}{2\tau_M}\right\}^{1/2}$$

$$\begin{aligned} &= \left\{\frac{\dfrac{0.333}{\mathrm{day}} \times 0.2925\ \mathrm{cm}^2 \times \left[\dfrac{1\ \mathrm{m}^2}{10^4\ \mathrm{cm}^2}\right] \times 100^2\ \mathrm{km}^4 \times \left[\dfrac{10^{12}\ \mathrm{m}^4}{\mathrm{km}^4}\right]}{2 \times 2\ \mathrm{days}}\right\}^{1/2} \\ &\quad \times \left[\frac{\mathrm{day}}{86400\ \mathrm{s}}\right] \\ &= 1.806\ \mathrm{m}^3/\mathrm{s} \end{aligned}$$

The base flow of 0.25 m^3/s increases the mean total flow to $E[Q] = 1.156 + 0.25 = 1.406$ m^3/s, but has no effect on the standard deviation, $\text{Std.Dev.}[Q] = \text{Std.Dev.}[Q_r] = 1.806$ m^3/s.

A one-year simulation of the flow profile for the gamma IUH filtered Poisson process model for this basin is shown in Figure 10.8*a*. The mean and standard deviation of this simulation, 1.422 m^3/s and 1.619 m^3/s, respectively, are similar to the respective theoretical values computed above (1.406 m^3/s and 1.806 m^3/s). Figure 10.8*b* compares the sample cdf of the simulated flow to a three-parameter gamma distribution with its location parameter equal to the minimum flow, $Q_b = 0.25$ m^3/s, and the remaining two parameters computed from the mean and standard deviation of Q_r, so that $Q \sim$ 3-Par-$\gamma(\theta = 0.25, a = 0.4097, b = 2.821)$ (using the same procedure as demonstrated in Example 7.1). The simulated and theoretically derived gamma distributions are very similar. Figure 10.8*c* compares the theoretical autocorrelation function from Table 10.1 (for $m = 2$) to that computed for the simulated flow; again a close correspondence is indicated.

The statistics above are derived for the stream flow model based on instantaneous flow rates; however, for many water quality calculations flow rates averaged over discrete time intervals are needed. For example, water quality standards for

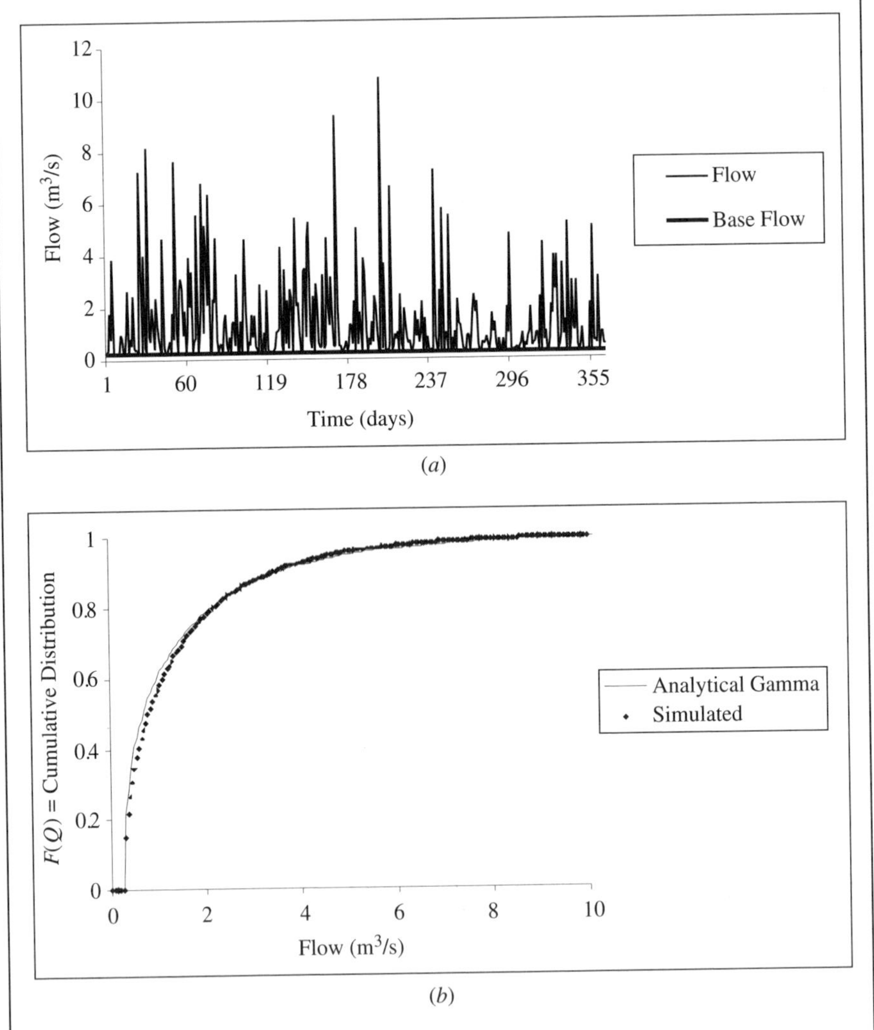

Figure 10.8 (*a*) One-year simulation of the flow profile for the γ IUH filtered Poisson process model in Example 10.2. (*b*) Comparison of simulated and analytical distributions for γ IUH filtered Poisson process for stream flow problem: instantaneous stream flows. *(continued)*

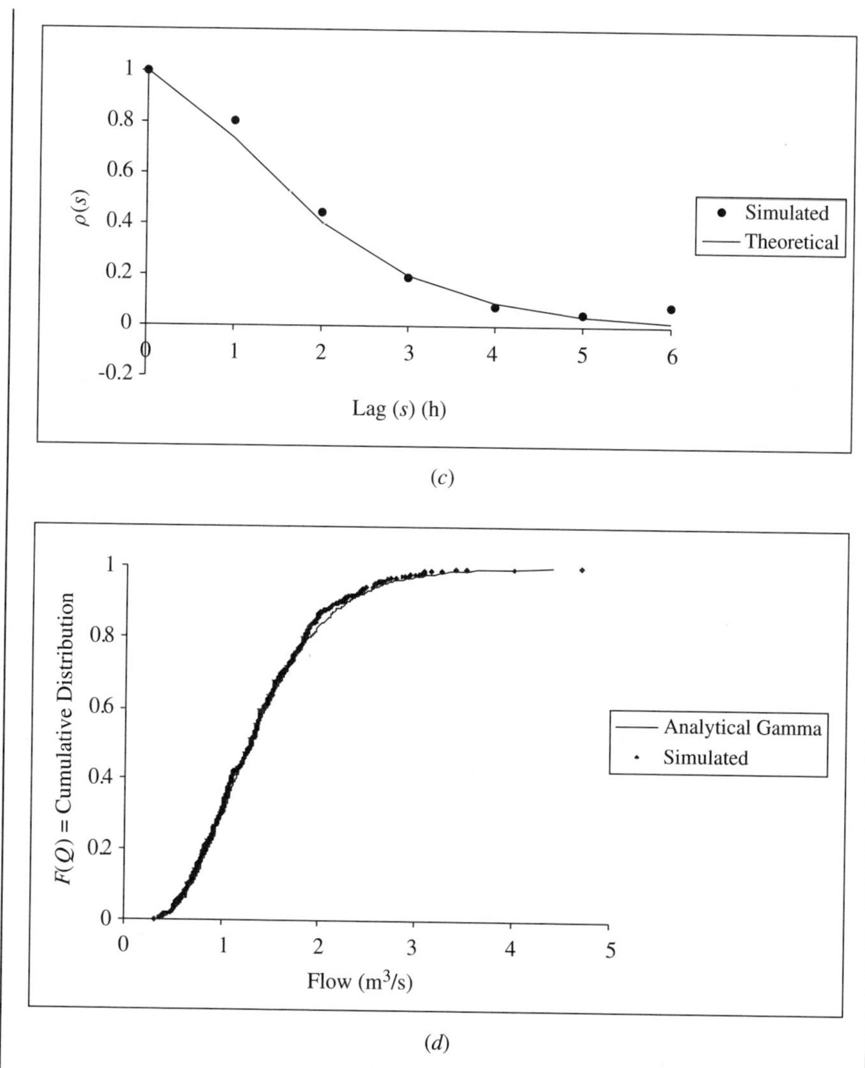

Figure 10.8 (*Continued*) (c) Comparison of autocorrelation function for simulated flow and theoretical autocorrelation function from Table 10.1 (for $m = 2$) for a gamma IUH filtered Poisson process for stream flow problem: instantaneous stream flows. (d) Comparison of simulated and derived gamma distribution for 30-day averaged flow.

aquatic toxicity and wildlife effects are often calculated using design flows for periods ranging from 1 to 90 days (US EPA, 1991, 1994). As demonstrated in Section 7.5.4, longer averaging periods for a random process yield a variable with the same mean but a reduced variance; and Eq. (7.48) can be used to calculate the variance reduction based on the autocorrelation function. This approach is used to evaluate the statistical properties of monthly (30-day) averaged flows and to estimate a 5th percentile (low flow) monthly flow rate that might be used to evaluate water quality for a low-flow condition. The calculation for the example stream flow model proceeds as follows:

1. Compute the reduction in the variance of stream flow resulting from averaging over a time period T_A, by substituting the autocorrelation function in Table 10.1 (for $m = 2$) into Eq. (7.48) to yield

$$\frac{\text{Var}\,[Q_{rA}(t)]}{\text{Var}\,[Q_r(t)]} = \frac{1}{\alpha_{TA}^2}\left[\left(\alpha_{TA} + \frac{3}{2}\right)\exp(-2\alpha_{TA}) + 2\alpha_{TA} - \frac{3}{2}\right]$$

where $Q_{rA}(t)$ is the time-averaged runoff flow rate, $Q_r(t)$ is the instantaneous runoff flow rate, and $\alpha_{TA} = T_A/\tau_M$, that is, the averaging time divided by the mean residence time of the runoff. Since the desired averaging time is 30 days, $\alpha_{TA} = 30$ days/2 days $= 15$, and the variance reduction resulting from time averaging is computed as:

$$\frac{\text{Var}\,[Q_{rA}(t)]}{\text{Var}\,[Q_r(t)]} = 0.1267$$

The variance of the 30-day averaged runoff flow rate is thus

$$\begin{aligned}\text{Var}[Q_{rA}(t)] &= 0.1267(1.806\ \text{m}^3/\text{s})^2\\ &= 0.413\ (\text{m}^3/\text{s})^2\end{aligned}$$

and the standard deviation of the 30-day averaged runoff flow rate is

$$\begin{aligned}\text{Std.Dev.}[Q_{rA}(t)] &= (0.413)^{1/2}\\ &= 0.643\ \text{m}^3/\text{s (compared to } 1.806\ \text{m}^3\text{/s for the instantaneous flow rate)}\end{aligned}$$

2. The 30-day averaged flow is once again assumed to follow a three-parameter gamma distribution with the same minimum flow ($= Q_b = 0.25\ \text{m}^3/\text{s}$), the same mean flow ($= 1.156 + 0.25 = 1.406\ \text{m}^3/\text{s}$), but the reduced standard deviation calculated above. The parameters of the gamma distribution for the 30-day averaged flow are computed with the revised moments to yield $Q_{30-\text{day}} \sim$ 3-Par-$\gamma(\theta = 0.25, a = 3.2346, b = 0.3574)$.

3. The 5th percentile of the 30-day flow is computed using the inverse cdf function for the gamma distribution in Excel, yielding

$$\begin{aligned}Q_{r30\text{-day},0.05} &= F^{-1}_{Qr30\text{-day}}[0.05]\\ &= \text{GAMMAINV}(0.05, 3.2346, 0.3574) = 0.336\ \text{m}^3/\text{s}\\ Q_{30\text{-day},0.05} &= Q_b + Q_{r30\text{-day},0.05} = 0.25 + 0.336 = 0.586\ \text{m}^3/\text{s}\end{aligned}$$

The derived gamma distribution for 30-day averaged flow is compared to that simulated over a 50-year time period in Figure 10.8*d*, indicating that the gamma distribution employed to estimate the statistics for the time-averaged flow provides a very good approximation.

Low-flow statistics are sometimes desired for extreme, critical low-flow conditions. A common critical design flow for wasteload allocation in streams is the "7Q10," the 7-day average low flow with a return period of 10 years. The 7Q10 is generally computed from observed long-term stream flow records (e.g., Riggs, 1965, 1980; Chapra, 1997, pp. 243–244) or estimated based on correlations with regional watershed characteristics (Kroll et al., 2004), but the methods used to derive the statistics for 30-day flow in Example 10.2 could also be used. However, the 7Q10 occurs at such a low percentile (approximately 0.002) that its value is determined more by the selection of Q_b than by the statistics computed for 7-day runoff. Indeed, the reader may wish to verify that the 7Q10 computed using the methods presented above yields a total flow only marginally above the base flow. A statistical model for the base flow itself is thus needed to gain insight on extremely low flows with very long return periods. Such a model is likely to be controlled by the longer-term features of climate and drought (e.g., Tasker, 1987; Cayan et al., 1998, 1999; Kroll and Vogel, 2002).

10.4.5 Water Quality Implications of the IUH Model

As indicated above, the IUH model can be used to identify probability-based design flows for water quality evaluations. In addition, a number of other interesting implications of the IUH model for water quality can be derived. The IUH, such as the gamma form in Eq. (10.13), is equivalent to the residence time distribution of fluid elements impacting the watershed to the point of discharge in the stream. It is also the impulse-response function of the concentration, $C(t)$, to an instantaneous spike of mass M, either in the incoming precipitation or mobilized at the surface of the watershed [with the first term in Eq. (10.13), $1/[\Gamma(m)\,\tau_F^m]$, replaced by $(M/V_i)/[\Gamma(m)\,\tau_F^{m-1}]$ to yield concentration]. As described below, the residence time distribution can be used to derive average water quality concentrations for chemicals undergoing linear chemical kinetic processes in the watershed, while the impulse–response function can be used to derive statistical properties of instream concentrations resulting from non-point-source pollution.

Consider a pollutant initially generated or entering a watershed that is modeled using a gamma IUH. The pollutant is assumed to have an initial concentration C_0, and undergoes first-order decay with rate coefficient $k(T^{-1})$. The average concentration at the discharge point in the stream (in the conceptual model, this is the average concentration in and exiting the mth linear reservoir) is computed as:

$$E[C] = E_\tau \lfloor C_0 e^{-k\tau} \rfloor$$
$$= \int_{\tau=0}^{\infty} \frac{C_0 e^{-k\tau}}{\Gamma(m)\tau_F^m} \tau^{m-1} \exp\left[-\frac{\tau}{\tau_F}\right] d\tau \qquad (10.21)$$
$$= C_0 \left(\frac{m}{m + k\tau_m}\right)^m$$

where again $\tau_m = m\tau_F$ is the mean residence time of the system. The expression $[m/(m + k\tau_m)]^m$ is simply the efficiency (fraction remaining) of a tanks-in-series set of reactors.

The solution in Eq. (10.21) can be adapted for chemicals that accumulate as fluid elements travel through the watershed, for example, due to ion exchange or chemical weathering. If the accumulation function is given by $C(\tau) = C_L[1 - e^{-k\tau}]$, where C_L is the limiting concentration for long-residence fluid elements (this might also be the concentration associated with the base flow, if, for example, the base flow is assumed to have been in contact with the watershed soils for a very long time), then:

$$E[C] = C_L\left[1 - \left(\frac{m}{m + k\tau_m}\right)^m\right] \qquad (10.22)$$

This solution could be applied to base cations or alkalinity, which accumulate as waters are in contact with the watershed soils (Paces, 1983; Velbel, 1985). In practice, such a solution might be implemented by identifying multiple rainfall–runoff pathways, for example, a "quick-flow" portion that reaches the stream directly via overland flow, and "delayed-flow" portions that travel through different soil layers (Cosby, et al., 1985; Hooper et al., 1990). The quick-flow portion has shorter residence time, as well as lower limiting concentrations and accumulation rates of base cations and alkalinity, while the delayed-flow fractions exhibit longer residence times and greater chemical accumulation rates.[11] Mean concentrations are calculated from the flow-weighted average of the contributing pathways.

10.5 HYDRAULIC MODELING AND FLOW ROUTING

In Chapter 5 we introduced the continuity and momentum balance (Navier–Stokes) equations for fluid flow and noted the simplifications that are commonly used to predict the relationship between water flow rates, velocities, and depth in a stream

[11] Measured chemical concentrations of base cations, alkalinity, or other naturally occurring constituents that accumulate with different rates in different compartments of a watershed are often used to trace and delineate the flow contributions of these different compartments, using an inverse-modeling/parameter estimation approach referred to as "hydrograph separation," often with conceptual models of the type considered here. See, for example, Hooper and Shoemaker (1986), Jakeman et al. (1991), and Weiler et al. (2003).

channel (see Section 5.2.3). Here we provide a more in-depth summary of the physical processes that affect water flow in rivers and the available analytical and numerical methods that are used for flow routing and prediction. An introduction to hydrodynamic models, used to characterize more complex, multidimensional currents in lakes and coastal waterways, is provided in Section 10.7.3.

A significant body of knowledge has accumulated to describe one-dimensional, longitudinal flow along a stream or river channel. As when describing pollutant concentrations, models for stream flow can be distinguished between those that assume constant, or *steady*, flow over time versus those that characterize time-varying, or *unsteady*, flow rates. Variations in flow conditions along the length of the stream channel determine whether the flow is *uniform*, that is, exhibiting constant cross-sectional area, velocity, and depth, or *nonuniform*, with changing conditions along the length of the stream segment. In the latter case, the flow may be *gradually varying* with distance, allowing use of a continuous solution along the stream channel, or rapidly varying, with a significant discontinuity in the conditions and mathematical descriptions that apply at a point in the stream, for example, where changes in channel geometry or slope induce a change from critical to subcritical flow (at the point of a hydraulic jump).

10.5.1 Hydraulic Flow Routing

Hydraulic flow routing methods combine the continuity assumptions employed in hydrologic flow routing with various components of the momentum balance. Allowing for the changes in flow rate and cross-sectional area that result from flow additions or subtractions along a stream segment, the continuity requirement is given by the Saint-Venant equation:

$$\frac{\partial A}{\partial t} + \frac{\partial Q}{\partial x} = q_{\mathrm{LIN}} \tag{10.23}$$

where A is the cross-sectional area of the flow (L^2), which is a function of the channel shape and depth of flow, Q is the flow rate ($L^3\,T^{-1}$), and q_{LIN} is the lateral inflow rate per unit length of stream channel ($L^3\,T^{-1}/L = L^2\,T^{-1}$). The Saint-Venant equation can be combined with the momentum equations [see Eq. (5.1)] and simplified for a one-dimensional channel to yield (Martin and McCutcheon, 1999):

$$\frac{\partial u}{\partial t} + u\frac{\partial u}{\partial x} + \frac{u q_{\mathrm{LIN}}}{A} = -g\frac{\partial h}{\partial x} - \left(\frac{g m^2}{C_0^2 R_h^{4/3}}\right) u^2 \tag{10.24}$$

where u is the average longitudinal velocity, h is the water surface elevation relative to a reference datum, g is gravity, and m, C_0, and R_h correspond, respectively, to the roughness coefficient, dimensional constant, and hydraulic radius defined for the Manning equation (5.6a) in Section 5.2.3. The first term on the rhs of Eq. (10.24) accounts for the effect of gravity forces on the water surface gradient while the second term accounts for frictional forces. Equation (10.24) allows a particular implementation of the Saint-Venant equations assuming (i) gradually varying flow, (ii) a mildly sloping channel, and (iii) friction forces are the same as determined for steady uniform flow. Full solution of the equation requires numerical hydraulic routing methods that utilize finite difference or finite element methods such as those described in Chapter 6. Many of the models currently available for solving the one-dimensional flow equation use implicit finite difference methods to ensure stability and accuracy. These include the U.S. Army Corps of Engineers CE-QUAL-RIV1 model (U.S. Army Engineers, 1990, 1995; see also http://www.wes.army.mil/el/elmodels/), the RIVMOD-H River Hydrodynamics Model available from the U.S. EPA (Hosseinipour, 1995; Warwick and Heim, 1995), and the U.S. Geological Survey BRANCH-Network Dynamic Flow Model (Schaffranek et al., 1981; Schaffranek, 1987; see also http://water.usgs.gov/software/branch.html).

Equation (10.24) can be simplified by assuming that the inertial term $(u\,\partial u/\partial x)$ is negligible compared to the pressure, friction, and gravity terms. When the resulting equation is combined with continuity, an equation of the following form is obtained (Bedient and Huber, 1992, p. 293):

$$\frac{\partial Q}{\partial t} + c_e \frac{\partial Q}{\partial x} = -D_Q \frac{\partial^2 Q}{\partial x^2} \tag{10.25}$$

where c_e is the celerity of a gravity wave $(= \sqrt{g\,\Delta h})$, Δh is the depth of the wave, and D_Q is a diffusion coefficient for flow $(L^2\,T^{-1})$. Note that the form of Eq. (10.25) is similar to that of the advection–dispersion equation for contaminant transport, and this noninertial model for dynamic flow routing is thus often referred to as the *diffusion wave model*. The flow diffusion coefficient can be estimated for a stream as (Bedient and Huber, 1992, p. 293):

$$D_Q = \frac{Q_p}{2BS_0} \tag{10.26}$$

where Q_p is the peak flow during the routing interval, B is the width at the top of the stream channel, and S_0 is the channel slope. This leads to modification of the Muskingum hydrologic flow-routing procedure [Eq. (10.18)], adding a fourth term so that:

$$Q_{\text{out}}(t+\Delta t) = G_1 Q_{\text{in}}(t+\Delta t) + G_2 Q_{\text{in}}(t) + G_3 Q_{\text{out}}(t) + G_4 \tag{10.27}$$

where G_1, G_2, and G_3 are as defined in Eq. (10.18), and

$$G_4 = \frac{q_{\text{LIN}}\,\Delta t\,L_e}{g_\tau(1-g_x) + 0.5\Delta t}$$

where L_e is the length of the stream reach. This modification of the Muskingum method, developed by Cunge (1969), is referred to as the Muskingum–Cunge method. With this method the travel time constant is given by $g_\tau = L_e/c_e$ and the weighting coefficient is given by $g_x = 0.5 - (D_Q/c_e L_e)$. Further guidance on parameter estimation for this method is found in Bravo et al. (1994).

Further simplifications of the Saint-Venant equations are employed to model steady-flow conditions. Here the time derivatives in Eqs. (10.23) and (10.24) are set equal to zero, so that stream flow and depth are expressed only as a function of length along the stream channel:

$$u\frac{\partial u}{\partial x} = -g\frac{\partial h}{\partial x} - \left(\frac{gm^2}{C_0^2 R_h^{4/3}}\right)u^2 \tag{10.28}$$

This equation can be solved numerically in an iterative manner under a number of conditions, serving as the basis for steady-flow backwater calculations (e.g., Martin and McCutcheon, 1999, Chapter 5, Section II). If sufficient time has ensued under steady-flow conditions for the gravitational and frictional forces to equilibrate, then the convective acceleration term on the lhs of Eq. (10.28) becomes zero, resulting in the kinematic wave equation:

$$0 = -g\frac{\partial h}{\partial x} - \left(\frac{gm^2}{C_0^2 R_h^{4/3}}\right)u^2 \tag{10.29}$$

Flow routing methods based on the kinematic wave equation can be derived assuming the evolution of successive quasi-steady-flow conditions (Bedient and Huber, 1992).

The kinematic wave model is also commonly used for routing overland flow as a part of surface runoff model calculations in non-point-source hydrology and pollution models (Morris and Woolhiser, 1980). Non-point-source pollution models that incorporate kinematic wave flow routing for both overland and channel flow include the U.S. Department of Agriculture—Agricultural Research Service (ARS) Kinematic Runoff and Erosion model, KINEROS (Woolhiser et al., 1990; see also http://www.tucson.ars.ag.gov/kineros/), and the U.S. EPA Storm Water Management Model, SWMM (Huber and Dickinson, 1988; see also http://www.ccee.orst.edu/swmm/ or http://www.epa.gov/ceampubl/swater/swmm/index.htm). The SWMM family also includes the EXTRAN module for fully dynamic hydraulic simulation in sewer system collection pipes and culverts using the Saint-Venant equations (Roesner et al., 1988).

The final simplification to Eq. (10.29) arises for the case of steady, uniform flow, where $\partial h/\partial x = S_0$, the channel slope. The formulation then becomes equivalent to the Chezy–Manning equation [Eq. (5.5)] relating the flow velocity and hydraulic radius.

10.6 WATER QUALITY IN RIVER SYSTEMS

Having introduced the basic elements of hydraulic flow modeling in streams, we are ready to consider the transport processes that determine water quality in riverine systems. The simplest water quality models for a stream environment involve application of the 1D steady-state plug flow equations, introduced in Section 1.6.2. This formulation still provides the basis for many assessments of water quality involving biochemical oxygen demand, microbial pathogens, ammonia toxicity, and other traditional environmental problems. The use of this model is demonstrated for predicting microbial pathogen dieoff with distance from a source, as well as dissolved oxygen concentrations in a stream when multiple-point, nonpoint, and instream processes affect the oxygen balance. Before this, however, a review of current approaches for characterizing stream mixing and dispersion is presented. This knowledge is necessary for dynamic models of stream response to intermittent stormwater loadings and also provides insight into the contribution of the main channel versus transient storage zones in river systems.

As shown in Figure 10.9, wastewater discharges that enter a river from an outfall pipe require some travel time and distance before the complete mix assumption, inherent in the use of a 1D model, becomes valid. An order-of-magnitude estimate of the travel distance, L_m, to the zone of complete mixing in a river, for a discharge in the middle of the stream, is given by (Thomann and Mueller, 1987):

$$L_m(m) = 4.3\,(\mathrm{s/m})\,u\frac{B^2}{H} \tag{10.30}$$

where u is the average stream velocity (m/s), B is the stream width (m), and H is the stream depth (m). When the discharge is to the side bank of the river, this distance is doubled. A more refined estimate of mixing zone distances, considering the effects of discharge configuration, velocity, and temperature, can be obtained using the CORMIX model introduced in Section 10.7.3.

When intermittent or highly time-variable discharges are considered, it is usually necessary to model the river as an advective–dispersive system. An approximate

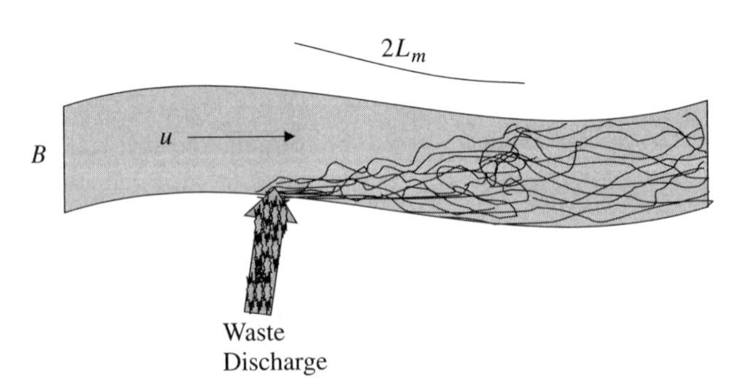

Figure 10.9 Schematic for mixing zone determination in a river with width B and velocity u (discharge to side bank of river).

estimate of the longitudinal dispersion coefficient for streams is given by (Fischer et al., 1979):

$$D_l(\text{m}^2/\text{s}) = 0.011\frac{u^2B^2}{U^*H} \quad (10.31)$$

where the shear velocity, U^* (m/s), is computed from the stream depth (H) and slope (S) as:

$$U^* = (gHS)^{1/2} \quad (10.32)$$

This equation does not consider the effects of pooling and dead zones in the stream. Expressions for estimating the transverse and vertical dispersion coefficients in a river are given by (adapted from Martin and McCutcheon, 1999):

$$D_t(\text{m}^2/\text{s}) = U^*H$$
$$D_v(\text{m}^2/\text{s}) = 0.1U^*H \quad (10.33)$$

These values are very approximate; significantly higher values of the dispersion coefficients can occur in rough streams with frequent meanders, flow blockages, and pooling. In large, slowly moving rivers, the vertical disperision coefficient can be reduced by the occurrence of thermal stratification.

EXAMPLE 10.3 INITIAL ESTIMATE OF STREAM DISPERSION COEFFICIENTS AND MIXING ZONE LENGTH

Consider a river of width $B = 20$ m, depth $H = 1$ m, and average velocity $u = 0.5$ m/s. The slope of the river is $S = 0.0004$. Estimate the longitudinal, transverse, and vertical dispersion coefficients, as well as the length of the mixing zone for a discharge to the side of the channel.

First, the shear velocity for the stream is estimated from Eq. (10.32) as $U^* = (gHS)^{1/2} = (9.81 \times 1 \times 0.0004)^{1/2} = 0.0626$ m/s. The three dispersion coefficients are then determined from Eqs. (10.31) and (10.33) as:

$$D_l(\text{m}^2/\text{s}) = 0.011\frac{u^2B^2}{U^*H} = 0.011\frac{(0.5)^2(20)^2}{(0.0626)(1)} = 18 \text{ m}^2/\text{s}$$

$$D_t(\text{m}^2/\text{s}) = U^*H = (0.0626)(1) = 0.06 \text{ m}^2/\text{s}$$

$$D_v(\text{m}^2/\text{s}) = 0.1U^*H = 0.1(0.0626)(1) = 0.006 \text{ m}^2/\text{s}$$

These estimates indicate that the longitudinal dispersion coefficient is 300 times larger than the transverse dispersion coefficient, which in turn is a factor of 10 larger than the vertical dispersion coefficient. The mixing zone length for a discharge to the side of the channel (as shown in Fig. 10.9) is computed from Eq. (10.30) as:

$$\text{Mixing zone length} = 2 \times L_m(\text{m}) = 2 \times 4.3(\text{s/m})u\frac{B^2}{H}$$
$$= 2 \times 4.3 \times 0.5 \times \frac{20^2}{1}$$
$$= 1720 \text{ m}$$

More accurate estimates of dispersion coefficients are often obtained for a specific river or stream section using dye studies. For a conservative (nonreactive) dye, the concentration response at distance x downstream at time t after it is introduced into the stream is, from Eq. (5.23) (with the first-order reaction rate, k, set equal to zero):

$$C(x,t) = \frac{m}{2A\sqrt{\pi D_x t}} \exp\left[\frac{-(x-ut)^2}{4D_x t}\right] \tag{10.34}$$

where m is the mass of the slug of dye introduced to the river, A is the cross-sectional area, and D_x here is equivalent to the longitudinal dispersion coefficient, D_l. Assuming u and A are known for the stream reach, a first, simple estimate for D_x can be obtained by measuring $C(x,t)$ at a single downstream location, x_1, and recording the peak concentration, C_p. This peak occurs at a travel time $t = x_1/u$, so that the argument of the exponential term in Eq. (10.34) is zero, and the peak concentration is given by:

$$C_p = \frac{m}{2A\sqrt{\pi D_x \left(\frac{x_1}{u}\right)}} \tag{10.35}$$

Rearranging this equation, the value of the longitudinal dispersion coefficient is estimated as:

$$\widehat{D}_x = \frac{u}{\pi x_1}\left(\frac{m}{2AC_p}\right)^2 \tag{10.36}$$

Note that the point where the dye is measured should be beyond the length of the mixing zone, so that the dye is well mixed across the stream cross-sectional area.

A more accurate estimate for D_x can be obtained using the full set of observations at two locations, x_1 and x_2, downstream of the point where the dye is introduced. Denote these measured concentrations as $C_{x1}(t_j)$, $j = 1, n_{x1}$ for the first location x_1; and $C_{x2}(t_j)$, $j = 1, n_{x2}$ for the second location x_2. The dispersion coefficient is determined by estimating the temporal moments of the dye concentration response at each location:

$$t_{c,x} = \widehat{E[t]} = \frac{\sum_{j=1}^{n_x}[C_x(t_j) \times t_j]}{\sum_{j=1}^{n_x} C_x(t_j)} \tag{10.37}$$

$$s_{t,x} = \widehat{\text{Std.Dev.}}[t] = \left(\frac{\sum_{j=1}^{n_x} \left[C_x(t_j) \times (t_j - t_{c,x})^2 \right]}{\sum_{j=1}^{n_x} C_x(t_j)} \right)^{1/2} \tag{10.38}$$

The difference in the mean time between the two stations is first used to estimate the velocity (assuming x_2 is downstream of x_1):

$$\widehat{u} = \frac{x_2 - x_1}{t_{c,x2} - t_{c,x1}} \tag{10.39}$$

The longitudinal dispersion coefficient is then estimated as:

$$\widehat{D}_x = \frac{\widehat{u}^3 \left(s_{t,x2}^2 - s_{t,x1}^2 \right)}{2(x_2 - x_1)} \tag{10.40}$$

Note that this method does not require a priori knowledge of either u or the cross-sectional area A.

EXAMPLE 10.4 ESTIMATION OF LONGITUDINAL DISPERSION FROM AN INSTREAM TRACER STUDY

For the stream described in Example 10.3, a tracer study is conducted by introducing 10 kg of a tracer dye, and the concentration is measured at locations 2 and 4 km downstream (see Table 10.2; monitoring results shown in columns 1 to 4). Estimate the dispersion coefficient in this stretch of the river.

We first utilize Eq. (10.36) to estimate the dispersion coefficient from the peak concentrations at the two locations. For this method, we use the assumed velocity and cross-sectional area for the stream: $u = 0.5$ m/s, and $A = B \times H = 20 \times 1 = 20$ m^2. From Eq. (10.36), the dispersion coefficient is computed from the peak concentration at location 1, $C_p = 0.457$ mg/L, as follows:

$$\begin{aligned} \widehat{D}_x &= \frac{u}{\pi x_1} \left(\frac{m}{2AC_p} \right)^2 \\ &= \frac{0.5 \text{ m/s}}{\pi \times 2000 \text{ m}} \left(\frac{10 \text{ kg} \times 10^6 \text{ mg/kg}}{2 \times 20 \text{ m}^2 \times 0.457 \text{ mg/L} \times 1000 \text{ L/m}^3} \right)^2 \\ &= 23.8 \text{ m}^2/\text{s} \end{aligned}$$

The reader may verify that if the same calculation is made for location 2 using the peak concentration observed at that location, $C_p = 0.271$ mg/L, then a higher dispersion coefficient ($\widehat{D}_x = 33.9$ m^2/s) is obtained. Recall a key lesson from

TABLE 10.2 Tracer Concentrations Measured at Location 1 ($x_1 = 2000$ m) and Location 2 ($x_2 = 4000$ m)[a]

1	2	3	4	5	6	7	8
	t_j	C_{x1}	C_{x2}	$t_j C_{x1}$	$t_j C_{x2}$	$[(t_j - 72.7)^2]C_{x1}$	$[(t_j - 144.0)^2]C_{x2}$
j	(min)	(mg/L)	(mg/L)	min* (mg/L)	min* (mg/L)	min^2(mg/L)	min^2(mg/L)
1	20	0	0	0	0	0	0
2	40	0.023	0	0.92	0	24.58	0
3	60	0.457	0	27.42	0	73.63	0
4	80	0.301	0.004	24.08	0.32	16.06	16.36
5	100	0.111	0.015	11.1	1.5	82.76	29.00
6	120	0.018	0.255	2.16	30.6	40.28	146.52
7	140	0.007	0.271	0.98	37.94	31.71	4.27
8	160	0	0.152	0	24.32	0	39.05
9	180	0	0.059	0	10.62	0	76.58
10	200	0	0.051	0	10.2	0	160.10
11	220	0	0.009	0	1.98	0	52.02
12	240	0	0	0	0	0	0
	Sum	0.917	0.816	66.66	117.48	267.82	523.88

[a]Calculations in columns 5 to 8 shown for dispersion estimate.

Chapter 5 that the apparent dispersion coefficient does tend to increase at larger spatial and temporal scales, and this appears to be the case in this example.

The second approach for computing the dispersion coefficient uses the temporal moments at the two locations, with Eqs. (10.37) to (10.40). Only a single estimate of the dispersion coefficient is obtained using this method. The first temporal moment at each location is calculated by implementing Eqs. (10.37) and (10.38) with the appropriate sums in columns 3 to 6 of Table 10.2, as follows:

$$t_{c,x1} = \frac{\sum_{j=1}^{n_x} [C_{x1}(t_j) \times t_j]}{\sum_{j=1}^{n_x} C_{x1}(t_j)} = \frac{\text{Sum column 5}}{\text{Sum column 3}} = \frac{66.66}{0.917} = 72.7 \text{ min}$$

$$t_{c,x2} = \frac{\sum_{j=1}^{n_x} [C_{x2}(t_j) \times t_j]}{\sum_{j=1}^{n_x} C_{x2}(t_j)} = \frac{\text{Sum column 6}}{\text{Sum column 4}} = \frac{117.48}{0.816} = 144.0 \text{ min}$$

The velocity in the stream is then calculated from Eq. (10.39) as:

$$\hat{u} = \frac{x_2 - x_1}{t_{c,x2} - t_{c,x1}} = \frac{4000 \text{ m} - 2000 \text{ m}}{144.0 \text{ min} - 72.7 \text{ min}} \times 1 \text{ min/60 s}$$

$$= 0.47 \text{ m/s}$$

which is very similar to the a priori estimate of $u = 0.5$ m/s. The second temporal moments are computed using the first moments determined above and the resulting sums in the last two columns of Table 10.2:

$$s_{t,x1} = \left(\frac{\sum_{j=1}^{n_x}\left[C_{x1}(t_j) \times (t_j - t_{cx1})^2\right]}{\sum_{j=1}^{n_x} C_{x1}(t_j)}\right)^{1/2} = \left(\frac{\text{Sum column 7}}{\text{Sum column 3}}\right)^{1/2} = \left(\frac{267.82}{0.917}\right)^{1/2}$$

$$= 17.1 \text{ min}$$

$$s_{t,x2} = \left(\frac{\sum_{j=1}^{n_x}\left[C_{x2}(t_j) \times (t_j - t_{cx2})^2\right]}{\sum_{j=1}^{n_x} C_{x2}(t_j)}\right)^{1/2} = \left(\frac{\text{Sum column 8}}{\text{Sum column 4}}\right)^{1/2} = \left(\frac{523.88}{0.816}\right)^{1/2}$$

$$= 25.3 \text{ min}$$

The dispersion coefficient is then estimated from Eq. (10.40) as:

$$\widehat{D}_x = \frac{\widehat{u}^3\left(s_{t,x2}^2 - s_{t,x1}^2\right)}{2(x_2 - x_1)}$$

$$= \frac{(0.47 \text{ m/s})^3\left[(25.3 \text{ min} \times 60 \text{ s/min})^2 - (17.1 \text{ min} \times 60 \text{ s/min})^2\right]}{2 \times (4000 - 2000) \text{ m}}$$

$$= 32.5 \text{ m}^2\text{/s}$$

This value is similar in magnitude to the two estimates of the dispersion coefficient determined using the peak concentration method, as well as the a priori estimate of $D_l = 18$ m^2/s (from Example 10.3).

Once a dispersion coefficient is estimated, it may be used to calculate the response of the stream to different types of transient loading patterns, such as those that might occur from non-point-source pollution during a storm event. If the storm is short, the load may be approximated as occurring as an instantaneous input, and Eq. (10.34) [or Eq. (5.23), when first-order decay of the pollutant is also considered] can be used to simulate the stream response. In the general case, numerical methods presented in Chapter 6, such as the finite difference or finite element techniques, are used to solve for the dynamic response to an arbitrary loading profile, with time-varying flow (using methods such as those described in Section 10.5) and concentration.

In systems that are highly advective, with low dispersion coefficients and high Peclet numbers, numerical accuracy and stability problems often arise in simulating the steep concentration gradients that arise from intermittent loads. Numerical methods that utilize a Lagrangrian framework can address these problems by shifting the coordinates with the fluid velocity so that the advection term is eliminated and only the dispersion and reaction terms need be included (McBride and Rutherford, 1984). Lagrangian methods, or mixed Lagrangian–Eulerian methods, where only the advection term is treated from a Lagrangian perspective and the other terms are treated in an Eulerian manner, are also often used for groundwater problems where steep gradients can lead to numerical problems (Cheng et al., 1984; Neuman, 1984; Yeh, 1990).

10.6.1 Transient Storage Model

In recent years, it has been observed that the simple advective–dispersive equation does not adequately describe the concentration response in many streams, especially those with stagnant zones occurring in side pools and channels, backwater areas behind natural or manmade barriers, and active side and bottom porous sediment zones that exchange water with the main channel flow. For these systems, some of the pollutant mass from a discharge may bypass these transient storage zones and reach a downstream monitoring point more rapidly than would be calculated based on the net flow rate through the entire system, while other portions of the chemical input may take much longer to reach the downstream location due to their passage through, and residence time within, the transient storage zones of the stream. When a tracer test is conducted, it thus appears that there is a greater spread (and variance) in the concentration response than would be expected from the processes of main channel advection and dispersion alone. The functional form and shape of the concentration curve is difficult to match using only the advective–dispersive equation. Furthermore, different chemical processes and interactions often occur in the transient storage zones. Effective representation of these processes thus dictates the explicit identification and modeling of a storage zone compartment. This has led to a class of transient storage models for solute transport in streams.

The early development of the transient storage model is found in Thackston and Schnelle (1970), Sabol and Nordin (1978), and Bencala and Walters (1983). The approach has now been incorporated into the U.S. Geological Survey (USGS) One-Dimensional Transport with Inflow and Storage (OTIS) model (Runkel et al., 1996; Runkel, 1998), and we describe the method using the OTIS formulation. Figure 10.10 illustrates the types of physical processes that lead to transient storage, while Figure 10.11 shows the principal transport mechanisms included in the model. As indicated, the OTIS model includes mass transfer from the main channel to and from an aggregate storage zone compartment, as well as lateral inflow and outflow from the main channel to the surrounding soils and groundwater. A simultaneous mass balance is computed for the solute concentration in the main channel C and in the storage zone C_S:

$$\frac{\partial C}{\partial t} = -\frac{Q}{A}\frac{\partial C}{\partial x} + \frac{1}{A}\frac{\partial}{\partial x}\left\langle AD\frac{\partial C}{\partial x}\right\rangle + \frac{q_{\mathrm{LIN}}}{A}(C_L - C) + \alpha_s(C_S - C) - kC \quad (10.41)$$

$$\frac{dC_S}{dt} = \alpha_s\frac{A}{A_S}(C - C_S) - k_{sz}C_S \quad (10.42)$$

where A = main channel cross-sectional area (L^2)
A_S = storage zone cross-sectional area (L^2)
D = dispersion coefficient in the main channel ($L^2\,T^{-1}$)
Q = flow rate in main channel ($L^3\,T^{-1}$)
q_{LIN} = lateral inflow rate ($L^3\,T^{-1}\,L^{-1}$)
α_s = storage zone exchange coefficient (T^{-1})

k = instream first-order decay coefficient (T^{-1})

k_{sz} = storage zone first-order decay coefficient (T^{-1})

The equation is solved in OTIS with appropriate boundary conditions and loading terms, using an implicit finite difference method for $C(x, t)$ that allows for a decoupled calculation of $C_S(t)$ (Runkel and Chapra, 1993). The transient storage model has been interpreted in terms of the stochastic residence time of fluid and contaminant elements in the main channel and the storage zone (Hart, 1995). In addition, it has many features that are similar to a statistical aggregate dead zone model that predicts the evolution of downstream concentrations using time-series equations that are physically related to the flow routing methods presented in Section 10.5 (Wallis, et al., 1989; Rutherford, 1994; Lees et al., 2000).

The transient storage model has been applied to estimate solute transport in many streams and rivers, and the OTIS model allows for sorption and reaction in the stream bed, equilibrium chemical reactions for pH and precipitating metals, and nutrient cycling and retention processes in the storage and sediment zones. The importance of the transient storage zone in affecting solute transport is determined by the relative cross-sectional area of the storage zone and the main channel, A_S/A, which indicates the physical size of the storage zone, and the exchange coefficient α_s,

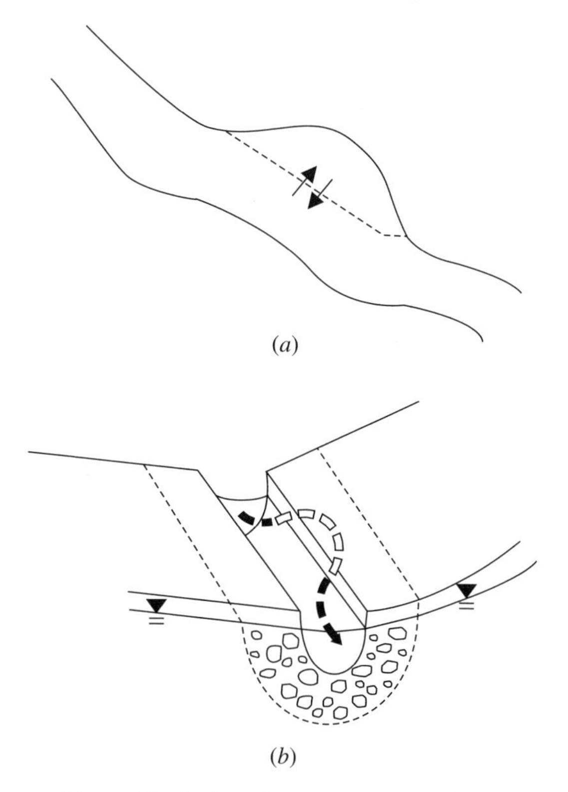

Figure 10.10 Transient storage mechanisms.

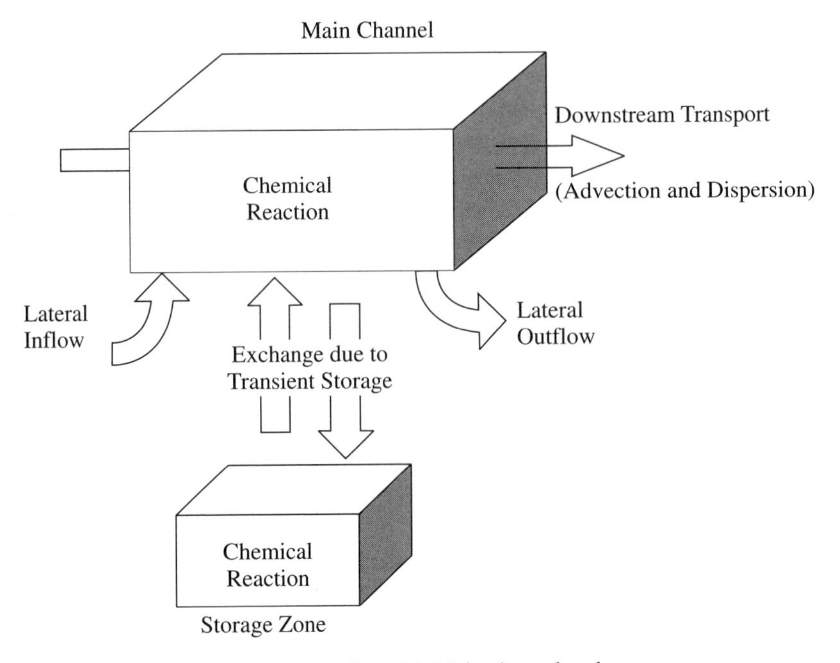

Figure 10.11 Conceptual model: Main channel and storage zone.

which indicates how quickly solutes move back and forth between the main channel and the storage zone. Typical values of the relative cross-sectional area range from $A_S/A \sim 0.03 - 0.35$ for reaches in large rivers, such as the Upper Willamette in western Oregon (Fernald et al., 2001) to $A_S/A \sim 1 - 3$ for the small, pool-and-riffle Uvas Creek in northern California (Bencala and Walters, 1983). The exchange coefficient is generally found to be in the range of $\alpha_s = 10^{-6}$ to $10^{-3}\ \mathrm{s}^{-1}$, with values of 1.7 to $2.0 \times 10^{-4}\ \mathrm{s}^{-1}$ reported for the mid-sized Clackamas River in Oregon, where A_S/A is estimated to be 0.23 (Runkel, 2000).

The effect of the transient storage zone on the model predictions is characterized in an overall manner by the Dahmkohler 1 number. The Dahmkohler 1 number provides a dimensionless measure of the ratio of mass transfer to bulk transport. For a stream reach of length L, the Da1 is given by:

$$\mathrm{Da1} = \alpha_s \frac{L}{u}\left[1 + \left(\frac{A}{A_S}\right)\right] \tag{10.43}$$

When Da1 is very large, for example, greater than 100, the exchange between the main channel and the storage zone is so rapid that the compartments can be assumed to be in equilibrium. When Da1 is very small, then solutes introduced to the main channel have little interaction with the storage zone. For either of these cases, the parameters of the transient zone model become difficult to estimate from observed field studies; a value of Da1 between 0.1 and 10 has been indicated as required for reliable parameter estimation of the transient storage zone model (Wagner and Harvey, 1997; Fernald et al., 2001).

Parameter estimation for the transient storage model is usually made using tracer studies in conjunction with a numerical parameter estimation method, and the OTIS model includes this capability (see Chapter 14 for a discussion of general parameter estimation methods). Approximate estimates can also be made by fitting tracer study concentration profiles to the first three temporal moments to estimate D, A_S/A, and α_s, assuming the stream velocity u is known—similar in approach to the use of the first two moments to estimate D and u in the standard advection–dispersion model, using Eqs. (10.39) and (10.40). The three temporal moments of the concentration profile resulting from the instantaneous release of a nonreactive tracer that yields a uniform concentration at the beginning of a stream reach are given by (Czernuszenko and Rowinski, 1997; Lees et al., 2000):

$$t_{c,\text{TSM}} = E[t_c] = \frac{2D}{u^2} + \frac{x}{u}\left(1 + \frac{A_S}{A}\right) \tag{10.44}$$

$$s^2_{t,\text{TSM}} = \text{Var}[t_c] = \frac{8D^2}{u^4} + \frac{2xD}{u^3}\left(1 + \frac{A_S}{A}\right) + \frac{2x(A_S/A)^2}{u\alpha_s} \tag{10.45}$$

$$\begin{aligned} g_{t,\text{TSM}} &= E\left[\left(t_c - t_{c,\text{TSM}}\right)^3\right] \\ &= \frac{2x^2 D}{u^4}\left(\frac{A_S}{A}\right)\left(1 + \frac{A_S}{A}\right)^2 + \frac{64D^3}{u^6} \\ &\quad + \frac{x}{u}\left[\frac{12D^2}{u^4}\left(1 + \frac{A_S}{A}\right)^2 + \frac{4D(A_S/A)^2}{u^2\alpha_s}\left(\frac{A_S}{A} + 2\right) + \frac{6(A_S/A)^3}{\alpha_s^2}\right] \end{aligned} \tag{10.46}$$

Note that the skewness, $g_{t,\text{TSM}}$, differs from the skewness *coefficient* defined in Chapter 7, Eq. (7.9), by a factor of $s^3_{t,\text{TSM}}$. The skewness coefficient is dimensionless, while $g_{t,\text{TSM}}$, as defined in Eq. (10.46), has units of (T^3). The first two moments, $t_{c,\text{TSM}}$ and $s_{c,\text{TSM}}$, are determined from the tracer concentration data using Eqs. (10.37) and (10.38). The third moment is determined in a similar manner as:

$$\hat{g}_{t,\text{TSM}} = \frac{\sum_{j=1}^{n_x}\left[C_x(t_j) \times (t_j - t_c)^3\right]}{\sum_{j=1}^{n_x} C_x(t_j)} \tag{10.47}$$

The three computed moments from the concentration data are set equal to the three moment equations (10.44) to (10.46), and the values of D, A_S/A, and α_s determined by simultaneous solution of these equations.

EXAMPLE 10.5 ESTIMATION OF TRANSIENT STORAGE MODEL PARAMETERS FROM TRACER DYE STUDY RESULTS

Using the results from the tracer study in Example 10.4 and the prior estimate of the stream velocity, $u = 0.5$ m/s, estimate the three parameters of the transient storage model (assuming a conservative tracer and no lateral inflow or outflow): D, A_S/A, and α_s.

As in Example 10.4, we use the results from both locations; the first provides an estimate for the stream reach from 0 to 2 km and the second from 0 to 4 km. The first two moments at each downstream location have already been calculated in Example 10.4 using the information in Table 10.2:

$$t_{c,x1} = 72.7 \text{ min} \qquad t_{c,x2} = 144.0 \text{ min}$$
$$s_{t,x1}^2 = (17.1 \text{ min})^2 \qquad s_{t,x1}^2 = (25.3 \text{ min})^2$$
$$= 292 \text{ min}^2 \qquad = 640 \text{ min}^2$$

The third moment is calculated from the results in Table 10.2, using Eq. (10.47), yielding:

$$g_{c,x1} = 5103 \text{ min}^3 \qquad g_{c,x2} = 12{,}819 \text{ min}^3$$

The three transient storage model parameters are estimated by simultaneous solution of Eqs. (10.44) to (10.46), using the Excel Solver. The resulting estimates are

- Using the concentration data for first location (0–2 km):

$$D = 26.6 \text{ m}^2/\text{s} \qquad A_S/A = 0.037 \qquad \alpha_s = 6.8 \times 10^{-5}\text{s}^{-1}$$

- Using the concentration data for the second location (0–4 km):

$$D = 29.4 \text{ m}^2/\text{s} \qquad A_S/A = 0.043 \qquad \alpha_s = 8.8 \times 10^{-5}\text{s}^{-1}$$

The values of D are very similar too, though slightly smaller than, that calculated for these data using the standard advective–dispersive model without a transient storage zone (in Example 10.4, $D = 32.5$ m^2/s). This indicates that some of the apparent dispersion computed for the standard model is attributable to the effects of a storage zone. However, the relatively small values for the relative storage zone areas and the exchange coefficients, as well as the low values for the Da1 numbers computed using 10.43 (for the first location, Da1 = 0.28, for the second Da1 = 0.73), suggest that the effects of the storage zone on the main channel transport in this case are only modest. Further insight into these effects is now obtained by examining the implications of the storage zone for computed water quality.

10.6.2 Implications of Transient Storage Model for Water Quality

Transient storage zones provide additional residence time for contaminants as they are transported downstream in a river system. Depending on the type of sorption and chemical reaction processes that take place in these zones, pollutant transport can be significantly delayed and attenuated. However, the storage zone might then also serve as a long-term source of the pollutant or its reaction products back to the main channel. This is very similar in concept to the attenuating effect that absorption and transport to isolated pore water zones have on the flux of contaminants in a groundwater plume (Goltz and Roberts, 1986, 1987).

A first analysis of transient zone effects on concentration profiles for primary and secondary pollutants (e.g., BOD and dissolved oxygen deficit) is provided by Chapra and Runkel (1999), assuming a plug flow system at steady state with a continuous discharge. Recall from Chapter 1 that the steady-state profile for a primary pollutant downstream of a continuous point source that results in a concentration, C_0, at the point of discharge is given by [Eq. (1.14)]:

$$C(x) = C_0 \exp\left(\frac{-kx}{u}\right) \tag{10.48}$$

Assume that k still denotes the first-order pollutant degradation rate coefficient for the main channel, but that the first-order reaction rate coefficient in the storage zone is k_{sz}. The exponential profile predicted by Eq. (10.48) still applies for the concentration as a function of distance along the main channel, but with k replaced by:

$$k_{\text{eff}} = k\left\{1 + \left[\frac{\dfrac{\alpha_s}{k}\left(\dfrac{A_S}{A}\right)}{\dfrac{\alpha_s}{k_{sz}} + \left(\dfrac{A_S}{A}\right)}\right]\right\} \tag{10.49}$$

so that

$$C(x) = C_0 \exp\left(\frac{-k_{\text{eff}}x}{u}\right) \tag{10.50}$$

The value of k_{eff} is greater than that of k due to the longer residence time provided by the transient storage compartment (especially when $k_{sz} > k$). The rate of exponential decay of the pollutant concentration along the stream channel is thus increased when a transient storage zone is present. Chapra and Runkel (1999) provide examples where, even with k_{sz} assumed equal to k, the value of k_{eff} is greater than k by a factor ranging from 1.5, for medium-sized rivers to 2 to 3 for small streams.

EXAMPLE 10.6 EFFECT OF TRANSIENT STORAGE ZONE ON STEADY-STATE CONCENTRATION PROFILE OF A PRIMARY POLLUTANT, FECAL COLIFORM, IN A STREAM

Microbial pathogens are an important primary pollutant in surface water systems. While a variety of bacteria, viruses, and protozoa can pose a threat to public health, coliform bacteria have traditionally been used to characterize the extent of sewage and sanitary waste contamination in streams, lakes, and beach waters. Common measures of microbial pollution include concentrations of total coliform (TC) and fecal coliform (FC) bacteria. The latter are a subset of TC that originate in the intestines of humans and other warm-blooded animals. FC concentrations are typically ~20% of TC, though this ratio can vary widely. Both TC and FC are measured as a number per 100 mL; this value is usually reported as a *most probable number*, or MPN/100 mL, since there is uncertainty in the actual total count of bacteria in an analyzed sample. FC concentrations in raw sewage generally range from 10^5 to 10^8 MPN/100 mL, with values in urban runoff and combined sewer overflows often about 10^4 to 10^5 MPN/100 mL, depending on the amount of dilution of the raw waste. Effective sewage treatment with solids removal and disinfection aims to result in an ~4 order-of-magnitude reduction in coliform concentrations in treated municipal discharges, though this level of treatment is not always achieved, especially when there are highly variable (e.g., storm-affected) flow rates.

Coliform bacteria undergo first-order decay in surface waters due to dieoff and settling, though resuspension can reintroduce settled bacteria to the water column during high flow periods. Coliform morbidity rates tend to increase with exposure to solar radiation, salinity, and high temperature. In surface waters used for contact recreation, ambient FC concentrations of 100 MPN/100 mL or lower are usually sought, with lower values required in waters that support shellfishing.

In the stream described in Examples 10.3 to 10.5, a waste discharge is introduced with a flow rate of $Q_W = 0.1$ m^3/s and an FC concentration $C_W =$ 40,000 MPN/100 mL. An FC removal rate of $k = 2.0$ day^{-1} is estimated for the main channel flow. Determine the steady-state concentration profile downstream from the discharge for a distance of 50 km, contrasting results obtained from the standard (main channel-only) model versus one that includes a transient storage zone. For the latter, first assume that $k_{sz} = k = 2.0$ day^{-1}, then consider the further affects of a higher mortality rate in the storage zone, $k_{sz} = 20.0$ day^{-1}. This higher dieoff rate may occur due to the warmer, shallower waters that are found in transient side pools—these areas are also more strongly affected by solar radiation—or as a result of filtration of the bacteria as water flows through channel sediments. Assume that the transient storage zone parameters determined using the second monitoring station in Example 10.5: $A_S/A = 0.043$; and $\alpha_s = 8.8 \times 10^{-5}$ s^{-1} = 7.60 day^{-1}; apply along a 50-km section of the river (note that the estimated value of D is not needed since longitudinal dispersion is ignored in the steady-state plug flow solution).

The FC loading rate from the waste discharge is first calculated as:

$$S_{\text{FC}} = Q_W C_W = 0.1\ \text{m}^3/\text{s} \times 40{,}000\ \text{MPN}/100\ \text{mL} \times \frac{10^3\ \text{L}}{\text{m}^3} \times \frac{10^3\ \text{mL}}{\text{L}}$$

$$= 4 \times 10^7\ \text{MPN/s}$$

The flow rate of the river upstream of the discharge is $Q_s = uA = 0.5$ m/s × 20 m^2 = 10 m^3/s. The water flow rate from the waste discharge, $Q_W = 0.1$ m^3/s, is a factor of 100 smaller and can thus be ignored it terms of its contribution to the streamflow, with the river flow rate and velocity remaining unchanged downstream of the discharge: $Q_s = 10$ m^3/s and $u = 0.5$ m/s. The FC concentration at the head of the reach is then given by:

$$C_0 = \frac{S_{\text{FC}}}{Q_s} = \frac{4 \times 10^7 \text{ MPN/s}}{10 \text{ m}^3\text{/s}} \times \frac{1 \text{ m}^3}{10^6 \text{ mL}}$$

$$= 400 \text{ MPN/100 mL}$$

This result, 100 times lower than the waste discharge concentration, is expected since the flow in the river provides a factor of 100 dilution relative to the flow rate of the waste discharge.

Figure 10.12 plots the steady-state concentration profile for each of the three cases:

1. Assuming fecal coliform removal occurs only in the main channel with $k = 2.0$ day^{-1}, so that

$$C(x) = 400 \times \exp\left[\left(\frac{-2.0 \text{ day}^{-1}}{0.5 \text{ m/s}} \times \frac{1 \text{ day}}{86{,}400 \text{ s}}\right) * x \text{ (m)}\right]$$

$$= 400 \exp\left[-4.63 \times 10^{-5} \text{ m}^{-1} * x \text{ (m)}\right]$$

2. Assuming fecal coliform removal also occurs in the transient storage zone with $k_{sz} = k = 2.0$ day^{-1}. Eq. (10.49) is first used to compute the effective first-order loss rate:

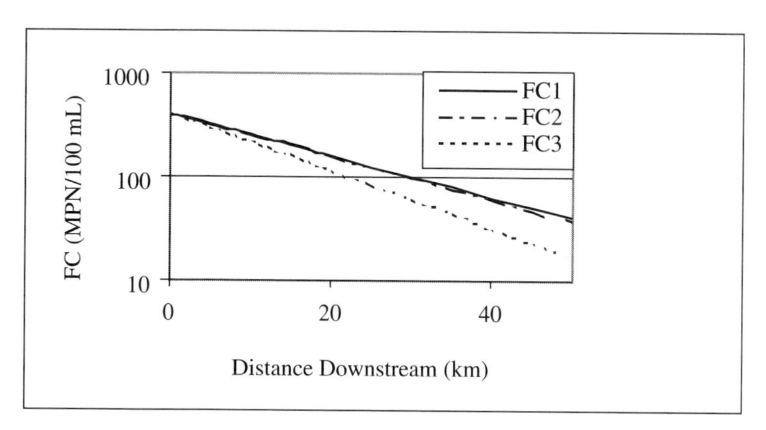

Figure 10.12 Steady-state fecal coliform profile for three cases examined in Example 10.6. (FC1 is case 1, main channel transport only; FC2 is case 2, transient storage zone added; FC3 is case 3, transient storage zone with enhanced dieoff.)

$$k_{\text{eff}} = k\left\{1 + \left[\frac{\frac{\alpha_s}{k}\left(\frac{A_S}{A}\right)}{\frac{\alpha_s}{k_{sz}} + \left(\frac{A_S}{A}\right)}\right]\right\} = 2.0\left\{1 + \left[\frac{\frac{7.60}{2.0}(0.043)}{\frac{7.60}{2.0} + 0.043}\right]\right\}$$

$$= 2.0 \times 1.043 = 2.086 \text{ day}^{-1}$$

and the steady-state FC concentration profile is given by:

$$C(x) = 400 \exp\left(-4.83 \times 10^{-5} \text{ m}^{-1} * x \text{ (m)}\right)$$

3. Assuming fecal coliform removal occurs in the transient storage zone at the higher rate, $k_{sz} = 20 \text{ day}^{-1}$. The effective first-order loss rate is then given by:

$$k_{\text{eff}} = k\left\{1 + \left[\frac{\frac{\alpha_s}{k}\left(\frac{A_S}{A}\right)}{\frac{\alpha_s}{k_{sz}} + \left(\frac{A_S}{A}\right)}\right]\right\} = 2.0\left\{1 + \left[\frac{\frac{7.60}{2.0}(0.043)}{\frac{7.60}{20.0} + 0.043}\right]\right\}$$

$$= 2.0 \times 1.386 = 2.772 \text{ day}^{-1}$$

and the steady-state FC concentration profile is

$$C(x) = 400 \exp\left[-6.42 \times 10^{-5}/\text{m}^{-1} * x(\text{m})\right]$$

As indicated in Figure 10.12, very similar steady-state fecal coliform profiles are predicted downstream of the discharge for the first two cases, since in this example the transient storage zone is small and has a low rate of exchange with the main channel. Only when a much higher rate of dieoff is included in the transient storage zone (case 3: plotted as FC3) is a significant difference apparent.

Chapra and Runkel (1999) also illustrate the effect of a transient storage zone on the steady-state concentration profile of a secondary pollutant, the dissolved oxygen deficit. Assuming that reaeration is negligible in the storage zone (due to the stagnant nature of side pools, or, in the case of a channel sediment storage zone, its isolation from the atmosphere), the peak dissolved oxygen deficit is computed to be larger and to occur closer to the source. Furthermore, dissolved oxygen concentrations are predicted to be even lower in the storage zone. We now consider other stream processes that affect the oxygen balance in a river to provide a more complete steady-state model for stream DO.

10.6.3 Models for the Steady-State Profile of Dissolved Oxygen in a Stream

In Section 1.6.2, we introduced the Streeter–Phelps model for dissolved oxygen in a river, considering only a single point source of BOD at the upstream end of a stream section. Here we add other important processes that can affect dissolved concentrations in a typical stream modeling application, including: (i) the DO deficit in the waste discharge, (ii) distributed, nonpoint sources of BOD entering the stream channel, (iii) bottom sediment oxygen demand, and (iv) algal photosynthesis, respiration, and other water column sources or sinks of dissolved oxygen. A standard stream channel is assumed (without a transient storage zone), however, the adjustments made for transient storage effects that are presented in the previous section could also be introduced for this full DO model.

The different types of distributed loads that can affect stream water quality are shown in Figure 10.13 (from Chapra, 1997). Point sources that we have thus far addressed have been defined by their mass loading rate, $S\ (M\ T^{-1})$, and have been assumed to occur at the beginning of a stream reach. In contrast, Figure 10.13 depicts (a) a line source, $S_d''(M\ L^{-1}\ T^{-1})$, that occurs along the length of the stream reach; (b) an areal source, $S_d'(M\ L^{-2}\ T^{-1})$, that acts either through the top or bottom surface of the stream; and (c) a volumetric source, $S_d(M\ L^{-3}\ T^{-1})$, that acts throughout the water column. For a primary contaminant undergoing first-order decay in the stream channel, the steady-state concentration profile for each of the distributed loading types shown in Figure 10.13 is determined by first calculating the equivalent volumetric loading rate, S_d, as follows:

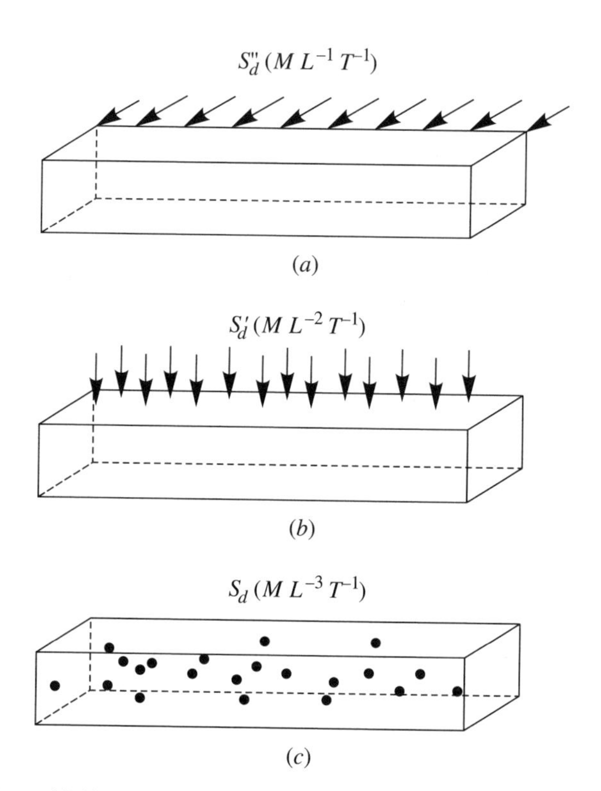

Figure 10.13 (*a*) Line load, (*b*) areal load, and (*c*) volumetric load.

a. For a line source,

$$S_d = S''_d \frac{L}{V} = \frac{S''_d}{A} \tag{10.51}$$

where L is the length of the stream reach, V is the volume of the water in the stream reach, and A is the cross-sectional area of the stream.

b. For an areal source,

$$S_d = S'_d \frac{A_b}{V} = \frac{S'_d}{H} \tag{10.52}$$

where A_b is the surface area of the bottom (or top, for an atmospheric load) of the stream channel through which the load enters and H is the depth of the stream.

Then, the solution for a volumetric distributed load is determined by solving the following mass-balance equation [adapted from Eq. (1.12)]:

$$\frac{\partial C}{\partial t} = -u\frac{\partial C}{\partial x} - kC + S_d \tag{10.53}$$

for steady-state conditions, that is, with $\partial C/\partial t = 0$, so that

$$0 = -u\frac{dC}{dx} - kC + S_d \tag{10.54}$$

For an upstream boundary condition, $C(x = 0) = C_0$, the solution to Eq. (10.54) is given by:

$$C(x) = C_0 \exp\left(\frac{-kx}{u}\right) + \frac{S_d}{k}\left[1 - \exp\left(\frac{-kx}{u}\right)\right] \tag{10.55}$$

The first term in Eq. (10.55) is equivalent to the result in Eq. (10.48) and represents the exponential decline of the upstream boundary concentration. The second term characterizes an exponential approach to a limiting concentration, $C(\infty) = S_d/k$, that balances the distributed volumetric loading rate of a pollutant with its first-order decay. As noted above, solutions for line or areal sources can be calculated using the equivalent volumetric loading rate from Eq. (10.51) or (10.52), and the concentration profile from individual loadings added, through use of superposition (since the model is linear) to obtain a total response.

The result in Eq. (10.55) is used to describe the steady-state BOD profile in a stream receiving point discharges of BOD (that result in new values of C_0 at the point of discharge), as well as distributed loadings along a reach, described by their equivalent $S_{\mathrm{BOD},d}$ values. Likewise, point and distributed discharges of dissolved oxygen deficit can be represented with this equation since for these cases, the dissolved oxygen deficit is the primary pollutant. All that remains is the solution of the (secondary) DOD (dissolved oxygen deficit) profile that results from the distributed loading of BOD [the DOD response to point sources of BOD is given by Eq. (1.18)]. The steady-state mass-balance equation is given by:

$$0 = -u\frac{d\,\text{DOD}}{dx} - k_a\,\text{DOD} + k_d C(x) \tag{10.56}$$

where k_d is the deoxygenation rate of the BOD, k_a is the reaeration rate of the dissolved oxygen, and $C(x)$ is the solution for BOD(x) in Eq. (10.55). The deoxygenation rate is typically estimated using BOD bottle tests for the targeted waste, while the total removal rate, k, for the BOD is fit to the observed stream data since it includes the effects of deoxygenation and other BOD removal processes, such as settling. The value of k_a is usually estimated independent of the observed field data, using a film- or surface-renewal theory model, such as the O'Connor–Dobbins equation presented in Section 4.3.1 [Eq. (4.13)].

Substituting the result for $C(x)$ in Eq. (10.55) into Eq. (10.56) and solving yields

$$\begin{aligned}\text{DOD}(x) &= \text{DOD}_0 \exp\left(\frac{-k_a x}{u}\right) + \frac{k_d C_0}{k_a - k}\left[\exp\left(\frac{-kx}{u}\right) - \exp\left(\frac{-k_a x}{u}\right)\right]\\ &\quad + \frac{k_d S_{\text{BOD},d}}{k_a k}\left[1 - \exp\left(\frac{-k_a x}{u}\right)\right] + \frac{k_d S_{\text{BOD},d}}{(k_a - k)k}\left[\exp\left(\frac{-kx}{u}\right) - \exp\left(\frac{-k_a x}{u}\right)\right]\end{aligned} \tag{10.57}$$

The first line on the rhs of Eq. (10.57) is the same as the result presented in Eq. (1.18), including the decay of the initial upstream deficit (first term) and the response to the upstream BOD point-source loading (second term). The two terms on the second line of Eq. (10.57) represent the contribution from the distributed sources of BOD. With the addition of distributed loadings of DOD, a full description of the steady-state DOD is obtained:

$$\begin{aligned}\text{DOD}(x) &= \text{DOD}_0 \exp\left(\frac{-k_a x}{u}\right) + \frac{k_d C_0}{k_a - k}\left[\exp\left(\frac{-kx}{u}\right) - \exp\left(\frac{-k_a x}{u}\right)\right]\\ &\quad + \frac{k_d S_{\text{BOD},d}}{k_a k}\left[1 - \exp\left(\frac{-k_a x}{u}\right)\right] + \frac{k_d S_{\text{BOD},d}}{(k_a - k)k}\left[\exp\left(\frac{-kx}{u}\right) - \exp\left(\frac{-k_a x}{u}\right)\right]\\ &\quad + \frac{S_{\text{DOD},d}}{k_a}\left[1 - \exp\left(\frac{-k_a x}{u}\right)\right]\end{aligned} \tag{10.58}$$

where the last added term is the contribution due to the distributed DOD loading.

Distributed loadings of DO deficit that are included in standard stream dissolved oxygen models include:

Volumetric Sources

- Algal photosynthesis, P (mg/L day), which is the amount of dissolved oxygen generated by photosynthesis:

$$S_{\text{DOD},d} = -P$$

- Algal respiration, R (mg/L day), which is the amount of dissolved oxygen consumed by respiration:

$$S_{\mathrm{DOD},d} = R$$

- (For long-term, steady-state DO models, the net value of P-R is often estimated.[12])

Areal Sources

- Sediment oxygen demand, SOD (g/m^2 day), which is the rate at which dissolved oxygen is consumed by biochemical reactions at the stream-bottom sediment interface:

$$S_{\mathrm{DOD},d} = \mathrm{SOD}/H$$

- A comprehensive review of methods for estimating SOD is provided by Di Toro (2001). A recent model for dissolved oxygen concentration in the Seneca River, New York, also included an areal source term for the oxygen demand exerted by zebra mussels, termed ZOD (Gelda et al., 2001).

Additional inputs of dissolved oxygen are sometimes included at points along a stream with significant physical reaeration, such as hydraulic jumps or flow over a dam or waterfall.

For a multisegment river with multiple point and distributed sources of BOD and DOD, Eqs. (10.55) and (10.58) are implemented sequentially down the stream channel, with the concentrations computed at the end of each reach serving (with appropriate mixing and dilution adjustments made for new waste or tributary loadings) as the initial concentrations (C_0 and DOD$_0$), for the next downstream reach. As before, the dissolved oxygen concentration profile, DO(x), is determined by subtracting the dissolved oxygen deficit from the saturation value of dissolved oxygen (the value that would be present in the stream in equilibrium with the atmosphere), DO$_{\mathrm{sat}}$: $\mathrm{DO}(x) = \mathrm{DO}_{\mathrm{sat}} - \mathrm{DOD}(x)$. The value of the dissolved oxygen saturation concentration can be computed from empirical equations that relate DO$_{\mathrm{sat}}$ to the water temperature and (for coastal or brackish waters) the salinity, for example (APHA, 1985; Thomann and Mueller, 1987):

$$\begin{aligned}\ln[\mathrm{DO}_{\mathrm{sat}}\ (\mathrm{mg/L})] = & -139.34411 + \frac{1.575701 \times 10^5}{T} - \frac{6.642308 \times 10^7}{T^2} \\ & + \frac{1.24380 \times 10^{10}}{T^3} - \frac{8.621949 \times 10^{11}}{T^4} \\ & - S\left(1.7674 \times 10^{-2} - \frac{10.754}{T} + \frac{2.1407 \times 10^3}{T^2}\right) \end{aligned} \qquad (10.59)$$

[12] In dynamic models for dissolved oxygen, the photosynthetic source of DO occurs during the day, so that supersaturated conditions (with DO greater than DO$_{\mathrm{sat}}$) may occur by late afternoon, whereas respiration is exerted at night, so that very low concentrations of dissolved oxygen can be experienced before sunrise.

where T is the temperature in degrees Kelvin and S is the salinity in parts per thousand (ppt). Further adjustment can be made for reduced dissolved oxygen saturation at high elevation (Thomann and Mueller, 1987):

$$DO_{sat}(Z)\ (\text{mg/L}) \cong DO_{sat}(Z = 0)[1 - 0.000115\ Z] \tag{10.60}$$

where Z is the elevation above sea level (m).

An extended example problem illustrating dissolved oxygen modeling for a river with point and distributed loadings of BOD and DOD is presented in the web appendix for this chapter at www.wiley.com/college/ramaswami.

10.7 AVAILABLE WATER QUALITY MODELS[13]

Numerous computer models are available to predict non-point-source loadings and receiving water response to both point and non-point-source discharges. Many models have similar overall capabilities but operate at different time and spatial scales and were developed for different purposes. The available models range between empirical and physically based. However, all simplify the relevant physical processes and often include empirical components for complex underlying mechanisms.

10.7.1 Non-Point-Source Pollution Loading Models

Watershed or loading models can be divided into categories based on their complexity, operation, time step, and simulation technique. In its review of available models for conducting allowable Total Maximum Daily Load TMDL studies, the U.S. EPA has grouped existing watershed-scale models into three categories based on the number of processes it incorporates and the level of detail it provides (U.S. EPA, 1997). Simple models implement empirical relationships between physiographic characteristics of the watershed and pollutant runoff. These relationships are fit using regression methods, and usually apply to aggregate watershed loadings over larger (e.g., monthly or annual) time scales. A number use the Universal Soil Loss Equation for solids loading rates, with pollutant concentrations associated with the solids. Examples include the EPA Screening Procedures developed by Mills et al. (1985), the USGS Regression Method described by Tasker and Driver (1988), the USGS Watershed Model (Walker et al., 1989), the U.S. Federal Highway Administration (FHWA) model (Driscoll et al., 1990a, 1990b), and the Watershed Management Model (WMM) developed for the State of Florida (Camp, Dresser and McKee, 1992).

The EPA also identifies watershed models of intermediate complexity. These use many of the same empirical relationships for estimating stormwater runoff rates and pollutant concentrations but allow for a higher degree of spatial and temporal disaggregation, simulating daily or event loading rates for multiple sub-basins. Examples of these include the Generalized Watershed Loading Functions (GWLF) model developed at Cornell University (Haith and Shoemaker, 1987; Haith et al., 1992), the

[13] Thanks and appreciation to Nicole Rowan who wrote the initial draft of Section 10.7.

USDA Agricultural Nonpoint Source Pollution Model (AGNPS) developed by Young et al. (1989) and linked to a GIS system by Tim and Jolly (1994), and the Source Loading and Management Model (SLAMM) (Ventura and Kim, 1993).

Detailed, mechanistic non-point-source pollution models use storm event or continuous simulation at hourly or shorter time scales to predict flow and pollutant concentrations for a range of flow conditions. These models explicitly simulate the physical processes of infiltration, runoff, pollutant accumulation, groundwater/surface water interaction, and hydraulic processes in conveyance channels and sewerage system pipes and overflow devices. They require a significant amount of site data and are rarely used at the initial planning stage for watershed evaluation and management (U.S. EPA 1997). However, they are very valuable at the detailed design stage, when specific non-point-source control options are located, sized, and evaluated. A list of mechanistic non-point-source pollution models and their capabilities is presented in the web appendix for this chapter at www.wiley.com/college/ramaswami.

10.7.2 Receiving Water Quality Models

Receiving water quality models differ in many ways, but some important dimensions of discrimination include their conceptual basis, input conditions, process characteristics, and output. Table 10.3 presents extremes of simplicity and complexity for each condition as a point of reference. Most receiving water quality models have some mix of simple and complex characteristics that reflect trade-offs made in optimizing performance for a particular task, and the state of the art in water quality modeling continues to evolve (Shanahan et al., 1998). Available steady-state surface water quality models are listed and described in the web appendix for this chapter. Table 10.4 summarizes the capabilities of four models that are often applied to evaluate dynamic water quality problems.

10.7.3 Multidimensional Hydrodynamic Models

In recent years a new class of water quality models has been developed that includes a full linkage between hydraulic and hydrodynamic flow prediction and water quality simulation. Water flow in multidimensional water systems is determined by a number of processes affecting their flow, heat, and momentum balances. Hydrodynamic models incorporating these processes are now used to compute water flow and circulation

TABLE 10.3 General Receiving Water Quality Model Characteristics

Model Characteristic	Simple Models	Complex Models
Conceptual basis	Empirical	Mechanistic
Input conditions	Steady state	Dynamic
Reaction process	Nonreactive, equilibrium, or linear kinetics	Nonlinear kinetics
Output conditions	Deterministic	Stochastic

TABLE 10.4 Descriptive List of Model Components—Dynamic Water Quality Models

Model	Water Body Type	Parameters Simulated	Process Simulated	
			Physical	Chemical/Biological
DYNTOX (Limno-Tech, 1994)	River	Conservative and nonconservative substances: probabilistic simulation	Dilution, advection	First-order decay
WASP5 (Ambrose, et al., 1993; Lung and Larson, 1995)	Estuary, river, (well-mixed/ shallow lake)	DO, CBOD, NBOD, ammonium, nitrate, nitrite, organic nitrogen, total phosphate, organic phosphorus, inorganic suspended solids, fecal coliform, conservative and nonconservative substances	Dilution, advection, dispersion, reaeration	First-order decay, process kinetics, daughter products, hydrolysis, oxidation, volatilization, photolysis, equilibrium adsorption. Settling, DO–CBOD, nutrient-algal cycle
CE-QUAL-RIV1 (U.S. Army Engineers, 1990, 1995)	Rivers	DO, CBOD, temperature, ammonia, nitrate, algae, coliform, phosphate, organic nitrogen	Dilution, advection, dispersion, heat balance	First-order decay, DO–CBOD, nutrient-algal cycle, carbon cycle
HSPF (Bicknell et al., 1993)	River (well-mixed/ shallow lakes)	DO, BOD, nutrients, pesticide, sediment, organic chemicals, and temperature	Dilution, advection, heat balance, particle fate, cohesive/ noncohesive sediment transport	First-order decay, process kinetics, daughter products, hydrolysis, oxidation, volatilization, photolysis, benthic demand, respiration, nutrient-algal cycle

Source: U.S. EPA (1997).

patterns in complex surface water systems, including lakes, estuaries, and coastal bays and harbors. A number of hydrodynamic models for surface waters are reviewed by Martin and McCutcheon (1999, see Chapter 15, Sections III–VI). The Water Environment Research Federation provides an overview of model accessibility and use for a number of these models, along with modeling selection software for narrowing the set of models that meet a particular specification (Fitzpatrick et al., 2001). Abbot

(1997) summarizes advances in integrating different types of hydrodynamic models with field data collection studies. Tetra Tech (2000) evaluates a number of approaches and models for characterizing ocean circulation and plume dispersion, with a particular focus on their applicability for use in modeling offshore ocean outfalls. Imhoff et al. (2003) compare the capabilities of a number of traditional water quality models with those of hydrodynamic water quality models for predicting the fate and transport of contaminated sediments. Here we provide a brief overview of the capabilities of a few of the more widely used models: the U.S. Army Engineers CE-QUAL-W2 model, the Virginia Institute of Marine Science EFDC/HEM-3D model, the computationally similar Princeton Ocean Model (POM), the FLOW hydrodynamic model, available as part of the Delft3D modeling system from Delft Hydraulics, and the EPA-approved Cornell Mixing Zone Expert System (CORMIX).

CE-QUAL-W2 is a 2D model that has been applied to rivers, lakes, reservoirs, and estuaries. The model assumes complete mixing in the lateral direction, allowing for longitudinal and vertical discretization. The model simulates water surface elevations, velocities, and temperature, with options for including temperature (buoyancy) and solute concentration effects on water density. Like its predecessor model, the Generalized Longitudinal-Hydrodynamics and Transport (GLVHT) model of Buchak and Edinger (1984, see also Edinger, 2002), CE-QUAL-W2 uses a zero-equation closure approach with the vertical eddy viscosity evaluated from a von Karman relationship (modified by a local Richardson number) with a mixing length that depends on layer depth and thickness (see Appendix 5A). Implicit finite difference solution methods are used and an automated time-step selection algorithm allows for efficient simulation while ensuring that numerical stability requirements are not violated. The model simulates sediment processes and multiple water quality constituents using modular algorithms. CE-QUAL-W2 is available from the U.S. Army Engineers Waterways Experiment Station (Cole and Buchak, 1995; see also http://www.wes.army.mil/el/elmodels/). Applications of CE-QUAL-W2 are described in Adams et al. (1997) and Wells and Cole (2000).

The Environmental Fluid Dynamics Code (EFDC) and the associated Hydrodynamic-Eutrophication Model 3-D (HEM-3D) are available as a package from the Virginia Institute of Marine Science (Hamrick, 1992; Park et al., 1995). The EFDC model solves the turbulent-averaged momentum and continuity equations for a variable density fluid with a free surface. Coupled transport equations are solved for turbulent kinetic energy, turbulent length scale, salinity, and temperature, using the Mellor-Yamada Level 2.5 turbulence closure method (Mellor and Yamada, 1982; Galperin et al., 1988). The equations are solved using finite difference methods with a semi-implicit mode splitting procedure that separates the baroclinic (internal shear) mode from the baratropic (external free surface gravity wave) mode.[14] The hydrodynamic and water quality portions of the HEM-3D model are internally linked, though their computations are decoupled. Applications of the model

[14] Baroclinic flow refers to circulation caused by horizontal density differences, while baratropic flow refers to circulation induced by differences in surface elevation.

are described in Park et al. (1998) and Shen et al. (1999). The EFDC is computationally similar to the Princeton Ocean Model (POM), a model often used for bay, continental shelf, and open-ocean applications, as both use methods derived from Blumberg and Mellor (1980, 1987). More information on POM is available at http://www.aos.princeton.edu/WWWPUBLIC/htdocs.pom/, and applications of the model are described in Ahsan et al. (1994) and Aikman and Wei (1995).

The Delft3D system is an integrated set of models for morphologic characterization and hydrodynamic and water quality simulation, available (for a fee, but with extensive user support) from Delft Hydraulics (see http://www.wldelft.nl/soft/index.html). The package includes utilities for bathymetry (bottom topography) generation, grid development, and visualization of model inputs and output. The FLOW hydrodynamics module of Delft3D provides a two- or three-dimensional simulation of nonsteady flow resulting from Coriolis, tidal, and/or meteorological forcing on a curvilinear, boundary-fitted grid that maintains a constant number of vertical segments. The model provides solution of the full Navier–Stokes equations with pressure gradients and density driven flows, with options for modeling vertical turbulent viscosity using zero- through higher-order closure methods. Horizontal turbulent exchange coefficients can be computed from the sum of 3D and 2D subgrid turbulence models. Equations are solved using various finite difference methods including higher-order central or backward differences and flux-corrected schemes to preserve steep gradients. The model and its applications are described in Casulli and Stelling (1998) and Davies and Gerritsen (1994).

As indicated, most of the available models for hydrodynamic simulation apply finite difference solution methods to some form of the Navier–Stokes and continuity equations, though models are also available using finite element methods.[15] The final model considered here, the Cornell Mixing Zone Expert System (CORMIX), uses a very different approach, based on a set of analytical solutions for buoyant and nonbuoyant plumes similar to the plume-rise equations for air pollution described in Section 8.3. CORMIX characterizes the near-field mixing of water and wastewater plumes discharged to surface waters using solutions for submerged single and multiport discharges and buoyant surface discharges. Buoyant discharge plumes commonly occur since discharged waters are typically warmer than ambient receiving waters. The CORMIX package also provides an expert-system interface to help users select among model configurations and assumptions. Dimensionless parameters involving discharge volume flux, momentum flux, buoyancy flux, and instream cross-flow velocity are used to identify length scales for the plume as it evolves from the near-field to the far-field, transitioning from jet spreading to buoyancy spreading and eventually to predominantly passive dispersion driven by ambient turbulence. The model predicts the resulting geometry and dilution characteristics of discharge plumes, considering also the effects of wind on thermal plume mixing and,

[15] For example, the Finite-Element Surface-Water Modeling System, FESWMS, available from the USGS (see http://water.usgs.gov/software/feswms.html), and the RMA family of models, available from the U.S. Army Engineers WES or Resource Management Associates (see http://hlnet.wes.army.mil/software/tabs/models.htp).

if applicable, tidal and other transient circulation patterns. The bases for the different CORMIX modules are described in Jirka and Donekaer (1991), Jirka and Akar (1991), and Jones et al. (1996). The model is available from the U.S. EPA (http://www.epa.gov/ceampubl/swater/cormix/index.htm) or Portland State University (http://www.cormix.info/index.php).

Homework problems for this chapter are provided at www.wiley.com/college/ramaswami.

11 Atmospheric Transformation and Loss Processes

The air quality models described in Chapter 8 focus on pollutant transport and thus are suitable for primary pollutants. The analytical models in particular are derived using the assumption that reaction and decay processes are negligible or can be treated as first-order losses. However, many harmful air pollutants, such as ozone, sulfates, and nitrates, are secondary pollutants which are not emitted directly but rather are formed in the atmosphere through chemical reactions. This chapter discusses *photochemical* models of gas-phase chemical reactions that form secondary air pollutants and models of secondary aerosol formation and growth.

Figure 11.1 illustrates major elements of photochemical air quality models, including the inputs they require and the main processes they track. Along with chemical reactions, dry deposition and wet scavenging are important loss processes for both gases and particles, so their treatment is also described here. Aqueous-phase chemistry in cloud and fog droplets is also important in air pollution. This topic is covered briefly in this chapter, supplementing the broader discussion of aqueous chemistry in Chapter 12. Although secondary air pollutants and atmospheric chemistry are modeled on scales ranging from those of aircraft or power plant plumes to the global scale, this chapter focuses on models developed for urban or regional-scale applications.

The next section discusses the gas-phase atmospheric chemistry involved in formation of tropospheric ozone, along with the treatment of gas-phase chemical kinetic mechanisms in air quality models. Section 11.2 introduces the atmospheric chemistry of sulfur and nitrogen oxides, which involves both gas- and aqueous-phase reactions. Sections 11.3 and 11.4, respectively, address aerosol dynamics and wet and dry deposition processes. Section 11.5 compares some of the currently available photochemical models and presents a case study of a photochemical model application.

11.1 GAS-PHASE ATMOSPHERIC CHEMISTRY

11.1.1 Photolysis Reactions

The gas-phase chemistry that leads to formation of ozone and to the oxidation of organic compounds, sulfur oxides, and nitrogen oxides is driven by sunlight and so is known as photochemistry. Two main types of reactions are involved in photochemistry: *photolysis* reactions in which gas molecules interact directly with sunlight and *thermal* reactions between two or more gas molecules.

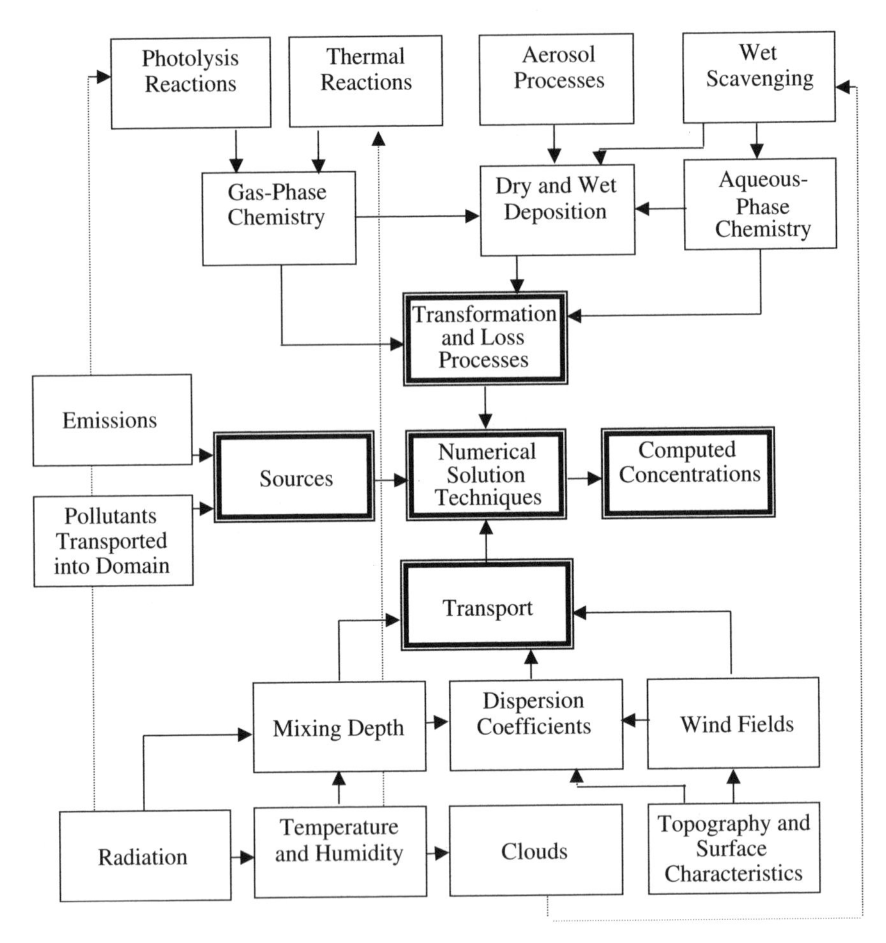

Figure 11.1 Inputs, outputs, and processes included in photochemical air quality models. (Adapted from Demerjian, 1975.)

In the troposphere, sunlight at wavelengths ranging from the near ultraviolet at about 290 nm through the visible region up to about 730 nm provides energy for dissociation reactions that lead to the formation of free radicals (molecules or atoms with an unpaired electron) and molecules in excited states. These radicals and molecules participate in chain reactions that drive carbon, nitrogen, and sulfur compounds in the atmosphere to their most oxidized states; for example, breaking down and oxidizing hydrocarbons to form CO_2. Sunlight at wavelengths below 290 nm is important for driving stratospheric chemistry, but almost all of it is absorbed by O_2 and O_3 in the stratosphere[1] and so is not available to participate in reactions that occur at lower altitudes.

[1] While this chapter focuses on processes that occur in the troposphere, Seinfeld and Pandis (1998) provide an excellent introduction to stratospheric chemistry.

Two of the most important photolysis reactions that occur in the troposphere are the dissociation of O_3 and NO_2:

$$O_3 + h\nu \rightarrow O^1D + O_2 \qquad \lambda < 320\ \text{nm} \tag{11.1}$$

$$NO_2 + h\nu \rightarrow NO + O^3P \qquad 290\ \text{nm} < \lambda < 430\ \text{nm} \tag{11.2}$$

where O^1D represents an oxygen atom in an excited electronic state (one in which absorbed energy has promoted electrons into high-energy orbitals) and O^3P a ground-state oxygen atom. In the presence of water vapor, O^1D can react to form hydroxyl radicals, which are the single most important oxidizing agent in the atmosphere:

$$O^1D + H_2O \rightarrow 2HO \tag{11.3}$$

The O^3P atom from reaction (11.2) reacts with molecular oxygen to form ozone:

$$O^3P + O_2 + M \rightarrow O_3 + M \tag{11.4}$$

where M represents a third molecule that is needed to absorb excess energy from the reaction. Photodissociation of NO_2 is the only reaction that leads to O_3 formation in the troposphere. In the stratosphere, ozone is also formed from photodissociation of O_2.

In general, interactions between a photolyzing gas molecule A and sunlight produce an excited-state molecule A^*,

$$A + h\nu \rightarrow A^* \tag{11.5}$$

which further reacts to produce one of four outcomes:

$$\text{Dissociation: } A^* \rightarrow B1 + B2 + \cdots \tag{11.6a}$$

$$\text{Direct reaction: } A^* + C \rightarrow D1 + D2 + \cdots \tag{11.6b}$$

$$\text{Fluorescence: } A^* \rightarrow A + h\nu \tag{11.6c}$$

$$\text{Collisional deactivation: } A^* + M \rightarrow A + M \tag{11.6d}$$

The probability that molecules of A^* will react by the ith pathway is known as the quantum yield, Φ_i, for that reaction route. The sum of the quantum yields of reactions (11.6a) to (11.6d) must equal 1. Of the four pathways, dissociation is the most important for atmospheric chemistry.

The tendency for molecule A to interact with sunlight of a given wavelength is given by its absorption cross section, $\sigma(\lambda, T)$. The overall rate at which A reacts with sunlight to produce new products by dissociation depends on the product of sunlight intensity, $I(\lambda)$, the quantum yield for pathway (11.6a), Φ_a, and the absorption cross section for the molecule, integrated across the relevant wavelength spectrum:

$$\frac{d[\mathrm{A}]}{dt} = -k_\lambda^{\mathrm{A}}[\mathrm{A}] \tag{11.7}$$

$$k_\lambda^{\mathrm{A}} = \int_{280\,\mathrm{nm}}^{730\,\mathrm{nm}} I(\lambda)\sigma(\lambda)\phi_a(\lambda)\,d\lambda \tag{11.8}$$

Absorption cross sections and quantum yields are usually determined in laboratory studies. The intensity and spectral distribution of sunlight is a function of the date, latitude, altitude, solar zenith angle, overhead ozone column, and presence of clouds or particles. Sunlight spectra can be measured for a given time and location or modeled using radiative transfer codes (e.g., Madronich et al., 1996).

11.1.2 Chemistry of Ozone Formation in the Troposphere

Ozone is produced in the troposphere through coupled reactions involving nitrogen oxides (NO_x), volatile organic compounds (VOCs), water vapor, and sunlight. Nitrogen oxides are emitted into the atmosphere from high-temperature combustion sources, primarily in the form of nitric oxide. Vapor-phase organic compounds are emitted from a wide variety of sources, including incomplete combustion, evaporation of solvents and liquid fuels, and vegetation. The presence of hundreds of different organic compounds in urban air significantly complicates the chemistry of ozone formation. However, the basic mechanism can be illustrated by considering methane as a model organic compound.

As noted above, ozone is formed from NO_2 photodissociation (11.2) followed by the reaction of ground-state oxygen atoms with O_2 (11.4). Ozone itself can undergo photolysis (11.1) or react with NO:

$$\mathrm{NO} + \mathrm{O_3} \rightarrow \mathrm{NO_2} + \mathrm{O_2} \tag{11.9}$$

In the absence of organic compounds, reactions (11.2), (11.4), and (11.9) rapidly reach a condition known as the photostationary state. When photostationary conditions are achieved, the rate of ozone production by reaction (11.4) is equal to the rate of ozone consumption by reaction (11.9):

$$k_4[\mathrm{O^3P}][\mathrm{O_2}][\mathrm{M}] = k_9[\mathrm{NO}][\mathrm{O_3}] \tag{11.10a}$$

where k_4 and k_9 represent the rate constants of reactions (11.4) and (11.9), respectively. Rearranging Eq. (11.10a) to solve for the O_3 concentration gives

$$[\mathrm{O_3}] = \frac{k_4[\mathrm{O^3P}][\mathrm{O_2}][\mathrm{M}]}{k_9[\mathrm{NO}]} \tag{11.10b}$$

Equation (11.10b) can be simplified because the ground-state oxygen atom, O^3P, is very short-lived. It can be assumed to react via reaction (11.4) as quickly as it is formed from reaction (11.2), that is:

$$k_2[NO_2] = k_4[O^3P][O_2][M] \tag{11.10c}$$

Rearranging Eq. (11.10c) to solve for $[O^3P]$ and substituting this expression into (11.10b) gives the photostationary state expression for ozone:

$$[O_3] = \frac{k_2[NO_2]}{k_9[NO]} \tag{11.10d}$$

In this case, no buildup of ozone occurs because once formed it quickly reacts with NO to regenerate NO_2.

EXAMPLE 11.1

Calculate the ozone concentration given by the photostationary state equation for polluted urban rush-hour concentrations of $[NO_2] = 0.02$ ppm and $[NO] = 0.10$ ppm. Use $k_2 = 0.3\ \text{min}^{-1}$, characteristic of a summer day at midlatitudes, and $k_9 = 26\ \text{ppm}^{-1}\ \text{min}^{-1}$.

Solution 11.1 The photostationary state equation gives

$$[O_3] = \frac{0.3\ \text{min}^{-1}\ (0.02\,\text{ppm})}{26\,\text{ppm}^{-1}\ \text{min}^{-1}\ (0.10\,\text{ppm})} = 0.002\,\text{ppm}$$

This concentration is very low compared to the 1-h average NAAQS for ozone of 0.12 ppm. So why is ground-level ozone a problem?

Hydrocarbons disrupt the sequence of reactions (11.2), (11.4), and (11.9) by providing an alternative means of oxidizing NO to NO_2 without consuming O_3. Their involvement is initiated by reaction with the hydroxyl radical, for example, for methane:

$$CH_4 + HO\bullet \rightarrow CH_3\bullet + H_2O \tag{11.11}$$

The methyl radical reacts very quickly with O_2 to produce a methyl peroxy radical:

$$CH_3\bullet + O_2 + M \rightarrow CH_3O_2\bullet + M \tag{11.12}$$

The methyl peroxy radical can react with NO to produce NO_2 and a methoxy radical, which then reacts with O_2 to produce formaldehyde and a hydroperoxy radical:

$$CH_3O_2 + NO \rightarrow NO_2 + CH_3O\bullet \tag{11.13}$$

$$CH_3O\bullet + O_2 \rightarrow HCHO + HO_2\bullet \tag{11.14}$$

The HO_2 radical continues the chain by reacting with a NO molecule to produce NO_2 and HO•:

$$HO_2\bullet + NO \rightarrow NO_2 + HO\bullet \tag{11.15}$$

The NO_2 molecules produced in reactions (11.13) and (11.15) can photolyze, leading to O_3 production. The HO• radical produced in reaction (11.15) is available to attack another organic compound molecule [e.g., through reaction (11.11)]. Alternatively, the radical chain may be terminated by the reaction of HO• with NO_2 to produce nitric acid:

$$HO\bullet + NO_2 \rightarrow HNO_3 \tag{11.16}$$

or by interactions between peroxy radicals, for example:

$$HO_2\bullet + HO_2\bullet \rightarrow H_2O_2 + O_2 \tag{11.17}$$

Nitric acid formation (11.16) is the most important chain termination reaction under the high NO_x conditions found in most urban areas, whereas radical–radical termination reactions such as reaction (11.17) are important in areas that are far from NO_x sources.

Formaldehyde is an important participant in O_3 photochemistry because it can undergo photolysis, with one of its photodissociation pathways producing two hydroperoxy radicals:

$$HCHO + h\nu \rightarrow 2HO_2\bullet + CO \tag{11.18}$$

$$HCHO + h\nu \rightarrow H_2 + CO \tag{11.19}$$

HCHO and its product CO also react with HO• to produce $HO_2\bullet$, with the potential of converting more NO to NO_2 through reaction (11.15):

$$HCHO + HO\bullet \rightarrow HO_2\bullet + CO + H_2O \tag{11.20}$$

$$CO + HO\bullet \rightarrow CO_2 + HO_2\bullet \tag{11.21}$$

Formaldehyde photolysis is an important source of radicals, especially in polluted areas.

As illustrated in Figure 11.2, reactions such as (11.1) to (11.6), (11.9), and (11.11) to (11.21) lead to a highly nonlinear relationship between ozone and its precursors, NO_x and VOCs. At point *A*, ozone concentrations could be reduced effectively by reducing VOC emissions, but would be increased by reducing NO_x emissions. In contrast, at point *B*, reducing VOC emissions would have little impact, but reducing NO_x emissions would be effective. Because of the nonlinearity in the system, photochemical air quality models with accurate emissions, chemistry, and transport are required to predict the consequences of proposed emissions control strategies.

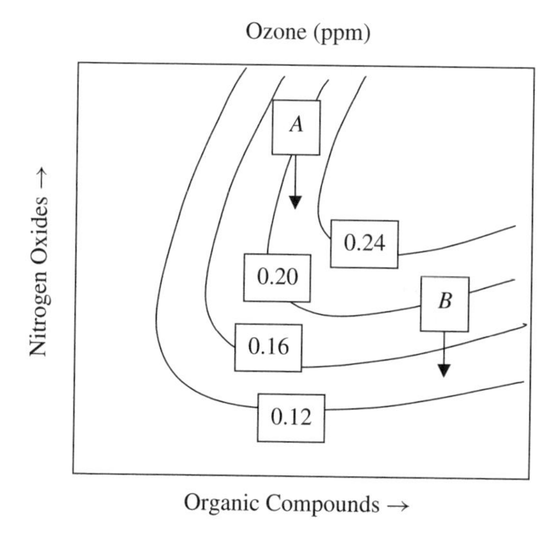

Figure 11.2 Schematic ozone isopleth diagram showing dependence of ozone concentrations on NO_x and organic compound emissions levels.

Although methane and formaldehyde illustrate the role that organic compounds play in ozone formation, these relatively simple compounds provide a somewhat misleading example. Oxidation mechanisms for larger molecules are more complex, as Figure 11.3 shows for propane and acetaldehyde. Adding further complexity, nonsymmetric hydrocarbon bonds or functional groups involving oxygen or halogens can present alternative sites for the addition of HO• or for hydrogen atom abstraction, resulting in competing reaction pathways (Finlayson-Pitts and Pitts, 1986). In addition, although alkanes and aromatic compounds react primarily with HO•, alkenes also react with O_3, NO_3, and O, and many carbonyl compounds photodissociate at significant rates. As discussed by Atkinson (1994), understanding of many of these reactions is still evolving.

Derwent and co-workers (Jenkin et al., 1997) have developed a "master" mechanism for tropospheric O_3 chemistry that includes more than 7000 reactions. Tracking that level of detail in a three-dimensional air quality model is not computationally feasible, nor is it justified by the state of understanding of many of the hypothesized reactions. Reactions of aromatic compounds are a good case in point. Toluene, xylenes, and trimethylbenzenes are widespread in the atmosphere, comprising 30% or more of the hydrocarbons in many urban areas (Jeffries, 1995). However, their oxidation mechanisms are not known. Laboratory studies have identified products accounting for less than 50% of the carbon in the aromatic molecules (Andino et al., 1996; Kwok et al., 1997). Air quality models generally use parameterized mechanisms for aromatic compound oxidation, estimated from ozone yields in laboratory experiments (Wang et al., 2000). In the end, most urban or regional-scale photochemical air quality models are simplified to include only about 150 to 250 chemical reactions (Carter, 1990, 2000; Gery et al., 1989; Stockwell et al., 1990, 1997).

Propane

$CH_3CH_2CH_3 + HO\cdot \rightarrow CH_3C(\cdot)HCH_3 + H_2O$

$CH_3C(\cdot)HCH_3 + O_2 \rightarrow CH_3CH(O_2\cdot)CH_3$

$CH_3CH(O_2\cdot)CH_3 + NO \rightarrow NO_2 + CH_3CH(O\cdot)CH_3$

$CH_3CH(O\cdot)CH_3 + O2 \rightarrow CH_3C(O)CH_3$ (acetone) $+ HO_2\cdot$

Acetaldehyde

$CH_3CHO + h\nu \rightarrow CH_3\cdot + HCO\cdot$

$CH_3CHO + HO\cdot \rightarrow CH_3C(\cdot)O + H_2O$

$CH_3\cdot + O_2 \rightarrow CH_3O_2\cdot$

$HCO\cdot + O_2 \rightarrow HO_2\cdot + CO$

$CH_3C(\cdot)O + O_2 \rightarrow CH_3C(O)O_2\cdot$

$CH_3O_2\cdot + NO \rightarrow NO_2 + CH_3O\cdot$

$CH_3O\cdot + O_2 \rightarrow HCHO + HO_2\cdot$

$CH_3C(O)O_2\cdot + NO \rightarrow CH_3C(O)O\cdot + NO_2$

$CH_3C(O)O\cdot + O_2 \rightarrow CH_3O_2\cdot + CO_2$

$CH_3C(O)O_2\cdot + NO_2 \rightarrow CH_3C(O)O_2NO_2$ (peroxy acetyl nitrate or PAN)

$CH_3C(O)O_2NO_2 \rightarrow CH_3C(O)O_2\cdot + NO_2$

Figure 11.3 Oxidation mechanisms for propane and acetaldehyde (Seinfeld, 1986).

Gas-phase mechanisms for urban atmospheric chemistry are tested primarily by comparison against controlled laboratory experiments. So-called smog chambers are large bags or rigid-walled structures in which mixtures of gases can be irradiated with either real or artificial sunlight. Indoor chambers typically use blacklights or xenon arc lights to approximate the solar spectrum. Outdoor smog chambers use real sunlight but are subject to variations in light intensity when clouds pass overhead. As illustrated in Figure 11.4, chamber experiments are also used to estimate lumped parameters for reaction mechanisms that are poorly known, such as those for aromatic compounds. Chamber experiments provide the only feasible means to test the mechanisms developed for air quality modeling because chemistry in the real atmosphere is affected by too many variables that are uncontrolled or poorly characterized. However, chamber experiments are themselves subject to artifacts that are difficult to account for in mechanism tests. Losses to the chamber walls and off-gassing from the walls have proven to be especially tricky. These artifacts are known to be significant but are not understood well enough to model using first principles (Carter et al., 1995).

11.1.3 Mathematical Modeling of Gas-Phase Chemical Kinetics Equations

The simplest modeling framework available for considering atmospheric chemistry is a closed, well-mixed box or batch reactor with specified initial conditions. Batch reactor models are widely used for testing chemical mechanisms and for exploratory analyses of atmospheric chemistry (e.g., Dodge et al., 1989; Gao et al., 1996). In a

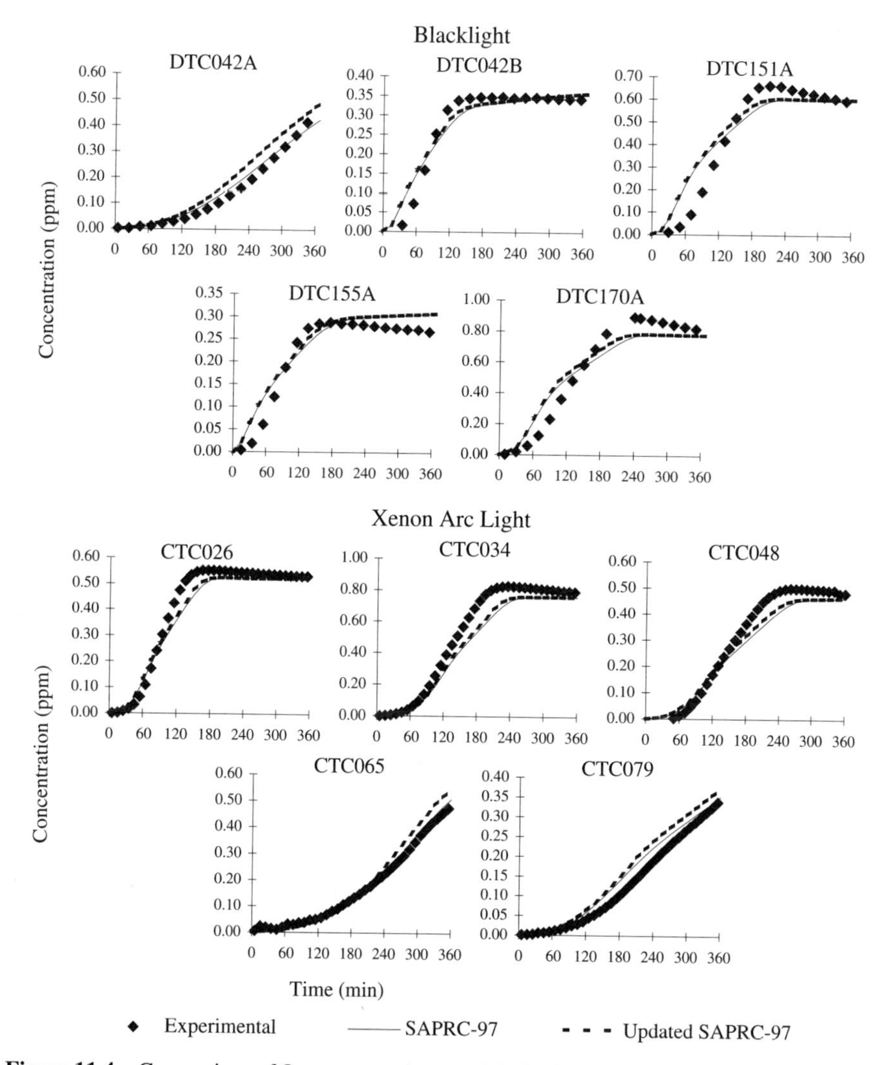

Figure 11.4 Comparison of O_3 concentrations modeled using the SAPRC-97 chemical mechanism with concentrations measured in an indoor smog chamber at the University of California at Riverside (Wang et al., 2000). In this series of experiments, toluene, NO_x, and air were injected into the chamber and the mixtures were irradiated with artificial lights. The experiments were used to adjust key parameters of the SAPRC-97 aromatics mechanism to achieve the best fit with the measured O_3 concentrations.

batch reactor, with no spatial dependence, mass-balance equations are formulated as a system of coupled, ordinary differential equations for the suite of N species that participate in the chemistry, as:

$$\left.\frac{dC^i}{dt}\right|_{\text{Rxn}} = \text{Prod}^i(\mathbf{C}, t) - \text{Loss}^i(\mathbf{C}, t)C^i \qquad i = 1, \ldots, N \qquad (11.22)$$

where $\mathbf{C} = \{C^1, C^2, \ldots, C^N\}^T$ is the transposed vector of species concentrations. In Eq. (11.22), Prod^i denotes the rate of production of species i through chemical reactions and $\text{Loss}^i C^i$ its loss rate. For example, based on reactions (11.1) to (11.6), (11.9), and (11.11) to (11.21), the rate expression for NO is

$$\left.\frac{d[\text{NO}]}{dt}\right|_{\text{Rxn}} = k_2[\text{NO}_2] - k_9[\text{NO}][\text{O}_3] - k_{13}[\text{CH}_3\text{O}_2\cdot][\text{NO}] - k_{15}[\text{HO}_2\cdot][\text{NO}] \tag{11.23}$$

The initial conditions for the system of equations are $C^i(t = 0) = C_0^i$, $i = 1, \ldots, N$.

The rate constants of nonphotolytic reactions are usually expressed as a function of temperature in the Arrhenius form:

$$k_j(T) = A_j T^{B_j} \exp\left(\frac{-E_{aj}}{RT}\right) \tag{11.24}$$

where A_j is the pre-exponential factor, B_j is the temperature exponent, and E_{aj} is the activation energy for the reaction. The values of each of these three reaction-dependent constants are either determined experimentally or estimated using structure–reactivity relationships. Rate constants for photolysis reactions are specified by Eq. (11.8).

Because emissions of organic compounds and NO_x vary significantly across urban areas, actual airsheds cannot be modeled accurately as batch reactors. To investigate spatial variations in secondary pollutant concentrations due to advection and dispersion, 3D photochemical models are required. These models provide numerical solutions to the 3D advection–dispersion–reaction equation:

$$\frac{\partial C^i(\mathbf{x}, t)}{\partial t} + \mathbf{u} \cdot \nabla C^i = \nabla \cdot D \nabla C^i \pm \text{Rxn}^i(\mathbf{C}) + \mathbf{S^i(x, t)} \qquad \mathbf{i = 1, \ldots, N} \tag{11.25}$$

When coupled photochemistry is being considered, Eq. (11.25) must be solved numerically using discrete cells or grids to approximate the spatial variations. Most 3D photochemical air quality models employ operator splitting to separate out the chemical integration steps from the transport steps. Using this approach, the numerical algorithms in the model alternate between transport steps and solving the chemical operator equations, which are just the rate equations for a batch reactor [Eq. (11.22)]. Operator splitting allows the solution techniques that are best suited to each process to be used. Numerical solution techniques that are employed to integrate the chemical rate equations are described in the supplemental information at www.wiley.com/college/ramaswami.

11.2 OXIDATION OF SULFUR AND NITROGEN OXIDES

In the late 1950s and the 1960s, scientists increasingly recognized the contributions that anthropogenic emissions of SO_2, NO_x, and organic compounds were making to the acidification of rainwater and in turn of lakes and streams in some areas. This trend was especially apparent in northern Europe, the northeastern United States, and southeastern Canada. Research conducted through the 1970s and 1980s clarified the importance of gas- and aqueous-phase oxidation of sulfur dioxide to sulfate, gas-phase oxidation of nitrogen oxides to HNO_3, and the relative significance of dry versus wet deposition of sulfur and nitrogen species.

Nitric acid is estimated to contribute about one-third of the excess hydrogen ion concentration (i.e., acidity) in precipitation in the eastern United States (Golumb, 1983). Nitric acid is produced in the gas phase through reaction of NO_2 with HO• [reaction (11.16)], which is its main formation route during daytime. At night, HO• concentrations are relatively low because the photolytic formation pathway [reaction (11.3)] shuts down. Nitric acid is formed then through the heterogeneous reaction of N_2O_5 with water:

$$NO_2 + O_3 \rightarrow NO_3 + O_2 \tag{11.26}$$

$$NO_3 + NO_2 \rightarrow N_2O_5 \tag{11.27}$$

$$N_2O_5 + H_2O(aq) \rightarrow 2HNO_3(aq) \tag{11.28}$$

The Henry's law coefficient for nitric acid[2] has a value of 2.1×10^5 mol/L atm at 298 K (Schwartz and White, 1981), indicating that it partitions strongly to the aqueous phase when cloud droplets are present. Once in the aqueous phase, nitric acid dissociates almost completely into H^+ and NO_3^-. Also because of its high solubility, HNO_3 is efficiently deposited onto wetted surfaces (see Section 11.4). The other significant fate of atmospheric nitric acid is the formation of condensed-phase ammonium nitrate:

$$NH_3(g) + HNO_3(g) = NH_4NO_3(s) \tag{11.29}$$

Particulate ammonium nitrate is a significant contributor to visibility impairment, especially in the western United States (NRC, 1993).

Sulfur dioxide [S(IV)] can be oxidized to sulfuric acid [S(VI)] in the gas phase through the following sequence of reactions:

$$SO_2 + HO\bullet + M \rightarrow HOSO_2\bullet + M \tag{11.30}$$

$$HOSO_2\bullet + O_2 \rightarrow HO_2 + SO_3 \tag{11.31}$$

$$SO_3 + H_2O + M \rightarrow H_2SO_4 + M \tag{11.32}$$

Once formed, H_2SO_4 can undergo homogeneous nucleation to generate particles or condense on existing particles, as discussed in Section 11.3. In the atmosphere, sulfuric acid exists predominantly as an aerosol. SO_2 can also be oxidized to SO_4^{2-}

[2] Henry's law constants for inorganic species are conventionally defined with the aqueous-phase concentration in the numerator and the gas-phase concentration in the denominator, exactly the opposite of the convention for organic compounds.

in the aqueous phase. The Henry's law coefficient for SO_2 is relatively small, with a value of 1.2 mol/L atm at 298 K (Maahs, 1982). However, acid–base reactions that occur in aqueous solution have the effect of substantially increasing the effective solubility of S(IV) species:

$$SO_2\,(g) + H_2O = SO_2{\cdot}H_2O \tag{11.33}$$

$$SO_2{\cdot}H_2O = H^+ + HSO_3^- \tag{11.34}$$

$$HSO_3^- = H^+ + SO_3^{2-} \tag{11.35}$$

At pH = 5, a typical value for rainwater, the effective Henry's law coefficient for the three S(IV) species ($SO_2{\cdot}H_2O$, HSO_3^-, and SO_3^{2-}) is 1500 mol/L atm (Seinfeld and Pandis, 1998; p. 351). Thus, in the presence of liquid water, much more S(IV) is found in the aqueous phase than would be indicated by the Henry's law coefficient for SO_2 itself. For pH values ranging from about 2 to 7, the predominant S(IV) species is bisulfite, HSO_3^-. This ion can be very quickly oxidized to sulfate if hydrogen peroxide, H_2O_2, is present in the aqueous phase:

$$HSO_3^- + H_2O_2 = SO_2OOH^- + H_2O \tag{11.36}$$

$$SO_2OOH^- + H^+ \rightarrow H_2SO_4 \tag{11.37}$$

H_2O_2 is formed in the gas phase through reactions such as reaction (11.17), but with a high Henry's law constant of 7.5×10^4 mol/L atm at 298 K (Seinfeld and Pandis, 1998) is found almost entirely in the aqueous phase in the presence of liquid water. Other significant avenues for oxidizing S(IV) to S(VI) in the aqueous phase include reaction with dissolved O_3 or reaction with dissolved O_2 catalyzed by Mn^{2+} or Fe^{3+}. The O_2 reaction is negligibly slow unless the catalysts are present. For pH values less than about 5, H_2O_2 is likely to be the dominant oxidation pathway. For higher pH values, oxidation by reaction with O_3 or Fe^{3+}-catalyzed reaction with O_2 may dominate (Seinfeld and Pandis, 1998).

Modeling aqueous chemistry in cloud, fog, or raindrops requires tracking three steps: mass transfer of gaseous species into the water droplet, distribution of the species throughout the droplet, and chemical reactions within the droplet. Mass transfer to the droplet is usually the rate-limiting step. Aqueous-phase chemical reactions, for example, acid–base reactions, are usually fast enough to treat as equilibrium processes. Modeling of equilibrium aqueous chemistry is described in more detail in Chapter 12. An extensive tabulation of reactions that are important in atmospheric chemistry together with thermodynamic and kinetic data can be found in Seinfeld and Pandis (1998, pp. 391–398).

11.3 SECONDARY AEROSOL FORMATION

11.3.1 Atmospheric Aerosols: Composition and Size Distributions

Aerosols are collections of very fine particles that are suspended in a gas. The particles may be liquid, solid, or mixtures of both condensed phases. At concentrations that are widespread in populated areas, fine particles suspended in the atmosphere cause respiratory health problems and impair visibility. Particles that are less than about 10 μm in diameter (known as PM10) can remain suspended in the atmosphere for long periods of time and be transported over long distances. Particles less than 2.5 μm in diameter (i.e., PM2.5) are especially efficient at penetrating deep into human lungs and at scattering sunlight and thus reducing visibility.

Atmospheric aerosols are complex in their composition, size distribution, and morphology. Primary particles, emitted from combustion, soil dust, and a variety of industrial sources evolve in the atmosphere through chemical and physical processes (Fig. 11.5). In addition, products of gas-phase reactions can undergo homogeneous nucleation to form new particles. The major differences between secondary aerosol and ozone models are thus the addition of modules for aerosol dynamics and growth, aqueous chemistry, and particle removal via wet and dry deposition. In addition to tracking multiple chemical species, aerosol models must either track the parameters or moments of one or more size modes (Whitby et al., 1991; Binkowski and Shankar, 1995) or else track mass in multiple size bins (Gelbard and Seinfeld, 1980; Meng et al., 1998; Lurmann et al., 1997). Correspondingly, these models require particulate emissions information that is classified by size as well as composition. The lack of adequate primary particulate matter and ammonia emissions data is a significant limitation in many applications.

Urban PM2.5 concentrations are typically comprised of roughly 20 to 60% secondary species (Hidy et al., 1998), with sulfate dominating the secondary fraction in the eastern United States and nitrate playing a relatively larger role in western cities. Organic carbon, which may be either primary or secondary in origin, contributes up to

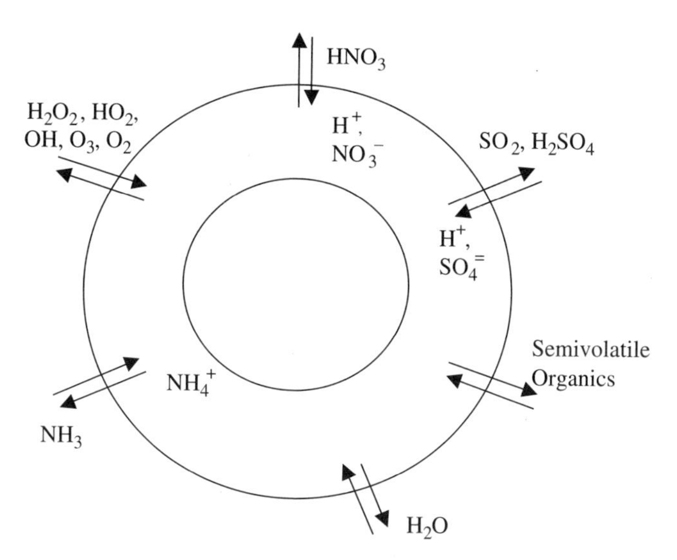

Figure 11.5 Selected species affecting particle composition. (Adapted from Seinfeld, 1989.) Species shown outside the outer circle undergo mass transfer to and from the condensed phases; those shown inside the circle are products of aqueous-phase dissociation.

40% of the fine aerosol in some areas (Gray et al., 1986). Many aerosol constituents are hygroscopic, so under humid conditions water may also comprise a significant fraction of aerosol mass. In addition to organic carbon, other components of primary particulate matter include elemental carbon (i.e., soot), sea salt, and crustal materials such as silicon, aluminum, and iron oxides. Primary particulate matter may also include toxic metals such as lead and cadmium.

Modeling the evolution of an aerosol requires characterizing and tracking its size distribution. Condensed matter exists in the atmosphere in sizes ranging from a few nanometers for nucleating molecular clusters to the size of coarse dust particles, which are on the order of tens of microns in diameter. Specific size ranges tend to be associated with particular processes of particle formation as well as particular chemical constituents, as shown in Figure 11.6.

Because particles in ambient air originate from a variety of sources, atmospheric size distributions are expected to be multimodal, as suggested by Figure 11.6. Size distributions can be expressed in terms of number, surface area, volume, or mass concentrations. Selection of a particular form for modeling depends on the effects of interest and the basis for particle measurements against which a model is being compared. In almost all ambient air quality models, particles are assumed to be spherical in shape. Further assuming uniform density, conversion of size distributions between these forms is straightforward, as shown below.

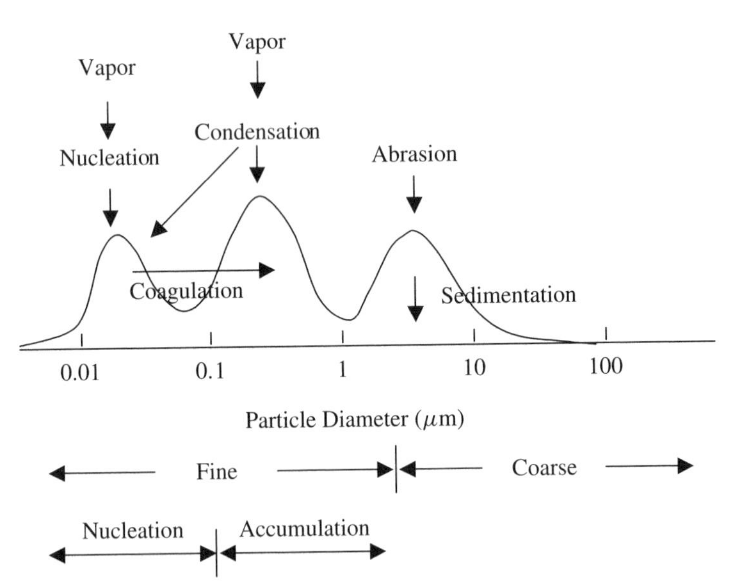

Figure 11.6 Schematic size distribution of an urban aerosol showing the three modes that typically contain much of the aerosol mass. The fine mode contains particles produced by nucleation and condensation of low-volatility gases. The midrange or *accumulation mode* results from coagulation of smaller particles and condensation of gases onto pre-existing particles. Coarse particles are usually generated mechanically.

Both discrete and continuous size distributions are used to characterize aerosols in air quality models. Discrete distributions used in air quality models may simply partition the mass concentration into coarse (e.g., > 2.5 μm) and fine (< 2.5 μm) size bins or at the other extreme may include tens of size intervals. Figure 11.7 shows a composite discrete size distribution for sulfate aerosols, which was compiled from many size distributions measured using cascade impactors having from 4 to 10 discrete stages. In this example, aerosol mass concentration is plotted as a function of the logarithm of the aerodynamic diameter of the particles, d_p, which is defined as the diameter of a spherical particle with unit density (1 g/cm^3) that would have the same aerodynamic behavior as the collected particles. Because of the wide range of particle sizes that must be represented, a logarithmic scale has been used in Figure 11.7. The value on the ordinate is the fraction of the total concentration that occurs in the size range from log d_p to $\log(d_p + \Delta d_p)$.

Lack of detailed information about size fractions of emitted particles and computational limitations may justify use of a small number of size classes. However, aerosol condensation, evaporation, coagulation, and deposition processes can be described with increasing accuracy as the number of size classes is expanded. Size classes used in models of aerosol evolution are usually fixed (Gelbard and Seinfeld, 1980), with mass transferred between classes as particles grow (or diminish) in size. Alternatively, Jacobson (1997) proposed a fixed boundary/moving center scheme in which the lower and upper boundaries of each size bin are fixed, but the average size within each bin can change as the aerosol evolves.

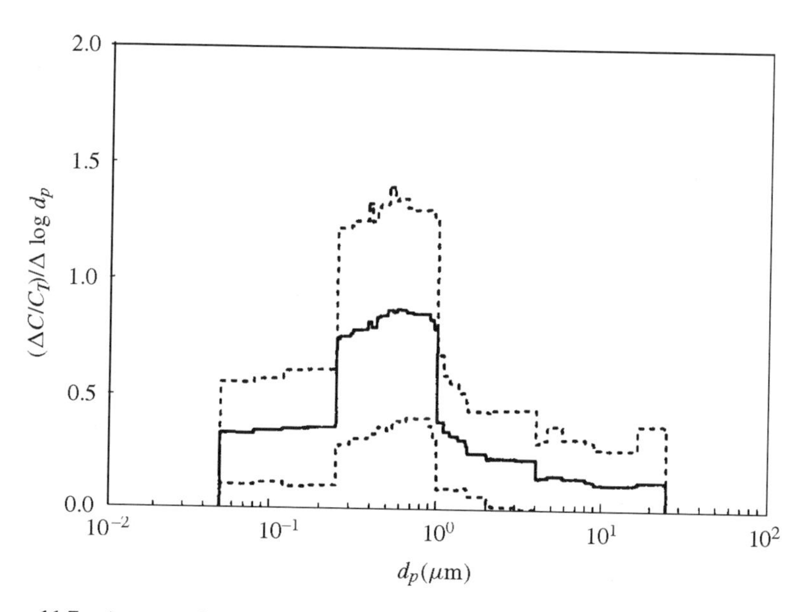

Figure 11.7 Average of 49 discrete size distributions for sulfate obtained using impactors. The heavy line shows the average, and the dashed lines one standard deviation above and below the average (Milford and Davidson, 1987).

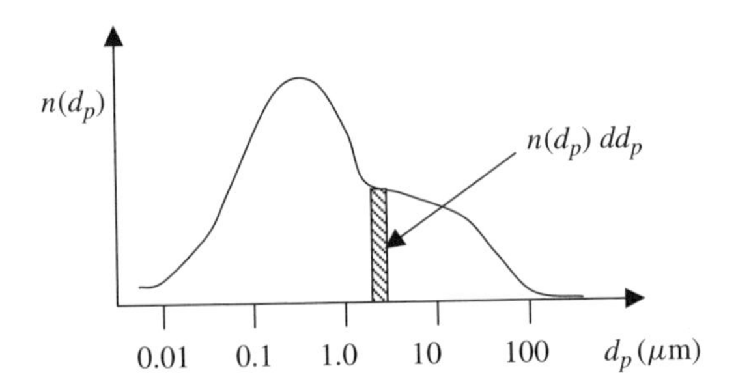

Figure 11.8 Continuous particle size distribution function, $n(d_p)$. The shaded area represents the number concentration of particles within the indicated size range.

Continuous size distribution functions reduce the length of the size classes into which aerosols are divided to infinitesimal increments. The continuous number distribution function, $n(d_p)$, illustrated in Figure 11.8, is defined such that $n(d_p)dd_p$ gives the number concentration of particles (e.g., particles per cm^3) with diameters ranging from d_p to $d_p + dd_p$. For a finite range of particle sizes from d_p^A to d_p^B, the number concentration is

$$N = \int_{d_p^A}^{d_p^B} n\left(d_p\right) dd_p \qquad (\text{number } L^{-3}) \tag{11.38}$$

For spherical particles, surface area, volume, and mass distributions can be obtained from the number distribution function as shown in Table 11.1.

TABLE 11.1 Total Number, Surface Area, Volume, and Mass for Spherical Particles

$$N_T = \int_0^{\infty} n\left(d_p\right) dd_p \qquad (\text{number } L^{-3}) \tag{11.39}$$

$$\mathrm{SA} = \pi \int_{d_p^A}^{d_p^B} d_p^2 n\left(d_p\right) dd_p \qquad (L^2\ L^{-3}) \tag{11.40}$$

$$V = \frac{\pi}{6} \int_{d_p^A}^{d_p^B} d_p^3 n\left(d_p\right) dd_p \qquad (L^3\ L^{-3}) \tag{11.41}$$

$$M = \frac{\pi}{6} \int_{d_p^A}^{d_p^B} \rho_p d_p^3 n\left(d_p\right) dd_p \qquad (M\ L^{-3}) \tag{11.42}$$

As with the discrete distribution shown in Figure 11.7, continuous distributions are commonly represented on a logarithmic scale. In this case a log-based distribution function is used, such that $n(\log d_p)\, d \log d_p$ gives the number of particles in the size range from $\log d_p$ to $\log d_p + d \log d_p$. The linear and log forms of the distribution functions have identical values: For a given size interval, $n(\log d_p) d \log d_p = n(d_p)\, dd_p$.

In many cases, size distributions of aerosols generated by a single process are described well by a lognormal distribution function (Seinfeld and Pandis, 1998):

$$\frac{n(d_p)}{N} = \frac{1}{d_p \ln \sigma_g \sqrt{2\pi}} \exp\left\{-\frac{1}{2}\left(\frac{\ln d_p - \ln \overline{d_{p_g}}}{\ln \sigma_g}\right)^2\right\} \tag{11.43}$$

In Eq. (11.43), σ_g is the geometric standard deviation of d_p and $\overline{d_{p_g}}$ is the median diameter, that is, 50% of the particles are smaller and 50% are larger than $\overline{d_{p_g}}$. Ambient aerosols in urban areas are produced from a variety of sources and secondary generation and growth processes. Their resulting multimodal distributions are reasonably well represented by a sum of $i = 1, \ldots, n$ lognormal distributions, each of which captures one of n modes, which is characterized by its own median and geometric standard deviation.

11.3.2 Aerosol Production and Growth

The evolution of aerosol size distributions in the atmosphere occurs through nucleation of particles from gas-phase species, condensation/evaporation of gases to and from particles, coagulation of smaller particles to form larger ones, and size-selective removal processes such as dry deposition. Homogeneous nucleation of H_2SO_4/H_2O particles is included in most models of atmospheric aerosol production and growth, along with condensation of H_2SO_4 onto existing particles. Condensation or dissolution and evaporation of volatile inorganics may either be treated as a kinetic mass-transfer process leading to a thermodynamically determined equilibrium state (e.g., Meng et al., 1998) or by assuming that equilibration between the gas and condensed or dissolved phases is instantaneous (e.g., Binkowski and Shankar, 1995). Thermodynamic treatments of gas/particle equilibria in various models differ in the number of species included in the equilibrium calculation and in the method used to calculate the activities of aqueous solutions.

Many aerosol models assume that all particles of a given size have the same composition, that is, that the aerosol population is "internally mixed." Kleeman et al. (1997) developed an externally mixed model in which particles emitted from different sources can have different compositions and then evolve differently. They found that the internally mixed aerosol assumption can distort size distributions if sulfate and sodium (e.g., from sea salt) are found in different particles of the same size or if material with different hygroscopic properties is externally mixed. However, Kleeman et al.'s approach is computationally intensive and has not been widely employed to date.

Neglecting spatial variations and transport, the general equation for the temporal evolution of species i through condensation, evaporation, and coagulation as well as internal sources and first-order losses is (Wexler et al., 1994; Meng et al., 1998):

$$\frac{\partial \chi^i(m,t)}{\partial t} = \underbrace{\Psi^i(m,t)\chi^i}_{\text{Evap.}} - \underbrace{\frac{\partial\left(m\chi^i\Psi\right)}{\partial m}}_{\text{Conden.}} + \underbrace{\int_0^m \vartheta\left(m', m-m'\right)\chi^i\left(m',t\right)\frac{\chi\left(m-m',t\right)}{m-m'}dm'}_{\text{Coagulation}}$$
$$\underbrace{-\chi^i(m,t)\int_0^\infty \vartheta\left(m',m\right)\frac{\chi\left(m',t\right)}{m'}dm'}_{\text{Coagulation}} + \underbrace{S^i(m,t)}_{\text{Sources}} - \underbrace{\text{Loss}(m,t)\chi^i(m,t)}_{\text{Deposition}} \tag{11.44}$$

In Eq. (11.44), χ^i (L^{-3}) is the mass concentration distribution of species i, which is defined such that $\chi^i(m,t)\,dm$ is the mass concentration of species i in the size range from mass m to $m + dm$. For a mixture of n species, the total mass concentration distribution is given by:

$$\chi(m,t) = \sum_{i=1}^{n} \chi^i(m,t) \tag{11.45}$$

For spherical particles, the mass concentration distribution is related to the number distribution through Eq. (11.42). The first two terms on the right-hand side of Eq. (11.44) represent evaporation and condensation; $\Psi^i(m,t)$ (T^{-1}) is the normalized growth or evaporation rate of species i in a particle of mass m:

$$\Psi^i(m,t) = \frac{1}{m}\frac{dm^i}{dt} \tag{11.46}$$

and Ψ is the sum of the normalized growth/evaporation rates over all of the species in the aerosol. The second pair of terms on the right represents coagulation into and out of the size range from m to $m + dm$, due to random collisions between particles. $\vartheta(m', m)$ represents the coagulation coefficient for pairs of particles of sizes m' and m. S^i ($L^{-3}T^{-1}$) and Loss (T^{-1}) represent particle sources including nucleation and the rate constant for deposition losses, respectively.

Coagulation Under ambient conditions, the most important coagulation mechanism for particles on the order of 1 μm or smaller is Brownian coagulation, by which particles are brought into contact through random motions resulting from bombardment by gas molecules. Coagulation by gravitational settling, where larger particles overtake smaller ones, can be important for larger particles. In modeling gas–particle interactions, the continuum regime is distinguished by a very small Knudsen number, $\text{Kn} = 2\lambda_g/d_p \ll 1$, where λ_g is the mean-free path length of the gas molecules. For air at 298 K and 1 atm, $\lambda_a = 0.0653\ \mu$m. The opposite case, with Kn $\gg 1$ is known as the kinetic or free molecular regime. For the continuum regime, the Brownian coagulation coefficient for particles of diameters d_p and d_p' is

$$\vartheta\left(d_p, d_p'\right) = \frac{2kT}{3\mu}\frac{\left(d_p + d_p'\right)^2}{d_p\, d_p'} \tag{11.47}$$

where $k = 1.38 \times 10^{-23}$ J/K is Boltzmann's constant and μ is the viscosity of air at temperature T. This expression can be modified for the free molecular regime through the use of a correction factor (Fuchs, 1964). Wexler et al. (1994) argue that except for extreme conditions of very high particle loadings, coagulation does not significantly affect the evolution of aerosol mass distributions. However, coagulation can rapidly alter number distributions of fine particles with diameters of $\sim$0.1 μm or smaller and may be important for describing overall mass distributions in concentrated plumes near air pollution sources. For a more detailed description of coagulation processes and approaches for estimating coagulation coefficients, the reader is referred to Seinfeld and Pandis (1998) and Jacobson (1999).

Evaporation and Condensation Evaporation and condensation or dissolution are often the most significant processes that alter the composition and mass size distribution of airborne particles. Evaporation and condensation can be treated dynamically by considering rates of mass transfer to or from a single particle of diameter d_p. For species i, Wexler et al. (1994) give

$$\Psi^i(m, t) = \frac{1}{m}\frac{dm^i}{dt} = \frac{2\pi\, d_p\, D_a^i\left(C_{g,\infty}^i - C_{g,I}^i\right)}{m\left(\dfrac{2\lambda_a}{\varsigma_i d_p} + 1\right)} \tag{11.48}$$

In this equation, transport to or away from the particle surface is assumed to occur through molecular diffusion, with the driving force for mass transfer expressed as the difference between the concentration of species i in the bulk gas phase, $C_{g,\infty}^i$, and the concentration at the particle surface, $C_{g,I}^i$; D_a^i is the molecular diffusion coefficient for species i in air. The concentration at the surface, $C_{g,I}^i$ is usually determined by thermodynamic considerations, assuming that the characteristic time for equilibration within the particle is short compared to the characteristic time for mass transfer to or from the particle. This assumption is most appropriate for high particle concentrations, high temperatures, and small particle diameters but is generally supported by field observations (Seinfeld and Pandis, 1998). The term in parentheses in the denominator accounts for noncontinuum effects, with ς_i representing the accommodation coefficient for species i on the particle. The accommodation coefficient expresses the probability that a condensing molecule will "stick" to a particle surface with which it collides. Values of ς are determined experimentally and are best known for interactions with dilute aqueous solutions (Seinfeld and Pandis, 1998); typical values for soluble species are about 0.1.

As with other chemical equilibria, phase equilibria in aerosols are determined by the condition that the Gibbs free energy of the system is minimized. For the case of a chemical reaction occurring within or between phases at constant temperature and pressure, for example, $a_1 A_1(\mathrm{g}) + a_2 A_2(\mathrm{s}) = a_3 A_3(\mathrm{s}) + a_4 A_4(\mathrm{g})$, the change in Gibbs free energy is given by:

$$\Delta G = \sum_{i=1}^{k} \kappa^i \, \Delta n^i \tag{11.49}$$

where κ^i is the chemical potential for the ith species participating in the reaction and Δn^i is the change in number of moles of the ith constituent that occurs in the reaction. At equilibrium, $\Delta G = 0$, which corresponds to the condition that

$$0 = \sum_{i=1}^{k} a^i \, \kappa^i \tag{11.50}$$

where a^i is the stoichiometric coefficient for the ith species (positive for reactants and negative for products).

For concentrated, nonideal aqueous solutions, the chemical potential of the ith constituent is related to temperature, pressure, and its activity, γ^i by:

$$\kappa^i = \kappa^{i,0}(T, P) + RT \ln \gamma^i \tag{11.51}$$

where $\kappa^{i,0}$ is the standard chemical potential of species i, which is defined as the chemical potential of the pure species. For ideal gases, the activity in Eq. (11.51) is equal to the partial pressure, and for solids it is equal to 1.

Combining Eqs. (11.50) and (11.51) the equilibrium condition can be written in the more common form:

$$\prod_{i=1}^{k} \kappa^i a^i = K = \exp\left(\frac{-1}{RT} \sum_{i=1}^{k} a^i \kappa^{i,0} \right) = K(T) \tag{11.52}$$

where $K(T)$ is the equilibrium constant. Combining equilibrium relationships for the phase changes and reactions that can occur in the aerosol with element balance and electroneutrality conditions (if applicable) produces a system of equations that can be solved for the activities of the aerosol constituents, including the surface partial pressures or concentrations of gases in equilibrium with the condensed phase.

Examples of key aerosol constituents whose evolution is governed by evaporation and condensation are ammonia, nitric acid, semivolatile organic compounds, and water. Important multiphase reactions include:

$$H_2O\,(\mathrm{g}) = H_2O\,(\mathrm{aq}) \tag{11.53}$$

$$NH_3\,(\mathrm{g}) + HNO_3\,(\mathrm{g}) = NH_4NO_3\,(\mathrm{s}) \tag{11.54}$$

$$NH_3\,(g) + HNO_3\,(g) = (NH_4NO_3\,(aq) \rightarrow)NH_4^+\,(aq) + NO_3^-\,(aq) \quad (11.55)$$

$$H_2SO_4\,(g) = (H_2SO_4\,(aq) \rightarrow)H^+\,(aq) + HSO_4^-\,(aq) \quad (11.56)$$

$$2NH_4^+\,(aq) + SO_4^{2-}\,(aq) = (NH_4)_2SO_4\,(s) \quad (11.57)$$

The notation $\rightarrow$ used for reactions (11.55) and (11.56) indicates that the species in parentheses rapidly dissociates in solution to the ionic products.

For trace gases such as ammonia and nitric acid, kinetic treatment of condensation and evaporation is recommended, so a mass-transfer rate between the bulk gas and aerosol would be calculated using an equation such as (11.48), with the gas-phase concentration at the particle surface determined by multicomponent equilibrium relationships. Because its vapor-phase concentrations are quite high under most circumstances, however, water is often assumed to instantaneously equilibrate between its vapor and condensed phases. For a liquid-phase aerosol, the dependence of the equilibrium aerosol water content (mass of water per volume of air) on the aqueous solution composition can be estimated using the ZSR (Zdanovskii, Stokes, and Robinson) relationship (Seinfeld and Pandis, 1998):

$$C_\ell^w = \sum_{i=1}^{n} \frac{C_\ell^i}{m_\ell^{i,0}(\gamma^w)} \quad (11.58)$$

where C_ℓ^i is the liquid-phase concentration of constituent i (mol L^{-3} air) and $m_\ell^{i,0}$ is the molality (mol M^{-1} water) that species i would have if it were the only solute present in solution with water at activity γ^w = RH/100. Assuming that water equilibrates instantaneously between the gas and particle phases, the rate at which water condenses onto or evaporates from an aqueous particle is then determined by the rate of change in the ambient relative humidity (RH) and the rate of change of the particle composition due to condensation or evaporation of other species.

Under conditions that are highly conducive to photochemical oxidant production, a significant fraction of fine particulate matter may be comprised of secondary organic aerosol (SOA). SOA is formed from the condensation of gas-phase oxidation products of primary reactive organic gases that have seven or more carbon atoms, such as aromatics and terpenes. Historically, aerosol models have used a somewhat simplified approach to represent SOA formation. For example, Pandis et al. (1992) modeled this process by assuming the gas-phase reaction of aromatics and other high-molecular-weight organic compounds produces a fixed yield of condensable products that are irreversibly added to the particle phase. The yield is defined as the fraction of the primary reactive organic gas that is converted to aerosol, on a mass basis. Yields incorporated into Pandis et al.'s model were estimated from chamber experiments, many of which were conducted at temperatures of 30°C or greater (Grosjean and Seinfeld, 1989). The dependence of the yields on temperature and seed-particle concentrations was not considered.

Odum et al. (1996) have developed a sorption-based approach to modeling SOA formation that appears to be more realistic than assuming fixed particle yields for organic compounds. They followed Pankow's (1994a, 1994b) suggestion that the fraction of the mass of condensable products found in the particle phase is governed by a temperature-dependent equilibrium partition coefficient:

$$K^i_{\mathrm{OM}} = \frac{C^i_{\mathrm{OM}}}{C^i_g C_{\mathrm{OM}}} \tag{11.59}$$

where C^i_{OM} is the mass concentration of product i in the condensed phase that is sorbed onto organic matter (OM); C^i_g is the gas-phase mass concentration of product i, and C_{OM} is the total mass concentration of aerosol organic matter. Since K^i_{OM} is a constant for a given temperature, the fraction of product i found in the condensed phase is assumed to increase as the total mass concentration of aerosol organic matter increases. For condensable product i, the definition of particle yield given above can be combined with the partition coefficient expression [Eq. (11.59)] and a mass balance for the total concentration of product i, $C^i = C^i_{\mathrm{OM}} + C^i_g$, to give

$$\mathrm{Yield}^i = C_{\mathrm{OM}} \left(\frac{w^i\, K^i_{\mathrm{OM}}}{1 + K^i_{\mathrm{OM}} C_{\mathrm{OM}}} \right) \tag{11.60}$$

where w^i is the mass of product i produced per mass of the primary organic compound reacted. The total particle yield from a given primary reactive organic is then

$$\mathrm{Yield} = C_{\mathrm{OM}} \sum_i \left(\frac{w^i\, K^i_{\mathrm{OM}}}{1 + K^i_{\mathrm{OM}} C_{\mathrm{OM}}} \right) \tag{11.61}$$

where the summation is taken over all of the condensable products of the primary reactive organic compound.

Nucleation Homogeneous nucleation of aerosols occurs when gas molecules cluster together to form stable nuclei. This requires that the partial pressure of the nucleating species exceed its vapor pressure, that is, that the vapor phase be supersaturated. Because nucleating species could alternatively condense on existing particles, homogeneous nucleation is most important when pre-existing particle concentrations are low and gas-phase production rates of nucleating species are high. For most condensing species, including water, homogeneous nucleation is a negligibly slow process that is overwhelmed by condensation on existing particles. The one exception that is commonly treated in aerosol dynamics models is binary homogeneous nucleation of sulfuric acid and water.

According to the classical theory of homogeneous nucleation (Wilemski, 1984; Kulmala et al., 1998), the formation of molecular clusters that will stay in the liquid phase and grow requires that the nucleating system surmount a critical free energy barrier associated with the formation of cluster surface area. At cluster radii larger than the metastable equilibrium point where the Gibbs free energy change reaches its maximum value, the reduction in chemical potential associated with the phase change from vapor to liquid dominates the system free energy change and the cluster grows quickly. The critical radius or number of molecules of compound A required for a viable cluster is related to the saturation ratio of the vapor phase, SR $= P^A/P_v^A(T)$, the surface tension and liquid-phase density of the nucleating species, and the temperature of the system (Seinfeld and Pandis, 1998). Given the critical cluster size, the instantaneous nucleation rate of viable nuclei for a homogeneous system can be described from the balance between molecular collisions that produce clusters of the critical size and evaporation of vapor molecules from those clusters. As the degree of supersaturation increases in a homogeneous system, the critical cluster size is reduced and the nucleation rate increases.

As mentioned above, for most species under atmospheric conditions homogeneous nucleation rates are negligible. Seinfeld and Pandis (1998) give the example of water vapor at a temperature of 20°C and relative humidity of 200% (SR = 2), in which case the homogeneous nucleation rate for water droplets is of the order 10^{-54} droplets/cm^3 s, which is extremely slow! The presence of liquid water in the atmosphere is due overwhelmingly to water condensation on existing particles, not homogeneous nucleation.

Although the process is not always important, many aerosol dynamics models do include homogeneous nucleation of the binary system of sulfuric acid and water. In binary systems, nucleation requires that each constituent be sufficiently supersaturated relative to its vapor pressure over the *solution* that the system is able to surmount the free energy barrier to stable cluster formation. In the H_2SO_4 and water case, partial pressures of the two components that are not supersaturated with respect to their pure-component vapor pressures may well be supersaturated with respect to their solution. As discussed above, sulfuric acid is formed through the gas-phase oxidation of SO_2 [reactions (11.30 to (11.32)]. In the presence of water vapor, H_2SO_4 forms hydrates, for example, $H_2SO_4 \cdot H_2O$, $H_2SO_4 \cdot 2\ H_2O$, with $H_2SO_4 \cdot H_2O$ being the dominant form. Homogeneous nucleation in this system thus involves cluster formation between pure molecules of H_2O and H_2SO_4 and hydrated sulfuric acid molecules.

While it is possible to predict the nucleation rate of the H_2SO_4/H_2O system from first principles, most models rely on simple parameterizations to reduce computing time. For example, Wexler et al. (1994) define a critical nucleation rate of 1 cm^{-3} s^{-1} as the rate required for nucleation to be significant under ambient atmospheric conditions. Based on theoretical calculations made by Jaecker-Voirol and Mirabel (1989), they developed a formula for the temperature and relative humidity dependence of the gas-phase H_2SO_4 concentration, $C_{\text{crit}}^{H_2SO_4}$, required to produce the critical nucleation rate. Wexler et al. (1994) evaluate $C_{\text{crit}}^{H_2SO_4}$ as time advances, and continually transfer any H_2SO_4 produced in the gas phase at concentrations above this value into their lowest particle size bin. The aerosol module of EPA's Models-3/CMAQ system (Byun and Ching, 1999) offers the choice of two parameterizations for treating H_2SO_4/H_2O nucleation. The first is a parameterization developed by Harrington and Kreidenweis (1998) from Jaecker-Voirol and Mirabel's (1989) study, which can be used to directly calculate nucleation rates. The second parameterization was developed by Kulmala et al. (1998), from theoretical calculations they performed based on Wilemski's (1984) formulation of binary homogeneous nucleation theory.

11.4 DEPOSITION PROCESSES

Air pollutants can be removed from the atmosphere by three mechanisms: the chemical transformations described previously and dry and wet deposition. Dry deposition is the term used for separation of particles or gases from the airflow with uptake at the surface. Wet deposition refers to the absorption of gases or particles into water droplets, followed by precipitation.

Deposition processes are generally represented in air quality models in terms of the flux $(M\,L^{-2}\,T^{-1})$ at a specified reference height:

$$J = v(z_r)C(z_r) \tag{11.62}$$

where v is the wet or dry deposition velocity, C is the contaminant concentration, and z_r is the reference height at which both are determined. Implicit in this scheme is the assumption that the reference height is located in the atmospheric surface layer, through which fluxes of heat, momentum, and chemical constituents are assumed to be constant. Moreover, the use of a deposition velocity parameterization is not valid in the case of species that react on time scales that are shorter than the time scale for transport through the surface layer, nor for species that are emitted from surface soils, water, or vegetation (Kramm et al., 1995).

Deposition fluxes are included in multilayered air quality models as surface boundary conditions, for example:

$$K_{zz} \left. \frac{\partial C^i}{\partial z} \right|_{z=0} = E^i - J^i \tag{11.63}$$

where K_{zz} is the vertical turbulent diffusion coefficient and E^i is the emissions flux of species i. Although particles and gases are treated identically for advection and dispersion, this is not true for deposition. Deposition velocities for particles and gases reflect the different mechanisms that govern their removal from the airflow that carries them.

11.4.1 Dry Deposition of Gases

Dry deposition of gases is viewed as occurring in three sequential stages (Fig. 11.9): turbulent transport down to the surface, diffusion across the quasi-laminar sublayer at the surface, and uptake by the surface, be it bare soil, vegetation, water, or engineered materials such as concrete or asphalt. As indicated in the figure, these three stages of transport are commonly modeled as a series of "resistances" $(T\,L^{-1})$, with the deposition velocity given by:

$$v_d = \frac{1}{r_a + r_d + r_c} \tag{11.64}$$

The first stage in Figure 11.9 is characterized by the aerodynamic resistance, r_a, a function of the wind speed, atmospheric stability, and surface roughness. The inverse of r_a is known as the momentum transfer velocity, v_m, and is given by:

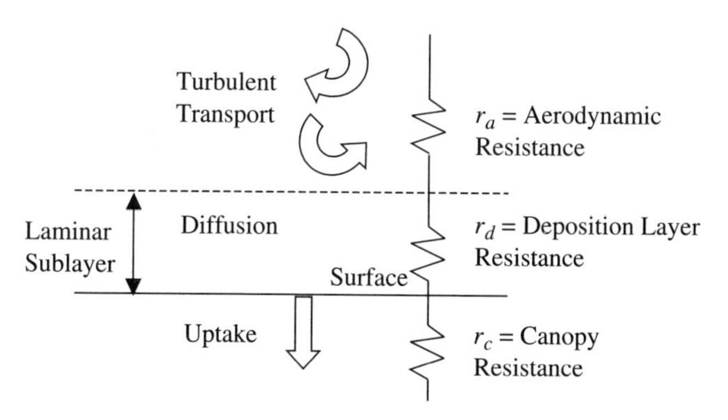

Figure 11.9 Resistance model for dry deposition of gases.

$$v_m = \frac{1}{r_a} = \frac{u_*^2}{u(z_r) - u(z_0)} \tag{11.65}$$

In Eq. (11.65), $u(z_r)$ is the horizontal wind speed at the reference height, u_* is the friction velocity, and z_0 is the surface roughness height, defined by $u(z_0) = 0$. The aerodynamic resistance is evaluated using wind profile expressions from similarity theory. For example, for adiabatic stability conditions:

$$\frac{u(z_r)}{u_*} = \frac{1}{\kappa} \ln\left(\frac{z_r}{z_0}\right) \tag{11.66}$$

where κ represents Von Karman's constant, which has a value of about 0.4. Combining Eqs. (11.65) and (11.66), the expression for r_a under adiabatic conditions is

$$r_a = \frac{1}{u_* \kappa} \ln\left(\frac{z_r}{z_0}\right) \tag{11.67}$$

The second resistance that must be overcome for a gas to be deposited from the atmosphere is the deposition layer resistance, r_d, which governs transport across the quasi-laminar sublayer at the surface. Because turbulence is assumed to disappear in this 0.1- to 1-mm thick layer, vertical transport of gases is modeled as occurring only by molecular diffusion. The deposition layer resistance thus depends on the molecular diffusion coefficient for the contaminant in air, $D_{m,\text{air}}$, and on the friction velocity, which controls the thickness of the laminar sublayer. The molecular diffusion coefficient is expressed in dimensionless form as the Schmidt number, which is the ratio of the kinematic viscosity of air, ν_{air}, to the diffusion coefficient:

$$\text{Sc} = \frac{\nu_{\text{air}}}{D_{m,\text{air}}} \tag{11.68}$$

Dimensional analysis suggests the commonly used expression for r_d:

$$r_d = \frac{a\,\mathrm{Sc}^b}{\kappa u_*} \tag{11.69}$$

where a and b are empirical constants with typical values of 5 and 0.66, respectively (Wesley, 1989).

The last resistance encountered by a depositing gas, r_c, the canopy or surface resistance, is by far the most complicated of the three. The canopy resistance depends on both surface characteristics and the properties of the contaminant, such as its solubility. Nitric acid reacts rapidly with most surfaces so its canopy resistance is close to zero. In contrast, CO has a comparatively high surface resistance. For land surfaces including soil and vegetation, a relatively simple treatment of r_c is depicted in Figure 11.10. Developed by Wesely (1989), this scheme is incorporated into several widely used air quality models, including the Regional Acid Deposition Model (RADM) (Chang et al., 1987) and the Urban Airshed Model (UAM) (Morris et al., 1991).

In Figure 11.10, the canopy resistance is shown as being comprised of parallel resistances through leaf stomata and their mesophyll layer, the vegetative canopy, and direct deposition to ground surfaces of soil, water, or other surfaces. Wesely (1989) provides estimates or expressions for these resistances for 11 land-use categories including urban land, coniferous and deciduous forests, and cropland and for 5 seasonal

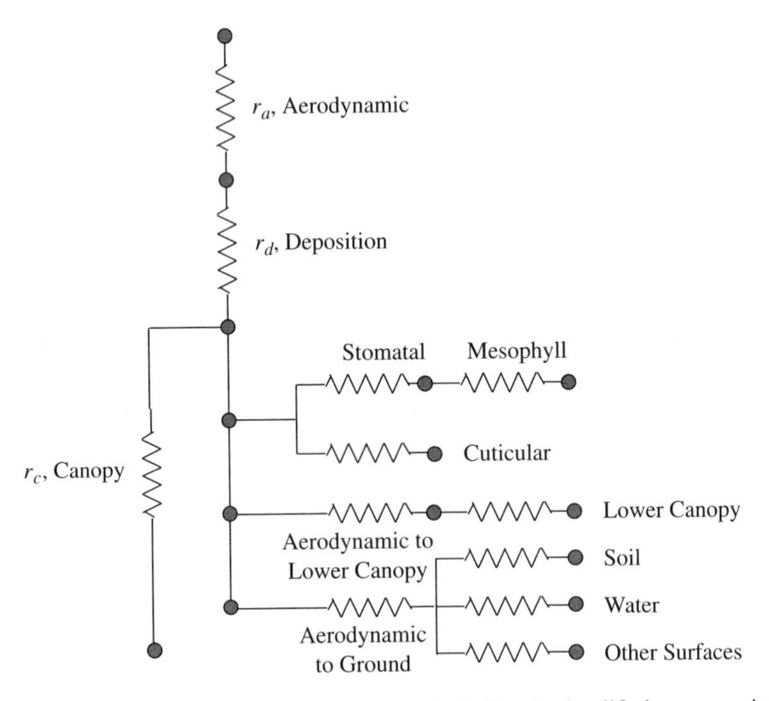

Figure 11.10 Resistance model for dry deposition, including the simplified canopy resistance model developed by Wesely (1989) for use in regional-scale air quality models.

categories ranging from midsummer with lush vegetation cover to winter with subfreezing temperatures and snow cover. The canopy resistances are parameterized in terms of the molecular diffusivity, effective Henry's law constant, and a reactivity factor for each species.

Although widely used, Wesely's (1989) scheme provides only a rough approximation of deposition rates, suitable for averaging over areas on the order of square kilometers and time scales of weeks. The limitations of the approach are characteristic of a common problem in environmental modeling: the need to parameterize processes that exhibit significant variability on scales that are much finer than the resolution of numerical model grids or the associated input data. One consequence of such parameterization schemes is that they are difficult to evaluate by comparison with measurements. The resistances in Wesely's canopy model, for example, are not directly measurable in field studies.

EXAMPLE 11.2

Calculate the dry deposition velocity of HNO_3 to the surface of a large lake with surface roughness $z_0 = 0.02$ m, using a reference height of $z_r = 2$ m and assuming conditions of neutral stability with $u(2\text{ m}) = 2$ m/s. For dry air at STP, $\nu = 0.15$ cm^2/s. The molecular diffusivity of HNO_3 in air is approximately 0.13 cm^2/s.

Solution 11.2 For the conditions given, the deposition velocity can be calculated from Eqs. (11.64) to (11.69), with the assumption that the canopy resistance $r_c = 0$ because HNO_3 is highly soluble.

For neutral stability conditions,

$$u_* = \frac{\kappa u\left(z_r\right)}{\ln\left(\dfrac{z_r}{z_0}\right)} = 0.17\text{ m/s}$$

The aerodynamic resistance r_a is then

$$r_a = \left(\frac{(u_*)^2}{u\left(z_r\right)}\right)^{-1} = 66\text{ s/m}$$

The Schmidt number for HNO_3 is $\text{Sc} = \nu/D = 1.15$, so the sublayer resistance is

$$r_d = \frac{5(\text{Sc})^{0.66}}{\kappa u_*} = 79\text{ s/m}$$

Finally, the dry deposition velocity for HNO_3 for a 2-m reference height is

$$v_d = \frac{1}{r_a + r_d} = 0.7 \text{ cm/s}$$

Dry deposition velocities for the same atmospheric conditions for gases with lower solubility or for unreactive gases would be lower than that for HNO_3 because their canopy resistances would be nonnegligible. Dry deposition velocities would generally be higher for less stable atmospheric conditions because the aerodynamic resistance would be smaller than for the case considered here.

11.4.2 Dry Deposition of Particles

Resistance models are also widely used for dry deposition of particles, but in this case the canopy resistance is commonly assumed to be negligible. In addition, for particles larger than about 10 μm in aerodynamic diameter, gravitational settling needs to be considered in parallel with turbulent transport down to the laminar sublayer and diffusional transport across that layer. Considering only the aerodynamic and deposition layer resistances along with gravitational settling, the dry deposition velocity for particles can be approximated as:

$$v_d = \frac{1}{r_a + r_d + r_a r_d v_g} + v_g \tag{11.70}$$

where v_g is the terminal settling velocity for the particle. The terminal settling velocity is the velocity that balances the drag, buoyancy, and gravitational forces acting on a particle such that its acceleration is zero. For Stokes flow conditions in which Re $<$ 0.1, where the Reynolds number is based on the particle diameter, v_g is given by:

$$v_g = \frac{d_p^2 (\rho_p - \rho_a) C_c g}{18 \mu} \tag{11.71}$$

In Eq. (11.71), g is the acceleration due to gravity, μ is the viscosity of air, ρ_a is the density of air, and C_c is the Cunningham correction factor. Important for particles smaller than about 1 μm in diameter C_c accounts for noncontinuum effects of fluid–particle interactions that occur when the particle diameter approaches or is smaller than the mean-free path length of air. It is given by:

$$C_c = 1 + \left(\frac{2\lambda_a}{d_p}\right)\left[1.257 + 0.40 \exp\left(\frac{-0.55\, d_p}{\lambda_a}\right)\right] \tag{11.72}$$

The aerodynamic resistance r_a is treated the same for particles as for gases. For particles, the deposition layer resistance r_d has a similar dependence on the Schmidt number and friction velocity as the deposition layer resistance for gases, but also accounts for a second mechanism of transport across the laminar sublayer. In addition to diffusive transport via Brownian motion, particles can also be transported across the sublayer due to their inertia. The mechanism of inertial impaction is characterized by the inertial relaxation time of the particle:

$$\tau = \frac{\rho_p d_p^2}{18\,\mu} \tag{11.73}$$

The relaxation time relates to the time required for a particle's velocity to return to that of the carrier airstream after an accelerating force has been removed. The relaxation time is represented in nondimensional form by the Stokes number, St $= \tau u_0/L$, where u_0 is a characteristic velocity and L a characteristic length. In the case of transport across the laminar sublayer, L is the thickness of this layer, which is proportional to the kinematic viscosity of the fluid and inversely proportional to the friction velocity. With u_* taken as the characteristic velocity, the Stokes number is given by St $= \tau u_*^2/\nu$.

The diffusion coefficient for a particle moving through a gas is its Brownian diffusivity, which characterizes random particle motions that occur due to bombardment by gas molecules:

$$D = \frac{kTC_c}{3\pi\,\mu\,d_p} \tag{11.74}$$

where k is the Boltzmann constant and T is the absolute temperature. Accounting for both Brownian motion and inertial impaction, Seinfeld and Pandis (1998) give the following empirical expression for r_d for particles:

$$r_d = \frac{1}{\left(\mathrm{Sc}^{-2/3} + 10^{-3/\mathrm{St}}\right) u_*} \tag{11.75}$$

Deposition velocities for particles show extremely strong dependence on particle diameter, as illustrated in Figure 11.11. As a consequence, particle diameter controls the distances over which airborne particles can be transported. For example, under dry conditions, sulfate particles with diameters on the order of 1 μm are transported over distances on the order of 1000 km because dry deposition velocities for particles in that size range are so small. The strong relationship between diameter and deposition velocities also governs the health effects of particles. Particles that are less than about 1 μm in diameter most readily travel through the nasopharyngeal and tracheobronchial regions of the human respiratory tract to reach the pulmonary region, where they are most harmful.

11.4.3 Wet Deposition

In the presence of precipitating clouds or fogs, wet deposition can be a very efficient removal mechanism for particles and for highly soluble gases. In addition to controlling the removal of some contaminants from the atmosphere, uptake of gases and particles into cloud and fog droplets is an important step in the aqueous-phase

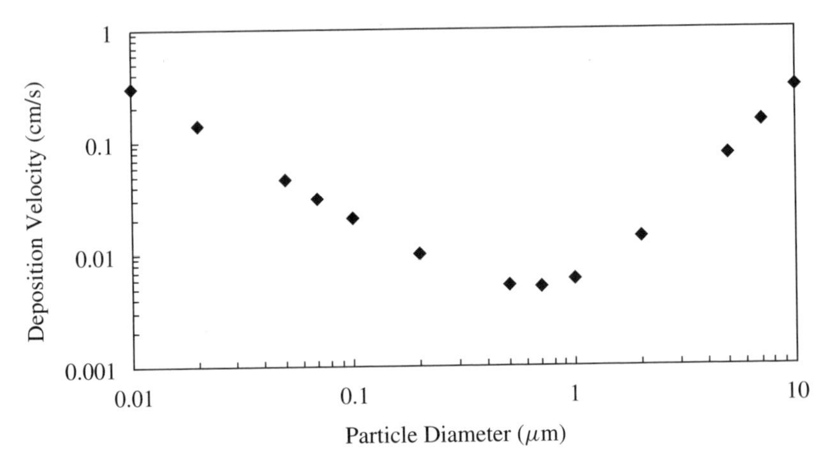

Figure 11.11 Deposition velocities calculated using Eqs. (11.70) to (11.75) for unit density particles with diameters ranging from 0.01 to 10 μm. The friction velocity and aerodynamic resistance were taken from Example 11.3. For particle diameters less than about 0.05 μm, the deposition velocity approaches $Sc^{-2/3}$. For particle diameters above about 5 μm, the deposition velocity approaches the settling velocity v_g.

oxidation of sulfur and some organic compounds, which leads to rainfall acidification or aerosol production.

Three steps are required in order for particles or gases to be removed from the atmosphere by wet deposition. First, the air mass containing the particle or gas must be brought into contact with water in the condensed phase (i.e., cloud or rain drops, snow, sleet, rime ice, or hail). Then the water droplets or other hydrometeors must scavenge the particle or gas. Finally, the hydrometeors must be carried to the surface to be deposited before the scavenging step is reversed by evaporation. Detailed modeling of wet deposition thus requires tracking the distribution of a contaminant between multiple forms of condensed water and the "dry" particle or gas phase. For warm cloud conditions, it is sufficient to account for contaminant transfer to cloud and rain drops; the discussion presented in this section is limited to this case. The microphysics of cloud formation and precipitation is a complex subject on its own. In the remainder of this section we describe how scavenging and wet deposition are typically handled in air quality models. The classic text by Pruppacher and Klett (1997) provides a more detailed discussion of the physics involved in these processes.

For particles, the scavenging step that dominates within clouds is *nucleation scavenging*, which is the activation and subsequent rapid growth of particles to form cloud droplets. Cloud formation and associated nucleation scavenging occurs when updrafts bring unsaturated air to a cooler altitude at which the water vapor concentration is suddenly supersaturated. Scavenging of interstitial aerosol within clouds by Brownian diffusion or impaction is a relatively slow process that is often neglected in air quality models. Below-cloud or *impaction scavenging* occurs through diffusion, interception, or impaction of particles into falling raindrops. Particle removal by wet deposition is thus the sum of contributions from impaction scavenging or "rainout" and nucleation scavenging accompanied by the conversion of cloud droplets to rain, which is referred to as "washout."

Gases are incorporated into cloud and raindrops by mass transfer followed by dissolution. Chemical reactions that then occur in the aqueous phase can have a significant impact on cloud or rainwater composition, or lead to new products when the cloud evaporates. Dissolution of sulfur dioxide into cloud droplets, followed by aqueous-phase oxidation to nonvolatile sulfate and water evaporation is an important mechanism for producing sulfate aerosol.

Like dry deposition, wet deposition is sometimes parameterized in terms of a deposition velocity, v_w, which relates the deposition flux to the concentration at some reference height [see Eq. (11.62)]. Alternatively, a vertically resolved scavenging coefficient, $\varphi(z)$ (T^{-1}) may be used, with the wet deposition flux, J_w, calculated as:

$$J_w = \int_0^h \varphi(z) C(z)\, dz \tag{11.76}$$

where h is the height of the layer through which rainout is occurring. Though widely employed, this parameterization is valid only if the scavenging process is independent of the aqueous-phase contaminant concentration and is irreversible.

11.4.4 Wet Deposition of Gases

Absorption of gases into liquid droplets is an interphase mass-transfer problem, as discussed in Chapter 4. For purposes of wet deposition calculations, the flux J_d $(M\, L^{-2}\, T^{-1})$ of a gas into a single water droplet can be expressed as:

$$J_d = k_m \left(C_g - C_s\right) \tag{11.77}$$

where k_m is a mass-transfer coefficient $(L\, T^{-1})$, C_g is the bulk gas concentration, and C_s is the concentration of the contaminant at the droplet surface that would exist in equilibrium with the aqueous-phase concentration.

As discussed in Chapter 4, mass-transfer coefficients are commonly expressed in terms of the dimensionless Sherwood number, $\mathrm{Sh} = k_m d_d / D_{m,\mathrm{air}}$, where $D_{m,\mathrm{air}}$ is the molecular diffusivity of the gas and d_d is the droplet diameter. For falling spherical droplets, the Sherwood number can be calculated as (Bird et al., 1960):

$$\mathrm{Sh} = 2 + 0.6\, \mathrm{Re}^{1/2} \mathrm{Sc}^{1/3} \tag{11.78}$$

where the Reynolds number uses the droplet's diameter as the characteristic length and its settling velocity as the characteristic velocity.

Assuming that water droplets with diameter d_d scavenge gases as they fall through a homogeneous layer of still air of height h, the mass scavenged per droplet will be

$$M_d = J_d \pi d_d^2 h v_{dt}^{-1} \tag{11.79}$$

where v_{dt} is the downward velocity of the droplet.[3] In the case of uniform droplet sizes, the number of drops falling per time over an area of dimensions $\Delta x \times \Delta y$ is

$$N_d' = \frac{p\,\Delta x\,\Delta y}{\frac{\pi}{6} d_d^3} \tag{11.80}$$

where $p\,(L\,T^{-1})$ is the precipitation rate (e.g., inches per hour). Combining Eqs. (11.77), (11.79), and (11.80), the instantaneous flux of the gas out of the homogeneous atmospheric layer is

$$J_w = \frac{6pk_m h}{d_d v_{dt}} \left(C_g - C_s\right) \tag{11.81}$$

For the case of uniform droplet diameters and irreversibly soluble gases such as nitric acid, for which $C_g \gg C_s$, the scavenging coefficient φ, defined in Eq. (11.76) above, is

$$\varphi = \frac{6pk_m}{v_{dt} d_d} \tag{11.82}$$

More generally, the scavenging coefficient integrates the product of droplet surface area and mass-transfer coefficient over the distribution of droplet sizes. For irreversibly soluble gases (Seinfeld and Pandis, 1998):

$$\varphi = \int_0^\infty \pi\, d_d^2 k_m\, n(d_d)\, dd_d \tag{11.83}$$

where $n(d_d)$ is the droplet number distribution.

Rather than explicitly modeling the detailed microphysics of cloud and raindrop evolution, scavenging calculations typically rely on the bulk physics assumption that droplet size distributions are fixed. The Marshall–Palmer distribution is typical of raindrop size distributions, which are parameterized in terms of precipitation intensity or rainfall rate (Pruppacher and Klett, 1997):

$$n(d_d) = n_0 \exp(-\overline{\omega}\, d_d) \tag{11.84}$$

where $\overline{\omega} = 4.1 \times 10^{-3}\, p^{-0.21}\ \mu\text{m}^{-1}$ with p in mm h^{-1} and $n_0 = 8 \times 10^{-6}$ drops cm^{-3} μm^{-1}.

[3] Note that the relatively large diameter of typical raindrops puts them outside the Stokes flow regime (Section 11.4.2). Their terminal velocities are significantly lower than Eq. (11.71) would predict. Seinfeld and Pandis (1998) discuss how to calculate terminal velocities for spherical droplets outside the Stokes flow regime.

11.4.5 Wet Deposition of Particles

As for gases, wet deposition calculations for particles generally need to account for both in-cloud and below-cloud scavenging processes. Within the cloud, nucleation scavenging is the dominant mechanism through which particles are incorporated into water droplets. Below precipitating clouds, impaction scavenging dominates.

Models of impaction scavenging start by defining a single droplet collection efficiency, which is the fraction of the particles in the volume swept by a falling raindrop that it collects. The assumption is made that particles are collected if they come into contact with the droplet due to diffusion out of the carrier airstream, inertial impaction, or interception (Fig. 11.12). The single droplet collection efficiency thus depends strongly on the particle diameter as well as on the droplet diameter. Slinn (1977, 1983) gives a semiempirical expression for the single droplet collection efficiency, η, which accounts for diffusion, interception, and impaction through three respective terms:

$$\eta\left(d_d, d_p\right) = \frac{4}{\mathrm{Pe}}\left(1 + 0.4\,\mathrm{Re}^{1/2}\,\mathrm{Sc}^{1/3}\right) + 4R_d\left(R_d + \frac{1 + 2R_\mu R_d}{1 + R_\mu \mathrm{Re}^{1/2}}\right) + \left(\frac{\mathrm{St} - \mathrm{St}^*}{\mathrm{St} + (2/3 - \mathrm{St}^*)}\right)^{3/2} \tag{11.85}$$

In Eq. (11.85), Pe is a Peclet number, defined as $d_d v_{dt}/2D$, where D is the Brownian diffusivity of the particle; Re is the Reynolds number of the droplet, defined using

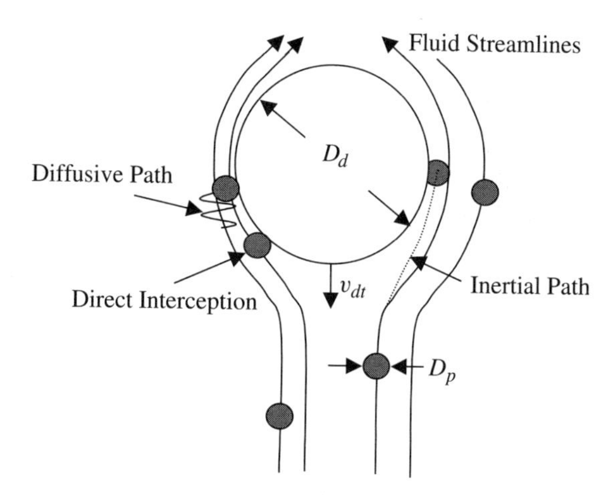

Figure 11.12 Particles are collected by rain drops falling with velocity v_{dt} through the mechanisms of inertial impaction, Brownian diffusion, and direct interception. Inertial impaction occurs when a particle's inertia prevents it from following the fluid streamline around the falling droplet. In direct interception, the fluid streamline comes within the particle's radius of the droplet. Brownian diffusion occurs when collisions with gas molecules impart a random deviation of the particle's path from the streamline.

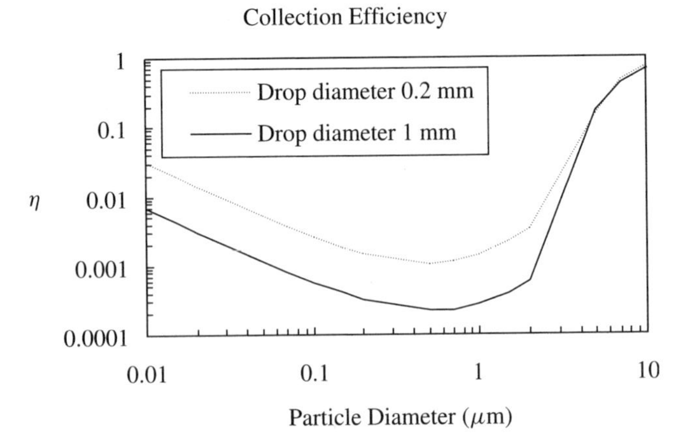

Figure 11.13 Particle collection efficiency of spherical rain drops of diameters $d_d = 0.2$ mm and 1 mm as a function of particle diameter, calculated using the semiempirical relationship of Slinn (1983) given in Eq. (11.85).

its radius as the characteristic length and its fall velocity as the characteristic velocity; Sc is the Schmidt number for the particle; R_d is the ratio of the particle and droplet diameters; R_μ is the ratio of viscosities of water and air, and St is a Stokes number, defined as $\tau v_d/d_d/2$, where τ is the particle relaxation time. The final parameter, St* is the critical Stokes number for the onset of impaction:

$$\mathrm{St}^* = \frac{1.2 + \frac{1}{12}\ln(1+\mathrm{Re})}{1+\ln(1+\mathrm{Re})} \tag{11.86}$$

The last term in Eq. (11.85) is set to zero except when St > St*. The relationship given in Eq. (11.85) is plotted in Figure 11.13 as a function of particle diameter, for two different collection droplet diameters. Like dry deposition velocities, single droplet collection efficiencies are very low for particles with diameters between about 0.1 and 1 μm.

Accounting for the single droplet collection efficiency, the flux of particles of size d_p scavenged by a single droplet of size d_d is

$$J_w = \frac{\pi d_d^2}{4} v_{dt} \eta\left(d_d, d_p\right) C \tag{11.87}$$

where $(\pi/4)d_d^2 v_{dt}(d_d)$ is the volume swept by the droplet per unit time.

For the general case of a distribution of droplet diameters (e.g., the Marshall–Palmer distribution), the scavenging coefficient is the integral over all droplet sizes:

$$\varphi\left(d_p\right) = \int_0^\infty \frac{\pi}{4} d_d^2 v_{dt}(d_d) \eta\left(d_d, d_p\right) n(d_d)\, dd_d \tag{11.88}$$

As mentioned above, use of a scavenging coefficient implies that the scavenging process is irreversible, that is, that evaporation of raindrops is negligible.

Nucleation scavenging of a particle occurs when water condenses rapidly and increases its diameter by several orders of magnitude, transforming it into a cloud droplet. Water condenses onto existing particles/solution droplets whenever the partial pressure of water vapor exceeds the *saturation* partial pressure that would exist at equilibrium over the droplet. This process happens relatively slowly until a critical particle radius is reached, beyond which further growth lowers the saturation partial pressure. At this point, the particle is *activated* and grows very quickly to the size of a cloud droplet.

Airborne particles capable of growing into cloud droplets under atmospheric conditions are known as cloud condensation nuclei, or CCN. Relative humidities measured inside clouds can range from 98 to 102%, but values of about 100.1% are typical. At these modest supersaturations, both models and measurements indicate that more than 50% of the accumulation mode (~0.5 to 1 μm diameter) aerosol act as CCN (Seinfeld and Pandis, 1998). However, the saturations required for activation increase sharply as the particle diameters decrease, so the fraction of very fine particles that are activated seldom exceeds a few percent.

Whether or not atmospheric conditions permit soluble inorganic particles, such as NaCl and $(NH_4)_2SO_4$ particles, to become CCN can be determined by comparing the ambient saturation ratio, SR = RH/100, to the equilibrium saturation ratio that would exist over an aqueous solution droplet. SR is the ratio of the ambient partial pressure of water vapor relative to the vapor pressure of water. (Recall that the vapor pressure of water is defined as the equilibrium partial pressure of water vapor over a flat surface of pure water.) The Köhler equation gives the equilibrium saturation ratio, $\mathrm{SR_{eq}}$, over an aqueous solution droplet of known composition as a function of the droplet radius r_d and temperature T. For a dilute aqueous solution and saturation ratio close to 1, the simplified form of the Köhler equation is (Pruppacher and Klett, 1997)

$$\mathrm{SR_{eq}} \cong 1 + \frac{A}{r_d} - \frac{B}{r_d^3}$$

$$A = \frac{2\,\mathrm{MW^w}\sigma^{\mathrm{w}}}{RT\rho^{\mathrm{w}}} \qquad B = \left(\frac{3\nu m^{\mathrm{s}}\,\mathrm{MW^w}}{4\pi\rho^{\mathrm{w}}\,\mathrm{MW^s}}\right) \tag{11.89}$$

In Eq. (11.89), $\mathrm{SR_{eq}}$ is the ratio of the *equilibrium* partial pressure of water vapor over the solution droplet relative to the vapor pressure of pure water; $\mathrm{MW^w}$, σ^{w}, and ρ^{w} are the molecular weight, surface tension, and density of water, R is the universal gas constant, T is the absolute temperature; $\mathrm{MW^s}$ is the molecular weight of the solute, ν is the number of ions into which the solute dissociates (e.g., 2 for NaCl), and m^{s} is the mass of solute in the droplet. The A term in Eq. (11.89) accounts for the effect of the curvature of the droplet on the saturation ratio. The B term accounts for the effect of the dissolved salt. When SR is greater than $\mathrm{SR_{eq}}$, the ambient conditions are supersaturated for a particle or droplet of radius r_d, so water will be transferred from the vapor phase to the droplet.

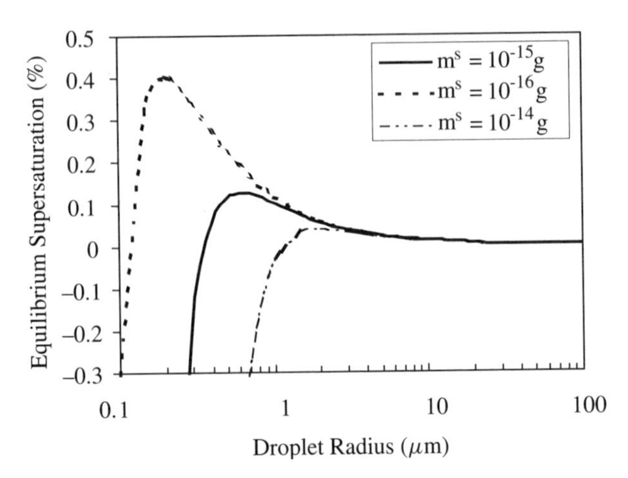

Figure 11.14 Example of Köhler curves for NaCl showing the equilibrium supersaturation ($SR_{eq} - 1$) (%) as a function of droplet radius and the mass of solute in the droplet.

Figure 11.14 shows the shape of the relationship given by the Köhler equation. The plot shows a maximum in SR_{eq}, which is known as the critical saturation, SR_c. The corresponding radius, r_{dc}, is the critical radius. Particles with radii larger than r_{dc} will be activated any time SR is greater than the value of SR_{eq} that corresponds to their original size because the equilibrium saturation ratio will decrease as they grow. Particles with radii smaller than r_{dc} will only activate if SR is greater than SR_c. For values of SR above SR_{eq} but below SR_c, any incremental increase in particle size will result in an increase in SR_{eq} that shuts down further growth. The critical radius and critical saturation ratio can be found by taking the derivative of Eq. (11.89) with respect to r_{dc} and setting it equal to zero. The results (for a dilute solution) are:

$$r_{dc} = \left(\frac{3B}{A}\right)^{1/2} \qquad SR_c \cong 1 + \left(\frac{4A^3}{27B}\right)^{1/2} \tag{11.90}$$

The Köhler equation can be modified to account for the effect of a wettable but insoluble particle core (Seinfeld and Pandis, 1998). In addition, Gorbunov et al. (1998) recently proposed a modification of the theory that accounts for the effect of soluble, insoluble, and surface-active organic compounds on particle activation. However, these detailed microphysical descriptions of droplet nucleation require more information about particle sizes, composition, and geometry than is often available for urban or regional-scale air quality model applications. Consequently, the treatment of nucleation scavenging in many air quality models is highly simplified (e.g., Taylor, 1989; Byun and Ching, 1999).

11.5 AVAILABLE PHOTOCHEMICAL AIR QUALITY MODELS

Integration of gas- and aqueous-phase chemistry, aerosol growth, and gas and particle deposition processes into spatially resolved models of air pollutant chemistry and transport is a significant challenge. Such photochemical air quality models have been under development since the 1970s, for application to urban and regional-scale ozone, acid deposition, and aerosol problems. Significant advances have been made over this time period in scientific descriptions of the transport and transformation processes that affect air pollutants and in numerical methods for solving the relevant governing equations. In this section, we describe and compare characteristics of state-of-the-art photochemical air quality models. We then present a case study that emphasizes a major source of uncertainty in many model applications: not the formulation of the model but rather its input data. A list of current photochemical models and their attributes is provided at www.wiley.com/college/ramaswami.

11.5.1 Features of State-of-the-Art Photochemical Models

Most state-of-the-art photochemical models employ a three-dimensional, Eulerian grid structure. Because of their computational efficiency, single-cell or vertically resolved trajectory model formulations are also used, but primarily for exploratory analyses (e.g., Kleeman et al., 1997; Bergin et al., 1999; Bergin and Milford, 2000). Recent Eulerian models such as CMAQ/Models-3 have incorporated grid nesting capabilities to allow finer spatial resolution to be used where emissions or concentration gradients are expected to be large. Both one-way and two-way nesting are used. In one-way nesting a complete simulation is first run in the coarser domain and then the outputs from that run are used as boundary conditions for simulations in the fine domain. In two-way nesting, air and contaminants flow from coarsely gridded regions into the finely gridded domain and back out into the coarse domain in a single simulation. In current models, the grid structure is fixed prior to beginning the simulation, but use of adaptive grids that move along with a plume of air pollutants is under investigation.

Most photochemical models currently use prognostic meteorological models to provide wind fields and other meteorological inputs. However, some advanced models, such as the Gas and Aerosol Transport and Reaction model (GATOR) (Lu et al., 1997) couple the solution of the advection–diffusion reaction equations for gases and aerosols with solution of the fluid dynamics and radiative transfer equations that govern atmospheric circulations. Coupled models allow investigation of feedbacks such as the effects of particles on radiative transfer, which in turn affects photolysis rates and the evolution of the planetary boundary layer. Coupled meteorology air quality models are expected to be an important area for ongoing research (Russell and Dennis, 2000).

Depending on their intended application, some photochemical models include only gas-phase chemistry, while others include aqueous chemistry and/or aerosol dynamics. Models used for urban-scale ozone, for example, often neglect aqueous chemistry as well as precipitation scavenging. The software structure of many recent models is modular so that specific processes can be turned on or off depending on whether they are significant for a particular application. Modular design also facilitates updating of components such as chemical mechanisms. The Models-3/CMAQ system is a good example of this flexible approach. Models-3/CMAQ can be used to model tropospheric ozone, acid deposition, fine particulates, and visibility on urban and regional scales. The Models-3/CMAQ system also gives users options of process parameterizations and solution algorithms; for example, a choice of three different gas-phase chemical mechanisms is provided.

11.5.2 Application of a Photochemical Air Quality Model

In their review of urban and regional-scale ozone models, Russell and Dennis (2000) conclude that current models generally do a good job of representing most of the processes that are important for ozone during typical high-concentration episodes.[4] Compared to gas-phase chemistry and transport models for ozone, descriptions of the processes involved in describing secondary aerosol formation and evolution are less advanced. Moreover, relatively few studies have evaluated size-resolved aerosol models because of lack of data needed to carry out such evaluations (Seigneur et al., 1997).

Despite the fact that ozone models are viewed as generally sound in their formulation, Russell and Dennis (2000) note that they have not performed as well as expected in many studies. Normalized gross errors of 25 to 35% for modeled ozone concentrations compared to observations were found to be typical, with little indication of improvement over time. The inability of the models to match observed concentrations is thought to be largely due to inaccurate model *inputs*, for example, emissions and meteorology. However, another factor may be a mismatch between the spatial resolution of the models and the point observations that are used to evaluate them (McNair et al., 1996).

An application of the CIT airshed model illustrates the importance of model inputs and their associated uncertainties. Harley et al. (1997) applied the CIT model to predict ozone concentrations in the South Coast Air Basin (SoCAB) for August 27–28, 1987, a period when intensive air quality monitoring was conducted as part of the Southern California Air Quality Study (SCAQS). A principal objective of the study was to examine the sensitivity of the model results to assumptions about the input motor vehicle emissions inventory. The effect of using different chemical mechanisms was also evaluated. Results obtained using the SAPRC93 mechanism (Carter, 1995) were compared to those produced with the older LCC mechanism (Lurmann et al., 1987). The SAPRC93 mechanism consists of 186 reactions involving 83 species. It is significantly more detailed than the LCC mechanism, which has only 38 species that participate in 95 reactions.

The CIT airshed model solves the 3D atmospheric diffusion equation for the concentrations of multiple species, C^i:

$$\frac{\partial C^i(\mathbf{x}, t)}{\partial t} + \mathbf{u} \cdot \nabla C^i = \nabla \cdot D \nabla C^i \pm \mathrm{Rxn}^i(\mathbf{C}) \qquad i = 1, 2, \ldots, n \tag{11.91}$$

The equations for the n species are coupled through the reaction terms, Rxn^i. Emissions are accounted for in the surface boundary condition:

[4] When important, cloud processing and point source plumes could be exceptions because they are not well-developed in many models. Atmospheric chemistry of aromatic compounds is also an important source of uncertainty.

$$- D_z \frac{\partial C^i}{\partial z} = E^i - v_g^i C^i \tag{11.92}$$

Concentrations are specified around the sides of the modeling domain. A no-flux boundary condition is specified for the top of the modeling region. Typical of high ozone episodes, weather conditions during the August 1987 episode were dry and sunny, so only gas-phase chemistry and dry deposition processes were modeled. The modeling domain used is shown in Figure 11.15. A fixed grid was used with 5 km × 5 km horizontal resolution and five vertical layers, with layer depths ranging from 38 m at the surface to 429 m for the highest level.

Meteorological inputs for the CIT model were interpolated from measurements made during SCAQS. Wind fields, for example, were interpolated from hourly average surface-level winds measured at 50 stations, and from upper-air soundings made 6 times per day at 8 sites. Data from the upper-air soundings were used to estimate time-varying mixing depths. Boundary and initial concentrations of ozone, nitrogen oxides, and hydrocarbons were estimated from surface monitors and limited aircraft measurements. Surface-level ozone concentrations were measured at more than 50 sites during SCAQS; however, only 12 sites reported hydrocarbon concentrations. As with many air quality modeling studies, the CIT model application would have benefited from more upper-air data for both meteorological variables and pollutant concentrations.

The baseline emissions inventory used with the CIT model was a spatially (25 km^2) and temporally (hourly) resolved inventory developed for the SoCAB in conjunction with SCAQS. Harley et al. (1993) note that the detailed inventory includes emissions from more than 800 source types and that the inventory for organic compounds includes more than 280 chemical species. Specifically, the baseline mobile source inventory was developed using a travel demand model to estimate vehicle activity together with California's model for emissions factors, EMFAC. EMFAC utilizes

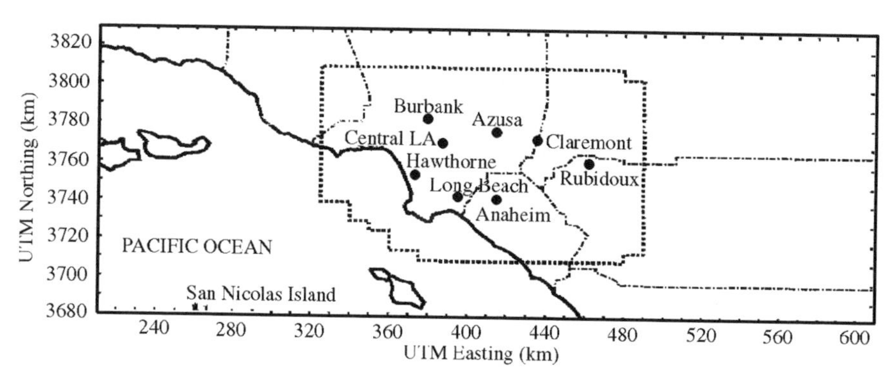

Figure 11.15 Map of the computational region used by Harley et al. (1997) indicating the locations of nine intensive monitoring sites from the Southern California Air Quality Study. The dashed line shows county boundaries; the dotted line shows the modeling region.

TABLE 11.2 Comparison of Baseline (MVEI 7E[a]), Current Official (MVEI 7G[a]), and Revised (Fuel-Based) Motor Vehicle Emissions Estimates for the South Coast Air Basin in Summer 1987

Emission Category	MVEI 7E Emissions 10^3 kg/day	MVEI 7G Emissions 10^3 kg/day	Fuel-based Emissions 10^3 kg/day	Ratio: Fuel-Based/ MVEI 7E	Ratio: Fuel-based/ MVEI 7G
CO	3966	6926	8100	2.0	1.3
NO_x	603	740	710	1.2	0.96
NMOC					
Exhaust	369	476	1240	3.4	2.6
Evaporative[b]	180	259	559	3.1	2.2
Total NMOC	549	735	1800	3.3	2.4

[a] Motor Vehicle Emissions Inventory based on EMFAC 7E and Motor Vehicle Emissions Inventory based on EMFAC 7G.

[b] Includes diurnal and hot soak evaporative emissions, running losses, and resting losses.

Source: Adapted from Harley et al. (1997).

laboratory data on emissions from vehicles representing a wide range of models, ages, and maintenance conditions and extrapolates these results to on-road vehicle populations and driving conditions. Two versions of EMFAC, "7E" and "7G" were available for comparison at the time the study was performed.

Because field experiments conducted earlier in the decade suggested that the official inventories underestimated on-road motor vehicle emissions of carbon monoxide and nonmethane organic compounds (NMOC) (Fujita et al., 1992; Pierson et al., 1990), Harley et al. (1997) replaced the official on-road vehicle emissions inventory with new emissions estimates developed from gasoline sales and infrared remote-sensing data on exhaust emissions from on-road vehicles (Stedman et al., 1994; Singer and Harley, 1996). Table 11.2 compares the fuel sales and remote-sensing-based emissions estimates with those based on EMFAC 7E and EMFAC 7G. The fuel sales and remote-sensing-based estimates for CO and NMOC are higher than the official estimates; the differences are consistent with the degree of underestimation in the official estimates that was found in the field studies.

Model results for ozone concentrations at selected locations are shown in Figure 11.16. Results across the modeling domain are shown in the form of contour plots for peak ozone concentrations in Figure 11.17. Comparison of model results obtained using the LCC and SAPRC93 mechanisms indicated that SAPRC93 gives peak ozone concentrations that are about 20% higher on average than the concentrations obtained with LCC. (Of note, subsequent revisions of the SAPRC mechanism have gone in the opposite direction, tending to reduce the amount of ozone produced for equivalent emissions of reactive organic gases.) Replacing the EMFAC 7E-based inventory with the inventory developed from fuel sales and remote-sensing data led to an increase of more than 80%, on average, in predicted peak ozone concentrations.

The performance of the CIT model for predicting ozone concentrations at the monitoring sites included in Figure 11.16 is good, compared to the typical errors

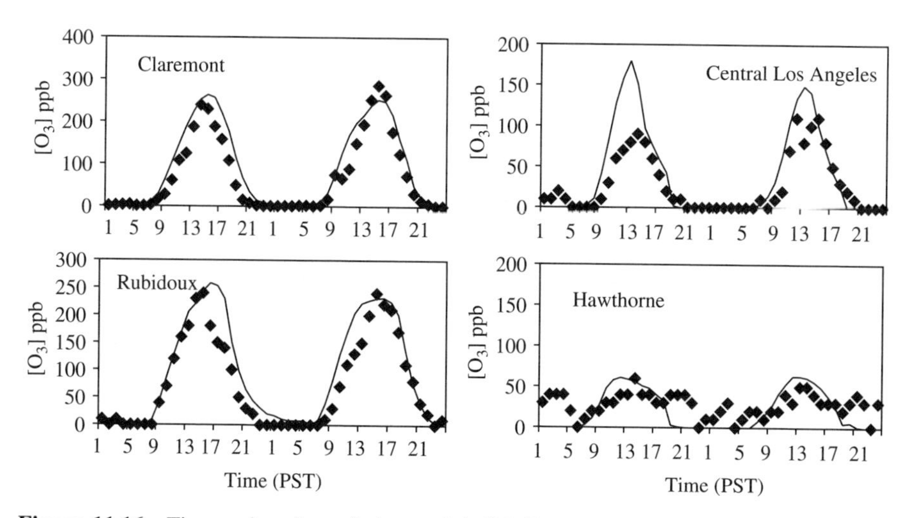

Figure 11.16 Time-series plots of observed (solid diamonds) and predicted (lines) ozone concentrations. Model predictions shown used the CIT model with SAPRC93 chemistry and the revised motor vehicle emissions inventory. (Adapted from Harley et al., 1997.)

in modeled ozone concentrations of 25 to 35% that were cited by Russell and Dennis (2000). However, considering all of the ozone monitoring data from all times and locations with ozone concentrations above 60 ppb, the version of the model with SAPRC93 and the revised motor vehicle emissions inventory exhibits an overall normalized bias of 34% and a gross error of 50%. (See Table 14.1 for mathematical definitions of these performance statistics.) Peak ozone concentrations at measurement stations are overpredicted by 20% on average. The basinwide peak is underpredicted by 4%.

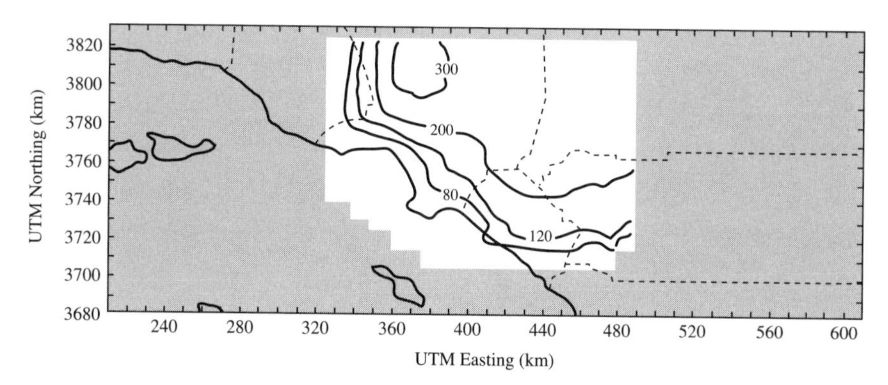

Figure 11.17 Contours of peak 1-h average ozone concentrations for August 28, 1987, modeled using the CIT model with SAPRC93 chemistry and the revised motor vehicle emissions inventory (Harley et al., 1997).

Was the performance of the CIT model satisfactory in the Harley et al. (1997) study? This question is discussed further in Chapter 14, which points out that both model performance in comparison to observations and *face validity,* or consistency with current scientific understanding, are important factors to consider when evaluating environmental models. Several lines of argument support the conclusion that the revised motor vehicle emissions inventory reduced one of the major uncertainties in input data that has compromised many photochemical air quality model applications. On the other hand, the fact that the CIT model's performance was worse than the norm reported by Russell and Dennis (2000) suggests that there is room for improvement. Likely avenues for improvement include further updates to the chemical mechanism and refining the model's meteorological inputs through the acquisition of more upper air meteorological data and the use of prognostic meteorological models.

12 Modeling Chemical Transformations in Water

Chapters 9 and 10 described contaminant transport in groundwater and surface water systems, respectively. A brief overview of chemical transformations occurring in the aqueous environment was presented in Section 2.3. The objective of this chapter is to couple chemical transformations with transport processes in order to describe the overall transport and fate of pollutants in aqueous environments.

Chemical reactions transform the molecular structure of pollutants, yielding products with altered reactivity, toxicity, and mobility characteristics. Chemical reactions can therefore result in removal of primary pollutants from water, while sometimes also creating secondary pollutants. Chemical reactions occurring in aqueous systems are often mediated by enzymes found in biota, for example, microbes and plants; such reactions are termed biochemical reactions. Still others may involve the interaction of mineral phases and enzymatic systems and are termed biogeochemical reactions. Reactions in which both reactants and products are dissolved in the aqueous phase are termed homogeneous aqueous-phase chemical reactions, while those in which either reactant or product exists as a separate phase (e.g., a solid-phase mineral) are termed heterogeneous reactions.

Chemical reactions can broadly be subdivided into two categories: (a) irreversible reactions that proceed from reactants to products only in the forward direction until exhaustion of the reactants occurs and (b) reversible chemical reactions wherein equilibrium between rapid forward and backward reactions controls the final chemical concentrations in the environment. See Table 2.4. Although presented separately in Table 2.4, reversible and irreversible physicochemical processes frequently co-occur in aqueous systems. For example, although acid–base dissociation reactions occur rapidly and reversibly in water enabling the use of instantaneous equilibrium models, the slow rate of dissolution of limestone and other minerals into water can function as a rate-limiting step that controls the overall rate of change in acidity or alkalinity of aqueous solutions. Likewise, irreversible reactions such as microbial degradation of organic pollutants can be affected by phase equilibria that control the availability of oxygen and pollutants in water. Because of the large number of physicochemical processes that co-occur in aqueous environments and, together, control the overall rate of chemical transformations, the process of developing water quality models must include three important steps:

1. Identifying the most relevant physicochemical processes affecting contaminant fate in water

2. Choosing appropriate equilibrium or rate-limited mathematical models for these processes based on analyses of Damkohler numbers (see Section 4.7)
3. Implementing numerical methods for incorporating the relevant reversible and/or irreversible reaction processes into chemical transport models in surface and groundwater systems

The process of developing water quality models is illustrated in this chapter by analyzing five case studies in which a wide spectrum of chemical reaction process models, ranging from instantaneous equilibrium models to kinetic irreversible reaction models, are coupled with transport processes occurring in surface and groundwater systems. Section 12.1 presents a study of Lake Constance, Germany (Sigg, 1985), focusing on equilibrium surface-sorption of metal ions on particles, followed by settling of particulates within the lake. Section 12.2 describes the evolution of models that describe the response of lake waters to acid rain, starting with an exclusively equilibrium representation of mineral dissolution and acid–base chemistry and culminating in a "trickle-down" model (Lin and Schnoor, 1986) that considers a combination of equilibrium and rate-limited processes occurring in multiple watershed compartments draining into a lake. The process of combining equilibrium and rate-limited transformation processes is further elaborated upon in Section 12.3 with a study of alkalinity changes down a mountain stream in Colorado in response to acid mine drainage (Broshears et al., 1996). Models for irreversible chemical transformations in water are presented in Section 12.4, with a focus on in situ biodegradation of organic pollutants is groundwater, described using two case studies. The first case study (Section 12.5) at a creosote-contaminated site in Texas utilizes a biodegradation kinetic model wherein degradation of organic pollutants is assumed to occur instantaneously, the overall degradation rate in groundwater being controlled by oxygen availability (Borden et al., 1986). The second case study at Dover Airforce Base, Delaware (Section 12.6), examines a sequence of aerobic and anaerobic reactions that result in degradation of chlorinated solvents yielding multiple by-products in groundwater (Clement et al., 2000). Thus, reaction models ranging from equilibrium to instantaneous kinetic models are presented in Sections 12.1 to 12.6.

12.1 CHEMICAL EQUILIBRIA IN ZERO-DIMENSIONAL TRANSPORT SYSTEMS

Important chemical reactions involving inorganic species in water are identified in Section 2.3, including acid–base dissociation, oxidation–reduction, complex formation–destabilization, and mineral dissolution–precipitation. Inorganic ions can also interact chemically with soil surfaces present in water, effectively resulting in sorption and desorption equilibria in solid–water systems. In the next few pages, we present an introduction to equilibrium surface complexation reactions that describe the interactions of aqueous metal ions with particulate matter.

12.1.1 Surface Complexation of Metals

Amphoteric minerals, for example, oxides of iron, aluminum, and silicate that exhibit variable surface charge conditions, are common constituents of Earth's crust. Particles composed of these minerals are frequently encountered in surface and groundwater. The surface of these metal or metalloid oxides is typically covered with surface hydroxyl groups formed by the interaction of the metal oxide mineral with water. Consider a metal oxide mineral MeO, where Me represents the metal or metalloid element within the mineral. When exposed to water, the electron-donating metal, Me, forms a coordination bond with a water molecule through the electronegative oxygen atom in water. This often yields a charged surface on soil particles with the Me^+ ions on the solid surface linked first with a layer of OH^- ions, forming a metal hydroxide surface, which in turn attracts H^+ ions (Stumm and Morgan, 1996). The metal oxide surface coordinated in this manner with a water molecule is represented as $\exists MeOH_2{}^+$ and is shown schematically in Figure 12.1*a*, with the symbol ∃ representing linkage to solid surfaces. The reactive surface can become deprotonated, that is, dissociate to release H^+ ions. This dissociation may be described, similar to that of a diprotic acid [see Chapter 2, Eq. (2.17) for monoprotic acid dissociation], as:

$$\exists MeOH_2{}^+ \Leftrightarrow H^+ + \exists MeOH \qquad K_{a1} = \frac{\{H^+\}\,[\exists MeOH]}{[\exists MeOH_2{}^+]} \tag{12.1a}$$

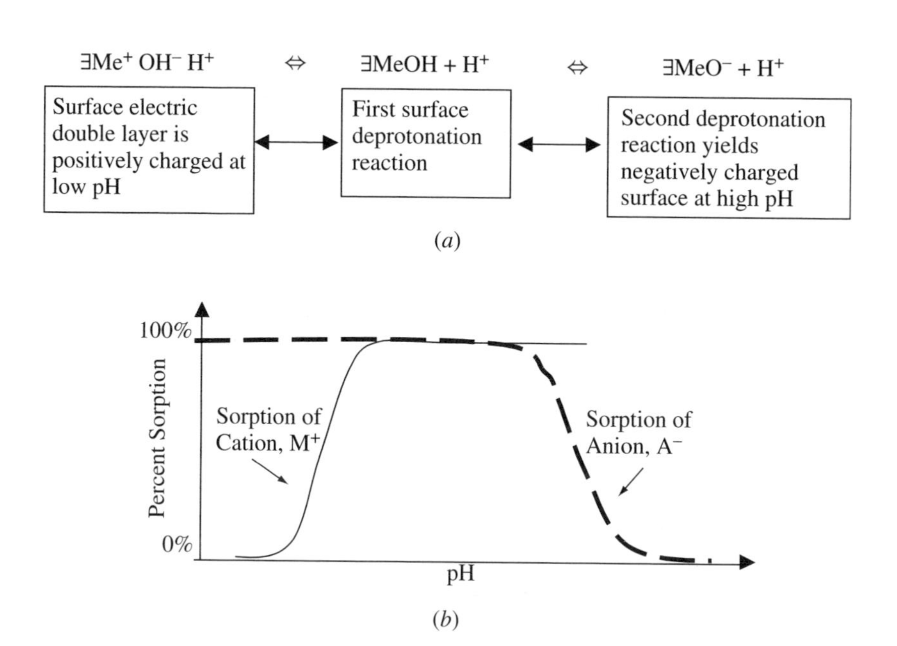

Figure 12.1 Generalized schematic showing (*a*) the amphoteric nature of mineral oxide surfaces and (*b*) the impact of pH on the "sorption" edge of cations (solid line) and anions (broken line).

$$\exists\text{MeOH} \Leftrightarrow \text{H}^+ + \exists\text{MeO}^- \qquad K_{a2} = \frac{\{\text{H}^+\}\,[\exists\text{MeO}^-]}{[\exists\text{MeOH}]} \tag{12.1b}$$

where the { } and [ ] symbols represent activity and molar concentration,[1] respectively, of the various species in aqueous solution, and K_{a1} and K_{a2} are termed the first and second apparent surface acid dissociation constants, respectively.[2] As indicated previously, the symbol ∃ represents species linked to solid surfaces.

From the above reactions one can see (by analogy with acid dissociation) that when pH $\ll$ pK_{a1}, positively charged surface species dominate, and when pH $\gg$ pK_{a2}, negatively charged surface species dominate. Thus, the metal oxide surface can exhibit variable charge at different pH conditions and is able to attract and "sorb" dissolved anions and cations present in water depending on the charge at the surface, as influenced by the solution pH. The sorption of dissolved cations to these surfaces is inhibited at low pH (due to repulsion between the incoming cation, M^+ and the positive surface charge) but is dramatically increased to 100% sorption at higher pH conditions. The reverse is true of dissolved anions in solution as shown in Figure 12.1*b*. The interaction of a divalent metal ion, M^{2+}, with the deprotonated surface, ∃MeOH, under favorable pH conditions for sorption can be written as:

$$\exists\text{MeOH} + \text{M}^{2+} \Leftrightarrow \exists\text{MeOM}^+ + \text{H}^+ \qquad K_{s1} = \frac{\{\text{H}^+\}\,[\exists\text{MeOM}^+]}{[\exists\text{MeOH}]\,\{\text{M}^{2+}\}} \tag{12.2a}$$

where K_{s1} represents the equilibrium constant for the formation of the metal–surface complex. Other metal–surface complexes may also be formed, for example:

$$2\exists\text{MeOH} + \text{M}^{2+} \Leftrightarrow (\exists\text{MeO})_2\text{M} + 2\text{H}^+ \qquad K_{s2} = \frac{\{\text{H}^+\}^2\,[(\exists\text{MeO})_2\text{M}]}{[\exists\text{MeOH}]^2\,\{\text{M}^{2+}\}} \tag{12.2b}$$

As indicated from the equilibrium representations in Eqs. (12.2a) and (12.2b), the concentrations of both surface–metal complexes are sensitive to the hydrogen ion concentration $[H^+]$ and hence pH.

A pH-dependent partition coefficient, K_d, can be defined for the sorption of metal ions from aqueous solution to the surface of metal–oxide particles. Following the definition of K_d in Chapter 3, the partition coefficient for metals can be defined as:

$$K_{d^\text{M}} = \frac{\sum \text{Conc. of surface complexed metal species (mol/kg}_\text{particles})}{\sum \text{Concentration of aqueous-phase metal species (mol/L}_\text{water})} \tag{12.3a}$$

[1] Activity and molar concentration are related through the activity coefficient, γ, which typically depends upon the ionic strength of the solution. At low ionic strength, { } = [ ]. Generally, [ ] = $\gamma \times$ { }.

[2] These dissociation constants are termed apparent dissociation constants as they relate to H^+ ion activity in bulk solution instead of H^+ ions on the surface. Between the bulk water and the charged surface lies a diffuse layer of counterions, often modeled as the Gouy–Chapman layer, which creates an electrostatic barrier affecting the entry of incoming ions to the surface. The apparent equilibrium constants incorporate the electrostatic effect by applying a Coulombic correction factor to an intrinsic equilibrium constant. See Eq. (12.4a).

and K_d has units of $L_w/kg_{particles}$, consistent with that of the soil–water partition coefficient for organic chemicals introduced in Chapter 3; however, K_d for metal ions is strongly pH sensitive. In the simple illustration presented in Eqs. (12.2a) and (12.2b) there is only one aqueous-phase species, M^{2+}, along with two surface-complexed metal species, in which case K_d is computed as:

$$K_d^{\mathrm{M}} = \frac{\sum \text{Conc. of surface complexed metal species}}{\sum \text{Concentration of aqueous-phase metal species}} = \frac{[\exists \mathrm{MeOM}^+] + [(\exists \mathrm{MeO})_2\mathrm{M}]}{[\mathrm{M}^{2+}]} \tag{12.3b}$$

The equilibrium concentration of the three metal species at a specified pH can be computed by simultaneously solving three equilibrium equations [(12.1a), (12.2a), and (12.2b)] along with mass-balance and charge-balance constraints, as was illustrated in Chapter 2. Available computer packages such as MINTEQ (Allison et al., 1991; Gustaffson, 2004) readily solve these types of problems employing a large built-in thermodynamic database that provides numerical values for the various equilibrium constants.

Need for Surface Charge Correction Before working on computer modeling, it is important to understand the correction for surface charge as it is applied to the equilibrium constants for the surface reactions shown in Eqs. (12.1) and (12.2). The equilibrium constants in Eqs. (12.1a), (12.1b), (12.2a), and (12.2b) are often referred to as apparent equilibrium constants as they relate to metal and hydrogen ion concentrations in bulk aqueous solution instead of at the metal oxide surface. The ions in bulk aqueous solution must overcome an energy barrier before they are attracted to the charged metal oxide–hydroxide surface. In the diffuse double-layer model often used to describe surface complexation reactions (Dzombak and Morel, 1990; Stumm and Morgan, 1996), the charged metal oxide surface is visualized to be surrounded by a second layer of diffuse counterions in solution, with a charge opposite to that of the surface. For example, when the metal–oxide surface is negatively charged, the diffuse layer is dominated by positively charged counterions in solution. An incoming metal ion (M^+) or a hydrogen ion (H^+), must overcome the electrostatic barrier thus created by the charged surface and its associated counterions in the diffuse layer. The energy barrier offered by the diffuse layer decreases in intensity with distance away from the surface and contributes to a coulombic correction factor $e^{-zF\Psi/RT}$, where z is the net change in surface charge as a result of the reaction, Ψ is the surface electric potential, F is Farraday's constant, R is the universal Gas constant, and T is temperature. The apparent acid dissociation constants in Eq. (12.1a and 12.1b) are obtained from corresponding intrinsic dissociation constants, that are independent of surface charge by applying the coulombic correction factor. For example:

$$K_{a1} = K_{a1}^{\mathrm{int}} \times e^{-zF\Psi/RT} \tag{12.4a}$$

The intrinsic dissociation constant for Eq. (12.1a), K_{a1}^{int}, may be now rewritten as:

$$K_{a1}^{\text{int}} = \frac{K_{a1}}{e^{-zF\Psi/RT}} = \frac{\{\text{H}^+\}\,[\exists\text{MeOH}](e^{-F\Psi/RT})}{[\exists\text{MeOH}_2{}^+]} \quad \text{for } z = -1 \qquad (12.4\text{b})$$

wherein the coulombic correction factor now effectively functions as a component that would appear as a product on the rhs of Eq. (12.1a), that is, the surface species $\exists MeOH_2{}^+$ can be envisioned to dissociate to yield the *three* components shown in the numerator in Eq. (12.4b) including the coulombic factor. Thus, surface charge effects are incorporated into equilibrium surface complexation models by employing intrinsic equilibrium constants along with the coulombic factor as a reactant or a product based upon a reaction-specific change in surface charge. A necessarily brief overview of surface complexation reactions at metal–oxide surfaces has been presented here, with focus on the diffuse double-layer model for describing the electrostatic barrier surrounding a charged surface. The reader is directed to classic textbooks in this area (Stumm and Morgan, 1996; Dzombak and Morel, 1990) for more in-depth learning of alternate models for describing surface charge and solution electrostatic effects.

EXAMPLE 12.1

Determine K_d for sorption of Pb^{2+} to hematite surfaces at aqueous solution pH conditions of 3, 4, 5, 6, and 7. Assume a total of 10^{-6} M lead is in contact with solid hematite present at a concentration of 8.6 mg/L of water. The ionic strength is 0.005 M. The hematite is characterized by a specific surface area of 40 m^2/g with site density of functional groups of 0.32 mol/kg (after Stumm and Morgan, 1996, pp. 573). The intrinsic equilibrium constants are $K_{a1}^{\text{int}} = 10^{-7.25}$; $K_{a2}^{\text{int}} = 10^{-9.75}$; and $K_{s1}^{\text{int}} = 10^{4.0}$.

Solution 12.1 We solve this problem first at a pH of 5 using Visual MINTEQ (http://www.lwr.kth.se/English/OurSoftware/vminteq/) and obtain the output shown in Table 12.1. Details of the solution technique are available in the supplementary materials for chapter 12 at www.wiley.com/college/ramaswami. At a pH of 5, the surface-sorbed lead species $>FeOPb^+$ is found to be at an aqueous concentration of $10^{-6.3}$ M. The total lead in the system was specified at 10^{-6} M. Thus, the fraction of lead sorbed to hematite at a pH of 5 is 0.5. Visual MINTEQ can be operated in a pH sweep mode to generate output for the other pH conditions specified in the problem, yielding the data for the sorption "edge" shown in Figure 12.2.

We now compute the K_d for lead at a pH of 5. Note, the concentration of lead

TABLE 12.1 Output from Visual MINTEQ for Sorption of 10^{-6} M Pb on to Hematite Solid Present at a Concentration of 8.6 mg/L of Water (2.8 x 10^{-6} M) Solution pH is Set at 5

	Concentration	Activity	Log Activity
>FeO– (1)	2.458E-07	2.458E-07	−6.609
>FeOH2+ (1)	2.6553E-07	2.6553E-07	−6.576
>FeOPb+ (1)	5.0078E-07	5.0078E-07	−6.3
>SOH(1)	1.6879E-06	1.6879E-06	−5.773
H+1	0.000010789	0.00001	−5
OH–	1.0864E-09	1.0069E-09	−8.997
Pb(OH)2 (aq)	2.9583E-14	2.9617E-14	−13.528
Pb(OH)3–	3.2174E-20	2.9822E-20	−19.525
Pb+2	4.9822E-07	3.6774E-07	−6.434
Pb2OH+3	1.0735E-14	5.4209E-15	−14.266
Pb3(OH)4+2	8.7194E-24	6.4359E-24	−23.191
Pb4(OH)4+4	7.9738E-27	2.3667E-27	−26.626
PbOH+	1.0035E-09	9.3011E-10	−9.031

Note: Attachment to solid surfaces is represented by the > symbol in Visual MINTEQ.

in the solid media is computed from the aqueous concentration of the metal oxide complex as:

$$C\frac{>\mathrm{FeOPb}^{+}}{\mathrm{solid}} = \frac{[>\mathrm{FeOPb}^{+}]}{8.6\ \mathrm{mg\ hematite/L_{water}}} = 0.058\frac{\mathrm{mol} > \mathrm{FeOPb}^{+}}{\mathrm{kg\ hematite}}$$

The total lead concentration in water is effectively $10^{-6.434}$ M as Pb^{2+}, since all other aqueous lead species concentrations shown in the output file are much smaller in comparison. The K_d for lead sorption to hematite can be computed from Eq. (12.3) to yield 157,550 L_w/kg hematite at a pH of 5.

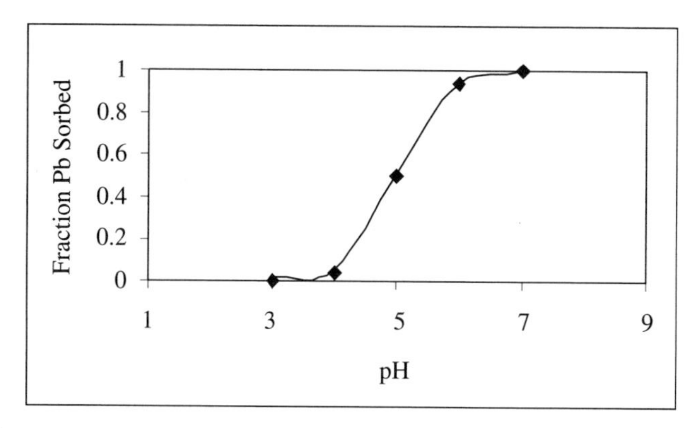

Figure 12.2 Sorption edge for lead complexation with hematite surfaces.

Example 12.1 shows that the percent lead sorbed from aqueous solution is strongly affected by solution pH, increasing from nearly zero at a pH of 3 to almost 1 at a pH of 7. A closed system was assumed in the simplified illustration presented in Example 12.1. The magnitude of the fraction sorbed, and hence the partition coefficient, is expected to change as more solid-phase and aqueous-phase complexation reactions are included in the equilibrium tableau, including the formation of aqueous-phase lead carbonate complex, $PbCO_3^0$ in an open system exposed to atmospheric CO_2. Computer programs such as MINEQL and MINTEQ allow the user to specify all relevant reacting species in water, such that all possible equilibrium complexation reactions are included in the equilibrium computations of the partition coefficient, K_d.

12.1.2 Case Study of Metals in Lake Constance, Germany

Sorption of metals to particles present in water bodies can effectively remove the metal from the aqueous phase, causing a corresponding decrease in the metal concentration in water. Furthermore, particles themselves undergo gravity settling and are removed from water by the process of sedimentation, thereby contributing to the removal of sorbed metals from water flow streams. The sorption–sedimentation effect of particles in lake waters is believed to result in the relatively low metal concentrations observed in some lakes, despite heavy inputs of metals through atmospheric deposition (Sigg, 1985).

Sigg (1985) presents a case study of Lake Constance, a prealpine lake in a calcite-rich region of the former Federal Republic of Germany. The hypolimnion of the lake is aerobic throughout the year. Lake water samples were collected during spring and summer of 1981 and 1982 at the deepest point in the lake. The samples were analyzed to determine pH, ionic strength, major cations, and anion concentrations, as well as the concentrations of heavy metals such as lead, zinc, and cadmium. Total aqueous concentrations of the heavy metals were measured since selective metal ions and complexes could not be distinguished by the electrodes used for analyses. The speciation of the aqueous metal ions was inferred from knowledge of the major ligands (e.g., OH^-, CO_3^{2-}, Cl^-) present in solution. Sediment traps were also placed in the lake, and data on particle sedimentation in the water column was collected at the same location and times as the aqueous phase was sampled. The pH of lake waters was fairly alkaline, ranging from 7.9 to 8.7, which is generally favorable for sorption of metals to particles. The heavy-metal concentrations in sediment were also measured.

Inorganic particles in the lake were assumed to include primarily oxides of silica, manganese-oxide-coated particles, and aluminum oxides. A range of partition coefficients for sorption of heavy metals to these inorganic surfaces was computed by Sigg (1985) using MINEQL and is shown in Table 12.2a. The MINEQL computations assume the particle concentration in water is 1 mg/L_w, the solution pH is 8 and the alkalinity is 2.5 meq/L_w. As seen in Table 12.2a, the computed K_d (L_w/$kg_{particles}$) for lead ranges from 30 for sorption to silicon oxides to 100,000 for sorption to magnesium oxides; the computed K_d for copper ranges from 1.6 for sorption onto silicon oxides to 280 for sorption to aluminum oxide surfaces. The actual mineral

TABLE 12.2 Computed and Measured Partition Coefficients (K_d) for Sorption of Lead and Copper to Particles in Lake Constance[a]

(a) **Theoretical computation of K_d** employing MINEQL to represent surface complexation reactions of metals with various pure mineral phase particles present at a particle concentration in water of 1 mg/L. Solution pH = 8, Alk = 2.5 meq/L.

Particle type	K_d for lead	K_d for copper
Silicon oxide	30	1.6
Aluminum oxide	240	390
Magnesium oxide	100,000	—

(b) **Observed K_d in Lake Constance** based on measured metal concentrations in lake water and sediments.

Particle type	K_d for lead	K_d for copper
Lake particles	1000	30

[a] Units of K_d are $L_{water}/kg_{particles}$.
Source: Adapted from Sigg (1985).

composition of the particles in Lake Constance was not known. Furthermore, in addition to mineral particles, particles of biogenic origin (e.g., algae) can also strongly sorb metals and subsequently undergo sedimentation. Analysis of sediments in Lake Constance showed a fairly constant heavy-metal composition ratio relative to phosphorus, suggesting removal with settling biota, with a hypothetical composition of $C_{113}N_{15}P_1Cu_{0.008}Zn_{0.06}Pb_{0.004}Cd_{0.00005}$. Based on a linear sorption model for partitioning of metals to both mineral and biological surfaces (at a fixed pH), a comprehensive constant partition coefficient K_d was deemed appropriate for use in the whole lake. The comprehensive partition coefficient K_d was computed from Eq. (12.3a) employing measured solid-phase and aqueous-phase metal concentrations, yielding the estimates shown in Table 12.2b. The measured K_d values are in the range of those computed using MINEQL assuming surface complexation equilibria (Table 12.2a), suggesting that equilibrium chemical sorption may be controlling metal interactions with particle surfaces.

Particle concentrations in the lake waters ranged from 0.1 to 5 mg/L, with a measured sedimentation rate of 10^3 g/m^{-2} yr. Metal concentrations in precipitation inputs and surface water inflows to the lake were measured, yielding metal input data to the lake. Data on sorption coefficients, particle concentrations, and sedimentation rates were coupled with metal input rates by employing a well-mixed lake model shown schematically in Figure 12.3. The model formulation involves the water balance and metal balance equations shown below:

Water Balance Equation $Q_{\text{inflow}} + Q_{\text{precip}} = Q_{\text{outflow}} = Q$, where Q_{inflow} is the average volumetric water flow rate entering the lake from surface and subsurface flows, and, Q_{precip} is the volumetric flow rate of direct precipitation water entering the lake.

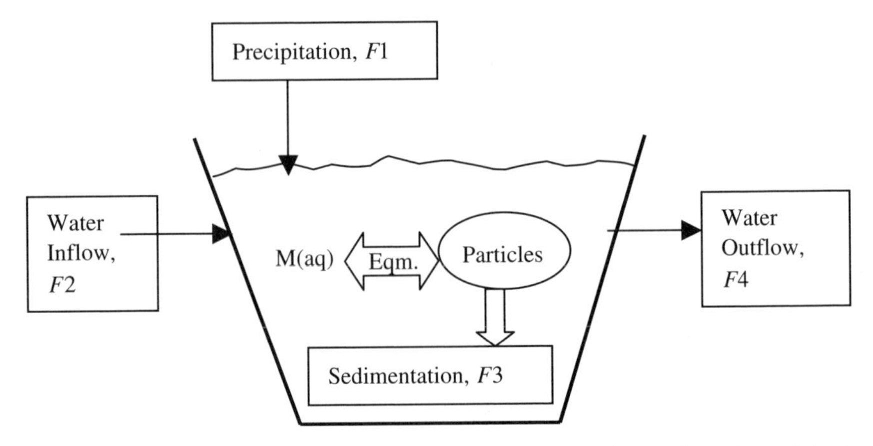

Figure 12.3 Schematic showing a well-mixed lake with metal inputs through precipitation and water inflows. Equilibrium metal sorption to particles and subsequent particle sedimentation contribute to metal removal from lake waters.

Metals Mass Balance

- *Metal mass input rates with precipitation* $= F1 = Q_{\text{precip}} \times C_{\text{precip}}$, where C_{precip} (mg/L_w) is the metal concentration measured in precipitation water.
- *Metal mass input rates with water flow* $= F2 = Q_{\text{inflow}} \times C_{\text{inflow}}$, where C_{inflow} (mg/L_w) is the metal concentration measured in influent waters.
- *Metal mass removal rates by sedimentation* $= F3 = K_d \times C_w \times S_{\text{particle}} \times A$, where K_d ($\text{L}_w/\text{kg}_{\text{particles}}$) is the particle-water partition coefficient for the metal of interest, C_w (mg/L_w) is the metal concentration in the well-mixed aqueous lake compartment, S_{particle} is the measured particle sedimentation rate in the lake waters [kg/m^{-2} day], and A (m^2) represents the area of the lake.
- *Metal mass outflow rates* $= F4 = Q_{\text{outflow}} \times C_w$.

At steady state, $F1+F2 = F3+F4$, from which the steady-state metal concentration in the lake waters, C_w, can be computed for various metals as:

$$C_w = \frac{F1 + F2}{Q + AS_{\text{particle}}K_d}$$

Using measured values of K_d shown in Table 12.2b, the computed steady-state concentrations for lead and copper in lake waters ranged from 0.02 to 0.5 mg/L and 0.02 to 1 mg/L, respectively. Lake measurements indicated lead and copper concentration ranging from 0.04 to 0.1 mg/L and 0.5 to 1 mg/L, respectively. Thus, an equilibrium sorption model for metal partitioning to particles, coupled with particle sedimentation in a well-mixed lake, is able to describe metal concentrations in the lake fairly well. More importantly, the steady-state modeling approach described above can be used to quantify the relative contribution of various physicochemical processes to heavy-metal fluxes in the lake. For example, more than 50% of the input of lead to the lake occurred from atmospheric deposition with precipitation, and, more than 90% of the total lead input to the lake is retained within lake sediment, resulting in low lead concentrations in lake waters. An acidic, alpine lake in the same region, which could be assumed to receive similar inputs of lead through precipitation from the atmosphere, showed much higher lead concentrations in water due to the low biological activity of the lake and low pH levels that inhibited metal sorption to mineral particles. Thus, this case study demonstrates that pH-sensitive equilibrium metal complexation with particle surfaces plays an important role in metal removal from surface waters.

The transport of dissolved cations and anions in groundwater can likewise be "retarded" by sorption to the reactive and charged mineral oxide soil surfaces encountered in the subsurface, provided appropriate pH conditions are maintained. Knowledge of surface chemistry is very important in determining the conditions that favor sorption of inorganic pollutants to such soil surfaces. While the mechanisms for this process are fairly well-understood, the difficulty lies in quantifying the concentration of surface-charged species in various soils. Soil suspensions in water undergo titration in water to determine the surface concentrations and pK_a values for the equilibrium expressions shown in Eqs. (12.1a) and (12.1b). Methods for determining these parameter values are detailed in Dzombak and Morel (1990).

12.2 COUPLED EQUILIBRIUM AND RATE-LIMITED CHEMICAL PROCESSES IN ZERO-DIMENSIONAL TRANSPORT SYSTEMS

Section 12.1 presented a simple, single-compartment well-mixed box model to describe metal concentrations in lakes by employing an exclusively equilibrium representation for metal sorption to particles. Other processes occurring in lakes, such as lake acidification, require a combination of equilibrium and rate-limited process models, best understood by studying the evolution of models describing the response of lakes to acid rain, illustrated schematically in Figures 12.4*a* to 12.4*d*. As shown in Figure 12.4*a*, the simplest model for acid rain response represents a lake as an impervious bowl open to the atmosphere, with only precipitation and evapotranspiration of water occurring. Cumulative inputs of acid species (e.g., sulfate) with precipitation would overwhelm any alkalinity initially offered by the lake water, after which lake water pH should drop steadily with a corresponding increase in SO_4^{2-} concentration. Although simple in concept, such a rudimentary model predicts an initial nominal equilibrium pH of 5.6 based on equilibration with atmospheric CO_2, followed by a sharp drop in lake pH in response to acid deposition since the role of mineral dissolution and cation exchange in neutralizing acid inputs is ignored. Minerals lining the lake bottom, or those found in surrounding soils, can dissolve into lake waters and buffer the atmospheric acid inputs to the lake. Equilibrium dissolution of carbonate minerals (e.g., limestone) is illustrated in the second conceptual model for lake acidification

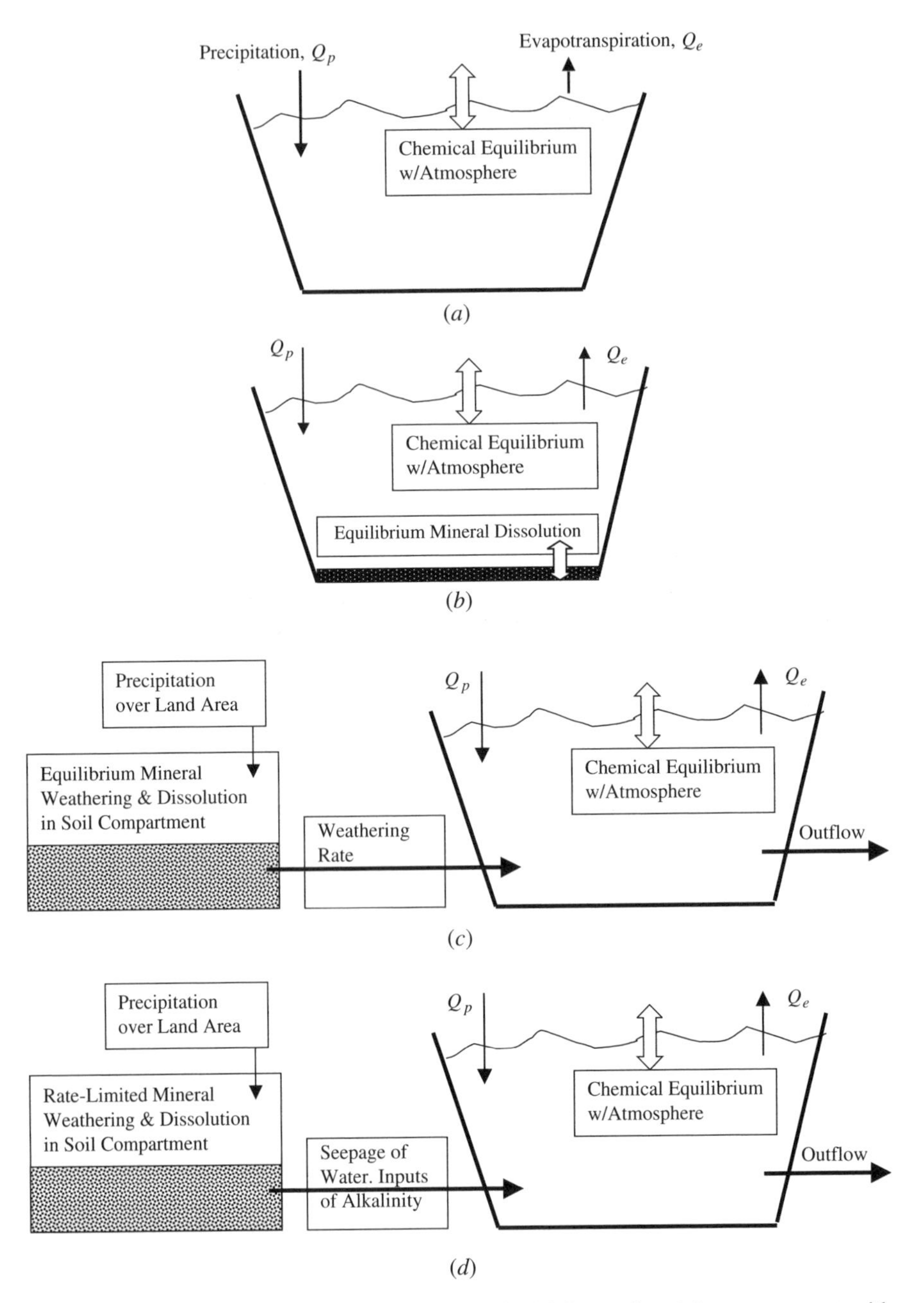

Figure 12.4 Evolution of lake acidification models: (*a*) Impervious lake compartment with precipitation and evaporation, (*b*) equilibrium dissolution of carbonate minerals contributes alkalinity to the mixed lake compartment, (*c*) equilibrium mineral dissolution and weathering from surrounding soils with limited availability of base cations, and (*d*) rate-limited weathering from surrounding soils controls alkalinity input through seepage water that enters the lake compartment.

shown in Figure 12.4*b*. In this case, an unlimited amount of limestone is assumed to line the lake bottoms. The limestone is assumed to instantaneously dissolve into the lake waters yielding conditions of instantaneous equilibrium mineral dissolution and saturation, which will neutralize any acid inputs to the lake, yielding the other extreme condition of little or no change in lake water pH in response to acid input.

More realistic models for lake acidification include some quantitative upper limit on the amount of acid neutralization capacity offered by mineral dissolution, cation exchange, and other geochemical processes occurring in the soils surrounding a lake or stream. Cosby et al. (1985a) present an equilibrium model for many of the geochemical processes contributing to the alkalinity of lake waters, including sorption of sulfate to soil, exchange of base cations and ions on mineral surfaces, dissolution of aluminum hydroxide and subsequent complexation reactions. A two-compartment model is used in which equilibrium processes contribute to weathering in the soil compartment, releasing a net flux of base cations and anions into the surface water compartment. The surface water compartment reequilibrates with a lower CO_2 partial pressure in the open atmosphere (compared to soil compartments); such equilibration determines the concentration of key constituents (including pH) leaving the lake. This third approach to modeling stream water response to acid inputs resulted in the development of the MAGIC class of models (Modeling Acidification of Groundwater In Catchments; Cosby et al., 1985a,b), illustrated schematically in the two-compartment equilibrium box models in Figure 12.4*c*. The primary disadvantage of this third modeling approach is that the weathering flux between the soil box and the surface water box is an unknown parameter since the total number of available minerals and base and acid exchange sites in soil is not known a priori. The weathering parameters pertaining to total ion exchange capacity must therefore be fit to the field observations.

A fourth approach is to explicitly model the weathering and dissolution of minerals in the soil "box" as a kinetic process, coupled with instantaneous equilibrium with ambient atmospheric CO_2 occurring in the linked surface water box. Experimental studies (e.g., Grandstaff, 1977; Busenberg and Plummer, 1982; Stumm et al., 1983; Schnoor and Stumm, 1985) have shown that the rate of mineral weathering depends upon the pH of the solution in contact with the minerals; likewise cation exchange kinetics are also dependent on the solution pH. The kinetics of mineral weathering and cation exchange in the soil zone can be used to explicitly estimate the flux of alkalinity entering the lake "box", where instantaneous equilibration with the open atmosphere occurs, as shown schematically in Figure 12.4*d* in the coupled kinetic-equilibrium model representation. The coupled kinetic-equilibrium model can be extended to multiple soil zones, water from all of which can trickle into a lake where equilibration with the atmosphere occurs. Other biochemical reaction kinetics may also be included, leading to a family of trickle down models developed by Lin and Schnoor (1986) and Nikoloaidis et al. (1988), further enhanced by incorporation of sulfate biodegradation and nitrogen reaction process models (Rees and Schnoor, 1994). Before presenting a case study on the calibration and application of the trickle-down model of Lin and Schnoor (1986), we present a series of simple exercises that will enable the reader to implement the various modeling approaches outlined in Figures 12.4*a* to 12.4*d*.

12.2.1 Exercises in Modeling Lake Response to Acid Inputs

We begin with the fundamentals by reviewing the case of carbonic acid present in water in a closed system, that is, there is no contact with CO_2 in the atmosphere and no contact with any carbonate containing minerals (solid phases). Carbonic acid is a di-protic acid that dissociates twice, releasing a hydrogen ion in each step. The dissociation reactions are shown below, with K_{a1} and K_{a2} representing the first and second acid dissociation constants, respectively:

Dissociation of carbonic acid (H_2CO_3):

$$H_2CO_3 \Leftrightarrow H^+ + HCO_3^- \qquad K_{a1} = \frac{[H^+][HCO_3^-]}{[H_2CO_3]} \tag{12.5a}$$

Dissociation of bicarbonate ion (HCO_3^-):

$$HCO_3^- \Leftrightarrow H^+ + CO_3^{2-} \qquad K_{a2} = \frac{[H^+][CO_3^{2-}]}{[HCO_3^-]} \tag{12.5b}$$

At typical ambient environmental temperatures, K_{a1} and K_{a2} are approximately of the order of $10^{-6.3}$ and $10^{-10.3}$, respectively. The total carbonate concentration in an aqueous system, C_{T,CO_3}, must equal the sum of the concentrations of the three species shown in Eqs. (12.5a) and (12.5b):

Mass Bal

$$C_{T,CO_3} = [H_2CO_3] + [HCO_3^-] + [CO_3^{2-}] \tag{12.6}$$

The charge-balance condition (see Section 2.3.2) for this system is written as:

Charge Bal

$$[H^+] = [HCO_3^-] + 2[CO_3^{2-}] + [OH^-] \tag{12.7}$$

In a closed system with 10^{-5} mol/L of C_{T,CO_3}, the equilibrium pH of water and the concentrations of the five species present in water can be determined by simultaneously solving the dissociation equations for carbonate [Eqs. (12.5a) and (12.5b)], the dissociation equation for water [Eqs. (2.15) and (2.16)] and the mass- and charge-balance equations shown in Eqs. (12.6) and (12.7), respectively. The graphical solution is shown on point *A* in Figure 12.5*a*, which matches the output from Visual MINTEQ shown in Table 12.3a.

In the open system shown in Figure 12.4*a*, the "impervious bowl" of water is at equilibrium with a typical atmosphere containing CO_2 at a partial pressure of $10^{-3.5}$ atm (=0.000316 atm). In this case, the concentration of dissolved carbonic acid is constant for all pH conditions and is determined solely by equilibration with the atmospheric CO_2 as:[3]

$$[H_2CO_3] = K_h^{CO_2} \times p_{CO_2} = 10^{-1.5}(\text{mol/L atm}) \times 10^{-3.5}\ \text{atm} = 10^{-5}\ \text{mol/L} \tag{12.8}$$

[3] Note, the Henry's constant K_h for inorganic chemicals is the reciprocal of the definition conventionally applied for organic chemicals presented in Eq. (3.7).

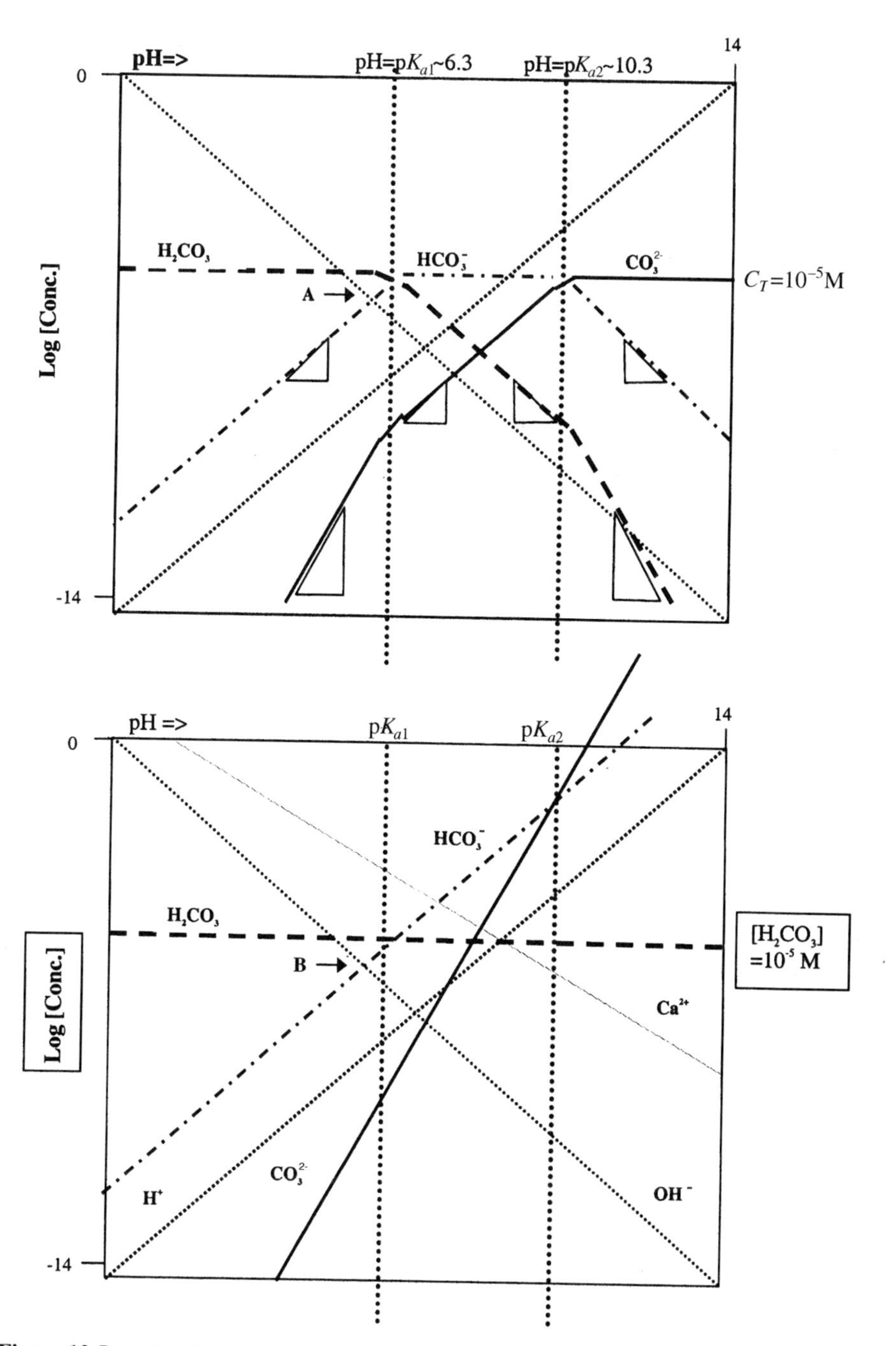

Figure 12.5 *pC*–pH plots of aqueous carbonate systems. (*a*) Closed system containing 10^{-5}M total carbonate and (*b*) aqueous system open to the atmosphere at a partial pressure of CO_2 of $10^{-3.5}$ atm (light grayscale line indicates Ca^{2+} in solution at equilibrium with an infinite supply of $CaCO_3$).

TABLE 12.3 Equilibrium Species Concentrations in Three Different Aqueous Systems Containing Carbonate[a]

(a) Equilibrium in Aqueous Carbonate Systems. No Sulfate

BASE CASE: Closed System With Total Conc of Carbonate = 0.00001 M

Species	Concentration	Activity	Log activity
CO3-2	4.7052E-11	4.6748E-11	−10.33
H+1	1.9037E-06	1.9007E-06	−5.721
H2CO3* (aq)	8.1016E-06	8.1016E-06	−5.091
HCO3−	1.8983E-06	1.8953E-06	−5.722
OH−	5.3064E-09	5.2978E-09	−8.276

CASE 1: Open System in Equilibrium with CO2 at a partial pressure of 0.000316 atm

Species	Concentration	Activity	Log activity
CO3-2	4.7107E-11	4.6781E-11	−10.33
H+1	2.1931E-06	2.1893E-06	−5.66
H2CO3* (aq)	0.000010757	0.000010757	−4.968
HCO3−	2.1884E-06	2.1846E-06	−5.661
OH−	4.6073E-09	4.5993E-09	−8.337

CASE 2: Same as Case 1, with the addition of an infinite amount of CaCO3(s)

Species	Concentration	Activity	Log activity
Ca+2	0.0014669	0.0010974	−2.96
CaCO3 (aq)	0.0001201	0.00012023	−3.92
CaHCO3+	0.000038997	0.000036268	−4.44
CaOH+	1.2863E-07	1.1963E-07	−6.922
CO3-2	0.000088247	0.000066016	−4.18
H+1	1.9817E-09	1.843E-09	−8.734
H2CO3* (aq)	0.000010746	0.000010757	−4.968
HCO3−	0.0027905	0.0025952	−2.586
OH−	5.8748E-06	5.4636E-06	−5.263

(b) Equilibrium in Aqueous Carbonate Systems. Addition of 10^{-5} M Sulfate

CASE 1 WITH ADDITION OF 0.00001 M SULFATE

Species	Concentration	Activity	Log activity
CO3-2	5.6992E-13	5.5547E-13	−12.255
H+1	0.000020221	0.000020092	−4.697
H2CO3* (aq)	0.000010757	0.000010757	−4.968
HCO3−	2.3959E-07	2.3805E-07	−6.623
HSO4−	1.9223E-08	1.91E-08	−7.719
OH−	5.044E-10	5.0117E-10	−9.3
SO4-2	9.9808E-06	9.7278E-06	−5.012

CASE 2 WITH ADDITION OF 0.00001 M SULFATE

Species	Concentration	Activity	Log activity
Ca+2	0.0014663	0.0010964	−2.96
CaCO3 (aq)	0.0001201	0.00012023	−3.92

(*continued*)

TABLE 12.3 (*Continued*)

CASE 2 WITH ADDITION OF 0.00001 M SULFATE (*continued*)			
Species	Concentration	Activity	Log activity
CaHCO3+	0.000038984	0.000036251	−4.441
CaOH+	1.2859E-07	1.1957E-07	−6.922
CaSO4 (aq)	1.5797E-06	1.5814E-06	−5.801
CO3-2	0.000088371	0.000066076	−4.18
H+1	1.981E-09	1.8421E-09	−8.735
H2CO3* (aq)	0.000010746	0.000010757	−4.968
HCO3−	0.0027921	0.0025964	−2.586
HSO4−	1.2189E-12	1.1335E-12	−11.946
OH−	5.8782E-06	5.4661E-06	−5.262
SO4−2	8.4207E-06	6.2963E-06	−5.201

[a]Output derived from Visual MINTEQ.

Since the system is open to the atmosphere, the total mass of carbonate in the system, C_{T,CO_3}, is no longer constant; however, the bicarbonate and carbonate concentrations are related to H_2CO_3 through Eqs. (12.5a) and (12.5b). When the charge balance in Eq. (12.7) is applied, the equilibrium pH in an aqueous system exposed to $10^{-3.5}$ atm CO_2 is found to be 5.66 (see Table 12.3a and point B in Fig. 12.5*b*).

Let us now assume that sulfate ions are being added to the open lake system at a concentration of 25×10^{-3} mol SO_4^{2-} per m^3 of water input as precipitation (an average precipitation rate of 1 m/yr is assumed). Let us further assume a lake of area 1,000,000 m^2 and depth 2.5 m, yielding a lake water volume of 2,500,000 m^3, which remains fairly constant, that is, water added through precipitation is lost through evapotranspiration. Thus, 25,000 mol of sulfate are added per year to the lake, yielding a sulfate concentration of 0.00001 mol/L in the lake waters at the end of one year, assuming no neutralization, sorption, or biological transformation occurs. The addition of 10^{-5} M sulfate causes the pH in the open lake system to drop from 5.66 to 4.69, a rapid decline over 1 year since no neutralization reactions are assumed to occur in the lake modeled as an impervious bowl. The corresponding output from Visual MINTEQ is shown in Table 12.3b. The reader should confirm this result by attempting a graphical solution using Figure 12.5*b*, applying a charge-balance equation that incorporates sulfate, yielding $[H^+] = [HCO_3^-] + 2[CO_3^{2-}] + 2[SO_4^{2-}] + [OH^-]$, satisfied at pH = 4.7 where $[H^+] \approx 2[SO_4^{2-}]$.

We now include an infinite supply of limestone (solid $CaCO_3$) to simulate the lake illustrated in Figure 12.4*b*, wherein lake waters are in equilibrium with the open atmosphere as well as with an infinite supply of solid limestone lining the lake bottom. A new component, Ca^{2+}, enters the system of equations, the concentration of which is controlled by the dissolution equilibrium of limestone, represented as:

$$CaCO_3\,(s) \Leftrightarrow Ca^{2+} + CO_3^{2-} \qquad K_{sp}^{CaCO_3} = 10^{-8.3} = [Ca^{2+}] \times [CO_3^{2-}] \qquad (12.9a)$$

The concentrations of carbonic acid (H_2CO_3), bicarbonate (HCO_3^-) and carbonate (CO_3^{2-}) are controlled by Eqs. (12.8), (12.5a), and (12.5b). The concentration of Ca^{2+}

is inversely related to that of CO_3^{2-}, through application of Eq. (12.9a), generating the complementary graph of Ca^{2+} shown in light grayscale on Figure 12.5*b*. In the absence of sulfate addition, the equilibrium pH is obtained by solving the charge-balance equation:

$$[H^+] + 2\lfloor Ca^{2+} \rfloor = [HCO_3^-] + 2[CO_3^{2-}] + 2[SO_4^{2-}] + [OH^-] \qquad (12.9b)$$

yielding an equilibrium pH = 8.74 (see Table 12.3a, Case 2). With the addition of 0.00001 M of sulfate over the 1-year acid precipitation input period, the equilibrium pH changes very little, as illustrated in Table 12.3b (Case 2 with sulfate). Thus, lake water pH is not impacted significantly if instantaneous equilibrium dissolution with an unlimited supply of limestone is assumed to occur. More realistic lake water quality models consider either the equilibrium dissolution and weathering of a finite mass of minerals from surrounding soils (Fig. 12.4*c*) or that mineral dissolution and weathering from surrounding soils occurs slowly, functioning as the rate-limiting process that controls alkalinity inputs to the lake (Fig. 12.4*d*).

Instead of separately considering all the different reactions that control the pH of lake waters, it is more convenient to consider alkalinity as the master variable that controls solution pH. Alkalinity, that is, the acid-neutralizing capacity of water, is defined as the net excess strong base over strong acid added to the solution. In waters that contain carbonic acid (in various stages of dissociation), addition of C_b equivalents of a strong base (yielding C_b equivalents of positive ions, e.g., Na^+) and/or C_a equivalents of a strong acid (yielding C_a equivalents of anions, e.g., Cl^-), results in the following electroneutrality condition:

$$[H^+] + C_b = [HCO_3^-] + 2[CO_3^{2-}] + [OH^-] + C_a \qquad (12.10a)$$

which may be rewritten as:

$$\text{Alkalinity} = \text{Alk} = C_b - C_a = [HCO_3^-] + 2[CO_3^{2-}] + [OH^-] - [H^+] \qquad (12.10b)$$

Equation (12.10b) indicates that the alkalinity of water, that is, the excess base equivalents in solution, is effectively given by the sum of equivalents of the carbonate and bicarbonate species that exist in the water. Hydroxyl and hydrogen ion concentrations in Eq. (12.10b) are typically much smaller than the two carbonate species and are often neglected in comparison. Another way of thinking about alkalinity in weak acid systems is to remember that the bicarbonate species (HCO_3^-) has the capacity to neutralize acid addition by reverting to H_2CO_3 by consuming one H^+ ion. Likewise, the carbonate species (CO_3^{2-}) can revert to H_2CO_3 by consuming two H^+ ions. When pH is lowered by addition of foreign agents such as, for example, a strong acid, the weak acid–base system is able to compensate by consuming the excess H^+ ions by reversing Eqs. (12.5a) and (12.5b). On the other hand, when pH is increased, resulting in a drop in H^+ ion concentrations, the acid dissociation proceeds in the forward direction producing more H^+ ions. The reversible nature of weak acid–base systems thus provides water with alkalinity and buffering capacity. If other weak acid–base components, for example, organic acids, are also present in solution, they must be included on the rhs of Eq. (12.10b).

Recognizing that aqueous concentrations of HCO^-_3 and CO^{2-}_3 are controlled by solution pH Eq. (12.5a) and (12.5b) and atmospheric CO_2 Eq. (12.8) and (12.10) enables us to relate the net addition into water of conservative base cations over conservative anions to the final pH condition of such a water, when it is exposed to the atmosphere. The sequential steps in lake water quality modeling, using alkalinity as the master variable, are illustrated below for a lake of volume V receiving volumetric inputs of water and sulfate through precipitation Q_p, and losing water at a volumetric rate of Q_e, due to evapotranspiration. We assume that sulfate addition is the only reaction contributing to an alkalinity decrease. The different steps in lake water quality modeling are:

1. Water flow modeling: Over a certain time interval Δt, a water balance equation can be written for this lake as:

$$V(t + \Delta t) = V(t) + (Q_p - Q_e)(\Delta t) \tag{12.11a}$$

   where V is the volume of water in the lake and Q_p and Q_e are volumetric water precipitation and evaporation rates, respectively.

2. Alkalinity balance: Considering sulfate species to be the major acid species entering the lake via precipitation, the lake water alkalinity is expected to change as:

$$V(t + \Delta t) \times \text{Alk}(t + \Delta t) = V(t) \times \text{Alk}(t) - 2 \times Q_p[\text{SO}_4^{2-}]_p \tag{12.11b}$$

   where the sulfate concentration in precipitation, $[SO_4{}^{2-}]$, is assumed to be constant.

3. An equilibrium aqueous chemistry submodel is solved next with alkalinity changes in water coupled with equilibrium with CO_2 in the ambient atmosphere. Equation (12.10b) is utilized with the carbonate species concentrations determined from equilibration with atmospheric CO_2 [Eq. (12.8)], dissociation of carbonic acid [Eq. (12.5a)], and dissociation of bicarbonate [Eq. (12.5b)], and the hydroxyl species $[OH^-]$ concentration determined from the equilibrium expression for the dissociation of water [Eq. (2.15)], yielding:

$$\text{Alk}(t + \Delta t) = \frac{K_{a1} K_h P_{\text{CO}_2}}{[\text{H}^+(t + \Delta t)]} + \frac{2K_{a1} K_{a2} K_h P_{\text{CO}_2}}{[\text{H}^+(t + \Delta t)]^2} - [\text{H}^+(t + \Delta t)] + \frac{K_w}{[\text{H}^+(t + \Delta t)]} \tag{12.11c}$$

With a known partial pressure of CO_2 in the atmosphere and an alkalinity computed at time $(t + \Delta t)$ from Eq. (12.11b), Eq. (12.11c) can be solved to yield the hydrogen ion concentration and hence the pH of lake waters at time $(t + \Delta t)$. Available numerical packages such as MINTEQ can be used to solve the equilibrium submodel, which computes the pH of the lake waters at the end of each time step. Alternatively, search routines such as the Excel Solver function, as well as numerical methods discussed in Chapter 6 for nonlinear systems of equations, may be applied directly to solve Eq. (12.11c).

A single reaction process, that is, the addition of sulfate to water, was assumed to consume alkalinity in Eq. (12.11b). Furthermore, a single soil compartment was considered, as shown in Figure 12.3*d*. The range of reactions that can consume or release base cations to water in soil and aquifer compartments are shown in Table 12.4,

TABLE 12.4 Processes That Modify the H^+ Balance in Waters

Processes	Changes in Alkalinity $\Delta[Alk]^a = -\Delta[H\text{-}Acy]^b$ [Equivalents per Mole Reacted (Reactant Is Underlined)]
1. Weathering reactions:	
$\underline{CaCO_3(s)} + 2H^+ \leftrightarrow Ca^{2+} + CO_2 + H_2O$	+2
$\underline{CaAl_2Si_2O_8(s)} + 2H^+ \leftrightarrow Ca^{2+} + H_2O + Al_2Si_2O_5(OH)_4(s)$	+2
$\underline{KAlSi_3O_8(s)} + H^+ + 4.5H_2O \leftrightarrow K^+ + 2H_4SiO_4 + 0.5Al_2Si_2O_5(OH)_4(s)$	+1
$\underline{Al_2O_3 \cdot 3H_2O} + 6H^+ \leftrightarrow 2Al^{3+} + 6H_2O$	+6
2. Ion exchange:	
$2ROH + \underline{SO_4^{2-}} \leftrightarrow R_2SO_4 + 2OH^-$	+2
$NaR + \underline{H^+} \leftrightarrow HR + Na^+$	+1
$CaR_2 + \underline{2H^+} \leftrightarrow 2HR + Ca^{2+}$	+2
3. Redox processes (microbial mediation):	
Nitrification: $\underline{NH_4^+} + 2O_2 \leftrightarrow NO_3^- + H_2O + 2H^+$	−2
Denitrification: $1.25CH_2O + \underline{0.25NO_3^-} + H^+ \rightarrow 1.25CO_2 + 0.5N_2 + 1.75H_2O$	+1
Oxidation of H_2S: $\underline{H_2S} + 2O_2 \rightarrow SO_4^{2-} + 2H^+$	−2
SO_4^{2-} reduction: $\underline{SO_4^{2-}} + 2CH_2O + 2H^+ \rightarrow 2CO_2 + H_2S + H_2O$	+2
Pyrite oxidation: $\underline{FeS_2(s)} + 3.75O_2 + 3.5H_2O \rightarrow Fe(OH)_3 + 2SO_4^{2-} + 4H^+$	−4
4. Synthesis and decomposition of biomass and humus:	Variable impact depending on specific reaction

[a][Alk] = alkalinity = acid-neutralizing capacity.
[b][H-Acy] = mineral acidity.
Adapted from Schnoor and Stumm, 1985, and Schnoor, 1996.

including mineral weathering and dissolution, sulfate sorption through ion exchange, and mineralization reactions. These reactions can be modeled as kinetic processes occurring in soil and aquifer systems that drain into a lake, contributing to alkalinity changes in the lake. Several linked compartments in a watershed can thus be evaluated for the net production of alkalinity by balancing acidity inputs with mineral weathering, sorption, and biomineralization reactions. Watcr and alkalinity inputs from the various linked compartments enter the lake, resulting in an effective trickle-down model. Alkalinity is treated as a master variable, with water flow balance equations, alkalinity balance equations, and the equilibrium lake water chemistry submodel coupled together as was illustrated for a single compartment in Eqs. (12.11a) to (12.11c). The next section presents a case study of the trickle-down model developed by Lin and Schnoor (1986), applied to describe the acidification of Vandercook Lake in northern Wisconsin.

12.2.2 Case Study: Acidification of Seepage Lakes

Lake Vandercook is a seepage lake, that is, a lake without any significant outlets or tributaries. Much of the water gain in Lake Vandercook occurs from precipitation events over the lake surface, and water losses occur largely due to surface evaporation, and, to a lesser extent seepage to groundwater. Thus, in addition to demonstrating chemical process integration into transport models, this case study also illustrates the linkages between surface water and groundwater subsystems.

The first step in modeling water quality in the lake is to model and calibrate water flow in the system. Water flow to and from Lake Vandercook is modeled to occur through a system of indexed compartments ($i = 1$ to 9) as shown in Figure 12.6 (Lin and Schnoor, 1986). The time-varying water volume of each compartment, $i = 2$ through $i = 6$ is represented as V_i. Following the notation used by Lin and Schnoor (1986), volumetric water flow rate from compartment j to compartment i is represented as $Q_{i,j}$. Five water balance equations can be written for the compartments $i = 2$ through $i = 6$, as shown below:

Volume of snow compartment ($i = 2$) = Precipitation − Evaporation − Snowmelt to soil

$$\frac{dV_2}{dt} = Q_{2,1} - Q_{1,2} - Q_{3,2}$$

Water volume in soil compartment ($i = 3$) = Precipitation + Snowmelt − Evapotranspiration − Vertical percolation to unsat. zone − Lateral flow to lake (overland flow)

$$\frac{dV_3}{dt} = Q_{3,1} + Q_{3,2} - Q_{4,3} - Q_{5,3}$$

Water volume in unsaturated zone ($i = 4$) = Vertical percolation from soil − Lateral flow to lake − Flow to satd. zone GW − Evapotranspiration

$$\frac{dV_4}{dt} = Q_{4,3} - Q_{5,4} - Q_{6,4} - Q_{1,4}$$

Volume of lake body ($i = 5$) = Precipitation + Lateral flow from soil + Lateral flow from unsat. zone + Inflow from groundwater − Evapotranspiration − Groundwater export − Surface water outflow + Overland flow

$$\frac{dV_5}{dt} = Q_{5,1} + Q_{5,3} + Q_{5,4} + Q_{5,6} - Q_{1,5} - Q_{6,5} - Q_{8,5} + \text{Overland}$$

Water volume in saturated groundwater ($i = 6$) = Percolation from unsat. zone + Export from lake + Groundwater inflow − Groundwater outflow − Evapotranspiration

$$\frac{dV_6}{dt} = Q_{6,4} + Q_{6,5} + Q_{6,9} - Q_{9,6} - Q_{1,6}$$

In the above equations, precipitation refers to the aerial input of water from the atmosphere into a compartment, i, while evapotranspiration (ET) refers to direct transfer of water from compartment i to the atmosphere by a combination of evaporation and plant transpiration. Hence, precipitation flow rates are indexed as $Q_{i,1}$, while evapotranspiration rates are indexed as $Q_{1,i}$. Measured monthly precipitation data (in mm) multiplied by compartment-specific surface areas yield precipitation inputs in the model simulation. Evaporation rates were estimated from measured pan evaporation rates corrected by compartment-specific pan evaporation coefficients and surface areas for evaporation. Empirical equations developed by the Army Corps of Engineers were used to describe the snowmelt rate on fair weather and rainy days. Both lateral and vertical water movement in the soil zone and unsaturated zone were modeled to be driven by differences in water content in soil compared to saturation conditions. The lateral movement of water in these zones depends upon a rate constant

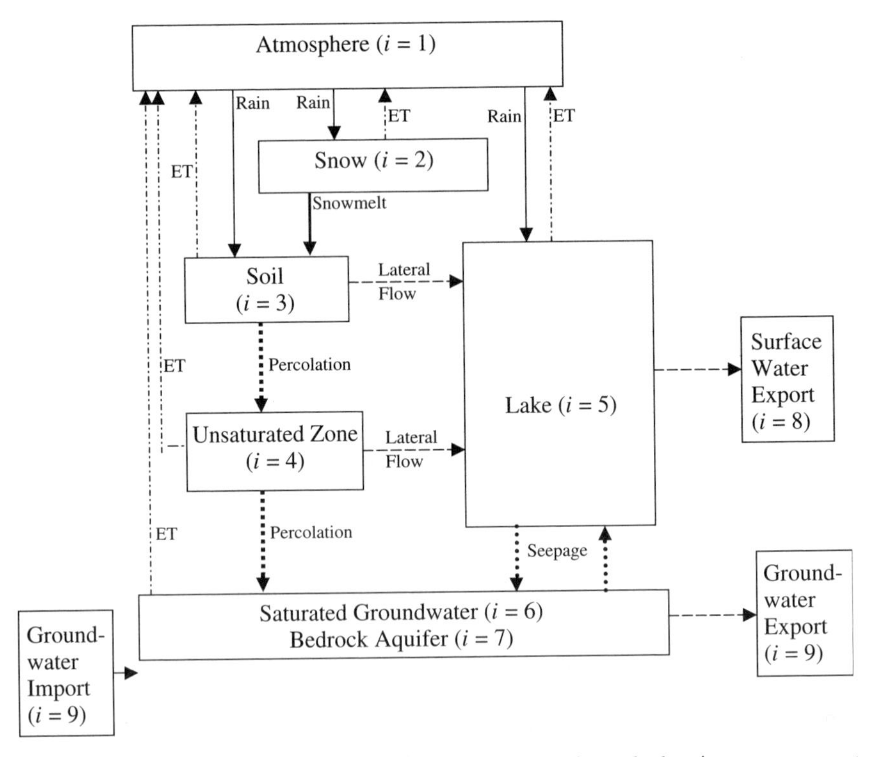

Figure 12.6 Schematic of hydrologic model for Lake Vandercook showing compartments and flows ($i = 1$ to 9). (Adapted from Lin and Schnoor, 1986).

TABLE 12.5 Parameters Used in Simulations at Lake Vandercook

	Compartment			
Input Parameters	Soil	Unsaturated Zone	Lake	Groundwater
a. Parameters Used in Water Flow Balance Equations				
Porosity	0.35	0.35		0.3
Moisture content	0.12	0.05		
Pan coefficient (ET)	0.5	0.1	0.7	0.1
Lateral flow rate constant (1/day)	10^{-5}	10^{-4}		
Vertical conductivity (m/day)	0.4	0.3		
Lake conductivity (m/day)			0.05	
b. Parameters Used in Alkalinity Balance Equations				
Weathering rate constant, $k_{\text{weathering}}$ (1/day)	10^{-4}	10^{-6}		5×10^{-6}
Sediment alkalinity transport rate, P(eq/m^2/day)			6×10^{-5}	
Sum of bases, B_{total} (eq/kg)	0.005	0.006		0.001

Source: Lin and Schnoor (1986).

for recession (see Chapter 10), specific for each of the two compartments. The vertical percolation rates were proportional to compartment-specific vertical hydraulic conductivity. Darcy's law (see Section 9.2) was used to model groundwater flow from and to the lake. The process models used in flow simulations are detailed as supplementary materials for Chapter 12 at www.wiley.com/college/ramaswami. The hydrologic model was calibrated by varying the vertical permeability and lateral recession flow rate constants within a reasonable range to achieve agreement between observations and model simulations of groundwater and lake stage. The parameters used in the calibrated flow model are shown in Table 12.5a, while the fit between model predictions and field observations is shown in Figure 12.7.

Once the flow model is calibrated, alkalinity balance equations are written for the five compartments of interest as shown below.

Snow Compartment ($i = 2$) Alkalinity input from precipitation and dry deposition minus alkalinity removed with evapotranspiration and snowmelt to soil:

$$\frac{\partial(V_2\text{Alk}_2)}{\partial t} = \text{Alk}_1 Q_{2,1} + \text{Alk}_d A_{\text{land}} - \text{Alk}_{10} Q_{1,2} - \text{Alk}_2 Q_{3,2}$$

Soil Compartment ($i = 3$) Alkalinity input from precipitation, snowmelt, and weathering minus alkalinity removed with evapotanspiration, vertical percolation to the unsaturated zone, and interflow to lake:

$$\frac{\partial(V_3\text{Alk}_3)}{\partial t} = \text{Alk}_1 Q_{3,1} + \text{Alk}_2 Q_{3,2} + W_3 - \text{Alk}_{10} Q_{1,3} - \text{Alk}_3 Q_{4,3} - \text{Alk}_3 Q_{5,3}$$

Unsaturated Zone ($i = 4$) Alkalinity input from vertical water percolation from soil and weathering minus alkalinity removed with lateral water flow to lake, water flow to the saturated zone groundwater, and through evapotranspiration:

$$\frac{\partial(V_4\text{Alk}_4)}{\partial t} = \text{Alk}_3 Q_{4,3} + W_4 - \text{Alk}_4 Q_{5,4} - \text{Alk}_4 Q_{6,4} - \text{Alk}_{10} Q_{1,4}$$

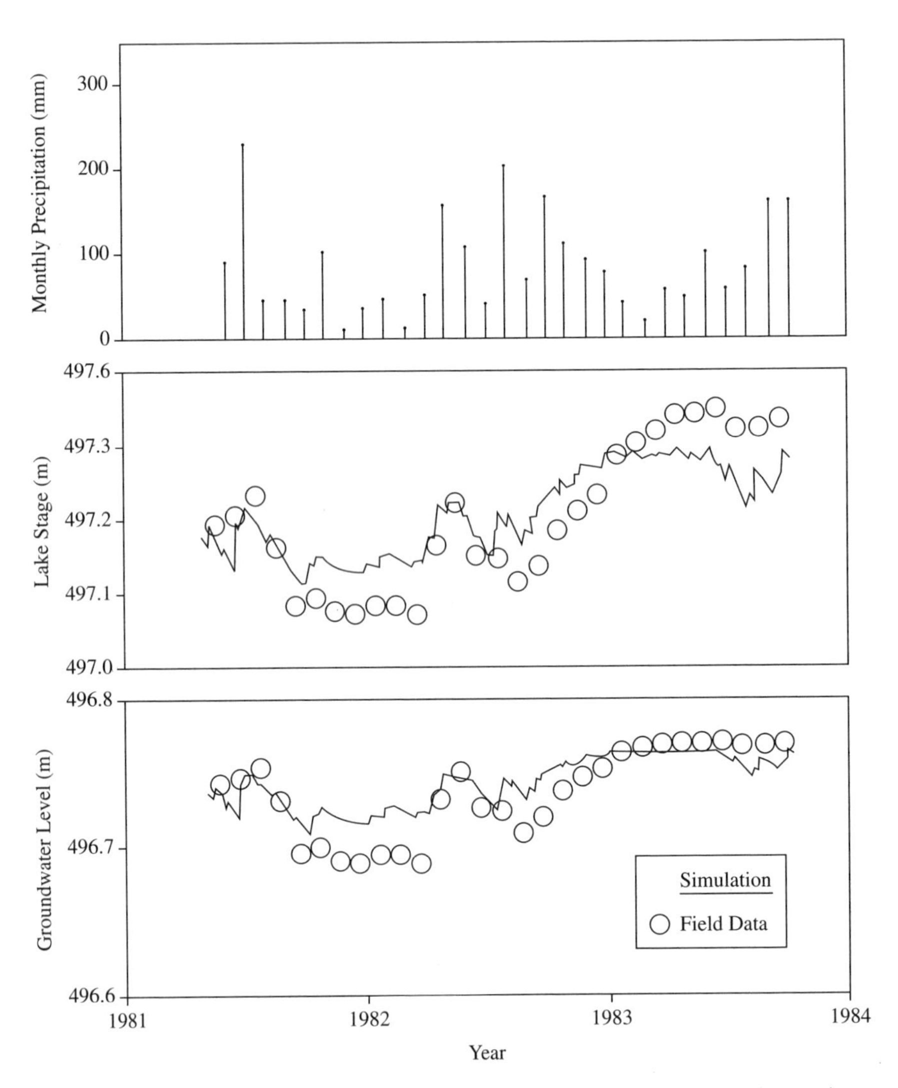

Figure 12.7 Monthly precipitation data and simulation results for lake stage and groundwater level at near-shore piezometer (from Lin and Schnoor, 1986).

Lake Body ($i = 5$) Alkalinity input from precipitation, dry deposition, lateral water flow from soil, inflow from groundwater and from lake alkalinity production (P), minus alkalinity removed with evapotranpiration, groundwater export, surface water outflow, and overland flow:

$$\frac{\partial(V_5 \text{Alk}_5)}{\partial t} = \text{Alk}_1 Q_{5,1} + \text{Alk}_d A_{\text{lake}} + \text{Alk}_3 Q_{5,3} + \text{Alk}_4 Q_{5,4} + \text{Alk}_9 Q_{5,9} + P - \text{Alk}_{10} Q_{1,5} - \text{Alk}_5 Q_{9,5} - \text{Alk}_5 Q_{8,5}$$

Groundwater—saturated ($i = 6$) Alkalinity input from water percolating from unsaturated zone, lake, and groundwater inflow and from weathering minus alkalinity removed with groundwater outflow and evapotranspiration:

$$\frac{\partial(V_6 \text{Alk}_6)}{\partial t} = \text{Alk}_4 Q_{6,4} + \text{Alk}_5 Q_{6,5} + \text{Alk}_9 Q_{6,9} + W_6 - \text{Alk}_6 Q_{9,6} - \text{Alk}_{10} Q_{1,6}$$

where Alk_i represents the alkalinity of compartment i. Alk_d represents the alkalinity of the dry deposition, which occurs over terrestrial areas, represented by A_{land}, and over the lake surface area, represented by A_{lake}. Alk_{10} represents alkalinity of the evapotranspiration stream and is set to zero. W_i represents the weathering rate in compartment i. The rate of weathering employs a first-order weathering rate constant k and is modeled to be proportional to the difference between the total base cations available in the porous media compartments and the base cation depletion in those compartments. Thus, the weathering rate W_i is represented as:

$$W_i = k_{\text{weathering},i} V_{\text{total},i} \rho_b [B_{\text{total}} - B_{\text{depleted}}]$$

The total number of equivalents of base cations in the porous matrix is unknown and is a fitting parameter in the model, as is the weathering rate constant. The base depletion is computed from alkalinity changes. Weathering expressions and an alkalinity balance equation are also applied to the aquifer bedrock compartment ($i = 7$). In addition to weathering of minerals in porous media, alkalinity in watersheds is also produced by mineral dissolution and weathering occurring within the lake itself, represented by the sediment alkalinity production rate or transport rate, P.

The alkalinity balance equations were solved by using the fourth-order Runge–Kutta technique for numerical integration (see Chapter 6). The solution converged to within $\pm 1\%$ when the time step for integration, Δt, was varied from 0.01 to 0.5 days, indicating stability of the numerical method. A time step of 0.1 days was used for the simulations reported by Lin and Schnoor (1986). The alkalinity of the different water compartments in the watershed, as well as wet deposition (precipitation) and dry deposition alkalinity were measured from May 1981 to September 1983. Calibration of the alkalinity submodel was achieved by varying the weathering rate, sediment transport rate, and total base cations in the porous matrices until a good fit was obtained between the alkalinity simulated in the lake and that measured for the lake over the simulation time period. The fitted model results are shown in Figure 12.8, with the best-fit weathering parameters shown in Table 12.5b.

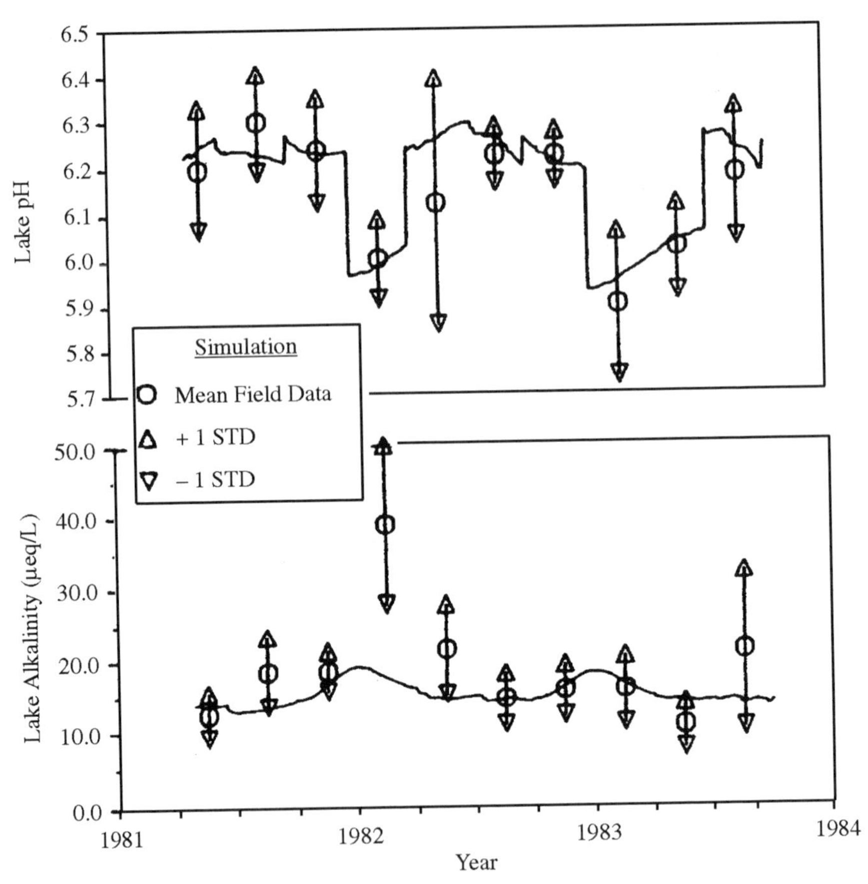

Figure 12.8 Simulation results and field data for lake alkalinity and pH (from Lin and Schnoor, 1986).

Once alkalinity computations were completed, the pH of the lake was determined by applying the carbonate equilibrium model shown previously in Eq. (12.11c). Lin and Schnoor (1986) use a trial-and-error method to solve Eq. (12.11c), since only the positive root for pH in the range of 0 to 14 would apply as a solution. Model results for lake pH and alkalinity matched field observations quite well, as shown in Figure 12.8. The calibrated trickle-down model was then applied in a long-term simulation to forecast future trends in lake acidification occurring with varying loads of acid deposition. The lake was found to be quite sensitive to acid precipitation, and the lake alkalinity was predicted to become negative within a period of 5 years if the rate of acid precipitation inputs were to double. Thus, this case study demonstrates the utility of a fully calibrated complex reactive flow model in assessing future water quality trends. The steps used in water quality modeling of Lake Vandercook are summarized in Figure 12.9, emphasizing the order and linkage between the water flow-balance module, the kinetically based alkalinity module, and the carbonate equilibrium pH submodel.

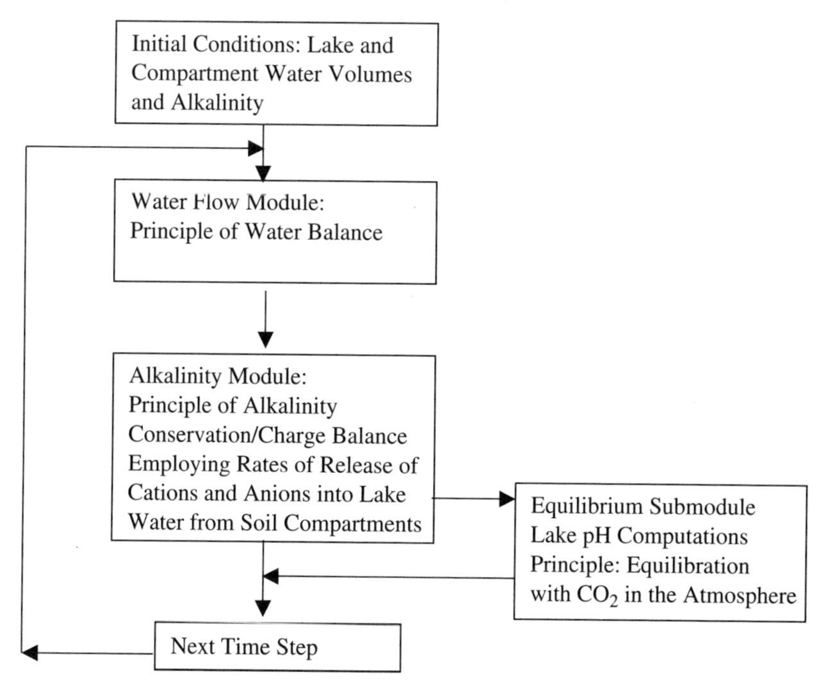

Figure 12.9 Schematic showing various modules used in water quality modeling at Lake Vandercook. (Adapted from Lin and Schnoor, 1986.)

12.3 COUPLED EQUILIBRIUM AND RATE-LIMITED CHEMICAL PROCESSES IN ONE-DIMENSIONAL TRANSPORT SYSTEMS

Section 12.2 presented a case study in which equilibrium and kinetic reaction models were coupled with water flow occurring in well-mixed lake and soil compartments. A similar approach has been taken by many modelers to describe water quality in streams and rivers. As shown in Figure 1.4, the dimensions across which well-mixed conditions are assumed determines the dimensionality of the water flow model. In many simulations of stream flow, a one-dimensional advection–dispersion model is used to represent transport occurring along the longitidinal (x) axis, with the cross section assumed to be well-mixed. Section 10.6.1 presented OTIS (Runkel, 1998), an improved one-dimensional solute transport model with inflow and transient storage of water, effective in describing flow in streams and rivers. In addition to advection and dispersion in streams, the model incorporates the impact of lateral inflows, for example, overland flow, groundwater flow, and so forth, as well as stagnant pools and backwater areas. In this section, we describe the coupling of this 1D advection–dispersion, lateral flow transient storage water flow model with a coupled equilibrium–kinetic water chemistry model. The model is applied to describe water quality changes occurring along an acid mine drainage stream in Colorado, in response to treatment with sodium carbonate. This case study is based upon the modeling and fieldwork described in Runkel et al. (1996) and Broshears et al. (1996).

Waters that drain coal and metal mining areas are typically acidic (pH < 4) and contain a high concentration of heavy metals. Such waters are called acid mine drainage and pose a hazard to aquatic life. A key step in the formation of acid mine drainage is the dissolution of pyrite (iron sulfide), which is found at enriched concentrations in the soil and rocks found at mining sites. The oxidation of iron pyrite initiates acid mine drainage through a sequence of reactions shown below. FeS_2 dissolves under oxidizing conditions to release ferrous (Fe^{2+}) iron to water accompanied by the release of acidity (H^+ ions) to water. Ferrous iron (Fe^{2+}) is oxidized to ferric iron (Fe^{3+}), which reacts with water to form ferric hydroxide, an orange–red precipitate. Because this material is a solid (precipitate), it remains in the source pyrite zone and functions as a reservoir of ferric iron that can now oxidize iron pyrite, further releasing acidity.

Pyrite oxidation by oxic waters: $FeS_2(s) + 7/2O_2 + H_2O \Leftrightarrow Fe^{2+} + 2SO_4^{2-} + 2H^+$

Oxidation of ferrous iron: $Fe^{2+} + \frac{1}{4}O_2 + H^+ \Leftrightarrow Fe^{3+} + \frac{1}{2}H_2O$

Precipitation of ferric iron: $Fe^{3+} + 3H_2O \Leftrightarrow Fe(OH)_3(s) + 3H^+$

Pyrite oxidation by ferric iron: $FeS_2(s) + 14Fe^{3+} + 8H_2O \Leftrightarrow 15Fe^{2+} + 2SO_4^{2-} + 16H^+$

Acid waters typically enhance the rate of dissolution of minerals, thereby rapidly releasing heavy metals into water. Furthermore, the sorption of these metal ions to soil is inhibited by the low pH of the acid waters due to the surface sorption phenomena described in Section 12.1 (see Figs. 12.1 and 12.2). Thus, high acidity promotes rapid metal dissolution as well as increased mobility of metal ions in flowing water. Because of the deleterious effects of heavy metals on human and aquatic health, treatment of acid mine drainage waters is a high-priority issue. The case study described in this section assesses the impact of addition of a neutralizing agent to acid surface waters in a 498-m reach of the St. Kevin Gulch, a small tributary of the Tennessee Creek in the headwaters of the Upper Arkansas River near Leadville, Colorado (Broshears et al., 1996).

The initial stream chemistry in St. Kevin Gulch was monitored before the injection of sodium carbonate into the stream. The initial background pH of the stream waters was 3.5. The pH modification at the start of the reach ($x = 0$) involved injecting a concentrated solution of sodium carbonate so that pH increased in a stepwise manner from 3.5 to 5.8 at the injection location. The injection began at $t = 9.3$ h and continued at the same rate until $t = 11.9$ h, after which the injection rate was increased and sustained until 14.9 h, when the injection was stopped. The transport of the pulse of increased pH was traced as it moved downstream, with sampling locations located at 24, 70, 251, and 498 m from the injection site. The pulse was impacted by water flow, mixing, and dilution, as well as chemical reactions occurring in the water and between water and stream sediments. These coupled processes were modeled as described next. As in Lake Vandercook, the water flow field was established first.

OTIS, a one-dimensional solute transport model was used to characterize advection, dispersion, inflow, and transient storage of water in the stream using parameters estimated from conservative tracer transport data (Runkel, 1998; Runkel et al., 1996). The advection–dispersion and transient storage equations [(10.41) and (10.42)] were solved numerically using finite difference methods, applying the Crank–Nicolson algorithm (Runkel, 1998; see also Chapter 6). Tracers (lithium and sodium) injection conducted in the stream prior to the pH modification experiment yielded stream flow parameters at each sampling location (see Example 10.5). In-stream flow ranged from 12.3 L/s at site 24 to 13.2 L/s at sites 251 and 498. The dispersion coefficient was set at 0.05 m^2/s for all four subreaches modeled within the stream. The stream exchange coefficient, α, was estimated to be 10^{-4} s^{-1} for the first three subreaches and 3×10^{-5} s^{-1} in the last reach.

With the flow parameters established, each stream cell within OTIS was linked with MINTEQA2 (Allison et al., 1991) to perform equilibrium water chemistry computations. The resulting combination of solute transport with chemical equilibria yields the OTEQ model (Runkel and Kimbell, 2002). Chemical reaction equilibria were represented and solved within each cell at each time step and the mobile products formed were transported to the surrounding stream cells in the next time step. Solids formed were tracked as well; any solids formed were retained in the cells that they are formed in and participated in dissolution and sorption–desorption equilibria in future time steps. The specific chemical processes simulated in the model include equilibration with atmospheric CO_2, equilibrium dissolution and precipitation of ferric hydroxide and aluminum hydroxide, and sunlight-driven redox reactions that controlled the amount of dissolved ferrous and ferric iron in solution under variable equilibrium and kinetic conditions. Sorption and desorption of aqueous species on to the stream bed solids was initially modeled as an equilibrium process. The stream bed was assumed to contain ferric oxide. The depth of sediments interacting with the water above was estimated as 0.1 to 0.5 mm, with a porosity ranging from 0.7 to 0.9, yielding effective iron oxide concentrations in the overlying waters in the four subreaches. The diffuse double-layer model (described in Section 12.1) was applied to describe all sorption–desorption equilibria. Equilibrium sorption–desorption models did not predict the general trends in stream chemistry observed in the field. Diffusion-limited sorption–desorption kinetics were therefore incorporated into the model to represent the departure from equilibrium conditions and were found to successfully describe the trends observed in the field.

The mixed equilibrium–kinetic OTEQ model simulations of pH, iron, and aluminum concentrations in stream waters, in response to the base injection, were compared with field observations, as shown in Figure 12.10. Note the pH change designed for a step increase from 3.4 to 5.8 at the injection location ($x = 0$) produces diminishing increases in pH with increasing downstream distance because of the cumulative buffering effect of the stream solids as one moves downstream. For example at location site 498 (farthest downstream, $x = 498$ m), the pH peaks at only 4.6 compared to 5.8 at the injection site. The general trends in chemical species concentration were nicely predicted by the model, although some deviations were noted. These results indicate that for fast-moving waters interacting with stream bed solids, a combination

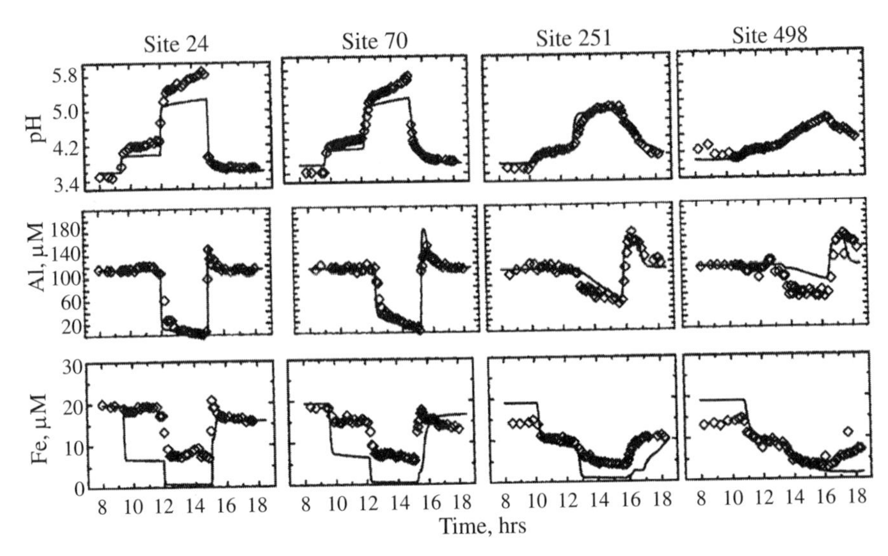

Figure 12.10 Observed (open diamonds) and simulated (solid lines) pH and concentrations of dissolved aluminum and iron during the pH modification experiment (from Broshears et al., 1996).

of equilibrium and rate-limited chemical processes, coupled with transport, may be required to most effectively describe the spatial and temporal distribution of inorganic ions. The reader is referred to Runkel and Kimball (2002) for more case studies on modeling inorganic chemical reactions coupled with transport processes occurring in streams. Exclusively equilibrium models of inorganic reactions are sometimes effective in describing groundwater quality because of the much slower flow rates of water in the subsurface that facilitate an approach toward equilibrium conditions (e.g., EPA, 1999). However, even in the case of groundwater chemistry, diffusion limitations to mass transfer often play an important role in controlling the rates of heterogeneous reactions, for example, mineral dissolution and sorption–diffusion in soil. Diffusion limitation may likewise limit microbial oxidation processes occurring in aqueous systems, particularly in groundwater. A general review of microbial kinetic processes is provided in the next section, followed by incorporation of these into groundwater transport models as demonstrated through two case studies on situ bioremediation.

12.4 KINETICS OF BIOCHEMICAL REACTIONS

The word *microorganisms,* or microbes, refers to small cellular organisms that are not visible to the human eye. Microbes can be unicellular organisms or multicellular organisms, belonging to both the plant kingdom, for example, algae, as well as the animal kingdom, for example, fungi and bacteria. Microbes in the plant kingdom process carbon dioxide to generate oxygen and organic matter and are also termed autotrophs, while those in the animal kingdom typically respire by oxidizing organic matter to carbon dioxide and water and are termed heterotrophs. The food and energy sources for microbes lead to further classifications such as chemotrophs, lithotrophs, phototrophs, that is, microbes using chemicals, minerals, and light, respectively. Microbes may also be classified as eukaryotes and prokaryotes, the prokaryotes being more "primitive" than the eukaryotes, which have a well-defined nucleus.

Among microorganisms, algae, fungi, and protozoa belong to the category of eukaryotes, while bacteria belong to the class of prokaryotes. Bacterial activity is primarily responsible for the transformation of many pollutants of interest and will be the primary focus of this section. Bacterial processes pertaining to contaminant oxidation are represented simplistically in a box diagram presented in Figure 12.11.

Bacteria have been found to degrade a wide range of organic pollutants, under different environmental conditions. Most bacteria fall under the category of Eubacteria, for example, the vast number of bacterial species in the *Pseudomonas* genus, that readily facilitate pollutant transformation under moderate environmental conditions (pH ranging from 6 to 8, and the moderate range of temperature from 10 to 45°C). Specialized bacteria termed Archaebacteria can operate in extreme environments with high temperature (thermophiles), high salinity (halophiles), and no oxygen (methanogens). Bacterial processes may also be categorized based on the electron acceptors used in oxidizing the organic substrate. Processes occurring in oxygenated waters are termed aerobic, those occurring in the absence of oxygen are termed anoxic or anaerobic, which may include the use of nitrates, sulfates, and even

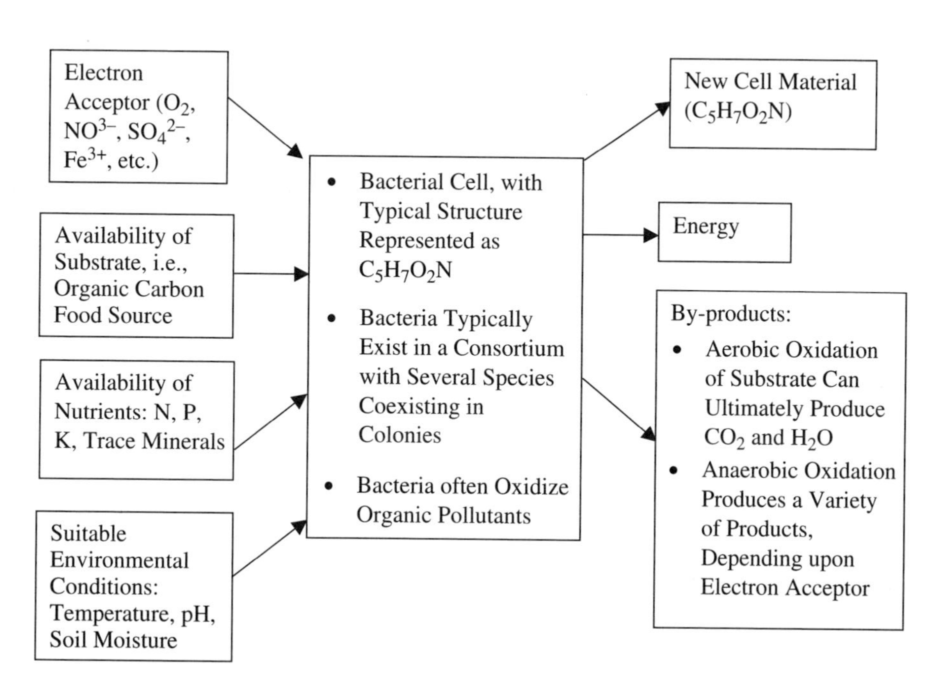

Figure 12.11 Schematic showing oxidation of organic pollutants by bacteria.

iron as oxidizing agents (electron acceptors). The feasibility and spontaneity of these reactions may be determined by assessing the net free energy of formation of the redox reaction pair, as was illustrated in Example 2.7. For example, Table 2.5 showed the following preference for oxidizing agents during the oxidation of an organic pollutant: oxygen first, followed by nitrates, ferric iron, sulfates, and eventually the use of CO_2 itself to produce methane (methanogenesis).

$$\text{Preference of electron acceptors:} \quad O_2 > NO_3^- > Fe_3^+ > SO_4^{2-} > CO_2$$

In the case of certain chlorinated aliphatic compounds, for example, TCE, reductive dehalogenation is energetically favored when oxygen and nitrates are unavailable, that is, the organic chemical undergoes *reduction* and releases a chloride ion (Vogel et al., 1987). A more detailed description of reductive dehalogenation is presented in Section 12.6. In this section we describe the subsurface biodegradation of typical organic contaminants that are readily oxidized, for example, fuel hydrocarbons, benzene, and so forth.

An aquifer contaminated with organic pollutants typically shows zoning effects, that is, aerobic zones along the leading edge of the plume with lower pollutant concentrations, and more anoxic zones in areas where organic chemical concentrations are very high thereby depleting the oxygen dissolved in water during microbially mediated oxidation reactions. In the typical case of pollutant oxidation, one may also expect to see depletion of electron acceptors in the anoxic zones in the order of their preferential use, as is shown in Figure 12.12. Figure 12.12 assumes that a variety of bacteria capable of aerobic and anoxic processes are readily available in

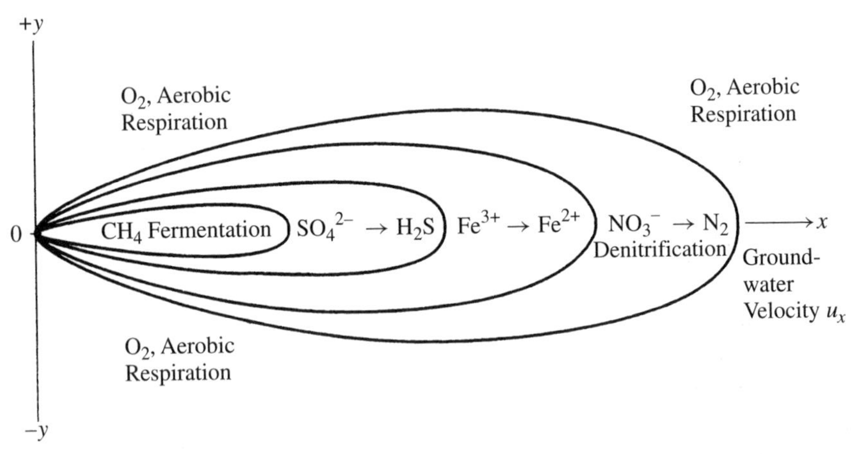

Figure 12.12 A typical hydrocarbon plume undergoing oxidation by various electron acceptors. The peripheral areas of the plume exhibit low hydrocarbon concentrations and are aerobic. Anaerobic conditions are found at the core of the plume where high concentrations of hydrocarbons have depleted the oxygen; a sequence utilization of preferred electron acceptors is shown in this zone (from Schnoor, 1996).

the subsurface. Bacteria that only grow in the presence of oxygen are called obligate aerobes, those that only grow in anaerobic environments are termed obligate anaerobes, while those that can adapt, if need be, from anaerobic to aerobic conditions are called facultative anaerobes. Field assays confirm the presence of a wide range of bacterial capabilities, ranging from aerobic to anaerobic, in the subsurface. In addition to bacteria, other microbes may also participate in pollutant transformations. One specific fungus—the white rot fungus—has been found to survive harsh environments and transform a variety of contaminants (Anderson et al., 1993). The rate at which microbes degrade pollutants in surface water and groundwater systems is important in determining "natural attenuation" of pollution in these systems, as well as in the design of engineered bioremediation schemes.

Bioavailability is an important factor controlling the overall biotransformation rate of pollutants in environmental systems. Bioavailability refers to the ready supply of pollutants, that is, substrates, to the microbes. While a few microbes have been known to directly degrade hydrocarbon liquids and solid substrates (Atlas, 1991), in general, pollutants are readily degraded by microbes only when present in aqueous solution. As a result, the overall rate of pollutant biotransformation has been observed to be linked to the rates of physicochemical mass-transfer processes that make the pollutant "available" for microbial degradation such as desorption from soil, diffusion from micropores, and dissolution from NAPLs (Ogram et al., 1985; Rijnaarts et al., 1990; Michelcic and Luthy, 1991; Scow and Alexander, 1992; Efroymson and Alexander, 1994; Ghoshal et al., 1995; Ramaswami and Luthy, 1997, 2002). Because of bioavailability constraints, several slightly different terms are used to represent rates of biodegradation of pollutants in environmental systems. The *overall biotransformation rate* refers to the overall rate at which a pollutant can be transformed to a product (not necessarily the end product) through a combination of microbial phenomena and physicochemical mass-transfer processes (advection, diffusion, dissolution) that control pollutant availability to microbes. The *biomineralization rate* refers to the overall rate at which a pollutant is converted to inorganic CO_2 and, once again, is a combination of intrinsic microbial degradation rates and physicochemical processes. The biotransformation rate may be equal to or, more typically, greater than the biomineralization rate because pollutants are not fully converted to CO_2, either because of conversion of some portion of the substrate to new cell biomass or into intermediate products that are slow to mineralize. The *intrinsic biodegradation rate* of a pollutant focuses on the microbial process alone and quantifies the rate at which microbes transform a pollutant under optimum conditions, without being constrained by limited electron acceptor availability or contaminant availability in aqueous solution. In summary, one can view overall biotransformation rates (somewhat simplistically) to be controlled by the following three main factors:

- Intrinsic microbial degradation rates
- Bioavailability of pollutants in the aqueous phase, controlled by equilibrium partitioning to water from various media (Chapter 3), as well as rates of advection, diffusion, dispersion, sorption–desorption, and NAPL dissolution (Chapters 4, 9, and 10)

- Availability of electron acceptors and nutrients at locations where microbes and substrate coexist.

The three processes are integrated as described in the next section.

12.4.1 Modeling Intrinsic Biodegradation of Pollutants

The intrinsic biodegradation rate of an aqueous-phase organic substrate can be described by several mathematical models, the two most commonly used models being:

1. The Michaelis–Menten model assumes the rate of biodegradation of aqueous-phase substrate is limited by the availability of enzymes.
2. The Monod model assumes substrate degradation contributes to cell growth; hence, microbial populations control the rate of degradation of the aqueous-phase substrate.

The classic model of Michaelis–Menten has proven useful with many kinds of enzymes. It is represented by the following multistep reaction:

$$\mathrm{E} + \mathrm{S} \underset{k_{-1}}{\overset{k_1}{\rightleftarrows}} \mathrm{ES} \xrightarrow{k_2} \mathrm{E} + \mathrm{P} \tag{12.12}$$

In the first step, the substrate (S) binds reversibly to the enzyme (E) to form an enzyme–substrate complex (ES), which then yields the product P. The rate constants of the various reaction steps are represented by k_1, k_{-1}, and k_2, as shown in Eq. (12.12). The product (P) can readily be measured in most systems, while the enzyme–substrate complex (ES) is much harder to identify and measure in most cases. However, at steady state, the rate of ES formation must equal the rate of ES dissociation either back to E + S or forward to E + P. Applying the law of mass action at steady-state conditions yields

$$k_1[\mathrm{S}][\mathrm{E}] = k_{-1}[\mathrm{ES}] + k_2[\mathrm{ES}] \tag{12.13}$$

Defining $\mathrm{E_{total}}$ as the total enzyme concentration, that is, $[\mathrm{E_{total}}] = [\mathrm{ES}] + [\mathrm{E}]$, the above equation can be rearranged to obtain an expression for the concentration of the enzyme–substrate complex (ES):

$$[\mathrm{ES}] = \frac{[\mathrm{E_{total}}]\,[\mathrm{S}]}{[\mathrm{S}] + K_m} \quad \text{where} \quad K_m = \frac{k_2 + k_{-1}}{k_1} \tag{12.14}$$

and K_m is called the Michaelis–Menten constant. From the law of mass action, the rate of product formation (also termed the enzyme velocity) is given by $k_2[\mathrm{ES}]$, from which the maximum enzyme velocity is defined as $V_{\max} = k_2[\mathrm{E_{total}}]$. Thus, by combining Eqs. (12.13) and (12.14), the rate of product formation can be written in terms of the substrate concentration, the maximum enzyme velocity, and the Michaelis–Menten constant, as:

$$\text{Enzyme velocity} = \frac{dP}{dt} = V = \frac{V_{max}[S]}{[S] + K_m} \tag{12.15}$$

The Monod representation for intrinsic biodegradation rates is similar in form to the Michaelis–Menten model shown above in Eq. (12.15) but is based on a different premise. In the Monod model, bacterial growth is directly related to intrinsic microbial degradation rates of the pollutant, when the pollutant serves as the sole carbon source for microbial growth. When a microbial population is first exposed to a new substrate, it may exhibit a lag period—that is, a period of no growth during which the microbial consortium is developing specific enzymatic capabilities to degrade the specific pollutant. The lag period is, hence, the acclimation time required for the microbes to become accustomed to their new environment. Acclimation periods can be as short as a few hours for readily degradable organics and as long as several hundred days when a recalcitrant compound is subject to microbial processes. In the case of recalcitrant pollutants, genetically engineered microbes can be designed to induce suitable enzymatic activity. In some cases, microbes at specific hazardous waste sites may have gone through the acclimation phase, and these or other genetically engineered microbes can be introduced at other waste sites to speed up bioremediation in a process called bioaugmentation.

Microbes reproduce by binary fission and, hence, after the lag period, if there is a sufficient supply of food (substrate), microbial growth occurs exponentially. When the microbial population has grown beyond what the food source or the environmental conditions can sustain, microbial growth enters a stationary phase during which the net growth is zero, that is, the rate at which new cells are formed equals the rate at which cell death occurs, and, hence, substrate consumption still continues but at a slower rate. When all substrate has been consumed, the cells stop growing and enter the final phase of death. The living microbes now consume the carbon produced from dead microbes at a rate termed the endogeneous decay rate. The entire microbial life cycle is captured in the most basic form of the Monod equation that represents the specific microbial growth rate, μ, as (Simkins and Alexander, 1984):

$$\mu = \frac{1}{X}\frac{dX}{dt} = \frac{\mu_m C}{K_s + C} \tag{12.16}$$

where X (mass/volume) is the microbial concentration at any time t; C (mass/volume) is the pollutant (substrate) concentration at time t; μ_m is the maximum specific growth rate (1/time), and K_s (mass/volume) is the half saturation coefficient. Equation (12.16) indicates that, when $C \gg K_s$, that is, an abundant supply of substrate is available, the microbial growth rate will be equal to the maximum specific growth rate with $\mu = \mu_m$, that is, the maximum possible microbial growth rate possible is sustained. On the other hand, when a limited amount of substrate (food) is available such that $K_s \gg C$, the microbial growth rate is proportional to the substrate concentration with $\mu = \mu_m C/K_s$. A stoichiometric relationship is used to relate changes in substrate concentration with microbial growth rates, as:

$$\frac{dC}{dt} = -\frac{1}{Y}\frac{dX}{dt}. \tag{12.17}$$

where Y is termed the yield coefficient and represents the milligrams of biomass formed for every milligram of substrate consumed. Combining the above two equations yields the general from of the Monod equation for substrate utilization:

$$\frac{dC}{dt} = -\frac{\mu_m}{Y}\left(\frac{C}{K_s + C}\right)X \tag{12.18a}$$

Note the similarity in mathematical form between the Monod model [Eq. (12.18)] and the Michaelis–Menten model [Eq. (12.15)]. Simkins and Alexander (1984) derive six models for substrate degradation kinetics from the general Monod model including: (1) zero order ($C \gg K_s$), (2) first order ($C \ll K_s$), (3) Monod with no growth, (4) logistic, (5) logarithmic, and (6) Monod with growth. (See supplementary materials for Chapter 12 at www.wiley.com/college/ramaswami.) Models (1) through (3) are of the no-growth type and assume a constant biomass concentration over time. The assumption of a "constant" biomass is applicable when the microbial concentration, X, is so large that any additional growth causes little change in the population. Microbial death, also termed endogenous decay, may also be included in the Monod equation as (Schirmer et al., 1999; Knightes and Peters, 1999; Guha et al, 1999):

$$\frac{dX}{dt} = -Y\left(\frac{dC}{dt}\right) - bX \tag{12.18b}$$

where b is a first-order decay constant (1/time).

The two Monod parameters, K_s and μ_m, and the yield, Y, can be estimated by assessing substrate degradation data, that is, a plot of $C(t)$ versus t, in batch or flow-through reactors. In many cases, when a certain substrate functions as the sole carbon source for the microbes, oxygen depletion can be measured and stoichiometrically related to substrate depletion (Bielefeldt and Stenzel, 1999). For the no-growth assumption, the Monod parameters can be determined graphically from their definitions. When microbial growth is specifically considered, numerical integration of Eqs. (12.17) and (12.18) is required with a best-fit procedure to determine the three unknown parameters from the two equations (Knightes and Peters, 2000). Software packages are currently available to fit a variety of biokinetic models to substrate depletion data, with some input required on the initial microbial concentration in the batch system. Note, measuring microbial numbers is not a very accurate process, and uncertainties in microbial concentrations can produce large uncertainties in the estimated parameter values. The parameter set is also nonunique and care must be exercised in parameter estimation from batch biodegradation data. See supplementary materials for Chapter 12 at www.wiley.com/college/ramaswami.

Variations of the Monod model incorporate the effect of toxicity/inhibition and competitive substrate utilization. Toxicity and inhibition occurs when either the substrate or an intermediate product of susbstrate degradation inhibits microbial activity. When the substrate itself inhibits microbial activity, a modified Monod model, referred to as the Haldane model or the Andrews equation (Andrews, 1968; Haldane, 1930), is often used in which:

$$\mu = \frac{\mu_m C}{K_s + C + \dfrac{C^2}{K_I}} \tag{12.19}$$

where K_I is an inhibition coefficient that must be estimated from substrate degradation data. When substrate concentrations are below the toxicity level (i.e., when $C^2 \ll K_I$), we revert to the basic Monod model. When multiple organic substrates are present the preference for one or more food source (substrate) can create competition, inhibiting degradation of the other substrates. Incorporating such competitive effects, the degradation rate of a substrate i in a system on n substrates is given as:

$$\frac{dC_i}{dt} = \left[\frac{\mu_{\max} C_i}{K_{s,i} + \sum_{j=1}^{n} \dfrac{K_{s,i}}{K_{s,j}} C_j}\right] \frac{X}{Y_i} \tag{12.20}$$

where the summation term in the denominator represents inhibition of degradation of substrate i due to the presence of a competing substrate, j, at concentration C_j, When $C_j \ll K_{s,j}$, the competitive effect is minimal and we revert to the basic Monod model shown in Eq. (12.17). The biomass concentration, X, in Eq. (12.20) refers to the total biomass concentration in the system obtained from the summation of the biomass growth contributed by each of the substrates.

12.4.2 Incorporating Biokinetics into Pollutant Transport Models

Coupling intrinsic biokinetics with the availability of substrate and of electron acceptors in surface and groundwaters is the primary challenge in modeling contaminant transport and fate in these systems. The substrate removal rate (dC/dt), represented by either the Monod or Michaelis–Menten models shown in Section 12.4.1, functions as the sink term in the advective–dispersive–reactive (ADR) equation written for the substrate chemical (see Chapters 5 to 9). For example, incorporating the full Monod model for contaminant biodegradation into a one-dimensional groundwater transport model for that chemical yields

$$\frac{\partial C(x,t)}{\partial t} = -v\frac{\partial C(x,t)}{\partial x} + D_L \frac{\partial^2 C(x,t)}{\partial x^2} + D_T \frac{\partial^2 C}{\partial y^2} - \frac{\mu_m C(x,t) X(x,t)}{Y[K_s + C(x,t)]} \tag{12.21}$$

Note, when contaminant concentrations are very low such that $C(x,t) \ll K_s$, we revert to first-order removal with respect to substrate concentrations, as long as the microbial concentration, X, is uniform and stable, that is, does not change significantly due to substrate degradation ($X_0 \gg YC_0$). When contaminant concentrations in water are high, that is, $C(t) \gg K_s$, Eq. (12.21) becomes zeroth order with respect to substrate concentration, although significant microbial growth can occur in this regime. Analytical solutions are available for the ADR equation with zeroth- and first-order contaminant removal; some of these analytical solutions have been presented previously in Chapters 9 and 10, in the context of groundwater and surface water systems, respectively. In the intermediate range of substrate concentrations $[0 < C(t) < K_s]$, the full Monod formulation must be employed. Although analytical solutions are unavailable to solve the Monod representation shown in Eq. (12.21), numerical models can readily be used. Since hazardous pollutants typically occur at very low ($C \ll K_s$) to moderately low ($0 < C < K_s$) concentrations in water, both the first-order as well as the Monod-kinetic representation for biodegradation are made available in most groundwater contaminant transport and fate models (e.g., MT3D, RT3D, BIOPLUME III, etc). In addition to the intrinsic rate of biodegradation of the contaminant chemical, overall biotransformation rates in the subsurface are also impacted by the availability of electron acceptors that enable microbially mediated oxidation of the contaminant. Methods for incorporating availability of electron acceptors into subsurface biodegradation models are discussed next.

12.4.3 Availability of Electron Acceptors: The Instantaneous Biodegradation Model

In the early days of bioremediation modeling aerobic processes were considered most important in biodegradation and hence substrate biodegradation was assumed to occur instantaneously, being limited only by oxygen supply. The rationale for employing an instantaneous reaction model is as follows: For many readily biodegradable compounds such as BTEX and naphthalene, field observations indicate a pseudo-first-order biotic decay rate constant of the order of days (Howard et al., 1991). In contrast, groundwater flow rates at most sites are typically of the order of tens to hundred meters per year. Thus, in a contaminated zone with size $\Delta x = 10$ to 50 m, the fluid residence time is of the order of years, while biodegradation is predicted to occur with a half-life of days. Hence, the readily degradable organic chemicals may be assumed to be instantaneously and fully degraded, the extent of degradation being limited only by the available electron acceptors present within the cell. These assumptions underlie the instantaneous biodegradation model. The model treats the subsurface system as a large "chemical reactor," focusing on availability of stoichiometric amounts of electron acceptors. The distribution and growth of microbes in the porous media matrix is not considered to be significant in affecting the patterns of biodegradation observed in the field.

Contaminant transport and fate with the instantaneous biodegradation assumption is modeled by simultaneously tracking the movement of the organic substrate, contaminant A, as well as the primary electron acceptor (oxygen). The chemical transport equations for the two species are written as:

Contaminant, A:

$$\frac{\partial C^{\mathrm{A}}(x,y,t)}{\partial t} = -v\frac{\partial C^{\mathrm{A}}(x,y,t)}{\partial x} + D_L\frac{\partial^2 C^{\mathrm{A}}(x,y,t)}{\partial x^2} + D_T\frac{\partial^2 C^{\mathrm{A}}(x,y,t)}{\partial y^2} \quad (12.22a)$$

Oxygen, O_2:

$$\frac{\partial C^{O_2}(x,y,t)}{\partial t} = -v\frac{\partial C^{O_2}(x,y,t)}{\partial x} + D_L\frac{\partial^2 C^{O_2}(x,y,t)}{\partial x^2} + D_T\frac{\partial^2 C^{O_2}(x,y,t)}{\partial y^2} \tag{12.22b}$$

Eqs. (12.22a) and (12.22b) enable modeling of the simultaneous transport of oxygen as well as substrate across a spatial grid. At each time step, within each cell, the substrate mass that is stoichiometrically able to react with the oxygen present in that cell is modeled to be consumed "instantaneously." The algorithm for computing substrate and oxygen depletion is shown below in Eqs. (12.23a) to (12.23c). A stoichiometric factor, F, is used to determine the change in concentrations of substrate and oxygen at any time in any spatial cell, as (Bedient et al., 1999):

$$\text{Oxygen depletion conditions: } \Delta C^{\mathrm{A}} = -C^{O_2}/F \text{ and } C^{O_2} = 0 \text{ when } C^{\mathrm{A}} > C^{O_2}/F \tag{12.23a}$$

$$\text{Substrate depletion conditions: } \Delta C^{O_2} = -C^{\mathrm{A}}F \text{ and } C^{\mathrm{A}} = 0 \text{ when } C^{\mathrm{A}} < C^{O_2}/F \tag{12.23b}$$

where

$$F = \frac{\text{Mass of } O_2 \text{ required for stoichiometric oxidation of 1 mol of substrate}}{\text{Molecular mass of substrate}} \tag{12.23c}$$

Equation (12.23a) indicates that when the pollutant concentration is greater than the equivalent availability of oxygen in water, all the oxygen in the cell is consumed "instantaneously," and the pollutant concentration drops by an amount equivalent to that stoichiometrically degraded by the available oxygen. In contrast, when the available oxygen is in excess of that required by the pollutant in water, the pollutant is completely degraded (i.e., goes to zero), while the oxygen concentration is depleted by that amount needed to degrade the pollutant. In this manner, the pollutant and oxygen plumes are superimposed and the aqueous concentration of both species at each time step is computed by applying Eqs. (12.23a) and (12.23b). The two species move forward in the next time step by advection–dispersion as described by Eq. (12.22), and the pollutant concentrations within each cell at the subsequent time step are computed once again from Eqs. (12.23a) to (12.23c). Thus, plumes of both species are developed over time and space as illustrated in a hypothetical example shown in Example 12.2.

EXAMPLE 12.2

In a 1D transport system (assume advection only, no dispersion), depict pictorially the fate of a benzene plume over a period of 4 years, assuming the fluid residence time in each cell is 1 year. The initial benzene concentration in the aquifer is shown in Figure 12.13*a*.

Consider first pseudo-first-order degradation of benzene with a half-life of 3 months, irrespective of oxygen availability. Next, consider instantaneous degradation of benzene governed by oxygen availability. Assume uniform initial oxygen concentration in the aquifer representing high saturation conditions (9 mg/L); no further oxygen is assumed to be available in the water.

Solution 12.2 *Pseudo-first-order benzene degradation, irrespective of oxygen availability*, yields essentially no benzene remaining in the aquifer after 4 years. After 4 years, the initial contents of cell $i = 1$ move forward to cell $i = 5$, but in that time period the concentration decreases by a factor of e^{-kt}, where k is the pseudo-first-order degradation constant estimated from half-life data as $k = 0.693/t_{1/2}$, where $t_{1/2}$ is given to be 0.25 years (3 months). Thus, at $t = 4$ years the attenuation factor is $e^{-0.693\times 16}$, as shown below, yielding essentially zero benzene concentrations (< 1 μg/L) across the aquifer at the end of 4 years, as shown in Figure 12.13*b*.

Instantaneous biodegradation coupled with O_2 availability is considered next: To implement the algorithm presented in Eq. (12.23), the stoichiometric factor F is first calculated by writing a balanced equation for benzene oxidation by O_2:

$$C_6H_6 + 7.5O_2 \rightarrow 6CO_2 + 3H_2O$$

where 7.5 mol of O_2 are needed to degrade 1 mol of benzene, and hence the factor F is computed as:

$$F = \frac{7.5 \text{ mol/O}_2 \times 32 \text{ g O}_2\text{/mol}}{78 \text{ g benzene}} \approx 3 \frac{\text{g O}_2}{\text{g benzene}}$$

Superposing the initial benzene and oxygen concentrations in the aquifer and computing the impact of instantaneous oxidation, the following results are obtained at $t = 0$: In cell $i = 1$, the high benzene concentration of 11 mg/L requires 33 mg O_2/L (since $F \simeq 3$). But only 9 mg O_2/L is available, which is consumed entirely to degrade 3 mg/L of benzene, yielding a final benzene concentration of 8 mg/L and a final O_2 concentration of 0 mg/L in cell 1. In cell 2, the benzene concentration of 2 mg/L requires 6 mg/L oxygen, which is available at a concentration of 9 mg/L. Hence, all the benzene in cell 2 is instantaneously degraded leaving 0 mg/L benzene and 3 mg O_2/L.

In the absence of dispersion, the benzene and oxygen are advected forward—no further reaction can take place since no additional oxygen or substrate available in each cell. At $t = 4$ yr, unlike the zero benzene concentration predicted over the entire aquifer in the first-order model, the instantaneous reaction model predicts a hot spot of benzene in cell 5, corresponding to the travel position of the initially high benzene concentration zone wherein oxygen depletion has occurred and no further aerobic oxidation reaction can take place. See Figure 12.13*c*. Note, if dispersion were to be incorporated, the benzene and oxygen would disperse forward and backward in adjacent cells and undergo further "instantaneous reactions" in time steps $t = 2$ years and onward.

The algorithm for instantaneous chemical degradation shown in Eq. (12.23) and demonstrated in Example 12.2 and Figure 12.13*c* provides the basis for one of the first in situ biodegradation models, BIOPLUME, that specifically incorporated oxygen availability in the subsurface (Borden and Bedient, 1986). Since the 1980s, many modifications have been made to the BIOPLUME models to include the variety of electron acceptors available in the subsurface, such as nitrates, sulfates, and the like. The current version of BIOPLUME (BIOPLUME III) considers simultaneous transport of substrate along with six electron acceptors found in the subsurface to enable switching from aerobic to anaerobic conditions when depletion of dissolved oxygen occurs (Bedient et al., 1999). The model allows the user to choose between instantaneous, first-order, or Monod representations for the contaminant biodegradation rate. Note, unless specifically linked with electron acceptor transport, the Monod model (or the simplified Monod pseudo-first-order model) assumes the contaminant degrades irrespective of the availability of electron acceptors. Hence, in oxygen-limiting conditions, overall contaminant transformation can occur quite slowly using the "instantaneous" reaction model, compared with the first-order (or Monod) representation. See Example 12.2. Thus, despite its name, the instantaneous degradation model does not imply that the chemical biodegrades "instantaneously" across the entire aquifer. The use of the instantaneous reactor model is indicated when (Bedient et al., 1999): (a) Rapid depletion of oxygen and other anaerobic electron acceptors such as nitrate and sulfate is observed in the source zone, accompanied by (b) production of anaerobic degradation by-products such as methane and ferrous iron is documented.

Other factors that affect overall contaminant biotransformation rates, such as bioavailability constraints caused by slow diffusion–desorption from soil or slow dissolution from NAPLs, can be incorporated into the contaminant transport equations by including appropriate concentration boundary conditions and mass-transfer kinetic terms (see Chapter 4). Bioavailability constraints have been measured and modeled in batch and column systems, often employing simplified first-order representation of intrinsic biokinetics (e.g., Seagren et al., 1993, 1994; Ghoshal et al., 1996; Ramaswami and Luthy, 1997). At field sites, the spatial variability in microbial number densities in soil can also cause significant spatial variation in degradation rates. Excessive microbial growth in highly polluted zones can cause biofilm growth (Rittmann, 1993) and result in changes in aquifer dispersivity and porosity (Bielefeldt et al., 2002). The growth, transport, and attachment of microbes, with consequent changes in aquifer properties, has been modeled by a few researchers (Taylor and Jaffe, 1990). Heterogeneities in microbial populations and availability of electron acceptors has been assessed in a stochastic analysis of susbsurface biodegradation (Miralles-Wilhen and Gelhar, 1996; Miralles-Wilhen et al., 1997); variation in microbial population densities can affect susbstrate degradation rates in the subsurface far more than aquifer advection–dispersion phenomena.

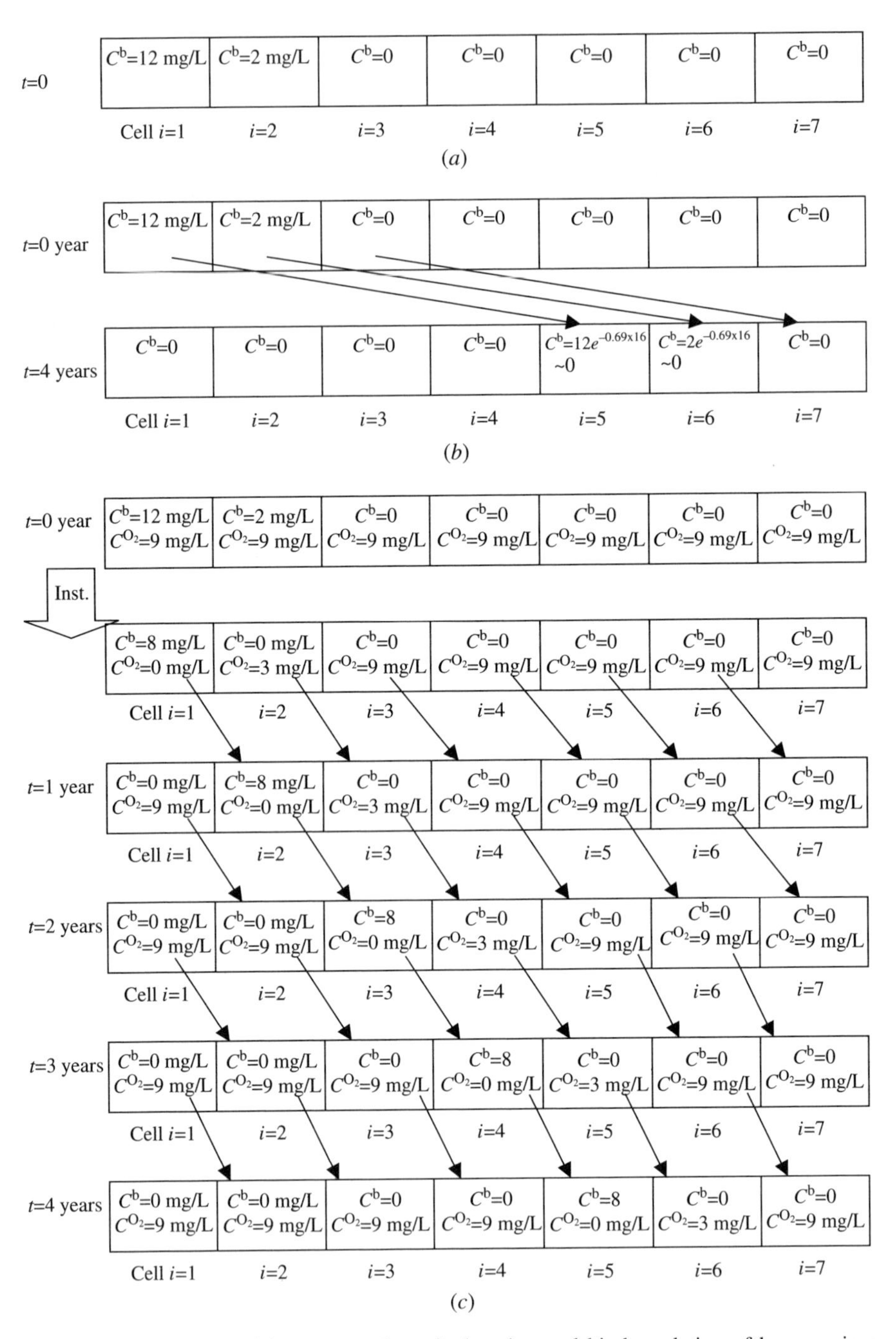

Figure 12.13 A pictorial representation of advection and biodegradation of benzene in a hypothetical 1-d aquifer: (*a*) Initial conditions; (*b*) First order biodegradation at $t = 4$ yrs with no oxygen limitations; (*c*) Instantaneous biodegradation with oxygen limitation over 4 yrs.

Efforts to quantify natural spatial distribution of microbial population densities is further complicated by the presence of vegetation, which exhibits a "rhizosphere effect" (Anderson et al., 1993). Plants and grass have been found to have several orders of magnitude larger microbial densities in their rhizospheres than in surrounding soils (Jordahl et al., 1997; Joner et al., 2001), a phenomenon used to advantage in phytodegradation of organic pollutants in soil (Anderson et al., 1993).

Notwithstanding the challenges in fully quantifying a natural and diverse process such as biodegradation, existing models have been used at several contaminated sites to predict plume shrinkage in response to biodegradation. Two case studies of natural plume attenuation due to microbial degradation processes are presented next. The first case study employs an instantaneous biodegradation model to describe attenuation of a naphthalene and PCP (pentachlorophenol) plume at a creosote-contaminated site in Conroe, Texas. The second case study represents a series of biodegradation reactions, each modeled as a first-order process, occurring at a site with a TCE DNAPL spill. The two different biokinetic models employed at the two sites highlight different approaches to biodegradation modeling used in the field.

12.5 INSTANTANEOUS BIOKINETICS COUPLED WITH TWO-DIMENSIONAL TRANSPORT SYSTEMS

As described in the previous section, the instantaneous biodegradation model focuses entirely on the availability of oxygen (and other electron acceptors) as the primary factor controlling the rates of organic chemical biotransformation in the subsurface—the microbial biodegradation reaction itself is assumed to occur instantaneously. This reaction model was first developed in the 1980s and has since been applied to describe PAH and petroleum hydrocarbon attenuation at several sites, for example, Conroe site, Texas (Borden et al., 1986), Hill Air Force Base, Utah (Parsons Engineering-Science, 1994; Newell et al., 1996), Cliff-Dow Superfund site, Michigan (Klecka et al., 1990), and Gas Plant Site, Michigan (Chiang et al., 1989). We present here an application of the instantaneous biodegradation model to describe groundwater plume development at a creosote-contaminated site in Conroe, Texas (Borden et al., 1986). The presentation on biodegradation modeling follows site conceptualization and groundwater modeling described previously for this site in Section 9.7.

12.5.1 Case Study of the Conroe Site, Texas

Overview Unknown quantities of wastes from wood-preserving operations were disposed from 1946 to 1972 into two unlined ponds on the Conroe site as shown by the hatched regions in Figure 9.14*a*. Chloride ion, naphthalene, and PCP were chosen as the principal representative chemicals within the contaminated plume emerging from the source zones (Borden et al., 1986). Field measurements showed concentrations of all organic chemicals decreasing along the water flow direction on the site. The objective of the modeling exercise was to characterize the transport and attenuation, attributed to biodegradation, of the organic chemicals found in groundwater at the site. An early version of the BIOPLUME (Borden et al., 1986) model was used to simulate the biodegradation of naphthalene and total hydrocarbons at the site.

Water Flow and Chloride Plume Modeling The reader is directed to Section 9.7 (Case Study 1), for a more detailed description of water flow modeling at the site. We summarize here the key results that the water flow at this site is toward the south, with a hydraulic gradient of 0.6%, an average porosity of 29%, and an average conductivity of $K = 0.74$ m/day. Plume dispersion parameters were determined from those fit to describe the existing chloride plume on the site and were $\alpha_L = 9.1$ m and $\alpha_T = 1.8$ m. The above aquifer advection and dispersion parameters were successful in simulating the existing chloride plume on the site, as shown in Table 9.8b.

Contaminant Retardation and Biodegradation Retardation and degradation parameters were determined by conducting a controlled pulse injection and recovery test at an uncontaminated area of the site—see Figure 9.14*c*. The peaks of all three curves in Figure 9.14*c* appear at approximately the same time, suggesting that retardation is not significant in this aquifer, confirmed by the very low f_{oc} ($< 0.1\%$) of the aquifer materials yielding low K_d values. Thus, a retardation factor of 1 is appropriate for use for the contaminants at the Conroe site. However, the percent mass recovered for the organic pollutants is much less than that recovered for chloride ion, suggesting that significant degradation of these chemicals is occurring at the site.

Evidence of Oxygen-Limited Biodegradation A combination of laboratory microcosm tests, along with the observed attenuation of these chemicals, both in the existing plume as well as in the on-site injection and recovery test, indicated that significant microbial degradation of these chemicals could be occurring in the aquifer. Microcosm samples of aquifer material were studied in laboratory tests. Microbial populations at the site were well-characterized with about 10^6 to 10^7 cells per gram of aquifer material. Laboratory studies of microbes obtained from the vicinity of the waste plume indicated that they were acclimated to the range of organic compounds found in groundwater, being able to degrade them fairly rapidly (to less than 20 μg/L within 20 days) in aerated laboratory microcosms. Systems maintained under nitrogen showed greatly lowered degradation rates suggesting that oxygen supply in the subsurface may be the primary rate-limiting step controlling in situ biodegradation of these chemicals. Furthermore, analysis of field data showed that naphthalene concentrations were consistently very low in areas with high dissolved oxygen, highlighting the importance of oxygen in degradation of naphthalene at the site. Borden et al. (1986) therefore hypothesized that oxygen availability controls the rate of biodegradation of organic chemicals at the site.

Biodegradation Modeling Borden et al. (1986) used an early version of the BIOPLUME model to describe the "instantaneous" oxygen-limited biodegradation of naphthalene at the Conroe site. Naphthalene was used as the representative chemical that typified the behavior of all the hydrocarbons present at the site. Regression analysis employing site-monitoring data showed that the aqueous total hydrocarbon concentration was related to the naphthalene concentration through a log–log regression equation ($C^{\text{total HC}} = 0.8342 \times [C^{\text{naphthalene}}]^{1.146}$) with an $r^2 = 0.99$. A stoichiometric factor, $F = 3$, was applied to describe the stoichiometric requirement of oxygen by the hydrocarbons present at the site. Some loss of hydrocarbons was observed to occur due to volatilization to the unsaturated zone; likewise reaeration of oxygen also occurred due to unsaturated zone interactions. The hydrocarbon loss was represented as a first-order process, the rate constant for which depended on the thickness of the unconfined aquifer ($b = 1$ to 3.5 m) and the vertical dispersion coefficient $\alpha_z = 3.5$ cm). Because of the losses to the unsaturated zone, strict superposition as shown in Eqs. (12.22) and (12.23) could not be applied. However, a subtraction routine was used to represent the decrease in hydrocarbon concentration based on the stoichiometric factor F and the oxygen concentration in the aquifer [Eqs. (12.23a) to (12.23c)]. In this manner, the simultaneous transport of oxygen and the organic chemical through the aquifer matrix was simulated using the measured initial (background) oxygen concentrations in the aquifer of 3 mg/L and the assumed source zone hydrocarbon concentrations as shown in Table 9.8. The parameters used in the simulations are summarized in Table 12.6a. The simulations were able to capture the main features of the naphthalene and oxygen concentrations within the aquifer as the plume developed over a period of 20 years (see Table 12.6b). Neither a first-order nor the Monod biodegradation models were successful in representing field observations, indicating the need to use the instantaneous biodegradation model linked to oxygen availability. Once the biodegradation model was calibrated and verified with present-day conditions, the model was applied to evaluate alternative remedial operations at the Conroe site after the source zone was contained (i.e., the wastes in the ponds were removed from the site). Natural attenuation was found to be sufficient in controlling the contaminated plumes and was expected to shrink the plume within a period of 60 years. Thus, modeling efforts showed that no further engineered treatment was required at the site once the source wastes were excavated and removed.

A similar conclusion regarding natural attenuation was reached at a different site in Dover, Delaware, contaminated with chlorinated solvents. At this site, sequential biodegradation of chlorinated ethenes was described using pseudo-first-order biokinetic models specified within delineated aerobic and anaerobic zones observed in the aquifer. Although transport of electron acceptors was not explicitly considered, the biodegradation modeling at this site involved coupling the transport of multiple reactive organic substratess in aerobic and anaerobic zones of the aquifer. The site study is described in more detail in the next section.

12.6 SEQUENTIAL BIOKINETICS IN A TWO-DIMENSIONAL TRANSPORT SYSTEM

Overview Chlorinated compounds can undergo a variety of abiotic as well as microbially mediated biochemical transformations, under both aerobic and anaerobic conditions (Vogel et al., 1987). The appropriate sequence of reactions at a certain

TABLE 12.6 Details of BIOPLUME Modeling at the Conroe Site, TX

a. Parameter Description and Values Used in Simulation

Parameter and Description	Estimated Parameter Value Used in Simulation
Hydraulic conductivity, K	0.74 m/day
Aquifer porosity, n	0.29
Longitudinal dispersivity, α_L	9.1 m
Transverse dispersivity, α_T	1.8 m
Vertical dispersivity, α_z	3.5 cm
Aquifer thickness, b	1 to 3.5 m
Retardation factor, R	1
Background O_2 conc. in water	3 mg/L
Stoichiometric factor, F	3

b. Comparison of Simulated and Observed Hydrocarbon and Oxygen Concentrations

Location (Well)	Comparison of Simulated and Observed Total Hydrocarbon Conc. in Groundwater (μg/L)			Comparison of Simulated and Observed Oxygen Conc. in Groundwater (mg/L)		
	Observed	Simulated	Error	Observed	Simulated	Error
SW-1	nd	0	0	2.9	2.9	0.0
SW-2	62[a]	285	223	—	0	—
SW-3	nd	0	0	0.7	0.6	−0.1
SW-4	108	40	−68	1.2	0	−1.2
SW-5	46	87	41	0.3	0	−0.3
SW-7	nd	0	0	–	1.4	—
SW-8	nd	0	0	–	3.0	—
SW-9	84	0	−84	0.7	2.1	1.4
RU-2	5352	5334	−18	0.2	0	−0.2
RU-3	14	0	−14	1.4	1.8	0.4
	Root-mean-squared error = 80			Root-mean-squared error = 0.7		

[a]Low accuracy of measurement.
Source: After Borden et al. (1986).

site depends strongly upon the types of electron acceptors and electron donors available at that site. Under anaerobic conditions, chlorinated aliphatic compounds such as trichloroethene (TCE) and trichloroethane (TCA) can undergo reductive dechlorination, wherein hydrogen replaces a chlorine group on the aliphatic compound. The hydrogen in this case functions as the electron donor, while the chlorinated compound acts as the electron acceptor. Reductive dechlorination occurs in anaerobic zones of the aquifer where both dissolved oxygen and nitrates are depleted, most often consumed by biodegradation of other organic compounds, for example, fuel hydrocarbons. See Example 12.4. The rate of dechlorination is often modeled as a first-order process; reductive dechlorination rates increase with the number of chlorines attached to the parent compound. Some typical pathways for reductive dechlorination are shown in Figure 12.14*a*, resulting in a sequence of reactions where more chlorinated

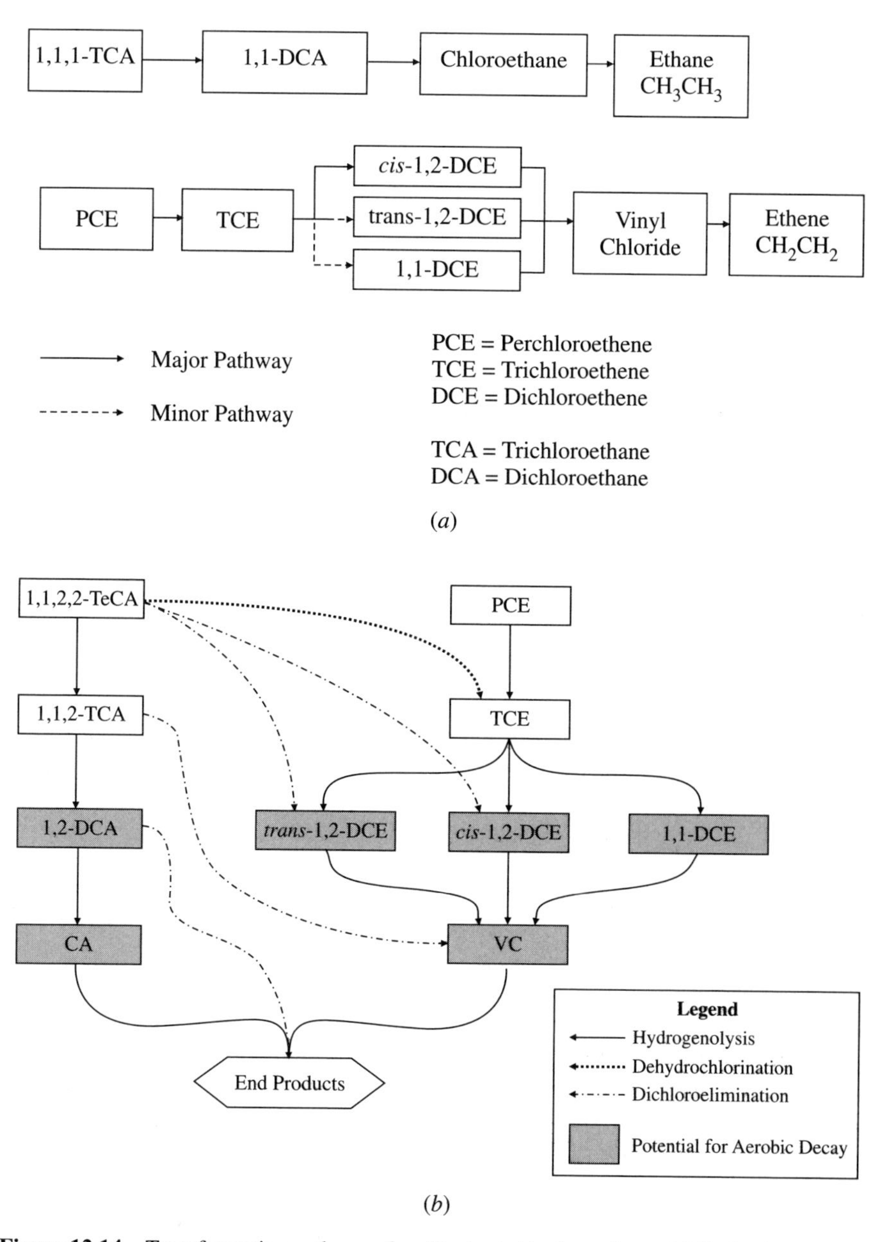

Figure 12.14 Transformation pathways for chlorinated hydrocarbons. (*a*) Reductive dechlorination pathways for chlorinated hydrocarbons (from Vogel et al., 1987) and (*b*) aerobic and anaerobic transformation pathways for the chlorinated ethane and ethane compounds. Note, TeCA = Tetrachloroethane, CA = Chloroethane, and VC = Vinyl Chloride. (Reprinted from *Journal of Contaminant Hydrology,* Vol. 59, Nos. 1–2, T. P. Clement, M. J. Truex, and P. Lee; A case study for demonstrating the application of U.S. EPA's monitored natural attenuation screening protocol at a hazardous waste site, pp. 133–162, © 2002, with permission from Elsevier.)

compounds react to yield less chlorinated hydrocarbons. Because the dechlorination rate typically decreases with the degree of chlorination of the reactant molecule, some of the intermediate products may accumulate in groundwater, being produced faster than they can themselves can be dechlorinated. Simultaneously, as the plume moves into more oxygenated regions of the aquifer, aerobic co-metabolic reactions can transform the chlorinated compounds (e.g., Alvarez-Cohen and McCarty, 1991). Clement et al. (2002) show the combination of anaerobic and aerobic biotransformations that can occur at a solvent-contaminated site (see Fig. 12.14*b*).

12.6.1 Case Study of Chlorinated Compounds at Dover AFB, Delaware

In our second case study on subsurface biodegradation, we study natural attenuation of chlorinated aliphatic compounds at Area 6 of the Dover Air Force Base, Delaware. The summary presented here is based upon model development and calibration with data as reported by Clement et al. (2000). The conceptual model for chemical transport and fate at the site was developed after synthesis of geologic, hydrologic, and biochemical information gathered from a variety of sources, including literature reports, field borings, piezometer data, laboratory tests, and field tracer tests. The conceptual model is described briefly first.

Location and Site Hydrogeology The Area 6 site located at Dover, Delaware, is about 3 km long (9000 ft) in the north–south direction and about 2 km wide (6000 ft) in the east–west direction. Surface features of the site are shown in Figure 12.15*a*. The St. Jones River loops around the site along the southern and western boundaries. Water flow is predominantly from the north toward the river. Details on the vertical aspects of the site were determined from drilling logs and are summarized in Figure 12.15*b*. Subsurface contamination by fuel and solvents occurs in an unconfined aquifer, bounded below by a low-permeability clayey layer called the Calvert formation. The unconfined aquifer is composed of a mixture of sandy, silty, and gravelly material. Two distinct regions, an upper shallow zone and a deeper zone, are discernible within the unconfined aquifer. The shallow zone has a thickness ranging from 5 to 10 ft and is composed of low-permeability silty-sandy material. The deep zone consists of sandy-gravelly material with thickness ranging from 10 to 15 ft. Average hydraulic conductivity was estimated at 35 and 85 ft/day for the shallow and deep zones, respectively, from slug tests and pumping tests. A contour map of groundwater elevations measured at the site during a sampling event in 1997 is shown in Figure 12.15*c*. The measured groundwater elevations range from 14 to 3 ft above sea level. Water flows from the north toward the river perpendicular to the equipotential lines seen in the figure. The head gradient is fairly shallow in the northern regions and becomes steeper closer to the river. The head at the river itself was measured at about 2 to 3 ft above sea level.

Modeling Groundwater Flow The primary area of contamination is the deep zone of the unconfined aquifer. Steady-state groundwater flow is assumed to occur within this zone. A 100 ft $\times$ 100 ft finite difference grid was used in the flow simulations with the USGS MODFLOW code (McDonald and Harbaugh, 1988). A uniform thickness

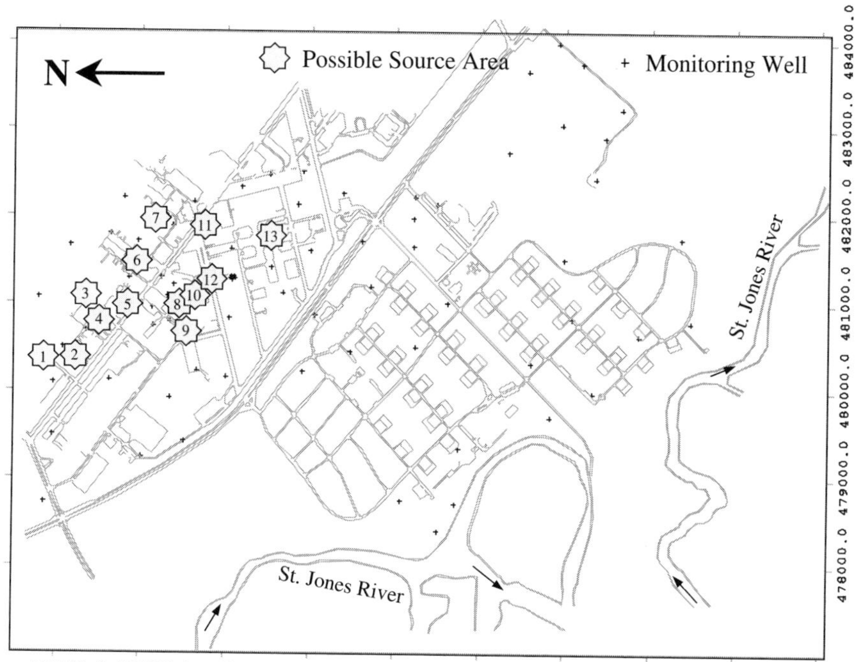

(*a*)

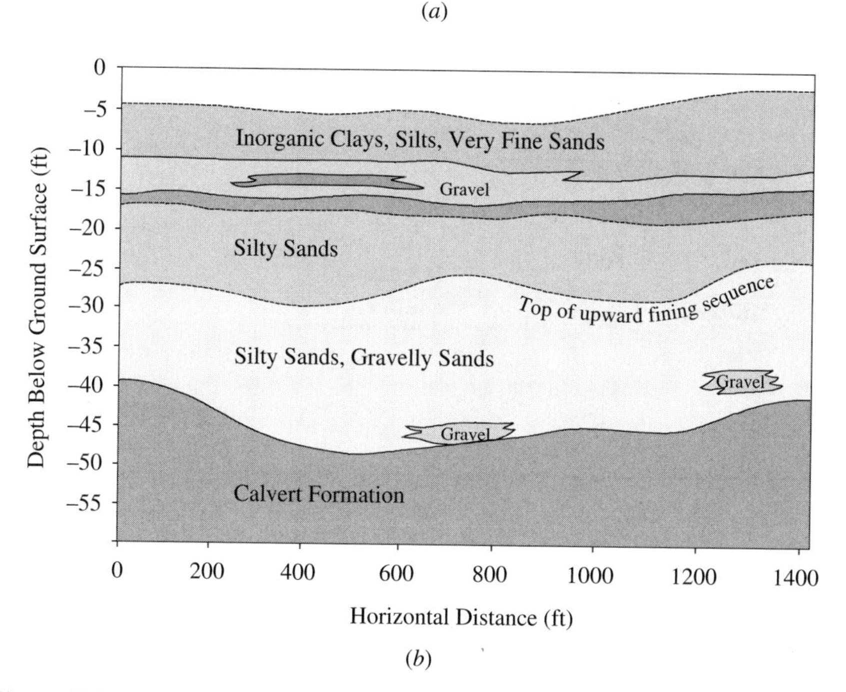

(*b*)

Figure 12.15 (*a*) Details of area 6 site, Dover Air Force Base, Deleware. (*b*) Conceptual model for site geology.

(continued)

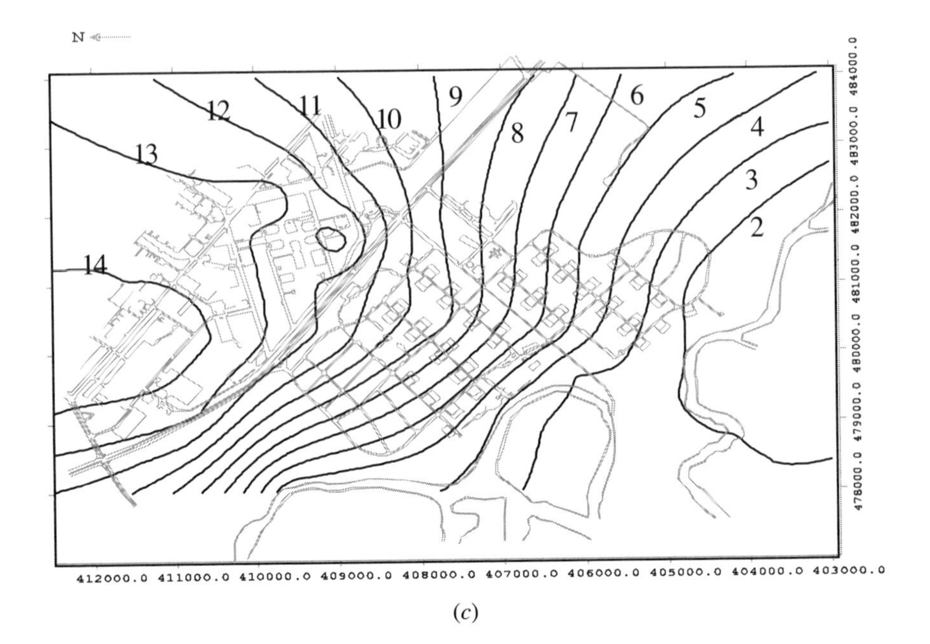

(*c*)

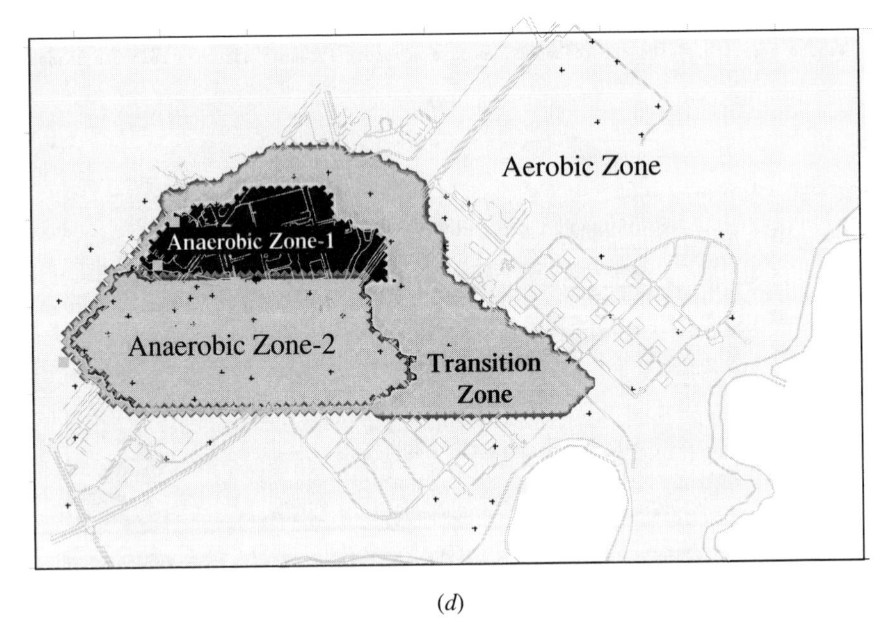

(*d*)

Figure 12.15 (*c*) Groundwater contours based on field-measured data, feet above MSL. (*d*) Conceptual model used for defining the anaerobic and aerobic reaction zones. (Reprinted from *Journal of Contaminant Hydrology,* Vol. 42, T. P. Clement, C. D. Johnson, Y. Sun, G. M. Klecka, and C. Bartlett, Natural attenuation of chlorinated ethene compounds: Model development and field-scale application at the Dover site, pp. 113–140, © 2000, with permission from Elsevier.)

of 12 ft was used to represent the deep zone of the aquifer. Trial-and-error methods were used to adjust the aquifer conductivity within the range of observed values measured at the site (25 to 100 ft/day). An anisotropy factor of 0.75 was used in the transverse directions to simulate the observed preferential flow in the southerly direction. Leakage of water from the upper low-permeability shallow zone to the deep zone was modeled by applying steady recharge rates of 0.0006 ft/day, which is also within the range measured at the site. Constant head boundaries were set up along the river and at the northern boundary based on measured head data shown in Figure 12.15*c*. Steady-state head distributions obtained from MODFLOW simulations (not shown) compared quite well with the observations shown in Figure 12.15*c*.

Source Characterization Much uncertainty in contaminant transport modeling at field sites arises from our lack of knowledge about the quantity, location, and time of release of contaminants at the specific site. It is speculated that hydrocarbon and chlorinated solvent releases occurred at the Dover site over a period of many years. TCE and PCE are the two primary chlorinated compounds originally released in Area 6, generating a larger TCE plume that dominates the site, as well as a smaller PCE-derived plume. The fuel hydrocarbons and solvent chemicals are believed to originate from the source areas marked with stars in Figure 12.15*a*, where solvent DNAPL spills were released into the subsurface and migrated downward into the deep aquifer. While no free-phase DNAPL exists currently at the site, DNAPL at residual saturation is believed to be a continuing source of chlorinated contaminants in the deep aquifer zone. As was indicated in Section 4.4, the rate of mass transfer from subsurface NAPL bodies is difficult to predict due to uncertainty about the total DNAPL mass released as well as the interfacial area of contact between water and NAPL bodies trapped in the subsurface. As a result, the source zone at the Dover site was approximated by modeling TCE- and PCE-contaminated water discharges into groundwater from hypothetical wells located at the source locations shown in Figure 12.15*a*. The well water discharge rates were set at low levels so as not to significantly perturb the steady-state groundwater flow field. Four stress periods were used, each simulating different levels of contaminant releases into the aquifer as shown in Table 12.7*a*. Mass-balance constraints were applied to ensure that more contaminant mass was not modeled for release into the water than that present during the 1987 sampling event (accounting for the PCE mass sorbed as well as estimated to be degraded). A 40-year time frame was established from the time of initial source release to the current 1997 sampling event.

Tracer Transport Parameters Before modeling the transport and fate of TCE and PCE in the aquifer, tracer advection and dispersion parameters are required. Advection is characterized by groundwater flow rates that were successfully modeled using the MODFLOW code, as explained above. Field-scale dispersivity is more difficult to estimate and, just as in the case of source release rates, contributes significantly to model uncertainty. Dispersion parameters were first estimated based upon the guidelines provided in Gelhar et al. (1982). Tracer simulations were then conducted to confirm that the assumed dispersivity values produce reasonably sized plumes. The

TABLE 12.7 Source Parameters and Reaction Rate Constants Used in Multispecies Simulation of Chlorinated Compounds at the Dover Airforce Base, DE

Source Number	Stress Period 1		Stress Period 2		Stress Period 3		Stress Period 4	
	PCE	TCE	PCE	TCE	PCE	TCE	PCE	TCE
a. Source Zone Modeling at the Dover Site. Assumed Mass Release Rates: $\dot{m}$ (kg/yr)								
1	1	10	2	10	1	8	1	1
2	1	2	1	2	1	1	1	0
3	1	52	1	1	0	1	0	1
4	1	19	2	19	2	8	2	2
5	25	413	25	165	17	74	0	17
6	1	9	0	1	0	1	0	1
7	1	1	1	1	1	1	0	0
8	0	5	0	5	0	5	0	5
9	10	1	31	517	10	1	1	1
10	0	41	0	41	0	4	0	2
11	2	0	0	413	0	0	0	0
12	0	21	0	21	0	17	0	2
13	0	0	0	0	0	0	0	0

b. Calibrated Degradation Rate Constants (day^{-1})

First-Order Rate Constant	Associated Contaminant	Anaerobic Zone 1	Anaerobic Zone 2	Transition Zone	Aerobic Zone
k_P (anaerobic)	PCE	3.2×10^{-4}	4.0×10^{-4}	1.0×10^{-4}	0.0
k_{T1} (anaerobic)	TCE	9.0×10^{-4}	4.5×10^{-4}	1.125×10^{-4}	0.0
k_{T2} (aerobic)	TCE	0.0	0.0	0.4×10^{-5}	1.0×10^{-5}
k_{D1} (anaerobic)	DCE	8.45×10^{-4}	6.5×10^{-4}	1.625×10^{-4}	0.0
k_{D2} (aerobic)	DCE	0.0	0.0	1.6×10^{-3}	4.0×10^{-3}
k_{V1} (anaerobic)	VC	8.0×10^{-3}	4.0×10^{-3}	1.0×10^{-3}	0.0
k_{V2} (aerobic)	VC	0.0	0.0	0.8×10^{-3}	2.0×10^{-3}
k_{E1} (anaerobic)	ETH	2.4×10^{-2}	1.2×10^{-2}	0.3×10^{-2}	0.0
k_{E2} (aerobic)	ETH	0.0	0.0	0.4×10^{-2}	1.0×10^{-2}

Source: Adapted from Clement et al. (2000).

assumed longitudinal dispersivity value of 40 ft and transverse dispersivity of 8 ft produced simulated tracer plumes similar in shape to the observed concentration plumes. Larger dispersivitiy values produced more smear than observed at this site, while smaller dispersivity values produced simulated plumes that were much narrower than those observed at the site. Hence, all contaminant transport simulations were conducted with these dispersivity values.

Retardation Factors Laboratory and field tests indicated an average soil porosity of 0.38 and retardation factors of 1.3, 1.2, and 1.1 for PCE, TCE, and DCE, respectively.

Evidence of Biological Transformations The presence of vinyl chloride, DCE, ethanol, and excess chloride ions at the site indicated that PCE and TCE were undergoing degradation reactions consistent with those shown in Figure 12.14*a*. Based on literature reports, PCE primarily undergoes anaerobic degradation, while both aerobic and anaerobic microbial processes can contribute to TCE degradation. Laboratory microcosm studies with Dover soils indicated that reductive dechlorination was a dominant pathway for breakdown of chlorinated compounds under *anaerobic* conditions. Laboratory microcosm tests indicated a first-order anaerobic degradation rate of approximately $\sim 3 \times 10^{-4}$ day^{-1} for PCE. Anaerobic decay rates (1/day) were of the order of 10^{-5} for TCE, 10^{-6} for DCE, and 10^{-2} for VC. Based on the pathways shown in Figure 12.14*a*, and the detection of DCE, VC, and elevated chloride on-site, the following pathway was assumed to apply: PCE $\rightarrow$ TCE $\rightarrow$ DCE $\rightarrow$ VC $\rightarrow$ ETH (Ethanol).

In addition to the PCE pathway, the TCE plume could itself undergo *aerobic* biotransformations, as has been reported in the literature on co-metabolism by mixed methanotrophic cultures (Alvarez-Cohen and McCarty, 1991). McCarty and Semprini (1994) indicate that both DCE and VC have the potential to degrade via both direct and co-metabolic pathways. Field studies conducted by Bradley et al. (1998) and Klier et al. (1999) also demonstrate aerobic breakdown pathways for DCE and VC. Lab microcosm studies with the Dover site soils indicated measured direct aerobic decay rates (1/day) of the order of 10^{-3} for DCE and 10^{-4} for VC. Field data on oxygen and methane measurements at the Dover site indicated that the plume area could be characterized by a central anoxic zone surrounded by an aerobic zone (see Fig. 12.15*d*). Putting together the available information on reaction pathways, the laboratory test data and the observation of distinct anoxic and aerobic zones at the Dover site, a combination of aerobic and anaerobic breakdown pathways was used to model the fate of TCE and PCE released from the source zones at the Dover site. A schematic representing the reaction pathway is shown in Figure 12.16. PCE degradation was modeled exclusively as an anaerobic reductive dechlorination process, while all other chemicals were modeled to undergo both aerobic and anaerobic transformations.

All the reactions shown in Figure 12.16 were modeled as first-order processes, that is, the microbial concentrations at the site are assumed to be fairly stable, neither increasing nor decreasing over time, such that a pseudo-first-order reaction model can be used wherein the rate of reaction depends primarily upon the concentration of the reactants. Note that such a first-order representation does not simulate the availability

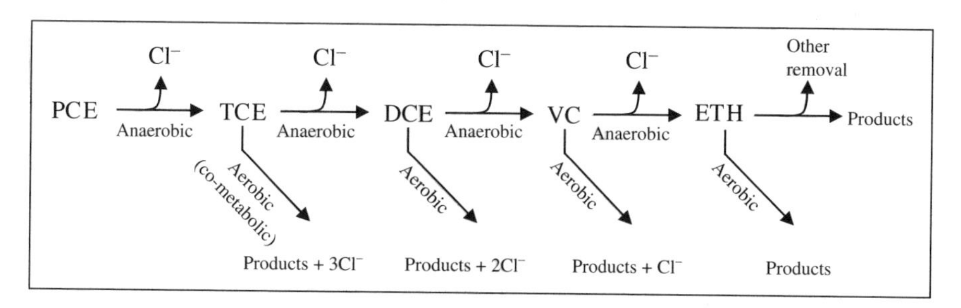

Figure 12.16 Conceptual model for biological reactions.

and transport of electron acceptors nor microbial growth and decay processes. The first-order assumption is considered to be valid when no active remediation measures are being taken at the site either by bioaugmenting the site with microbes, biostimulating the site with nutrients to promote microbial growth, or by biosparging and supplying oxygen to the site of interest (Wiedemeier et al., 1996). As discussed in Section 12.3.1, the first-order representation of the Monod model is also appropriate at low pollutant concentrations. Different first-order rate constants for biotransformation were assumed for each of the different chlorinated constituents within four geochemically separate zones shown in Figure 12.15*d*—two anaerobic zones close to the NAPL spill site (zones 1 and 2), where hydrocarbon fuel spills are believed to have depleted the oxygen, a transition zone (zone 3), and a surrounding aerobic zone, zone 4. However, within each zone, the rate constants are assumed to be spatially and temporally constant.

The advection–dispersion sorption–reaction equation (see Chapter 9) is written for each of the eight species shown in Figure 12.16. In the set of equations shown below, the reaction terms on the rhs represent the rates of production and depletion of the chemical due to the sequence of reactions shown in Figure 12.16. Assuming first-order biodegradation kinetics, transport, and transformation of PCE, TCE, DCE, VC, ETH, and Cl can be simulated by solving the following set of partial differential equations:

$$R_P \frac{\partial[\mathrm{PCE}]}{\partial t} = \frac{\partial}{\partial x_i}\left(D_{ij}\frac{\partial[\mathrm{PCE}]}{\partial x_j}\right) - \frac{\partial(v_i[\mathrm{PCE}])}{\partial x_i} + \frac{q_s}{\phi}[\mathrm{PCE}]_s - k_P[\mathrm{PCE}]$$

$$R_T \frac{\partial[\mathrm{TCE}]}{\partial t} = \frac{\partial}{\partial x_i}\left(D_{ij}\frac{\partial[\mathrm{TCE}]}{\partial x_j}\right) - \frac{\partial(v_i[\mathrm{TCE}])}{\partial x_i} + \frac{q_s}{\phi}[\mathrm{TCE}]_s + Y_{T/P}k_P[\mathrm{PCE}]$$
$$- k_{T1}[\mathrm{TCE}] - k_{T2}[\mathrm{TCE}]$$

$$R_D \frac{\partial[\mathrm{DCE}]}{\partial t} = \frac{\partial}{\partial x_i}\left(D_{ij}\frac{\partial[\mathrm{DCE}]}{\partial x_j}\right) - \frac{\partial(v_i[\mathrm{DCE}])}{\partial x_i} + \frac{q_s}{\phi}[\mathrm{DCE}]_s + Y_{D/T}k_{T1}[\mathrm{TCE}]$$
$$- k_{D1}[\mathrm{DCE}] - k_{D2}[\mathrm{DCE}]$$

$$R_V \frac{\partial[\mathrm{VC}]}{\partial t} = \frac{\partial}{\partial x_i}\left(D_{ij}\frac{\partial[\mathrm{VC}]}{\partial x_j}\right) - \frac{\partial(v_i[\mathrm{VC}])}{\partial x_i} + \frac{q_s}{\phi}[\mathrm{VC}]_s + Y_{V/D}k_{D1}[\mathrm{DCE}]$$
$$- k_{V1}[\mathrm{VC}] - k_{V2}[\mathrm{VC}]$$

$$R_E \frac{\partial[\mathrm{ETH}]}{\partial t} = \frac{\partial}{\partial x_i}\left(D_{ij}\frac{\partial[\mathrm{ETH}]}{\partial x_j}\right) - \frac{\partial(v_i[\mathrm{ETH}])}{\partial x_i} + \frac{q_s}{n}[\mathrm{ETH}]_s + Y_{E/V}k_{V1}[\mathrm{VC}]$$
$$- k_{E1}[\mathrm{ETH}] - k_{E2}[\mathrm{ETH}]$$

$$R_C \frac{\partial[\text{Cl}]}{\partial t} = \frac{\partial}{\partial x_i}\left(D_{ij}\frac{\partial[\text{Cl}]}{\partial x_j}\right) - \frac{\partial(v_i[\text{Cl}])}{\partial x_i} + \frac{q_s}{n}[\text{Cl}]_s + Y1_{C/P}k_{P1}[\text{PCE}]$$
$$+ Y1_{C/T}k_{T1}[\text{TCE}] + Y1_{C/D}k_{D1}[\text{DCE}] + Y1_{C/V}k_{V1}[\text{VC}]$$
$$+ Y2_{C/T}k_{T2}[\text{TCE}] + Y2_{C/D}k_{D2}[\text{DCE}] + Y2_{C/V}k_{V2}[\text{VC}]$$

where [PCE], [TCE], [DCE], [VC], [ETH], and [Cl] represent contaminant concentrations of various species (mg/L); k_P, k_{T1}, k_{D1}, and k_{V1}, and k_{E1} are first-order anaerobic degradation rate constants (day^{-1}); k_{T2}, k_{D2}, k_{V2}, and k_{E2} are first-order aerobic degradation rate constants (day^{-1}); and R_P, R_T, R_D, R_V, R_E, and R_C are retardation factors. Stoichiometric yield values, Y, can be calculated from the reaction stoichiometry and molecular weights. For example, anaerobic degradation of one mole of PCE would yield one mole of TCE; therefore, $Y_{T/P} =$ molecular weight of TCE/molecular weight of PCE (131.4/165.8 = 0.79). Based on similar calculations, values of $Y_{D/T}$, $Y_{V/D}$, and $Y_{E/V}$ (which are chlorinated compound yields under anaerobic reductive dechlorination conditions) were calculated to be 0.74, 0.64, and 0.45, respectively; values of $Y1_{C/P}$, $Y1_{C/T}$, $Y1_{C/D}$, and $Y1_{C/V}$ (which are yield values for chloride under anaerobic conditions) were calculated to be 0.21, 0.27, 0.37, and 0.57, respectively; and finally values of $Y2_{C/T}$, $Y2_{C/D}$, and $Y2_{C/V}$ (yield values for chloride under aerobic conditions) were calculated to be 0.81, 0.74, and 0.57, respectively. Note that the reaction models used assume that the biological degradation reactions only occur in the aqueous phase, which is a conservative assumption.

With advection, dispersion, retardation, and reaction parameters in place, the multidimensional reactive transport simulator RT3D (Clement, 1997) was used to simulate the development of the PCE and TCE plumes, as well as that of the degradation intermediate products, DCE, VC, EtOH, and excess chloride. RT3D is a general-purpose reactive transport solver that can be used to solve any number of transport equations coupled through biodegradation reactions (see equations above). RT3D was developed from the EPA transport code MT3D (Zheng, 1990) and utilizes several numerical methods to obtain a solution to the problem, including an operator-split strategy that allows definition of different reaction models through a plug-in user-defined reaction module (Clement, 1997). In implementing the RT3D simulations for the Dover site, both the source zone release rates of TCE and PCE, as well as the presumed first-order rate constants for biodegradation of the eight species were unknown. A general boundary on these values was determined by applying mass-balance constraints, that is, the mass of TCE and PCE currently found in water and soil at the site was used to provide a lower bound for the mass of these chemicals released from the source zone. The mass presumed to be biodegraded could be added on to provide an upper bound for source release rates. The model was calibrated by varying the source loadings and the degradation rate constants until a fairly good match was observed between the simulated and observed plumes for the five chemicals observed in the field. The model parameters used in the best-fit calibration are shown in Table 12.7*b*. A sensitivity analysis was performed to assess the change in simulation results caused by a change in model input parameters. As at the Conroe site, the calibrated model was used to demonstrate the potential for natural attenuation occurring in the subsurface.

Model results indicated success in describing the complex series of reactions

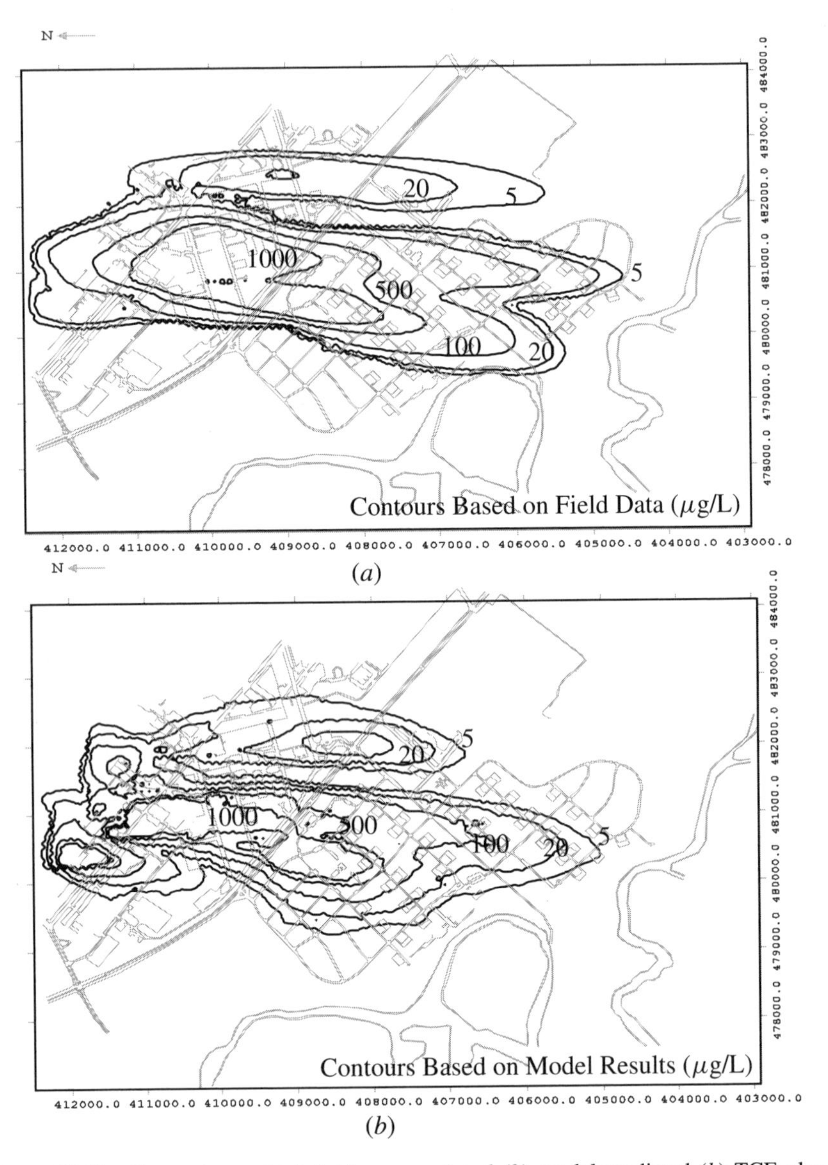

Figure 12.17 Comparison of (*a*) field-measured and (*b*) model-predicted (*b*) TCE plume contours at the Dover site. (Reprinted from *Journal of Contaminant Hydrology*, Vol. 42, T. P. Clement, C. D. Johnson, Y. Sun, G. M. Klecka, and C. Bartlett, Natural attenuation of chlorinated ethene compounds: Model development and field-scale application at the Dover site, pp. 113–140, © 2000, with permission from Elsevier.)

occurring at the Dover site. Simulated PCE and TCE plumes matched fairly well with plume data observed over the 1997 sampling event (TCE observations and simulations are shown in Fig. 12.17*a,b*). The total mass of PCE and TCE simulated to be present at the site after 40 years also matched fairly well with that observed during the sampling even. The first-order degradation rate constants used in the simulations were similar in magnitude both to those reported in the literature and to those observed in the lab microcosm tests. The simulated plumes of the intermediate reaction products (DCE, VC, and ETH) matched fairly well with the general trends observed at the site for these plumes. Furthermore, the mass of "excess chloride" observed at the site was consistent with the original release of TCE and PCE used in the simulations. Thus, the modeling effort was successful in describing a complex problem, both in terms of chemistry, as well as the integration of chemistry with flow and dispersion occurring in the subsurface.

The five case studies presented in this chapter have provided an overview of the different techniques used to couple chemical and biochemical transformations with transport processes occurring in aqueous systems. Chapter 13 addresses the use of groundwater and surface water contaminant transport and fate models in human exposure and risk assessment. Chapter 14 presents formal methods for model evaluation and verification.

13 Exposure and Risk Assessment

Most of this text has focused on the processes affecting the transport and fate of chemicals in the environment. However, the attention given to pollutant concentrations in the environment is usually motivated by concern over their impacts on human health and ecosystem well-being. To cause these effects, humans or other target organisms must be exposed to the pollutants and absorb or ingest them in a manner that leads to disease or impairment. Since such exposure is rarely certain to lead to a health or ecological impact, but instead imparts a *risk* of harm, the study of chemical impacts on human health and the environment is referred to as *risk assessment*. This chapter provides an overview of methods and models for exposure and risk assessment and shows how they may be linked to models for pollutant transport and fate.

Risk and exposure assessments are conducted for many regulatory applications, such as the determination of whether a chemical or waste is "hazardous" or whether a site remediation plan or emission control program is adequate. Procedures for exposure and risk assessment have become standardized for some of these applications. Despite this standardization, exposure and risk estimates are usually highly uncertain. This chapter provides an introduction to standard approaches to exposure and risk assessment. At the same time, new and emerging methods and needs for further research are identified. The chapter also introduces physiologically based pharmacokinetic (PBPK) models that predict chemical transport and fate in the body and multimedia environmental models that incorporate modules for predicting human exposure and risk.

13.1 RISK ASSESSMENT PARADIGM

The 1983 National Research Council Report, *Risk Assessment in the Federal Government: Managing the Process* (NRC, 1983, often referred to as the Red Book, the color of its cover), had a major impact in establishing and promoting the risk assessment paradigm for regulatory decision making. While a number of the tenets of the report have been challenged in recent years (NRC, 1996a), much of the systematic approach to risk assessment put forth by the Red Book remains central to current practice. Other, more recent texts providing detailed descriptions of the current practice and state-of-the-art of risk assessment include Crawford-Brown (1997), Louvar and Louvar (1998), Kammen and Hassenzahl (1999), Paustenbach (2002), and McDaniels and Small (2004).

The major steps in risk assessment include:

1. *Hazard Identification*: Does the agent cause an adverse effect?
2. *Exposure Assessment*: What exposures are currently experienced or anticipated under different conditions?
3. *Dose–Response Assessment*: What is the relationship between dose and incidence of the adverse effect in humans?
4. *Risk Characterization*: What is the estimated incidence of the effect in a given population?

This list is slightly reordered from that presented in the Red Book, which places the dose–response assessment before the exposure assessment. This reordering is

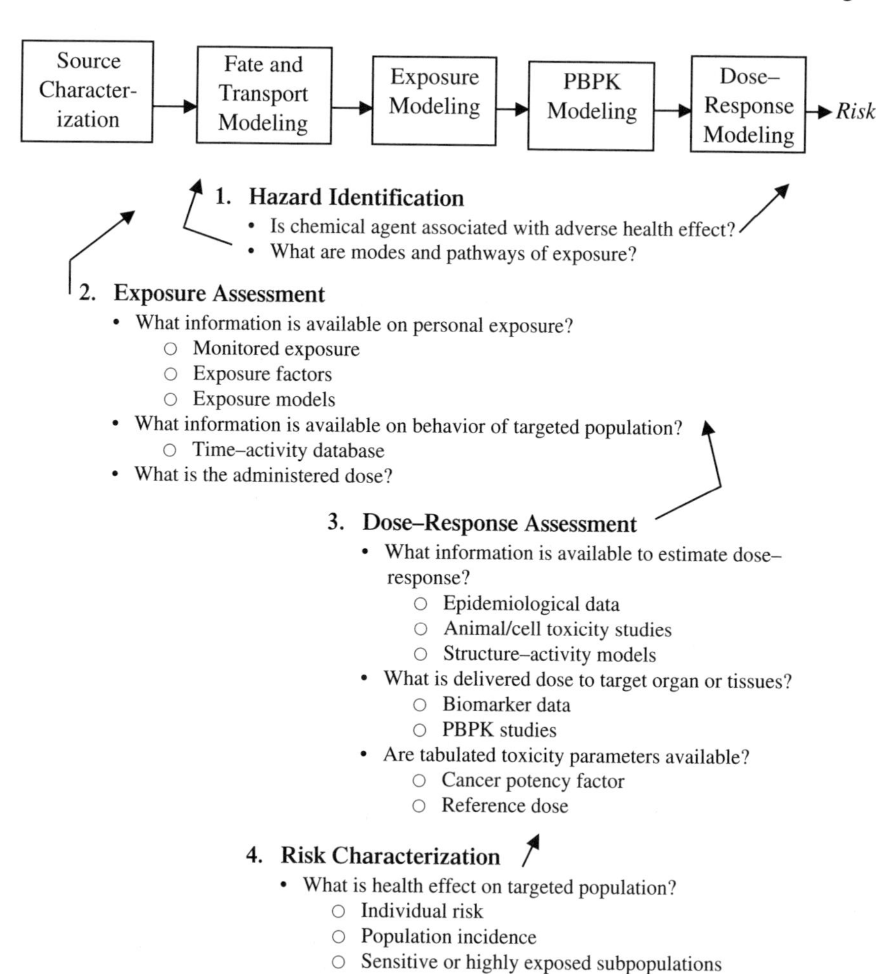

Figure 13.1 Risk assessment paradigm and the major steps in a risk assessment.

consistent with the sequential modeling of emissions, ambient concentrations, exposures, doses, and risks, first discussed in Chapter 1 and reiterated in Figure 13.1. The major steps and assumptions undertaken when performing a risk assessment are addressed below, relating these to the needs of a sequential model of the type shown in Figure 13.1.

13.2 HAZARD IDENTIFICATION

The first step in a risk assessment is the hazard identification, which asks, "What, if any, adverse effects are caused by a chemical agent? Is there a significant hazard?" The hazard identification is generally qualitative, seeking to identify the presence of health impacts, their strength or magnitude, and the weight of evidence establishing the causative link to the pollutant(s) of concern.

As discussed in Chapter 2, the major human health effects of concern for pollutants include:

- Carcinogenicity: cancer
- Mutagenicity: genetic or inherited defects
- Teratogenicity: birth defects
- Neurotoxicity: damage to the nervous system
- Cardiopulmonary impact: damage to the respiratory or circulatory system
- Systemic toxicity: damage to a particular organ or tissue (e.g., kidney, liver, skin)

For each of these categories, effects can be *chronic*, occurring over many years or a lifetime, or *acute*, resulting in short-term effects over days or weeks. Some effects are fatal, resulting in *mortality*, while others result only in *morbidity*, a nonfatal disease or illness.

The hazard identification relies on evidence from:

1. *Epidemiological studies*, showing a positive statistical association between exposure to the chemical in the target population and the incidence of the health endpoint
2. *Animal toxicity studies*, showing a positive association between exposure and incidence among laboratory test animals
3. *In vitro studies* on test cell cultures
4. *Structure–activity relationships*, which correlate toxicity to molecular structure

The evidence from these various sources is weighed and combined, identifying key assumptions and uncertainties.

As a prelude to the subsequent exposure assessment, the hazard identification also attempts to identify the likely media and pathways of exposure for chemicals from their source emissions to the targeted population. Examples of major modes and pathways of exposure include:

- The inhalation of chemicals
  - Emitted directly to the outdoor or indoor environment
  - Resuspended from soil, dust, building, or plant surfaces
  - Volatilized from a contaminated water supply during showering, bathing, or other water-use activities
- The ingestion of chemicals
  - On soils, dust, or building surfaces due to pica (hand-to-mouth) activity
  - In drinking water from the community water supply or private wells
  - In surface waters during swimming
  - In foods such as fish from contaminated surface waters, fruits, or vegetables exposed to atmospheric deposition or contaminated waters, or meat or dairy products from farm animals exposed via air, water, or food-chain pathways
- Dermal uptake of chemicals
  - In the water supply during washing or bathing
  - In surface waters during swimming or boating
  - In contaminated soils during gardening or childhood play

As indicated, a wide variety of routes and pathways can lead to human exposure, and many of these pathways are related and interact. The hazard identification attempts to identify those routes of exposure most likely to be important, prior to implementation of a more detailed exposure assessment.

13.3 EXPOSURE ASSESSMENT

The field of exposure assessment encompasses evaluation of the sources, timing, and magnitude of human exposure to chemicals in the environment, through both measurement and modeling (Ott, 1995; Lioy et al., 1998; McKone, 1999; Asante-Duah, 2002; Nieuwenhuijsen, 2003). When measurements are available, they can provide direct estimates of exposure. In many cases, however, adequate data are lacking. Exposure estimates are needed to:

1. *Reconstruct past exposures*: This may be necessary to establish a dose–response relationship for a chronic exposure in a population. It may also be necessary to establish a legal argument for probable causation for individuals with the observed health effect. If good historical data are available, they are used. However, this is rarely the case, and modeling is generally required for historical exposure reconstruction, hopefully in combination with data that can be used to calibrate or corroborate some aspects of the model (Esman and Marsh, 1996). The use of human biological markers, or *biomarkers*, is an important area of current research, using changes in the exposed individual's body tissue, cells, or DNA to infer past exposure levels (Sexton, et. al., 1995a, 2004; Mendelsohn et al., 1998; Aylward and Hays, 2002).

2. *Estimate current exposure*: A variety of new and innovative approaches are available for measuring exposure. Examples include passive monitors for airborne gases and particles (Spengler et al., 1994; Janssen et. al., 1998), intensive studies of chemical concentrations in soil, dust, water, and food (Sexton et al., 1995b; Berry, 1997; Jantunen et al., 1998), and measurement of chemical concentrations in body tissues (Sexton et al., 1995a; Pirkle et al., 1995), fluids (Polissar et al., 1990), or exhaled air (Wallace et al., 1991; Roy and Georgopoulos, 1998). Again, fully effective measurements may not be possible in all cases and some modeling is usually necessary. In particular, models for current conditions must be developed, calibrated, and verified using available data so that they can be subsequently used for prediction.

3. *Predict future exposure under alternative risk management strategies*: Since we are rarely able to test and measure the effects of proposed exposure reduction options (e.g., source removal or control) prior to implementation, we must invariably rely upon model projections to predict their effects.

13.3.1 Metrics of Exposure and Dose

A number of metrics are needed to define the relationship between ambient concentration, exposure, and dose. Consider an individual exposed to chemical A, with time varying concentration $C_A(t)$ over the time interval $0 - T$. The *exposure*, E_A, over this interval, with units of [concentration—time], for example, (mg/L) $\times$ days, is computed as:

$$E_A = \int_0^T C_A(t)\, dt \tag{13.1}$$

The *average exposure concentration* over this time interval is

$$C_{E,A} = E_A/T \tag{13.2}$$

The exposure is subsequently converted to dose D_A through consideration of the rate R of uptake of, or contact with, the exposure medium:

$$D_A = E_A R \tag{13.3}$$

For air inhalation or drinking water ingestion, R is given in units of volume/time, generally m^3/day for air inhalation and liters/day for water ingestion. For exposures to chemicals in soil or food, the units of R are mass/time, for example, grams/day of soil or food ingestion (with mass/mass concentration units used for C_A when dealing with these media). For the dermal exposure mode, R is expressed as:

$$R_{derm} = A_S P_{derm} \tag{13.4}$$

where A_S (L^2) is the surface area of the exposed skin and P_{derm} (L/T) is the dermal permeability (mass transfer) coefficient for the chemical through the skin.

The dose computed in Eq. (13.3) is referred to as a *potential*, or *administered dose*, since it is the dose of the chemical that crosses the outermost envelope of the body. However, not all of the potential dose is retained in the body and transported to target organs or tissues to become an *effective* or *delivered dose*, causing the toxicological response. A simple, "fraction-absorbed" model relating the effective and potential dose is presented in Section 13.3.4; mechanistic models for this same purpose are discussed in Section 13.6.

13.3.2 Statistical Models for Exposure from Observed Data

A number of monitoring studies of pollutants in different environmental media have been conducted over the past few decades. Examples include:

- The Total Exposure Assessment Methodology (TEAM) studies for VOCs (Wallace et al., 1986) and airborne particulate matter (Ozkaynak et al., 1996)
- The National Human Exposure Assessment Survey (NHEXAS) for multiple chemical exposures in the United States (Sexton et al., 1995b; Robertson et al., 1999)
- The Air Pollution Exposure Distributions of Adult Urban Populations in Europe (EXPOLIS) study (Jantunen et. al., 1998; Kousa et al., 2002)

With the observed chemical concentration data from this type of study, exposure estimates can be developed using the *microenvironment* approach. With this approach, the exposure for individual i is computed using a discrete sum to approximate Eq. (13.1):

$$E_i = \sum_{j=1}^{m} t_{i,j} C_{i,j} \tag{13.5}$$

where $t_{i,j}$ is the amount of time individual i spends in microenvironment j, and $C_{i,j}$ is the concentration of the chemical in microenvironment j when individual i is present. The average exposure concentration is then calculated as:

$$C_{E,i} = \sum_{j=1}^{m} f_{i,j} C_{i,j} \tag{13.6}$$

where $f_{i,j}$ is the fraction of time individual i spends in microenvironment $j(=t_{i,j}/T)$.

Typical microenvironments that are considered for computing airborne inhalation exposure include (see, e.g., Ott, 1995):

- Indoor in the home
- Indoor office/factory
- Indoor bar/restaurant
- Indoor other (shopping center, library, etc.)

- Outdoors
- In vehicle

Other microenvironments are included when they are thought to be a significant source of exposure, such as the shower or bathroom for inhalation exposure to volatile chemicals from a contaminated water supply, or the backyard for childhood exposure to contaminants in soils. In the study of Ott (1984), the indoor bar/restaurant was included because of the potential of this location to be a source of high exposure to environmental tobacco smoke (ETS). For exposure modes such as drinking water ingestion, food ingestion, or dermal contact, the exposure estimate must consider the individual's distribution of ingestion or contact amounts from different sources (e.g., their diet), and the chemical concentrations associated with each of these sources.

The exposures or average concentrations given by Eqs. (13.5) and (13.6) can be modeled across a population by simulating different individuals assumed to spend different amounts of time in contact with each of the microenvironments and by sampling different concentrations for each microenvironment contact. Concentration distributions for a microenvironment are defined across individuals as well as over time, to simulate both individual-to-individual variation in the population and day-to-day variation for a given individual.

The amount of time that individuals spend in different microenvironments is described by their *time–activity pattern*. A common method for collecting information on time–activity patterns is through telephone surveys, asking people what they did the previous day. Study participants may also be asked to keep a diary of their activities as they perform them. Important time–activity databases have been generated in this manner for California (Wiley et. al., 1991), the United States (Robinson and Thomas, 1991; Glen et al., 1997), and other countries (Freijer et al., 1998; Leech et al., 2002). These data are subject to errors due to recollection or recording error or bias; however, efforts are made to correct for the errors and validate the estimates. The U.S. EPA's Comprehensive Human Activity Database (CHAD) is one of the largest and up-to-date source of information of this type, including nearly 17,000 person-days of activity from 114 different locations in the United States (EPA, 1999).

An illustration of the time–location distribution for adults sampled in the California database is shown in Figure 13.2 (Ott, 1995). As indicated, most California adults are indoors, no matter what time of day it is. This is true of adults in most industrialized countries. However, Americans, including Californians, may be unique in how large a proportion of them are (at any time during the day) in their vehicles.

More intensive exposure studies can be performed using subjects wearing passive air pollution monitors who also keep a written or electronic diary of their locations and activities. These studies can help estimate both exposures and pollutant concentrations associated with different microenvironments. Examples of these activities include water use, smoking, and the use of household chemicals, such as cleaners, paint strippers, or insecticides (Wallace et al., 1989; Ott and Roberts, 1998; Riley et al., 2000).

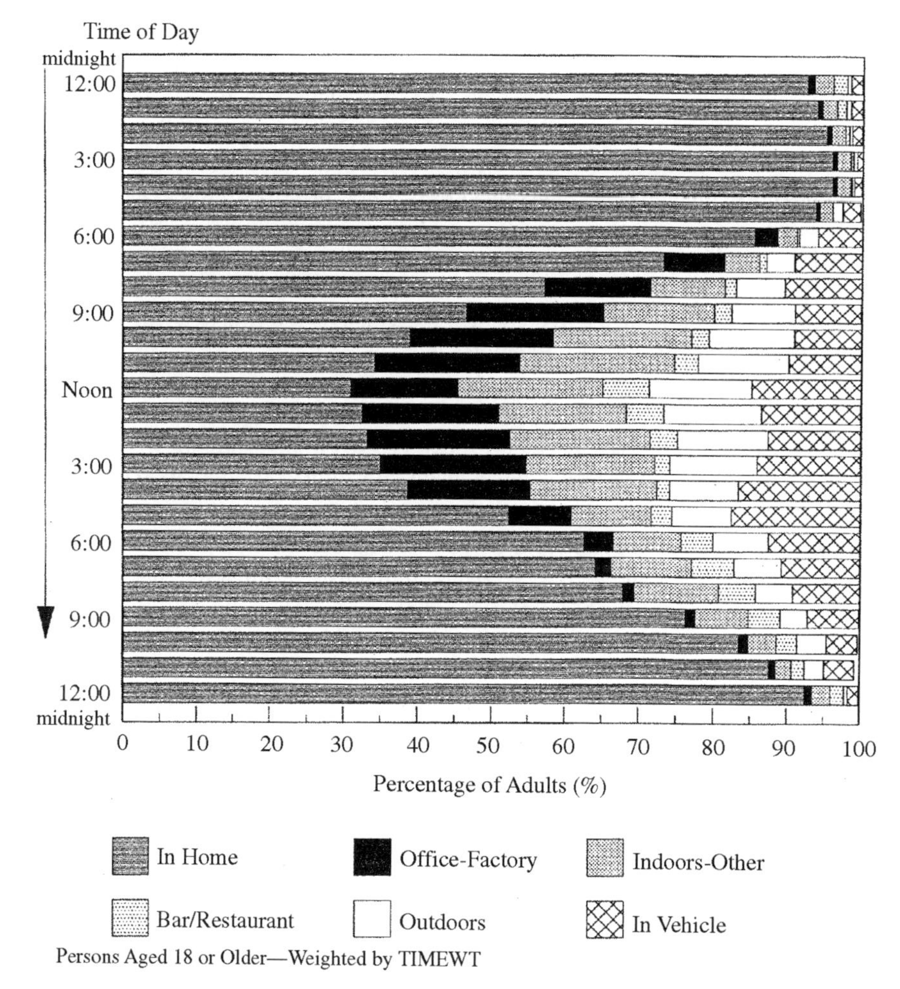

Figure 13.2 Proportion of the California adult population present in each of six location categories graphed by time of day in 1-h intervals (Ott, 1995).

13.3.3 Exposure Factors for Potential Dose

The total exposure and potential dose to an individual depend upon chemical concentrations in their contact media, uptake rates, and their frequency and duration of exposure. Combining Eqs. (13.2) to (13.6), the potential dose can be expressed as:

$$D_P = f_E T C_E R \tag{13.7}$$

where f_E is the frequency of exposure, T is the duration of exposure, and R is the uptake rate from the media with average exposure concentration C_E. For long-term exposures, T is generally measured in years, with f_E measured in days/year or events/year. Population-wide exposure and risk estimates are needed to determine the incidence of the health effect in the target population. However, there is also often a need to model *individual risk*, particularly when such an individual is thought to be representative of a highly exposed or especially vulnerable segment of the population. If it can be shown that these individuals are protected with a high degree of safety and confidence, then it is also likely that efforts to protect the overall population will be effective.

Exposure and risk assessments often consider hypothetical individuals who are described by very conservative exposure and behavioral assumptions that lead to high-end estimates of dose. Such a hypothetical individual is referred to as the *maximally exposed individual*, or the *MEI*. Examples include

- Individuals chained to the fence at the boundary of a waste disposal site all of their life, breathing air at this (or a nearby) point of maximum concentration, drinking water from a well at the boundary location, and eating food grown next to the fence
- A subsistence fisherman for a barge disposal scenario that follows the barge around, catching and eating fish that continuously swim at the maximum concentration point of the barge plume

The problem with the MEI approach is that we have no way of knowing how conservative and extreme the exposure factor assumptions are—do the assumptions result in a dose to an individual at the 99th percentile of the population, at the 99.99th percentile, or simply much higher than any dose ever experienced by anyone?

In response, exposure analysts have adopted more realistic, though still conservative, assumptions, leading to the *reasonably exposed individual (REI)*, or a *high-end exposure Estimate (HEEE)* (NRC, 1994). As described in Section 13.5, methods are still needed to determine where this individual lies on the overall exposure distribution.

A set of standard or default exposure factors has emerged for use in exposure assessment. These are presented in reports such as *Risk Assessment Guidance for Superfund* (EPA, 1989) and the *Exposure Factors Handbook* (EPA, 1997a). Early versions of the *Handbook* presented single-point estimates for exposure factors; more recent issues identify distributions that describe variability across the population and the uncertainty of these estimates. Typical exposure factors are presented in Table 13.1.

A number of the values presented in Table 13.1 and similar references are based upon limited or outdated surveys, and many are highly uncertain. This can result in significant controversy, for example, in the selection of soil ingestion rates for children (Binder et al., 1991; Thompson and Burmaster, 1991; Calabrese and Stanek, 1995, 1996; Calabrese et al., 1997; Stanek et al., 1998). Site-specific data should be collected and used where feasible, for both exposure factors and exposure concentrations. Special attention should be given to vulnerable or highly exposed subgroups. Such data are often difficult to obtain but are essential as a basis for model estimation

TABLE 13.1 Exposure Factors Commonly Used to Compute Potential Dose

Residential Exposures

$T = 70$ years for lifetime chronic exposure
30 years for upper 90th percentile at one residence
9 years for 50th percentile at one residence

Inhalation Exposure

$R = 20$ m^3/day, average value for adult; 30 m^3/day upper bound
(higher during outdoor work, exercise, etc.)
$f_E = 365.25$ days/year for ambient inhalation
(corrected as needed for fraction of time spent outdoors or at the location of exposure, e.g., for exposure in vehicles, showers, or workroom)

Ingestion of Drinking Water

$R = 1.4$ L/day for average adult; 2 L/day upper 90th percentile
$f_E = 365.25$ days/yr (corrected as needed for use of other sources of drinking water)

Ingestion while Swimming

$R = 0.05$ L/h
$f_E = 7$ days/yr national average $\times$ 2.6 h swimming/day
(each depends on climate and age)

Dermal Uptake while Swimming, Bathing or Showering [see Eq. (13.4)]

$R = A_S \times 8.4 \times 10^{-4}$ cm/h (for water)
where $A_S = 0.728$ m^2 for male child ages 3–6; 0.711 m^2 for female child ages 3–6
$= 1.16$ m^2 for male or female ages 9–12
$= 1.94$ m^2 for adult male; 1.69 m^2 for adult female
$f_E = 1$ shower/day; see above for swimming;
reported shower duration of 7 min (median) to 12 min (90th percentile)

Soil Ingestion

$R = 200$ mg/day for children 1–6 years old
$= 100$ mg/day for adults and children 7 years and older
$f_E = 365.25$ day/yr (corrected as needed for days spent away from contaminated soil)

Food Ingestion

R given in kg/meal or kg/day for different vegetables, milk, eggs, meat, and fish
f_E given in meals/year or days/year using the given source for each type of food

Sources: EPA (1997a), Berry (1997), and Louvar and Louvar (1998).

and validation. The selection of the level of individual to protect is a risk management one (e.g., should extreme behavior be considered, such as "soil eating" among children?), but the best scientific characterization is still needed to describe just where these protected individuals lie on the variability and uncertainty distributions for the targeted population.

EXAMPLE 13.1 COMBINING TIME–ACTIVITY PATTERNS WITH POLLUTANT FATE AND TRANSPORT MODELS

To illustrate the linkage of exposure assessment to environmental models, consider the Model for Analysis of Volatiles and Residential Indoor-Air Quality (MAVRIQ) developed by Wilkes et al. (1992). MAVRIQ simulates chemical volatilization from a domestic water supply from various water-use activities and calculates the resulting indoor air concentrations and inhalation exposure. To demonstrate the model, Wilkes et al. assumed a three-person household with two adults and one young (~6-month-old) child living in the one-bedroom apartment illustrated in Figure 13.3. The water supply was assumed to be contaminated with trichloroethylene (TCE) at a concentration of 20 mg/m^3.

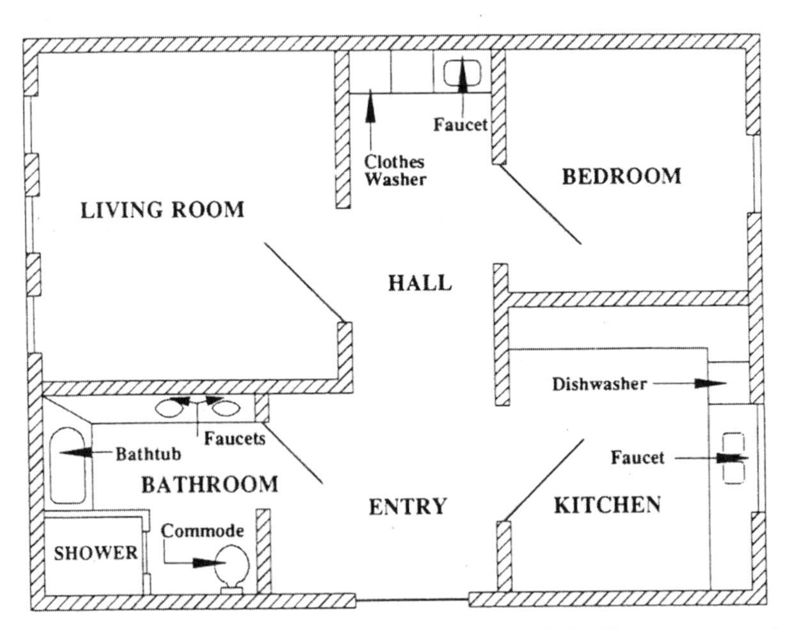

Figure 13.3 Modified five-room test house (after Axley, 1988). The water sources have been added for this application (Wilkes et al., 1992).

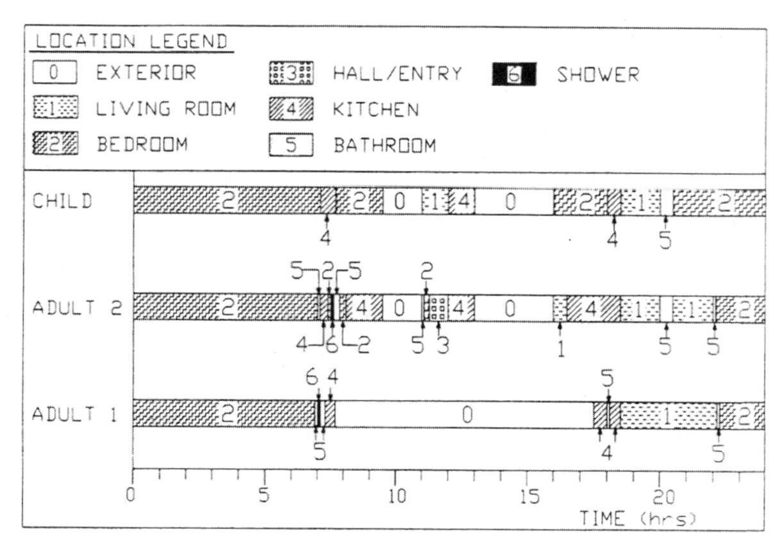

Figure 13.4 Assumed daily household activity patterns for the three occupants (Wilkes et al., 1992).

In the model volatilization occurs to different extents from the shower, bath, toilet, dishwasher, clothes washer, and kitchen, bathroom, and laundry faucets, and the airflow rates between rooms are defined with and without the bathroom door open [see Table 1 and Fig. 3 of Wilkes et al. (1992)].[1] A daily time–activity profile is assumed for each member of the household, as shown in Figure 13.4, with water use occurring during a number of these activities. Adult 1 works outside of the home and is away between 7:45 a.m. and 5:30 p.m. Adult 2 has primary child care responsibility and performs associated tasks as well as other household work (though adult 1 does stay with the child in the kitchen for breakfast while adult 2 showers). Adult 2 and the child are at home for the majority of the day.

The model simulates TCE emissions and computes the resulting indoor concentrations for each room of the apartment, as shown in Figure 13.5. Combining these with the time–location patterns in Figure 13.4 and assuming a 20-m^3/day breathing rate for the adults and 2 m^3/day for the child, yields the cumulative exposure and potential inhalation doses shown in Figure 13.6. For reference, the potential dose associated with drinking 1.4 L of the contaminated water (at 20 mg/m^{-3}) is 0.028 mg, slightly less than the daily inhalation dose received by adult 1, and less than half of that received by adult 2. Note that even adult 2's shower, ostensibly identical to that of adult 1, results in more exposure than adult 1's since adult 1 showers first, and as a result, TCE concentrations in the bathroom and shower stall are already elevated by the time adult 2's shower begins.

[1] See Howard-Reed et al. (1999) and Moya et al. (1999) for more recent estimates of chemical volatilization rates from household water uses.

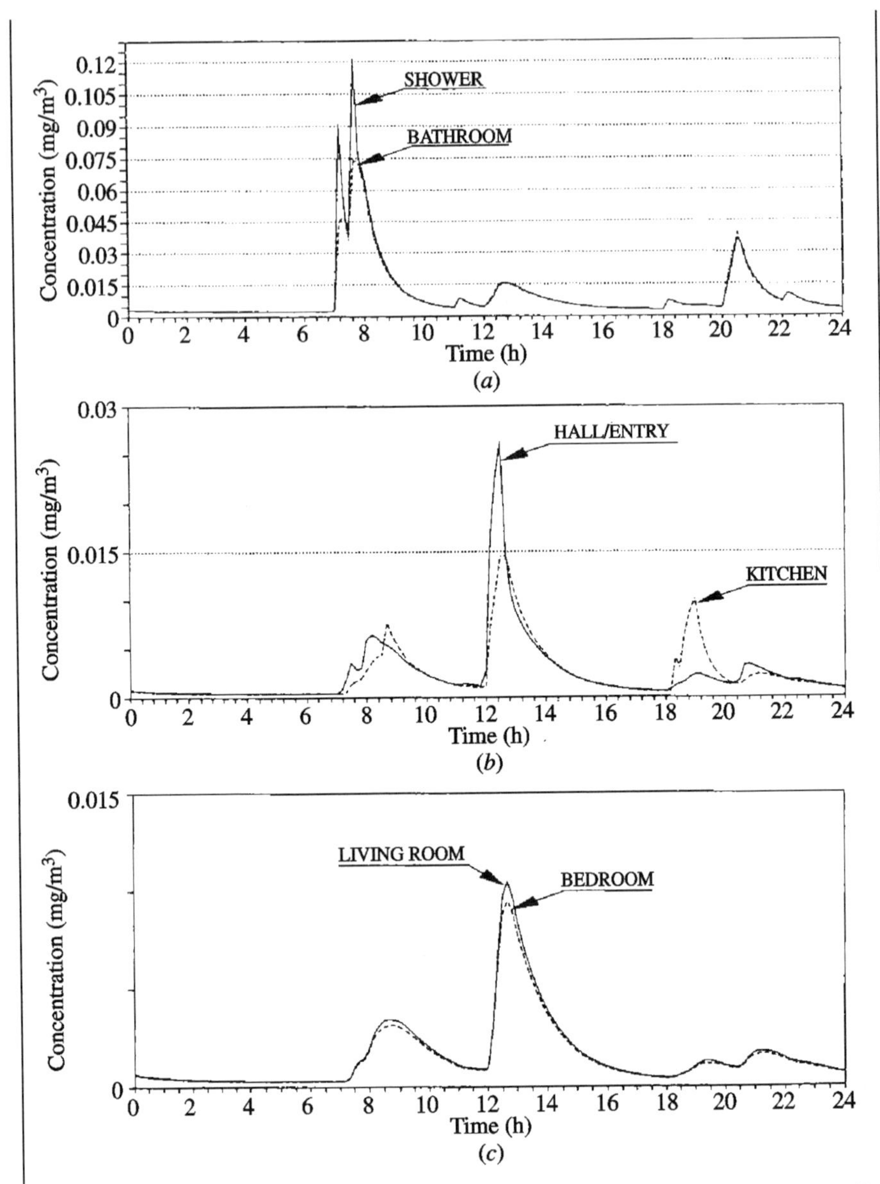

Figure 13.5 Modeled indoor air concentrations of TCE: (*a*) Bathroom and shower, (*b*) kitchen and hall/entry, and (*c*) living room and bedroom. Note the different scales of each. (Wilkes et al., 1992).

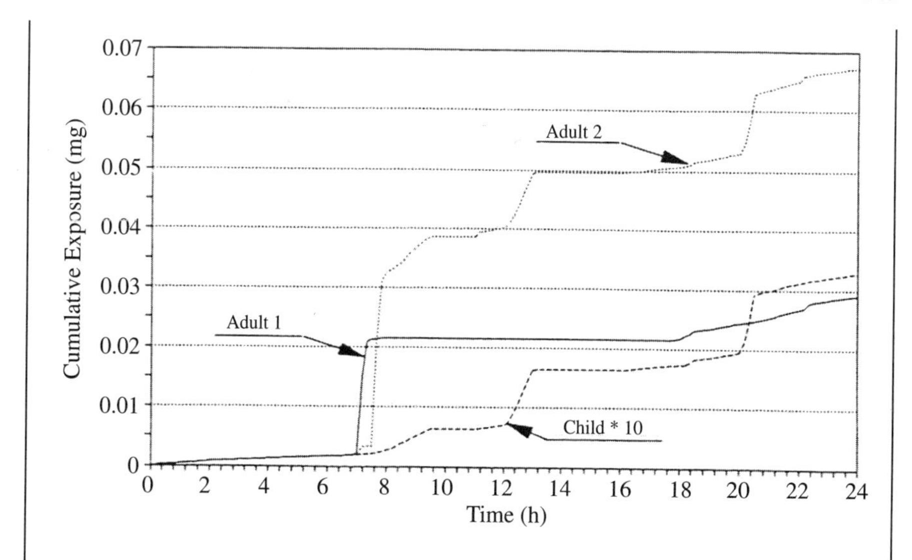

Figure 13.6 Modeled TCE inhalation exposures (= potential inhalation dose) of the three occupants (Wilkes et al., 1992).

What can be learned from this example?

- Inhalation exposures from VOCs, such as TCE, resulting from household use of contaminated waters can be as high, or higher, than those resulting from ingestion of the same water. Public health and regulatory officials now recognize this in the guidance they give during water contamination events—no longer do they routinely advise, "Don't drink the water, but other domestic uses are OK." Furthermore, inhalation exposure from volatilization of contaminated household water is now generally incorporated in risk assessments for site cleanup (e.g., Labieniec et al., 1996) and those aimed at determining maximum allowable contaminant levels (MCLs) for chemicals in the water supply (ATSDR, 1997; Hopke et al., 2000; Williams et al., 2002).
- Differences in time activity patterns can significantly affect exposure.
- Take the first shower!

The example also illustrates how human behavior, time–activity patterns, and exposure factors can be linked with environmental models in a systematic manner to predict exposure and potential dose.

13.3.4 Effective Dose

Once a chemical enters the body, there are a number of physiological factors that determine how much of the chemical reaches target organs or tissues for toxicological effect. As a first approximation, a *fraction absorbed* or retained by the body, f_a, can be defined to distinguish between that portion of the potential dose that remains in the body for long-term effect and that which is excreted over a relatively short time period (e.g., through the breath or urine). For the subsequent toxicological calculation, it is customary to express the effective dose, D_e, on a daily basis, relative to the body weight of the individual:

$$D_e = \frac{D_p f_a}{\text{BW} \times T} \tag{13.8}$$

where BW is the body weight, typically assigned the value of 70 kg (=154 lb) for a "representative adult" (this value is actually intended to represent a male's average weight over his lifetime) and T is once again the exposure duration, assumed to be 70 years for a chronic lifetime exposure or less as appropriate for acute or "subchronic" exposures.

The effective dose can also be expressed as the product of the exposure E [typically in (mg/m^3) × days], and the dosimetry factors, which include the exposure factors and the relative dose factors (fraction absorbed f_a, and body weight BW):

$$D_e = E\left[(\text{mg/m}^3) \times \text{days}\right] \times \left[\frac{R(\text{m}^3/\text{day}) \times f_a[\,]}{\text{BW(kg)} \times T(\text{days})}\right] \tag{13.9}$$

The resulting units of D_e are mg/(kg · day). The effective dose is computed as a *lifetime average daily dose*, or *LADD*, for chronic effects such as cancer (by setting $T = 70$ years = 25,567.5 days). A *maximum daily dose*, or *MDD*, is used for evaluating acute toxicity (with $T = 1$ day). A conservative estimate of the fraction absorbed, $f_a = 1.0$, is often used in standard risk assessment. More rigorous estimates of f_a can be obtained from PBPK models, as described in Section 13.6.

EXAMPLE 13.2 EXPOSURE AND DOSE IN A MULTIMEDIA UNIT WORLD

In Example 3.9 the steady-state concentrations of benzene in multiple environmental compartments are computed using a level II model, assuming a constant loading rate to the environment of $W_{\text{total}}^{\text{benzene}} = 1000$ kg/hr. What is the total exposure and LADD of benzene for a "unit person" in this unit world? Assume an exposure duration of $T = 70$ years and a body weight of BW = 70 kg. Consider the following pathways and exposure factors:

- Inhalation exposure to the air with:
  $R = 20$ m^3/day
  $f_E = 365.25$ days/yr
- Ingestion of water with
  $R = 1.4$ L/day
  $f_E = 365.25$ days/yr

- Ingestion of soil with
  $R = 100$ mg/day
  $f_E = 365.25$ days/yr
- Consumption of fish with
  $R = 0.2$ kg/meal
  $f_E = 180$ meals/yr

The computed exposures and doses are shown in Table 13.2. Columns 2 and 3 of the table show the steady-state concentrations and distributions of the emitted benzene in each of the media, as computed by the level II model in Example 3.9. (Note that the mass distribution percentages do not sum to 100 since a small amount of the benzene was computed to reside in the water sediments, which are not included in the exposure calculation—the individual is assumed to drink filtered water.)

To illustrate the dose calculations, the potential dose for air inhalation is computed as:

$$D_p = f_E T C_E R$$
$$= 365.25 \text{ day/yr} \times 70 \text{ years} \times (0.197\ \mu\text{g/m}^3 \times 1 \text{ mg}/1000\ \mu\text{g}) \times 20 \text{ m}^3\text{/day}$$
$$= 100.7 \text{ mg}$$

and the effective dose (LADD) is

$$\text{LADD } (= D_e) = \frac{D_p f_a}{\text{BW} \times T}$$
$$= \frac{100.7 \text{ mg} \times 1.0}{70 \text{ kg} \times (70 \text{ years} \times 365.25 \text{ days/yr})}$$
$$= 5.63 \times 10^{-5} \text{ mg/kg} \cdot \text{day}$$

The fraction absorbed for all exposures and doses is assumed equal to 1.0.

TABLE 13.2 Benzene Exposure and Dose in the Level II Unit World

Exposure Media	Level II Results: Concentration	Level II Results: Mass Distribution (%)	Potential Dose [Eq. (13.7)] (mg)	Effective Dose (LADD) [Eq. (13.8) or (13.9)] (mg/kg day)	Dose Distribution (%)
Air	$0.197\ \mu\text{g/m}^3$	99.0	100.7	5.63×10^{-5}	99.95
Water	$0.877\ \mu\text{g/m}^3$	0.882	0.0314	1.76×10^{-8}	0.031
Soil	$9.7 \times 10^{-7}\ \mu\text{g/g}$	0.105	2.48×10^{-6}	1.39×10^{-12}	~0
Fish	$5.9 \times 10^{-6}\ \mu\text{g/g}$	5.95×10^{-6}	0.0149	8.33×10^{-9}	0.015
TOTAL	—	~100	~100.746	5.633×10^{-5}	100

As indicated in Table 13.2, the benzene dose is dominated by the air inhalation pathway. While the concentration in the water is somewhat higher than that in the air, the significantly larger uptake rate for air inhalation versus drinking water ingestion results in a much greater potential dose through the air. Since the fraction absorbed is assumed to be the same (=1.0) for all modes of exposure (which, in reality, may not be the case), the relative distribution of the effective dose between different pathways is the same as the relative distribution of the potential dose. In this example, the dose distribution percentages follow a similar pattern to the partition percentages of benzene in the unit environment, except that fish consumption results in a greater dose than soil consumption, even though considerably less of the benzene is distributed in the fish as compared to the soil. This occurs because our assumed unit person's fish consumption is much higher than his or her soil ingestion, and the benzene concentration in the fish is also somewhat higher than that in the soil.

This example is highly idealized, with each environmental media represented by a single, fully mixed compartment, each exposure assumed to derive directly from contact with the respective environmental media (i.e., exposure concentrations assumed equal to ambient concentrations), and the use of standard, conservative exposure factors. However, simple applications of this type can provide a number of useful insights. With the addition of similarly idealized risk estimates (demonstrated in the following section), unit-world concentration and exposure estimates can provide a first basis for screening chemicals based on their likely extent of transport and persistence, and the resulting potential for human exposure and risk (e.g., Eisenberg and McKone, 1998; Hertwich et al., 1998; Rodan et al., 1999).

13.4 DOSE–RESPONSE ASSESSMENT

Dose–response relationships are used to compute the risk or incidence of adverse health impacts for a given dose. The development of these relationships is the main goal of health scientists such as epidemiologists and toxicologists, though they generally go about their work in very different ways. Dose–response relationships can be derived from human health studies conducted during or following a known historic exposure. This exposure may have occurred as the result of an accident, or in an occupational setting, often prior to recognition of the chemical's harmful effects. For some constituents that occur naturally or commonly in the environment, a considerable database of human exposure and recorded health effects may be available for study, for example, for arsenic in drinking water (Chiou et al., 1995; Andrew et al., 2003; Yu et al., 2003) or ambient PM (Daniels et al., 2000; Pope et al., 2002).

While historic exposure and health effects data are available and acceptable for use[2] for some problems, for most chemical exposures, reliable data of this type are

[2] In certain (fortunately rare) cases, harmful exposures have been forced upon human subjects, such as prisoners or concentration camp inmates during World War II, unknowingly or without their consent. While

unavailable. In many cases, the data were collected for conditions with doses much higher than those applicable to current environmental exposures. Examples include workers involved in the direct manufacture, use, or mining of a product, such as chemical workers exposed to benzene (Brief et al., 1980; White et al., 1982; Crump, 1996), uranium mine workers exposed to high levels of radon (NRC, 1988; Brill et al., 1994), workers exposed to beryllium (Kolanz et al., 2001), and workers at chromate production facilities exposed to hexavalent chromium (Crump et al., 2003). Similarly, exposure and health outcome data are available for a few major chemical spill or release incidents, such as the release of large quantities of mercury to Minamata Bay, Japan, and spills in Iraq (Ellis, 1989); the 1984 disaster involving the release of methyl isocyanate following an explosion at the Union Carbide pesticide plant in Bhopal, India (Andersson et al., 1988; Cullinan et al., 1997); and more localized, acute affects from smaller spills of toxic chemicals (Rubin, 1988). For certain, less severe effects, including respiratory irritation from pollutants such as ozone, or to study uptake rates of compounds such as nickel, perchlorate, or household chemicals or pesticides, clinical toxicology studies have been conducted with volunteer subjects (Sunderman et al., 1989; Gong et al., 1998; Greer et al., 2002), although once again, important safety and ethical issues arise, and the conduct of these studies must be subject to strict evaluation by human subject institutional review boards (Dourson et al., 2001; Weinhold, 2001).

Even if a statistically significant relationship can be identified for highly exposed populations, how can this relationship be extrapolated to the lower exposures and doses usually experienced by the general population? Epidemiologists may attempt to address this question by studying the effects of exposures across a wider set of the population, controlling for other factors that correlate with the health endpoint. Often, however, background rates of the disease are such that, with the low incremental risks that are expected, it is hard to separate the epidemiological signal from the noise.

Toxicologists study the physical mechanisms of health effects in humans but are usually limited in their investigations to laboratory animal studies, most commonly with rats or mice, though sometimes with dogs, monkeys, or other primates.[3] These studies are conducted at high dosage rates since a high dose is usually needed to yield effects in a high enough proportion of the test animals to achieve statistical

---

all agree that this is immoral and criminal, an ethical dilemma does arise when experiments of this type have been performed in the past and the perpetrators brought to justice, but the results are now available for use by health scientists. Some argue that such data, though immorally obtained, can and should be put to good use to help save lives today; others argue that no good use should come of past acts of this nature (Beecher, 1966; Moe, 1984; Cohen, 1990). The U.S. EPA did rule in 1988 that data obtained during human subject studies by Nazi physicians during World War II could not be used as part of a health risk assessment (Sun, 1988).

[3] Animal toxicology is also faced with ethical issues related to the extent to which test animals can be subjected to captivity and harmful exposures and effects to provide information that may help save human lives (NRC, 1996b; Barnard and Kaufman, 1997; Botting and Morrison, 1997; Mukerjee, 1997). Animal studies remain an important component of modern toxicology, though efforts are made to avoid poor conditions and treatment of the test animals and to limit studies to those demonstrated to be effective and essential for advancing the state of knowledge (Festing et al. 1998).

significance, while limiting the study to a feasible sample size. Extrapolations are thus once again needed from high dose to low dose.

Both high-dose human health data and laboratory animal results can be represented with a variety of different mechanistic or empirical functions, and a number of different statistical procedures can be used to fit these relationships (Crump, 1981; Anderson et al., 1993; Kammen and Hassenzahl, 1999). These different methods can yield large differences when the results are extrapolated back to the low-dose region of concern.

When toxicity factors are based on animal studies, another extrapolation is needed to predict the human health effect. Different toxicological and pharmacokinetic responses in animals and humans lead to great uncertainty in the animal-to-human extrapolation (Crouch and Wilson, 1978; Gold et al., 1992; Infante, 1993; Kuo et al., 2002; Munns and MacPhail, 2002). When this is combined with the uncertainty in the high- to low-dose extrapolation, the overall procedure for estimating human health toxicity factors from animal studies results in very high, often order-of-magnitude, uncertainties in the dose–response assessment, and some question the most fundamental assumptions of the approach (Ames et al., 1996). Health scientists now increasingly attempt to combine the evidence from animal toxicology and human epidemiological studies to provide the most informed estimate of dose–response relationships.

The study of human health and toxicology is a broad discipline with a long history and rapid, ongoing progress. The following sections attempt only to provide a brief introduction to how these methods are used as part of an integrated exposure, dose, and risk assessment.

13.4.1 Dose–Response Relationship

Examples of alternative dose–response relationships for human health risk are shown in Figure 13.7. The dashed curve is the most general case of a dose–response relationship, exhibiting nonlinearity and a threshold below which no effects occur. Often, however, a linear, no-threshold relationship is assumed (the solid line in Fig. 13.7), as is the standard procedure for cancer risk assessment. The linear, no-threshold dose–response relationship is fully specified by its *slope*, or *potency factor* β, with units of 1/dose, usually $(\text{mg/kg} \cdot \text{day})^{-1}$. The higher the slope or potency factor for a chemical, the greater its toxicity.

When a linear, no-threshold toxicity function is applied to estimate chronic (lifetime) risk, the lifetime individual risk (LIR) is computed from the lifetime average daily dose as:

$$\begin{aligned} \text{LIR} &= \beta \times \text{LADD} \\ (\,) &= \left((\text{mg/kg} \cdot \text{day})^{-1}\right) \times (\text{mg/kg} \cdot \text{day}) \end{aligned} \tag{13.10}$$

This is the probability that an individual will contract the disease at some time during his or her lifetime. Equation 13.10 is clearly only approximate since high values of

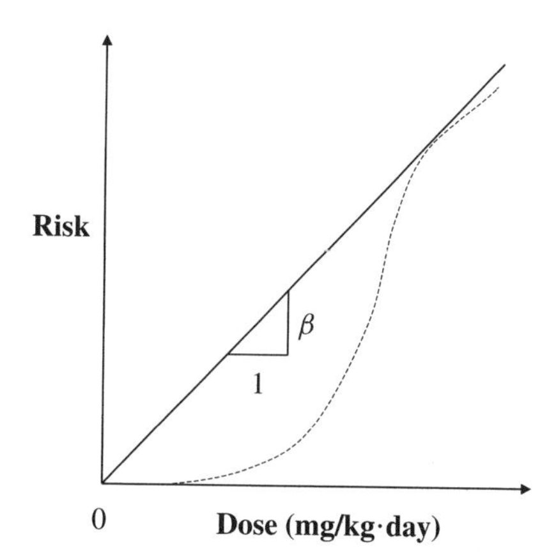

Figure 13.7 Alternative dose–response relationships for human health risk: dashed line is general nonlinear relationship with threshold and solid line is linear no-threshold relationship.

LADD will eventually cause the LIR to approach and exceed 1.0—since the LIR is a probability, this is impossible. The linear dose–response equation is only valid when the computed LIR is below 0.01.

For higher risks, a more fundamental risk–probability relationship must be used, such as the "one-hit" model equation (Louvar and Louvar, 1998):

$$\text{LIR} = 1 - \exp(-\beta \times \text{LADD}) \tag{13.11}$$

The one-hit model derives from a probabilistic representation in which the probability of a damaging transition, or a "hit," at a target cell site is proportional to the dose. "Multihit" models can also be employed, assuming that multiple cell transitions are necessary for the onset of the disease (Crawford-Brown, 1997). Equations (13.10) and (13.11) are equivalent for the very low values of LADD and computed LIR common in environmental risk assessment.

Epidemiologists may also express the dose–response relationship in terms of the *relative risk*, RR, relative to the background incidence rate of the disease in the population (Aunan, 1996):

$$\text{RR} = \exp(\beta \times \text{dose}) \tag{13.12}$$

As the dose goes to zero, the RR approaches 1.0. This dose–response relationship has no-threshold, and, like Eq. (13.10), is usually only valid in the low-dose, linear range.

13.4.2 Cancer Slope Factors

Despite the uncertainties in human and animal health studies, consensus values of cancer slope factors are available. These values are based on the deliberations of expert panels reviewing the available studies and overall weight of evidence. Upper 95% confidence level estimates are often used for slope factors to provide a degree

TABLE 13.3 Cancer Slope Factors for Selected Chemicals

		Slope Factor $(mg/kg \cdot day)^{-1}$		
Chemical	CAS#	Dermal	Inhalation	Oral
Benzene	71432	2.99×10^{-2}	2.90×10^{-2}	2.90×10^{-2}
Benzo[*a*]pyrene	50328	2.35×10^{1}	3.10	7.30
Chlordane	57749	7.00×10^{-1}	1.30	3.50×10^{-1}
Chloroform	67663	3.05×10^{-2}	8.10×10^{-2}	6.10×10^{-3}
DDT	50293	4.86×10^{-1}	3.40×10^{-1}	3.40×10^{-1}
Dichloroethylene, 1,1-	75354	6.00×10^{-1}	1.20	6.00×10^{-1}
Methylene chloride	75092	7.89×10^{-3}	1.65×10^{-3}	7.50×10^{-3}
TCDD, 2,3,7,8-	1746016	3.00×10^{5}	1.50×10^{5}	1.50×10^{5}
Trichloroethylene	79016	7.33×10^{-2}	6.00×10^{-3}	1.10×10^{-2}
Vinyl chloride	75014	1.90	3.00×10^{-1}	1.90

Source: DOE Risk Assessment Information System [updated October 1999]: http://risk.lsd.ornl.gov/tox/tox_values.shtml.

of conservatism in the evaluation. Examples include the EPA Cancer Assessment Group (CAG) and Human Health Assessment Group (HHAG) slope factor estimates. These are tabulated in a variety of toxicity databases, including the EPA Integrated Risk Information System (IRIS) database (EPA, 1997b); the EPA's Health Effects Assessment Summary Tables (HEAST) (EPA, 1995), and the International Agency for Research on Cancer (IARC) estimates (IARC, 1987). The IRIS system provides online documentation of chemicals listed alphabetically and by CASRN (Chemical Abstracts Service Registry Number), summaries of methods for estimating slope and other toxicity factors; and background documents and references that provide support for the estimates (see http://www.epa.gov/iriswebp/iris/index.html). The DOE Risk Assessment Information System provides an online summary of toxicity factors from a number of sources and guidance on the use of these factors for risk assessment (see http://risk.lsd.ornl.gov/rap_hp.shtml).

A summary of cancer slope factors for selected chemicals listed in the DOE Risk Assessment Information System is provided in Table 13.3.[4] Since different absorption fractions and mechanisms of toxicity can apply to chemical doses through the inhalation and oral pathways, slope factors determined from studies using the potential (administered) dose metric are listed separately for these pathways. For chemicals for which the inhalation and oral slope factors differ, exposures via these different routes should be tracked separately. Dermal exposures are usually converted to an equivalent oral dose using the appropriate value of the fraction absorbed for the dermal pathway. Separate slope factors are needed for estimates of skin cancer occurring via the dermal exposure pathway.

[4] Note that the values in the DOE database, as well as the EPA IRIS database, are subject to periodic update. So, for example, the latest value listed for the inhalation slope factor for benzene (accessed on January 1, 2004) is 2.73×10^{2} $(mg/kg \cdot day)^{-1}$, rather than the value 2.90×10^{2} $(mg/kg \cdot day)^{-1}$ listed in Table 13.3.

EXAMPLE 13.3 CANCER RISK FROM BENZENE EXPOSURE IN THE MULTIMEDIA UNIT WORLD

In Example 13.2, a LADD of 5.633×10^{-5} mg/kg · day is computed for benzene exposure for a unit person in a steady-state (level II) unit world. What is the resulting cancer risk?

As indicated in Table 13.3, the cancer slope factor for benzene is 0.029 (mg/kg · day)$^{-1}$ for both the inhalation and oral pathways (the only ones considered in this example). The resulting lifetime individual risk can thus be computed as:

$$\text{LIR} = 0.029(\text{mg/kg} \cdot \text{day})^{-1} \times 5.633 \times 10^{-5}\ \text{mg/kg} \cdot \text{day} = 1.6 \times 10^{-6}$$

This estimated risk, slightly greater than one in a million, is just at the level at which regulatory concern begins to occur. Depending on the regulatory program and the type and source of exposure (e.g., natural vs. anthropogenic, voluntary vs. involuntary, occupational vs. residential), risk management efforts are usually triggered at individual risks between 10^{-6} and 10^{-3}. Thus, while the computed risk is well below the background rate for cancer (which in the United States is about 20%) and undoubtedly undetectable in any epidemiological study, it may trigger regulatory efforts to reduce emissions and/or exposure.

Such a conservative approach to risk-based regulation for cancer is usually justified by the high degree of uncertainty in the overall exposure and risk assessment methodology and by the desire to have a high level of safety assurance in the protection of public health from controllable (especially involuntary) exposures. It also recognizes that regulations based on health risks estimated for single chemicals must consider the cumulative effects of exposures to multiple chemicals from the regulated source, as well as present in the general environment (Wilkinson et al., 2000; EPA, 2003).

13.4.3 Noncarcinogenic Effects Models

For most health effects other than cancer, the *reference dose* (RfD) method is used. An RfD is that level below which no health effects are expected. This value corresponds to the threshold value in the dashed curve in Figure 13.7, with safety factors added to provide a conservative level of protection. Reference doses may be determined for chronic effects (between 7 years and a lifetime) or subchronic effects (between 2 weeks and 7 years). Reference doses have been developed for various health endpoints, including neurological and neurobehavioral effects, liver damage, and reproductive effects.

The reference dose is estimated based on human data when possible, though often only animal studies are available. The RfD is computed from the no-observed-adverse-effect-level (NOAEL) with appropriate uncertainty and modifying factors:

$$\text{RfD} = \frac{\text{NOAEL}}{\text{UF}_1 \times \text{UF}_2 \times \cdots \times \text{MF}} \tag{13.13}$$

Uncertainty factors (UF) are used to account for different sensitivities in target populations (when human health studies are used), extrapolation from animals to humans (when laboratory studies form the basis for the NOAEL), or when different administration rates or dose delivery periods (e.g., subchronic vs. chronic) are used. Values of 10 are typically assumed for each of the UF_i's. Since laboratory studies must be conducted at fixed values of the dose and researchers have no way of knowing the threshold a priori, the RfD is often computed from the lowest-observed-adverse-effect level, or LOAEL—that value of the dose at which the toxic effect is first observed in the laboratory studies. When the RfD is computed from the LOAEL instead of the NOAEL, an additional uncertainty factor (again, usually assumed to equal 10) is used in Eq. (13.13).[5] The modifying factor (MF) in Eq. (13.13) is the final safety factor to account for any additional uncertainty based on professional judgment, and is usually between 1 and 10 (Louvar and Louvar, 1998).

The risk from noncancer exposures is expressed in terms of the ratio of the effective dose to the reference dose, known as the *hazard quotient*, HQ:

$$\text{HQ} = \frac{D_e}{\text{RfD}} \tag{13.14}$$

When the hazard quotient is below 1, the exposure is considered nontoxic; an HQ value greater than 1 implies a risk of harmful effects. While the health risk is believed to be proportional to the HQ, the hazard quotient is not used to compute individual risk or the incidence of the disease in an exposed population, rather it provides only a binary indication of health risk. The hazard quotient for subchronic toxicity effects is usually computed using the maximum daily dose, MDD, for the effective dose in Eq. (13.14). When multiple chemicals affect the same target organ or target organ systems with the same mode of action, the cumulative effect of exposure to these multiple chemicals is sometimes represented by a *hazard index, HI*, which is equal to the sum of the individual chemical hazard quotients (Chen et al., 2001):

$$\text{HI}_m = \sum_k \text{HQ}_{k,m} \tag{13.15}$$

where k indexes the compounds that affect target organ m, with the same mode of action. This assumes that the effects of the individual chemicals in the mixture on the target organ are independent and additive, without synergistic or antagonistic interactions (Mumtaz et al., 1997; Hertzberg and MacDonell, 2002).

[5] In order to avoid the uncertainty that arises in determining NOAELs and LOAELs for reference doses, an alternative approach to determining thresholds for noncancer effects has been developed involving the estimation of a *benchmark dose,* or BMD. The BMD is determined from a fitted dose–response curve, and is the dose that results in a fixed amount of predicted increase in the health endpoint (e.g., 10%) beyond background levels (Crump, 1984; Faustman and Bartell, 1997).

Reference doses, like cancer slope factors, are found in various EPA, DOE, and other health or environmental organization databases such as IRIS, HEAST, and the DOE Risk Assessment Information System (BMDs are not yet commonly reported). Selected chronic and subchronic RfDs are listed in Table 13.4. As noted by Goldhaber and Chessin (1997), differences between toxicity factors given by different agencies may exist due to the use of different studies, uncertainty, and modifying factors, though the differences are usually within an order of magnitude. Also, ongoing research and new data may result in frequent revisions, and many reference doses, slope factors, and related health benchmarks remain controversial (e.g., De Rosa, 1998; Renner, 1999).

TABLE 13.4 Chronic and Subchronic Reference Doses (mg/kg · day) for Selected Chemicals

		Dermal		Inhalation		Oral	
Chemical	CAS#	Chronic	Subchronic	Chronic	Subchronic	Chronic	Subchronic
Chloroform	67663	2×10^{-3}	2×10^{-3}			1×10^{-2}	1×10^{-2}
DDT	50293	3.5×10^{-4}	3.5×10^{-4}			5×10^{-4}	5×10^{-4}
Dichloroethylene, 1,1-	75354	9×10^{-3}	9×10^{-3}			9×10^{-3}	9×10^{-3}
Trichloroethylene	79016	4.5×10^{-5}		1.14×10^{-2}		3×10^{-4}	
Methylene chloride	75092	5.7×10^{-2}	5.7×10^{-2}	8.57×10^{-1}	8.6×10^{-1}	6×10^{-2}	6×10^{-2}
Arsenic (inorganic)	7440382	1.23×10^{-4}	1.23×10^{-4}			3×10^{-4}	3×10^{-4}
Barium	7440393	4.9×10^{-3}	4.9×10^{-3}	1.43×10^{-4}	1.4×10^{-3}	7×10^{-2}	7×10^{-2}
Beryllium and Compounds	7440417	2×10^{-5}	5×10^{-5}	5.71×10^{-6}		2×10^{-3}	5×10^{-3}
Chromium (III) (insoluble salts)	16065831	7.5×10^{-3}	5×10^{-3}			1.5	1.0
Cyanide (CN−)	57125	3.4×10^{-3}	3.4×10^{-3}			2×10^{-2}	2×10^{-2}
Hydrogen sulfide	7783064	6×10^{-4}	6×10^{-3}	2.86×10^{-4}	2.9×10^{-3}	3×10^{-3}	3×10^{-2}
Mercuric chloride	7487947	2.1×10^{-5}	2.1×10^{-4}			3×10^{-4}	3×10^{-3}
Methyl mercury	22967926	9×10^{-5}	9×10^{-5}			1×10^{-4}	1×10^{-4}
Nickel (soluble salts)	7440020	5.4×10^{-3}	5.4×10^{-3}			2×10^{-2}	2×10^{-2}
Nitrate	14797558	8×10^{-1}				1.6	
Nitrite	14797650	5×10^{-2}	5×10^{-2}			1×10^{-1}	1×10^{-1}
Phosphine	7803512	6×10^{-5}	6×10^{-5}	8.75×10^{-5}	8.6×10^{-4}	3×10^{-4}	3×10^{-4}
Selenium	7782492	2.2×10^{-3}	2.2×10^{-3}			5×10^{-3}	5×10^{-3}
Silver	7440224	9×10^{-4}	9×10^{-4}			5×10^{-3}	5×10^{-3}
Zinc (metallic)	7440666	6×10^{-2}	6×10^{-2}			3×10^{-1}	3×10^{-1}

Source: DOE Risk Assessment Information System [updated October 1999]: http://risk.lsd.ornl.gov/tox/tox_values.shtml.

13.5 RISK CHARACTERIZATION

The risk characterization provides a summary of the risk assessment, estimating the incidence of the health effect in the target population. In recent years the concept of risk characterization has been expanded to include consideration of a broad range of issues associated with environmental risk, including socioeconomic impacts and implications for legal rights, informed consent, and democratic principles (NRC, 1996a). Issues such as environmental justice and equity are considered as part of the overall assessment of impacts and alternatives, as is the fairness of the process by which the public and affected parties participate in the framing and deliberation of the pertinent issues. This chapter is limited primarily to the technical steps involved in estimating and summarizing health risks and incidence, though many of the tools for characterizing subgroup risk distributions in a population can and should be used to evaluate issues related to rights and equity. The reader is encouraged to read the 1996 NRC *Understanding Risk* report and to be sensitive to this broader landscape of concerns when framing and implementing environmental risk assessments and models.

The risk characterization includes a summary of:

1. Point estimates for individual exposure and risk (for the MEI, HEEE, etc.)
2. Population estimates for the pollutants of concern and various health endpoints, including the estimated risk distribution and incidence for the overall population and different subgroups of the population
3. Analysis of key uncertainties, the effort needed to resolve them, and the likelihood that this will occur in the near future

This chapter focuses on the first two of these tasks, providing methods for characterizing the *variability* of exposure and risk across affected populations. Methods for characterizing the *uncertainty* of risk estimates are presented in Chapter 14.

Figure 13.8, from Sexton et al. (1995b), provides a framework for summarizing exposure distributions across an affected population. In Figure 13.8a, exposure distributions are shown for various subgroups of the population, then aggregated to yield the overall exposure distribution for the general population. The calculus for this aggregation follows rules for *mixtures of distributions.* If an overall population is comprised of N subgroups: $j = 1, N$, each with exposure probability density function $f_j(E)$, then the probability density function for the general population, $f(E)$, is simply the weighted mixture of the subgroup distributions:

$$f(E) = \sum_{j=1}^{N} p_j f_j(E) \tag{13.16}$$

where p_j is the fraction of the population in subgroup j. The same weighting procedure applies to the cumulative distribution function:

$$F(E) = \sum_{j=1}^{N} p_j F_j(E) \tag{13.17}$$

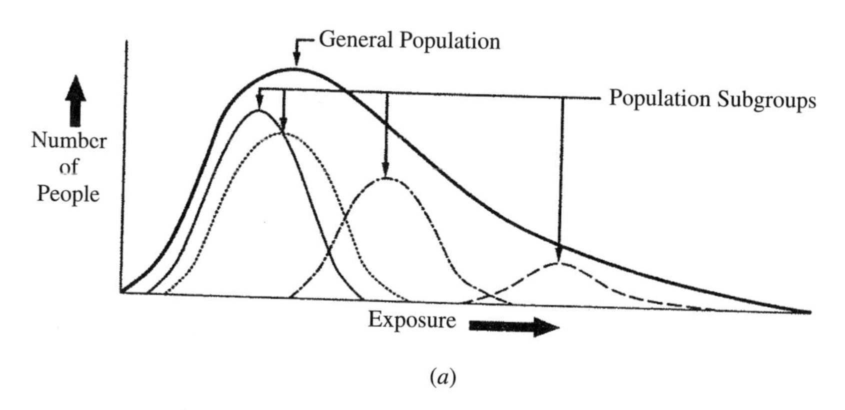

(*a*)

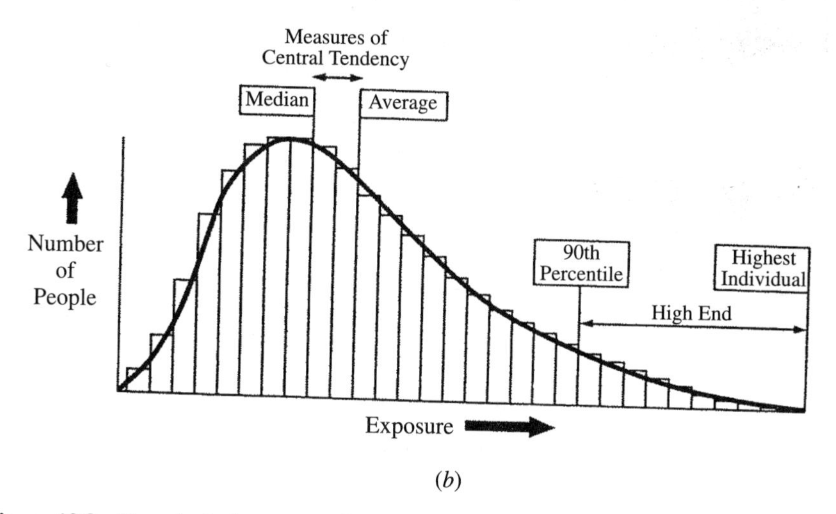

(*b*)

Figure 13.8 Hypothetical exposure distribution illustrating (*a*) the relationship between the general population and important subgroups and (*b*) common descriptors of distribution characteristics (Sexton et al., 1995b).

and the various moments of the exposure distribution:

$$\text{Expected value}\,[E^m] = \sum_{j=1}^{N} p_j \,\text{Expected value}\,[E_j^m] \tag{13.18}$$

The resulting exposure distribution is characterized by the common descriptors in Figure 13.8*b*, including measures of central tendency (the median and mean) and measures of high-end exposure, such as the 90th percentile or other percentiles near the upper tail of the distribution.

The implications for dose, risk, and incidence are summarized in Figure 13.9. For noncancer effects where a reference dose is determined, the fraction of the population with doses above the reference dose is computed (or, as shown in Fig. 13.9*a*, a reference *concentration* can be back-calculated from the reference dose, and the fraction with exposure concentrations above this value determined). Assuming that the distribution of dose is available, the fraction above the reference dose is simply

$$\text{Fraction above Rfd} = 1 - F_{D_e}(\text{RfD}) \tag{13.19}$$

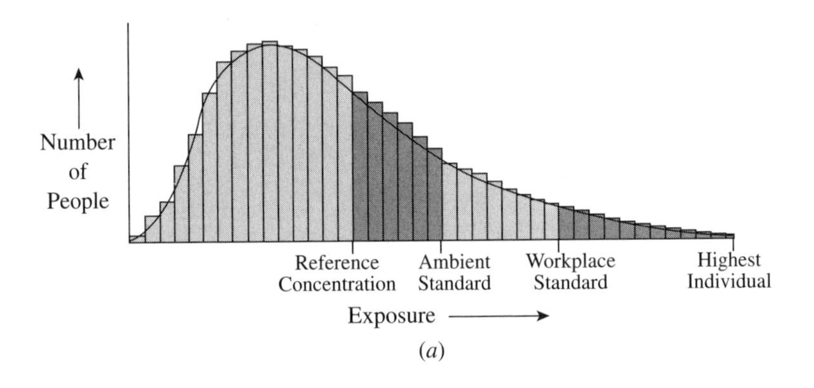

(*a*)

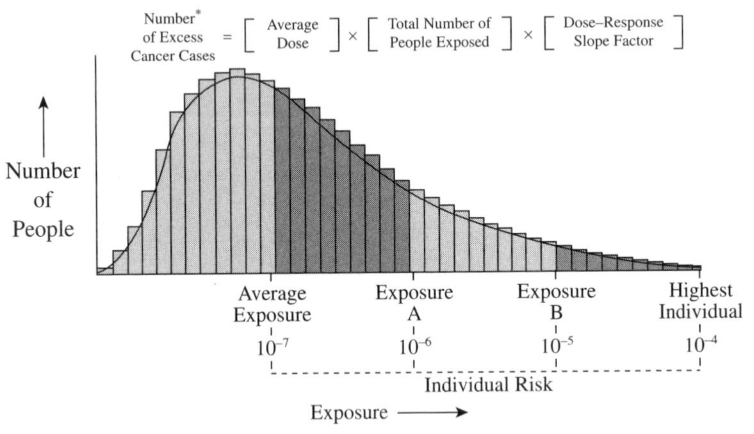

(*b*)

Figure 13.9 Hypothetical exposure distribution for (*a*) a noncancer toxicant and (*b*) a carcinogen, illustrating calculation of individual risk and population incidence (Sexton et al., 1995b).

For cancer effects, the exposure and dose distributions are converted to risk estimates and the overall distribution summarized as in Figure 13.9*b*. The cancer incidence rate in the population is computed as the average value of the lifetime individual risk times the total number of people P in the population, divided by the lifetime exposure period (i.e., $T_L = 70$ years):

$$\text{Incidence (cancers/year)} = \left[\int_{D_e} \text{LIR}(D_e) f_{D_e}(D_e)\, dD_e \right] \times \frac{P}{T_L} \qquad (13.20)$$

The lifetime average daily dose is used for the effective exposure (D_e = LADD) in Eq. (13.9). Assuming the linear, no-threshold dose–response relationship of Eq. (13.10), with slope factor β (i.e., LIR $= \beta \times D_e$), the incidence is determined directly from the average dose, $E[D_e]$:

$$\text{Incidence (cancers/year)} = \beta \times E[D_e] \times \frac{P}{T_L} \qquad (13.21)$$

Similar calculations can be conducted for each subgroup to estimate the fraction above the reference dose or the incidence of cancer within that subgroup.

EXAMPLE 13.4 LOGNORMAL DOSE AND RISK DISTRIBUTION FOR BENZENE IN THE UNIT WORLD

As mentioned previously, pollutant concentrations often follow a lognormal distribution. A lognormal model is also often used to describe exposure, dose and risk variability across a population (e.g., Crouch et al., 1983; Brand and Small, 1995; Rai and Krewski, 1998; Hattis et al., 1999). The genesis for this can be seen in Eq. (13.9), which shows the effective dose as the product of the exposure, the uptake rate, and the fraction absorbed—divided by the body weight and the exposure period. If each of these factors varies lognormally across the population, then the effective dose is likewise lognormal. Even if some of these distributions deviate from lognormality (as, e.g., the distribution of the fraction absorbed must, since it is constrained to the range 0–1), the central limit theorem, when applied to the logarithm of the effective dose:

$$\ln(D_e) = \ln(E) + \ln(R) + \ln(f_a) - \ln(\text{BW}) - \ln(T) \qquad (13.22)$$

dictates that the distribution of $\ln(D_e)$ approaches normality, and D_e thus approaches the lognormal.

To illustrate this type of model for the unit world, assume that the computed values of the air exposure ($E = 1.97 \times 10^{-4}$ mg/m$^3 \times$ 25,567.5 day $= 5.0$ mg/m^3

days), air inhalation rate ($R = 20$ m^3/day) and body weight (BW = 70 kg) in Examples 13.2 and 13.3 are actually median values, and that these factors vary as lognormal distributions across the population. The following parameters describe the lognormal distribution for the three variable exposure parameters.[6] Recall also that the inhalation exposure and dose are the dominant component of the total exposure and dose in Examples 13.2 and 13.3.

X = Exposure Parameter	Median = X_{50} = GM(X)	$\mu_{\ln(X)}$	$\sigma_{\ln(X)}$	ν_X = Coeff Var. (X)
E (mg/m^3 days)	5	1.609	0.833	1.0
R (m^3/day)	20	2.996	0.010	0.1
BW (kg)	70	4.248	0.198	0.2

These values are typical of the magnitude of the variation one might expect across an exposed population (see, e.g., Taylor, 1993; Finley et al., 1994; Thompson and Graham, 1996; Burmaster and Crouch, 1997; Burmaster and Murray, 1998; Cullen and Frey, 1999; Hattis et al., 1999). Variation in the fraction absorbed and the exposure period is ignored, with each assumed equal to their nominal values, $f_a = 1.0$ and $T = 25{,}567.5$ days. It is also assumed that the modeled factors (E, R, and BW) vary independently across the population (there is in fact often correlation in the exposure factors across the population, e.g., positive correlation is expected between an individual's body weight and his or her inhalation rate).

The resulting distribution of dose is lognormal with:

$$\begin{aligned}\mu_{\ln(\text{LADD})} &= \mu_{\ln(E)} + \mu_{\ln(R)} + \ln(f_a) - \mu_{\ln(\text{BW})} - \ln(T)\\ &= 1.609 + 2.996 + 0 - 4.248 - 10.149\\ &= -9.792\end{aligned}$$

and

$$\begin{aligned}\sigma^2_{\ln(\text{LADD})} &= \sigma^2_{\ln(E)} + \sigma^2_{\ln(R)} + \sigma^2_{\ln(\text{BW})}\\ &= (0.833)^2 + (0.010)^2 + (0.198)^2 = 0.733\\ \sigma_{\ln(\text{LADD})} &= (0.733)^{1/2} = 0.856\end{aligned}$$

The distribution of lifetime individual risk is similarly lognormal with:

$$\begin{aligned}\mu_{\ln(\text{LIR})} &= \ln(\beta) + \mu_{\ln(\text{LADD})}\\ &= \ln(0.029) - 9.729 = -13.332\\ \sigma_{\ln(\text{LIR})} &= \sigma_{\ln(\text{LADD})} = 0.856\end{aligned}$$

[6] See Section 7.1.1 for the equations used to relate the lognormal parameters, $\mu_{\ln(X)}$ and $\sigma_{\ln(X)}$, to the distribution median, X_{50}, and coefficient of variation, ν_X.

This assumes that the slope factor [$\beta = 0.029$ (mg/kg · day)$^{-1}$] is the same for all individuals in the population. Any interindividual variability in the susceptibility to cancer from benzene is thus ignored.[7] The coefficient of variation for both the dose and the risk distribution is $\nu_{\text{LADD}} = \nu_{\text{LIR}} = 1.04$, corresponding to $\sigma_{\ln X} = 0.856$. The variability in dose and risk for this example is clearly dominated by the variability in the exposure, with the variability in R and BW contributing only a very small amount. In other situations with less variable exposure concentrations and more variable uptake rates, this situation could be reversed [see MacIntosh et al. (1999) for an example where this occurs involving exposure to pesticides in drinking water].

The population distribution of benzene dose and associated cancer risk is shown in Figure 13.10. Figure 13.10*a* presents the result as a lognormal probability plot; Figure 13.10*b* shows the pdf (since LIR is just the slope factor times the LADD, the same curves suffice for each, only a change in scale is needed to translate from one to the other). Approximately 71% of the population are estimated to have an LIR above 10^{-6} (corresponding to a standard normal deviate of $Z = -0.56$), while ~2% have an LIR greater than 10^{-5} (corresponding to $Z = 2.13$). If the HEEE is designated as the 95th percentile of the exposure–dose–risk distribution ($Z = 1.645$), then the HEEE LIR is 6.6×10^{-6}.

To calculate the cancer incidence rate, assume an exposed population of $P = 10$ million people. To apply Eq. (13.21), first compute the average dose (see Table 7.1 for the expression for the mean of a lognormal distribution):

$$\begin{aligned} E[\text{LADD}] &= \exp\left(\mu_{\ln(\text{LADD})} + 0.5 \times \sigma^2_{\ln(\text{LADD})}\right) \\ &= \exp(-9.792 + 0.5 \times 0.733) \\ &= 8.06 \times 10^{-5}\ \text{mg/kg} \cdot \text{day} \end{aligned}$$

The cancer incidence is then given by:

$$\begin{aligned} \text{Incidence} &= 8.06 \times 10^{-5}\ \text{mg/kg} \cdot \text{day} \times 10^{7}\ \text{people} \times 0.029\ (\text{mg/kg} \cdot \text{day})^{-1}/70\ \text{years} \\ &= 0.33\ \text{cancer cases/year} \end{aligned}$$

This result could have also been computed by first calculating the *average risk* in the population, $E[\text{LIR}] = 8.06 \times 10^{-5}$ mg/kg · day × 0.029 (mg/kg · day)$^{-1}$ = 2.3×10^{-6}.

Example 13.4 shows some of simple calculations that are performed to provide a first-, or screening-level, estimate of risk distributions and incidence across a

[7] The probabilistic treatment of dose–response parameters is a matter of controversy (Hattis, 2004). Some argue that our ability to characterize interindividual variability and uncertainty in toxicity parameters is so poor that we are better off using a single, consensus number for cancer potency factors, reference doses, etc. U.S. EPA policy has thus far supported this view. Others argue that ignoring the variability and uncertainty in these factors, while treating emissions, fate-and-transport, and exposure in a probabilistic manner, is inappropriate and has held back advancement in the field.

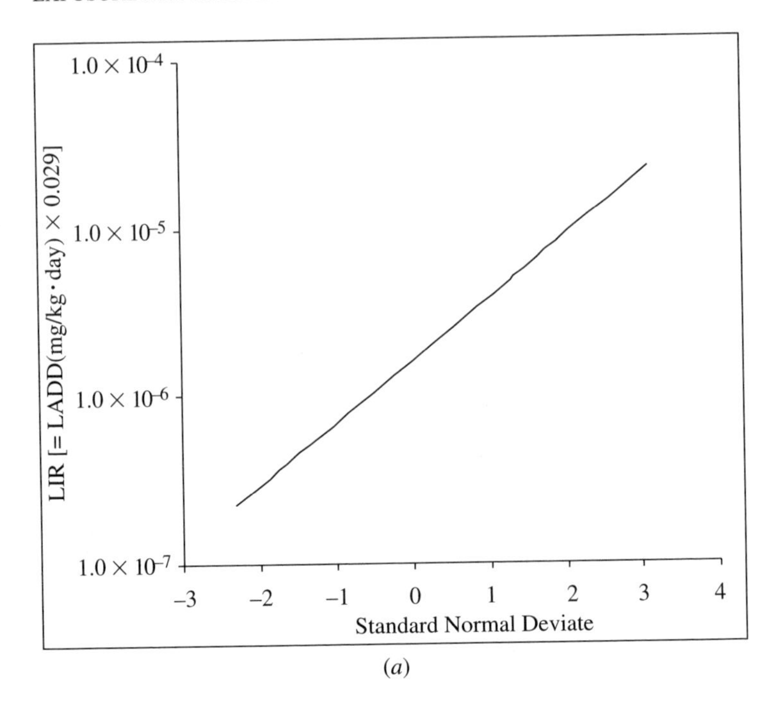

(*a*)

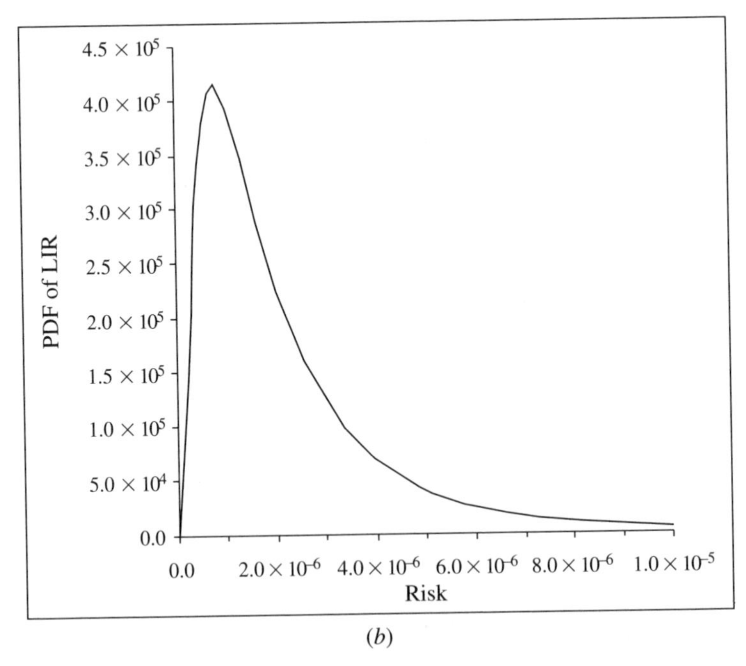

(*b*)

Figure 13.10 Population distribution of cancer risk (LIR) from benzene exposure in Example 13.5: (*a*) lognormal probability plot and (*b*) pdf of risk.

population. To illustrate a more rigorous exposure and dose assessment for a population based on a combination of microenvironmental sampling, modeling, and time activity patterns, the study of MacIntosh et al. (1995) is presented, characterizing the exposure and dosage rates of benzene for residents of six midwestern U.S. states.

EXAMPLE 13.5 POPULATION-BASED EXPOSURE MODEL FOR BENZENE

MacIntosh et al. (1995) develop and present the Benzene Exposure and Absorbed Dose Simulation (BEADS) model to predict the distribution of personal exposure and effective dosage rate of benzene, based on a combination of microenvironment data and simulation models. The model includes:

- Exposure to benzene in the air from inhalation
  - Of outdoor ambient air
  - At gasoline stations
  - In the automobile cabin
  - Indoors at home
  - Indoors at work or school
  - Indoors at other locations
- Exposure to benzene in water
  - From ingestion of tap water
  - From dermal uptake while bathing
  - From inhalation of volatilized benzene during and after showering

Exposure to benzene from both active smoking and environmental tobacco smoke (ETS) is included in the air inhalation calculation. The fraction absorbed, f_a, is assumed equal to 1.0 for water ingestion and 0.47 for air inhalation. A constant dermal permeability coefficient is assumed for dermal exposure.

MacIntosh et al. (1995) apply the model to residents of U.S. EPA Region 5 (Illinois, Indiana, Michigan, Minnesota, Ohio, and Wisconsin) and Arizona; only results for EPA Region 5 are shown here. A two-dimensional simulation approach is used to model the variability across the population and the uncertainty in this estimate. Best estimates for the distributions of benzene concentration in the air and benzene dose across the population are shown in Figure 13.11. Figure 13.11 is presented as a lognormal probability plot. The straightness of the lines indicates that the population distributions are approximately lognormal, except that there is an

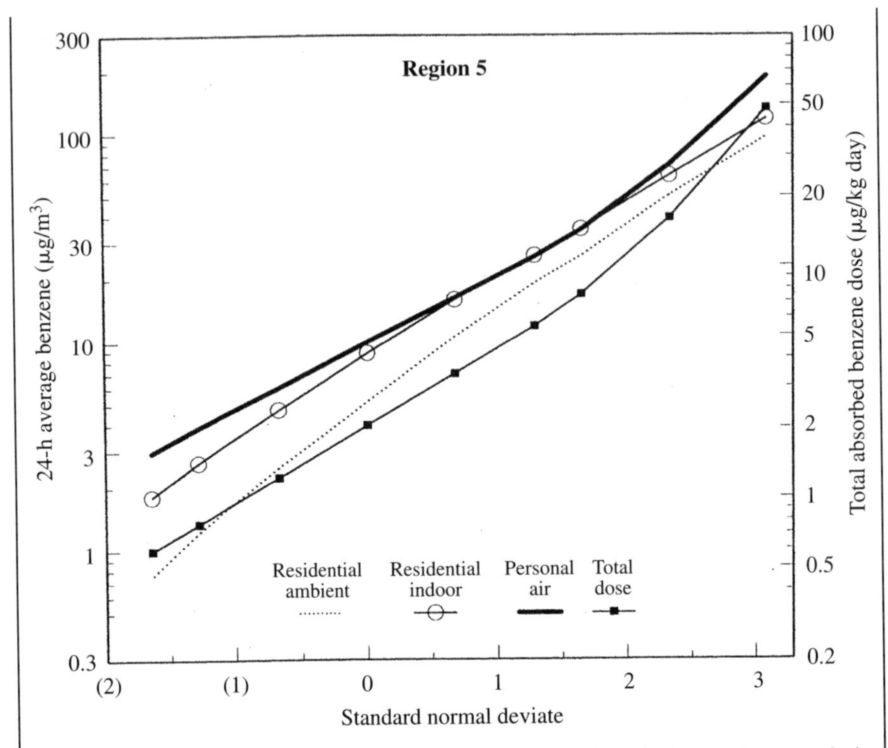

Figure 13.11 Distribution of predicted 24-h average outdoor, indoor, and personal air benzene concentrations and total effective benzene dose for EPA Region 5, as predicted by the BEADS model (MacIntosh et al., 1995).

upward curvature at the upper end of the personal-air and total-dose distributions. MacIntosh et al. attribute this heavy upper tail to individuals with high occupational exposures. To help differentiate the exposure and dose among subgroups of the population, they compare distributions for rural versus nonrural residents, smokers versus nonsmokers, ETS- versus non-ETS-exposed individuals, and those with and without occupational exposure. The results for the latter are shown in Figure 13.12. Individuals with an occupational exposure to benzene clearly have a much higher personal exposure and total dose than those without.

The occupational versus nonoccupational exposure subgroups can be compared using the framework for mixture distributions presented above. Each line in Figure 13.12 is nearly straight and the corresponding distributions are thus approximately lognormal. The following parameters are estimated for the total dose distributions of the occupationally and nonoccupationally exposed subgroups. As indicated, the parameters are approximate, based solely on the information presented in MacIntosh et al. (1995). Approximately 0.4% of the total population is estimated to have occupational exposure to benzene.

Subgroup	$\mu_{\ln(D_e)}$ [ln(μg/kg · day)]	$\sigma_{\ln(D_e)}$ [ln(μg/kg · day)]	p_j	$E[D_e]$ (μg/kg · day)
$j = 1$: Nonoccupationally exposed	0.693[a]	0.742[b]	0.996[c]	2.6
$j = 2$: Occupationally exposed	2.715[a]	0.956[b]	0.004[c]	23.9

[a]Based on reported median doses of 2.0 and 15.1 μg/kg · day for $j = 1$ and 2, respectively. Assumes lognormal distribution with median = geometric mean (GM).

[b]Based on reported geometric standard deviations (GSDs) of 2.1 and 2.6 for *personal exposure* for $j = 1$ and 2, respectively. Based on Figure 13.11, the slopes of the total dose and personal exposure lines are nearly equal to each other for each subgroup, therefore the GSDs of dose can be approximated to be the same as those for exposure.

[c]Estimated using Eq. (13.16) to match the 99th percentile value (standard normal deviate = 2.33) of dose for the total population from Figure 13.11 (≅ 15 μg/kg · day).

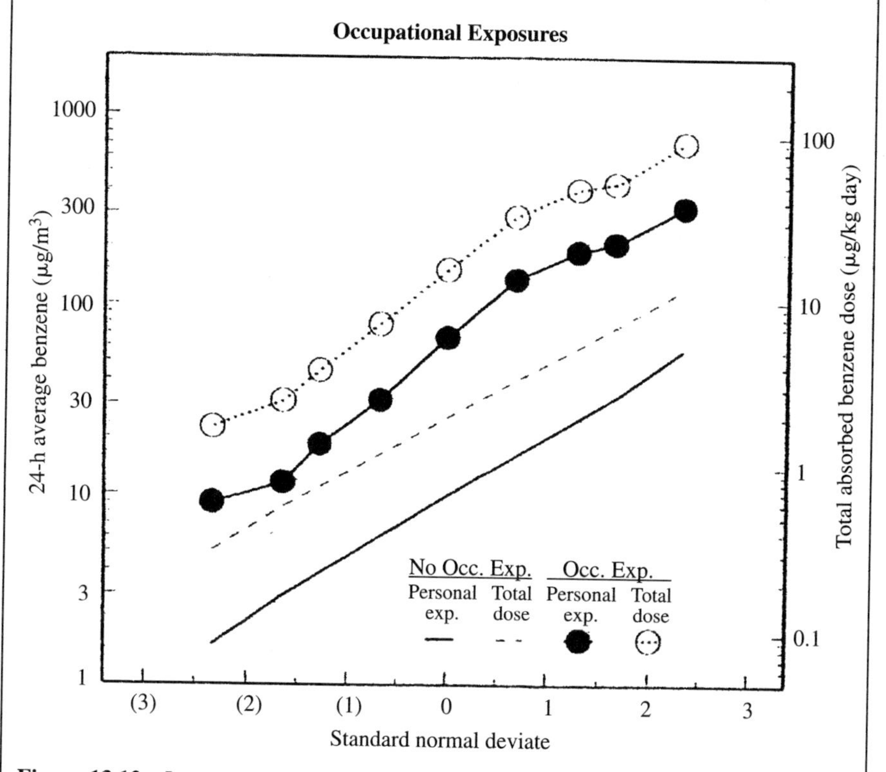

Figure 13.12 Lognormal probability plot of estimated personal benzene exposures and total absorbed doses by occupational category (occupational exposure vs. no occupational exposure) for EPA Region 5, as predicted by the BEADS model (MacIntosh et al. 1995).

The cancer incidence associated with benzene exposure to residents of U.S. EPA Region 5 can now be estimated from the mean dose. For the total population, the mixture model indicates a mean dose of $0.996 \times 2.6 + 0.004 \times 23.9 = 2.7$ μg/kg · day. Assuming the cancer slope factor of 0.029 (mg/kg · day)$^{-1}$ [$= 2.9 \times 10^{-5}$ (μg/kg · day)$^{-1}$] and a total population of $P = 48{,}920{,}000$ people in the six states comprising Region 5 (U.S. Census estimate for 1998: http://www.census.gov/), the following incidence rates are estimated using Eq. (13.21):

Subgroup	$E[D_e]$ (μg/kg · day)	E[LIR]	P = Population	Cancer Incidence (cases/year)
$j = 1$: Nonoccupationally exposed	2.6	7.5×10^{-5}	48,724,320	52
$j = 2$: Occupationally exposed	23.9	6.9×10^{-4}	195,680	2
TOTAL	2.7	7.8×10^{-5}	48,920,000	54

The benzene exposures and cancer risks estimated for residents of EPA Region 5 in this example are much higher than those computed for the hypothetical population assumed in Example 13.4, where only a single, hypothetical source of benzene was considered. The overall mean risk of 7.8×10^{-5} and especially the mean LIR for those with occupational exposure (6.9×10^{-4}) are high enough to warrant serious attention. While those with occupational exposure constitute only 0.4% of the population, they are estimated to have nearly 4% of the approximately 54 expected cancer incidences per year associated with benzene exposure in Region 5. This is consistent with the result that their estimated exposures and risks are nearly 10 times higher than those of the general population.

13.6 PHYSIOLOGICALLY BASED PHARMACOKINETIC (PBPK) MODELS

As the study of toxicology and physiology has progressed, health scientists have recognized that a more rigorous estimate of effective dose and toxic impact can be achieved by studying the way in which chemicals move about and through the body to reach target tissues or organs. This has led to the development of physiologically based pharmacokinetic (PBPK) models that provide a mechanistic representation of chemical fate and transport in the body. These models, originally developed in the field of pharmacology to study how medicines reach target organs and cells, are remarkably similar in approach and structure to multimedia environmental compartment models. PBPK models are developed to predict retention and clearance times for chemicals in the human body (e.g., Sunderman et al., 1989), to allow incorporation of tissue and other biomarker data in exposure and dose assessment (Sexton et al., 2004), and to compute the variation of effective dose for dose–response calculations across a target population (Pelekis et al., 2003). They are also developed for different species (e.g., rats and mice) to allow interpretation of laboratory studies and to provide a basis for cross-species (animal-to-animal or animal-to-human) extrapolation (Andersen et al., 1991; Barton et al., 1998).

The structure of a typical PBPK model for a chemical inhaled through the lungs, absorbed into blood, and transported to target organs and tissues via the circulatory system is shown in Figure 13.13 (Ramsey and Andersen, 1984; Bogen and McKone, 1988). The model variables and parameters associated with the figure are summarized in Table 13.5. Also listed are typical values of these parameters for a reference male (Bogen and McKone, 1988; Ward et al. 1988).

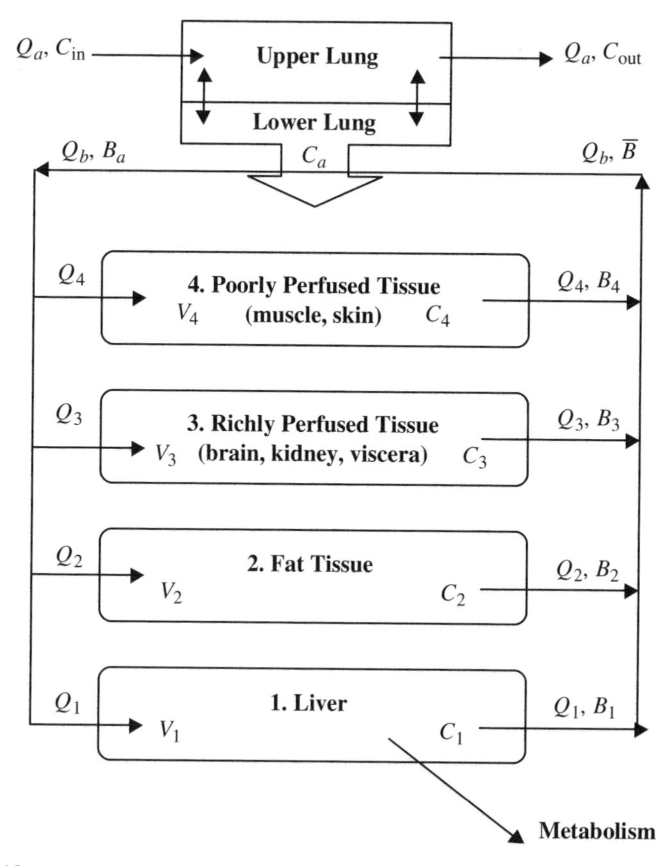

Figure 13.13 Structural diagram of PBPK model for inhaled contaminant partitioning to blood and body tissues (from Bogen and McKone, 1988). See Table 13.5 for definition of model variables.

TABLE 13.5 Compartment and Parameter Values for Ramsey–Andersen PBPK Model[a]

Parameter	Definition	Units and Assumed Values
C_{in}	Concentration in inhaled air	mg/L air (input)
C_a	Concentration in alveolar air	mg/L air[b]
C_{out}	Concentration in exhaled air	mg/L air[b] (assumed = C_a)
Q_a	Air inhalation rate	L air/h (353.5)
Q_b	Total blood flow rate	L blood/h (371.6)
$Q_i/Q_b(i = 1, 4)$	Fraction blood flow by compartment	(0.25/0.05/0.51/0.19)
P_b	Blood–air partition coefficient	L air/ L blood (10.3)
B_a	Concentration in arterial blood	mg/L blood[b]
$\bar{B}$	Concentration in venous blood	mg/L blood[b]
A_m	Pollutant mass metabolized in liver	mg[b]
W	Body weight	kg (70)
$V_i/W(i = 1, 4)$	Fraction body weight by compartment	(0.04/0.20/0.05/0.62)
$B_i(i = 1, 4)$	Concentration in blood in compartment	mg/L blood[b]
$C_i(i = 1, 4)$	Concentration in compartment tissue	mg/L tissue[b]
$A_i(i = 1, 4)$	Pollutant mass in compartment	mg[b]
$P_i(i = 1, 4)$	Tissue/blood partition coefficient	L blood/L tissue (6.82/159./6.82/7.77)
$V_{\max}$	Maximum metabolic rate in liver	mg/h (4.1)
K_m	Metabolic rate half-saturation constant	mg/L (0.19)

[a]Physiological parameter values from Ward et al. (1988), partition coefficients from gas uptake experiments.

Compartment subscripts: i = 1; liver (metabolizing tissue group)
= 2; fat tissue (very poorly perfused)
= 3; richly perfused tissues (brain, kidney, viscera)
= 4; poorly perfused tissues (muscle, skin)

[b]Computed by model.

Source: From Bogen and McKone (1988).

The modeled contaminant enters the body through inhalation and undergoes air–liquid exchange in the lung alveoli to enter the arterial bloodstream. From there it is distributed to four principal types of tissue in the body:

1. The liver, where the chemical is metabolized
2. Fat tissue, which is very poorly perfused by blood vessels
3. Richly perfused tissues, such as the brain, kidney, and viscera
4. Moderately to poorly perfused tissue, including muscle and skin

Each of these compartments is assumed to be well mixed; and the chemical is partitioned in an equilibrium manner between the tissue and the blood within (and exiting) each compartment. Chemical metabolism occurs in the liver as a kinetic process following a Michaelis–Menten rate equation. The blood flows exiting each compartment are recombined in the venous bloodstream for recirculation through the lungs, where equilibrium is once again achieved with the air in the alveoli.

When inhalation exposure to a chemical begins and ensues, concentrations in the blood and body tissue compartments build up over time and can eventually reach a value in steady state with a fixed exposure concentration. Once the exposure is terminated (or reduced), the blood and tissue concentrations decrease and eventually approach zero (or the new steady-state value).

The PBPK model is formulated and solved by first defining the mass balance of the chemical in the respiratory system. Assuming no accumulation in the lungs:

$$\text{Mass in} = \text{Mass out}$$

$$(Q_a C_{in} + Q_b \bar{B})\, dt = (Q_a C_a + Q_b B_a)\, dt \tag{13.23}$$

As indicated, the air enters the respiratory system with concentration C_{in}, but following equilibration with the blood, exits with concentration C_a. Similarly, the blood enters the respiratory system with concentration $\bar{B}$, but exits (following equilibration with C_a) at B_a. As such, we require

$$C_a = B_a / P_b \tag{13.24}$$

The blood–air partition coefficient, P_b, is similar to the dimensionless air–water partition coefficient for chemicals in a multimedia fate and transport model. Equations (13.23) and (13.24) have two unknowns (C_a and B_a), which can be solved simultaneously to yield

$$B_a = \frac{Q_a C_{\text{in}} + Q_b \bar{\text{B}}}{(Q_a / P_b + Q_b)} \tag{13.25}$$

Equations (13.25) and (13.24) are thus used to compute B_a and C_a from model inputs Q_a, C_{in}, Q_b, and P_b and from the computed value of $\bar{B}$ at each time step. The concentration in the venous blood reentering the pulmonary system is simply the flow-weighted average of the concentrations exiting each tissue compartment:

$$\bar{B} = \frac{1}{Q_b} \sum Q_i B_i \tag{13.26}$$

The mass balances in the nonmetabolizing compartments ($i = 2$, 3, and 4) are computed as:

$$\text{Mass in} = \text{Mass accumulation} + \text{Mass out}$$

$$Q_i B_a dt = \qquad dA_i \qquad + Q_i B_i\, dt \tag{13.27}$$

As such, A_i is the mass of the chemical in the tissue of compartment i, given by:

$$A_i = V_i C_i = V_i B_i P_i \tag{13.28}$$

so that $dA_i = V_i P_i \, dB_i$. Dividing Eq. (13.27) by dt and substituting for dA_i, yields

$$\frac{dB_i}{dt} = \frac{Q_i(B_a - B_i)}{V_i \, P_i} \tag{13.29}$$

In the liver ($i = 1$), mass is lost from the tissue–blood system via chemical metabolism, so Eq. (13.29) is modified to

$$\frac{dB_1}{dt} = \frac{Q_1(B_a - B_1)}{V_1 \, P_1} - \frac{1}{V_1 \, P_1} \frac{dA_m}{dt} \tag{13.30}$$

where dA_m/dt is the rate of chemical metabolism. The Michaelis–Menten kinetic formula is used to compute this rate from the concentration of the chemical in the blood of the liver:

$$\frac{dA_m}{dt} = \frac{V_{\max} B_1}{K_m + B_1} \tag{13.31}$$

The rate of chemical metabolism increases with B_1, is equal to one-half of its maximum value when $B_1 = K_m$, and approaches its maximum value of $V_{\max}$ as B_1 becomes larger (see Section 12.4.1 for other examples of Michaelis–Menten kinetics). Since, in this case, it is the metabolites of chemicals absorbed in the body that result in the toxic effect, the cumulative fraction metabolized, relative to the potential dose, provides an estimate of the retention fraction, or fraction absorbed (over the time period $0 - T$):

$$f_a = \int_0^T \left[\frac{dA_m(t)}{dt} \times \frac{1}{Q_a(t) C_{\text{in}}(t)} \right] dt \tag{13.32}$$

The PBPK model is specified and initiated by:

1. Inputting the physiological parameters for an individual:

   P_b, $P_1 - P_4$, K_m, $V_{\max}$, $Q_1 - Q_4$, and $V_1 - V_4$

2. Specifying the initial conditions:

   $C_a(0)$, $B_a(0)$, $B_1(t) - B_4(0)$ (these are assumed equal to zero for many problems)

3. Inputting the individual's breathing rate and personal exposure concentration, which force the model:

   $Q_a(t)$ and $C_{\text{in}}(t)$

The model may be readily modified to add other exposure-intake routes, including ingestion or dermal uptake. The model is solved numerically by simultaneously calculating $B_a(t+\Delta t)$ and $C_a(t+\Delta t)$ from Eqs. (13.24) and (13.25), then computing $B_1 - B_4(t+\Delta t)$, $C_1 - C_4(t+\Delta t)$, and $dA_m/dt|_{t\to t+\Delta t}$ from Eqs. (13.28) to (13.31). A numerical method, such as the fourth-order Runge–Kutta, is utilized.

A direct steady-state solution can be computed for the case of very long exposure to a constant C_{in} (with constant Q_a), by equating the mass rate of entry into the body ($Q_a C_{\text{in}}$) to that exiting the body by exhalation ($Q_a C_a$) plus that which is metabolized in the liver (dA_m/dt). This results in two equations:

$$B_a = B_2 = B_3 = B_4 = \frac{Q_a C_{\text{in}} + Q_1 B_1}{Q_a/P_b + Q_1} \tag{13.33}$$

$$Q_1 \left(B_a - B_1\right) = \frac{V_{\max} B_1}{K_m + B_1} \tag{13.34}$$

which can be solved simultaneously for the two unknowns, B_a and B_1. This also results in constant values of the tissue concentrations, $C_1 - C_4$ [from Eq. (13.28)], chemical metabolism rate [Eq. (13.31)], and fraction absorbed, f_a.

EXAMPLE 13.6 IMPLEMENTATION OF A PBPK EQUILIBRIUM MODEL

To illustrate the PBPK model's prediction of physiological response to changes in chemical exposure, the internal compartments of a "reference male human" characterized in Bogen and McKone (1988) are studied as inhaled tetrachloroethylene (PCE) levels are varied according to a simple step function. The subject breathes in a steady concentration, C_{in}, of 10 μg/L for 8 h, followed by a 16-h period with no exposure. A fourth-order Runga–Kutta method with a time step of 1 min is used to numerically integrate the set of equations described in Eqs. (13.23) to (13.32).

Figure 13.14 shows the buildup and subsequent flushing of PCE in the different bloodstreams. For blood in (and leaving) the lung, liver, and viscera compartments, relatively rapid increases in their respective PCE concentrations occur during the period of exposure, followed by correspondingly rapid loss after the exposure is completed. The response time for the blood concentration in the muscle and skin compartment is somewhat slower than that for the lung, liver, and viscera, due to this compartment's large volume and mass (62% of the total body weight) and resulting capacity for removing PCE from the incoming bloodstream. The fat tissue concentration responds very slowly and continues to increase even after inhalation of PCE stops, since with the low blood flow and resulting low PCE delivery rate to these tissues, the fat cell concentration is still well below equilibrium

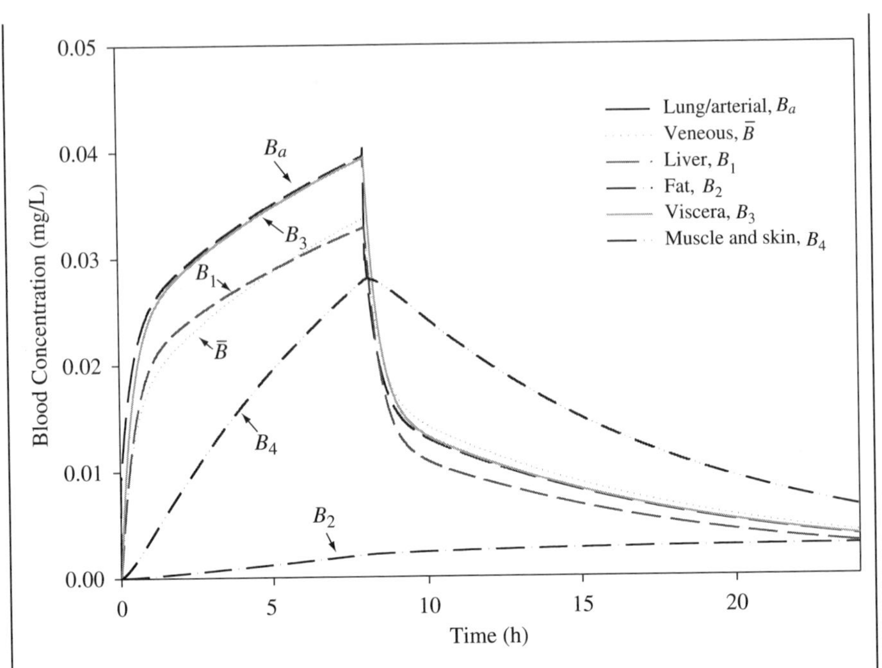

Figure 13.14 Response of blood concentrations to a step-function PCE input (for blood in and leaving the indicated compartments).

with the blood when the inhalation period ends. As PCE concentrations decrease in the blood, the fat tissue concentration remains below that which would be in equilibrium with even these lower blood concentrations for a considerable time, and thus continues to increase. Therefore, even after PCE inhalation is terminated, PCE is redistributed throughout the compartments and continues to be taken up by the fat tissue. The veneous blood concentration is the flow-weighted average of the return blood from all of the compartments, so its behavior is not greatly influenced by that of the fat (which has only 5% of the body's blood flow), instead reflecting the general trend of the other compartments.

Figure 13.15 shows the accumulated mass in each tissue compartment, which appropriately mirrors their respective blood concentrations. These masses are proportional to the blood concentrations by a factor that accounts for the compounded effect of compartment volume and affinity for PCE. As Figure 13.15 shows, the extremely large volume of muscle and skin accommodates more PCE, and the high permeability and relatively high volume (20% of the total) of fat attracts a significant quantity of PCE as well.

Figure 13.15 also includes a monotonically increasing curve that indicates the cumulative mass metabolized in the liver. Equation (13.32) allows us to estimate that 14% of the PCE entering the lungs will have been metabolized by the liver at the end of the initial inhalation period. The liver continues to consume the PCE retained in the body, however, and by the conclusion of the 24-h simulation

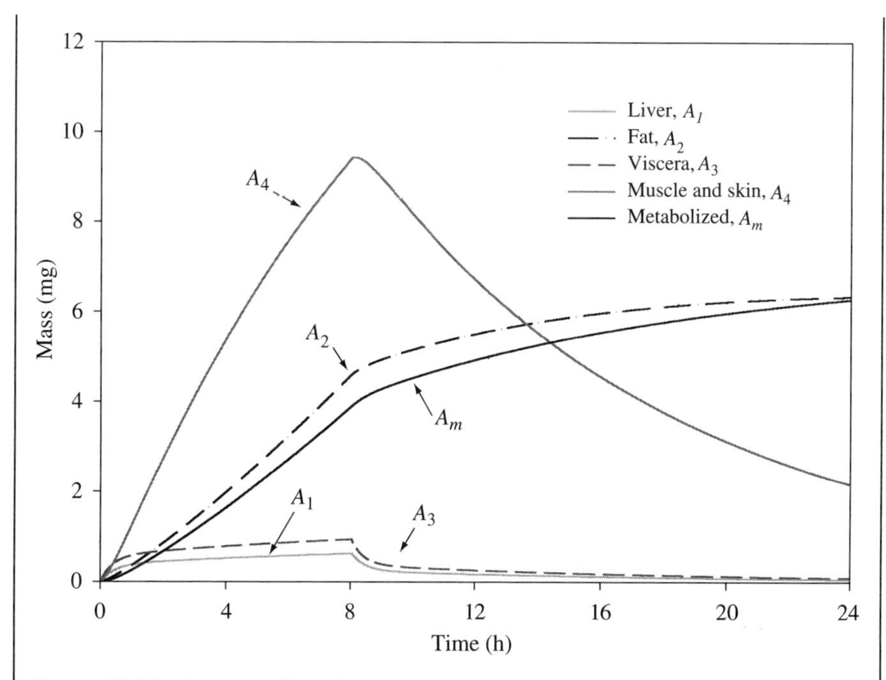

Figure 13.15 Accumulation of PCE mass in tissue in response to airborne PCE inhalation.

period, 22% of the entire 28 mg of toxin inhaled will be metabolized. Since the metabolites of PCE in the liver are considered to be the major source of toxicity, this is the portion of the potential dose that results in toxic impact, and the PBPK model (under the particular scenario considered) thus suggests that the fraction absorbed, f_a, relating the effective dose to the potential dose is ~0.20 to 0.25. Figure 13.16 compares the concentration of PCE in the inhaled air to that predicted in the exhaled air. This figure effectively illustrates the collective dynamic response and accumulation behavior of the body to a step-function input of PCE. The model results suggest that even a 16-h period free of exposure does not fully purge the body of the PCE burden.

The PBPK model illustrated here is very similar to the level II multimedia environmental model presented in Chapter 3, with equilibrium partitioning between compartments, advection from compartments (from the lungs via the arterial blood and exhalation; from the tissue compartments via the venous blood), and kinetic transformations within compartments (in this case, only within the liver). The model is driven by a dynamic loading function, yielding a dynamic response. PBPK models like the one illustrated here continue to be developed and applied for various applications in environmental health and risk assessment. In the next section, we return to multimedia environmental models and show how modules for human exposure and risk assessment are incorporated.

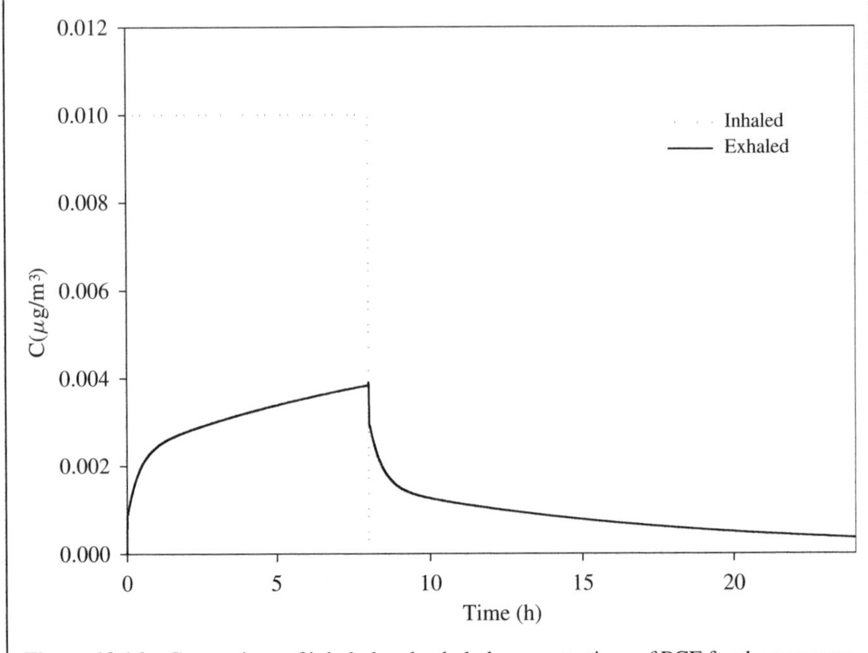

Figure 13.16 Comparison of inhaled and exhaled concentrations of PCE for the exposure scenario.

13.7 MULTIMEDIA ENVIRONMENTAL FATE, TRANSPORT, EXPOSURE, AND RISK MODELS

A number of models for predicting ambient pollutant concentrations in a multimedia environment have in recent years been adapted or extended to compute human exposure and risk. Examples include the RESRAD (Residual Radiation) model (Yu et. al., 1993); the Multimedia Contaminant Fate, Transport, and Exposure Model (MMSOILS) (EPA, 1992a); and the Multimedia Environmental Pollutant Assessment System (MEPAS) model (Buck et al., 1995; Whelan et al., 1992). These models are based on a similar representation of the environment—each using a series of linked, single-media models. Additional information on RESRAD, MMSOILS and MEPAS is provided in the web site section for this chapter, and those models have been evaluated and compared in a number of benchmark studies (Mills et al., 1997; Laniak et al., 1997).

Multimedia exposure and risk calculations have also been added to multicompartment equilibrium models of the type demonstrated in this text, such as CalTOX: a multimedia total exposure model for hazardous-waste sites developed by the University of California, Davis, and the Lawrence Livermore National Laboratory, for the California Environmental Protection Agency, Department of Toxic Substances

Control (CalTOX, 1994). A more detailed overview and illustration of the CalTOX model is presented in the following section.

13.7.1 The CalTOX Model[8]

CalTOX was developed as a set of spreadsheet models and spreadsheet data sets for assessing human exposures from continuous releases to air, soil, and water (McKone, 1993a, 1993b, 1993c). It has also been used for waste classification and for setting soil cleanup levels at uncontrolled hazardous waste sites (McKone et al., 1996). The components of CalTOX include a multimedia transport and transformation model, multipathway exposure scenario models, and add-ins to quantify and evaluate variability and uncertainty. The multimedia transport and transformation model is dynamic so it can be used to assess time-varying concentrations of contaminants introduced initially into soil layers or released continuously to air, soil, or water. The exposure model encompasses 23 exposure pathways, which are used to estimate average daily doses (intake and uptake) within a human population linked geographically to a release region. The exposure assessment process consists of relating contaminant concentrations in the environmental compartments to concentrations in the media with which a human population has direct contact (personal air, tap water, foods, household dusts/soils, etc.). The explicit treatment of the link between pollutant concentrations in different environmental media and the concentrations in human microenvironments distinguishes CalTOX from many other exposure models. All parameter values used as inputs to CalTOX are distributions, described in terms of mean values and a coefficient of variation, instead of point estimates or plausible upper values as employed by most other models. This stochastic approach allows sensitivity and uncertainty analyses to be conducted directly.

In CalTOX, the distribution of individual lifetime risk, $H(t)$, at some time t after a release begins is calculated by summing the dose and effect over exposure routes, environmental media, and exposure pathways:

$$H(t) = C_s(0) \times \left\{ \sum_{j\text{ routes}} \sum_{\substack{k\text{ environ-}\\ \text{mental}\\ \text{media}}} \sum_{\substack{i\text{ exposure}\\ \text{media}}} \left[Q_j \left(\mathrm{ADD}_{ijk}\right) \times \left(\frac{\mathrm{ADD}_{ijk}}{C_k}\right) \times \Phi\left[S(0) \rightarrow C_k, t\right] \right] \right\} \tag{13.35}$$

In this expression, $\Phi[S(0) \rightarrow C_k, t]$ is the multimedia dispersion function that converts the pollutant release rate $S(0)$ (mol/day) and exposure duration, ED, into a contaminant concentration C_k in environmental medium k for a time t after the release begins. (Units of C_k are mg/kg for soil, mg/m^3 for air, and mg/L for water). ADD_{ijk}/C_k is the unit dose factor, which is the average daily potential dose over a

[8] This section on the CalTOX model was written by Thomas McKone with the help of Edward Butler.

specified averaging time from exposure medium i by route j (inhalation, ingestion, dermal uptake) attributable to environmental compartment k, divided by C_k, where C_k is constant over the duration ED. The exposure is summed over the number of media that link potential dose by route j to contaminants in compartment k. Function $Q_j(\text{ADD}_{ijk})$ is the dose–response function that relates the potential dose, ADD_{ijk} (mg/kg · d), by route j to the lifetime probability of detriment per individual, for example, $Q_j(\text{ADD}_{ijk}) = \beta_j \times \text{ADD}_{ijk}$, where β_j $(\text{mg/kg} \cdot \text{d})^{-1}$ is the cancer potency factor for exposure route j.

Two major CalTOX components are required by Eq. (13.35). Each component can also operate independently. The first component is the multimedia transport and transformation model, which is used to determine the dispersion of soil contaminants among soil, water, and air media. The second component is the human exposure model, which translates environmental concentrations into estimates of human contact and potential dose.

Multimedia Transport and Transformation Model The current version of CalTOX (CalTOX4) is an eight-compartment regional and dynamic multimedia fugacity model. CalTOX derives environmental concentrations by determining the likelihood of competing processes by which chemicals (a) accumulate within the compartment of origin, (b) are physically, chemically, or biologically transformed within this compartment (i.e., by hydrolysis, oxidation, etc.), or (c) are transported to other compartments by cross-media transfers that involve dispersion or advection (i.e., volatilization, precipitation, etc.). CalTOX uses a level III multimedia chemical partitioning model to characterize mass-transfer processes between compartments and transformation within compartments. While the multimedia chemical partitioning model approach is best suited to nonionic organic chemicals for which partitioning is strongly related to chemical properties, such as vapor pressure, solubility, and the octanol–water partition coefficient, CalTOX has been designed to also handle ionic organic contaminants, inorganic contaminants, radionuclides, and metals, with a modified fugacity approach. For all species, fugacity and fugacity capacities are used to represent chemical potential and mass storage within compartments.

The compartment structure of CalTOX is illustrated in Figure 13.17. The eight CalTOX4 compartments are (1) air, (2) plant surfaces (cuticle) (3) plant leaf biomass (leaves), (4) ground-surface soil, (5) root zone soil, (6) the vadose zone soil below the root zone, (7) surface water, and (8) sediments. The air, surface water, cuticle, leaves, ground-surface soil, and sediment compartments are assumed to be in quasi-steady-state with the root zone soil and vadose zone soil compartments. Contaminant inventories in the root zone soil and vadose zone soil are treated as time-varying state variables. Contaminant concentrations in groundwater are based on the leachate from the vadose zone soil.

Quantities or concentrations within compartments are described by a set of linear, coupled, first-order differential equations. A compartment is described by its total mass, total volume, solid-phase mass, liquid-phase mass, and gas-phase mass. Contaminants are moved among and lost from each compartment through a series of transport and transformation processes that can be represented mathematically as

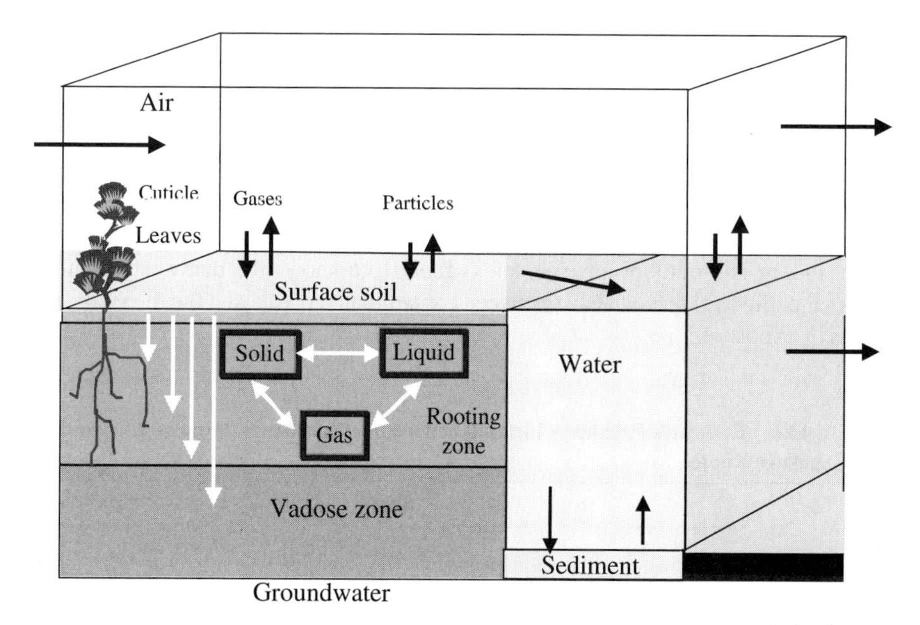

Figure 13.17 Mass-exchange processes in the CalTOX 4.0 eight-compartment environmental transport and transformation model.

first-order losses. Thus, the transport and transformation equations solved in CalTOX have the form:

$$\frac{d}{dt}N_i(t) = -k_i N_i(t) - \sum_{j=1, j\neq i}^{m} T_{ij} N_i(t) + \sum_{j=1, j\neq i}^{m} T_{ji} N_j(t) + S_i(t) - T_{io} N_i(t) \tag{13.36}$$

where $N_i(t)$ is the time-varying inventory of a chemical species in compartment i (mol); k_i is the first-order rate constant for removal of the species from compartment i by transformation (1/day); T_{ij} is the rate constant for the transfer of the species from compartment i to compartment j (1/day); and T_{ji} is the rate constant for the transfer of the species from compartment j to compartment i (1/day); T_{io} is the rate constant for the transfer of the species from compartment i to a point outside of the defined landscape system (1/day); S_i is the source term for the species into compartment i (mol/day); and m is the total number of compartments within the landscape system. Equation (13.36) is solved for the compartments shown in Figure 13.17. Mass flows among compartments include solid-phase flows, such as dust suspension or deposition, and liquid-phase flows, such as surface runoff and groundwater recharge. CalTOX simulates all decay and transformation processes (such as radioactive decay, photolysis, biodegradation, etc.) as first-order, irreversible removals. Direct emissions are considered from sources in the air, ground-surface soil, root zone soil, vadose zone soil, and surface water compartments but not in the leaf or sediment layers.

Human Exposure Model Dose models used in CalTOX are based on those described by the U.S. EPA (EPA, 1989, 1992b) and by the California Department of Toxic Substances Control (DTSC, 1992). The nature and extent of multimedia exposures depends largely on human factors and the concentrations of a chemical substance in the contact media. Human factors include contact rates with food, air, water, soils, drugs, and the like. Activity patterns are also significant because they directly affect the magnitude of inhalation and dermal exposures to substances present in different indoor and outdoor environments. Table 13.6 shows the many interrelationships (or pathways) that can exist between contaminated media and the three possible routes of exposure.

TABLE 13.6 Exposure Pathways Linking Environmental Media, Exposure Scenarios, and Exposure Routes

	Media		
Exposure Routes	Air (gases and particles)	Soil (ground surface soil; root zone soil)	Water (surface water and groundwater)
Inhalation	• Gases and particles in outdoor air • Gases and particles transferred from outdoor air to indoor air	• Soil vapors that migrate to indoor air • Soil particles transferred to indoor air	• Contaminants transferred from tap water to indoor air
Ingestion	• Fruits, vegetables, and grains contaminated by transfer of atmospheric chemicals to plant tissues • Meat, milk, and eggs contaminated by transfer of contaminants from air to plants to animals • Meat, milk, and eggs contaminated through inhalation by animals • Mother's milk	• Human soil ingestion • Fruits, vegetables, and grains contaminated by transfer from soil • Meat, milk, and eggs contaminated by transfer from soil to plants to animals • Meat, milk, and eggs contaminated through soil ingestion by animals • Mother's milk	• Ingestion of tap water • Irrigated fruits, vegetables, and grains • Meat, milk, and eggs from animals consuming contaminated water • Fish and sea food • Surface water during swimming or other water recreation • Mother's milk
Dermal contact	• Not included	• Dermal contact with soil	• Dermal contact in baths and showers • Dermal contact while swimming

Multimedia, multiple pathway exposure equations are used in CalTOX to estimate average daily doses within a human population geographically linked to a pollutant release. The overall model encompasses 23 potential exposure pathway scenarios. The end product of these exposure assessments is an estimate of the distribution of potential dose among the population at risk. Each of the exposure scenarios relating an average daily dose to an environmental medium concentration for a specific pathway is developed in the following form:

$$\text{ADD} = \text{Intake}_i = \text{TF}(k \rightarrow i) \times \left[\frac{\text{IU}_i}{\text{BW}}\right] \times \frac{\text{EF} \times \text{ED}}{\text{AT}} \times C_k \qquad (13.37)$$

In this expression, the concentration of the contaminant in environmental medium k is assumed constant over the exposure duration. ADD is the average daily potential dose rate in mg/kg · day, which is the intake of a contaminant from exposure medium i; TF($k \rightarrow i$) is the intermedia transfer factor, $[C_i/C_k]$, which expresses the ratio of contaminant concentration in the *exposure* medium i to the concentration in an environmental medium k; and [IU$_i$/BW] is the intake or uptake factor per unit body weight associated with the exposure medium i. The exposure frequency, EF, for the exposed individual is the number of days per year that an individual contacts the contaminated medium k; ED is the exposure duration for the exposed individual; and AT is the averaging time for the exposed individual (days). The formulation of Eq. (13.37) is directly parallel to that of Eqs (13.7) to (13.9).

Uncertainty and Sensitivity Analyses with CalTOX As discussed in the next chapter, uncertainty analysis of mathematical models involves the determination of the variation in an output function due to the collective variation of model inputs, whereas sensitivity analysis involves the determination of the changes in model response as a result of changes in individual model parameters (Morgan and Henrion, 1990). CalTOX has been designed to provide users with the options of: (1) making point estimates of average exposure, (2) applying a sensitivity analysis to these point estimates, (3) using a Monte Carlo add-in to make stochastic estimates of the distribution of exposure uncertainty or variability, and (4) determining the contributions of model inputs to the overall variance in the estimated distribution of exposure and risk.

EXAMPLE 13.7 APPLICATIONS OF CALTOX

To illustrate the use of CalTOX for comparative exposure assessment, we apply the model to two compounds: hexachlorobenzene and pentachlorophenol, which are structurally similar but differ significantly in chemical properties and exposure pathways. This calculation is applied to a landscape of 1 million square kilometers. For each chemical, we consider a continuous release for a period of 15 years. The multimedia releases used for both chemicals in these examples are

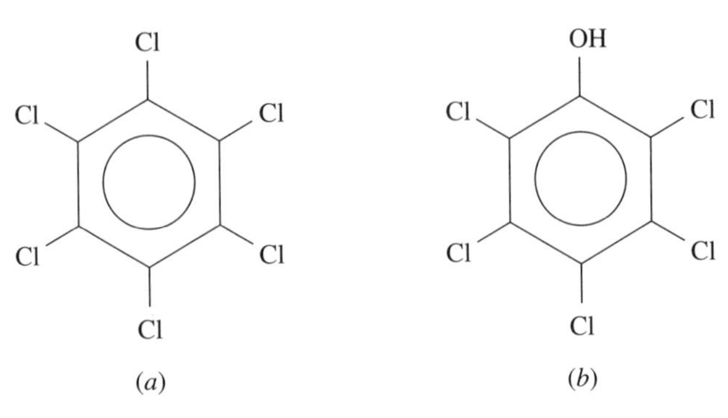

Figure 13.18 Chemical structures of (*a*) hexachlorobenzene and (*b*) pentachlorophenol.

100,000 mol/day to air, 10,000 mol/day applied to the soil surface, and 1000 mol/day to surface water.

Hexachlorobenzene (HCB) is a fat-soluble, persistent, and ubiquitous chlorinated compound. HCB is formed as a waste product during the production of several chlorinated hydrocarbons and is a contaminant in some pesticides. HCB

TABLE 13.7 Chemical Properties for Hexachlorobenzene (HCB) and Pentachlorophenol (PCP)

Description	HCB	PCP
CAS ID number	118-74-1	87-86-5
Chemical formula	C_6Cl_6	C_6Cl_5OH
Molecular weight (g/mol)	285	267
Octanol–water partition coefficient	350,000	130,000
Melting point (K)	501	447
Vapor pressure in (Pa)	0.0017	0.0042
Solubility (mol/m^3)	5.3×10^{-5}	0.052
Henry's law constant (Pa · m^3/mol)	110	0.0025
Diffusion coefficient in pure air (m^2/day)	0.47	0.46
Diffusion coefficient; pure water (m^2/day)	5.8×10^{-5}	4.2×10^{-5}
Organic carbon partition coefficient [kg(water)/kg(organic carbon)]	46,000	900
Octanol/air partition coefficient [m^3(water)/m^3 (octanoal)]	7.7×10^6	1.2×10^{11}
Soil–water partition coefficient in root zone soil layer [L(water)/kg(soil-solids)]	770	1.2
Sediment–water partition coefficient in surface water sediments [L(water)/kg(solid)]	920	18
Bioconcentration factor fish/water (L/kg)	17,000	640
Reaction half-life in air (day)	850	11
Reaction half-life in surface/root soil (day)	470	50
Reaction half-life in deep soil (day)	5,000	100
Reaction half life in surface water (day)	1,500	0.4
Reaction half-life in sediments (day)	4,700	230

is released to air as a fugitive emission from these chemical facilities and is also released to air in flue gases and fly ash from waste incineration. HCB is released to soil as a contaminant in pesticides. HCB is persistent in the environment due to its chemical stability and resistance to biodegradation (Howard, 1989). The chemical structure of HCB is shown in Figure 13.18.

Pentachlorophenol (PCP) is a manufactured chemical not found naturally in the environment. Over the last several decades PCP has been used as a biocide for preserving wood and wood products and in the manufacture of sodium pentachlorophenate. The chemical structure of PCP is shown in Figure 13.18; as indicated PCP is very similar to HCB, except that an OH group is substituted for one of the six Cl atoms on the benzene ring. The chemical properties of HCB and PCP used in CalTOX are listed in Table 13.7. As indicated, the polarity introduced

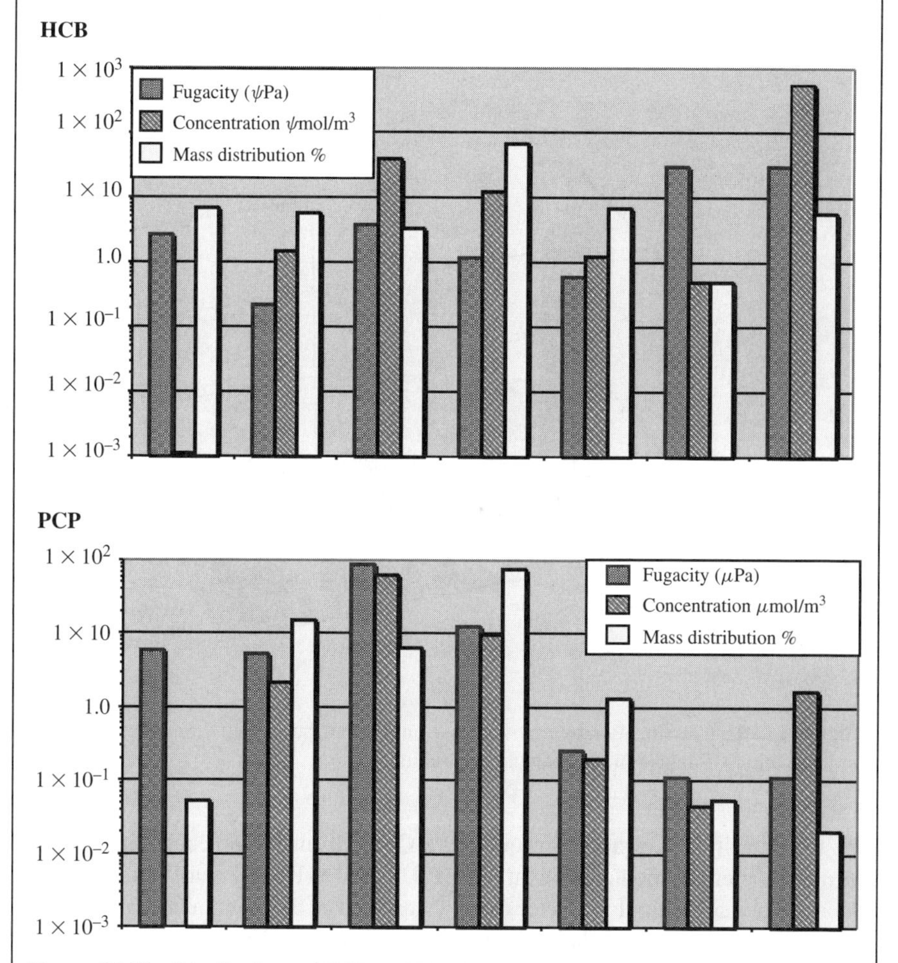

Figure 13.19 Distributions of HCB and PCP in a multimedia environment averaged over 15 years based on continuous emissions to air, surface soil, and surface water.

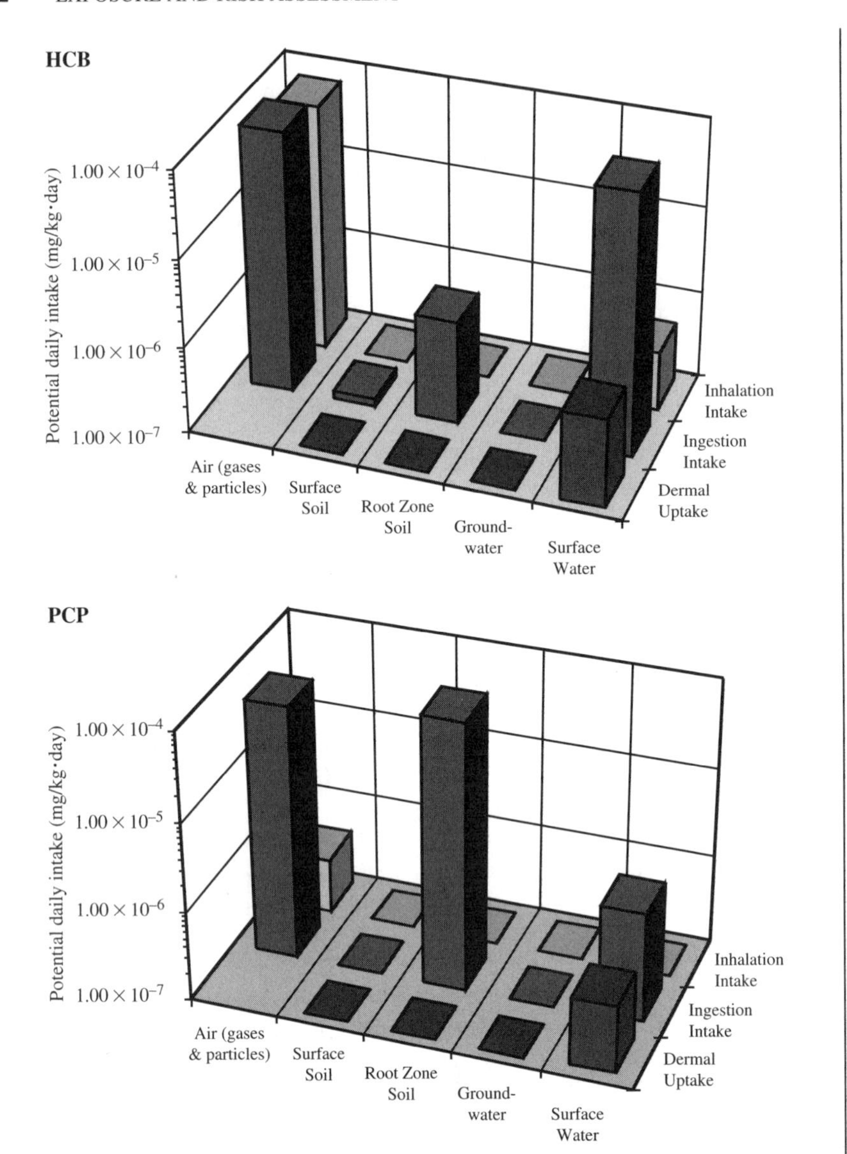

Figure 13.20 Classification by exposure route and by attributable environmental pathway of HCB and PCP intake in the example calculations.

by the substitution of an OH group for a Cl atom in the molecule reduces the octanol–water partition coefficient of PCP relative to HCB by nearly a factor of 3, increases its solubility by a factor of 1000, and makes the compound significantly more susceptible to degradation in the air, water, soil, and sediments.

Figure 13.19 summarizes and compares the environmental fate of HCB and PCP as calculated by CalTOX. Shown here are the fugacity, concentration, and mass distribution in seven compartments. While both compounds exhibit significant partitioning to the root soil compartment, PCP is found at lower levels in the air (due to its high solubility and low Henry's constant) and the water (due to its higher degradation rate).

Figure 13.20 compares the relative contribution of HCB and PCP exposures calculated with CalTOX from the mass distributions in Figure 13.19. Both compounds have a majority of their exposure associated with the ingestion of foods and water, though as shown, HCB results in a significant inhalation intake via the air route as well, while PCP does not.

13.8 ECOLOGICAL RISK ASSESSMENT

This chapter has focused primarily on human exposure and risk. However, in many applications, threats to the natural environment and its flora and fauna are of equal or greater concern. While the modeling of ecological systems has a history paralleling that of models for pollutant fate and transport (and ecological impacts are explicitly or implicitly considered in a number of fate and transport models, especially those for surface waters) formal methods for ecological risk assessment have remained less developed and less standardized.A number of the methods for human health risk assessment can and have been adapted to ecological risk assessment. For example, hazard quotients or indices can be computed based on the ratio of a calculated dose to the reference dose for important species (Newman, 1995; Hanson and Solomon, 2002), and such approaches have been linked to multimedia fate and transport models similar to CalTOX (Fenner et al., 2002). However, the inherent complexity of ecological systems and the many possible endpoints of ecosystem harm and impact that might be considered have caused many analysts to shy away from such simplifications. Much of the most recent guidance on ecological risk assessment (e.g., EPA, 1998; Suter, 1999) has emphasized the need for conceptual models that consider:

- The effects of multiple factors that impact individual species and ecosystems, including climate, habitat, pollution, and competition (Caughley and Gunn, 1996; Power and Adams, 1997)
- The occurrence of multiple, population-level impacts, some of which occur through indirect effects in ecosystems (Walthall and Stark, 1997; Tanaka and Nakanishi, 2000)
- The interaction of ecological impacts with economic, cultural, and societal effects (Lande, 1998).

No simple, cookbook approach can be adequate for addressing the full range of complex system interactions and endpoints of concern in ecological risk assessment. Indeed, consideration of cumulative effects of chemical mixtures and the interaction of exposure and health effects with a broader range of "habitat" factors, is now also recognized as necessary for improved human health risk assessment, and this is a part of the motivation for the field of human ecology (Marten, 2001). The systems-based approach of ecology could thus help to provide a framework for broader assessments and characterizations of risk for both human health and ecological risk assessment.

14 Tools for Evaluation, Analysis, and Optimization of Environmental Models

As first discussed in Chapter 1, environmental fate and transport models are developed both for scientific purposes and as applied tools for environmental management. In the context of developing models to advance scientific understanding, model evaluation can be viewed as an example of the scientific process of *hypothesis testing*, where the model represents a collection of hypotheses about the nature and significance of physical and chemical processes that act on a contaminant. When models are developed as predictive tools for environmental management applications, model evaluation is instructively viewed as a stage in an iterative *design* process (Fig. 14.1). From this perspective, model design/development begins with careful specification of the objectives and constraints that are important for the intended application. Alternative formulations are evaluated against specified criteria such as model performance and ease of use. A successful design/development process ends, often after many iterative rounds of testing and modification, with a model that is judged to be adequate for its intended purpose.

The model evaluation and analysis tools discussed in this chapter include model calibration and parameter estimation, statistical performance evaluation, sensitivity and uncertainty analysis, and optimization. A comprehensive modeling study directed at a site remediation problem, for example, might utilize all of these tools at various stages, beginning with model calibration to represent the hydrologic conditions at the remediation site and culminating in optimization of pumping rates for groundwater capture and treatment. This chapter begins with an overview of the model evaluation process in Section 14.1. Section 14.2 discusses statistical and graphical techniques used to evaluate model performance in comparison to field observations. The formulation and solution of model calibration and parameter estimation problems is discussed in Section 14.3. Sensitivity and uncertainty analysis methods are described in Section 14.4, which includes an introduction to Bayesian Monte Carlo analysis, a new approach that combines information from uncertainty analysis and performance evaluations to allow quantitative characterization of overall confidence levels for model results.

14.1 MODEL EVALUATION

In many environmental management applications, a primary objective of model development is to maximize the accuracy with which environmental concentrations of

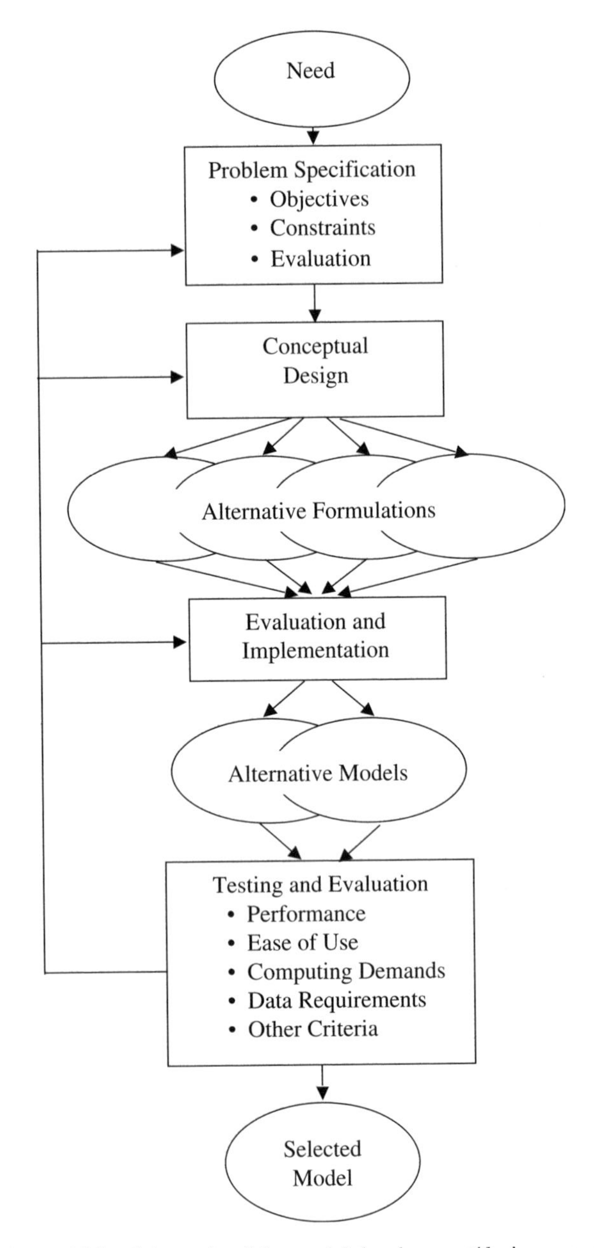

Figure 14.1 Schematic of the model development/design process.

a particular contaminant can be predicted. This objective is usually tempered by constraints on expenditures for input data collection or computing power, or on the level of expertise needed to run the model. The range of conditions to which the model will be applied is also a critical issue. For example, designing an air quality model for application to episodes of high pressure and sunny skies is a different problem than designing a model that handles pollutant scavenging during storms.

Not all model applications have the objective of maximizing predictive accuracy (Beck et al., 1997; Beck and Chen, 2000). For regulatory screening models, the objective may be to ensure that the model gives conservative predictions of environmental concentrations or exposures. In such situations, processes such as biodegradation or chemical decay that are critical for accurately predicting environmental fate and transport may be deliberately omitted. Other models may be intentionally simplified or idealized to maximize the insight they can give to novice users or the lay public. The objectives for which models are developed thus bear strongly on the question of how they should be evaluated.

Since all models employ a simplified representation of the real world, no model can ever be validated in the pure sense of the word (Oreskes et al., 1994). If a statistical approach to model validation is adopted, that is, we decide whether or not to reject the null hypothesis that the observed data were sampled from a world in which the model is true (Reckhow et al., 1990; Reckhow, 1993), then as a more accurate and larger observed data set is collected for the test, the null hypothesis should eventually be rejected for all models. Indeed, since a null hypothesis is never "accepted," only "rejected" or "not rejected," a model can never be proven correct, only proven (or not proven) to be incorrect (Konikow and Bredehoeft, 1992). For this reason, many modelers prefer to say that their model is confirmed or corroborated by the observed data, rather than "validated" or even "verified" (Chapra and Reckhow, 1983; Reckhow, 1993; Small et al., 1995).

An alternative approach to model validation is to address the issue in a relative, rather than an absolute, manner. Alternative models—with different formulations and/or parameter values—can be compared in terms of their relative consistency and goodness of fit with the data, as well as their relative strengths in terms of credibility and standing as state-of-the-art representations in the scientific community, responsiveness to decision-making needs, ease and transparency of use, and so forth. This is the approach that is introduced in Chapter 7, where the relative likelihoods or relative posterior probabilities are compared among alternative statistical models (Section 7.1). The spirit of this approach is continued in this chapter, where relative measures of goodness of fit and overall utility are emphasized for alternative models.

For almost all intended uses, model evaluation includes review of what has been called *face validity*, that is, whether the model formulation is consistent with current scientific understanding of the processes it needs to describe. Any conscientious model developer will evaluate the face validity of her or his own model, but in many cases external peer review is also conducted.

For models that utilize numerical solution techniques, a second evaluation step is testing the numerical implementation of the model. This step generally includes *benchmarking* the code against known analytical solutions for idealized cases and/or against highly accurate (but perhaps less efficient) computational techniques for more realistic cases, and testing the sensitivity of model results to the grid resolution or time steps used in the numerical solution. Tests of numerical implementation frequently assess mass conservation during transport of a nonreacting tracer and examine stability and numerical dispersion problems, as discussed in Chapter 6.

The next step, *performance evaluation*, involves graphical or statistical comparison of model results with field observations for one or more historical cases. Common measures of model performance are defined in Section 14.2. For many model applications, performance evaluation is considered the heart of the evaluation process. However, in other cases this step may not be possible because appropriate observations do not exist, or it may not be relevant if conservatism rather than accuracy is the objective. In some applications, performance evaluation may be preceded by *calibration* in which values of selected parameters are "tuned" so the model outputs match the observations. The mathematical programming techniques used for model calibration are discussed in more detail in Section 14.3. Ideally, the calibrated model would then be tested against an independent data set, without further calibration. Little confidence in a model's predictive capabilities can be gained from performance evaluation if too many degrees of freedom are exploited to calibrate the model (Gupta and Sorooshian,1983; Beck, 1987; Kleissen et al., 1990; Jakeman and Hornberger, 1993; Reichert and Omlin, 1997).

In real-world applications, environmental model results always carry some uncertainty due to imprecisely known input or parameter values. *Sensitivity analysis* explores how model outputs respond to changes in inputs or parameter values. *Uncertainty analysis* combines this information with estimates of the uncertainties in the inputs or parameters, producing estimates of uncertainty in model results. This uncertainty propagation step is often accomplished using Monte Carlo simulation methods. Quantifying uncertainties in model predictions is an important element of the evaluation process, especially when the predictions are to be considered in decision making. Sensitivity and uncertainty analysis methods are presented in Section 14.4.

To make an overall assessment of a model's suitability for a given application, model developers, users, and peer reviewers combine information from each applicable stage of the evaluation process. Over time, retrospective assessments of how the model performed in earlier predictive applications may become part of a model's resume, building (or eroding) confidence in its value as a predictive tool. A *postaudit* of model performance—comparing the observed response of an environmental system to major modifications in emissions or other engineering controls, to that predicted by the model *before* the modifications were implemented (Di Toro et al., 1987; Thomann, 1987; Long, 1996) can yield especially valuable insight into how accurate future predictions made with current models are likely to be. Traditionally, overall judgments about a model's validity for use in a particular application have been expressed in qualitative terms, for example, state-of-the-art or acceptable. As described in Section 14.4, Bayesian Monte Carlo analysis is a new approach that combines information from uncertainty analysis and performance evaluations to allow quantitative characterization of overall confidence levels for model results, as well as relative comparisons among alternative model formulations.

14.2 PERFORMANCE EVALUATION

In a performance evaluation, model results are compared against observations for one or more historical cases. To distinguish it from other model evaluation steps, performance evaluation is sometimes known as history matching. Although performance evaluation and model calibration are often conducted iteratively, the most instructive performance evaluations are carried out after calibration has been completed, with observational data sets that are independent of those used to calibrate the model. However, this is difficult to do in many cases, especially for systems (such as groundwater aquifers) that only respond to change very slowly over time (Konikow and Bredehoeft, 1992).

In this section, we present a variety of statistical and graphical measures of model performance that are widely used for fate and transport models. However, the choice of performance measures to use is less important than the more fundamental questions of what cases and observations to use in the evaluation. A fortunate model developer may have the chance to help design a field study to collect data for model evaluation purposes. In other situations, she or he may have to use whatever data are available. Either way, the best data sets for model evaluation:

- Encompass a range of conditions similar to those of interest in the ultimate model application that include conditions that are likely to *challenge* the model, to help identify potential limitations
- Allow individual model components to be evaluated separately, usually by providing data on intermediate variables in the model, as well as final outputs
- Allow *direct* comparison between model results and observations for the spatial resolution and averaging times of interest, minimizing the need to interpolate or extrapolate data or estimate secondary variables from those that are observable
- Provide information on potential errors or uncertainties in model inputs and in the observations against which the model is being compared.

Many of the models discussed in this book give time series of contaminant concentrations at one or more receptor locations as their primary outputs. In this case, model performance can be evaluated graphically by plotting both the modeled concentrations, C^i_{pr}, and the corresponding observed concentrations, C^i_{ob}, versus time, as illustrated for ozone concentrations in the Los Angeles area in Figure 14.2*a*. Scatter plots of modeled versus observed concentrations paired in time and/or space are also useful (Fig. 14.2*b*) although any information about temporal or spatial patterns in the differences is lost in this format. *Residuals* between model results and observations are defined as $w_i = C^i_{\text{pr}} - C^i_{\text{ob}}$ and are commonly plotted versus time or as histograms showing their distribution across discrete bins (Fig. 14.2*c*).

Though somewhat less informative to some than graphical presentations, a variety of statistics are useful for summarizing the comparison between numerous pairs of model results and observations. Statistical comparisons do provide an objective, replicable means for evaluating goodness of fit (see Section 7.1) and for testing hypotheses about whether the observed data are consistent with environmental conditions and responses predicted by alternative models. Still, the statistical measures must be chosen carefully to capture the important features of these comparisons.

Some of the most commonly used statistics for model performance are defined in Table 14.1, with n giving the number of prediction/observation pairs. A related way to

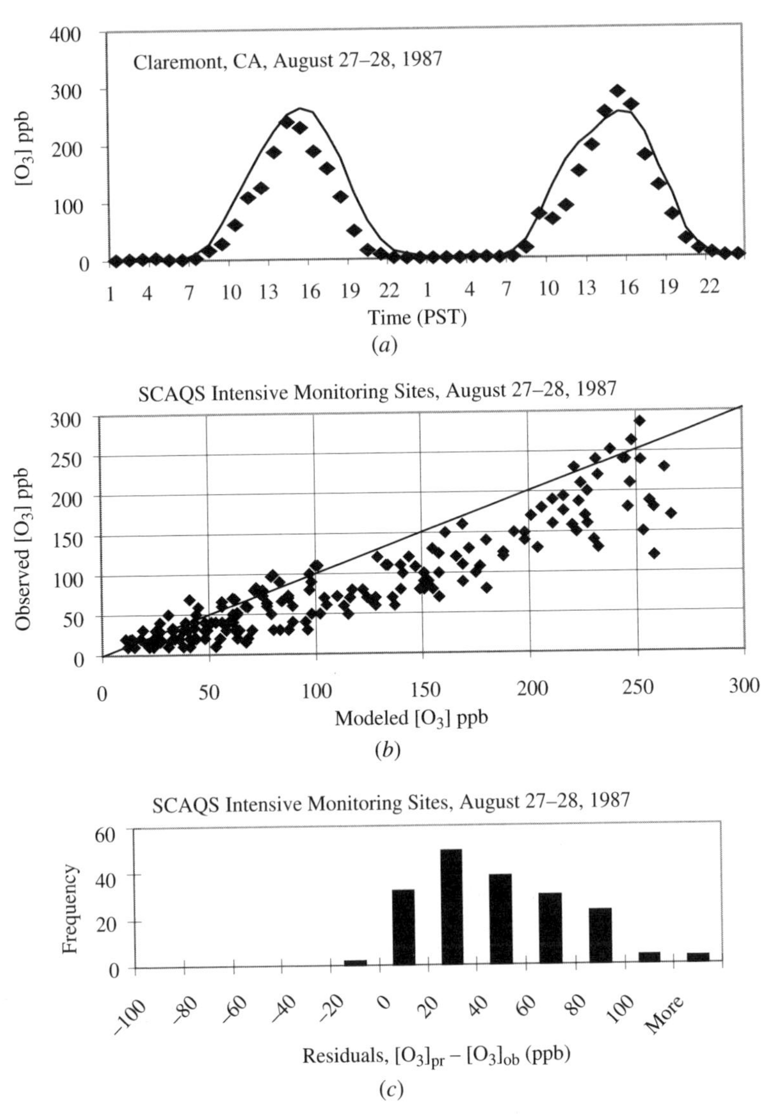

Figure 14.2 Graphical performance evaluation results for the CIT photochemical airshed model applied to the August 27–28, 1987, episode of the Southern California Air Quality Study (SCAQS) by Harley et al. (1997). The model and simulation conditions are described in Section 11.5. (*a*) Time-series plot of modeled (–) and observed (♦) concentrations at the Claremont, California, monitoring location. (*b*) Scatter plot of observations vs. model results for eight intensive monitoring locations, omitting concentrations less than 10 ppb. (*c*) Histogram of residuals between predicted and observed ozone concentrations at the eight intensive monitoring locations. Model performance at Claremont is considered good, and model results and observations are well correlated over the 2-day period and eight intensive monitoring sites. However, the histogram plot of residuals shows a tendency for the model to overpredict ozone concentrations.

TABLE 14.1 Statistics Widely Used to Summarize Model Performance[a]

Name	Definition	Comments	CIT $[O_3]$ Result
Bias	$\mu = \frac{1}{n}\sum_{i=1}^{n} w_i$	Mean of residuals	29 ppb
Normalized bias	$\frac{1}{n}\sum_{i=1}^{n}\left(\frac{w_i}{C_{ob}^i}\right) \times 100\%$		54%
Root-mean-squared error (RMSE)	$\left(\frac{\sum_{i=1}^{n} w_i^2}{n}\right)^{1/2}$		42 ppb
Standard deviation of residuals	$\left[\frac{1}{n}\sum_{i=1}^{n}(w_i - \mu)^2\right]^{0.5}$	Measure of spread, insensitive to bias	30 ppb
Absolute average gross error	$\frac{1}{n}\sum_{i=1}^{n}\lvert w_i\rvert$	Less sensitive to outliers than RMSE	32 ppb
Correlation coefficient	$\frac{\sum_{i=1}^{n}\left(C_{pr}^i - \overline{C_{pr}}\right)\left(C_{ob}^i - \overline{C_{ob}}\right)}{\left[\sum_{i=1}^{n}\left(C_{pr}^i - \overline{C_{pr}}\right)^2 \sum_{i=1}^{n}\left(C_{ob}^i - \overline{C_{ob}}\right)^2\right]^{1/2}}$	Degree to which spatial or temporal changes in observations are followed by those in predictions. Overbar indicates the mean.	0.91

Sources: Thomann (1980), Fox (1981), and McRae and Seinfeld (1983).

[a] Illustrative values are given for O_3 concentrations from the CIT photochemical airshed model applied to the August 27–28, 1987, SCAQS episode. The values given here are calculated for eight intensive monitoring locations, omitting concentrations less than 10 ppb.

summarize performance evaluation results is to use regression analysis to estimate the relationship between the observations and model outputs shown graphically in a scatter plot such as Figure 14.2*b*. Assuming a linear relationship between the explanatory variable (model result) and the observations, linear least-squares regression analysis estimates the scalar coefficients, b_0 and b_1 of the relationship: $C_{ob} = b_0 + b_1 C_{pr} + \epsilon$, where ϵ represents a random variable with expectation $E(\epsilon) = 0$ and constant variance, $\mathrm{Var}(\sigma^2)$. Ideally, the relationship between the model results and observations is such that $b_o = 0$ and $b_1 = 1$. The coefficient of the regression, R^2, indicates the fraction of the variance in the observations that can be explained by the model and has an ideal value of 1. For the photochemical model application for which results are shown in Figure 14.2, the least-squares linear relationship between observed and predicted ozone concentrations is given by $[O_3]_{ob} = -5.2 + 0.79[O_3]_{pr}$, with concentrations in ppb units. The coefficient of the regression, R^2, has a value of 0.83, indicating that the regression model explains 83% of the variance in the observed concentrations. However, the negative value of the intercept ($b_0 = -5.2$) and a slope value less than 1.0 ($b_1 = 0.79$) both indicate that the model systematically overestimates observed concentrations.

When the observations used in a regression analysis or in computing performance statistics constitute a sample of independent, normally distributed random variables, standard statistical methods that are covered in elementary statistics texts can be used to calculate confidence intervals on the mean or variance of residuals, the correlation between predictions and observations, and the regression coefficients, b_0 and b_1. If n is large, and the observations are independent but not normally distributed, approximate confidence intervals for the mean of the residuals (i.e., the bias) can be derived from the standard formulas because the mean of any random variable calculated from a large enough sample is approximately normally distributed. Alternatively, if the observations are independent but not normally distributed, confidence intervals for any of the performance statistics can be estimated using bootstrap or jackknife approaches (Efron and Gong, 1983; Hanna, 1989; Frey and Burmaster, 1999).

In the bootstrap technique, repeated random samples are drawn with replacement from the whole data set to choose a subset of n' observations from which performance statistics will be estimated. Quantiles of the cumulative distribution function derived from hundreds of repeated evaluations of the performance measure are interpreted as confidence limits for the measure corresponding to sample size n'. While assymtotically correct, the bootstrap method can have problems estimating the tails of a distribution. This problem can be addressed by blocking the data or by using bootstrap methods only to estimate the mean and variance of the performance measure. In the latter case, confidence limits on the mean can be estimated from the bootstrapped standard deviation estimate, using the classical student's t-distribution formula.

The jackknife approach entails calculating performance measures with each observation omitted in turn. Confidence limits are calculated by defining $x_j^* = n x_{all} - (n-1)x_j$, where x_j is the value of the measure calculated with the jth observation omitted and x_{all} is its value calculated from all of the observations. The Z% confidence interval on the performance measure x is $\lfloor \overline{x^*} - t_{Z_{n-1}} \sigma_{x*}, \overline{x^*} + t_{Z_{n-1}} \sigma_{x*} \rfloor$, where $\overline{x^*}$ is the mean value of the x_j^* and σ_x^* is its standard deviation, and $t_{Z,n-1}$ is the value of the student's t distribution corresponding to Z, with $n - 1$ degrees of freedom. The jackknife method is efficient but tends to slightly overestimate the width of confidence intervals (Hanna, 1989).

Both standard confidence interval estimators and jackknife and bootstrap techniques assume that performance statistics are calculated from independent observations. However, observations used to evaluate environmental models are often serially correlated, in which case the effective sample size used in confidence interval estimates or significance tests must be reduced (see Chapter 7).

The statistics listed in Table 14.1 allow model performance to be quantified. This facilitates comparison of different models or alternative parameterizations. In particular, such statistics are useful for parameter estimation and model calibration, when parameter values are adjusted to achieve the best fit between model results and observations. However, caution should be exercised before making summary judgments about the superiority of one model versus another from top-down performance evaluations alone. Models can appear to perform well due to compensating errors. Statistical performance measures give little or no diagnostic information about causes of lack of agreement between model results and observations. Errors in observations and incomparability between observations and model outputs also need to be considered (e.g., Stow and Reckhow, 1996; Sohn and Small, 1999; Sohn et al., 2000). For example, urban-scale photochemical air quality models typically produce ozone concentrations that represent averages over a 25-km^2 grid cell. McNair et al. (1996) interpolated observed concentrations from 37 ozone monitoring stations in the Los Angeles network onto a 25-km^2 grid and found a normalized root-mean-squared (rms) error of about 30% between the interpolated and observed concentrations. They argued that error associated with spatial averaging sets a floor for the error that can be expected between photochemical model results and point observations. Thus, in drawing conclusions about model performance, the suitability, rigor, and robustness of the evaluation need to be considered, along with the potential that errors in the observations could obscure the model's predictive skill.

14.3 MODEL CALIBRATION AND PARAMETER ESTIMATION

For some modelers, the ideal environmental model is formulated entirely from fundamental principles that are universally applicable, with independent laboratory studies or theoretical calculations providing values for all of its parameters, and direct measurements supplying boundary and initial conditions and contaminant loadings. Although this ideal is sometimes realized, in many cases important processes are not understood or cannot be described in detail, necessitating the use of site-specific parameter values that can only be estimated in the field. If, furthermore, these parameters are not directly observable, indirect parameter estimation or calibration techniques may be required.

Figure 14.3 illustrates a generic model calibration process in which one or more model parameters are adjusted within reasonable limits to minimize the difference between model results and observations for a historical case. The fit of the model to

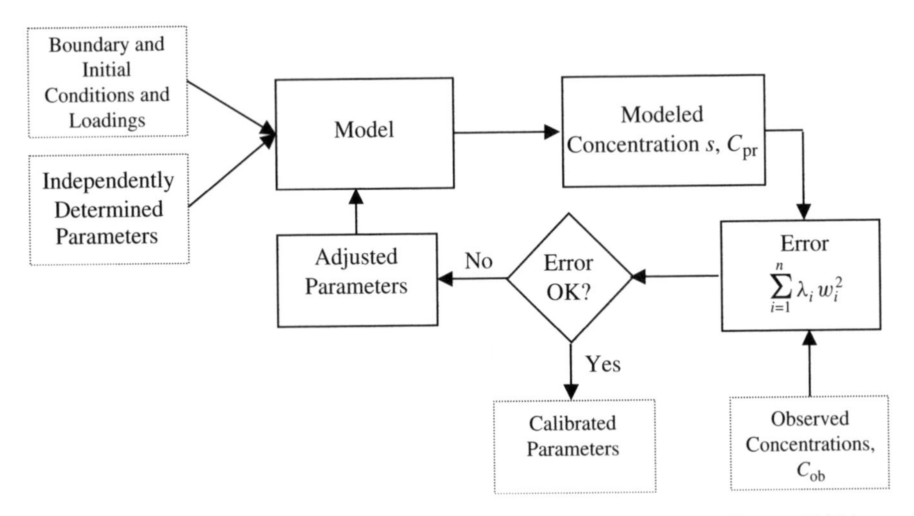

Figure 14.3 Schematic diagram of model calibration. (Adapted from Chapra, 1997.)

the observations is usually quantified by the sum of the squared residuals $\sum_{i=1}^{n} w_i^2$ or by a weighted sum $\sum_{i=1}^{n} \lambda_i w_i^2$, where λ_i is the weight assigned to the ith observation/prediction pair and typically reflects the estimated error in the observation, as discussed below. The match between the observations and model results can be optimized by trial-and-error adjustments of the parameters or by more formal least-squares minimization techniques.

An important problem in some model calibration studies is that models are overparameterized relative to the amount of information that is available to calibrate them (Beck, 1987; Reichert and Omlin, 1997). This problem exists when the model output being observed is not uniquely responsive to the parameter being estimated, for example, because errors in two different parameters cancel each other out in affecting the model's response. Model sensitivity analysis (discussed in Section 14.4) should be performed as part of the process of designing a field study for calibration purposes, to ensure that the data that will be collected do correspond to variables that respond uniquely to the parameters being estimated.

In a formal approach to parameter estimation, mathematical programming methods for optimization are used, together with principles of statistical inference. Thorough parameter estimation studies produce not only estimates of parameter values but also estimates of the error in these values and some assessment of the goodness of fit of the optimal model. In the optimization problem, the unknown model parameters or inputs are treated as control variables, and the objective function is related to the degree of fit (or misfit) between the model predictions and the observed data. Constraints enter the problem through physical restrictions on the values that can be considered for the fitted (i.e., optimal) parameters. For example, reaction rates, dispersion coefficients, and unspecified non-point-source loadings (and almost every other input or parameter that might be calibrated as part of a parameter estimation exercise) cannot be negative. Because the relationship between model parameters and model state variables is rarely linear, parameter estimation generally relies upon methods of *nonlinear programming*. However, in special cases, unconstrained linear regression can be used.

The application of simple linear regression to estimate a first-order decay rate constant from log-transformed concentration data shown in Figure 14.4 is a familiar problem that illustrates the process of parameter estimation. The concentration of a substance undergoing first-order decay in a batch reactor is specified by the following differential equation and initial condition:

$$\frac{dC}{dt} = -kC \qquad C(t = 0) = C_0 \tag{14.1}$$

The solution to this equation is:

$$C(t) = C_0 \exp(-kt) \tag{14.2}$$

If the rate constant k is unknown, it can be determined from an experiment in which the concentration is measured as the substance decays. Taking the natural logarithm of both sides of Eq. (14.2) gives

$$\ln C(t) = -kt + \ln C_0 \quad \text{or} \quad -\ln\left[\frac{C(t)}{C_0}\right] = kt \tag{14.3}$$

A plot of $-\ln[C(t)/C_0]$ versus time is thus expected to be a straight line with a slope equal to k. The concentration measurements are likely to have errors associated with them, leading to some scatter about the line, as suggested in Figure 14.4.

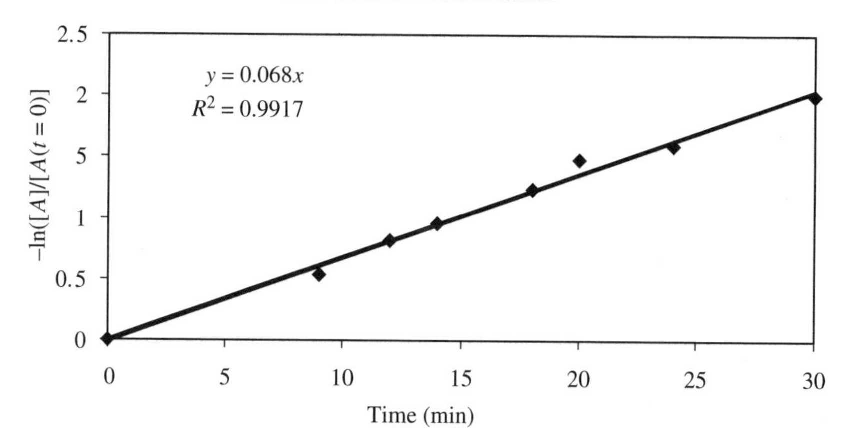

Figure 14.4 Use of linear regression to determine the rate constant for a first-order decay reaction. In this case, the rate constant is 0.068 min^{-1}.

In general, simple linear regression applies when observations can be described by an equation of the form:

$$Y_i = \beta_0 + \beta_1 X_i + \varepsilon_i \tag{14.4}$$

Linear regression gives the "best-fit" equation for the line, which includes an estimate of β_1, its slope, and β_0, its intercept. But what is meant by "best fit"?

Ordinary least-squares (OLS) linear regression seeks to minimize the sum of the squared residuals between the observations, the Y_i's, and the model results, denoted by $\hat{Y}_i$:

$$\text{Minimize} \sum_{i=1}^{n} \left(Y_i - \hat{Y}_i\right)^2 \tag{14.5}$$

The solution to this *unconstrained* optimization problem is $\hat{Y} = b_0 + b_1 X$, where:

$$b_1 = \frac{\sum_{i=1}^{n} \left(X_i - \overline{X}\right)\left(Y_i - \overline{Y}\right)}{\sum_{i=1}^{n} \left(X_i - \overline{X}\right)^2} \quad \text{and} \quad b_0 = \overline{Y} - b_1 \overline{X} \tag{14.6}$$

The overbar in Eq. (14.6) denotes the arithmetic mean of the sample values. It is important to realize that no statistical assumptions are used to obtain the expressions for b_1 and b_0. They are derived simply by substituting Eq. (14.6) into the objective function [Eq. (14.5)] and then setting the derivatives with respect to X and Y equal to zero. However, certain error characteristics allow confidence limits for the values of b_1 and b_0 to be estimated, and statistical tests of significance and goodness of fit to be applied to the linear regression model. The standard statistical assumptions are that the errors in the observations are independent and normally distributed random variables with a mean of zero and a constant variance.

Use of least-squares minimization as the objective function in linear regression is intuitively appealing because it accounts for both positive and negative residuals without letting them cancel each other out, as would be the case if a simple sum of differences were used. Furthermore, with the assumptions that the errors in all of the observations have the same probability distribution, it makes sense to weight equally the residuals that correspond to each observation. The field of statistics offers an additional rationale for the use of OLS when the assumptions stated in the previous paragraph apply. As discussed in Chapter 7, the *likelihood function* gives the probability that a particular sample of observations, Y_i's, would be observed given that the estimated parameter values used in the model are correct. Using this definition, the "best" parameters should be those that maximize the likelihood function. If the ε_i's, are in fact independent and normally distributed with mean of zero and constant variance, then OLS gives the maximum-likelihood (ML) parameter estimates.

Multiple linear regression is used to estimate the constant coefficients (or parameters) of a linear model relating a dependent variable to a set of independent variables that are thought to affect it, that is, for the ith observation:

$$Y_i = \beta_0 + \sum_{j=1}^{J} \beta_j X_{ji} + \varepsilon_i \tag{14.7}$$

With the objective function given by Eq. (14.5), the OLS solution is given by:

$$\hat{Y} = b_0 + b_1 X_1 + \cdots + b_j X_j + \cdots + b_J X_J \tag{14.8a}$$

where the coefficients are given by:

$$\mathbf{b} = \left(X^{\mathrm{T}} X\right)^{-1} X^{\mathrm{T}} \mathbf{Y} \tag{14.8b}$$

In Eq. (14.8b), matrix notation is used with X representing an augmented $n \times (J+1)$ matrix of observations of the independent variables:

$$X = \begin{bmatrix} 1 & X_{11} & \cdots & X_{J1} \\ 1 & X_{12} & \cdots & X_{J2} \\ \vdots & \vdots & \cdots & \vdots \\ 1 & X_{1n} & \cdots & X_{Jn} \end{bmatrix} \tag{14.8c}$$

Again, this solution follows from the objective of least-squares minimization [Eq. (14.5)] without relying on any assumptions about the distribution of the errors in Eq. (14.7). However, as with simple linear regression, if the ε_i's are independent and normally distributed with a mean of zero and a constant variance, then the OLS solution gives the maximum-likelihood estimators. If the ε_i's are independent, normal, have a mean of zero, and have known but not necessarily constant variances, the ML solution is given by weighted least squares:

$$\text{Minimize} \sum_{i=1}^{n} w_i \left(Y_i - \hat{Y}_i\right)^2 = (\mathbf{Y} - X\boldsymbol{\beta})^{\mathrm{T}} w (\mathbf{Y} - X\boldsymbol{\beta}) \tag{14.9}$$

where $w_i = 1/\sigma_i^2$. For multiple linear regression, the weighted least-squares (WLS) solution is

$$\mathbf{b} = \left(X^{\mathrm{T}} w X\right)^{-1} X^{\mathrm{T}} w \mathbf{Y} \tag{14.10}$$

where

$$w = \begin{bmatrix} \sigma_1^2 & 0 & \cdots & 0 \\ 0 & \sigma_2^2 & \cdots & 0 \\ \vdots & \vdots & \cdots & \vdots \\ 0 & 0 & \cdots & \sigma_n^2 \end{bmatrix}$$

Solutions for other error distributions (e.g., nonzero covariances) are available in textbooks on regression analysis and parameter estimation (Beck and Arnold, 1977).

Linear regression is a powerful parameter estimation method, but its applicability is limited to special cases in which the state variables are linear functions of the parameters to be estimated (or can be transformed to this condition). Moreover, even when the relationship between observable state variables and unspecified parameters is linear, the values that can be assumed by the parameters may be constrained, for example, to be positive. This changes a least-squares linear regression problem into a constrained optimization problem with a quadratic objective function. In environmental models, the relationship between the state variables and parameters is often not linear and, in fact, may not even be able to be expressed in closed algebraic form. This would be the case, for example, if numerical solution techniques are required to integrate the ordinary or partial differential equations that comprise the model.

As in linear regression, the most common approaches to nonlinear parameter estimation utilize OLS, WLS, or ML objective functions. For example, using weighted least squares, the objective function for a nonlinear parameter estimation problem can be expressed as:

$$\text{Minimize} \sum_{i=1}^{n} w_i \left[Y_i - \hat{Y}_i(\mathbf{k}, \mathbf{p}; \mathbf{x}, t) \right]^2 = \left[\mathbf{Y} - \hat{\mathbf{Y}}(\mathbf{k}, \mathbf{p}; \mathbf{x}, t) \right]^T w \left[\mathbf{Y} - \hat{\mathbf{Y}}(\mathbf{k}, \mathbf{p}; \mathbf{x}, t) \right] \tag{14.11}$$

where $\mathbf{p}$ is the vector of parameters to be estimated, $\mathbf{k}$ denotes the other parameters in the model, $\mathbf{x}$ is the position vector (for spatially variable models), and t is time. In the objective function, $\hat{Y}$ might be an explicit function of $\mathbf{p}$. However, it might also be an output from a numerical model that uses finite differences or finite elements to solve a system of differential equations. In this case, the numerical model is often treated as a "black box" that delivers an output value (or set of values) for a trial set of parameter values.

Parameter estimation problems may also have constraints on the values of $\mathbf{p}$ that may be linear or nonlinear and may be either explicit or implicit functions of $\mathbf{p}$. Often, the constraints are simple bounds on the values that $\mathbf{p}$ can assume. For these cases, constrained optimization methods must be used. A brief introduction to the theory and nomenclature of constrained optimization is provided in the supplementary information at www.wiley.com/college/ramaswami.

A wide variety of solution techniques are available for nonlinear optimization problems. The most common methods include quasi-Newton methods and steepest ascent/descent methods such as Levenberg–Marquardt, which is briefly described in Example 14.1. Bard (1974) and Beck and Arnold (1977) provide detailed discussions of nonlinear programming algorithms with applications to parameter estimation problems. Press et al. (1994) outline algorithms for several popular methods. Many environmental computer models for air, water, groundwater, ecosystem, or pharmacokinetic systems now imbed one of these routines to allow for automated parameter estimation. A number of general parameter estimation software packages are also available, including the DAKOTA (Design Analysis Kit for Optimization and Terascale Applications) toolkit (http://endo.sandia.gov/DAKOTA/software.html), the Parameter Estimation, Inc. (PEST) package (http://www.parameter-estimation.com/), the freeware PEST package (http://www.sspa.com/pest/), the UCODE (a computer code for universal inverse modeling) (http://typhoon.mines.edu/freewarc/ucode/), and the PopTools package for use with Excel and Solver (http://sunsite.univie.ac.at/Spreadsite/poptools/). These packages can be used as a shell around any general model with accessible input and output files, given a set of data against which the model output is compared and optimized.

EXAMPLE 14.1

In an investigation of the effectiveness of engineering techniques for controlling exposure to environmental tobacco smoke, Miller et al. (1997) used nonlinear parameter estimation to estimate airflow rates in a two-room building. They modeled the building as two well-mixed zones that exchange air with each other and with the outside. Figure 14.5 shows the basic configuration of the rooms and the airflows that were considered in the model. Miller and her colleagues released two different tracer gases, one in each room of the building, and then tracked their concentrations in each room.

Assuming complete mixing and constant flow rates, the four tracer concentrations (two tracers in two rooms) are described by a system of four differential equations:

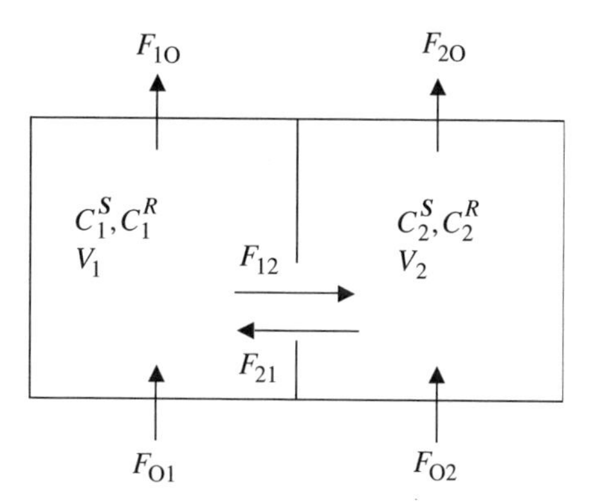

Figure 14.5 Schematic diagram of the two-zone model. Tracer gas S is injected into zone 1; gas R into zone 2.

$$
\begin{aligned}
\frac{dC_1(t)}{dt} &= \frac{dC_1^S(t)}{dt} = \frac{1}{V_1}\left[F_{21}C_2^S(t) - (F_{12} + F_{1O})C_1^S(t)\right] \\
\frac{dC_2(t)}{dt} &= \frac{dC_2^S(t)}{dt} = \frac{1}{V_2}\left[F_{12}C_1^S(t) - (F_{21} + F_{2O})C_2^S(t)\right] \\
\frac{dC_3(t)}{dt} &= \frac{dC_1^R(t)}{dt} = \frac{1}{V_1}\left[F_{21}C_2^R(t) - (F_{12} + F_{1O})C_1^R(t)\right] \\
\frac{dC_4(t)}{dt} &= \frac{dC_2^R(t)}{dt} = \frac{1}{V_2}\left[F_{12}C_1^R(t) - (F_{21} + F_{2O})C_2^R(t)\right]
\end{aligned}
\tag{14.12}
$$

and initial conditions:

$$
\begin{aligned}
&C_1(t=0) = C_1^S(t=0) = C_{SO};\ C_2(t=0) = C_2^S(t=0) = 0 \\
&C_3(t=0) = C_1^R(t=0) = 0;\ C_4(t=0) = C_2^R(t=0) = C_{RO}
\end{aligned}
$$

In Eq. (14.12), $C_k^j(t)$ [ppb(v)] represents the mole fraction of the jth tracer in the kth room, V_k (m^3) is the volume of the kth room, and F_{kl} represents the volumetric flow rate (m^3/h) from room k to room l, where O indicates the outdoors. The linear, first-order system of ordinary differential equations (14.12) can be solved analytically using eigenvalues.

Two airflow balance equations also apply:

$$
\begin{aligned}
F_{21} + F_{O1} &= F_{12} + F_{1O} \\
F_{12} + F_{O2} &= F_{21} + F_{2O}
\end{aligned}
\tag{14.13}
$$

In addition, all of the flow rates must remain nonnegative.

Miller et al. (1997) used a weighted least-squares objective function with nonnegativity constraints to estimate F_{12}, F_{21}, F_{1O}, and F_{2O}:

$$
\text{Minimize} \sum_{m=1}^{4} \frac{1}{C_{m,\max}} \sum_{i=1}^{N} \frac{\left[C_{m,i} - \hat{C}_m(t_i, \mathbf{F})\right]^2}{\sigma_i^2}
\tag{14.14}
$$

$$
\text{such that } \mathbf{F} > \mathbf{0}
$$

In Eq. (14.14), $C_{m,\max}$ represents the maximum measured concentration of the mth tracer/room combination, $C_{m,i}$ is the concentration measured at time t_i, and $\hat{C}_m(t_i, \mathbf{F})$ is the concentration estimated from the analytical solution to Eq. (14.12) using the trial airflow rates; σ_i is the standard deviation associated with the error in the measurements. Given optimal estimates of F_{12}, F_{21}, F_{1O}, and F_{2O}, they then calculated F_{O1} and F_{O2} from Eq. (14.13).

Miller et al. (1997) used the Levenberg–Marquardt (LM) method (Press et al., 1994) to perform the weighted least-squares optimization. Starting with an initial guess, successive iterations $\mathbf{F}^{q+1} = \mathbf{F}^q + \delta\mathbf{F}^{q+1}$ move the parameter estimates toward the values that optimize Eq. (14.14). The iterations are continued until acceptable convergence is achieved with minimal changes in either the value of the objective function or the decision variables. In the LM method, the correction vector, $\delta\mathbf{F}^{q+1}$ is based on the gradient of the objective function (method of steepest descent) when the trial solution is still far from the optimum. As the optimum value is approached, the size of the iterative step is reduced and the correction vector determined using a Gauss–Newton technique (Beck and Arnold, 1977).

Miller et al. (1997) initially assumed that the errors in their observations would be independent, normally distributed random variables with a mean of zero and a constant variance of σ^2. Figure 14.6 shows the normalized residuals, $[C_{m,i} - \hat{C}_m(t_i, \mathbf{F})]/C_{m,i}$, obtained after the data in one of their experiments were fit using a constant value of σ in Eq. (14.14). The figure shows that in this case, the assumptions of independent, random errors with constant variance are not met. To correct for this problem, the time dependence of the error distribution was modeled using a power function of the form αt^β, which was fit to the absolute value of the residuals. This has the effect of downweighting the fit during the first hour of the experiments, when incomplete mixing is thought to have introduced relatively large errors. The final concentration results for the same experiment are shown in Figure 14.7. The corresponding optimal parameter values are listed in the caption. Sohn and Small (1999) provide an evaluation of alternative optimization methods for parameter estimation for a simulated system similar to that of Miller et al. (1997), but with open compartments assumed to have effective (unknown) volumes. Their analysis simulates the parameter estimation errors that result from limited, imprecise measurements of the trace gas concentrations and

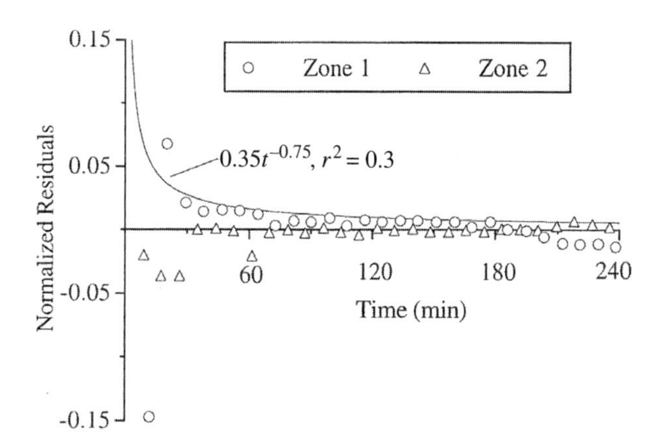

Figure 14.6 Normalized tracer concentration residuals as a function of time, assuming constant measurement errors. Open circles or triangles represent measurements; the solid line is the power function fit to the residuals. The magnitude of the residuals indicates the size of the deviation relative to the peak normalized concentration, e.g., a residual of 0.15 indicates a 15% deviation. (Reprinted by permission from Miller et al., 1997.)

compares the performance of steepest descent and simulated annealing optimization methods for such poorly specified systems that exhibit parameter identification problems.

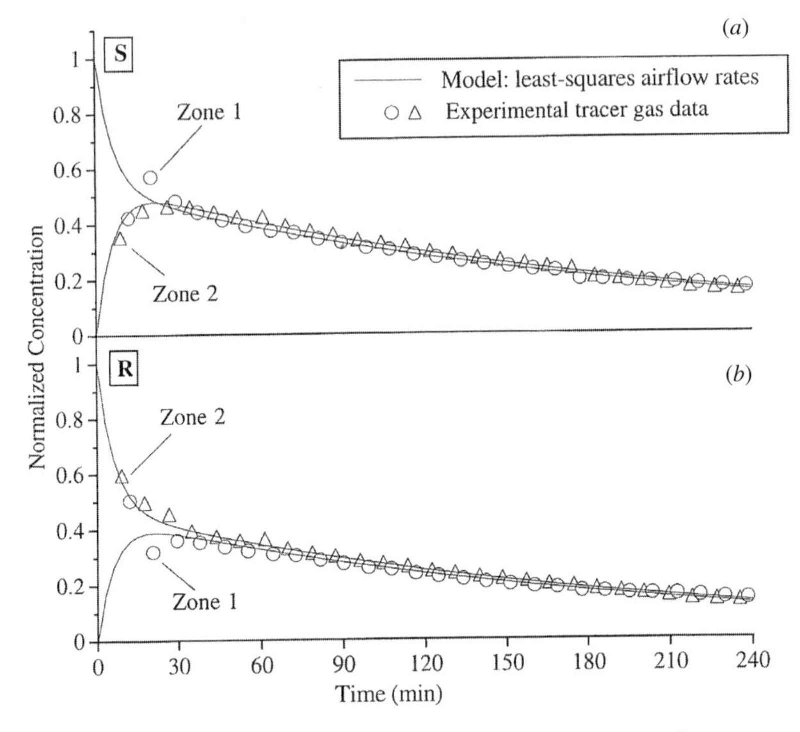

Figure 14.7 Fitted tracer concentrations of two tracers (denoted R and S) in two zones. Open circles or triangles represent measurements; the solid lines show the model results. The optimal parameter values used in the model are: F_{21} (m^3/h) $= 154 \pm 17$; F_{12} (m^3/h) $= 163 \pm 18$; F_{10} (m^3/h) $= 10 \pm 5$; F_{20} (m^3/h) $= 11 \pm 5$; F_{01} (m^3/h) $= 19 \pm 1.4$; F_{02} (m^3/h) $= 2 \pm 1.3$ (m^3/h); C_{S0} (ppbv) $= 129 \pm 2$; C_{R0} (ppbv) $= 365 \pm 8$.

14.4 SENSITIVITY AND UNCERTAINTY ANALYSIS

The point estimate outputs produced from a standard model run give an incomplete and often misleading representation of the modeler's understanding of the system he or she is modeling. Two types of information are missing. First, point estimates give no indication of the range of *variability* that might be expected in the system. Second, they provide no information about the *uncertainty* (or conversely level of confidence) in the model results.

Sources of uncertainty in environmental models include model structure, for example, missing chemical reactions or inadequate spatial resolution, inherent uncertainty associated with the stochastic nature of future weather and/or turbulent flows (Fox, 1984; Small and Mular, 1987; Venkatram, 1988; Jury and Gruber, 1989; Weil et al., 1992), and lack of knowledge concerning the appropriate values for empirical parameters and inputs such as rate constants or emissions rates. Although they are difficult to treat probabilistically, uncertainties in model structure can be explored by comparing results from alternative models, for example, by comparing wind fields estimated by objective interpolation schemes versus prognostic meteorological models, or groundwater contaminant predictions generated by a model with single versus multiple aquifer layers (NRC, 1990; Sohn et al., 2000). Venkatram (1988) and Weil et al. (1992) have suggested that inherent uncertainties in atmospheric transport models may best be explored through laboratory studies of turbulence or by resolving more of the turbulent motion in environmental models using large eddy or direct numerical simulation techniques. Beck (1987) notes that critical sources of uncertainty in surface water quality modeling involve issues of complexity and aggregation in representing multiscale, multiprocess hydrologic systems. In groundwater systems, the greatest uncertainty is usually associated with the stochastic nature of subsurface deposits and the resulting spatial field of hydraulic conductivity (Dagan, 1987; Gelhar, 1993; Gross and Small, 1998; Rubin, 2003), although random variations and uncertainty in soil chemical properties must also be considered (Robin et al., 1991). Important issues of uncertainty in environmental exposure and human health risk models arise from the fundamental variation in response to chemical and other stresses at the cellular, tissue, organ, and individual levels and the inability to predict the patterns of individual and societal behavior necessary to construct future scenarios for exposure and risk (Rhomberg, 2004; Hattis, 2004; Cullen and Small, 2004).

Uncertainty and variability associated with fixed model parameters such as partition coefficients and application-dependent inputs such as initial concentrations or loadings can often be quantified probabilistically. In this case, single-valued, best-estimate model results can be supplemented with information about the probability that a range of output values could occur. Uncertainties in model parameters/inputs may be due to: (1) both random and systematic errors in measurements, (2) assumptions required to estimate parameter values from theory and indirect measurements, or (3) approximation errors entailed in abstracting from reality to make models tractable. Variability over space, over time, or between subjects (e.g., variability in fish consumption between individuals in an exposure study) is a distinct issue that can also be treated probabilistically. In many studies, variability is considered together with other sources of uncertainty, but it is usually best to treat it separately, especially in exposure and risk analyses (Hattis and Burmaster, 1994; Hoffman and Hammonds, 1994; Cullen and Frey, 1999; Cullen and Small, 2004). The important distinction is that, in principle, obtaining better information will reduce uncertainty, whereas variability is inherent to the system and cannot be reduced. Two-dimensional probabilistic analyses that separate uncertainty and variability are discussed below.

A systematic approach to quantitative analysis of uncertainty (and/or variability) for any mathematical model begins by specifying the conditions to which the model will be applied and identifying potential sources of uncertainty (Alcamo and Bartnicki, 1987). Often, sensitivity analysis is performed next, to determine which parameters or inputs are likely to have a significant effect on model outputs (e.g., Dakins et al., 1996). Experience has shown that in most cases, a relatively small subset of the variables is influential. Estimates of uncertainty/variability in the influential parameters and inputs are then propagated through the model for the specified simulation conditions. The results are analyzed to assess the uncertainty in the model outputs and to determine how much of it is due to each source that was considered. Sensitivity analysis methods for quantifying a model's response to changes in its inputs or parameters are presented in the next subsection. Methods for quantifying uncertainty in model outputs due to input and parameter uncertainties are described in Sections 14.4.2 to 14.4.4. Much of our presentation in Sections 14.4.1 to 14.4.3 follows that of Morgan and Henrion (1990), who provide additional detail on these topics. Additional discussion of available methods is found in Saltelli et al. (2000) and Frey and Patil (2002).

14.4.1 Sensitivity Analysis

Sensitivity analysis examines the effect of changes in a model's structure, inputs, or parameter values on its predictions. Here we focus on analysis of sensitivity to model parameters or input variables. Figure 14.8 illustrates the concept of sensitivity analysis for a model that can be specified by the equation $y = f(x_1, x_2)$, where y is an

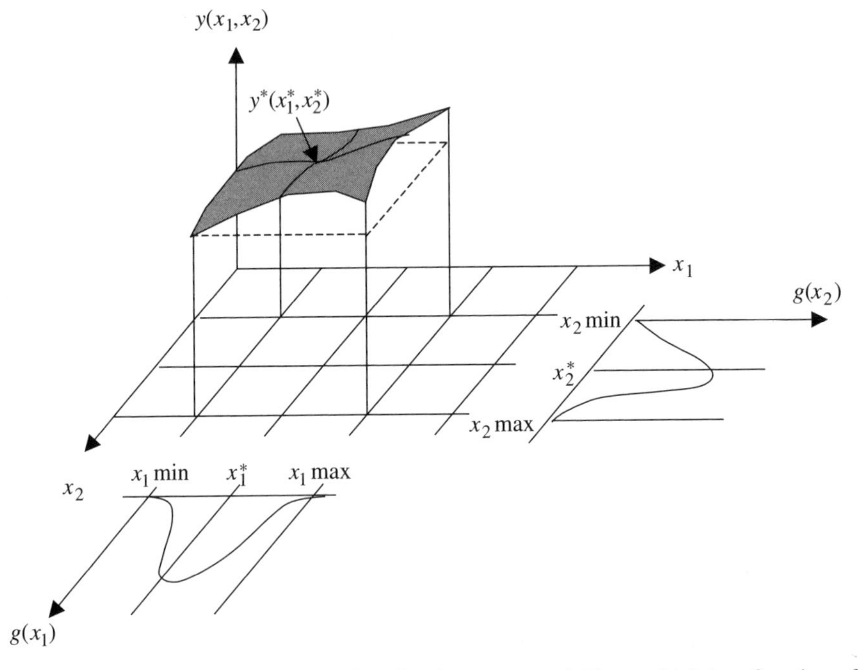

Figure 14.8 Schematic response surface for the output variable y, which is a function of the parameters x_1 and x_2. (Adapted from McRae, 1986.). The response surface results from solving for y as x_1 and x_2 are varied over their full domains.

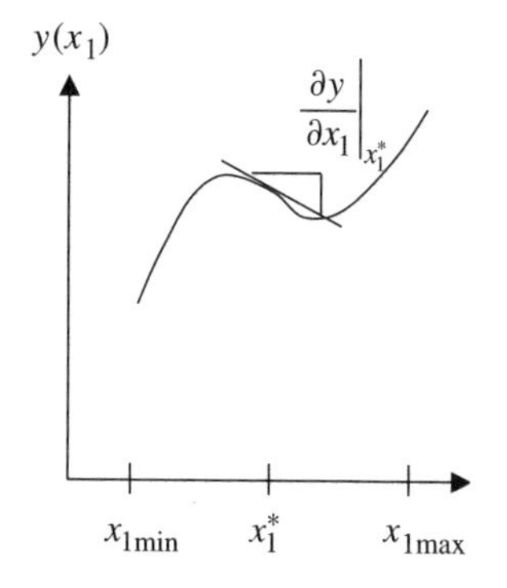

Figure 14.9 Graphical definition of a local sensitivity coefficient showing the response of an output variable, y, to a small change in an input variable in the neighborhood of its nominal value, x^*.

output variable, x_1 and x_2 are model inputs, and f is the continuously differentiable function (or model) that describes their relationship. The figure presents a response surface, which shows directly how y varies with x_1 and x_2. The nominal or base case scenario, $y^* = f(x_1^*, x_2^*)$ gives the best estimate value of the output, y^*, corresponding to the best estimate values of the inputs. A *local* sensitivity analysis determines the rate of change of the output y associated with small changes in the input values in the neighborhood about their nominal values:

$$S_1 = \left.\frac{\partial y}{\partial x_1}\right|_{x_1^*, x_2^*} \qquad S_2 = \left.\frac{\partial y}{\partial x_2}\right|_{x_1^*, x_2^*} \tag{14.15}$$

The sensitivity coefficients, S_1 and S_2, are graphically interpreted as the slopes of the tangents to the response surface as shown in Figure 14.9, evaluated at the nominal input values. For models with multiple outputs and/or more than two inputs or parameters, sensitivity coefficients for all of the output/input pairs can be evaluated from partial derivatives analogous to those given in Eq. (14.15).

The values of the sensitivity coefficients given by Eq. (14.15) depend on the units of x_i and y and so are not directly comparable across input parameters with different units. To circumvent this problem, normalized sensitivities or *elasticities* are commonly used:

$$\begin{aligned} S_1' &= \left.\frac{\partial y}{\partial x_1}\right|_{x_1^*, x_2^*} \times \frac{x_1^*}{y^*} = \left.\frac{\partial \ln y}{\partial \ln x_1}\right|_{x_1^*, x_2^*} \\ S_2' &= \left.\frac{\partial y}{\partial x_2}\right|_{x_1^*, x_2^*} \times \frac{x_2^*}{y^*} = \left.\frac{\partial \ln y}{\partial \ln x_2}\right|_{x_1^*, x_2^*} \end{aligned} \tag{14.16}$$

Local sensitivity coefficients are often approximated by finite differences calculated from repeated function evaluations or model runs, for example, using a forward difference approximation:

$$S_1 \cong \left.\frac{\Delta y}{\Delta x_1}\right|_{x_1^*, x_2^*} = \frac{f(x_1^* + \Delta x_1, x_2^*) - f(x_1^*, x_2^*)}{\Delta x_1} \tag{14.17}$$

Backward or central difference approximations can also be utilized for the derivatives. This approach, in which the model is run repeatedly while one parameter at a time is varied a small amount from its nominal value, is sometimes called *brute force* sensitivity analysis. More efficient methods, which use symbolic differentiation and allow evaluation of many sensitivity coefficients simultaneously, are also available (Hwang et al., 1978; Dunker, 1984; Bischof et al., 1992; Horwedel et al., 1992; Piasecki and Katopodes, 1997; Dunker et al., 2002; Hakami et al., 2003).

Figures 14.8 and 14.9 illustrate the fact that local sensitivity coefficients depend on the point in input parameter space at which they are evaluated. If the parameter values are uncertain, sensitivities for more than the nominal point may be of interest. This issue can be examined graphically through parametric analysis, in which the model is run repeatedly for a range of values of the ith parameter, x_i, with other parameters held at their nominal values. Graphs of the results for y of parametric analyses in which x_1 or x_2 alone is varied would correspond to slices through the response surface shown in Figure 14.8. The response surface itself shows the results of a parametric analysis in which both input parameters are varied.

14.4.2 Uncertainty Analysis

Uncertainty analysis adds important information beyond sensitivity analysis by considering the degree of uncertainty in model inputs and parameters and then estimating the effect of these uncertainties on model outputs. Two basic approaches are commonly used to propagate uncertainty estimates through environmental fate, transport, and risk models. The first approach uses a Taylor series approximation to represent the model and estimates the variance in an output variable from this approximation. The Taylor series is commonly truncated after the first term, in which case this approach is known as first-order uncertainty analysis. The second popular approach to uncertainty propagation is Monte Carlo analysis, a combinatorial method in which repeated model runs are completed using sample vectors of input parameter values that are drawn from probability distribution functions assigned to the parameters.

14.4.3 First-Order Uncertainty Analysis (FOUA)

Consider a model that can be represented as $y = f(\mathbf{x})$, where $\mathbf{x} = \{x_1, x_2, \ldots, x_n\}^T$ is a vector of uncertain parameters or model inputs and the function f is continuously differentiable in the neighborhood about $\mathbf{x}$. The Taylor series expansion for y about its nominal value, y^*, is

$$y = y^* + \sum_{i=1}^{n} (x_i - x_i^*) \left[\frac{\partial y}{\partial x_i} \right]\Bigg|_{\mathbf{x}^*} + \frac{1}{2} \sum_{i=1}^{n} \sum_{j=1}^{n} (x_i - x_i^*)(x_j - x_j^*) \left[\frac{\partial^2 y}{\partial x_i\, \partial x_j} \right]\Bigg|_{\mathbf{x}^*} + \text{HOT} \tag{14.18}$$

If $x_i^* = E[x_i]$ is the expected value of the parameter x_i, then the variance of y, $E[(y - y^*)^2]$, resulting from uncertainty in the x_i values, can be approximately evaluated from the first-order term of the Taylor series as:

$$E\left[(y-y^*)^2\right] \cong E\left[\left(\sum_{i=1}^{n}(x_i-x_i^*)\left[\frac{\partial y}{\partial x_i}\right]\bigg|_{\mathbf{x}^*}\right)^2\right]$$
$$= \sum_{i=1}^{n}\sum_{j=1}^{n}\mathrm{Cov}[x_i,x_j]\left[\frac{\partial y}{\partial x_i}\right]\bigg|_{\mathbf{x}^*}\left[\frac{\partial y}{\partial x_j}\right]\bigg|_{\mathbf{x}^*} \tag{14.19}$$

If the x_i are independent random variables ($\mathrm{Cov}[x_i, x_j] = 0$ for $i \neq j$) then Eq. (14.19) reduces to

$$E\left[(y-y^*)^2\right] \cong \sum_{i=1}^{n}\mathrm{Var}[x_i]\left(\left[\frac{\partial y}{\partial x_i}\right]\bigg|_{\mathbf{x}^*}\right)^2 \tag{14.20}$$

The individual terms of the sum on the rhs of Eq. (14.20) represent the contributions of each of the n parameters to the variance in the output variable.

Because second-order and higher terms are neglected in deriving Eqs. (14.19) and (14.20), the accuracy of the approximation of the output uncertainty may be poor if the uncertainties in the input parameters are large or if y is an especially nonlinear function of $\mathbf{x}$. Morgan and Henrion (1990) show the exact analytical formulas that can be derived from the first-order Taylor series expansion in the special cases of linear models and multiplicative models with log-transformed parameters. They also discuss higher-order Taylor series approximations.

For models with multiple outputs of concern, for example, concentrations of a given chemical predicted at multiple locations and times, or concentrations and exposures predicted for multiple chemicals and receptors, the uncertainties in these outputs are likely to be correlated. For example, in the multimedia exposure and risk model for benzene that is presented in Chapter 13, if the lifetime average daily dose (LADD) is higher than its nominal estimate, then the lifetime individual risk (LIR) is likely to be higher than its nominal estimate as well. A positive covariance (and a positive correlation coefficient) is thus expected for the uncertainties in the LADD and the LIR. For other model outputs, it may be more difficult to predict whether a positive, negative, or approximately zero association in the uncertainties will occur. Consider, for example, the ambient concentrations of benzene in the water, C_w, and the air, C_a, for the integrated exposure and risk model presented in Chapter 13. If the uncertainties in these model state variables are dominated by the effects of the uncertain K_{aw}, with C_a increasing as K_{aw} increases and C_w decreasing as K_{aw} increases, then a negative correlation coefficient is expected for the uncertainties in the air and water concentrations. However, if the uncertainties in the air and water concentrations are dominated by the joint effects of the uncertain half-lives, $t_{1/2,\mathrm{air}}$ and $t_{1/2,\mathrm{water}}$, since these uncertain inputs affect both outputs in the same manner (increasing the half-life in either the air or water will cause both C_a and C_w to increase), a positive correlation should be predicted. A method is thus needed to assess both the sign and the magnitude of the uncertainty covariance among model outputs.

A matrix-based extension of FOUA is available for estimating the covariance of model outputs, given the matrix of model sensitivities (also known as the Jacobian of the input–output system) and the covariance matrix representing the joint uncertainty of the model inputs. For n_x model inputs, $x_i, i = 1, \ldots, n_x$; and n_y model outputs, $y_k, k = 1, \ldots, n_y$; the n_y by n_x model sensitivity matrix, A_{yx}, is given by:

$$[A_{yx}] = \left[\left. \frac{\partial y_k}{\partial x_i} \right|_{\mathbf{x}^*} \right] \tag{14.21}$$

where $\mathbf{x}^*$ denotes that the local sensitivity coefficients are computed with all inputs at their nominal values. The $n_x \times n_x$ covariance matrix of the inputs, S_x, is defined as:

$$[S_x] = \left[\ \mathrm{Cov}[x_i, x_j] \ \right] \tag{14.22}$$

The $n_y \times n_y$ covariance matrix of the outputs, S_y, is then computed as:

$$[S_y] = [A_{yx}] \times [S_x] \times [A_{yx}]^{\mathrm{T}} \tag{14.23}$$

where $[\ ]^{\mathrm{T}}$ denotes the matrix transpose operation. Application of Eq. (14.23) is illustrated for the integrated benzene exposure and risk model in the following example.

EXAMPLE 14.2

Based on the multicompartment benzene concentration from Example 3.9, level II multimedia model, and cancer risk estimation from Examples 13.2 and 13.3, perform an uncertainty analysis of the following model outputs: the benzene concentration in water, the benzene concentration in air, the benzene dose to a representative individual, and the lifetime individual risk (LIR) for cancer.[1] Assume that the following model inputs are uncertain with the indicated uncertainty distributions, and that the input uncertainties are independent:

Air–water partition coefficient	$\sim U(0.15, 0.30)$
Soil–water partition coefficient	$\sim U(2.0, 3.31)$
Reaction half-life in air (h)	$\sim U(10, 24)$
Reaction half-life in water (h)	$\sim U(100, 240)$
Water ingestion rate (L/day)	$\sim U(1, 1.8)$
Air inhalation rate (m^3/day)	$\sim U(15, 25)$
Body weight (kg)	$\sim U(60, 80)$
Cancer slope factor $(\text{mg/kg}\cdot\text{day})^{-1}$	$\sim U(0.005, 0.053)$

[1] This example, along with Examples 14.3 and 14.4, are based on a student project conducted by Win Trivitayanurak, for the Carnegie Mellon University course Mathematical Modeling of Environmental Systems. His contribution is gratefully acknowledged.

Use the FOUA method to compute the uncertainty of the four targeted model outputs, including the correlation among them. Also, demonstrate the relative importance of the eight uncertain model inputs to the uncertainty in the predicted risk.

Solution 14.2 First construct a model that links the level II model with the exposure and risk calculations. This model will receive eight uncertain inputs and produce four outputs. Since the inputs are all independent, the variance of each output is determined by Eq. (14.20). The derivative of an output with respect to each input, x_i, can be calculated either analytically or numerically. The mean and variances of the inputs are computed for a uniform distribution $\sim U(a, b)$ by:

$$x^* = \frac{a+b}{2} \qquad \text{Var}[x] = \frac{(b-a)^2}{12}$$

The mean and variance of the uncertain inputs are

Inputs	$\sim U(a,b)$		Mean	Variance
	a	b		
K_{aw}	0.15	0.3	0.225	0.0019
K_d	2	3.31	2.655	0.1430
$t_{1/2,\text{air}}$ (h)	10	24	17	16.33
$t_{1/2,\text{water}}$ (h)	100	240	170	1633
R_{water} (L/day)	1	1.8	1.4	0.053
R_{air} (m^3/day)	15	25	20	8.33
BW (kg)	60	80	70	33.33
β $(\text{mg/kg}\cdot\text{day})^{-1}$	0.005	0.053	0.029	0.00019

The output derivatives are estimated numerically by central differencing, using $\Delta x = 0.05\%$ of x^*;

$$\frac{\partial y}{\partial x_i} \cong \frac{\Delta y}{\Delta x_i} = \frac{y|_{\mathbf{x}^*, x_i^* + \Delta x} - y|_{\mathbf{x}^*, x_i^* - \Delta x}}{2\Delta x}$$

First-order uncertainty analysis is then performed on each output with respect to the eight inputs. The partial derivatives of the outputs with respect to each input are shown in Table 14.2, and the contributions to the variances of each output from each input are shown in Table 14.3.

As indicated, the variance in the benzene water concentration (C_w) is due to nearly equal contributions from the uncertainty in K_{aw} and $t_{1/2,\text{ air}}$, while $t_{1/2,\text{ air}}$ is responsible for virtually all of the uncertainty in C_a and more than half (57%) of the variance in the dose. The remaining portion of the dose variance is due to the assumed inhalation rate, R_{air} (32%), and body weight, BW (11%). The same three inputs affect the uncertainty in the lifetime individual risk, however, their effects are all relatively small compared to those caused by uncertainty in the cancer slope factor, β, which contributes 78% of the total variance in the LIR.

TABLE 14.2 First-Order Derivatives of Outputs with Respect to Each Input

Input	$\frac{\partial C_w}{\partial x}$	$\frac{\partial C_a}{\partial x}$	$\frac{\partial (\text{LADD})}{\partial x}$	$\frac{\partial (\text{LIR})}{\partial x}$
K_{aw}	-3.88×10^{-3}	7.99×10^{-7}	1.14×10^{-7}	3.30×10^{-9}
K_d	-8.47×10^{-9}	-1.91×10^{-9}	-5.44×10^{-10}	-1.58×10^{-11}
$t_{1/2,\text{air}}$	4.13×10^{-5}	9.29×10^{-6}	2.65×10^{-6}	7.70×10^{-8}
$t_{1/2,\text{water}}$	3.67×10^{-9}	8.26×10^{-10}	2.36×10^{-10}	6.84×10^{-12}
R_{water}	0	0	1.25×10^{-8}	3.62×10^{-10}
R_{air}	0	0	2.81×10^{-6}	8.15×10^{-8}
BW	0	0	-8.04×10^{-7}	-2.33×10^{-8}
β	0	0	0	5.63×10^{-5}

The covariances of the outputs are computed using Eq. (14.23). In this case, the covariance matrix S_x used in Eq. (14.23) has the variances of the x_i on the diagonal, and the off-diagonal elements are zeros because the inputs are assumed to be independent. Correlation coefficients are calculated from the covariances using the relation:

$$\rho_{Y_1,Y_2} = \frac{\text{Cov}(Y_1, Y_2)}{\sigma_{Y_1} \cdot \sigma_{Y_2}} \tag{14.24}$$

TABLE 14.3 Variance of Indicated Model Output Contributed by Each Input, as Calculated by FOUA[a]

Input	Var[C_w] (mg/m^3)^2	Var[C_a] (mg/m^3)^2	Var[LADD] (mg/kg $\cdot$ day)^2	Var[LIR]
K_{aw}	2.83×10^{-8} (50%)	1.20×10^{-15} ($\ll$1%)	2.43×10^{-17} ($\ll$1%)	2.04×10^{-20} ($\ll$1%)
K_d	1.03×10^{-17} ($\ll$1%)	5.20×10^{-19} ($\ll$1%)	4.24×10^{-20} ($\ll$1%)	3.57×10^{-23} ($\ll$1%)
$t_{1/2,\text{air}}$	2.78×10^{-8} (50%)	1.41×10^{-9} (100%)	1.15×10^{-10} (57%)	9.68×10^{-14} (12%)
$t_{1/2,\text{water}}$	2.20×10^{-14} ($\ll$1%)	1.11×10^{-15} ($\ll$1%)	9.10×10^{-17} ($\ll$1%)	7.65×10^{-20} ($\ll$1%)
R_{water}	0 —	0 —	8.33×10^{-18} ($\ll$1%)	7.00×10^{-21} ($\ll$1%)
R_{air}	0 —	0 —	6.59×10^{-11} (32%)	5.54×10^{-14} (7%)
BW	0 —	0 —	2.15×10^{-11} (11%)	1.81×10^{-14} (2%)
β	0 —	0 —	0 —	6.08×10^{-13} (78%)
Total variance	5.61×10^{-8} (100%)	1.41×10^{-9} (100%)	2.03×10^{-10} (100%)	7.78×10^{-13} (100%)

[a] The percentage of the total variance of the output contributed by the given input uncertainty is indicated in parentheses.

TABLE 14.4 Correlation Coefficients Between Outputs

	C_w	C_a	LADD	LIR
C_w	1.00			
C_a	0.70	1.00		
LADD	0.53	0.75	1.00	
LIR	0.25	0.35	0.47	1.00

Computed correlation coefficients of the benzene model outputs are shown in Table 14.4. As expected, positive correlation coefficients are computed for the uncertainties in the LIR and the LADD, as well as between these variables and both C_a and C_w. The higher correlations between both LIR and LADD with C_a, as compared to C_w, are indicative of the result that the exposure and risk for benzene is dominated by the air exposure pathway. The positive correlation between C_a and C_w suggests that the negative-correlation-inducing effect of the uncertainty in the value of K_{aw} is less important than the positive-correlation-inducing effects of the uncertainties in the half-lives, especially $t_{1/2,\text{air}}$.

14.4.4 Monte Carlo Uncertainty Analysis

In FOUA, uncertainties in model outputs are estimated by using a linear approximation for their response to variations in input parameters. For environmental models that exhibit nonlinear behavior, this approximation may not be satisfactory. Moreover, only the second moment of the output uncertainty distribution (the variance) is estimated. Monte Carlo analysis is a combinatorial uncertainty propagation technique that can provide estimates of full probability distribution functions for model outputs and can accurately reflect any nonlinearity in a model. The downside of Monte Carlo analysis is that it can be computationally intensive, requiring hundreds or thousands of model runs in many cases.

In Section 7.4, we introduced Monte Carlo methods for simulating random variables and deriving distributions from simple bivariate relationships. Here, we extend this to the case of a general computational model with multiple inputs and outputs. As in the previous section, consider a model that can be represented by the function $y = f(\mathbf{x})$ where $\mathbf{x} = \{x_1, x_2, \ldots, x_n\}$. The first step of a Monte Carlo analysis is to estimate the probability distribution function, $g(\mathbf{x})$, that characterizes the joint uncertainty in the inputs $\mathbf{x}$. If the inputs are assumed to be independent, then the joint distribution is characterized by the individual, marginal probability distribution functions for each, $g_1(x_1), g_2(x_2), \ldots, g_n(x_n)$. For example, if x_1 is thought to be normally distributed, its probability distribution function $N(\mu_{x1}, \sigma_{x1})$ is specified by its mean, μ_{x1}, and standard deviation, σ_{x1}. As described in Chapter 7, these distributions can be developed based on statistical analysis of available data or on subjective judgment.

Once the model input uncertainty distribution is specified, Monte Carlo analysis proceeds by sampling a vector of inputs at random—for independent inputs these are generated by sampling from the distributions of each of the n variables—producing a sample vector, $\mathbf{x}^j = \{x_1^j, x_2^j, \ldots, x_n^j\}$, which defines the jth sample or *scenario*. The model is run using the jth vector of input parameter values and the resulting output value y^j is recorded. A new input sample vector is drawn and the model run again for this scenario. After this process has been repeated m times, a sample of m output values, $y^1, y^2, \ldots, y^m$ is produced, from which the distribution of the output can be estimated. Output distribution functions from Monte Carlo analysis are commonly presented graphically as histograms or cumulative distribution curves.

Confidence limits for the mean, standard deviation, or fractiles of the distribution of an output estimated using a given Monte Carlo sample size, m, can be determined using standard statistical techniques, assuming the sample vectors are random and independent. Conversely, the same statistical relationships can be used to determine the number of Monte Carlo scenarios needed to produce a confidence interval with a specified width.

Monte Carlo analysis with random sampling requires large numbers of model runs, especially if accurate estimation of the tails of the output distribution is of interest. This is often true in environmental modeling applications, if regulatory attention and/or significant risk are associated with high concentrations. Compared to random sampling, Latin hypercube sampling (LHS; Iman and Shortencarier, 1984) is a relatively efficient scheme for selecting combinations of input parameter values for uncertainty propagation. The goal of LHS and other stratified sampling techniques is to ensure that the full range of the input parameter space is sampled more uniformly, for a given number of samples, than is achieved using completely random sampling. The idea of LHS is illustrated in Figure 14.10. The technique starts by dividing the probability distribution function for each input parameter into m intervals of equal probability p. For each uncertain parameter, LHS draws a value at random from within each of its m intervals. The samples representing the m equally probable intervals for each parameter are then combined into m scenarios by randomly selecting one value for each parameter, without replacement.

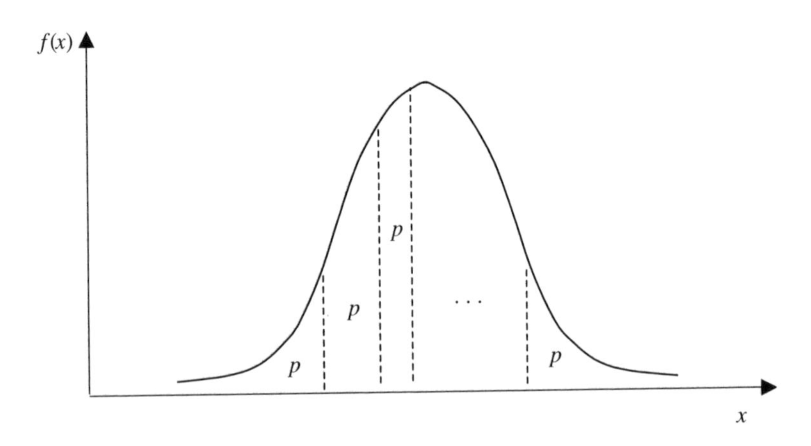

Figure 14.10 Equal probability intervals for x as used in Latin hypercube sampling.

LHS is more efficient than purely random Monte Carlo analysis in that it provides more accurate estimates of the mean, standard deviation, and fractiles of the output distribution for a given sample size (Helton et al., 2002). However, LHS has the disadvantage that standard methods for estimating confidence limits as a function of sample size do not apply because the scenarios used in LHS are not independent. Furthermore, the uniformity properties of LHS sampling tend to break down when sampling an input space with a large number of input variables, n. Alternative methods for generating stratified, multidimensional input samples include the Hammersley sequence (Hammersley, 1960; Kalagnanam and Diwekar, 1997; Reichert et al., 2002), the related Halton sequence (Halton, 1960), and the Sobol method (Sobol, 1988; Bratley and Fox, 1988). For n_x uncertain inputs, these methods generate a set of points on the n_x-dimensional unit cube, $[0, 1]^{nx}$, that is very nearly uniformly distributed, even for small sample sizes. While these methods are most often used for numerical integration, they have also been used in statistical applications (Shaw, 1988). The value for the ith input for the jth model scenario is determined by applying the inverse cdf function for x_i to the computed 0,1 variable [see Eq. (7.28)].

Although so far in the discussion of Monte Carlo analysis we have assumed that the uncertain inputs or parameters are independent, many environmental models involve either positively or negatively correlated inputs or parameters. Correlation commonly arises from physical interactions that are not explicitly modeled (e.g., when measured flow velocities and temperatures are used as inputs to a water quality model, but the dependence of temperature on flow velocity is not described). Correlation can also occur when two input variables depend on a third (e.g., emissions of CO and NO_x from motor vehicles both depend on the level of motor vehicle activity.) When feasible, Morgan and Henrion (1990) recommend modeling such relationships explicitly in uncertainty analyses, by introducing the common factor or dependence into the model. When such treatment is not possible or unduly complicates a model, correlation can be introduced into Monte Carlo or LHS sampling. The LHS code of Iman and Shortencarier (1984) provides this capability. Morgan and Henrion (1990) describe how to generate pairs of correlated variables, while Smith et al. (1992) and Haas (1999) illustrate the possible implications of parameter correlation on risk and uncertainty analysis results.

Beyond estimating the overall uncertainty in model outputs that results from uncertain inputs or parameters, modelers are often interested in knowing which parameters contribute most to the output uncertainty. Following the use of Monte Carlo methods for uncertainty propagation, a useful graphical approach for exploring uncertainty contributions is to examine scatter plots of the output values in the Monte Carlo scenarios versus the corresponding value of each uncertain input. If the relationships generally appear monotonic and linear, uncertainty contributions can be quantified by calculating pairwise correlation coefficients between the output variable and each of the uncertain inputs, or by performing a multivariate linear regression analysis of the output against a set of inputs. If significant nonlinear but still monotonic relationships are apparent in the scatter plots, rank-order correlations between the output and inputs provide a meaningful measure of uncertainty contributions (Jaffe and Ferrara, 1984; Helton et al., 2002; Saltelli, 2002).

EXAMPLE 14.3

In Example 14.2 the uncertainty analysis of the multimedia model for benzene fate, exposure, and risk was performed using first-order uncertainty analysis. For the same model, use Monte Carlo analysis to estimate the uncertainty of the targeted outputs and the correlation among them. Show the relative importance of the eight uncertain inputs and the uncertainty in the predicted risk. Compare the results from the two methods.

Solution 14.3 From the given input distributions, generate uniformly distributed random values for j values, which lead to j sets of inputs for j scenarios. In this example, random sampling is used with 1000 scenarios. The model is run with each sample scenario to produce output values for each scenario, as illustrated in Figure 14.11. The mean and variance of the outputs are shown in Table 14.5 and compared to the values obtained from the first-order uncertainty analysis method. The simulated cdf's of the four output variables are plotted in Figure 14.12.

Comparison of the means of the outputs from the Monte Carlo method with those estimated using first-order uncertainty analysis shows very close agreement, with the biggest difference occurring between the mean values of C_{water} (~4%). The variances of the outputs also generally agree, although the variances of C_{water} differ by about 20%. The S-shaped curves of the cdf plots for the log-scale outputs in Figure 14.12 (for all outputs, except C_a) suggest that the output uncertainty distributions tend to have a lognormal shape, even though each of the input uncertainties to the model is assumed to follow a uniform distribution. As noted in Chapter 7, this result is not unexpected for model outputs that result from multiplicative-like model transformations involving a number of uncertain inputs, due to the central limit theorem.

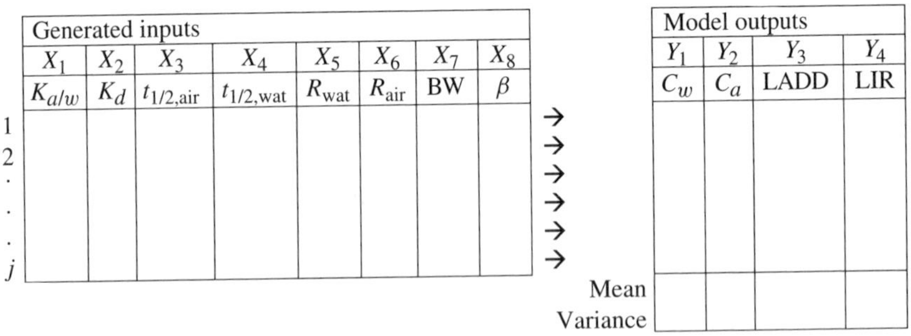

	Generated inputs									Model outputs			
	X_1	X_2	X_3	X_4	X_5	X_6	X_7	X_8		Y_1	Y_2	Y_3	Y_4
	$K_{a/w}$	K_d	$t_{1/2,\text{air}}$	$t_{1/2,\text{wat}}$	R_{wat}	R_{air}	BW	β		C_w	C_a	LADD	LIR
1									→				
2									→				
.									→				
.									→				
.									→				
j									→				
									Mean Variance				

Figure 14.11 Schematic of the Monte Carlo analysis procedure for estimating uncertainties in the outputs of a benzene risk model.

TABLE 14.5 Mean and Variance of Model Outputs from Monte Carlo and First-Order Uncertainty Analysis

Method	Parameter	C_w (mg/m^3)	C_a (mg/m^3)	LADD (mg/kg · day)	LIR
Monte Carlo	Mean	9.08×10^{-4}	1.97×10^{-4}	5.64×10^{-5}	1.68×10^{-6}
	Variance	6.83×10^{-8}	1.45×10^{-9}	2.14×10^{-10}	8.37×10^{-13}
FOUA	Mean	8.75×10^{-4}	1.97×10^{-4}	5.62×10^{-5}	1.63×10^{-6}
	Variance	5.61×10^{-8}	1.41×10^{-9}	2.03×10^{-10}	7.78×10^{-13}

The correlation coefficients between outputs and inputs are estimated and tabulated in Table 14.6. Correlation coefficients among the outputs are shown in Table 14.7. Figure 14.13 provides scatter plots illustrating the correlation among selected inputs and the predicted LIR (those shown in bold in Table 14.6), while Figure 14.14 shows the simulated relationship among selected model outputs (those

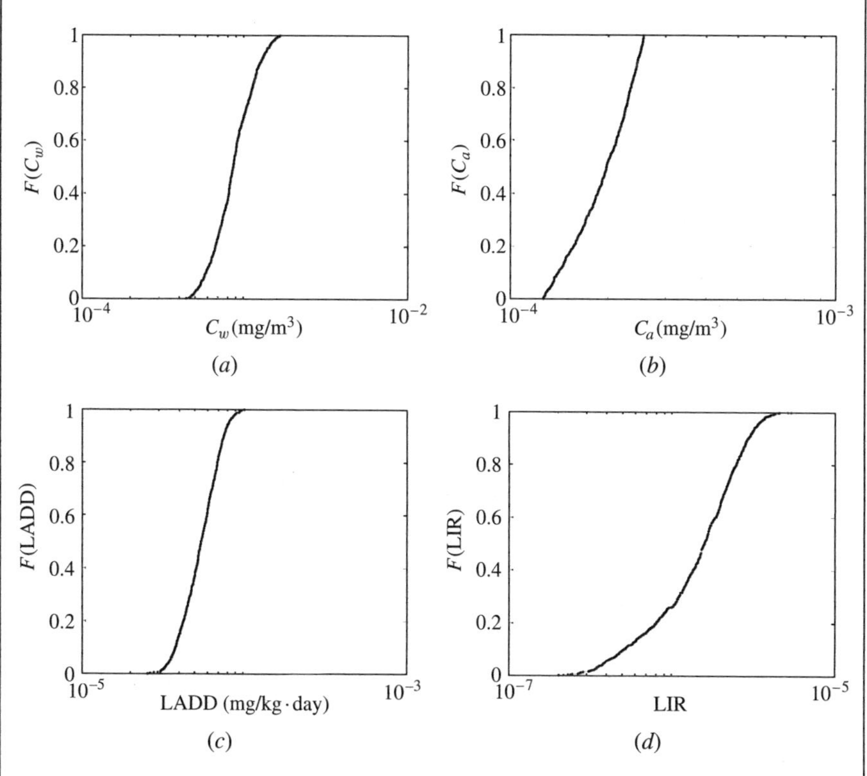

Figure 14.12 Empirical cumulative distribution function plots of uncertain model output variables (*a*) C_w, benzene concentration in the water, (*b*) C_a, benzene concentration in air, (*c*) LADD, benzene dose to a representative individual, and (*d*) LIR, lifetime individual risk for cancer. Note that a logarithmic scale is used for each model output, with a one order of magnitude range for C_a, a two order of magnitude range for C_w and LADD, and a three order of magnitude range for the LIR.

TABLE 14.6 Monte Carlo Correlation Coefficients Between Inputs and Outputs[a]

	C_w	C_a	LADD	LIR	% Variance of LIR from FOUA
K_{aw}	−0.710	−0.012	−0.012	−0.023	≪1%
K_d	−0.036	−0.001	−0.001	−0.022	≪1%
$t_{1/2,\text{air}}$	0.687	0.999	0.742	**0.353**	**12%**
$t_{1/2,\text{water}}$	0.025	−0.020	−0.038	0.019	≪1%
R_{water}	−0.018	−0.012	−0.012	0.010	≪1%
R_{air}	−0.033	−0.022	0.555	**0.277**	**7%**
BW	−0.033	0.007	−0.331	**−0.137**	**2%**
β	0.011	−0.008	−0.008	**0.849**	**78%**

[a]For comparison, the final column is the percent variance contribution to the LIR predicted by the FOUA (from Table 14.3).

shown in bold in Table 14.7). As expected, the scatter plots in Figure 14.13 and computed correlation coefficients in Table 14.6 yield similar information, indicating the strongest relationship between the predicted LIR and the cancer potency (β) input, followed by the benzene half-life in air ($t_{1/2,\text{ air}}$). Among the outputs (Fig. 14.14 and Table 14.7), the LIR is somewhat more closely associated with C_a than with C_w, but most closely associated with the LADD, which differs from the LIR by only the single factor of β.

The results of the Monte Carlo analysis yield very similar insights to those that were obtained using FOUA in Example 14.2. Comparing the last two columns of Table 14.6, the simulated correlation coefficients between the LIR and the inputs, and the FOUA-computed contributions to the LIR variance, suggest the same rank ordering for importance among the uncertain model inputs: (1) β; (2) $t_{1/2,\text{ air}}$; (3) R_{air}; and (4) BW, with the remaining four inputs (K_{aw}, K_d, $t_{1/2,\text{water}}$, and R_{water}) contributing a negligible amount to the uncertainty in the predicted LIR. Likewise the simulated and FOUA-predicted output variable correlation coefficients in Table 14.7 are essentially the same. The similarity in the Monte Carlo and FOUA results indicates that the governing equations in the benzene multimedia fate, exposure, and risk model are monotonic with respect to the inputs and fairly linear over the range of input uncertainties considered.

TABLE 14.7 Monte Carlo Correlation Coefficients of Outputs[a]

	C_w	C_a	LADD	LIR
C_w	1.00			
C_a	0.69 (0.70)	1.00		
LADD	0.51 (0.53)	0.74 (0.75)	1.00	
LIR	**0.26 (0.25)**	**0.35 (0.35)**	**0.47 (0.47)**	1.00

[a]For comparison, values in parentheses are the predicted correlation coefficients from the FOUA (from Table 14.4).

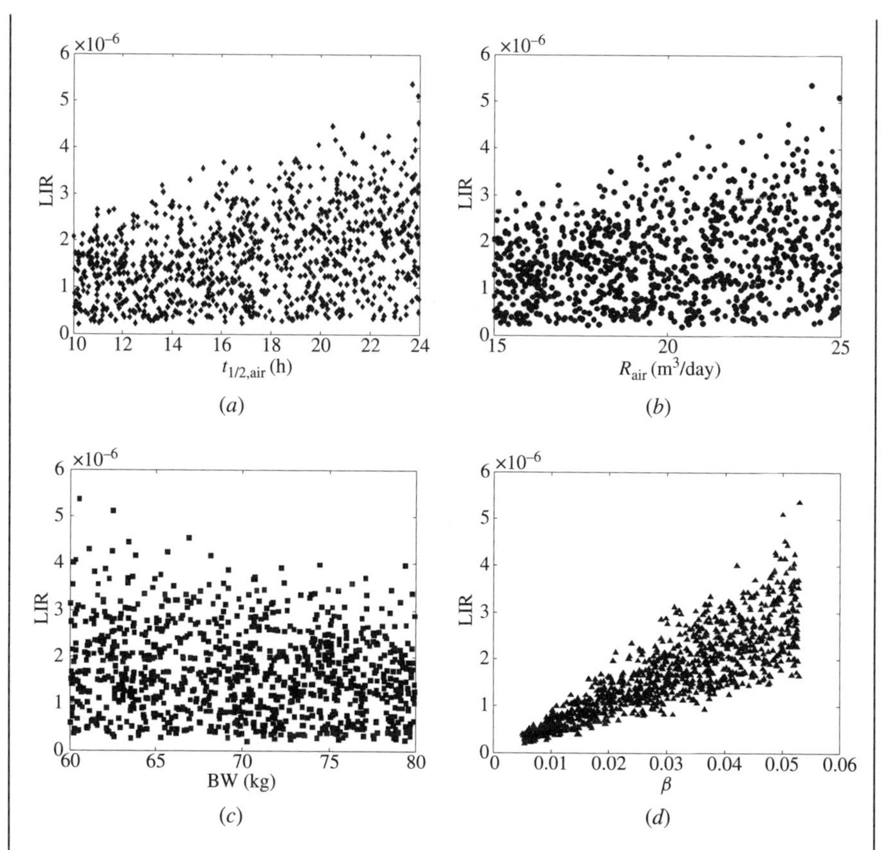

Figure 14.13 Scatter plots of predicted LIR vs. indicated model input. (*a*) LIR vs. $t_{1/2,\text{air}}$, reaction half-life in air, (*b*) LIR vs. R_{air}, air inhalation rate, (*c*) LIR vs. BW, body weight, and (*d*) LIR vs. β, cancer potency.

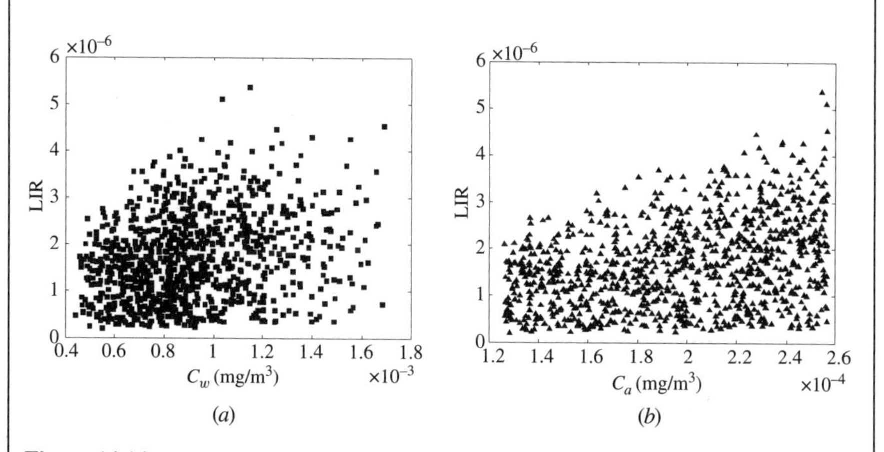

Figure 14.14 Scatter plots of each model output vs. LIR. (*a*) LIR vs. C_w, benzene concentration in water, (*b*) LIR vs. C_a, benzene concentration in air, and (*continued*)

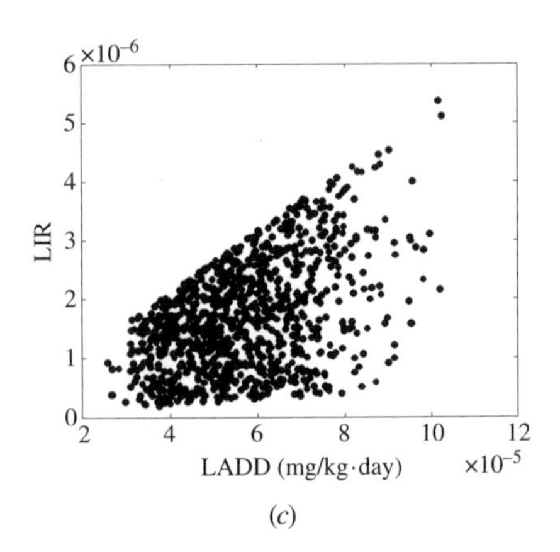

Figure 14.14 (*Continued*) (*c*) LIR vs. LADD, benzene dose to a representative individual.

14.4.5 Bayesian Monte Carlo Analysis

A disconnect often exists between uncertainty estimates produced by Monte Carlo analysis and other uncertainty propagation techniques versus the results of model performance evaluations. As separate steps in the process of model evaluation, performance evaluations are usually performed for previous conditions (for which observed data are available, i.e., history matching), while Monte Carlo uncertainty analysis is often performed for future scenarios with a set of assumed conditions and modified loading rates. However, in these standard Monte Carlo studies, model assumptions and input parameter values that generate model results that closely matched the observations are not necessarily given more weight than those that yielded poor agreement in the history matching exercise. If the joint distribution of the uncertain parameter values is determined as part of a classical parameter estimation procedure, with, for example, the mean and covariance matrix determined for the inputs, then the derived uncertainty will reflect the level of agreement with these observed data. However, this approach ignores the exogenous, expert knowledge that the analyst may possess regarding the uncertainty present in some of the parameters. A method that *combines* the use of (prior) expert knowledge as applied in traditional Monte Carlo analysis with the (posterior) "ground-truthing" provided by parameter estimation procedures is presented in this section. This method, referred to as Bayesian Monte Carlo (BMC) analysis, provides a means of adjusting uncertainty estimates to reflect model performance.

The neglect of model performance information in Monte Carlo analysis was first addressed in water quality modeling studies by discarding simulations that fell outside a specified limit for agreement with the observations (e.g., Hornberger and Spear, 1980; Beck, 1987). Dilks et al. (1992) then proposed Bayesian Monte Carlo analysis as an improvement over such binary schemes for acceptance or rejection. BMC has since been applied to estimate uncertainties in a wide variety of environmental models including surface water, groundwater, and air quality models (Patwardhan and Small, 1992; Brand and Small, 1995; Dakins et al., 1996; Sohn et al., 2000; Bergin and Milford, 2000).

The BMC method refines uncertainty estimates by using a continuous likelihood function to weight the results of individual Monte Carlo simulations. The likelihood function quantifies the probability of obtaining a specified difference between model results and observations, accounting for the errors in the observations. Through the likelihood function, the highest probability is assigned to those Monte Carlo runs that most closely match the observations, with relatively little weight given to runs that produce a poor fit. BMC can also be used to refine estimates for model parameters, identifying and accounting for covariances between them. Perhaps most importantly, because BMC provides a simultaneous update of the prior probabilities of all elements associated with the jth scenario of a Monte Carlo simulation—all model inputs and computed outputs, including predictions for new cases—it provides a means of accounting for model performance as well as input uncertainties in developing estimates of uncertainty in model *predictions*.

From Bayes's theorem, the posterior probability of model output Y_k from simulation k is defined as (Brand and Small, 1995):

$$p'_k = p(Y_k|O) = \frac{L(Y_k|O)p(Y_k)}{\sum_{j=1}^{N} L(Y_j|O)p(Y_j)} \tag{14.25}$$

where $p(Y_k) = p_k$ is the prior probability of the kth scenario out of $j = 1, \ldots, N$ runs. The likelihood, $L(Y_k|O) = p(O|Y_k)$ is the probability of observing value O given that the model output Y_k is the true value. The prior probability of each Monte Carlo simulation is equal to $1/N$. As such, Eq. (14.25) is simplified to:

$$p'_k = p(Y_k|O) = \frac{L(Y_k|O)}{\sum_{j=1}^{N} L(Y_j|O)} \tag{14.26}$$

That is, the posterior probability for each scenario is simply its likelihood given the data, divided by the sum of the likelihoods of all simulated scenarios given the data.

Using the posterior probabilities given by Eq. (14.26), the posterior (updated) distribution for output variable Y has an empirical cdf given by:

$$\hat{F}'_Y(y) = \sum_{\text{all } Y_j \leq y} p'_j(Y_j) \tag{14.27a}$$

a mean of:

$$\mu'_Y = \sum_{j=1}^{N} (Y_j \times p'_j) \tag{14.27b}$$

and standard deviation of:

$$\sigma'_Y = \sqrt{\sum_{j=1}^{N} \left(Y_j - \mu'_Y\right)^2 p'_j} \tag{14.27c}$$

The posterior probabilities, p', developed from the base case model outputs can also be applied to update estimates of sampled input parameter values or of model predictions such as control scenario results. This is done by substituting the values of the parameters, inputs, or predictions in each run for values of Y in Eqs. (14.27a), (14.27b), and (14.27c). Finally, the posterior correlation between output Y and input parameter X_i is

$$\rho'_{Y,X_i} = \frac{\sum_{j=1}^{N} \left(Y_j - \mu'_Y\right)\left(X_{i,j} - \mu'_{X_i}\right) p'_j}{\sigma'_Y \sigma'_{X_i}}. \tag{14.28}$$

Equation (14.28) can also be used to calculate posterior correlations between pairs of input parameters or pairs of output variables.

Equation (14.26) requires both a set of Monte Carlo simulations and a corresponding set of observations. Also, an estimate must be made of what the best possible agreement could be between the model results and observations, which is quantified by the likelihood function. As discussed in Chapter 7, the formulation of the likelihood function, $L(Y_k|O)$, must consider the error structure of the individual measurements that comprise O, including whether these measurements are independent. For the case where n_{meas} independent measurements are used for the update, each given by $O_{y,i}$, $i = 1, \ldots, n_{\text{meas}}$, the overall likelihood is [similar to Eq. (7.15)] given by:

$$L(Y_k|O) = \prod_{i=1}^{n_{\text{meas}}} L(Y_k|O_{y,i}) \tag{14.29}$$

Formulation of a reasonable likelihood function is a critical challenge for Bayesian Monte Carlo studies. Known functions for the distribution of field measurement errors provide a starting point for estimating the likelihood. However, further error could arise due to an imperfect association between the predicted model outputs and the observed data, resulting from subtle (or not so subtle) differences between what is being modeled and what is being measured (Patwardhan and Small, 1992). For example, the model might predict the total PCB concentration in the water column, fish, or sediments, as the sum of the predicted concentrations of the 10 PCB homologs represented in the model (each homolog corresponds to the sum of each of the PCB congeners, or unique molecules, with n_{Cl} chlorine atoms in the molecule, where n_{Cl} ranges from 1 to 10), while the observed data might represent the estimated sum of multiple PCB congeners, some of which appear together (or co-elute), followed by a statistical weighting scheme for estimating the total PCBs. The likelihood function should capture the modeling and measurement errors inherent in these different approaches to total PCB aggregation; however, these errors are very difficult to characterize.

Further errors in the association between model predictions and observed data could arise due to differences in the level of temporal and spatial averaging. For example, field samples are typically collected at a single location [unless sample compositing is used, see, e.g., Gilbert (1987)], while the model might predict concentrations averaged over a spatial grid (Sohn et al., 2000). Finally, determination of the impact of correlated errors among the multiple data points used for Bayesian updating is very difficult (Sohn, et al., 2000; Bergin and Milford, 2000). These effects, along with the tendency for scientists to underestimate errors in both their measurements and models, and to ignore the implications of model structure misspecification (Small and Fischbeck, 1999), suggest that a sensitivity analysis should be conducted to determine the effects of alternative error specification models for the likelihood function. As such, the sensitivity of Bayesian Monte Carlo results to both alternative prior distributions and alternative likelihood functions should be explored and reported.

EXAMPLE 14.4

Based on Examples 14.2 and 14.3, use a Bayesian Monte Carlo analysis to compute the posterior uncertainties, including correlations in the model inputs and the four model outputs induced by observations of the following model outputs. Which model inputs are changed the most as a result of the Bayesian update? Which model inputs have the strongest correlation in their uncertainties?

Assume each set of observations has a lognormal error structure with the indicated median, y_{50}, and coefficient of variation, ν;

Ambient air–benzene concentration: $y_{50} = 0.25\ \mu\text{g/m}^3,\ \nu = 0.2$

Ambient water–benzene concentration: $y_{50} = 0.80\ \mu\text{g/m}^3,\ \nu = 0.2$

Effective dose (LADD): $y_{50} = 7 \times 10^{-5}\ \text{mg/kg} \cdot \text{day},\ \nu = 0.4$

Lifetime individual risk (LIR): $y_{50} = 3 \times 10^{-6},\ \nu = 0.75$

Solution 14.4 From the given observations, the model outputs are updated with the likelihood computed by a lognormal probability density function with parameters a and b, which are given by (see Table 7.1):

$$a = \ln(y_{50}) \qquad b = \left[\ln(\nu^2 + 1)\right]^{1/2}$$

The likelihood can then be calculated for each observation compared to the model output, invoking the lognormal pdf formula as shown in the equation. The process of finding the likelihood is illustrated in Figure 14.15.

$$L(Y_k|O = y_{50}) = p(O = y_{50}|Y_k) = \frac{1}{b \cdot y_{50} \cdot \sqrt{2\pi}} \exp\left[\frac{-(a - \ln Y_k)^2}{2b^2}\right]$$

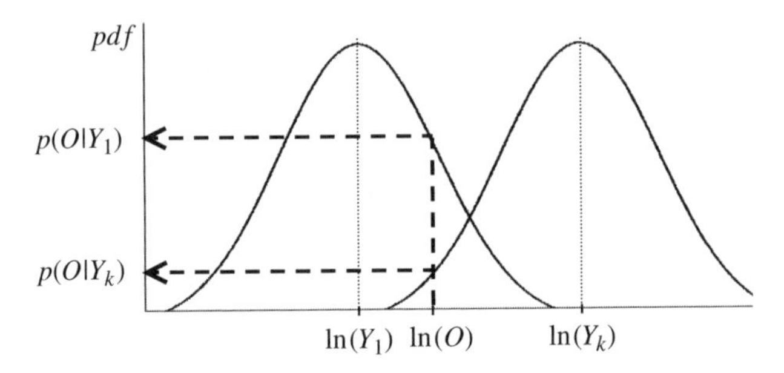

Figure 14.15 Illustration of probability $p(O|Y_k)$ of observing O with respect to Y_k.

For $N = 1000$, the posterior probability can be computed by Eq. (14.26), assuming the prior probabilities, $p(Y_k)$, are all equal to $1/N$ for all k. Since the likelihood is

$$L(Y_k|O) = L(Y_k|O_1, O_2, O_3, O_4) = \prod_{j=1}^{4} L(Y_k|O_j)$$

Eq. (14.26) becomes

$$p'_k = \frac{L(Y_k|O)}{\sum_{j=1}^{1000} L(Y_j|O)} = \frac{p'_{Cw,k} p'_{Ca,k} p'_{\text{LADD},k} p'_{\text{LIR},k}}{\sum_{j=1}^{1000} p'_{Cw,j} p'_{Ca,j} p'_{\text{LADD},j} p'_{\text{LIR},j}}$$

As outputs are updated with the estimated posterior probability, the posterior cdf's, means and standard deviations are then calculated according to Eqs. (14.27a), (14.27b), and (14.27c), respectively. The computed posterior means and standard deviations are shown and compared to prior values in Table 14.8 for outputs and Table 14.9 for inputs. Figure 14.16 compares the posterior cdf's for the four model outputs to the prior results from Figure 14.12. A close examination of Figure 14.16 reveals that the same points are plotted along the x axis for the prior and posterior cdf's (corresponding to the original 1000 Monte Carlo scenario results), but with

TABLE 14.8 Comparison of Prior and Posterior Mean and Variance of Outputs

	C_w (mg/m^3)	C_a (mg/m^3)	LADD (mg/kg · day)	LIR
Basecase	8.75×10^{-4}	1.97×10^{-4}	5.62×10^{-5}	1.63×10^{-6}
Observation	8.00×10^{-4}	2.50×10^{-4}	7.00×10^{-5}	3.00×10^{-6}
Monte Carlo mean	9.08×10^{-4}	1.97×10^{-4}	5.64×10^{-5}	1.68×10^{-6}
BMC posterior mean	8.29×10^{-4}	2.44×10^{-4}	7.21×10^{-5}	2.75×10^{-6}
Monte Carlo variance	6.83×10^{-8}	1.45×10^{-9}	2.14×10^{-10}	8.37×10^{-13}
BMC posterior variance	3.67×10^{-10}	3.18×10^{-11}	4.81×10^{-11}	4.52×10^{-13}

TABLE 14.9 Comparison of Prior and Posterior Mean and Variance of Inputs

	K_{aw}	K_d	$t_{1/2,\text{air}}$ (h)	$t_{1/2,\text{water}}$ (h)	R_{water} (L/day)	R_{air} (m³/day)	BW (kg)	β (mg/kg · day)$^{-1}$
Initial mean	0.225	2.655	17	170	1.4	20	70	0.029
Posterior mean	0.294	2.639	22.352	159.19	1.4	20.5	69.7	0.038
% change of mean	30.6	−0.6	31.5	−6.4	2.5	2.7	−0.5	31.8
Initial variance	1.88×10^{-3}	0.14	1.67	1633	0.05	8.33	33.33	0.00019
Posterior variance	5.38×10^{-5}	0.150	0.46	832	0.05	3.01	20.63	0.00007
% change of variance	−97.1	0.2	−72.2	−49.1	−0.5	−63.9	−38.1	−61.5

nonuniform increments along the y axis for the posterior cdf's, as determined by the posterior probabilities computed for each. As indicated, the posterior cdf's exhibit a decreased variance and are shifted toward somewhat higher individual risk, exposure, and concentration in the air, with somewhat lower concentrations of benzene predicted in the water, consistent with the observations incorporated

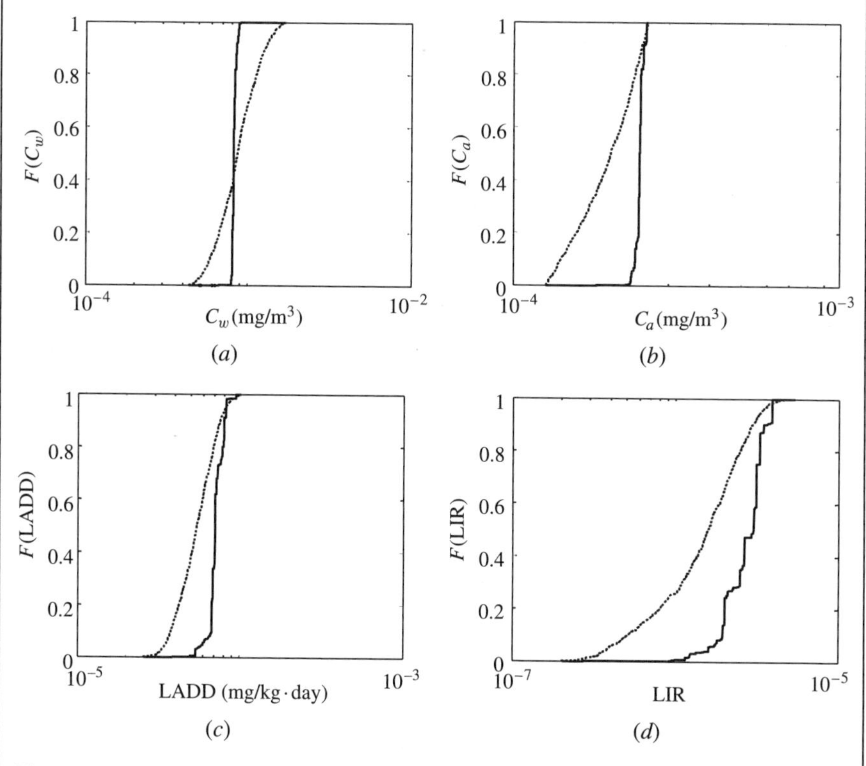

Figure 14.16 Posterior (solid lines) cdf plots of model output variables compared with prior cdf plots (dotted lines, reproduced from Fig. 14.12) for (*a*) C_w, benzene concentration in the water, (*b*) C_a, benzene concentration in air, (*c*) LADD, benzene dose to a representative individual, and (*d*) LIR, lifetime individual risk for cancer.

TABLE 14.10 Posterior Correlation Coefficient Between Inputs and Outputs

	C_w	C_a	LADD	LIR
K_{aw}	−0.54	0.57	0.18	0.21
K_d	−0.30	−0.23	−0.20	0.12
$t_{1/2,\text{air}}$	0.39	**1.00**	0.19	0.13
$t_{1/2,\text{water}}$	−0.03	0.11	−0.59	−0.54
R_{water}	0.16	0.21	0.21	0.04
R_{air}	0.12	0.14	0.75	0.01
BW	0.31	0.27	−0.43	−0.52
β	−0.10	0.06	0.01	**0.92**

for each model output. The greatest variance reductions for the inputs are achieved for the air–water partition coefficient of benzene (–97%), the half-life of benzene in the air (–72%), the air inhalation rate (–64%), and the cancer potency factor (–62%), though significant reductions in variance are also apparent for the half-life in water (–49%) and the BW (–38%). The largest shifts in the mean estimates for the input parameters occur for K_{aw}, $t_{1/2,\text{air}}$, and β, with their posterior means each increased by ~30%. This is consistent with the (assumed) fact that the observed air concentration and lifetime individual risk are somewhat higher than predicted by the prior model, while the observed water concentration is somewhat lower.

The posterior correlation coefficients between outputs and inputs are computed from Eq. (14.28) and shown in Table 14.10. Similarly, the correlation coefficients between the outputs are tabulated in Table 14.11 and those between the inputs in Table 14.12.

Comparison of posterior correlation coefficients between outputs and inputs shows significant correlation between the predicted benzene concentrations in air and the reaction half-life of benzene in air, while the LIR is most closely correlated with β, similar to that predicted under the prior conditions in Examples 14.2 and 14.3. As for the outputs, the LADD and the LIR are the most correlated, followed closely by the benzene concentrations in air and water. Finally, the posterior correlations between the model inputs can be evaluated to determine whether the updating has jointly constrained the possible values of the model inputs in a significant manner. These input correlations do not necessarily reflect physical associations among the parameters, rather they can arise due to the need for

TABLE 14.11 Posterior Correlation Coefficient Between Outputs

	C_w	C_a	LADD	LIR
C_w	1.00			
C_a	**0.39**	1.00		
LADD	−0.01	0.19	1.00	
LIR	−0.09	0.13	**0.40**	1.00

TABLE 14.12 Posterior Correlation Coefficient Between Inputs

	K_{aw}	K_d	$t_{1/2,\text{air}}$	$t_{1/2,\text{water}}$	R_{water}	R_{air}	BW	β
K_{aw}	1.00							
K_d	0.08	1.00						
$t_{1/2,\text{air}}$	0.56	−0.23	1.00					
$t_{1/2,\text{water}}$	0.13	−0.14	0.10	1.00				
R_{water}	0.04	−0.49	0.21	−0.10	1.00			
R_{air}	0.02	−0.42	0.15	−0.20	0.11	1.00		
BW	−0.04	−0.33	0.27	0.63	−0.08	0.23	1.00	
β	0.15	0.22	0.06	−0.34	−0.65	−0.30	−0.38	1.00

the model to counterbalance the joint effects of the parameters on the model outputs, which are now observed to vary over a more limited range (Sohn et al., 2000; Reichert et al., 2002). An example of this is the negative correlation induced between β and R_{air}, since if both are high (low), then the computed LIR is too high (low) relative to the observed data. Other input correlations induced by the BMC update are more difficult to interpret, perhaps due to the small number of Monte Carlo scenarios with significant posterior probability. Still, the shifts in the marginal posterior probability distributions of both the model inputs and the outputs are significant and serve to underscore the utility of the BMC method for simultaneous parameter estimation and uncertainty analysis, especially for cases where observed data sets for elements of the model are limited or lacking.

Alternative Methods for Bayesian Updating The Bayesian Monte Carlo method provides a conceptually simple approach for enabling numerical uncertainty analyses to be better informed by the available data. However, a number of problems have been observed in its implementation. In particular, if the prior distribution over the input parameter space is very broad relative to the posterior distribution, then only a few of the prior model scenario runs may be assigned significant posterior probability by application of Eq. (14.26), yielding a very coarse characterization of the posterior. This can be addressed by generating a much larger prior sample, but this leads to inefficiencies in the method. Alternative methods for implementing numerical Bayesian updating include the Markov chain Monte Carlo (MCMC) method (Gelman et al., 1995; Gamerman, 1997) and iterative methods for finding and characterizing the posterior probability space. An introduction to these methods is provided in Appendix 14.B in the electronic website for Chapter 14.

REFERENCES

Abbaspour, K.C., R. Schulin, E. Schlappi, and H. Fluhler (1996). A Bayesian approach for incorporating uncertainty and data worth in environmental projects. *Environ. Modeling Assessment* **1**:151–158.

Abbott, M.B. (1997). Range of tidal flow modeling. *J. Hydraulic Eng.* **123**(4):257–277.

Abernathy, C.O., R. Cantilli, J.T. Du, and O.A. Levander (1993). Essential Versus Toxicity: Some Consideration in the Risk Assessment of Essential Trace Elements. In *Hazard Assessment of Chemicals*, 8th ed. Taylor & Francis, Washington, DC., pp. 81–113.

Abramowitz, M. and I.A. Stegun (eds.) (1964). *Handbook of Mathematical Functions with Formulas, Graphs, and Mathematical Tables.* Number 55 in National Bureau of Standards Applied Mathematics Series. U.S. Government Printing Office, Washington, DC.

Abramowitz, M. and I.A. Stegun (1965). *Handbook of Mathematical Functions,* National Institute of Standards and Technology, Gaithersburg, MD, and Dover Publications, New York.

Abrioloa, L.M. and G.F. Pinder (1985). A multiphase approach to the modeling of porous media contamination by organic compounds: 1. Equation development. *Water Resources Res.* **21**(1):11–18.

Adams, B.J. and F. Papa (2000). *Analytical Probabilistic Models for Stormwater Management Planning.* Wiley, New York.

Adams, W.R., E.L. Thackston, and R.E. Speece (1997). Modeling CSO impacts from Nashville using EPA's demonstration approach. *J. Environ. Eng.* **123**(2):126–133.

Ahn, B.S. and W.K. Lee (1991). Mass transfer at liquid-liqud interfaces. *Ind. Eng. Chem. Res.* **29**:1927–1935.

Ahrens, L.H. (1954). The lognormal distribution of the elements. *Geochim. Cosmoschim. Acta* **5**:49–53.

Ahsan, A.K.M.Q., M.S. Bruno, L-Y. Oey, and R.I. Hires (1994). Wind-driven dispersion in New Jersey coastal waters. *J. Hydraulic Eng.* **120**(11):1264–1273.

Aikman, F. III and E. Wei (1995). A comparison of model-simulated trajectories and observed drifters in the Middle Atlantic Bight. *J. Marine Environ. Eng.* **2**(1–2): 123–139.

Alcamo, J. and J. Bartnicki (1987). A framework for error analysis of a long-range transport model with emphasis on parameter uncertainty. *Atmos. Environ.* **21**: 2121–2131.

Alexander, M. (1999). *Biodegradation and Bioremediation*, 2nd ed. Academic, San Diego.

Alley, W.M. and P.E. Smith (1982). *Multi-event Urban Runoff Quality Model*, Open File Report 82-764. U.S. Geological Survey, Denver, CO.

Allison, J.D., D.S. Brown, and K.J. Novo-Gradac (1991). *MINTEQA2/PRODEFA2, A geochemical assessment model for environmental systems: Version 3.0 User's Manual.* EPA/600/3-91/021, USEPA, Washington, DC.

Alvarez-Cohen, L.M. and P.L. McCarty (1991). Effects of toxicity, aeration, and reductant supply on trichloroethylene transformation by a mixed methanotrophic culture. *Appl. Environ. Microbiol.* **57**(1):228–235.

Ambrose, R.B., T.A. Wool, and J.L. Martin (1993). *The Water Quality Analysis Simulation Program, Part A: Model Documentation, WASP5 Version 5.10.* U.S. Environmental Protection Agency, Athens, GA; see http://www.epa.gov/ceampubl/swater/wasp/index.htm.

Ames, B.N., L.S. Gold, and M.K. Shigenaga (1996). Cancer prevention, rodent high-dose cancer tests, and risk assessment (Letter to the Editor). *Risk Anal.* **16**(5):613–617.

Amin, I.E. and M.E. Campana (1996). A general lumped parameter model for the interpretation of tracer data and transit-time calculation in hydrologic systems. *J. Hydro.* **179**:1–21.

Andersen, M.E., H.J. Clewell, M.L Gargas, M.G. MacNaughton, R.H. Reitz, R.J. Nolan, and M.J. McKenna (1991). Physiologically based toxiokinetic modeling with dichloromethane, its metabolite carbon monoxide, and blood carboxhemoglobin in rats and humans. *Toxicol. Appl. Pharmacol.* **108**:14–27.

Anderson, E., P. Deisler, D. McCallum, C. St. Hilaire, H. Spitzer, H. Strauss, J. Wilson, and R. Zimmerman (1993). Key issues in carcinogenic risk assessment guidelines, Society for Risk Analysis. *Risk Anal.* **13**(4):379–382.

Anderson, M., S.L. Miller, and J.B. Milford (2001). Source apportionment of exposure to toxic volatile organic compounds using positive matrix factorization. *Int. J. Exposure Anal. Environ. Epidemiol.* **11**:295–307.

Anderson, T.A., E.A. Guthrie, and B.T. Walton (1993). Bioremediation. *Environ. Sci. Technol.* **27**(13):2630–2636.

Andersson N., M.K. Muir, V. Mehra, and A.G. Salmon (1988). Exposure and response to methyl isocyanate: Results of community-based survey in Bhopal. *Brit. J. Ind. Med.* **45**: 469–475.

Andino, J.M., J.N. Smith, R.C. Flagan, W.A.Goddard, and J.H.Seinfeld (1996). Mechanism of atmospheric photooxidation of aromatics: A theoretical study. *J. Phys. Chem.* **100**:10967–10980.

Andrew, A.S., M.R. Karagas, and J.W. Hamilton (2003). Decreased DNA repair gene expression among individuals exposed to arsenic in United States drinking water. *Int. J. Cancer* **1094**:263–268.

Andrews, J.F. (1968). A mathematical model for the continuous culture of microorganisms utilizing inhibitory substrates. *Biotech. Bioeng.***10**:707–723.

APHA (Am. Public Health Association) (1985). *Standard Methods for the Examination of Water and Waste Water*, 16th ed. APHA, Washington, DC.

Aron, G. and E.L. White (1982). Fitting a gamma distribution over a synthetic unit hydrograph. *Water Resources Bull.* **17**(4):691–698.

Arrow, K., G. Daily, P. Dasgupta, S. Levin, K.G. Mäler, E. Maskin, D. Starrett, T. Sterner, and T. Tietenberg (2000). Managing ecosystem resources. *Environ. Sci. Tech.*, **34**(8):1401–1406.

Arya, S.P. (1999). *Air Pollution Meteorology and Dispersion.* Oxford University Press, New York.

Asante-Duah, K. (2002). *Public Health Risk Assessment for Human Exposure to Chemicals.* Kluwer Academic, Dordrecht.

Atkinson, R. (1994). Gas phase tropospheric chemistry of organic compounds. *J. Phys. Chem. Ref. Data*, Monograph No 2.

ATSDR (Agency for Toxic Substance and Disease Registry) (1997). *Toxicological Profile for Trichloroethylene*. U.S. Department of Health and Human Services, Washington, DC.

Aunan, K. (1996). Exposure-response functions for health effects of air pollutants based on epidemiological findings. *Risk Anal.* **16**(5):693–709.

Axley, J. (1988). *Progress Toward a General Analytical Method for Predicting Indoor Air Pollution in Buildings: Indoor Air Quality Modeling Phase III Report*. NBSIR 88-3814, National Bureau of Standards, National Eng. Laboratory, Center for Building Technology, Building Environment Division, Gaithersburg, MD.

Aylward, L.L. and S.M. Hays (2002). Temporal trends in human TCDD body burden: Decreases over three decades and implications for exposure levels. *J. Exposure Anal. Environ. Epidemiol.* **12**(5):319–328.

Bacci, E., D. Calamari, C. Gaggi, and M. Vighi (1990a). Bioconcentration of organic chemical vapors in plant leaves: Experimental measurements and correlation. *Environ. Sci. Tech.***24**:885–889.

Bacci, E., M.J. Cerejeira, C. Gaggi, G. Chemello, D. Calarmi, and M. Vighi (1990b). Bioconcentration of organic chemical vapors in plant leaves; the azalea model. *Chemosphere* **21**:525–535.

Baes, C.F. (1982). Prediction of radionuclide K_d values from soil-plant concentration ratios. *Trans. Am. Nuclear Soc.* **41**:53–54.

Baker, A.J.M. and R.R. Brooks (1989). Terrestrial higher plants which hyperaccumulate metallic elements—A review of their distribution, ecology and phytochemistry. *Biorecovery* **1**:81–126.

Ball, W.P. and P.V. Roberts (1989). Long-term sorption of halogenated organic chemicals by aquifer material. 1. Equilibrium. 2. Intraparticle diffusion. *Environ. Sci. Technol.* **25**(7): 1223–1249.

Banerjee, P., M.D. Piwoni, and K. Ebeid (1985). Sorption of organic contaminants to a low carbon subsurface core. *Chemosphere* **8**:1057–1067.

Barad, M.L. (1958). Project Prairie Grass, a field program in diffusion, Geophys. Res. Papers, No. 59, Vols. I and II, Report AFCRC-TR-58-235, U.S. Air Force Cambridge Res. Center.

Barancourt, C., J.D. Creutin, and J. Rivoirad (1992). A method for delineating and estimating rainfall fields. *Water Resources Res.*, **28**(4):1133–1144.

Bard, Y. (1974). *Nonlinear Parameter Estimation*. Academic, New York.

Barnard, N.D. and S.R. Kaufman, (1997). Animal research is wasteful and misleading. *Sci. Am.* **276**(2): 80–82.

Barton, H.A., M.E. Andersen, and H.J. Clewell III (1998). Harmonization: Developing consistent guidelines for applying mode of action and dosimetry information to cancer and noncancer risk assessment, *Human Ecolog. Risk Assess.* **4**:75–115.

Bastin, G., B. Lorent, C. Duque, and M. Gevers (1984). Optimal estimation of the average areal rainfall and optimal selection of rain gauge locations. *Water Resources Res.* 20:463–470 .

Baum, E.J. (1998). *Chemical Property Estimation*. Lewis Publishers, Chelsea, MI.

Bear, J. (1979). *Hydraulics of Groundwater*. McGraw-Hill, New York.

Bear, J. and A. Verruijt (1987). *Modeling Groundwater Flow and Pollution*. D. Reidel, Dordrecht, Netherlands.

Beck, M.B. (1987). Water quality modeling: A review of the analysis of uncertainty. *Water Resources Res.* **23**:1393–1442.

Beck, J.V. and K.J. Arnold (1977). *Parameter Estimation in Engineering and Science.* Wiley, New York.

Beck, M.B. and J. Chen (2000). Assuring the Quality of Models Designed for Predictive Tasks. In *Sensitivity Analysis*, A. Saltelli, K. Chan, and E.M. Scott (Eds.). Wiley, West Sussex, UK, pp. 401–420.

Beck, M.B., J.R. Ravetz, L.A. Mulkey and T.O. Barnwell (1997). On the problem of model validation for predictive exposure assessments. *Stochastic Hydrol. Hydraul.* **11**:229–254.

Bedient, P.B. and W.C. Huber (1992). *Hydrology and Floodplain Analysis*, 2nd ed. Addison-Wesley, Reading, MA.

Bedient, P.B., H.S. Rifai, and C. J. Newell (1999). *Groundwater Contamination: Transport and Remediation* 2nd Ed. Prentice Hall, Upper Saddle River, NJ.

Beecher, H.K. (1966). Ethics and clinical research. *N. Engl. J. Med.* June 16, 1966:1354–1360.

Bencala, K.E. and R.A. Walters (1983). Simulation of solute transport in a mountain pool-and-riffle stream: A transient storage model. *Water Resources Res.* **19**(3): 718–724.

Bennett, D.H., W.E. Kastenberg, and T.E. McKone (1999). General formulation of characteristic time for persistent chemicals in a multimedia environment. *Environ. Sci. Tech.* **33**(3):503–509.

Bergin, M. S. and J.B. Milford (2000). Application of Bayesian Monte Carlo analysis to a Lagrangian photochemical air quality model. *Atmos. Environ.* **34**(5):781–792.

Bergin, M.S., G.S. Noblet, K. Petrini, J. Dhieux, J.B. Milford, and R.A.Harley (1999). Formal uncertainty analysis of a Lagrangian photochemical air pollution model, Environ. Sci. Tech. **33**:1116–1126.

Berkenpas, M.B., E.S. Rubin, and B. Toole-O'Neil (1996). Evaluating trace chemical emissions from electric power plants. Proceedings of the International Symposium ECOS'96, Stockholm, Sweden, June 25–27.

Berry, M.R. (1997). Advances in dietary exposure Res. at the United States Environmental Protection Agency—National Exposure Res. Laboratory. *J. Exposure Anal. and Environ. Epidemiol.* **7**(1):3–16.

Berthouex, P.M. and L.C. Brown (1994). *Statistics for Environmental Engineers.* Lewis, Boca Raton, FL.

Beyer, A., D. Mackay, M. Matthies, F. Wania, and E. Webster (2000). Assessing long-range transport potential of persistent organic pollutants. *Environ. Sci. Tech.* **34**(4):699–703.

Biau, G., E. Zorita, H. Von Storch, and H. Wackernagel (1999). Estimation of precipitation by kriging in the EOF space of the sea level pressure field. *J. Climate* **12**:1070–1085.

Bicknell, B.R., J.C. Imhoff, J.L. Kittle, A.S. Donigan, and R.C. Johanson (1993). *Hydrological Simulation Program—FORTRAN (HSPF): User's Manual for Release 10.0.* EPA 600/3-84-066. U.S. Environmental Protection Agency, Environmental Res. Laboratory, Athens, GA; see http://www.epa.gov/ceampubl/swater/hspf/index.htm.

Bielefeldt, A.R. and H.D. Stensel (1999). Evaluation of biodegradation kinetic testing methods and longterm variability in biokinetics for BTEX metabolism. *Water Resources* **33**(3):733–740.

Bielefeldt, A.R., C. McEachern, and T. Illangasekare (2002). Hydrodynamic changes in sand due to biogrowth on naphthalene and decane. *J. of Environ. Eng.* **128**:51–59.

Binder, S., S. Sokal, and D. Maughan (1991). Estimating soil ingestion: The use of trace elements in estimating the amount of soil ingested by young children. *Arch. Environ. Health* **41**:341–345.

Binkowski, F.S. and U. Shankar (1995). The Regional Particulate Matter Model: 1. Model description and preliminary results. *J. Geophysi. Res.* **100**:26191–26209.

Bird, R.B., W.E. Stewart, and E.N Lightfoot (1960). *Transport Phenomena*. Wiley, New York.

Bird, W.L. (1999). Better Living: Advertising, Media, and the New Vocabulary of Business Leadership 1935–1955, Evanston.

Bischof, C., G. Corliss, L. Green, A. Griewank, K. Haigler, and P. Newman (1992). Automatic differentiation of advanced CFD codes for multidisciplinary design. *J. Comput. Sys. Eng.*, **3**(6):625–638.

Blumberg, A.F. and G.L. Mellor (1987). A Description of a Three-Dimensional Coastal Ocean Circulation Model. In *Three-Dimensional Coastal Ocean Models, Coastal and Estuarine Sciences*, Vol 4. N. Heaps (Ed.). Am. Geophysical Union, Washington, DC, pp. 1–16.

Bogen, K.T. and T.E. McKone (1988). Linking indoor air and pharmacokinetic models to assess tetrachloroethylene risk. *Risk Anal.* **8**(4):509–520.

Bogen, K.T. and R.C. Spear (1987). Integrating uncertainty and interindividual variability in environmental risk assessment. *Risk Anal.* **7**:427–436.

Borden, R.C. and P.B. Bedient (1987). In situ measurement of adsorption and biotransformation at a hazardous waste site. *Water Resources Bull. Am. Water Resources Assoc. (AWWRA)* **23**(4):629–636.

Borden, R.C., P.B. Bedient, M.D. Lee, C.H. Ward, and J.T. Wilson (1986). Transport of dissolved hydrocarbons influenced by oxygen-limited biodegradation 2. Field application. *Water Resources Res.* **22**(13):1983–1990.

Boris, J.P. and D.L. Book (1973). Flux-corrected transport, I. SHASTA, a fluid transport algorithm that works. *J. Comput. Phys.* **11**:38–69.

Borrazzo, J.E., C.I. Davidson, and M.J. Small (1992). A stochastic model for diurnal variations of CO, NO, and NO_2 concentrations in occupied residences. *Atmos. Environ.* **26B**:369–377.

Bothner, M.H., M. Buchholtz ten Brink, and Manheim, F.T. (1998). Metal concentrations in surface sediments of Boston Harbor—Changes with time. *Marine Environ. Res.* **45**(2):127–155.

Botting, J.H. and A.R. Morrison (1997). Animal research is vital to medicine. *Sci. Ame.* **276**(2):83–85.

Boussinesq, J. (1877). *Essay on the Theory of Flowing Water*. French Academy of Sci.s, Paris.

Boufadel, M.C. (1998). Unit hydrographs derived from the Nash model. *J. Am. Water Resources Assoc.* **34**(1):167–177.

Bowie, G.L., W.B. Mills, D.B. Porcella, C.L. Campbell, J.R. Pagenkopf., G.L.Rupp, K.M. Johnson, P.W.H. Chan, S.A. Gherni, and C.E. Chamberlin (1985). Rates, Constants and Kinetic Formulations in Surface Water Quality Modeling, USEPA, ORD, Athens, GA, ERL, EPA/600/3-85/040.

Box, G., G. Jenkins, and G. Reinsel (1994). *Time Series Analysis: Forecasting and Control,* 3rd ed. Prentice Hall, Englewood Cliffs, NJ.

Bradley, P.M. and F.H. Chapelle (1998). Effect of contaminant concentration on aerobic microbial mineralization of DCE and VC in stream-bed sediments. *Environ. Sci. Tech.* **32**(5):553–557.

Brand, K.P. and M.J. Small (1995). Updating uncertainty in an integrated risk assessment: Conceptual framework and methods. *Risk Anal.* **15**(6):719–731.

Bras, R.L. and I. Rodriguez-Iturbe (1994). *Random Functions and Hydrology*. Addison Wesley, new printing, Dover, New York.

Bratley, P. and B.L. Fox (1988). ALGORITHM 659: Implementing Sobol quasi-random sequence generator. *ACM Trans. Math. Software* **14**:88–100.

Bratley, P., B.L. Fox and L.E. Schrage (1987). *A Guide to Simulation*. Springer, New York.

Brattin, W.J., T.M. Barry, N. Chiu (1996). Monte Carlo modeling with uncertain probability density functions. *Human Ecolog. Risk Assessment* **2**(4):820–840.

Bravo, R., D.A. Dow, and J.R. Rogers (1994). Parameter determination for the Muskingum-Cunge flood routing method. *J. Am. Water Resources Assoc.* **30**(5):891–899.

Brief, R.S., J. Lynch, T. Bernath, and R.A. Scala (1980). Benzene in the workplace. *Am. Ind. Hyg. Assoc. J.* **41**:616–623.

Briggs, G.A. (1969). *Plume Rise.* U.S. Atomic Energy Commission Critical Review Series, U.S Atomic Energy Commission, Oak Ridge, TN.

Briggs, G.A. (1973). Diffusion Estimation for Small Emissions, Contribution No. 79, Atmospheric Turbulence and Diffusion Laboratory, Oak Ridge, TN.

Briggs, G.G., R.H. Bromilow, and A. Evans (1982). Relationship between lipophilicity and root uptake and translocation of non-ionized chemicals by barley. *Pesticide Sci.* **13:**495–504.

Brill, A.B., D.V. Becker, K. Donahoe, S.J. Goldsmith, B. Greenspan, K. Kase, H. Royal, E.B. Silberstein, and E.W. Webster (1994). Radon uptake: Facts concerning environmental radon levels, mitigation strategies, dosimetry, effects and guidelines—SNM committee on Radiobiological Effects of Ionizing Radiation. *J. Nuclear Med.* **35**(2):368–385.

Broshears, R.E., R.L. Runkel, B.A. Kimball, D.M. McKnight, and K.E. Bencala (1996). Reactive solute transport in an acidic stream: Experimental pH increase and simulation of controls on pH, aluminum, and iron. *Environ. Sci. and Tech.* **30**:3016–3024.

Brown, S.T. (1987). Concentration prediction for meandering plumes based on probability theory. *Atmos. Environ.* **21**(6):1321–1330.

Brooks, R.H. and A.T. Corey (1966). Properties of porous media affecting fluid flow. *Proceedings Am. Society of Civil Engineers, Irrigation and Drainage Division 92* no. IR2:61-87. ASCE, Washington.

Brooks, R.R. (1983). *Biological Methods of Prospecting for Minerals*. Wiley, New York.

Brusseau, M.L. (1991a). Rate-limited sorption and nonequilibrium transport of organic chemicals in low organic carbon aquifer materials. *Water Resources Res.* **27**(6):1137–1145.

Brusseau, M.L. (1991b). Application of a multiprocess nonequilibrium sorption model to solute transport in a stratified porous medium. *Water Resources Res.* **27**(4):589–595.

Buchack, E.M. and J.E. Edinger (1984). *Generalized Longitudinal-Vertical Hydrodynamics and Transport: Development, Programming and Applications*. Report No. 84–18-R, U.S. Army Corps of Engineers Waterways Experiment Station, Vicksburg, MS.

Buck, J.W., G. Whelan, J.G. Droppo, Jr., D.L. Strenge, K.J. Castleton, J.P. McDonald, C. Sato, and G.P. Streile (1995). *Multimedia Environmental Pollutant Assessment System (MEPAS) Application Guidance: Guidelines for Evaluating MEPAS Input Parameters for Version 3.1*. Report PNL-10395, Pacific Northwest Laboratory, Richland, WA.

Burken, J.G. and J. L. Schnoor (1997). Uptake and metabolism of atrazine by poplar trees. *Environ. Sci. Tech.* **31**(5):1399–1406.

Burmaster, D.E. and E.A. Crouch (1997). Lognormal distributions for body weight as a function of age for males and females in the United States, 1976–1980. *Risk Analy.* **17**:499–506.

Burmaster, D.E. and D.R. Murray (1998). A trivariate distribution for the height, weight, and fat of adult men. *Risk Anal.* **18**(4):385–389.

Burmaster, D.E. and K.M. Thompson (1998). Fitting second-order parametric distributions to data using maximum likelihood estimation. *Human Ecolog. Risk Assessment* **4**(2):319–339.

Burmaster, D.E. and A.M. Wilson (1996). An introduction to second-order random variables in human health risk assessments. *Human Ecolog. Risk Assessment*, **2**(4):892–919.

Busenberg, E. and L. Plummer (1982). The kinetics of dissolution of dolomite in CO_2–H_2O systems at 1.5 to 65°C and 0 to 1 atm pCO_2. *Am. J. Sci.* **282**:45–78.

Businger, J.A., J.C. Wyngaard, Y. Izumi, and E.F. Bradley (1971). Flux-profile relationships in the atmospheric surface layer. *J. Atmos. Sci.* **28**:181–189.

Byun, D.W. and J.K.S. Ching (Eds.) (1999). *Science Algorithms of the EPA Models-3 Community Multiscale Air Quality (CMAQ) Modeling System.* U.S. Environmental Protection Agency, EPA/600/R-99/030, Washington, DC.

Calabrese, E.J. and E.J. Stanek III (1995). Resolving intertracer inconsistencies in soil ingestion estimation, *Environ. Health Perspect.* **103**(5):454–457.

Calabrese, E.J. and E.J. Stanek III (1996). Methodology to estimate the amount and particle size of soil ingested by children: Implications for exposure assessment at waste sites. *Regul. Toxicol. Pharmacol.* **24**:264–268.

Calabrese, E.J., E.J. Stanek III, P. Pekow, and R.M. Barnes (1997). Soil ingestion estimates for children residing on a Superfund site. *Ecotoxicol. Environ. Safety* **36**:258–268.

CalTox: A Multimedia Tool Exposure Model for Hazardous Waste Site, prepared for the State of California, Toxic Substances Control, Lawrence Livermore National Laboratory, Livermore, CA, UCRL-CR-111456PEII, 1993.

CalTOX (1994). *CalTOX, A Multimedia Total Exposure Model for Hazardous-Waste Sites, Spreadsheet User's Guide.* Prepared by University of California, Davis, in Cooperation with Lawrence Livermore National Laboratory, for California Environmental Protection Agency, Department of Toxic Substances Control, Sacramento, CA.

C&EN:Chemical & Eng. News (2000). Climate treaty stalemate. CENEAR **78**(51):17–20.

Camp, Dresser and McKee (CDM) (1992). *Watershed Management Model User's Manual, Version 2.0.* Florida Department of Environmental Regulation, Tallahassee, FL.

Carey, G.F. (Ed.) (1995). *Finite Element Modelling of Environmental Problems: Surface and Subsurface Flow and Transport.* Wiley, New York.

Carson, R. (1962). *Silent Spring*. Houghton Mifflin, New York.

Carter, W.P.L. (1990). A detailed mechanism for the gas-phase atmospheric reactions of organic compounds. *Atmos. Environ.* **24A**:481–518.

Carter, W.P.L. (1995). Computer modeling of environmental chamber measurements of maximum incremental reactivities of volatile organic compounds. *Atmos. Environ.* **29**:2513–2527.

Carter, W.P.L. (2000). Documentation of the SAPRC-99 Chemical Mechanism for VOC Reactivity Assessment. Report to the California Air Resources Board, Contracts 92–329 and 95–308, May.

Carter, W.P.L., D. Luo, I. Malkina, and J.Pierce (1995). Environmental chamber studies of atmospheric reactivities of volatile organic compounds: Effects of varying chamber and light source, NREL/TP-425–7621. University of California, Riverside, CA.

Casada, L.S., S. Rouhani, C.A. Cardelino, and A.J. Ferrier (1994). Geostatistical analysis and visualization of hourly ozone data. *Atmos. Environ.* **28**:2105–2118.

Cassell, E.A. and J.C. Clausen (1993). Dynamic simulation modeling for evaluating water quality response to agricultural BMP implementation, *Water Sci. Tech.* **28**:635–648.

Casulli, V. and G.S. Stelling (1998). Numerical simulation of 3D quasi-hydrostatic free-surface flows. *J. Hydraulic Eng.* **124**(7):678–686.

Caughley G. and A. Gunn (1996). *Conservation Biology in Theory and Practice*. Blackwell Science, Cambridge, UK.

Cayan, D.R., M.D. Dettinger, H.F. Diaz, and N.E. Graham (1998). Decadal variability of precipitation over western North America. *J. Climate*, **11**:3148–3156.

Cayan, D.R., K.T. Redmond, and L.G. Riddle (1999). ENSO and hydrologic extremes in the western United States. *J.Climate*, **12**:2881–2893.

Celia, M.A., E.T. Bouloutas, and R.L. Zabra (1990). A general mass-conservative numerical solution for the unsaturated flow equation. *Water Resources Res.* **26**(7):1483–1496.

Chang, J.S., R.A. Brost, I.Isakson, S. Madronich, P. Middleton, W.R. Stockwell, and C.J Walcek (1987). A three-dimensional Eulerian acid deposition model: Physical concepts and formulation. *J. Geophys. Res.* **92**:14681–14700.

Chapra, S.C. (1997). *Surface Water-Quality Modeling*. McGraw-Hill, New York.

Chapra, S.C. and R.P. Canale (1988). *Numerical Methods for Engineers*, 2nd ed. McGraw-Hill, New York.

Chapra, S.C. and K.H. Reckhow (1983). *Engineering Approaches for Lake Management, Vol. 2: Mechanistic Modeling*. Butterworth, Boston.

Chapra, S.C. and R.L. Runkel (1999). Modeling impact of storage zones on stream dissolved oxygen. *J. Environ. Eng.* **125**(5):415–419.

Chatfield, C. (1996). *The Analysis of Time Series: An Introduction*, 5th ed. Chapman and Hall, London.

Chen, C.W., J. Herr, L. Ziemelis, R.A. Goldstein, and L. Olmsted (1999). Decision support system for total maximum daily load. *J. Environ. Eng.* **125**(7):653–659.

Chen, J.J., Y. Chen, and R.L. Kodell (2001). Using dose addition to estimate cumulative risks from exposures to multiple chemicals. *Regul. Toxicol. Pharmacol.* **34**:35–41.

Cheng, R.T., V. Casulli, and S.N. Milford (1984). Eurlerian-Lagrangian solution of the convection-dispersion equation in natural coordinates. *Water Resources Res.* **20**(7): 1541–1549.

Cheng, X. and M.P. Anderson (1994). Simulating the influence of lake position on groundwater. *Water Resources Res.* **30**(7):2041–2049.

Chiang, C.Y., J.P. Salanitro, E.Y. Chai, J.D. Colthart, and C.L. Klein (1989). Aerobic biodegradation of benzene, toluene, and xylene in a sandy aquifer—Data analysis and computer modeling. *Groundwater* **27**(6):823–833.

Chin, E.H. (1977). Modeling daily precipitation occurrence process with Markov Chain. *Water Resources Res.* **13**(6):949–959.

Chiou, H.Y., Y.M. Hsueh, K.F Liaw, S.F. Horng, M.H. Chiang, Y.S. Pu, J.S. Lin, C.H. Huang, and C.J. Chen (1995). Incidence of internal cancers and ingested inorganic arsenic: A seven-year follow-up study in Taiwan. *Cancer Res.* **55**(6):1296–1300.

Chou, S.T., L.T. Fan, and R. Nassar (1988). Modeling of complex chemical reactions in a continuous flow reactor: A Markov chain approach. *Chem. Eng. Sci.* **43**:2807–2815.

Chow, V.T. (1959). *Open-Channel Hydraulics*. McGraw-Hill, New York.

Christen, K. (2001). The arsenic threat worsens. *Environ. Sci. & Tech.* **35**(13):286A–291A.

Churchill, R.V. (1944). *Modern Operational Mathematics in Engineering*.McGraw-Hill, New York.

Cimorelli, A.J., S.G. Perry, A. Venkatram, J.C. Weil, R.J. Paine, R.B. Wilson, R.F. Lee, and W.D. Peters (1998). *AERMOD: Description of Model*. U.S. Environmental Protection Agency web site, http://www.epa.gov/scram001.

Chock, D.P. (1985). A comparison of numerical methods for solving the advection equation—II. *Atmos. Environ.* **19**:571–586.

Chock, D.P. (1991). A comparison of numerical methods for solving the advection equation—III, *Atmos. Environ.* **25A**:853–871.

Chock, D.P. and A.M. Dunker (1983). A comparison of numerical methods for solving the advection equation. *Atmos. Environ.* **17**:11–24.

Clement, T.P. (1997). RT3D—A modular computer code for simulating reactive multi-species transport in 3-dimensional groundwater aquifers. Pacific Northwest National Laboratory Report, PNNL-11720 (http://bioprocess.pnl.gov/rt3d.htm).

Clement, T.P., C.D. Johnson, Y. Sun, G.M. Klecka, and C. Bartlett (2000). Natural attenuation of chlorinated ethene compounds: Model development and field-scale application at the Dover Site. *J. Contaminant Hydrol.* **42**:113–140.

Clement, T.P., M.J. Truex, and P. Lee, (2002). A case study for demonstrating the application of U.S. EPA's monitored natural attenuation screening protocol at a hazardous waste site. *J. Contaminant Hydrol.* **59**(1–2):133–162.

Cohen, B. (1990). The ethics of using medical data from Nazi experiments. *J. Halacha Contemp. Soc.* **19**: 103–126.

Cole, T.M. and E.M. Buchak (1995). *CE-QUAL-W2: A Two-Dimensional, Laterally Averaged Hydrodynamic and Water Quality Model, Version 2.0 Users Manual*. Instruction Report EL-95-1, U.S. Army Corps of Engineers Waterways Experiment Station, Vicksburg, MI; see also http://www.coe.uncc.edu/~jdbowen/neem/w2_v2_manual.pdf.

Conradsen, K., A.A. Nielsen, and K. Windfeld (1992). Analysis of Geochemical Data Sampled on a Regional Scale. In *Statistics in the Environmental & Earth Sciences*. Wiley, New York, pp. 283–300.

Cooper, H.H., Jr., J.D. Bredehoeft, and I.S. Papadopoulos (1967). Response of a finite-diameter well to an instantaneous charge of water. *Water Resources Res.* **3**:263–269.

Corapcioglu, M.Y. and S. Jiang (1993). Colloid-facilitated groundwater contaminant transport. *Water Resources Res.* **29**(7):2215–2226.

Cordova, J. R. and R.L. Bras (1981). Physically based probabilistic models of infiltration, soil moisture, and actual evapotranspiration, *Water Resources Res.* **17**:93–106.

Cosby, B.J., G.M. Hornberger, R.B. Clapp, and T.R. Ginn (1984). A statistical exploration of the relationships of soil moisture characteristics to the physical properties of soils. *Water Resources Res.* **20**:682–690.

Cosby, B.J., R.F. Wright, G.M. Hornberger and J.N. Galloway (1985a). Modeling the effects of acid deposition: Estimation of long-term water quality responses in a small forested catchment. *Water Resources Res.* **21**(11):1591–1601.

Cosby B.J., G.M. Hornberger, and N.J. Galloway (1985b). Modeling the effects of acid deposition: Assessment of a lumped parameter model of soil water and streamwater chemistry. *Water Resources Res.* **21**:51–63.

Cowley, W.P. (1968). A global numerical ocean model, Part 1. *J. Comp. Phys.* **3**:111–147.

Cowpertwait, P.S.P. (1991). Further developments of the Neyman-Scott clustered point process for modeling rainfall. *Water Resources Res.* **27**(7):1431–1438.

Cowpertwait, P.S.P. and P.E. O'Connell (1993). A Neyman-Scott Shot Noise Model for the Generation of Daily Streamflow Time Series. In *Advances in Theoretical Hydrology, A Tribute to James Dooge*. J.P.J. O'Kane (Ed.). Elsevier, Amsterdam, pp. 75–94.

Crank, J. (1979). *The Mathematics of Diffusion*. Clarendon, Oxford.

Crawford-Brown, D.J. (1997). *Theoretical and Mathematical Foundations of Human Health Risk Analysis, Biophysical Theory of Environmental Health Science*. Kluwer Academic, Boston, MA.

Cressie, N. (1991). *Statistics for Spatial Data*. Wiley, New York.

Cronin, M.T.D. and J. C. Dearden (1995). Review: QSAR in toxicology. 1. Prediction of aquatic toxicity, *Quantitative Structure—Activity Relationships* **14**:1–7.

Crouch, E. and R. Wilson (1978). Interspecies comparison of carcinogenic potency. *J. Toxicol. Environ. Health* **5**:1095–1118.

Crouch, E.A.C., R. Wilson, and L. Zeise (1983). The risks of drinking water. *Water Resources Res.* **19**(6): 1359–1357.

Crump, C., K. Crump, E. Hack, R. Luippold, K. Mundt, E. Liebig, J. Panko, D. Paustenbach, and D. Proctor (2003). Dose-response and risk assessment of airborne hexavalent chromium and lung cancer mortality. *Risk Anal.* **23**(6):1147–1163.

Crump, K.S. (1981). An improved procedure for low-dose carcinogenic risk assessment from animal data. *J. Environ. Pathol. Toxicol.* **52**:675–684.

Crump, K.S. (1984). A new method for determining allowable daily intakes. *Funda. Appl. Toxicol.* **4**:854–871.

Crump, K. (1996). Risk of benzene-induced leukemia predicted from the Pliofilm cohort. *Environ. Health Perspect.* **104**(6):1437–1441.

Cullen, A.C. and H.C. Frey (1999). *Probabilistic Techniques in Exposure Assessment—A Handbook for Dealing with Variability and Uncertainty in Models and Inputs*. Plenum, New York.

Cullen, A. and M.J. Small (2004). Uncertain Risks: The Role and Limits of Quantitative Analysis. In *Risk Analysis and Society: Interdisciplinary Perspectives*. T. McDaniels and M. Small (Eds.). Cambridge University Press, Cambridge, UK.

Cullinan P., S.D. Acquilla, and V.R. Dhara (1997). Respiratory morbidity 10 years after the Union Carbide gas leak at Bhopal: A cross sectional survey. *Brit. Med. J.* **314**:338–342.

Culver, T.B. and C.A. Shoemaker (1992). Dynamic optimal control for groundwater remediation with flexible management period. *Water Resources Res.* **28**(3):629–641.

Cunge, K.A. (1969). On the subject of a flood propagation method (Muskingum method). *J. Hydro. Res.* **7**(2):205–230.

Cussler, E.L. (1983). *Diffusion Mass Transfer in Fluid Systems*. Cambridge University Press, New York.

Czernuszenko W. and P. Rowinski (1997). *J. of Hydraul. Res.* **35**(4):491–504.

Dagan, G. (1982). Stochastic modeling of groundwater flow by unconditional and conditional probabilities 2. The solute transport. *Water Resources Res.* **18**(4):835–848.

Dagan, G. (1986). Statistical theory of groundwater flow and transport: Pore to laboratory, laboratory to formation, and formation to regional scale. *Water Resources Res.* **22**:120S–134S.

Dagan, G. (1987). Theory of solute transport by groundwater. *Ann. Rev. Fluid Mech.* **19**:183–215.

Dakins, M.E., J.E. Toll, M.J. Small, and K.P Brand (1996). Risk-based environmental remediation: Bayesian Monte Carlo analysis and the expected value of sample information. *Risk Anal.* **16**:67–79.

Danckwerts, P.V. (1951). Significance of liquid-film coefficients in gas absorption, *Ind. Eng. Chem.* **43**(6):469.

Daniels, M.J., F. Dominici, J.M. Samet, and S.L. Zeger (2000). Estimating particulate matter-mortality dose-response curves and threshold levels: An analysis of daily time-series for the 20 largest US cities. *Am. J. of Epidemiol.* **152**(5):397–406.

Davies, A.M. and H. Gerritsen (1994). An intercomparison of three-dimensional tidal hydrodynamic models of the Irish Sea. *Tellus* **46A**:200–221.

Davis, J.C. (1986). *Statistics and Data Analysis in Geology,* 2nd ed. Wiley, New York.

Dawson, H.E. (1995). Screening-Level Tools for Modeling Fate and Transport of NAPLs and Trace Organic Chemicals in Soil and Groundwater: SOILMOD, TRANS1D, NAPLMOB, Colorado School of Mines Special Programs and Continuing Education, Golden, CO.

Deardorff, J.W. (1970). Convective velocity and temperature scales for the unstable planetary boundary layer and for Rayleigh convection. *J. Atmos. Sci.* **27**:1211–1213.

DeGroot, M.H. and M.J. Schervish (2002). *Probability and Statistics,* 3rd ed. Addison Wesley, Boston.

Demerjian, K.L. (1975). Photochemical diffusion models for air quality simulation: Current status. In Proceedings of the Conference on the State of the Art of Assessing Transportation-Related Air Quality Impacts, Oct. 22–24, 1975. Transportation Res. Board, Washington, DC.

De Roo, A.P.J., L. Hazelhoff, and G.B.M. Heuvelink (1992). Estimating the effects of spatial variability of infiltration of a distributed runoff and soil erosion model using Monte Carlo methods. *Hydrol. Processes* **6**:127–143.

De Rosa, C.T. (1998). Setting health risk benchmarks (letter). *Environ. Sci. Tech.* **32**(13):300A–301A.

Desai, C.S. (1979). *Elementary Finite Element Method.* Prentice Hall, Englewood Cliffs, NJ.

Desmond, A.F. and B.T. Guy (1991). Crossing theory for non-Gaussian stochastic processes with an application to hydrology. *Water Resources Res.* **27**(10):2791–2797.

Deutsch, C. and A. Journel (1992). *GSLIB: Geostatistical Software Library and User's Guide.* Oxford University Press.

Devillers, J., S., Bintein, and D. Domine (1996). Comparison of BCF models based on log P. *Chemosphere* **33**:1047–1065.

Diggle, P. (1983). *Statistical Analysis of Spatial Point Patterns.* Academic, London.

Dingman, S.L. (1993). *Physical Hydrology.* Prentice-Hall, Englewood Cliffs, NJ.

Di Toro, D.M. (1972). Line source distribution in two dimensions: Applications to water quality. *Water Resources Res.* **8**(6):1541–1546.

Di Toro, D.M. (1980). Statistics of Advective Dispersive System Response to Run-off. In *Proceedings Urban Stormwater and Combined Sewer Overflow Impact on Receiving Water Bodies*, Y. A. Yousef et al. (Eds.). Rep. EPA-600/9-80-056, EPA Municipal Environ. Res. Lab., Cincinnati, OH.

Di Toro, D.M. (1984). Probability model of stream quality due to runoff. *J. Environ. Eng.* **110**(3):607–628.

Di Toro, D.M. (2001). *Sediment Flux Modeling.* Wiley-InterScience, NewYork.

Di Toro, D.M. and M.J. Small (1979). Stormwater interception and storage. *J. Environ. Eng. Div. ASCE* **105**:43–54.

Di Toro, D.M., N.A. Thomas, C.E. Herdendorf, R.P. Winfeld, and J.P. Connolly (1987). A post audit of a Lake Erie eutrophication model. *J. Great Lakes Res.* **13**(4):801–825.

Dodge, M.C. (1989). A comparison of three photochemical oxidant mechanisms. *J. Geophys. Res.* **94**:5121–5136.

Dooge, J.C.I. (1973). *Theory of Hydrologic Systems*. Technical Bulletin 1468, U.S. Department of Agriculture. Washington, DC.

Dourson, M., M. Andersen, L. Erdreich, and J. MacGregor (2001). Using human data to protect the public's health, *Reg. Toxicol. Pharmacol.* **33**(2):234–256.

Draper, D. (1995). Assessment and propagation of model uncertainty (with discussion). *J. Roy. Statist. Soc. B.* **57**(1):45–97.

Driscoll, E.D., P.E. Shelley, and E.W. Strecker (1990a). *Polllutant Loadings and Impacts from Highway Stormwater Runoff, Volume I: Design Procedure.* U.S. Federal Highway Administration, Office of Eng. and Highway Operations R&D, Washington, D.C.

Driscoll, E.D., P.E. Shelley, and E.W. Strecker (1990b). *Polllutant Loadings and Impacts from Highway Stormwater Runoff, Volume II: Users Guide for Interactive Computer Implementation of Design Procedure.* U.S.Federal Highway Administration, Office of Eng. and Highway Operations R&D, Washington, DC.

Drost, W., D. Klotz, A. Koch, H. Moser, F. Neumaier, and W. Rauert (1968). Point dilution methods of investigating groundwater flow by means of radioisotopes. *Water Resources Res.* **4**:125–146.

DTSC (1992). *Guidance for Site Characterization and Multimedia Risk Assessment for Hazardous Substances Release Sites, Volume 2.* Department of Toxic Substances Control, State of California.

Duffy, C.J., L.W. Gelhar, and P.J. Wierenga (1984). Stochastic models in agricultural watersheds. *J. Hydrol.* **69**:145–162.

Dunker, A.M. (1984). The decoupled direct method for calculating sensitivity coefficients in chemical kinetics. *J. Chem. Phys.* **81**: 2385–2393.

Dunker, A.M., G. Yarwood, J.P. Ortmann, and G.M. Wilson (2002). The decoupled direct method for sensitivity analysis in a three-dimensional air quality model—implementation, accuracy and efficiency. *Environ. Sci. Technol.* **36**(13):2965–2976.

Dzombak, D.A. and F.M.M. Morel (1990). *Surface Complexation Modeling: Hydrous Ferric Oxide*. Wiley, New York, p. 393.

Eagleson, P.S. (1978). Climate, soil and vegetation 1–7, *Water Resources Res.* **14**:705–776.

Edinger, J.E. (2002). *Waterbody Hydrodynamic and Water Quality Modeling*. ASCE Press, Am. Society of Civil Engineers, Reston, VA.

Efron, B. and G. Gong (1983). A leisurely look at the bootstrap, the jackknife and cross validation. *Am. Statist.* **37**: 36–48.

Efroymson, R.A. and M. Alexander (1994). Biodegradation by an arthobacter species of hydrocarbons partitioned into organic solvents. *Envir. Sci. Technol.* **28**(9):1172–1179.

Eisenberg, J.N.S. and T. E. McKone (1998). Decision tree method for the classification of chemical pollutants: Incorporation of across-chemical variability and within-chemical uncertainty. *Environ. Sci. Tech.* **32**(21):3396–3404.

Eliassen, A. (1980). A review of long-range transport modeling. *J. Appl. Meteorol.* **19**:231–240.

Ellis, D. (1989). *Environments at Risk: Case Histories of Impact Assessment*. Springer, Berlin.

Ellison, A.M. (1996). An introduction to Bayesion inference for ecological research and environmental decision-making. *Ecolog. Appl.* **6**(4):1036–1046.

El-Shaarawi, A.H. (1989). Inference about the mean from censored water quality data. *Water Resources Res.* **25**(4):685–690.

Entekhabi, D., G.R. Asrar, A.K. Betts, K.J. Beven, R.L. Bras, C.J. Duffy, T. Dunne, R.D. Koster, D.P. Lettenmaier, D.B. McLaughlin, W.J. Shuttleworth, M.T. van Genuchten, M.Y. Wei, and E.F. Wood (1999). An agenda for land-surface hydrology and a call for the Second International Hydrologic Decade. *Bull. Am. Meteor. Soc.* **80**(10):2043–2058.

EPA (1989). *Risk Assessment Guidance for Superfund, Volume I, Human Health Evaluation Manual (Part A).* U.S. Environmental Protection Agency, EPA/540/1-89/002, Washington, DC.

EPA (1992a). *MMSOILS: Multimedia Contaminant Fate, Transport, and Exposure Model, Documentation and User's Manual.* U.S. Environmental Protection Agency, Office of Environmental Processes and Effects Res., Office of Res. and Development, Washington, DC.

EPA (1992b). Guidelines for exposure assessment; Notice, U.S. Environmental Protection Agency. 57 *Fed. Reg.*, 22888–22938, May 29.

EPA (1995). *Health Effects Assessment Summary Tables. Annual Update*. U.S. Environmental Protection Agency, Office of Res. and Development, EPA 540/R-95-036, Washington, DC.

EPA (1997a). *Exposure Factors Handbook.* U.S. Environmental Protection Agency, NTIS: PB98-124217, *http://www.epa.gov/ncea/exposfac.htm* , Washington, DC.

EPA (1997b). *Integrated Risk Information System.* U.S. Environmental Protection Agency, Environmental Criteria and Assessment Office, Washington, DC.

EPA (1998). *Guidelines for Ecological Risk Assessment.* EPA/630/R-95/002F, U.S. Environmental Protection Agency, Risk Assessment Forum, Washington, DC.

EPA (1999). TRIM.Expo: Technical Support Document (External Review Draft). U.S. Environmental Protection Agency, Office of Air Quality Planning and Standards, EPA-435/D-99-001, Res. Triangle Park, NC.

EPA (1999). *An In-Situ Permeable Reactive Barrier for the Treatment of Hexavalent Chromium and Trichloroethylene in Ground Water: Volume 3, Multicomponent Reactive Transport Modeling.* EPA/600/R-99/095c. Washington, DC.

EPA (2003). Framework for Cumulative Risk Assessment. U.S. Environmental Protection Agency, Risk Assessment Forum, Report EPA/630/P-02/001F, Washington, DC; see http://cfpub.epa.gov/ncea/raf/recordisplay.cfm?deid=54944; accessed January 1, 2004.

ES&T (1999). Production and Use of 12 Persistent Organic Pollutants. *Environ. Sci. Tech.* **33**(21):444A.

Esman, N.A. and G. M. Marsh (1996). Applications and limitations of air dispersion modeling in environmental epidemiology. *J. Exposure Anal. Environ. Epidemiol.* **6**(3):339–353.

Facemire, C.F. (1995). Mercury in Wildlife. In *National Forum on Mercury in Fish.* Proceedings, September 27–29, 1994, New Orleans, LA. EPA 823-R-95-002, U.S. EPA, Office of Water, Washington, DC, pp. 53–60.

Farman, J.C., B.G. Gardiner, and J.D. Shanklin (1985). Large losses of total ozone in Antarctica reveal seasonal ClO_x/NO_x interaction. *Nature* **315**: 207–210.

Farr, A.M., R.J. Houghtalen, and D.B. McWhorter (1990). Volume estimation of light non-aqueous phase liquids in porous media. *Ground Water* **28**(1):48–56.

Faustman, E.M. and S.M. Bartell (1997). Review of noncancer risk assessment: Applications of benchmark dose methods. *Human Ecolog. Risk Assess.* **3**:893–920.

Fay, J.A. and J.J. Rosenzweig (1980). An analytical diffusion model for long distance transport of air pollutants. *Atmos. Environ.* **14**:355–365.

Fay, J.A., D. Golomb, and S. Kumar (1985). Source apportionment of wet sulfate deposition in eastern North America. *Atmos. Environ.* **19**:1773–1778.

Fay, J.A., D. Golomb, and S. Kumar (1986). Erratum. *Atmos. Environ.* **20**(6):1315.

Fenner, K., C. Kooijman, M. Scheringer, and K. Hungerbuhler (2002). Including transformation products into the risk assessment for chemicals: The case of nonylphenol ethoxylate usage in Switzerland. *Environ. Sci. Technol.* **36**(6):1147–1154.

Fernald, A.G., P.J. Wigington, Jr. and D.H. Landers (2001). Transient storage and hyporheic flow along the Williamette River, Oregon: Field measurements and model estimates. *Water Resources Res.*, **37**(6):1681–1694.

Festing, M.F.W., V. Baumans, R.D. Combes, M. Halder, C.F.M. Hendriksen, B.R. Howard, D.P. Lovell, G.J. Moore, P. Overend, and M.S. Wilson (1998). Reducing the use of laboratory animals in biomedical Res.: Problems and possible solutions. *Alternatives Lab. Animals (ALTA)* **26**(3):283–301.

Fetter, C.W. (1993). *Contaminant Hydrogeology*. Macmillan, New York.

Finlayson-Pitts, B.J. and J. N. Pitts, Jr. (1986). *Atmospheric Chemistry*. Wiley, New York.

Finley, B., D. Proctor, P. Scott, N. Harrington, D. Paustenbach, and P. Price (1994). Recommended distributions for exposure factors frequently used in health risk assessment. *Risk Analy.* **14**:533–553.

Fischer, H.B., E.J. List, R.C. Koh, J. Imberger, and N.H. Brooks (1979). *Mixing in Inland and Coastal Waters*, Academic, New York.

Fisher, B.E.A. (1983). A review of the processes and models of long-range transport of air pollutants. *Atmos. Environ.* **17**:1865–1880.

Fitzpatrick, J., J. Imhoff, E. Burgess, and R. Brashear (2001). *Water Quality Models: A Survey and Assessment*. Water Environment Res. Foundation, Alexandria, VA.

Ford, D.E. and L.S. Johnson (1986). As Assessment of Reservoir Mixing Processes. Technical Report E-86-7, U.S. Army Corps of Engineers, Waterways Experiment Station, Vicksburg, MS.

Fox, D.G. (1981). Judging air quality model performance. *Bull. Am. Meteorol. Soc.* **62**:599–609.

Fox, D. (1984). Uncertainty in air quality modeling. *Bull. Am. Met. Soc.* **65**:27–36.

Freeze, A.R. and J.A. Cherry (1979). *Groundwater*. Prentice-Hall, Upper Saddle River, NJ.

Freijer, J.I., H.J.T.H. Bloemen, S. de Loos, M. Marra, P.J.A. Rombout, G.M. Steentjes, and M.P. van Veen (1998). Modelling exposure of the Dutch population to air pollution. *J. Haz. Mat.* **61**:107–114.

Frey, H.C. and D.E. Burmaster (1999). Methods for characterizing variability and uncertainty: Comparison of bootstrap simulation and likelihood-based approaches. *Risk Anal.* **19**(1):109–130.

Frey, H.C. and S.R. Patil (2002). Identification and review of sensitivity analysis methods. *Risk Anal.* **22**(3):553–578.

Fuchs, N.A. (1964). *Mechanics of Aerosols*. Pergamon, New York.

Fujita, E.M., B.E. Croes, C.L. Bennett, D.R. Lawson, F.W. Lurmann, and H.H. Main (1992).

Comparison of emission inventory and ambient concentration ratios of CO, NMOG and NO_x in California's South Coast Air Basin. *J. Air Waste Manag. Assoc.* **42**:264–276.

Gabrial, R. and J. Neuman (1962). A Markov chain model for daily rainfall occurrence at Tel Aviv, Israel. *J. Roy. Meteorol. Soc. London* **88**:90–95.

Gaganis, P. and L. Smith (2001). A Bayesian approach to the quantification of the effect of model error on the prediction of groundwater models. *Water Resources Res.*, **37**(9):2309–2322.

Galperin, B., L.H. Kantha, S. Hassid, and A. Rosati (1988). A quasi-equilibrium turbulent energy model for geophysical flows. *J. Atmos. Sci.* **45**:55–62.

Gao, D., W.R. Stockwell, and J.B. Milford (1996). Global uncertainty analysis of a regional-scale gas-phase chemical mechanism. *J. Geophys. Res.* **101**:C4, 9107–9119.

Gelbard, F. and J.H. Seinfeld (1980). Simulation of multicomponent aerosol dynamics. *J. Colloid Interface Sci.* **78**:485–501.

Gelda, R.K., S.W. Effler, and E.M. Owens (2001). River dissolved oxygen model with zebra mussel oxygen demand (ZOD). *J. Environ. Eng.* **127**(9):790–801.

Gelhar, L.W. (1986). Stochastic subsurface hydrology from theory to applications. *Water Resources Res.* **22**(9):135S–145S.

Gelhar, L.W. (1993). *Stochastic Subsurface Hydrology*. Prentice Hall, Englewood Cliffs, NJ.

Gelhar, L.W. and C.L. Axness (1983). Three-dimensional stochastic analysis of macrodispersion in aquifers. *Water Resources Res.* **19**(1):161–80.

Gelhar, L.W., A.L. Gutjahr, and R.L. Naff (1979). Stochastic analysis of macrodispersion in a stratified aquifer. *Water Resources Res.* **15**(6):1387–1391.

Geller, J. T. and J. R. Hunt (1993). Mass transfer from nonaqueous phase organic liquids in water saturated porous media. *Water Resources Res.* **29**(4):233–245.

Gelman, A., J.B. Carlin, H.S. Stern, and D.B. Rubin (1995). *Bayesian Data Analysis*. Chapman & Hall, London.

Georgopoulos, P.G. and J.H. Seinfeld (1982). Statistical distributions of air pollutant concentrations. *Environ. Sci. Tech.* **16**:401–416.

Gery, M.W., G.Z. Whitten, J.P. Killus, and M.C. Dodge (1989). A photochemical mechanism for urban and regional scale computer modeling. *J. Geophys. Res.* **94**:12925–12956.

Ghosh, U., J.S. Gillette, R.G. Luthy, and R.N. Zare (2000). Microscale location, characterization, and assoc. of polycyclic aromatic hydrocarbons on harbor sediment particles. *Environ. Sci. Technol.* **34**(9):1729–1736.

Ghoshal, S., A. Ramaswami, and R.G. Luthy, (1996). Biodegradation of naphthalene from coal tar and hepta-methylnonane in mixed batch systems. *Environ. Sci. Tech.* **30**(4):1282–1291.

Gifford, F.A. (1959). Statistical properties of a fluctuating plume dispersion model. *Adv. Geophys.* **6**:117–138.

Gifford, F.A. (1961). Use of routine meteorological observations for estimating atmospheric dispersion. *Nuclear Safety*, **2**:47–51.

Gifford, F.A., Jr. (1968). An Outline of Theories of Diffusion in the Lower Layers of the Atmosphere. In *Meteorology and Atomic Energy*. D.H. Slade (Ed.). U.S. Atomic Energy Commission, Technical Information Center, Oak Ridge, TN.

Gilbert, R.O. (1987). *Statistical Methods for Environmental Pollution*, Van Nostrand Reinhold, New York.

Glen, G., Y. Lakkadi, J.A. Tippett, and M. del Valle-Torres (1997). Development of NERL/

CHAD: The National Exposure Res. Laboratory Consolidated Human Activity Database. ManTech Environmental Technology for U.S. EPA, NERL Contract No. 68-D-0049.

Gobas, F.A.P.C., E.J. McNeil, L. Lovett-Doust, and G.D. Haffner (1991). Bioconcentration of clorinated aromatic hydrocarbons in aquatic macrophytes. *Environ. Sci. Tech.* **25**:924–929.

Gold, L.S., N.B. Manley, and B.N. Ames (1992). Extrapolation of carcinogenicity between species: Qualitative and quantitative factors. *Risk Anal.* **12**(4):579–588.

Goldhaber, S.B. and R.L. Chessin (1997). Comparison of hazardous air pollutant health risk benchmarks. *Environ. Sci. Technol.* **31**(12):568A–573A.

Goltz, M.N. and P.Y. Roberts (1986). Three-dimensional solutions for solute transport in an infinite medium with mobile and immobile zones. *Water Resources Res.* **22**(7):1139–1148.

Goltz, M.N. and P.Y. Roberts (1987). Using the method of moments to analyze three-dimensional diffusion-limited solute transport from temporal and spatial perspectives. *Water Resources Res.* **23**(8):1575–1585.

Golumb, D. (1983). Acid deposition-precursor emission relationship in the northeastern U.S.A.: The effectiveness of regional emission reduction. *Atmos. Environ.* **17**:1387–1390.

Gong, H. Jr., R. Wong, R.J. Sarma, W.S. Linn, E.D. Sullivan, D.A. Shamoo, K.R. Anderson, and S.B. Prasad (1998). Cardiovascular effects of ozone exposure in human volunteers, *Am. J. Respir. Crit. Care Med.* **158**(2):538–546.

Goodin, W.R., G.J. McRae, and J.H. Seinfeld (1979). An objective analysis technique for constructing three-dimensional, urban scale wind fields. *J. Appl. Meteor.* **19**:98–108.

Goolsby, D.A., W.A. Battaglin, G.B. Lawrence, R.S. Artz, B.T. Aulenbach and R.P. Hooper (1999). *Flux and Sources of Nutrients in the Mississippi-Atchafalaya River Basin: Topic 3 Report for the Integrated Assessment of Hypoxia in the Gulf of Mexico*. NOAA Coastal Ocean Program Decision Analysis Series No. 17. NOAA Coastal Ocean Program, Silver Spring, MD; see http://www.nos.noaa.gov/pdflibrary/hypox_t3final.pdf, accessed January 21, 2002.

Gorbunov, B., R. Hamilton, N. Clegg, and R. Toumi (1998). Water nucleation on aerosol particles containing both organic and soluble inorganic substances. *Atmos. Res.* **47–48**:271–283.

Gorelick, S.M., C.I. Voss, P.E. Gill, W. Murray, M.A. Saunders, and M.H. Wright (1984). Aquifer reclamation design: The use of contaminant transport simulation combined with nonlinear programming. *Water Resources Res.* **20**:415–427.

Gradshteyn, I.S. and I.M. Ryzhik (1980). *Table of Integrals, Series, and Products*. Academic, New York.

Grady, C.P.L., Jr., G.T. Daigger, and H.C. Lim (1999). *Biological Wastewater Treatment*, 2nd ed. Marcel Dekker, New York.

Grandstaff, D. (1977). Some kinetics of bronzite orthopyroxene dissolution. *Geochim. Cosmochim. Acta* **41**:1097.

Gray, H.A., G.R. Cass, J. Huntzicker, E. Heyerdahl, and J. Rau (1986). Characteristics of atmospheric organic and elemental carbon particle concentrations in Los Angeles. *Environ. Sci. Tech.* **20**:580–589.

Green, D.W. and J.O. Maloney (Eds.) (1997). *Perry's Chemical Engineers' Handbook*, 7th ed. McGraw-Hill, New York.

Green, J.R. (1964). A model for rainfall occurrence. *J. Roy. Statist. Soc., Series B* **26**:345–353.

Green, W.H. and G.A. Ampt (1911). Studies of soil physics: 1. The flow of air and water through soils. *J. Agricul. Sci.* **4**(1):1–24.

Greer M.A., G. Goodman, R.C. Pleus, and S.E. Greer (2002). Health effects assessment for environmental perchlorate contamination: The dose response for inhibition of thyroidal radioiodine uptake in humans, *Environ. Health Perspect.* **110**:927–937.

Grell, G., J. Dudhia, and D. Stauffer (1994). A Description of the Fifth Generation Penn State/NCAR Mesoscale Model (MM5), NCAR/TN-398+STR, National Center for Atmospheric Res., Boulder, CO.

Grolimund D., M. Elimelech, M. Borkovec, K. Bartmettler, R. Kretzschmar, and H. Sticher (1998). Transport of in situ mobilized colloidal particles in packed soil columns. *Environ. Sci. Technol.* **32**:3562–3569.

Grosjean, D. and J.H. Seinfeld (1989). Parameterization of the formation potential of secondary organic aerosols. *Atmos. Environ.* **23**:1733–1747.

Gross, L.J. and M.J. Small (1998). River and floodplain process simulation for subsurface characterization. *Water Resources Res.* **34**(9):2365–2376.

Grubler, A. (2000). Managing the global environment, *Environ. Sci. & Techn.* **34**(7):184A–187A.

Guha, S., C.A. Peters, and P.R. Jaffe (1999). Multisubstrate biodegradation kinetics of naphthalene, phenanthrene, and pyrene mixtures. *Biotech. BioEng.* **65**:491–499.

Gupta, S.C. and W.E. Larson (1979). Estimating soil water retention characteristics from particle size distribution, organic matter content, and bulk density. *Water Resources Res.* **15**:1633–1635.

Gupta, V.K. and S. Sorooshian (1983). Uniqueness and observability of conceptual rainfall runoff model parameters: The percolation process examined. *Water Resources Res.* **19**(1):269–276.

Gustafsson, J. (Last update: 2004). Visual MINTEQ 2.23. http://www.lwr.kth.se/English/OurSoftware/vminteq/.

Guymon, G.L. (1970). A finite element solution of the one-dimensional diffusion-convection equations. *Water Resources Res.* **6**:204–210.

Guymon, G.L., Scott, V.H., and Hermann, C.R. (1970). A general numerical solution of two-dimensional diffusion-convection equation by the finite element method, *Water Resources Res.* **6**:1611–1617.

Haas, C.N. (1999). On modeling correlated random variables in risk assessment. *Risk Anal.* **19**(6):1205–1214.

Haas, T.C. (1990). Kriging and automated variogram modeling within a moving window. *Atmos. Environ.* **24A**:1759–1769.

Haith, D.A. and L.L. Shoemaker (1987). Generalized watershed loading functions for stream flow nutrients. *Water Resources Bull.* **23**(3):471–478.

Haith, D.A., R. Mandel, and R.S. Wu (1992). *GWLF—Generalized Watershed Loading Functions, Version 2.0—User's Manual*. Department of Agricultural Eng., Cornell University, Ithaca, NY.

Hakami, A., M.T. Odman, and A.G. Russell (2003). High-order direct sensitivity analysis of multi-dimensional air quality models. *Environ. Sci. Technol.* **37**(11):2442–2452.

Haldane, J.B.S. (1930). *Enzymes*. Longmans, Green, UK. Republished by M.I.T. Press, Cambridge, MA (1965).

Halevy, E., H. Moser, O. Zellhofer, and A. Zuber (1967). Borehole Dilution Techniques: A Critical Review. In *Isotopes in Hydrology*. IAEA, Vienna, pp. 531–564.

Halton, J.H. (1960). On the efficiency of certain quasi-random sequences of points in evaluating multi-dimensional integrals. *Numerische Math.* **2**:84–90.

Hammersley, J.M. (1960). Monte Carlo methods for solving multidimensional problems. *Ann. NY Acad. Sci.* **86**:844–870.

Hammond, D., H.J. Simpson, and G. Mathieu (1975). Methane and Radon-222 as Tracers for Mechanisms of Exchange across the Sediment-Water Interface in the Hudson River Estuary. In *Marine Chemistry in the Coastal Environment*, Am. Chemical Society Symposium Series 18, T.M. Church (Ed.). Am.Chemical Society, Washington D.C.

Hamrick, J.M. (1992). *A Three-Dimensional Environmental Fluid Dynamics Computer Code: Theoretical and Computational Aspects*. Special Report in Applied Marine Sci. and Ocean Eng., Number 317, Virginia Institute of Marine Sci., Gloucester Point, VA.

Hanna, S.R. (1984). Concentration fluctuations in a smoke plume. *Atmos. Environ.* **18**:1091–1106.

Hanna, S.R. (1989). Confidence limits for air quality model evaluations, as estimated by bootstrap and jackknife resampling methods. *Atmos. Environ.* **23**:1385–1398.

Hanna, S.R., B.A. Egan, J. Purdum, and J. Wagler (2000). Comparison of AERMOD, ISC3 and ADMS Model Performance with Five Field Data Sets, Proceedings of the 93rd Annual Meeting of the Air & Waste Management Association, Salt Lake City, June 18–22.

Hansch, C. and A. Leo (1979). *Substituent Constants for Correlation Analysis in Chemistry and Biology*. Wiley, New York.

Hanson, M.L. and K. R. Solomon (2002). New technique for estimating thresholds of toxicity in ecological risk assessment. *Environ. Sci. Tech.* **36**(15):3257–3264.

Harley, R.A., A.G. Russell, G.J. McRae, G.R., Cass, and J.H. Seinfeld (1993). Photochemical modeling of the Southern California Air Quality Study. *Environ. Sci. Tech.* **27**:378–388.

Harley, R.A., R.F. Sawyer, and J.B. Milford (1997). Updated photochemical modeling for California's South Coast Air Basin: Comparison of chemical mechanisms and motor vehicle emission inventories. *Environ. Sci. Tech.* **31**:2829–2839.

Harrington, D.Y. and S.M. Kreidenweis (1998). Simulations of sulfate aerosol dynamics—I. Model description. *Atmos. Environ.* **32**:1691–1700.

Hart, D.R. (1995). Parameter estimation and stochastic interpretation of the transient storage model for solute transport in streams. *Water Resources Res.* **31**(2):323–328.

Harvey, J.W. and K.E. Bencala (1993). The effect of streambed topography on surface-subsurface water exchange in mountain catchments. *Water Resources Res.* **29**(1):89–98.

Harvey, L.D.D. and S.H. Schneider (1985). Transient climatic response to external forcing on 10^0 - 10^4 year time scales, 1: Experiments with globally averaged coupled atmosphere and ocean energy balance models. *J. Geophys. Res.* **90**:2191–2205.

Harvey, L.D.D., J. Gregory, M. Hoffert, A. Jain, M. Lal, R. Leemans, S.C.B. Raper, T.M.L. Wigley, and J.R. de Wolde (1997). *An Introduction to Simple Climate Models Used in the IPCC Second Assessment Report*. IPCC Technical Paper II. Houghton, J.T., L.G. Meira Filho, D.J. Griggs, and K. Maskell (Eds.) Intergovernmental Panel on Climate Change,

Geneva, Switzerland. See http://www.ipcc.ch/pub/techpap2.pdf, accessed January 21, 2001.

Hattis, D. (2004). The Conception of Variability in Risk Analyses—Developments Since 1980. In *Risk Analysis and Society: Interdisciplinary Perspectives*. T. McDaniels and M. Small (Eds.). Cambridge University Press, Cambridge, UK.

Hattis, D. and D.E. Burmaster (1994). Assessment of variability and uncertainty distributions for practical risk analyses. *Risk Anal.* **14**:713–730.

Hattis, D., P. Banati, R. Goble, and D.E. Burmaster (1999). Human interindividual variability in parameters related to health risks. *Risk Anal.* **19**(4):711–726.

Hay, L.E. and G.J. McCabe (2002). Spatial variability in water-balance model performance in the conterminous United States. *J. Am. Water Resources Assoc.* **38**(3):847–860.

Heinsohn, R. J., and R.L. Kabel (1999). *Sources and Control of Air Pollution*. Prentice Hall, Upper Saddle River, NJ.

Helsel, D.R. and R.M. Hirsch (1992). *Statistical Methods in Water Resources*. Elsevier, New York.

Helton, J.C., F.J. Davis, and J.C. Helton (2002). Illustration of sampling-based methods for uncertainty and sensitivity analysis. *Risk Anal.* **22**(3):591–622.

Hemond, H.F. and E. J. Fechner-Levy (2000). *Chemical Fate and Transport in the Environment*. Academic, San Diego.

Hertwich, E.G. and T. E. McKone (2001). Pollutant-specific scale of multimedia models and its implications for the potential dose. *Environ. Sci. Tech.* **35**(1):142–148.

Hertwich, E.G., W.S. Pease, and T.E. McKone (1998). Evaluating toxic impact assessment methods: What works best? *Environ. Sci. Technol.* **32**(5):138A–144A.

Hertzberg R.C. and M.M. MacDonell (2002). Synergy and other ineffective mixture risk definitions. *Sci. Total Environ.* **288**(1–2):31–42.

Hidy, G.M. (1984). Source-receptor relationships for acid deposition: Pure and simple? *J. Air Pollution Control Assoc.* **34**:518–531.

Hidy, G.M., P.M. Roth, J.M. Hales, and R. Scheffe (1998). Oxidant pollution and fine particles: Issues and needs, White paper prepared for the 1998 North Am. Res. Strategy for Tropospheric Ozone (NARSTO) Critical Review Series, June 30, 1998.

Higgins, J.J. and S. Keller-McNulty (1995). *Concepts in Probability and Stochastic Modeling*, Duxbury (Wadsworth), Belmont, CA.

Hine, J. and P.K. Mookerjee (1975). The intrinsic hydrophilic character of organic compounds. Correlations in terms of structural contributions. *J. Organic Chem.* **40**:292–298.

Hoffman, F.O. and J.S. Hammonds (1994). Propagation of uncertainty in risk assessments: The need to distinguish between uncertainty due to lack of knowledge and uncertainty due to variability. *Risk Anal.* **14**:707–712.

Hoffman, J.D. (1992). *Numerical Methods for Engineers and Scientist*. McGraw-Hill, New York.

Holdaway, M.R. (1996). Spatial modeling and interpolation of monthly temperature using kriging. *Climate Res.*, **6**:215–225.

Hooper, R.P. and C.A. Shoemaker (1986). A comparison of chemical and isotopic hydrograph separation. *Water Resources Res.* **22**(10):1444–1454.

Hooper, R.P., N. Christophersen, and N.E. Peters (1990). Modelling streamwater chemistry as a mixture of soilwater end-members—an application to the Panola Mountain Catchment, Georgia, USA. *J. Hydrol.* **116**:321–343.

Hopke, P.K. (1985). *Receptor Modeling in Environmental Chemistry*. Wiley, New York.

Hopke, P.K., T.B. Borak, J. Doull, J.E. Cleaver, K.F. Eckerman, L.C.S. Gundersen, N.H. Harley, C.T. Hess, N.E. Kinner, K.J. Kopecky, T.E. McKone, R.G. Sextro, and S.L. Simon (2000). Health risks due to radon in drinking water. *Environ. Sci. Tech.* **34**(6):921–926.

Hornberger, G.M. and R.C. Spear (1980). Eutrophication in Peel Inlet—I. The problem-defining behavior and a mathematical model for the phosphorus scenario. *Water Res.* **14**: 29–42.

Horne, A. and C.R. Goldman (1994). *Limnology*, 2nd ed. McGraw-Hill, New York.

Horwedel, J.E., R.J. Raridon, and R.Q. Wright (1992). Automated sensitivity analysis of an atmospheric dispersion model. *Atmos. Environ.* **26A**:1643–1649.

Hosseinipour, E.Z. (1995). *The One-Dimensional Riverine Hydrodynamic Model, RIVMOD-H —Model Documentation and Users' Manual.* U.S. EPA Office of Res. and Development, Environmental Res. Laboratory, Athens, GA.

Howard, P.H. (1989). *Handbook of Environmental Fate and Exposure Data for Organic Chemicals, Volume I.* Lewis, Chelsea, MI.

Howard, P.H., R.S. Boethling, W.F. Jarvis, W.M. Meylan, and E.M. Michalenko (1991). *Handbook of Environmental Degradation Rates*. Lewis Publishers, Chelsea, MI.

Howard-Reed, C., R.L. Corsi, and J. Moya (1999). Mass transfer of volatile organic compounds from drinking water to indoor air: The role of residential dishwashers. *Environ. Sci. Tech.* **33**(13):2266–2272.

Huang, C.H. (1979). A theory of dispersion in turbulent shear flow. *Atmos. Environ.* **13**:453–463.

Hubbert, M.K. (1940). The theory of groundwater motion. *J. Geol.* **48**:785–944.

Huber, W.C. and R.E. Dickinson (1988). *Storm Water Management Model User's Manual, Version 4.* EPA/600/3-88/001a, U.S. Environmental Protection Agency, Athens, GA; see http://www.epa.gov/ceampubl/swater/swmm/index.htm.

Hunt, B. (1978). Dispersive sources in uniform ground-water flow. *Am. Soc. Civil Eng. J. Hydraulics Div.* **104**(HY1):75–85.

Hunt, J.R., N. Sitar, and K.S. Udell (1988). Nonaqueous phase liquid transport and cleanup: Analysis of mechanisms. *Water Resources Res.* **24**:1247–1258.

Huyakorn, P.S. and G.F. Pinder (1983). *Computational Methods in Subsurface Flow*. Academic, San Diego.

Hvorslev, M.J. (1951). Time lag and soil permeability in groundwater observations. *U.S. Army Corps Engrs. Waterways Exp. Sta. Bull.36*, Vicksburg, MI.

Hwang, J.T., E.P. Dougherty, S. Rabitz, and H. Rabitz (1978). The Green's function method of sensitivity analysis in chemical kinetics. *J. Chem. Phys.* **69**:5180–5191.

HydroScience (1979). *A Statistical Method for the Assessment of Urban Storm Water*. EPA-440/3-79-023, U.S. Environmental Protection Agency (EPA), Washington, DC.

IARC (International Agency for Res. on Cancer) (1987). *IARC Monographs on the Evaluation of the Carcinogenic Risks to Humans: Overall Evaluation of Carcinogenicity: An Updating of IARC Monographs 1 to 42, Supplement 7.* World Health Organization, Lyon, France.

Illangasekare, T.H., J.L. Ramsey, K.H. Jensen, and M. Butts (1995). Experimental study of movement and distribution of dense organic contaminants in heterogeneous aquifers. *J. Contaminant Hydrology* **20**:1–25.

Iman, R. and S. Hora (1989). Bayesian methods for modeling recovery times with an application to the loss of off-site power at nuclear power plants. *Risk Anal.* **9**(1):25–36.

Iman, R.L. and M.J. Shortencarier (1984). A FORTRAN 77 Program and User's Guide for the Generation of Latin Hypercube and Random Samples for Use with Computer Models. NUREG/CR-3624, SAND83-2364, Sandia National Laboratory.

Imhoff, J.C., A. Stoddard, and E.M. Buchak (2003). *Evaluation of Contaminated Sediment Fate and Transport Models, Final Report*, U.S. Environmental Protection Agency, National Exposure Res. Laboratory, Athen, GA; see http://hspf.com/pdf/FinalReport110.pdf.

Infante, P.F. (1993). Use of rodent carcinogenicity test results for determining potential cancer risk to humans. *Environ. Health Perspect.* **101**(Suppl 5):143–148.

Inter-Governmental Panel on Climate Change (IPCC) (1996). *Climate Change: The Sci. of Climate Change, Contribution of WGI to the Second Assessment report of the IPCC.* Houghton et al. (eds.). Cambridge University Press, New York.

i Salau, J.S., R. Tauler, J.M. Bayona, and I. Tolosa (1997). Input characterization of sedimentary organic contaminants and molecular markers in the northwestern Mediterranean Sea by exploratory data analysis. *Environ. Sci. Tech.* **31**(12):3482–3490.

Isnard, P. and S. Lambert (1988). Estimating bioconcentration factors from octanol–water partition coefficient and aqueous solubility. *Chemosphere* **17**:21–34.

Jacobson, M.Z. (1997). Development and application of a new air pollution modeling system—II. Aerosol module structure and design. *Atmos. Environ.* **31**:131–144.

Jacobson, M.Z. (1999). *Fundamentals of Atmospheric Modeling*. Cambridge University Press, Cambridge, England.

Jaecker-Voirol, A. and P. Mirabel (1989). Heteromolecular nucleation in the sulfuric-acid–water system. *Atmos. Environ.* **23**:2053–2057.

Jaffe, P.R. and R.A. Ferrara (1984). Modeling sediment and water column interactions for hydrophobic pollutants. *Water Res.* **18**(9):1169–1174.

Jakeman, A.J. and G.M. Hornberger (1993). How much complexity is warranted in a rainfall-runoff model? *Water Resources Res.* **29**(8):2637–2649.

Jakeman, A.J, I.G. Littlewood, and P.G. Whitehead (1991). Computation of the instantaneous unit hydrograph and identifiable component flows with application to two upland catchments. *J. Hydrol.* **117**:275–300.

Janssen, N.A.H., G. Hoek, H. Harssema, and B. Brunekreef (1998). Personal sampling of airborne particles: Method performance and data quality. *J. Exposure Anal. Environ. Epidemiol.* **8**(1):37–49.

Jantunen, M.J., O. Hanninen, K. Katsouyanni, H. Knoppel, N. Kuenzli, E. Lebret, M. Maroni, R. Sram, and D. Zmirou (1998). Air pollution exposure in European cities: The "EXPOLIS" study. *J. Exposure Anal. Environ. Epidemiol.* **8**(4):495–518.

Javendel, I. and C.F. Tsang (1986). Capture-zone type curves—A tool for aquifer cleanup. *Groundwater* **24**(5):616–625.

Jeffries, H.E. (1995). Photochemical Air Pollution. In *Composition, Chemistry and Climate of the Atmosphere*. H.B. Singh (Ed.). Van Nostrand Rheinhold, New York.

Jenkin, M.E., S.M. Saunders, and M.J. Pilling (1997). The tropospheric degradation of volatile organic compounds: A protocol for mechanism development. *Atmos. Environ.* **31**:81.

Jirka, G.H. and P.J. Akar (1991). Hydrodynamic classification of submerged multiport diffuser discharges. *J. Hydraul. Eng.* **117**(9):1113–1128.

Jirka, G.H. and R.L. Doneker (1991). Hydrodynamic classification of submerged single port diffuser discharges. *J. Hydraul. Eng.* **117**(9):1095–1112.

Johnson, G.L., C.L. Hanson, S.P. Hardegree, and E.B. Ballard (1996). Stochastic weather simulation: Overview and analysis of two commonly used models. *J. Appl. Meteorol.* **35**:1878–1896.

Johnson P.R. and M. Elimelech (1995). Dynamics of colloid deposition in porous media: Blocking based on random sequential adsorption. *Langmuir* **11**(3):801–812.

Johnson P.R., N. Sun, and M. Elimelech (1996). Colloid transport in geochemically heterogeneous porous media: Modeling and measurements. *Environ. Sci. Technol.* **30**:3284–3293.

Johnson, W.B. (1983). Interregional exchange of air pollution: Model types and applications. *J. Air Pollution Control Assoc.* **33**:563–574.

Joner, E.J., A. Johansen, A.P. Loibnmer, M.A. de la Cruz, O.H. Szolar, J.M. Portal, and C. Leyval (2001). Rhizosphere effects on microbial community structure and dissipation and toxicity of polycyclic aromatic hydrocarbons (PAHs) in spiked soil. *Environ. Sci. Tech.* **35**(13):2773–2777.

Jones, G.R., J.D. Nash, and G.H. Jirka (1996). *CORMIX3: An Expert System for Mixing Zone Analysis and Prediction of Buoyant Surface Discharges*. Cornell University DeFrees Hydraulics Laboratory for U.S. EPA Office of Sci. and Technology, Cooperative Agreement No. CR 818527, Washington, DC.

Jordahl, J.L., L. Foster, J.L. Schnoor, and P.J.J. Alvarez (1997). Effect of hybrid poplar trees on microbial populations important to hazardous waste bioremediation. *Environ. Toxicol. Chem.* **16**(6):1318–1321.

Journel, A.G. and C.J. Huijbregts (1981). *Mining Geostatistics*, Academic, San Diego.

Junge, C.E. (1975). Basic Considerations about Trace Gas Constituents in the Atmosphere as Related to the Fate of Global Pollutants. In *Fate of Pollutants in the Air and Water Environments*, I.H. Suffet (Ed.). Wiley, New York, pp. 7–25.

Jury, W.A. and J. Gruber (1989). A stochastic analysis of the influence of soil and climatic variability on the estimate of pesticide ground water pollution. *Water Resources Res.* **25**:2465–2474.

Jury, W.A., Farmer, W.J., and Spencer, W.F. (1983). Behavior assessment model for trace organics in soil. 1. Model development, *J. Environ. Qual.* **12**:558–564.

Jury, W.A., Russo, D., Streile, G., and El Abd, H. (1990). Evaluation of volatilization by organic chemicals residing below the soil surface. *Water Resources Res.* **26**:13–20.

Kalagnanam, J.R. and U.M Diwekar (1997). An efficient sampling technique for off-line quality control. *Technometrics* **39**:308–319.

Kammen, D.M. and D.M. Hassenzahl (1999). *Should We Risk It? Exploring Environmental, Health, and Technological Problem Solving*. Princeton University Press, Princeton, NJ.

Kan, A.T. and M.B. Tomson (1996). UNIFAC prediction of aqueous and nonaqueous solubilities with environmental interest. *Environ. Sci. Tech.* **30**:1369–1376.

Karcher, W., and J. Devillers (1990). SAR and QSAR in Environmental Chemistry and Toxicology: Scientific Tool or Wishful Thinking? In *Practical Applications of Quantitative Structure-Activity Relationships (QSAR) in Environmental Chemistry and Toxicology.* W. Karcher and J. Devillers (Eds.). Kluwer Academic, Dordrecht, The Netherlands.

Karickhoff, S.W. (1984). Organic pollutant sorption in aquatic system. *J. Hydraulic Eng.* **110**:705–735.

Kim, K., M.P. Anderson, and C.J. Bowser (2000). Enhanced dispersion in groundwater caused by temporal changes in recharge rate and lake levels. *Adv. Water Resources* **23**:625–635.

Klecka, G.M., J.W. Davis, D.R. Gray, and S.S. Madsen (1990). Natural bioremediation of organic contaminants in ground water: Cliffs-Dow Superfund site. *Ground Water* **4**:534–543.

Kleeman, M.J., G.R. Cass, and A. Eldcring (1997). Modeling the airborne particle as a complex source-oriented external mixture. *J. Geophys. Res.* **102**:21355–21372.

Kleissen, F.M., M.B. Beck, and H.S. Wheater (1990). The identifiability of conceptual hydrochemical models. *Water Resources Res.* **26**(12):2979–2992.

Klier, N.J., R.J. West, and P.A. Donberg (1999). Aerobic biodegradation of dichloroethylenes in surface and subsurface soils. *Chemosphere* **38**:1175–1188.

Knightes, C.D. and C.A. Peters (2000). Statistical analysis of nonlinear parameter estimation for Monod biodegradation kinetics using bivariate data. *Biotech. BioEng.* **69**(20):160–170.

Koch, D.H. and T.A. Prickett (1993). *RAN3D, A Three-Dimensional Ground-Water Solute Transport Model, User's Manual*. Eng. Technologies Associates, Inc., Ellicot City, MD.

Kolanz, M.E., A.K. Madl, M.A. Kelsh, M.S. Kent, R.M. Kalmes, and D.J. Paustenbach (2001). A comparison of historical and current exposure assessment methods for beryllium: Implications for evaluating risk of chronic beryllium disease. *Appl. Occupat. Environ. Hyg.* **15**(5):593–614.

Konikow, L.F. and J.D. Bredehoeft (1992). Ground-water models cannot be validated. *Adv. Water Resources* **15**:75–83.

Kousa, A., L. Oglesby, K. Koistinen, N. Künzli, and M. Jantunen (2002). Exposure chain of urban air $PM_{2.5}$—Associations between ambient fixed site, residential outdoor, indoor, workplace and personal exposures in four European cities in the *EXPOLIS* study. *Atmos. Environ.* **36**(18):3031–3039.

Kramm, G., R. Dlugi, G.J. Dollard, T. Foken, N. Molders, H. Mueller, W. Seiler and H. Sieverin (1995) On the dry deposition of ozone and reactive nitrogen species. *Atmos. Environ.* **29**:3209–3231.

Kroll, C.N. and R.M. Vogel (2002). Probability distribution of low streamflow series in the United States. *J. Hydrol. Eng.* **7**(2):137–146.

Kroll, C., J. Luz, B. Allen, and R.M. Vogel (2004). Developing a watershed characteristics database to improve low streamflow prediction. *J. Hydrol. Eng.* **9**(2):116–125.

Kulmala, M., A. Laaksonen, and L. Pirjola (1998). Parameterizations for sulfuric acid/water nucleation rates. *J. Geophys. Res.* **103**:8301–8307.

Kuo, J., I. Linkov, L. Rhomberg, M. Polkanov, G. Gray, and R. Wilson (2002). Absolute risk or relative risk? A study of intraspecies and interspecies extrapolation of chemical-induced cancer risk. *Risk Anal.* **22**(1):141–157.

Kwok, E.S.C., S.M. Aschmann, R. Atkinson, And J. Arey (1997). Products of the gas-phase reactions of *o*-, *m*- and *p*-xylene with the OH radical in the presence and absence of NO_x. *J. Chem. Soc. Faraday Trans.* **93**(16):2847–2854.

Labieniec, P.A., D.A. Dzombak, and R.L. Siegrist (1996). Soil Risk: Risk assessment model for organic contaminants in soil. *J. Environ. Eng.* **122**(5):388–398.

Lande, R. (1998). Anthropogenic, ecological and genetic factors in extinction and conservation. *Res. Population Ecol.* **40**:259–269.

Lange, T.R., H.E. Royals, and L.L. Connor (1993). Influence of water on mercury concentrations in largemouth bass and Florida lakes, *Trans. Am. Fish. Soc.* **122**:74–84.

Laniak, G.F., J.G. Droppo, Jr., E.R. Faillace, E.K. Gnanapragasam, W.B. Mills, D.L. Strenge, G. Whelan, and C. Yu (1997). An overview of a multimedia benchmarking analysis for three risk assessment models: RESRAD, MMSOILS, and MEPAS. *Risk Anal.* **17**(2):203–214.

Leadbetter, M.R., G. Lindgren, and H. Rootzen (1982). *Extremes and Related Properties of Random Sequences and Processes*. Springer, New York.

Lee, L.S., P.S.C. Rao,, and I. Okuda (1992). Equilibrium partitioning of polycyclic aromatic hydrocarbons from coal tar into water. *Environ. Sci. Techn.* **26**:2110–2115.

Leech, J.A., W.C. Nelson, R.T. Burnett, S. Aaron, and M.E. Raizenne (2002). It's about time: A comparison of Canadian and Am. time-activity patterns. *J. Exposure Anal. Environ. Epidemiol.* **12**(6):427–432.

Lees, M.J., L.A. Camacho, and S. Chapra (2000). On the relationship of transient storage and aggregated dead zone models of longitudinal solute transport in streams. *Water Resources Res.* **36**(1):213–224.

Lin, C.J. and J.L. Schnoor (1986). An acid precipitation model for seepage lakes. *J. Environ. Eng.* **114**(4):677–694.

Linsley, R.K., M.A. Kohler, and J.L.H. Paulhus (1975). *Hydrology for Engineers*, 2nd ed. McGraw-Hill, New York.

Lioy, P., G. Foley, and J.M. Waldman (1998). Exposure analysis: Its place in the 21st Century. *J. Exposure Anal. Environ. Epidemiol.* **8**(3):279–285.

Liu, M.K. and J.H. Seinfeld (1975). On the validity of grid and trajectory models of urban air pollution. *Atmos. Environ.* **9**:553–574.

Lockwood, J.R., M.J. Schervish, P. Gurian, and M.J Small (2001). Characterization of arsenic occurrence in source waters of US community water systems. *J. Am. Statis. Assoc.* **96**(456):1184–1193.

Long, W.S. (1996). Post audit of the Upper Mississippi River BOD/DO model. *J. Environ. Eng., ASCE* **122**(5):350–358.

Louvar, J.F. and B.D. Louvar (1998). *Health and Environmental Risk Analysis: Fundamentals with Applications*. Prentice Hall, Upper Saddle River, NJ.

Lu, F.C. (1991). *Basic Toxicology: Fundamentals, Target Organs, and Risk Assessment.* Hemisphere, New York.

Lu, R., R.P. Turco, and M.Z. Jacobson (1997). An integrated air pollution modeling system for urban and regional scales. 1. Structure and performance. *J. Geophys. Res.* **102**:6063–6079.

Lung, W. and C.E. Larson (1995). Water quality modeling of the Upper Mississippi River and Lake Pepin. *J. Environ. Eng.* **121**(1):691–699.

Lurmann, F.W., W.P.L. Carter, and L.A. Coyner (1987). A Surrogate Species Chemical Reaction Mechanism for Urban-Scale Air Quality Simulation Models. ERT Inc., Newbury, CA, and Statewide Air Pollution Res. Center, Riverside, CA.

Lurmann, F., A. Wexler, S. Pandis, S. Musarra, N. Kumar, and J.H. Seinfeld (1997). Modeling urban and regional aerosols: II. Application to California's South Coast Air Basin. *Atmos. Environ.* **31**:2695–2715.

Luthy, R., D.A. Dzombak, C.A. Peters, S. Roy, A. Ramaswami, D.V. Nakles, and B.R. Nott (1994). Remediating tar contaminated soils and manufactured gas plant sites: Technological challenges. *Environ. Sci. Tech.* **28**(3):266A–276A.

Lyman, W.J., W.F. Reehl, and D.H. Rosenblatt (Eds.) (1982). *Handbook of Chemical Property Estimation Methods*. McGraw-Hill, New York.

Lyman, W.J., P.J. Reidy, and B. Levy (1992). *Mobility and Degradation of Organic Contaminants in Subsurface Environments*, C.K. Smoley, Chelsea, MI.

Maahs, H.G. (1982). Sulfur Dioxide/Water Equilibria between 0° and 50°C: An Examination of Data at Low Concentrations. In *Heteorgeneous Atmospheric Chemistry*. Am. Geophysical Union, Washington, DC.

MacIntosh, D.L., J. Xue, H. Ozkaynak, J.D. Spengler, and P.B. Ryan (1995). A population-based exposure model for benzene. *J. Exposure Anal. Environ. Epidemiol.* **5**(3):375–403.

MacIntosh, D.L., K. Hammerstrom, and P.B. Ryan (1999). Longitudinal exposure to selected pesticides in drinking water. *Human Ecolog. Risk Assess.* **5**(3):575–588.

Mackay, D. (1982). Correlation of bioconcentration factors. *Environ. Sci. Tech.* **16**:274–278.

Mackay, D. (1991). *Multimedia Environmental Models*. Lewis Publishers, Chelsea, MI.

Mackay, D. (2001). *Multimedia Environmental Models, the Fugacity Approach*, 2nd ed. Lewis Publishers, Chelsea, MI.

Mackay, D.E. and S. Paterson (1991). *Environ. Sci. Tech.* **25**:427–436.

Mackay, D.E. and A.T.K. Yuen. (1979). Mass transfer coefficients correlations for volatilization of organic solutes from water. *Environ. Sci. Tech.* **13**:333–337.

Mackay, D., W.Y. Shiu, A. Maijanen, and S. Feenstra (1991). Dissolution of non-aqueous phase liquids in groundwater. *J. Contaminant Hydrol.* **8**:23–42.

MacQuarrie, K.T.B. and E.A. Sudicky (1990). Simulation of biodegradable organic contaminants in groundwater. 2. Plume behavior in uniform and random flow fields. *Water Resources Res.* **26**(2):223–239.

Madronich, S., S.J. Flocke, J. Zeng, and I. Petropavlovskyh (1996). Tropospheric Ultraviolet-Visible Model (TUV). National Center for Atmospheric Res., Boulder, CO.

Mandal, B.K., T.R. Chowdhury, and G. Samanta (1996). Arsenic in groundwater in seven districts of West Bengal, India—The biggest arsenic calamity in the world, *Curr. Sci.* **70**(11):976–986.

Marten, G.G. (2001). *Human Ecology: Basic Concepts for Sustainable Development*. Earthscan, London.

Martin, D.O. (1976). Comment on the change of concentration standard deviation with distance. *J. Air Poll. Control Assoc.* **26**:145–146.

Martin, J.L. and S.T.C. McCutcheon (1999). *Hydrodynamics and Transport for Water Quality Modeling*. Lewis, Boca Raton, FL.

Mason, P.J. (1994). Large-eddy simulation: A critical review of the technique. *Quarterly J. Roy. Meteor. Soc.* **120**:1–26.

Massmann, J., R.A Freeze, L. Smith, T. Sperling, and B. James (1991). Hydrogeological decision analysis: 2. Applications to ground-water contamination. *Ground Water* **29**(4):536–548.

MathSoft, Inc. (1997). *Mathcad7 User's Guide*. MathSoft, Inc., Cambridge, MA.

MathWorks, Inc. (1995). *The Student Edition of MATLAB, Version 4, User's Guide*, Prentice Hall, Englewood Cliffs, NJ.

McBride, G. and J. Rutherford (1984). Accurate modeling of river pollutant transport, *J. Environ. Eng.* **110**(4):808–827.

McCarthy, J. and J. Zachara (1989). Subsurface transport of contaminants. *Environ. Sci. Technol.* **23**(5):496–502.

McCarty, P.L. and L. Semprini (1994). Groundwater Treatment for Chlorinated Solvents. In *Handbook of Bioremediation*. Norris et al. (Eds.). Lewis, Chelsea, MI, pp. 87–116.

McDaniels, T. and M.J. Small (Eds.) (2004). *Risk Analysis and Society: Interdisciplinary Perspectives*. Cambridge University Press, Cambridge, UK.

McDonald, M.G. and A.W. Harbaugh (1988). MODFLOW: *A Modular Three-Dimensional Finite-Difference Ground Water Flow Model,* Book 6 Modeling Techniques, Scientific Software Group, Washington, DC.

McKinney, D.C. and M.D. Lin (1995). Approximate mixed-integer nonlinear programming methods for optimal aquifer remediation design. *Water Resources Res.* **31**(3):731–740

McKone, T.E. (1993a). *CalTOX, A Multimedia Total-Exposure Model for Hazardous-Wastes Sites Part I: Executive Summary*. Lawrence Livermore National Laboratory, Livermore, CA, UCRL-CR-111456PtI.

McKone, T.E. (1993b). *CalTOX, A Multimedia Total-Exposure Model for Hazardous-Wastes Sites Part II: The Dynamic Multimedia Transport and Transformation Model*. Lawrence Livermore National Laboratory, Livermore, CA, UCRL-CR-111456PtII.

McKone, T.E. (1993c). *CalTOX, A Multimedia Total-Exposure Model for Hazardous-Wastes Sites Part III: The Multiple-Pathway Exposure Model*. Lawrence Livermore National Laboratory, Livermore, CA, UCRL-CR-111456PtIII.

McKone, T.E. (1996). *Reliab. Eng. Sys. Saf.* **54**:165–181.

McKone, T.E. (1999). The rise of exposure assessment among the risk sciences: An evaluation through case studies. *Inhalation Toxicol.* **11**:611–622.

McKone, T.E., D. Hall, and W.E. Kastenberg (1996). *Modifications of CalTOX to Assess the Potential Health Impacts of Hazardous Wastes Landfills*, Report prepared by the University of California, Berkeley for the Office of Scientific Affairs, Department of Toxic Substances Control, California Environmental Protection Agency. January.

McNair, L.A., R.A. Harley, and A.G. Russell (1996). Spatial inhomogeneity in pollutant concentrations and their implications for air quality model evaluation. *Atmos. Environ.* **30**:4291–4301.

McRae, G.J. (1986). Methods for Sensitivity and Uncertainty Analysis of Long Range Transport Models, In *Quantifying Uncertainty in Long Range Transport Models*. Am. Meteorological Society, Boston, MA.

McRae, G.J. and J.H. Seinfeld (1983). Development of a second-generation mathematical model for urban air pollution—II. Evaluation of model performance. *Atmos. Environ.* **17**:501–522.

McRae, G.J., W.R. Goodin, and J.H. Seinfeld (1982). Development of a second generation mathematical model for urban air pollution. I. Model formulation. *Atmos. Environ.* **16**:679–696.

Mein, R.G. and L.C. Larson (1973). Modeling infiltration during a steady rain. *Water Resources Res.* **9**(2):384–394.

Mellor, G.L. and T. Yamada (1982). Development of a turbulence closure model for geophysical fluid problems. *Rev. Geophys. Space Phys.* **20**:851–875.

Mendelsohn, M.L., L.C. Mohr, and J.P. Peeters (Eds.) (1998). *Biomarkers: Medical and Workplace Applications*. Joseph Henry Press, Washington, DC.

Meng, Z., D. Dabdub, and J.H. Seinfeld (1998). Size- and chemically resolved model of atmospheric aerosol dynamics. *J. Geophys. Res.* **103**(D3):3419–3435.

Mercer M.W. and R.M. Cohen (1990). A review of immiscible fluids in the subsurface: Properties, models, characterization and remediation. *J. Contaminant Hydrol.* **6**:107–163.

Meyers, H. (1899). Zur Theorie der Alkohol-narkose. I. Welche Eigenschaft der Anesthetica bedingt ihre narkotische Wirkung. *Arch. Exp. Pathol. Pharmakol.* **42**:109–118.

Mihelcic, J.R. and R.G. Luthy (1991). Sorption and microbial degradation of naphthalene in soil–water systems under denitrification conditions. *Environ. Sci. Tech.* **25**:169–177.

Milford, J.B. and C.I. Davidson (1987). The sizes of particulate sulfate and nitrate in the atmosphere—A review. *Air Pollution Control Assoc. J.* **37**:125–134.

Miller C.T., M.M. Poirier-McNeill, and A.S. Mayer (1990). Dissolution of trapped nonaqueous phase liquids: Mass transfer characteristics. *Water Resources Res.* **26**:2783–2796.

Miller, S.L., K. Leiserson, and W.W. Nazaroff (1997). Nonlinear least-squares minimization applied to tracer gas decay for determining airflow rates in a two-zone building. *Indoor Air* **7**:64–75.

Mills, A.F. (1995). *Heat and Mass Transfer*. Irwin, Chicago.

Mills, W.B., J.J. Cheng, J.G. Droppo, Jr., E.R. Faillace, E.K. Gnanapragasam, R.A. Johns, G.F. Laniak, C.S. Lew, D.L. Strenge, J.F. Sutherland, G. Whelan, and C. Yu (1997). Multimedia benchmarking analysis for three risk assessment models: RESRAD, MMSOILS, and MEPAS. *Risk Anal.* **17**(2):187–201.

Milly, P.C.D. (2001). A minimalist probabilistic description of root zone soil water. *Water Resources Res.* **37**(3):457–463.

Miralles-Wilhelm, F. and L.W. Gelhar (1996). Stochastic analysis of transport and decay of a solute in heterogeneous aquifers. *Water Resources Res.* **32**(12):3451–3459.

Miralles-Wilhelm, F., L.W. Gelhar, and V. Kapoor (1997). Stochastic analysis of oxygen-limited biodegradation in three-dimensionally heterogeneous aquifers. *Water Resources Res.* **33**(6):1251–1263.

Moe, K. (1984). Should the Nazi research data be cited? *Hastings Center Rep.* December: 5–7.

Moeng, C-H. (1984). A large-eddy simulation model for the study of planetary boundary-layer turbulence, *J. Atmos. Sci.* **41**:2052–2062.

Molina, M.J., and F.S. Rowland (1974). Stratospheric sink for chlorofluoromethanes: Chlorine atom-catalyzed destruction of ozone. *Nature* **249**:810–812.

Monahan, J.F. (2001). *Numerical Methods of Statistics*. Cambridge University Press, Cambridge, UK.

Monin, A.S., and A.M. Obukhov (1954). Basic turbulent mixing laws in the atmospheric surface layer. *Tr. Geofiz. Inst. Akad. Nauk. SSSR* **24**(151):163–187.

Monteith, J.L. (1965). Accomodation between transpiring vegetation and the convective boundary layer. *J. Hydrol.* **166**:251–263.

Morel-Seytoux, H.J. (1989). *Unsaturated Flow in Hydrologic Modeling: Theory and Practice*, Kluwer, Netherlands.

Morgan, G.M. and M. Henrion (1990). *Uncertainty: A Guide to Dealing with Uncertainty on Quantitative Risk and Policy Analysis*. Cambridge University Press, Cambridge, UK.

Morris, E.M. and D.A. Woolhiser (1980). Unsteady one-dimensional flow over a plane: Partial equilibrium and recession hydrographs. *Water Resources Res.* **16**(2):355–360.

Morris, K.R., R. Abramowitz, R. Pinal, P. Davis, and S.H. Yalkowsky (1988). Solubility of aromatic pollutants in mixed solvents. *Chemosphere* **17**:285–298.

Morris, R.E., T. Myers, S. Douglas, M. Yocke, and V. Mirabella (1991). Development of a nested-grid urban airshed model and application to Southern California, Paper No. 91-66.8, 84th Annual Meeting of the Air and Waste Management Assoc., June 16–21, 1991, Vancouver, British Columbia.

Moya, J., C. Howard-Reed, and R.L. Corsi (1999). Volatilization of chemicals from tap water to indoor air from contaminated water used for showering. *Environ. Sci. Tech.* **33**(14):2321–2327.

Mukerjee, M. (1997). Trends in animal research. *Sci. Am.* **276**(2):86–93.

Mumtaz, M.M., K.A. Poirier, and J.T. Colman (1997). Risk assessment for chemical mixtures: Fine-tuning the hazard index approach. *J. Clean Tech. Environ. Toxicol. Occupat. Med.* **6**(2):189–204.

Munns, W.R., Jr. and R. MacPhail (2002). Extrapolation in human health and ecological risk assessments: Proceedings of a symposium. *J. Human Ecolog Risk Assess.* **8**(1):1–6.

Munro, R.J., P.C. Chatwin, and N. Mole (2001). The high concentration tails of the PDF of a dispersing scalar in the atmosphere. *Boundary-Layer Meteorol.* **98**:315–339.

Murphy, A.H. and R.W. Katz (Eds.) (1985). *Probability, Statistics, and Decision Making in the Atmospheric Sci.s.* Westview, Boulder, CO.

Murthy, C.R. (1976). Horizontal diffusion characteristics in Lake Ontario. *J. Phys. Oceanogr.* **6**:76–84.

Najarian, T.O. and D.R.F. Harleman (1977). Real time simulation of nitrogen cycle in an estuary. *Am. Soc. Civil Eng., J. Environ. Eng. Div.* **103**(EE4):523–538.

Najarian, T.O., T.T. Griffin, and V.K. Gunawardana (1986). Development impacts on water quality: A case study. *J. Water Resources Planning Manage. ASCE* **112**(1):20–35.

Namkung, E. and B.E. Rittmann (1987). Estimating volatile organic compound emissions from publicly owned treatment works. *J. Water Pollution Control Federation* **59**:670–678.

Nash, J.E. (1958). Determining runoff from rainfall. *Proc. Inst. Civil Eng.* **10**:163–184, Dublin, Ireland.

Nash, J.E. (1959). Systematic determination of unit hydrograph parameters. *J. Geophys. Res.* **64**:1953–1956.

National Research Council (2000). *Clean Coastal Waters: Understanding and Reducing the Effects of Nutrient Pollution.* National Academy Press, Washington, DC.

Nernst, W. (1904). *Z. Phys. Chem.* **47**:52. Theorie der Reaktionsgesch eschwindigkeit in Heterogenen Systemen.

Neuman, S.P. (1984). Adaptive Eurlerian-Lagrangian finite element method for advection–dispersion. *Int. J. Numerical Methods Eng.* **20**:321–337.

Neuman, S.P. (1990). Universal scaling of hydraulic conductivities and dispersivities in geologic media. *Water Resources Res.* **26**(8):1749–1758.

Neuman, S.P., C.L. Winter, and C.N. Newman (1987). Stochastic theory of field-scale Fickian dispersion in anisotropic porous media. *Water Resources Res.* **23**(3):453–66.

Neumann, G. and W.J. Pierson (1966). *Principles of Physical Oceanography.* Prentice-Hall, Englewood Cliffs, NJ.

Newell, C.J., R.K. McLeod, and J.R. Gonzalez (1996). *BIOSCREEN Natural Attenuation*

Decision Support System User's Manual, Version 1.3, EPA/600/R-96/087. Robert S. Kerr Environmental Res. Center, Ada, OK.

Newton, H.J. (1988). *Timeslab: A Time Series Analysis Laboratory*, Wadsworth & Brooks/ Cole, Belmont, CA.

Nielsen, D.R., M.Th. Van Genuchten, and J.W. Biggar (1986). Water flow and solute transport processes in the unsaturated zone. *Water Resources Res.* **22**(9):89S–108S.

Nikolaidis, N.P., H. Rajaram, J.L. Schnoor, and K.P. Georgakakos (1988). A generalized soft water acidification model. *Water Resources Res.* **24**(12):1983–1996.

Nikolaidis, N.P., H. Shen, H. Heng, H.L. Hu, and J.C. Clausen (1993). Movement of nitrogen through an agricultural riparian zone: 2. Distributed modeling. *Water Sci. Techn.* **28**:613–623.

NRC (1983). *Protecting Visibility in National Parks and Wilderness Areas*. National Res. Council, National Academy Press, Washington, DC.

NRC (National Res. Council) (1983). *Risk Assessment in the Federal Government: Managing the Process*. National Academy Press, Washington, DC.

NRC (National Res. Council) (1988). *Health Effects of Radon and Other Internally Deposited Alpha Emitters, BEIR IV*. National Academy Press, Washington, DC.

NRC (National Res. Council) (1990). *Ground Water Models Scientific and Regulatory Applications*. Committee on Ground Water Modeling Assessment. National Academy Press, Washington, DC.

NRC (National Res. Council) (1994). *Science and Judgment in Risk Assessment*. National Academy Press, Washington, DC.

NRC (National Res. Council) (1996a). *Understanding Risk: Informing Decisions in a Democratic Society*, Stern, P.C., H.V. Fineberg (Eds.). Committee on Risk Characterization, National Academy Press, Washington, DC.

NRC (National Res. Council) (1996b). *Guide for the Care and Use of Laboratory Animals*, Institute of Laboratory Animal Resources, Commission on Life Sci.s, National Academy Press, Washington, DC.

NRC: National Res. Council (2001). *Assessing the TMDL Approach to Water Quality Management*. NRC Committee to Assess the Scientific Basis of the Total Maximum Daily Load Approach to Water Pollution Reduction, National Academy Press, Washington, DC.

NREL (2000). http://srrl.nrel.gov/bms. Solar Radiation Res. Laboratory, National Renewable Energy Laboratory, accessed June 28, 2000.

Obukhov, A.M. (1941). Distribution of energy in the spectrum of turbulent current. *Comptes Rendus de L'Academie des Sci.s de l'URSS,* 32: 19, and *Izvestiya Akademii Nauk SSSR, Seriya Geofizicheskaya,* **5**:453–466.

O'Connor, D.J. (1962). The bacterial distribution in a lake in the vicinity of a river discharge. Proceedings of Second Purdue Industrial Waste Conference, Purdue University, West Lafayette, IN.

O'Connor, D.J. and W.E. Dobbins. (1958). Mechanisms of re-aeration in natural streams. *Trans. Am. Soc. Civil Eng.* **123**:641–684.

Odman, M.T. and A.G. Russell (1993). A nonlinear filtering algorithm for multi-dimensional finite element pollutant advection schemes. *Atmos. Environ.* **27A**:793–799.

Odum, J.R., T. Hoffman, F. Bowman, D. Collins, R. Flagan, and J. Seinfeld (1996). Gas/particle partitioning and secondary organic aerosol yields. *Environ. Sci. Tech.* **30**:2580–2585.

Oeschger, H., U. Siegenthaler, U. Schotterer, and A. Gugelmann (1975). A box diffusion model to study the carbon dioxide exchange in nature, *Tellus* **27**:168–192.

Ogata, A. and R.B. Banks (1961). A solution of the differential equation of longitudinal dispersion in porous media. *U.S. Geol. Surv*. Prof. Paper 411-A.

Ogram, A.V., R.E. Jessup, L.T. Ou, and P.S.C. Rao (1985). Effects of sorption on biological degradation rates of acetic acid in soils. *App. Environ. Microbiol*. **49**(3):582–587.

Okubo, A. (1971). Oceanic diffusion diagrams. *Deep-Sea Res*. **18**:789–802.

Oliver, B.G. (1984). The relationship between bioconcentration factor in rainbow trout and physical-chemical properties for some halogenated compounds. In *QSAR in Environmental Toxicology*. K.L.E. Kaiser (Ed.). D. Reidel Publishing, Dordrecht, Holland, pp. 300–317.

Oreskes, N., K. Shrader-Frechette, and K. Belitz (1994). Verification, validation and confirmation of numerical models in the earth sciences. *Science* **263**:641–646.

Ortiz, E., M. Kraatz, and R.G. Luthy (1999). Organic phase resistance to dissolution of polycyclic aromatic hydrocarbon compounds. *Environ. Sci. Tech*. **33**:235–242.

Ostler N.K., T.E. Byrne, and M.J. Malachowski (1996). *Health Effects of Hazardous Materials*. Vol. 3. Prentice-Hall, Englewood Cliffs, NJ.

Ott, W. (1984). Exposure estimates based on computer generated activity patterns. *J. Toxicol. Clin. Toxicol*. **21**:97–128.

Ott, W.R. (1995). *Environmental Statistics and Data Analysis*. CRC Press, Boca Raton, FL.

Ott, W.R. (1995). Human exposure assessment: The birth of a new science. *J. Exposure Anal. Environ. Epidemiol*. **5**(4):449–472.

Ott, W.R. and J.W. Roberts (1998). Everyday exposure to toxic pollutants. *Sci. Am*. **278**:86–91.

Ott, W.R., L. Langan and P. Switzer (1992). A time series model for cigarette smoking activity patterns: Model validation for carbon monoxide and respirable particles in a chamber and an automobile. *J. Exposure Anal. Environ. Epidemiol*. **2**(S2):175–200.

Ott, W., J. Thomas, D. Mage, and L. Wallace (1998). Validation of the simulation of human activity and pollutant exposure (SHAPE) model using paired days from the Denver, CO, carbon monoxide field study. *Atmos. Environ*. **22**:2101–2113.

Overton, E. (1899). Ueber die allgemeinen osmotischen Eigenschaften der Zelle, ihre vermutlichen Ursachen und ihre Bedeutung fur die Physiologie. *Vierteljahresschr. Naturforsch. Ges. Zurich* **44**:88–135.

Ozkaynak, H., J. Xue, J. Spengler, L. Wallace, E. Pellizzari, and P. Jenkins (1996). Personal exposure to airborne particles and metals: Results from the Particle Team Study. *J. Exposure Anal. Environ. Epidemiol*. **6**(1):57–78.

Paces, T. (1983). Rate constants of dissolution derived from the measurement of mass balance in hydrologic catchments. *Geochim. Cosmochim. Acta* **47**:1855–1863.

Paine, R.J., R.F. Lee, R. Brode, R.B. Wilson, A.J. Cimorelli, S.G. Perry, J.C. Weil, A. Venkatram and W.D. Peters (1998). Model Evaluation Results for AERMOD. U.S. Environmental Protection Agency website, http://www.epa.gov/scram001.

Pan, L. and P.J. Wierenga (1995). A transformed pressure head-based approach to solve Richards' equation for variably saturated soils. *Water Resources Res*. **31**(4):925–931.

Pandis, S.N., R.A. Harley, G.R., Cass, and J.H. Seinfeld (1992). Secondary organic aerosol formation and transport. *Atmos. Environ*. **26A**:2269–2282.

Pankow, J.F. (1994a). An absorption model of gas/particle partitioning of organic compounds in the atmosphere. *Atmos. Environ*. **28A**:185–188.

Pankow, J.F. (1994b). An absorption model of the gas/aerosol partitioning involved in the formation of secondary organic aerosol. *Atmos. Environ.* **28A**:189–193.

Panofsky, H.A. and J.A. Dutton (1984). *Atmospheric Turbulence*. Wiley, New York.

Park, K., A.Y. Kuo, J. Shen, and J. Hamrick (1995). *A Three-Dimensional Hydrodynamic-Eutrophication Model (HEM-3D): Description of Water Quality and Sediment Process Submodels*. Special Report in Applied Marine Sci. and Ocean Eng., Number 327, Virginia Institute of Marine Sci., Gloucester Point, VA.

Park, K., J. Shen, and A.Y. Kuo (1998). Application of a multi-step computation scheme to an intratidal estuarine water quality model. *J. Ecolog. Modeling* **110**(3):281–292.

Parsons Engineering-Science, Inc. (1994). Intrinsic Remediation Eng. Evaluation/Cost Analysis for UST Site 870, Ogden, Utah, Hill Air Force Base. Parsons Eng. Sci., Inc., Denver, CO. Sept. 1994.

Parzen, E. (1962). *Stochastic Processes*. Holden-Day, San Francisco.

Pasquill, F. (1962). *Atmospheric Diffusion*. Van Nostrand, London.

Paterson, S., D. Mackay, E. Bacci, and D. Calamari (1991). Correlation of the equilibrium and kinetics of leaf-air exchange of hydrophobic organic chemicals. *Environ. Sci. Tech.* **25**:866–871.

Patterson, D.E., R.B. Husar, W.E. Wilson, and L.F. Smith. (1981). Monte Carlo simulation of daily regional sulfur distribution: Comparison with SURE sulfate data and visual range observation during August 1972. *J. App. Meteorol.* **20**:404–420.

Patwardhan, A. and M.J. Small (1992). Bayesian methods for model uncertainty analysis with application to future sea level rise. *Risk Analysis* **12**:513–523.

Paustenbach, D.J. (ed.) (2002). *Health Risk Assessment: Theory and Practice*. Wiley, New York

Pelekis, M., M.J. Nicolich, and J.S. Gautheir (2003). Probabilistic framework for the estimation of the adult and child toxicokinetic intraspecies uncertainty factors. *Risk Anal.* **23**(6):1239–1255.

Penman, H.L. (1948). Natural evaporation from open water, bare soil and grass. *Proc. Roy. Soc., London, Ser. A* **27**:779–787.

Penman, H.L. (1956). Estimating evaporation. *Trans. Am. Geophy. Union* **37**(1):43–50.

Penrose, W.R., W.L. Polzer, E.H. Essington, D.M. Nelson, and K.A. Orlandini (1990). Mobility of plutonium and americium through a shallow aquifer in a semiarid region; *Environ. Sci. Technol.* **24**:228–234.

Perry, R.H., D.W. Green, and J.O. Maloney (1997). *Perry's Chemical Eng. Handbook*, 7th ed. McGraw-Hill, New York.

Person, M., J. Raffensperger, S. Ge, and G. Garven (1996). Basin-scale hydrogeological modeling, *Rev. Geophys.* **34**:61–87.

Peters, C.A., C.D. Knightes, and D.G. Brown (1999). Long-term composition dynamics of PAH-containing NAPLs and implications for risk assessment. *Environ. Sci. Tech.* **33**:4499–4507.

Peters, C.A., S. Mukherji, and W.J. Weber, Jr. (1998). UNIFAC modeling of multicomponent NAPLs containing polycyclic aromatic hydrocarbons. *Environ. Toxicol. Chem.* **18**(3):426–429.

Piasecki, M. and N.D. Katopodes (1997). Control of contaminant releases in rivers and estuaries. Part I: Adjoint sensitivity analysis. *J. Hydraul. Eng., ASCE* **123**(6):486–492.

Piegorsch, W.W., E.P. Smith, D. Edwards, and R.L. Smith (1998). Statistical advances in environmental science. *Statistical Sci.*, 13(2): 186–208.

Pielke, R.A., W.R. Cotton, R.L.Walko, C.J. Tremback, W.A. Lyons, L.D. Grasso, M.E. Nichols, M.D. Moran, D.A. Wesley, T.J. Lee and J.H. Copelan (1992). A Comprehensive Meteorological Modeling System—RAMS. *Meteor. Atmos. Phys.* **49**:69–91.

Pierson, W.R., A. W. Gertler, and R.L. Bradow (1990). Comparison of the SCAQS tunnel study with other on-road vehicle emission data. *J. Air Waste Management Assoc.* **40**:1495–1504.

Pinder, G.F. and W.G. Gray (1977). *Finite Elements in Surface and Subsurface Hydrology.* Academic, San Diego.

Pirkle, J.L., L.L. Needham, and K. Sexton (1995). Improving exposure assessment by monitoring human tissues for toxic chemicals. *J. Exposure Anal. Environ. Epidemiol.* **5**(3):405–424.

Plotkin, J.B., M.D. Potts, N. Leslie, N. Manokaran, J. LaFrankie, and P.S. Ashton (2000). Species-area curves, spatial aggregation, and habitat specialization in tropical forests. *J. Theor. Biol.* **207**:81–99.

Poeter, E.P. and M.C. Hill (1998). Documentation of UCODE, a computer code for universal inverse modeling: U.S. Geological Survey Water-Resources Investigations Report 98-4080.

Polissar, A.V., P.K. Hopke, and L. Poirot (2001). Atmospheric aerosol over Vermont: Chemical composition and sources. *Environ. Sci. Tech.* **35**(23): 4604–4621.

Polissar, L., K. Lowry-Coble, D. Kalman, J. Hughes, G. Van Belle, D. Covert, T. Burbacher, D. Bolgiano, and N. Mottett (1990). Pathways of human exposure to arsenic in a community surrounding a copper smelter. *Environ. Res.* **53**:29–47.

Pope, C.A., R.T. Burnett, M.J. Thun, E.E. Calle, D. Krewski, K. Ito, and G.D. Thurston (2002). Lung cancer, cardiopulmonary mortality, and long-term exposure to fine particulate air pollution. *JAMA* **287**(9):1132–1141.

Power, M. and S.M. Adams (Eds.) (1997). Perspectives of the scientific community on the status of ecological risk assessment. *Environ. Manage.* **21**:803–830.

Powers, S.E., L.M. Abriola, and W.J. Weber (1992). An experimental investigation of nonaqueous phase liquid dissolution in saturated subsurface systems: Steady–State Mass Transfer Rates. *Water Resources Res.* **28**:2691–2705.

Prandtl, L. (1904). On fluid motion with very small friction, Proc. 3rd Intl. Math. Congr., Heidelberg.

Press, S.J. (1989). *Bayesian Statistics: Principles, Models, and Applications.* Wiley, New York.

Press, W.H., S.A. Teukolsky, W.T. Vetterling, and B.P. Flannery (1994). *Numerical Recipes in FORTRAN: The Art of Scientific Computing*, 2nd ed. Cambridge University Press, Cambridge, England.

Priestley, C.H.B. and Taylor, R.J. (1972). On the assessment of surface heat flux and evaporation using large scale parameters. *Monthly Weather Rev.* **100**:81–92.

Pruppacher, H.R. and J.D. Klett (1997). *Microphysics of Clouds and Precipitation*, 2nd ed. Kluwer Academic, Dordrecht.

Purves, D. (1985). *Trace Element Contamination of the Environment*. Elsevier, New York.

Rai, S.N. and D. Krewski (1998). Uncertainty and variability analysis in multiplicative risk models. *Risk Anal.* **18**(1):37–45.

Rajagopalan, R. and C. Tein (1976). Trajectory analysis of deep bed filtration with the sphere in cell porous media model. *AIchE J.* **22**:523.

Ramaswami, A. and R.G. Luthy (2002). Measuring and Modeling Physicochemical Limitation to Bioavailability and Biodegradation. In C.J. Hurst (Ed.). *Manual of Environmental Microbiology.* ASM Press, 2nd ed. Washington, DC.

Ramaswami, A. and R.G. Luthy, (1997). Mass transfer and bioavailability of PAH compounds in coal tar NAPL-slurry systems. 1. Model development. *Environ. Sci. Tech.* **31**(8):2260–2267.

Ramaswami, A. and M.J. Small (1994). Modeling the spatial variability of natural trace element concentrations in groundwater. *Water Resources Res.* **30**(2):269–282.

Ramaswami, A., S. Ghoshal, and R.G. Luthy (1997). Mass transfer and bioavailability of PAH compounds in coal tar (NAPL)-water slurry systems. 2. Experimental evaluations. *Environ. Sci. Tech.* **31**(8):2268–2276.

Ramsey, J.C. and M.E. Andersen (1984). A physiologically based description of the inhalation pharmacokinetics of styrene in rats and humans. *Toxicol. Appl. Pharmacol.* **73**:59–175.

Rawls, W., D.L. Brakensiek, and N. Miller (1983). Green-Ampt infiltration parameters from soils data. *J. Hydraul. Eng., ASCE* **109**(1):62–70.

Reckhow, K.H. (1993). Validation of Simulation Models: Philosphy and Statistical Methods of Confirmation. In *Concise Encyclopedia of Environmental Systems*, P.C. Young (Ed.). Pergamon Press, Oxford.

Reckhow, K.H., J.T. Clements, and R.C. Dodd (1990). Statistical evaluation of mechanistic water-quality models, *J. Environ. Eng.*, 116, 250–268.

Rees, T.H. and J.L. Schnoor (1994). Long-term simulation of decreased acid loading on forested watershed. *J. Environ. Eng.* **120**:291–312.

Reeves, A.L. (1981). *Toxicology: Principles and Practices.* Vol. 1. Wiley, New York.

Reichert, P. and M. Omlin (1997). On the usefulness of overparameterized ecological models. *Ecol. Model* **95**(2–3):289–299.

Reichert, P., M. Schervish, and M.J. Small (2002). An efficient sampling technique for Bayesian inference with computationally demanding models. *Technometrics* **44**(4):318–327.

Reid, R.C., J.M. Prausnitz, and B.E. Poling (1987). *The Properties of Gases and Liquids*, 4th ed. McGraw-Hill, New York.

Renner, R. (1999). Consensus on health risks from mercury exposure eludes federal agencies. *Environ. Sci. Tech.* **33**(13):269A–270A.

Resnick, S. I. (1987). *Extreme Values, Regular Variation and Point Processes.* Springer, New York.

Reynolds, O. (1883). An experimental investigation of the circumstances which determine whether the motion of water shall be direct or sinuous, and of the law of resistance in parallel channels. *Phil. Trans. Roy. Soc. London* **174**:935–982.

Reynolds, O. (1894). On the dynamical theory of incompressible viscous fluids and the determination of the criterion. *Phil. Trans. Roy. Soc. London* **186**:123–161.

Rhomberg, L.R. (2004). Mechanistic Considerations in the Harmonization of Dose-Response Methodology: The Role of Redundancy at Different Levels of Biological Organization. In *Risk Analysis and Society: Interdisciplinary Perspectives.* T. McDaniels and M. Small (Eds.). Cambridge University Press, Cambridge, UK.

Rice, S.O. (1944). Mathematical analysis of shot noise, *Bell Syst. Tech. J.* **23**:282–332.

Richardson, L.F. (1920). The supply of energy from and to atmospheric eddies. *Proc. Roy. Soc. London* **A97**:355–373.

Richardson, L.F. (1926). Atmospheric diffusion shown on a distance-neighborhood graph. *Proc. Roy. Soc. London, Series A* **110**:709–727.

Richardson, C.W. (1981). Stochastic simulation of daily precipitation, temperature, and solar radiation. *Water Resources Res.* **17**:182–190.

Rifai, H. and P.B. Bedient (1990). Comparison of biodegradation kinetics with an instantaneous reaction model for groundwater. *Water Resources Res.* **26**(4):637–645.

Riggs, H.C. (1965). Estimating probability distributions of drought flows. *Water Sewage Works* **112**(5):153–157.

Riggs, H.C. (1980). Characteristics of low flows. *J. Hydraul. Eng.* **106**(5):717–731.

Rijnaarts, H., A. Bachman, J. Jumelet, and A. Zehnder (1990). Effect of desorption and intraparticle mass transfer on the aerobic biomineralization of *a-hexachlorocyclohexane* in a contaminated calcareous soil. *Environ. Sci. Technol.* **24**(6):1349–1354.

Riley, D.M., M.J. Small, and B. Fischhoff (2000). Modeling methylene chloride exposure-reduction options for home paint-stripper users. *J. Exposure Anal. Environ. Epidemiol.* **10**(3):240–250.

Ripley, B.D. (1992). Stochastic Models for the Distribution of Rock Types in Petroleum Reservoirs. In *Statistics in the Environmental and Earth Sci.s.* A. Walden and P. Guttorp (Eds.). Edward Arnold, London, pp. 247–282.

Rittmann, B.E. (1993). The significance of biofilms in porous media. *Water Resources Res.* **29**(7):2195–2202.

Roberts, P.V., M.N. Goltz, and D.M. Mackay (1986). A natural gradient experiment on solute transport in a sand aquifer. 3. Retardation estimates and mass balances for organic solutes. *Water Resources Res.* **22**(13):2047–2058.

Robertson, G.L., M.D. Lebowitz, M.K. O'Rourke, S.Y. Gordon, and D. Moschandreas (1999). The National Human Exposure Assessment Survey (NHEXAS) study in Arizona: Introduction and preliminary results. *J. Exposure Anal. Environ. Epidemiol.* **9**(5):427–434.

Robeson, S.M. and D.G. Steyn (1990). Evaluation and comparison of statistical forecast models for daily maximum ozone concentrations. *Atmos. Environ.* **24B**:303–312.

Robin, M.J.L., E.A. Sudicky, R.W. Gillham, and R.G. Kachanoski (1991). Spatial variability of strontium distribution coefficients and their correlation with hydraulic conductivity in the CFB Borden aquifer, *Water Resources Res.* **27**(10):2619–2632.

Robinson, J.P. and J. Thomas (1991). *Time Spent in Activities, Locations, and Microenvironments: A California-National Comparison Project Report*. U.S. Environmental Protection Agency, Environmental Monitoring Systems Laboratory, Las Vegas.

Rodan, B.R., D.W. Pennington, N. Eckley, and R.S. Boethling (1999). Screening for persistent organic pollutants: Techniques to provide a scientific basis for POPs criteria in international negotiations. *Environ. Sci. Tech.* **33**(20):3482–3488.

Rodhe, H. and J. Grandell (1972). On the removal time of aerosol particles from the atmosphere by precipitation scavenging. *Tellus* **24**:442–454.

Rodriguez-Iturbe, I. (1986). Scale of fluctuation of rainfall models. *Water Resources Res.* **22**:15S–37S.

Rodriguez-Iturbe, I., D.R. Cox, and V. Isham (1988). A point process for rainfall: Further developments. *Proc. Roy. Soc. London* **A417**:283–298.

Roelke, M.E., D.P. Schultz, C.F. Facemire, S.F. Sundlof, and H.E. Royals (1991). Mercury Contamination in Florida Panthers. Prepared by the Technical Subcommittee of the Florida Panther Interagency Committee, Gainesville, FL.

Roesner, L.A., J.A. Aldrich, and R.E. Dickinson (1998). *Storm Water Management Model User's Manual, Version 4, Addendum I, EXTRAN*. EPA/600/3-88/001b, U.S. Environmental Protection Agency, Athens, GA.

Ross, S.A. (1989). *Introduction to Probability Models*, 3rd ed. Academic, San Diego.

Roy, A. and P.G. Georgopoulos (1998). Reconstructing week-long exposures to volatile organic compounds using physiologically based pharmacokinetic models. *J. Exposure Anal. Environ. Epidemiol.* **8**(3):407–422.

Rubin, D.F. (1988). Occupational health implications of a toxic spill of propylene dichloride. *Western J. Med.* **148**(1):78–79.

Rubin, E. and A. Ramaswami (2001). Phytoremediation of MTBE. *Water Res.* **35**(5):1348–1353.

Rubin, Y. (2003). *Applied Stochastic Hydrogeology*. Oxford University Press, New York.

Runkel, R.L. (1998). *One-dimensional Transport with Inflow and Storage (OTIS): A Solute Transport Model for Streams and Rivers*. US Geological Survey Water Resources Investigations Report 98-4018, http://webserver.cr.usgs.gov/otis/documentation/primary/wrir98-4018.pdf.

Runkel, R.L. (2000). *Using OTIS to Model Solute Transport in Streams and Rivers*. U.S. Geological Survey Fact Sheet FS-138-99, http://webserver.cr.usgs.gov/otis/documentation/primary/fs.pdf.

Runkel, R.L. and S.C. Chapra (1993). An efficient numerical solution of the transient storage equations for solute transport in small streams. *Water Resources Res.* **29**(1):211–215.

Runkel, R.L., B.A. Kimball (2002). Evaluating remedial alternatives for an acid mine drainage stream: Application of a reactive transport model. *Envir. Sci. and Technol.* **36**(5):1093–1101.

Runkel, R.L., K.E. Bencala, R.E. Broshears, and S.C. Chapra (1996). Reactive solute transport in streams: 1. Development of an equilibrium-based model. *Water Resources Res.* **32**(2):409–418.

Russell, A.G. and R. Dennis (2000). NARSTO critical review of photochemical models and modeling. *Atmos. Environ.* **34**:2283–2324.

Rutherford, J.C. (1994). *River Mixing*. Wiley, New York.

Ryan, J.N. and M. Elimelech (1996). Colloid mobilization and transport in groundwater. *Colloids Surfaces A: Physiochem. Eng. Aspects* **107**:1–56.

Ryan, J.N., M. Elimelech, R.A. Ard, R.W. Harvey, and P.R. Johnson (1999). Bacteriophage PRD1 and silica colloid transport and recovery in an iron oxide-coated sand aquifer. *Environ. Sci. Technol.* **33**:63–73.

Sabol, G.V. and C.F. Nordin, Jr. (1978). Dispersion in rivers as related to storage zones. *J. Hydraul. Div., ASCE* **104**:695–708.

Sakulyanontvittaya, T. (2003). Evaluation of ISCST3 and AERMOD for Modeling Benzene Dispersion in Commerce City, M.S. Thesis, Department of Mechanical Eng., University of Colorado at Boulder.

Saltelli, A. (2002). Sensitivity analysis for importance assessment. *Risk Anal.* **22**(3):579–590.

Saltelli, A., K. Chan, and E.M. Scott (2000). *Sensitivity Analysis*. Wiley, West Sussex, UK.

Samson, P.J. and M.J. Small (1984). Atmospheric Trajectory Models for Diagnosing the Sources of Acid Precipitation. In *Acid Precipitation, Vol. 9: Modeling of Total Acid Precipitation Impacts*. J.L. Schnoor (Ed.). Ann Arbor Sci., Butterworth, Boston, pp. 1–23.

Sawyer, R.F., R.A. Harley, S.H. Cadle, J.M. Norbeck, R. Slott, and H.A. Bravo (2000). Mobile sources critical review: 1998 NARSTO assessment. *Atmos. Environ.* **34**:2161–2181.

Saxton, K., W. Rawls, J. Romberger, and R. Papendick (1986). Estimating generalized soil water characteristics from texture. *Soil Sci. Society Am. J.* **50**:1031–1036.

Schaffranek, R.W. (1987). *Flow Model for Open-Channel Reach or Network.* U.S. Geological Survey Professional Paper 1384, USGS, Reston, VA.

Schaffranek, R.W., R.A. Balzer, and D.E. Goldberg (1981). *A Model for Simulation of Flow in Singular Interconnected Channels: US Geological Survey Techniques of Water-Resources Investigations.* Chapter C3, Book 7, USGS, Reston, VA.

Schaug, J., T. Iversen, and U. Pedersen (1993). Comparison of measurements and model results for airborne sulfur and nitrogen components with kriging. *Atmos. Environ.* **27A**:831–844.

Scheringer, M. (1997). Characterization of the environmental distribution behavior of organic chemicals by means of persistence and spatial range. *Environ. Sci. Tech.* **31**(10):2891–2897.

Schijven, J.F. and S.M. Hassanizadeh (2000). Removal of viruses by soil passage: Overview of modeling, processes, and parameters. *Crit. Rev. Environ. Sci. Tech.* **30**(1):49–127.

Schirmer, M., B.J. Butler, J.W. Roy, E.O. Frind, and J.F. Barker (1999). A relative-least-squares technique to determine unique Monod kinetic parameters of BTEX compounds using batch experiments. *J. Contaminant Hydrol.* **37**:69–86.

Schmugge, T.J. and J. André (Eds.) (1991). *Land Surface Evaporation: Measurement and Parameterization*, Springer, New York.

Schnoor, J. (1996). *Environmental Modeling: Fate and Transport of Pollutants in Water, Air and Soil.* Wiley, New York.

Schnoor, J.L. and W. Stumm (1985). Acidification of Aquatic and Terrestrial Systems. In *Chemical Processes in Lakes.* Wiley, New York, pp. 311–338.

Schnoor, J.L., L.A. Licht, S.C. McCutcheon, N.L. Wolfe, and L.H. Carreira (1995). Phytoremediation of organic and nutrient contaminants. *Environ. Sci. Tech.* **29**(7):318A–323A.

Schwartz, S.E. and W.H. White (1981). Solubility equilibria of the nitrogen oxides and oxyacids in dilute aqueous solution. *Adv. Environ. Sci. Eng.* **4**:1.

Schwartzenbach, R.P. and J. Westall (1981). Transport of nonpolar organic compounds from surface water to groundwater. *Environ. Sci. Tech.* **15**:1360:1367.

Schwarzenbach, R., P. Gschwend, and D. Imboden (1993). *Environmental Organic Chemistry.* Wiley, New York.

Schwille, F. (1988). *Dense Chlorinated Solvents in Porous and Fractured Media.* Translated by James F. Pankow. Lewis, Chelsea, MI.

Scire, J.S., D.G. Strimaitis, and R.J. Yamartino (2000). *A User's Guide for the CALPUFF Dispersion Model.* Earth Tech, Inc., Concord, MA.

Scow, K. and M. Alexander (1992). Effect of diffusion on the kinetics of biodegradation: Experimental results with synthetic aggregates. *J. Soil Sci. Soc. Am.* **51**:128–134.

Seagren, E.A., B.E. Rittmann, and J.A. Valochhi (1993). Critical evaluation of the local-equilibrium assumption in modeling NAPL-pool dissolution. *J. Contaminant Hydrol.* **12**: 103–132.

Seagren, E.A., B.E. Rittmann, and J.A. Valochhi (1994). Quantitative evaluation of the enhancement of NAPL-pool dissolution by flushing and biodegradation. *Environ. Sci. Tech.* **28**(5):833–839.

Seaman, N.L. (2000). Meteorological modeling for air quality assessments. *Atmos. Environ.* **34**:12–14, 2231–2259.

Seigneur, C., P. Pai, J-F. Louis, P. Hopke, and D. Grosjean (1997). *Review of Air Quality Models for Particulate Matter*. Publication No. 4669, Am. Petroleum Institute, Washington, DC.

Seinfeld, J.H. (1986). *Atmospheric Chemistry and Physics of Air Pollution*. Wiley, New York.

Seinfeld, J.H. (1989). Urban air pollution: state of the science. *Science* **243**:745–752.

Seinfeld, J.H. and S.N. Pandis (1998). *Atmospheric Chemistry and Physics: From Air Pollution to Climate Change*. Wiley, New York.

Sexton, K., S.G. Selevan, D.K. Wagener, amd J.A. Lybarger (1992). Estimating human exposures to environmental pollutants: Availability and utility of existing databases, *Arch. Environ. Health* **47**(6):398–407.

Sexton, K., D.K. Wagener, S.G. Selevan, T.O. Miller, and J.A. Lybarger (1994). An inventory of human exposure-related data bases. *J. Exposure Anal. Environ. Epidemiol.* **4**(1):95–109.

Sexton, K., M.A. Callahan, and E.F. Bryan (1995a). Estimating exposure and dose to characterize health risks: The role of human tissue monitoring in exposure assessment. *Environ. Health Perspect.* **103**(Suppl.3):13–29.

Sexton, K., D.E. Kleffman, and M.A. Callahan (1995b). An introduction to the National Human Exposure Assessment Survery (NHEXAS) and related Phase I field studies. *J. Exposure Anal. Environ. Epidemiol.* **5**(3):229–232.

Sexton, K., L.L. Needham, and J.L. Pirkle (2004). Human biomonitoring of environmental chemicals. *Am. Scient.* **92**(1):38–45.

Shanahan, P., M. Henze, L. Koncsos, W. Rauch, P. Reichert, L. Somlyódy, and P. Vanrolleghem (1998). River water quality modelling: II. Problems of the art. *Water Sci. Tech.* **38**(11):245–252.

Shanahan, P., D. Borchardt, M. Henze, W. Rauch, P. Reichert, L. Somlyódy, and P. Vanrolleghem (2001). River water quality model no. 1 (RWQM1): I. Modeling approach. *Water Sci. Tech.* **43**(5):1–9.

Shaw, J.E.H. (1988). A quasi-random approach to integration in Bayesian statistics. *Ann. Stat.* **16**:895–914.

Shen, J., J. Boon, and A.Y. Kuo (1999). A modeling study of a tidal intrusion front and its impact on larval dispersion in the James Estuary, Virginia. *Estuary* **22**(3A):127–135.

Sigg, L. (1985). Metal Transfer Mechanisms in Lakes: The Role of Settling Particles. In *Chemical Processes in Lakes*. Wiley, New York, pp. 283–310.

Simkins, S. and M. Alexander (1984). Models for mineralization kinetics with the variables of substrate concentration and population density. *Appl. Envir. Microbiol.* **47**:1299–1306.

Singer, B.C. and R.A. Harley (1996). A fuel-based motor vehicle emission inventory. *J. Air Waste Manage. Assoc.* **46**:581–593.

Singh, V.P. (1988). *Hydrologic Systems, Vol. 1, Rainfall-Runoff Modeling*. Prentice Hall, Englewood Cliffs, NJ.

Singh, V.P. (1992). *Elementary Hydrology*. Prentice Hall, Englewood Cliffs, NJ.

Singh, V.P. and D.A. Woolhiser (2002). Mathematical modeling of watershed hydrology. *J. Hydrol. Eng.* **7**(4):270–292.

Skeffington, R.A. (1999). The use of critical loads in environmental policy making: A critical appraisal. *Environ. Sci. Tech.* **33**(11):245A–252A.

Slinn, W.G. (1977). Some approximations for the wet and dry removal of particles and gases from the atmosphere. *Water, Air Soil Pollution* **7**:513–543.

Slinn, W.G.N. (1983). Air-to-Sea Transfer of Particles. In *Air-Sea Exchange of Gases and Particles*, P.S. Liss and W.G.N. Slinn (Eds). Reidel, Dordrecht.

Small, M.J. (1982). Stochastic model of atmospheric regional transport. Ph.D. thesis, Environmental and Water Resources Eng. Program, University of Michigan, Ann Arbor, MI.

Small, M.J. (1997). Groundwater detection monitoring using combined information from multiple constituents. *Water Resources Res.* **33**(5):957–969.

Small, M.J. and P.S. Fischbeck (1999). False precision in Bayesian updating with incomplete models. *Human Ecolog. Risk Assess.* **5**(2):291–304.

Small, M.J. and D.J. Morgan (1986). The relationship between a continuous time renewal model and a discrete Markov chain model of precipitation occurrence. *Water Resources Res.* **22**:1422–1430.

Small, M.J. and J.R. Mular (1987). Long-term pollutant degradation in the unsaturated zone with stochastic rainfall-infiltration. *Water Resources Res.* **23**:2246–2256.

Small, M.J. and M.C. Sutton (1986). A direct distribution model for regional aquatic acidification. *Water Resources Res.* **22**:1749–1758.

Small, M.J., M.C. Sutton, and M.W. Milke (1988). Parametric distributions of regional lake chemistry: Fitted and derived. *Environ. Sci. Tech.* **22**:196–204.

Small, M.J., C. Bloyd, G. Keeler, and R.J. Marnicio (1989). Stochastic simulation of meteorological variability for long-range atmospheric transport: 2. Long-term statistical models. *Atmos. Environ.* **23**:2825–2840.

Small, M.J., A.B. Nunn III, B.L. Forslund, and D.A. Daily (1995). Source attribution of elevated residential soil lead near a battery recycling site. *Environ. Sci. Tech.* **24**(4):883–895.

Smith, A.E, P.B. Ryan, and J.S. Evans (1992). The effect of neglecting correlations when propagating uncertainty and estimating the population distribution of risk. *Risk Anal.* **12**:467–474.

Smith, J.Q. and S. French (1993). Bayesian updating of atmospheric dispersion models for use after an accidental release of radiation. *Statistician* **42**:501–511.

Smith, R.L. and T.S. Shively (1995). Point process approach to modeling trends in tropospheric ozone based on exceedances of a high threshold. *Atmos. Environ.* **29**:3489–3499.

Smolarkiewicz, P.K. (1983). A simple positive definite advection scheme with small implicit diffusion. *Monthly Weather Rev.* **111**:479–486.

Snoeyink, V.L. and D. Jenkins (1980). *Water Chemistry*, Wiley, New York.

Sobol, I.M. (1988). On the distribution of points in a cube and the approximate evaluation of integrals. *USSR Comput. Math. Phys.* **7**:51–70.

Sohn, M.D. (1998). Updating Uncertainty in Site Characterization and Chemical Transport Using Bayesian Monte Carlo Methods. PhD Thesis, Department of Civil & Environmental Eng., Carnegie Mellon University, Pittsburgh, PA.

Sohn, M.D. and M.J. Small (1999). Parameter estimation of unknown air exchange rates and effective mixing volumes from tracer gas measurements for complex multi-zone indoor air models. *Build. Environ.* **34**:293–303.

Southworth, G.R. (1979). The role of volatilization in removing polycyclic aromatics from aquatic environments. *Bull. Environ. Contamin. Toxicol.* **21**:507–514.

Spalding M.G., R.D. Bjork, G.V.N. Powell, and S.F. Sundloff (1994). Mercury and cause of death in great white herons. *Wildlife Management* **58**(4):735–739.

Spengler, J., M. Schwab, P.B. Ryan, I.H. Billick, S. Colome, A.L. Wilson, and E. Becker (1994). Personal exposure to nitrogen dioxide in the Los Angeles basin. *J. Air Waste Manage. Assoc.* **44**:39–47.

Spielman, L.A. and S.K. Friedlander (1974). Role of the electrical double layer in particle deposition by convective diffusion. *J. Colloid Interface Sci.* **46**(1):22–31.

Srinivasan, R. and J.G. Arnold (1994). Integration of a basin-scale water quality model with GIS. *Water Resources Bull.* **30**(3):453–462.

Stanek, E.J. III, E.J. Calabrese, K. Mundt, P. Pekow, and K. Yeatts (1998). Prevalence of soil mouthing/ingestion among healthy children aged 1–6. *J. Soil Contamination* **7**:227–242.

Stauffer, D.R. and N.L. Seaman (1990). Use of four dimensional data assimilation in a limited area mesoscale model. Part 1. Experiments with synoptic data. *Monthly Weather Rev.* **118**:1250–1277.

Stedinger, J.R. (1980). Fitting lognormal distributions to hydrologic data. *Water Resources Res.* **16**:481–490.

Stedman, D.H., G. A. Bishop, S.P. Beaton, J.E. Peterson, P.L. Guenther, I.F. McVey, and Y. Zhang (1994). On-road remote sensing of CO and HC emissions in California. University of Denver, Denver, CO.

Stiber, N.A., M. Pantazidou, and M.J. Small (1999). Expert system methodology for evaluating reductive dechlorination at TCE sites. *Environ. Sci. Tech.* **33**(17):3012–3020.

Stockwell, W.R., P. Middleton, J.S. Chang, and X. Tang (1990). The RADM2 chemical mechanism for regional air quality modeling. *J. Geophys. Res.* **95**:16343–16367.

Stockwell, W.R., F. Kirchner, and M. Kuhn (1997). A new mechanism for regional atmospheric chemistry modeling. *J. Geophys. Res.* **102**, D22:25847–25879.

Stow, C. and K.H. Reckhow (1996). Estimator bias in a lake phosphorus model with observation error. *Water Resources Res.* **32**:165–170.

Stow, C.A. and S.S. Qian (1998). A size-based probabilistic assessment of PCB exposure from Lake Michigan fish consumption. *Environ. Sci. Tech.* **32**(15):2325–2330.

Streeter, H.W. and E.B. Phelps (1925). A Study of the Pollution and Natural Purification of the Ohio River, III. Factors Concerning the Phenomena of Oxidation and Reaeration, U.S. Public Health Service, Pub. Health Bulletin No. 146, February. Reprinted by U.S. Department of Health Education and Welfare, Public Health Administration, 1958.

Stull, R.B. (1988). *An Introduction to Boundary Layer Meteorology.* Kluwer Academic, Dordrecht.

Stumm, W. and J.M. Morgan (1981). *Aquatic Chemistry: An Introduction Emphasizing Chemical Equilibria in Natural Waters*, 2nd ed. Wiley, New York.

Stumm, W. and J.J. Morgan (1996). *Aquatic Chemistry*. Wiley-InterSci., New York, 1970; 2nd ed. 1981, 3rd ed. 1996.

Stumm, W., G. Furrer, and B. Kunz (1983). The role of surface coordination in precipitation and dissolution of mineral phases. *Croat. Chem. Acta* **56**:585–611.

Sullivan, P.P., J.C. McWilliams and C-H. Moeng (1996). A grid nesting method for large-eddy simulation of planetary boundary layer flows, *Boundary-Layer Meteorol.* **80**:167–202.

Sun, M. (1988). EPA bars use of Nazi data. *Sci.* **240**(4848):21.

Sunderman, F.W. Jr., S.M. Hopfer, K.R. Seeney, A.H. Marcus, B.M. Most, and J. Creason (1989). Nickel absorption and kinetics in human volunteers. *Exp. Biol. Med.* **191**(1):5–11.

Suter, G.W. II (1999). A framework for assessment of ecological risks from multiple activities. *Human Ecolog. Risk Assess.* **5**(2):398–413.

Sutton, O.G. (1953). *Micrometeorology*. McGraw-Hill, New York.

Swain, E.D. and E.J. Wexler (1996). *A Coupled Surface-Water and Ground-Water Flow Model (MODBRNCH) for Simulation of Stream-Aquifer Interaction*. U.S. Geological Survey Techniques of Water-Resources Investigations, Book 6, Chapter A6, Denver, CO.

Switzer, P. and W. Ott (1992). Derivation of an indoor air averaging time model from the mass balance equation for the case of independent source inputs and fixed air exchange rates. *J. Exposure Anal. Environ. Epidemiol.* **2**(Suppl. 2):113–135.

Tabios G.Q. and J.D. Salas (1985). A comparative analysis of techniques for spatial interpolation of precipitation. *Water Resources Bull.* **21**(3):365–380.

Takada, H., J.W. Farrington, M.H. Bothner, C.G. Johnson, and B.W. Tripp (1994). Transport of Sludge-Derived Organic Pollutants to Deep-Sea Sediments at Deep Water Dump Site 106. *Environ. Sci. Tech.* **28**(6):1062–1072.

Tanaka, Y. and J. Nakanishi (2000). Mean extinction time of populations under toxicant stress and ecological risk assessment. *Environ. Toxicol. Chemis.* **19**:2856–2862.

Tang, A.J., W.H. Chan, D.H.S. Chung, and M.A. Lusis (1986). Spatial and temporal pattern of sulfate and nitrate wet deposition in Ontario. *Water, Air Soil Pollution* **30**:263–273.

Tasker, G.D. (1987). Comparison of methods for estimating low flow characteristics of streams. *Water Resources Bull.* **23**(6):1077–1083.

Tasker, G.D. and N.E. Driver (1988). Nationwide regression models for predicting urban runoff water quality at unmonitored sites. *Water Resources Bull.* **24**(5):1091–1101.

Taylor, A.C. (1993). Using objective and subjective information to generate distributions for probabilistic exposure assessment. *J. Exposure Anal. Environ. Epidemiol.* **3**:285–298.

Taylor, G.R. (1989). Sulfate production and deposition in midlatitude continental cumulus clouds. Part II: Chemistry model formulation and sensitivity analysis. *J. Atmos. Sci.* **46**: 1991–2007.

Taylor, S.W. and P.R. Jaffe (1990). Substrate and biomass transport in a porous medium. *Water Resources Res.* **26**(9):2181–2194.

Taylor, S.W., P.R. Jaffe, (1990a) Biofilm growth and the related changes in the physical properties of a porous medium. 1. Experimental investigation. *Water Resour. Res.* 26(9):2161–2170.

Tetra Tech (2000). *Ocean Circulation and Plume Dispersion Modeling Review, with Emphasis on Orange County Sanitation District's Offshore Outfall and Wastewater Plume*. Prepared by Tetra Tech, Inc. for Orange County Sanitation District, Fountain Valley, California; see http://www.ocwatersheds.com/watersheds/pdfs/Ocean_Circulation_Plume_Dispersion_Modeling.pdf.

Thackston, E. and K. Schnelle (1970). Predicting effects of dead zones on stream mixing. *J. Sanitary Eng. Div., ASCE* **93**:319–331.

Theis, C.V. (1935). The relation between the lowering of the piezometric surface and the rate and duration of discharge of a well using groundwater storage. *Trans. Am. Geophys. Union* **2**:519–524.

Thomann, R.V. (1963). Mathematical model for dissolved oxygen, *J. Sanitary Eng. Division, ASCE*, **89**:1–30.

Thomann, R.V. (1972). *Systems Analysis and Water Quality Management*. McGraw-Hill, New York.

Thomann, R.V. (1980). Measures of Verification. In *Workshop on Verification of Water Quality Models*. R.V. Thomann and T.O. Barnwell, Jr. (Eds.). U.S. Environmental Protection Agency, EPA-600/9-80-016, Athens, GA, pp. 37–61.

Thomann, R.V. (1987). System Analysis in Water Quality Management—A 25-year Retrospective. In *Systems Analysis and Water Quality Management*. M.B. Beck (Ed.). Pergamon, London, pp. 1–14.

Thomann, R.V., and J.A. Mueller (1987). *Principles of Surface Water Quality Modeling and Control*. Harper & Row, New York.

Thompson, K.M. and D.E. Burmaster (1991). Parametric distributions for soil ingestion by children. *Risk Anal.* **11**:339–342.

Thompson, K.M. and J.D. Graham (1996). Going beyond the single number: Using probabilistic risk assessment to improve risk management. *Human Ecolog. Risk Assess.* **2**(4):1008–1034.

Tilman, D. and P. Kareiva (1997). *Spatial Ecology*. Princeton University Press, Princeton, N.J.

Tim, U.S. and R. Jolly (1994). Evaluating agricultural nonpoint-source pollution using integrated Geographic Information Systems and hydrologic/water quality model. *J. Environ. Qual.* **23**:25–35.

Tompson, A.F.B., R. Ababou, and L.W. Gelhar (1989). Implementation of the three-dimensional turning bands random field generator. *Water Resources Res.* **25**(10):2227–2243.

Trapp, S. (1995). In *Plant Contamination: Modeling and Simulation of Organic Chemical Processes*. S. Trapp and J. McFarlane(Eds.). Lewis Publishers, Boca Raton, FL, pp. 107–151.

Tremolada, P., V. Burnett, D. Calamari, and K.C. Jones (1996). Spatial distribution of PAHs in the U.K. atmosphere using pine needles. *Environ. Sci. Tech.* **30**(12):3570–3577.

Trijonis, J.C. (1974). Economic air pollution control model for Los Angeles County in 1975. *Environ. Sci. Techn.* **9**:811–826.

Troen, I. and E.L. Petersen (1989). *European Wind Atlas*. Risø National Laboratory, Risø, Denmark.

Turner, B. (1969). *Workbook of Atmospheric Dispersion Estimates*. U.S. Department of Health, Education and Welfare, Cincinnati, OH.

U.S. Department of Commerce (1995). DIPPR Data Compilation of Pure Compound Properties, 1995. Version 10.0. Project sponsored by the Design Institute for Physical Property Data. Am. Institute of Chemical Engineers, NIST Standard Reference Database 11, U.S. Department of Commerce, Gaithersburg, MD.

U.S. EPA (1986). *Guidelines for Carcinogen Risk Assessment, Federal Register*, **51**(185): 33992–34003.

U.S. EPA (1991). *Guidance for Water Quality-Based Decisions: The TMDL Process*. EPA 440/4-91-001, U.S. Environmental Protection Agency, Office of Water, Washington, DC.

U.S. EPA (1991). *Technical Support Document for Water Quality-based Toxics Control*. EPA-823-B-94-005, U.S. Environmental Protection Agency, Office of Water, Washington, DC.

U.S. EPA (1994). *Water Quality Standards Handbook*, 2nd ed. EPA-823-B-94-005, U.S. Environmental Protection Agency, Office of Water, Washington, DC; see http://www.epa.gov/waterSci./standards/handbook/.

USEPA (1994). Guidance for Evaluating the Technical Impracticability of Groundwater Restoration. Directive: 9234.2-25. Office of Solid Waste and Emergency Response, Washington DC.

U.S. EPA (1997a). *Technical Guidance Manual for Developing Total Maximum Daily Loads: Book 2, Rivers and Streams; Part 1—Biochemical Oxygen Demand/Dissolved Oxygen & Nutrient Eutrophication*. EPA 823/B-97-002, U.S. Environmental Protection Agency, Office of Water, Washington, DC.

U.S. EPA (1997b). *Compendium of Tools for Watershed Assessment and TMDL Development*. EPA 841B-97-006. U.S. Environmental Protection Agency, Office of Wetlands, Oceans, and Watersheds, Washington, DC.

U.S. EPA (1997c). *Guidelines for Carcinogen Risk Assessment: Review of ORD's Draft Guidelines*, An SAB Report, EPA-SAB-EHC-97-010, Washington, D.C.

Van Genuchten, M.Th. (1980). A closed-form equation for predicting the hydraulic conductivity of unsaturated soils. *Soil Sci. Soc. Am. J.* **44**:892–898.

Van Genuchten, M.Th. (1981). Analytical solutions for chemical transport with simultaneous adsorption, zero-order production, and first-order decay. *J. Hydrol.* **49**:213–233.

van Kampen, N.G. (1992). *Stochastic Processes in Physics and Chemistry*. North-Holland, Amsterdam.

Vanmarcke, E. (1983). *Random Fields: Analysis and Synthesis*. MIT Press, Cambridge, MA.

Van Pul, W.A.J., F.A.A. M. de Leeuw, J.A. van Jaarsveld, M.A. van der Gaag, and C.J. Sliggers (1998). The potential for long-range transboundary atmospheric transport. *Chemosphere,* **37**:113–141.

Veith, G.D., D.L. DeFoe, and B.V. Bergtedt (1979). Measuring and estimating the bioconcentration factor of chemicals in fish. *J. Fish Res. Boards Can.* **36**:1040–1048.

Veith, G.D., K.J. Macek, S.R. Petrocelli, and J. Carroll (1980). An evaluation of using partition coefficients and water solubility to estimate bioconcentration factors for organic chemical in fish. In *Aquatic Toxicology. Proceedings of the Third Annual Symposium on Aquatic Toxicology*, ASTM Special Technical Publication 707. J.G. Eaton, P.R. Parrish, and A.C. Hendricks (Eds.). Am. Society for Testing and Materials, Philadelphia, pp. 116–129.

Velbel, M.A. (1985). Geochemical mass balances and weathering rates in forested watersheds of the Southern Blue Ridge. *Am. J. Sci.* **285**:904–930.

Venkatram, A. (1980). Estimating the Monin–Obukhov length in the stable boundary layer for dispersion calculations. *Boundary Layer Meteorol.* **19**:481–485.

Venkatram, A. (1988). On the use of kriging in the spatial analysis of acid precipitation data. *Atmos. Environ.* **22**:1963–1979.

Venkatram, A. (1988). Inherent uncertainty in air quality modeling. *Atmos. Environ.* **22**:1221–1227.

Vennard, J.K. and Street, R.L. (1975). *Elementary Fluid Mechanics*, 5th ed. Wiley, New York.

Ventura, S.J. and K.H. Kim (1993). Modeling urban nonpoint source pollution with a Geographical Information System. *Water Resources Bull.* **29**(2):189–198.

Visser, A.W. (1997). Using random walk models to simulate the vertical distribution of particles in a turbulent water column. *Marine Ecology Progress Series* **158**:275–281.

Vogel, T.M., C.S. Criddle, and P.L. McCarty (1987). Transformations of halogenated aliphatic compounds. *Environ. Sci. Tech.* **21**(8):722–736.

Von Karman, T. (1936). Turbulence and skin friction. *J. Aero. Sci.* **1**:1.

Wagner, B.J. and J.W. Harvey (1997). Experimental design for estimating parameters of rate-limited mass transfer: Analysis of stream tracer studies. *Water Resources Res.* **33**(7):1731–1741.

Wait, R. and A.R. Mitchell (1985). *Finite Element Analysis and Applications.* Wiley, Chichester, England.

Wallace L.A., E.D. Pellizzari, T.D. Harwell, R.Whitmore, C. Sparacino, and H. Zelon (1986). Total Exposure Assessment Methodology (TEAM) Study: Personal exposures, indoor-outdoor relationships, and breath levels of volatile organic compounds in New Jersey. *Environ. Int.* **12**:369–387.

Wallace, L.A., E. Pellizzari, T.D. Hartwell, V. Davis, L.C. Michail, and R.W. Whitmore (1989). The influence of personal activities on exposure to volatile organic compounds. *Environ. Res.* **50**(1):37–55.

Wallace, L., W. Nelson, R. Ziegenfus, E. Pellizzari, L. Michael, R. Whitmore, H. Zelon, T. Hartwell, R. Perritt, and D. Westerdahl (1991). The Los Angeles TEAM study: Personal exposures, indoor-outdoor air concentrations, and breath concentrations of 25 volatile organic compounds. *J. Exposure Anal. Environ. Epidemiol.* **1**(2):157–192.

Wallis, S.G., P.C. Young, and K.J. Beven (1989). Experimental investigation of the aggregated dead-zone model for longitudinal solute transport in stream channels. *Proc. Inst. Civil Eng.* **87**, Part 2: 1–22.

Walker, J.F., S.A. Pickard, and W.C. Sonzogni (1989). Spreadsheet watershed modeling for nonpoint-source pollution management in a Wisconsin basin. *Water Resources Bull.* **25**(1): 139–147.

Walthall, W.K. and J.D. Stark (1997). A comparison of acute mortality and population growth rates as endpoints of toxicological effects. *Ecotoxicol. Environ. Safety* **37**:45–52.

Wang, H.F. and M.P. Anderson (1982). *Introduction to Groundwater Modeling.* W.H. Freeman, New York.

Wang, L. and J. B. Milford (2001). Reliability of optimal control strategies for photochemical air pollution. *Environ. Sci. Tech.* **35**:1173–1180.

Wang, L., J.B. Milford, and W.P.L. Carter (2000). Reactivity estimates for aromatic compounds, I. Uncertainty in chamber-derived parameters. *Atmos. Environ.* **34**4337–4348.

Ward, R.C., C.C. Travis, D.M. Hetrick, M.E. Andersen, and M.L. Gargas (1988). Pharmacokinetics of tetrachloroethylene. *Toxicol. Appl. Pharmacol.* **93**:108–117.

Warwick, J.J. and K.J. Heim (1995). Hydrodynamic modeling of the Carson River and Lahontan Reservoir, Nevada. *Water Resources Bull.* **31**(1):67–77.

Warwick, J.J. and P. Tadepalli (1991). Efficacy of SWMM application. *J. Water Resources Planning Manag.* **117**(3):352–366.

Waymire, E. and V.K. Gupta (1981). The mathematical structure of rainfall representations. 2. A review of the theory of point processes. *Water Resources Res.* **17**(5):1273–1286.

Webber, W.J. and W.A Huang (1996). Distributed reactivity model for sorption by soils and sediments. 4. Intraparticle heterogeneity and phase-distribution relationships under non-equilibrium conditions. *Env. Sci. Technol.* **30**:881–888.

Weber, W. and F. Di Giano (1996). Process Dynamics in Environmental Systems, Wiley, New York.

Weil, J.C., R.I. Sykes, and A. Venkatram (1992). Evaluating air-quality models: Review and outlook. *J. Appl. Meteorol.* **31**:1121–1145.

Weiler, M., B.L. McGlynn, K.J. McGuire, and J.J. McDonnell (2003). How does rainfall become runoff? A combined tracer and runoff transfer function approach. *Water Resources Res.* **39**(11):1315, doi:10.1029/2003WR002331.

Weinhold, B. (2001). Putting people to the test. *Environ. Health Perspect.* **109**(10):A482–A485.

Weiss, G. (1977). Shot noise models for the generation of synthetic streamflow data, *Water Resources Res.* **13**:101–108.

Wells, P.G., J.N. Butler, and J.S. Hughes (1995). Introduction, Overview, Issues. In *Exxon Valdez Oil Spill: Fate and Effects in Alaskan Waters*. ASTM STP 1219. P.G. Wells, J.N. Butler, and J.S. Hughes (Eds.). American Society for Testing and Materials, Philadelphia, pp. 3–38.

Wells, S.A. and T.M. Cole (2000). *CE-QUAL-W2, Version 3*, Water Quality Technical Notes Collection, ERDC WQTN-AM-09, U.S. Army Engineers Res. and Development Center, Vicksburg, MI; see http://www.wes.army.mil/el/elpubs/pdf/wqtnam09.pdf.

Wesely, M.L. (1989). Parameterization of surface resistances to gaseous dry deposition in regional-scale numerical models. *Atmos. Environ.* **23**:1293–1304.

Westall, J.C. (1986). *MICROQL: A Chemical Equilibrium Program in BASIC*. Department of Chemistry, Oregon State University, Report 86-02.

Wexler, A.S., F. W. Lurmann, and J.H. Seinfeld (1994). Modelling urban and regional aerosols —I. Model development. *Atmos. Environ.* **28**:431–546.

Whelan, G., J.W. Buck, D.L. Strenge, J.G. Droppo, Jr., B.L. Hoopes, and R.J. Aiken (1992). An overview of the Multimedia Assessment Methodology MEPAS. *Haz. Waste Haz. Mat.* **9**(2):191–208.

Whitby, E.R., P.H. McMurry, U. Shankar, and F.S. Binkowski (1991). Modal Aerosol Dynamics Modeling, EPA Report No. 600/3-91-020. U.S. Environmental Protection Agency, Res. Triangle Park, NC.

White, M.C, P.F. Infante, and K.C. Chu (1982). A quantitative estimate of leukemia mortality associated with occupational exposures to benzene. *Risk Anal.* **2**:195–204.

Wiedemeier, T.H., M.A. Swanson, D.E. Moutoux, E.K. Gordon, J.T.Wilson, B.H. Wilson, D.H. Kampbel, J. Hansen, and P. Haas (1996). Technical protocol for evaluating natural attenuation of chlorinated solvents in groundwater, Air Force Center for Environmental Excellence, Technology Transfer Division, Brooks AFB, San Antonio, TX.

Wiener, J.G. (1995). Bioaccumulation of mercury in fish. National Forum on Mercury in Fish: Proceedings; September 27–29, 1994; New Orleans, LA. EPA 823-R-95-002, U.S. EPA, Office of Water. Washington, DC, pp. 41–47.

Wilemski, G. (1984). Composition of the critical nucleus in multicomponent vapor nucleation. *J. Chem. Phys.* **80**:1370–1372.

Wiley, J.A., J.P. Robinson, T. Piazza, K. Garrett, K. Cirksena, Y. Cheng, and G. Martin (1991). Activity Patterns of California Residents. Final Report to Res. Division, California Air Resources Board, Contract No. A6–177-33, Sacramento, CA.

Wilhm, J. and L. King (1982). Morphometric characteristics of reservoirs of Oklahoma. *Proc. Oklahoma Acad. Sci.* **62**:14–17; see http://digital.library.okstate.edu/oas/oas_pdf/v62/p14_17.pdf.

Wilke, C.R. and P. Chang (1955). *A.I.Ch.E. J.* **1**:264–270.

Wilkes, C.R., M.J. Small, J.B. Andelman, N.J. Giardino, and J. Marshall (1992). Inhalation exposure model for volatile chemicals from indoor uses of water. *Atmos. Environ.* **26A**(12):2227–2236.

Wilkes, C.R., M.J. Small, C.I. Davidson, and J.B. Andelman (1996). Modeling the effects of water usage and co-behavior on inhalation exposures to contaminants volatilized from household water. *J. Exposure Anal. Environ. Epidemiol.* **6**(4):393–412.

Wilkinson, C.F., G.R. Christoph, E. Julien, J.M. Kelley, J. Kronenberg, J. McCarthy, and R. Reiss (2000). Assessing the risks of exposures to multiple chemicals with a common mechanism of toxicity: How to cumulate? *Regul. Toxicol. Pharmacol.* **31**:30–43.

Williams, P., L. Benton, J. Warmerdam, and P. Sheehan (2002). Comparative risk analysis of six volatile organic compounds in California drinking water. *Environ. Sci. Tech.* **36**(22):4721–4728.

Wilson, J.L. and P.J. Miller (1978). Two-dimensional plume in uniform ground-water flow. *Am. Soc. Civil Eng., J. Hydraulics Div.* **104**(HY4):503–514.

Wilson, J.M. and P.J. Miller (1978). Two-dimensional plume in uniform ground-water flow. *J. Hyd. Div* ASCE 104(HY4) Proc. Paper 13665, 503–514.

Whitman, W.G. (1923). The two-film theory of gas absorption. *Chem. Metal. Eng.* **29**:146–148.

Wolfram, S. (1991). *Mathematica: A System for Doing Mathematics by Computer*, 2nd ed. Addison-Wesley, Reading, MA.

Wolfson, L.J., J.B. Kadane, and M.J. Small (1996). Bayesian environmental policy decisions: Two case studies. *Ecolog. Appl.,* **6**(4):1056–1066.

Wood, E. and I. Rodriguez-Iturbe (1975). A Bayesian approach to analyzing uncertainty among flood frequency models. *Water Resources Res.* **11**(6):839–843.

Wood, E.F., D.P. Lettenmaier, and V.G. Zartarian (1992). A land-surface hydrology parameterization with subgrid variability for general circulation models, *J. Geophys. Res.* **97**:2717.

Woolhiser, D.A., R.E. Smith, and D.C. Goodrich (1990). *KINEROS, A Kinematic Runoff and Erosion Model: Documentation and User manual.* U.S. Department of Agriculture, Agricultural Res. Service, ARS-77, Washington, DC.

Wu, J.S., E.L. King, and M. Wang (1985). Optimal identification of Muskingum routing coefficients. *Water Resources Bull.* **21**(3):417–421.

Yalkowsky, S.H., S.C. Valvani, and G.L. Amidon (1976). Solubility of nonelectrolytes in polar solvents. IV. Nonpolar drugs in mixed solvents. *J. Pharmac. Sci.* **65**:1488–1494.

Yeh, G.T. (1990). A Lagrangian-Eulerian method with zoomable hidden fine-mesh approach to solving advection-dispersion equations. *Water Resources Res.* **26**(6):1133–1144.

Young, R.A., C.A. Onstad, D.D. Bosch, and W.P. Anderson (1989). AGNPS: A nonpoint-source pollution model for evaluating agricultural watershed. *J. Soil Water Conserv.* **44**: 168–173.

Yu, Z. and F.W. Schwartz (1998). Application of integrated Basin-Scale Hydrologic Model to simulate surface water and ground-water interactions in Big Darby Creek Watershed, Ohio. *J. Am. Water Resources Assoc.* **34**(2):1–17.

Yu, C., A.J. Zielen, J.J. Cheng, Y.C. Yuan, L.G. Jones, D.J. LePoire, Y.Y. Wang, C.O. Loureiro, E. Gnanapragasam, E. Faillace, A. Wallo III, W.A. Williams, and H. Peterson (1993). *Manual for Implementing Residual Radioactive Material Guidelines Using RESRAD, Version 5.0.* Argomne National Laboratory, Environmental Assessment Division, Distribution Category UC-902, Argonne, IL.

Yu, W.H., C.M. Harvey, and C.F. Harvey (2003). Arsenic in groundwater in Bangladesh: A geostatistical and epidemiological framework for evaluating health effects and potential remedies. *Water Resources Res.* **39**(6):1146.

Zannetti, P. (1990). *Air Pollution Modeling: Theories, Computational Methods and Available Software*. Van Nostrand Reinhold, New York.

Zaphiropoulos Y., P. Dellaportas, E. Morfiadakis, and G. Glinou (1999). Prediction of wind speed and direction in potential site. *Wind Eng.* **23**:167–175.

Zheng, C. (1990). MT3D—A modular three-dimensional transport model for simulation of advection, dispersion and chemical reactions of contaminants in groundwater systems, U.S. E.P.A Report, Washington, DC.

Zheng, C. and G.D. Bennett, (1995). *Applied Contaminant Transport Modeling: Theory and Practice*. Van Nostrand Reinhold, New York.

Zheng C. (ModIME V1.1) (2004). www.sspa.com/products/modime.htm.

Zienkiewicz, O.C. and R.L. Taylor (2000). *Finite Element Method: Volume 1. The Basis* and *Vol. 3 Fluid Dynamics*. Butterworth Heinemann, London.

Zison, S.W., W.B. Mills, D. Diemer, and C.W. Chen (1978). Rates, Constants, and Kinetic Formulations in Surface Water Quality Modeling. USEPA, ORD, Athens, GA, ERL, EPA/600/3-78-105.

INDEX